Remote Sensing of the Environment

An Earth Resource Perspective

Remote Sensing of the Environment

An Earth Resource Perspective

Second Edition

John R. Jensen

University of South Carolina

Upper Saddle River, NJ 07458

Library of Congress Cataloging-in-Publication Data

Jensen, John R.
  Remote sensing of the environment: an earth resource perspective / John R. Jensen. — 2nd ed.
    p.   cm. — (Prentice Hall series in geographic information science)
Includes bibliographical references and index.
ISBN 0-13-188950-8
1. Earth Sciences - Remote sensing.

QE33.2.R4J46 2007
550.28—dc22   2006041573

Acquisitions Editor: Jeff Howard
Editor in Chief, Science: Daniel Kaveney
Executive Managing Editor: Kathleen Schiaparelli
Production Editor: Edward Thomas
Manufacturing Buyer: Alan Fischer
Director of Marketing, Science: Patrick Lynch
Marketing Assistant: Renee Hanenberg
Copy Editor: Barbara Booth
Editorial Assistant: Margaret Ziegler
Production Assistant: Nancy Bauer
Art Director: Jayne Conte

Cover Credits: (top left) Dispersion of white light using a prism (courtesy of David Parker, Photo Researchers, Inc.); (top right) Artist's rendition of the *Terra* satellite (courtesy of NASA); (left center) High spatial resolution natural color image of Denver, CO, recorded by the *QuickBird* satellite on July 17, 2002 (courtesy DigitalGlobe, Inc.); (right center) Three Gorges Dam in China recorded by the *QuickBird* satellite on July 13, 2003 (courtesy DigitalGlobe, Inc.); (bottom left) Landsat Thematic Mapper 30 × 30 m color composite of Mount Kilimanjaro in Tanzania (courtesy of NASA and Earth Satellite, Inc.); (bottom right) Digital elevation model of Mount Kilimanjaro (30 × 30 m) viewed from an oblique vantage point looking W-NW derived from Space Shuttle Topography Mission (SRTM) interferometric SAR data (courtesy USGS EROS Data Center and NASA Jet Propulsion Lab).

© 2007, 2000 Pearson Education, Inc.
Pearson Prentice Hall
Pearson Education, Inc.
Upper Saddle River, NJ 07458

Pearson Prentice Hall™ is a trademark of Pearson Education, Inc.

Printed in the United States of America.

10 9 8 7 6 5 4 3 2 1

ISBN 0-13-188950-8

Pearson Education Ltd., *London*
Pearson Education Australia Pty., Ltd., *Sydney*
Pearson Education *Singapore*, Pte. Ltd.
Pearson Education North Asia Ltd., *Hong Kong*
Pearson Education Canada, Ltd., *Toronto*
Pearson Educacion de Mexico, S. A. de C. V.
Pearson Education–Japan, *Tokyo*
Pearson Education Malaysia, Pte. Ltd.

Brief Contents

Contents

Preface

Remote sensing is the art and science of obtaining information about an object without being in direct physical contact with the object. Remote sensing can be used to measure and monitor important biophysical characteristics and human activities on Earth. *Remote Sensing of the Environment: An Earth Resource Perspective, 2nd Edition,* was written to introduce the reader to:

- the characteristics of electromagnetic radiation and how the energy interacts with Earth surface materials such as vegetation, soil, rock, water, and urban infrastructure,

- how electromagnetic energy reflected or emitted from these materials is recorded using a variety of remote sensing instruments (e.g., cameras, multispectral scanners, hyperspectral instruments, RADAR, LIDAR), and

- how we can extract biophysical or land use/land cover information from the remote sensor data to solve important problems.

The book introduces the principles of visual photo-interpretation and image analysis. It is a companion volume to *Introductory Digital Image Processing: A Remote Sensing Perspective* (2005) published by Pearson Education, Inc., which introduces the fundamentals of digital image analysis.

This book was written for physical, natural, and social scientists interested in how remote sensing of the environment can be used to solve real-world problems. The reader should have education or practical experience in a systematic body of knowledge (e.g., physical geography, biogeography, geology, hydrology, urban planning, agriculture, forestry, marine science) to which the remote sensing science can be applied. The reader should understand college algebra. This book can be used in undergraduate or graduate, one- or two-semester courses in airphoto interpretation and/or remote sensing.

People responsible for managing the Earth's natural resources and planning future development recognize the importance of accurate, spatial information residing in a geographic information system (GIS). Many of the most important layers of biophysical, land use/land cover, and socioeconomic information in a GIS database are derived from an analysis of remotely sensed data. Consequently, we now see a significant increase in the demand for remote sensing data. Hundreds of public and commercial remote sensing systems throughout the world now collect remote sensor data. Some remote sensor data are available at a reasonable cost per km^2 and may be obtained by the user over the Internet. Some remote sensor data are very expensive.

To utilize the remotely sensed data to its full potential, it must be interpreted correctly. This requires an understanding of how energy interacts with the atmosphere and terrain, how remote sensing systems function, and how to extract useful information from the remote sensor data.

The following features hopefully make the information in this book easy to comprehend and apply:

- Each chapter includes illustrations that were specially designed to make complex principles easy to understand.

- Each chapter contains a substantive reference list.

- The large 8.5 × 11 in. format allows the images and diagrams to be readable and visually informative.

- The second edition contains forty-eight pages of color to display color images or biophysical information that may be extracted from remote sensor data.

- An appendix contains Internet addresses for many of the most important sources of remote sensing information and public and private sources of remotely sensed data.

Below is a brief summary of the topics discussed in each chapter and improvements made in the second edition.

Chapter 1. Remote Sensing of the Environment. The book begins by carefully defining *in situ* and remote sensing data collection. Elements of the remote sensing process are updated, including statement of the problem, remote sensor data collection, data analysis (visual and digital), and information presentation alternatives. Spatial, spectral, temporal, and radiometric resolution are defined along with trade-off considerations. A new remote sensing Earth observation economics section is included. The chapter reviews how remote sensing is used for a variety of national and international Earth resource monitoring programs.

Chapter 2. Electromagnetic Radiation Principles. Conduction, convection, and radiation are introduced as energy transfer mechanisms. Wave and particle models are used to describe how electromagnetic radiation is generated and transported through space. The reader is introduced to how electromagnetic energy interacts with the atmosphere, the terrain, and how the reflected or emitted energy is recorded by a remote sensing system. Important radiometric concepts are carefully defined. New spectral reflectance curves of selected materials are introduced. Emphasis is placed on minimizing the effects of atmospheric scattering to maximize the remote sensing signal from the terrain.

Chapter 3. History of Aerial Photography and Aerial Platforms. A history of photography is provided, including the invention of specific types of emulsions and fixative agents. A history of the collection of aerial photography from various platforms is then reviewed, including: balloons, kites, rockets, pigeons, gliders, aircraft, and spacecraft. A history of photo reconnaissance in the Civil War, World War I and II, the Cuban Missile Crisis, the Cold War, and war in the Middle East is presented. New developments in remote sensing platforms including unmanned aerial vehicles (UAV) and stealth technology are introduced.

Chapter 4. Aerial Photography – Vantage Point, Cameras, Filters, and Film. Vertical and oblique aerial photography are introduced using Hurricane Katrina examples. Aerial camera components, the most important types of aerial cameras, and band-pass filters used during data collection are introduced. New sections describe the characteristics of small-, medium-, and large-format digital cameras. Major black-and-white and color films (emulsions) used for aerial photography are identified, including color-infrared film. Methods of digitizing aerial photography are described.

Chapter 5. Elements of Visual Image Interpretation. Unique characteristics of remote sensor data are reviewed and demonstrated using new examples, including: the importance of the aerial regional perspective; the power of three-dimensional depth perception; the utility of obtaining knowledge beyond our human visual perception; and the significance of being able to collect a historical spatial record of Earth resources to document change. The fundamental elements of image interpretation are then introduced. The chapter concludes by summarizing various methods of search involving collateral information, convergence-of-evidence, and use of the multi-concept in image analysis.

Chapter 6. Photogrammetry. Methods of extracting quantitative information from vertical aerial photography are presented. The chapter defines the nomenclature and geometry of a single aerial photograph and how scale and height measurements are made. Stereoscopic principles are introduced. New sections describe interior and exterior orientation of stereoscopic images using soft-copy photogrammetry and the extraction of spot elevations, contours, digital elevation models and orthophotos. A new section describes how thematic features of interest are extracted based on the principle of the floating mark. The final section describes how area measurements are obtained from vertical remote sensor data.

Chapter 7. Multispectral Remote Sensing Systems. Multispectral remote sensing terminology is defined. The major types of remote sensing instruments are described along with significantly updated information on the most important remote sensing systems. Proposed sensor systems are discussed such as DigitalGlobe's WorldView I and II; NASA's Landsat Data Continuity Mission; and NOAA's NPOESS. The USGS *Global Visualization Viewer* is introduced to help students locate various types of remote sensor data.

Chapter 8. Thermal Infrared Remote Sensing. This chapter describes how surface temperature mapping is performed using thermal infrared remote sensing. The history of thermal infrared remote sensing includes more information on Herschel's experiment. Thermal infrared radiation properties, radiation laws, and atmospheric windows are summarized. The diurnal temperature characteristics of vegetation, soil, rock, water, and urban phenomena are introduced. The characteristics of thermal infrared scanning instruments and forward looking infrared (FLIR) sensors are presented. New aircraft thermal infrared remote sensing systems are introduced such as the Canadian Thermal Airborne Broadband Imager (TABI) and the Thermal Airborne Spectrographic Imagery (TASI). New case studies are provided, including: mapping stream temperature in Oregon, mapping septic tank leakage in SC, monitoring industrial roof moisture in NC, and monitoring agricultural plant stress degree days.

Chapter 9. Active and Passive Microwave Remote Sensing. The history of active microwave remote sensing is reviewed. Methods of sending and receiving a pulse of microwave energy are presented. A new table summarizes the characteristics of the most important Earth-orbiting Synthetic Aperture Radars. Important radar system parameters such as frequency, polarization, pulse length, depression angle, look direction, incident angle, and slant-range and ground-range image geometry are discussed. The major environmental factors influencing active microwave backscatter are presented, including: surface roughness, complex dielectric constant, and the cardinal effect. Sections on the response of vegetation, surface water, and urban phenomena to active microwave energy are expanded. New information is provided about the Canadian RADARSAT-1 and -2; the European Space Agency's ERS-1 and -2, and Envisat ASAR; and NASA's TRMM Microwave Imager and Shuttle Radar Topography Mission (SRTM). The section on active microwave interferometry includes new Utah and Tanzania SRTM examples. Passive microwave remote sensing is discussed with new data on NASA's Advanced Microwave Scanning Radiometer (AMSR-E).

Chapter 10. LIDAR Remote Sensing. This completely new chapter introduces the fundamental principles of remote sensing using Light Detection and Ranging (LIDAR) tech-

nology. Emphasis is placed on how LIDAR sensors function and on extracting useful information from first-, intermediate-, and last-return LIDAR data. A case study documents the utility of LIDAR to provide detailed digital surface and digital terrain models. Attention is also given to the usefulness of LIDAR intensity data.

Chapter 11. Remote Sensing of Vegetation. Photosynthesis fundamentals are introduced. The chapter then identifies the dominant factors controlling leaf reflectance in the visible, near-infrared, and middle-infrared portions of the spectrum. The bidirectional reflectance distribution function (BRDF) concept is discussed. The importance of understanding the temporal phenological cycle of the vegetation types under investigation is emphasized. The section on remote sensing-derived vegetation indices now includes more than twenty of the most widely adopted algorithms. There is new information on the nonlinear relationship between the Simple Ratio index and the Normalized Difference Vegetation Index (NDVI). The use of remotely sensed data for computing landscape ecology metrics and for assisting in the assessment of biodiversity are introduced. A new case study documents how South African agriculture can be monitored to predict agricultural water demand.

Chapter 12. Remote Sensing of Water. This chapter introduces how electromagnetic energy interacts with the water surface, subsurface, and the bottom as a function of wavelength. The impact of organic (e.g., chlorophyll) and/or inorganic (e.g., suspended sediment) material in the water column on spectral reflectance is discussed. Eight of the most important algorithms used to measure phytoplankton abundance in water are now provided. A new section reviews improvements in bathymetric mapping using passive optical (aerial photography) and active remote sensing (SONAR and LIDAR). Remote sensing methods for monitoring water surface temperature, precipitation, aerosols and clouds, water vapor, and snow are presented. The major satellite remote sensing systems used to measure ocean color are now summarized.

Chapter 13. Remote Sensing the Urban Landscape. Remote sensing extraction of urban information is now placed in the context of the National Spatial Data Infrastructure (NSDI). Urban spatial, spectral, and temporal resolution considerations are introduced. The most important land use and/or land cover classification schemes are now provided. Many new remotely-sensed images of residential, commercial, industrial, transportation, communications, and utilities land use are provided. The use of soft-copy photogrammetry and LIDAR for extracting three-dimensional information in urban areas is presented. There are new sections on impervious surface mapping and floodplain delineation.

Chapter 14. Remote Sensing of Soils, Minerals, and Geomorphology. New information on the influence of soil texture, moisture content, organic matter, iron oxide, salinity, biological crusts, and surface roughness on the spectral reflectance of soils is presented. Rock and mineral discrimination using imaging spectroscopy techniques is discussed. The chapter identifies how geologic lithology, structure, drainage density and pattern may be monitored using remote sensing. Drainage morphometrics that can be extracted from remote sensor data are now included. The chapter provides many new examples of how remote sensor data may be used to identify geomorphic features on the surface of the Earth.

Chapter 15. In Situ Reflectance Measurement. This new chapter introduces the fundamental principles of *in situ* spectral reflectance measurement using solar illumination in the field or artificial illumination in the laboratory. Handheld spectroradiometer characteristics are reviewed including: spectral resolution, sampling interval, design, data collection speed, viewing geometry and field of view.

Acknowledgments

I would like to thank Marsha, my wife, who provided patient encouragement. Judith Berglund, Jungho Im, Maria Garcia-Quijano, Shawn Callihan, Charles Huang, and Brian Hadley reviewed selected chapters. Pearson Education, Inc. personnel were very helpful including Dan Kaveney (Editor in Chief, Science), Jeff Howard (Editor), Barbara Booth (copyediting) and Ed Thomas (Production).

The American Society for Photogrammetry & Remote Sensing, Association of American Geographers, Geocarto International Centre, Inc. and Taylor & Francis, Inc. granted permission for the author to extract copyrighted material from articles published in *Photogrammetric Engineering & Remote Sensing,* the *Manual of Remote Sensing, Manual of Color Aerial Photography, Geocarto International,* and *International Journal of Remote Sensing.*

Elsevier Inc., granted permission to reprint figures from articles in *Remote Sensing of Environment* (*RSE*) and *Geoderma*: Torgersen, C. E. et al., 2001, "Airborne Thermal Remote Sensing for Water Temperature Assessment in Rivers and Streams," *RSE*, 76:386-398; Metternicht, G. I. and J. A. Zinck, 2003, "Remote Sensing of Soil Salinity: Potentials and Constraints," *RSE*, 85:1-20; Legleiter, C. J. et al., 2004, "Passive Optical Remote Sensing of River Channel Mor-

phology and In-stream Habitat: Physical Basis and Feasibility," *RSE*, 93:493-510; Chen, J. et al., 2005, "A New Index for Mapping Lichen-dominated Biological Soil Crusts in Desert Areas," *RSE*, 96:165-175; Farifteh, J. et al., 2006, "Assessing Salt-affected Soils Using Remote Sensing, Solute Modelling, and Geophysics," *Geoderma,* 130:191-206.

John Pike (Federation of American Scientists) made available several remote sensing images. Kite photography was provided by Dr. Cris Benton (University of California Berkeley). Dr. Stefan Sandmeier provided much of the BRDF material. Several historical photographs were made available by the Harry Ransom Humanities Research Center, University of Texas, Austin, TX, and the Deutsches Museum, Munich, Germany.

Several agencies were helpful, including: Smithsonian National Air & Space Museum, Library of Congress, NASA Observatorium, NASA Goddard Tropical Rainfall Measurement Mission Office (Alan Nelson), NASA Goddard Solar and Heliospheric Observatory (Joseph Gurman), NASA Johnson Space Center (Dr. Kamlesh Lulla), NASA Jet Propulsion Lab, USGS Desert Processes Working Group, USGS Western Region Coastal & Marine Geology, USGS Imaging Spectroscopy Lab (Roger Clark), National Optical Astronomy Observatory (Kathie Coil), NOAA, and the Geological Survey of Canada.

Commercial remote sensing data providers made available examples of their platforms, sensors, and image products. Special thanks to: DigitalGlobe Inc. (Chuck Herring), ImageSat Intl., Inc. (Karen Anisfeld), Space Imaging, Inc. (Linda Lidov), SPOT Image, Inc. (Clark Nelson), Lockheed Martin, Inc. (Jeannie Duisenberg and Eric Schulzinger), Litton Emerge Spatial, Inc. (Don Light), Positive Systems, Inc. (Kim Hickman), ORBIMAGE, Inc. (Sue Hale), Intermap, Inc., Aeromap USA, Inc., RADARSAT, Inc., Eastman Kodak, Inc. (Robert Lundquist and Mary Skerrett), and Hughes Santa Barbara Research Center, Inc. Photogrammetric engineering firms provided information and images of their sensors, including: Leica-GeoSystems, Inc. (Alistair Wells), L-H Systems, Inc. (Dr. A. Stewart Walker), Marconi Integrated Systems, Inc. (Bob Hayes), E. Coyote Enterprises and Z/I Imaging, Inc. (Marilyn O'Cuilinn), Analytical Spectral Devices, Inc., Stockton Infrared Thermographic Services, Inc. (Cindy Hawks), and Vexcel, Inc.

John R. Jensen
University of South Carolina

Remote Sensing of the Environment 1

S cientists observe nature, make measurements, and then attempt to accept or reject hypotheses concerning these phenomena. The data collection may take place directly in the field (referred to as *in situ* or *in-place* data collection), or at some remote distance from the subject matter (referred to as *remote sensing* of the environment).

 ### *In situ* Data Collection

One form of *in situ* data collection involves the scientist going out in the field and questioning the phenomena of interest. For example, a census enumerator may go door to door, asking people questions about their age, sex, education, income, etc. These data are recorded and used to document the demographic characteristics of the population.

Conversely, a scientist may use a *transducer* or other *in situ* measurement device at the study site to make measurements. Transducers are usually placed in direct physical contact with the object of interest. Many different types of transducers are available. For example, a scientist could use a thermometer to measure the temperature of the air, soil, or water; an anemometer to measure wind speed; or a psychrometer to measure air humidity. The data recorded by the transducers may be an analog electrical signal with voltage variations related to the intensity of the property being measured. Often these analog signals are transformed into digital values using analog-to-digital (A-to-D) conversion procedures. *In situ* data collection using transducers relieves the scientist of monotonous data collection often in inclement weather. Also, the scientist can distribute the transducers at important geographic locations throughout the study area, allowing the same type of measurement to be obtained at many locations at the same time. Sometimes data from the transducers are telemetered electronically to a central collection point for rapid evaluation and archiving (e.g., Teillet et al., 2002).

Two examples of *in situ* data collection are demonstrated in Figure 1-1. Leaf-area-index (LAI) measurements are being collected by a scientist at a study site using a handheld ceptometer in Figure 1-1a. Spectral reflectance measurements of vegetation are being obtained at a study site using a handheld spectroradiometer in Figure 1-1b. LAI and spectral reflectance measurements obtained in the field may be used to calibrate LAI and spectral reflectance measurements collected by a remote sensing system located on an aircraft or satellite (Jensen et al., 2005).

In situ Measurement

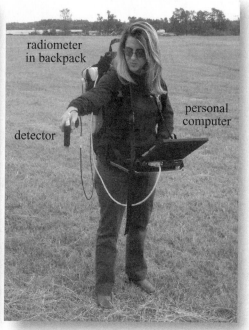

a. Leaf-area-index (LAI) measurement
using a ceptometer.

b. Spectral reflectance measurement
using a spectroradiometer.

Figure 1-1   *In situ* (in-place) data are collected in the field. a) A scientist is collecting leaf-area-index (LAI) measurements of soybeans (*Glycine max L. Merrill*) using a ceptometer that measures the number of "sunflecks" that pass through the vegetation canopy. The measurements are made just above the canopy and on the ground below the canopy. The *in situ* LAI measurements may be used to calibrate LAI estimates derived from remote sensor data. b) Spectral reflectance measurements from vegetation are being collected using a spectroradiometer located approximately 1 m above the canopy. The *in situ* spectral reflectance measurements may be used to calibrate the spectral reflectance measurements obtained from a remote sensing system.

Data collection by scientists in the field or by instruments placed in the field provide much of the data for physical, biological, and social science research. However, it is important to remember that no matter how careful the scientist is, error may be introduced during the *in situ* data-collection process. First, the scientist in the field can be *intrusive*. This means that unless great care is exercised, the scientist can actually change the characteristics of the phenomenon being measured during the data-collection process. For example, a scientist could lean out of a boat to obtain a surface-water sample from a lake. Unfortunately, the movement of the boat into the area may have stirred up the water column in the vicinity of the water sample, resulting in an unrepresentative, or *biased*, sample. Similarly, a scientist collecting a spectral reflectance reading could inadvertently step on the sample site, disturbing the vegetation canopy prior to data collection.

Scientists may also collect data in the field using biased procedures. This introduces *method-produced error*. It could involve the use of a biased sampling design or the systematic, improper use of a piece of equipment. Finally, the *in situ* data-collection measurement device may be calibrated incorrectly. This can result in serious measurement error.

Intrusive *in situ* data collection, coupled with human method-produced error and measurement-device miscalibration, all contribute to *in situ* data-collection error. Therefore, it is a misnomer to refer to *in situ* data as *ground truth data*. Instead, we should simply refer to it as *in situ ground reference data*, acknowledging that it contains error.

 Remote Sensing Data Collection

Fortunately, it is also possible to collect information about an object or geographic area from a distant vantage point using *remote sensing* instruments (Figure 1-2). Remote sens-

Remote Sensing Measurement

Orbital platform

Suborbital platform

Suborbital platform

Remote sensing instrument

H
altitude above ground level (AGL)

β

β
instantaneous-field-of-view (IFOV) of the sensor system

Object, area, or materials within the ground-projected IFOV

$\leftarrow D \rightarrow$
diameter of the ground-projected IFOV

Figure 1-2   A remote sensing instrument collects information about an object or phenomenon within the instantaneous-field-of-view (IFOV) of the sensor system without being in direct physical contact with it. The remote sensing instrument may be located just a few meters above the ground and/or onboard an aircraft or satellite platform.

ing data collection was originally performed using cameras mounted in suborbital aircraft. *Photogrammetry* was defined in the early editions of the *Manual of Photogrammetry* as:

> the art or science of obtaining reliable measurement by means of photography (American Society of Photogrammetry, 1952; 1966).

Photographic interpretation is defined as:

> the act of examining photographic images for the purpose of identifying objects and judging their significance (Colwell, 1960).

Remote sensing was formally defined by the American Society for Photogrammetry and Remote Sensing (ASPRS) as:

> the measurement or acquisition of information of some property of an object or phenomenon, by a recording device that is not in physical or intimate contact with the object or phenomenon under study (Colwell, 1983).

In 1988, ASPRS adopted a combined definition of photogrammetry and remote sensing:

> Photogrammetry and remote sensing are the art, science, and technology of obtaining reliable information about physical objects and the environment, through the process of recording, measuring and interpreting imagery and digital representations of energy patterns derived from non-contact sensor systems (Colwell, 1997).

But where did the term *remote sensing* come from? The actual coining of the term goes back to an unpublished paper in the early 1960s by the staff of the Office of Naval Research Geography Branch (Pruitt, 1979; Fussell et al., 1986). Evelyn L. Pruitt was the author of the paper. She was assisted by staff member Walter H. Bailey. Aerial photo interpretation had become very important in World War II. The space age was just getting under way with the 1957 launch of *Sputnik* (U.S.S.R.), the 1958 launch of *Explorer 1* (U.S.), and the collection of photography from the then secret CORONA program initiated in 1960 (Table 1-1). In addition, the Geography Branch of ONR was expanding its research using instruments other than cameras (e.g., scanners, radiometers) and into regions of the electromagnetic spectrum beyond the visible and near-infrared regions (e.g., thermal infrared, microwave). Thus, in the late 1950s it had become apparent that the prefix "photo" was being stretched too far in view of the fact that the root word, *photography*, literally means "to write with [visible] light" (Colwell, 1997). Evelyn Pruitt (1979) wrote:

> The whole field was in flux and it was difficult for the Geography Program to know which way to move. It was finally decided in 1960 to take the problem to the Advisory Committee. Walter H. Bailey and I pondered a long time on how to present the situation and on what to call the broader field that we felt should be encompassed in a program to replace the aerial photointerpretation project. The term 'photograph' was too limited because it did not cover the regions in the electromagnetic spectrum beyond the 'visible' range, and it was in these non-visible frequencies that the future of interpretation seemed to lie. 'Aerial' was also too limited in view of the potential for seeing the Earth from space.

The term *remote sensing* was promoted in a series of symposia sponsored by ONR at the Willow Run Laboratories of the University of Michigan in conjunction with the National Research Council throughout the 1960s and early 1970s, and has been in use ever since (Estes and Jensen, 1998).

Maximal/Minimal Definitions

Numerous other definitions of remote sensing have been proposed. In fact, Colwell (1984) suggests that "one measure of the newness of a science, or of the rapidity with which it is developing is to be found in the preoccupation of its scientists with matters of terminology." Some have proposed an all-encompassing *maximal definition*:

> Remote sensing is the acquiring of data about an object without touching it.

Such a definition is short, simple, general, and memorable. Unfortunately, it excludes little from the province of remote sensing (Fussell et al., 1986). It encompasses virtually all remote sensing devices, including cameras, optical-mechanical scanners, linear and area arrays, lasers, radar systems, sonar, seismographs, gravimeters, magnetometers, and scintillation counters.

Others have suggested a more focused, *minimal definition* of remote sensing that adds qualifier after qualifier in an attempt to make certain that only legitimate functions are included in the term's definition. For example:

> Remote sensing is the noncontact recording of information from the ultraviolet, visible, infrared,

and microwave regions of the electromagnetic spectrum by means of instruments such as cameras, scanners, lasers, linear arrays, and/or area arrays located on platforms such as aircraft or spacecraft, and the analysis of acquired information by means of visual and digital image processing.

Robert Green at NASA's Jet Propulsion Lab (JPL) suggests that the term *remote measurement* might be used instead of *remote sensing* because data obtained using the new hyperspectral remote sensing systems are so accurate (Robbins, 1999). Each of the definitions are correct in an appropriate context. It is useful to briefly discuss components of these remote sensing definitions.

Remote Sensing: Art and/or Science?

Science: A *science* is defined as a broad field of human knowledge concerned with facts held together by *principles* (rules). Scientists discover and test facts and principles by the scientific method, an orderly system of solving problems. Scientists generally feel that any subject that humans can study by using the scientific method and other special rules of thinking may be called a science. The sciences include 1) *mathematics* and *logic*, 2) *physical sciences*, such as physics and chemistry, 3) *biological sciences*, such as botany and zoology, and 4) the *social sciences*, such as geography, sociology, and anthropology (Figure 1-3). Interestingly, some persons do not consider mathematics and logic to be sciences. But the fields of knowledge associated with mathematics and logic are such valuable *tools* for science that we cannot ignore them. The human race's earliest questions were concerned with "how many" and "what belonged together." They struggled to count, to classify, to think systematically, and to describe exactly. In many respects, the state of development of a science is indicated by the use it makes of mathematics. A science seems to begin with simple mathematics to measure, then works toward more complex mathematics to explain.

Remote sensing is a tool or technique similar to mathematics. Using sophisticated sensors to measure the amount of electromagnetic energy exiting an object or geographic area from a distance and then extracting valuable information from the data using mathematically and statistically based algorithms is a *scientific* activity (Fussell et al., 1986). Remote sensing functions in harmony with other geographic information sciences (often referred to as *GIScience)*, including cartography, surveying, and geographic information systems (GIS) (Curran, 1987; Clarke, 2001; Jensen, 2005). Dahlberg and Jensen (1986) and Fisher and Lindenberg (1989) suggested a model where there is interaction

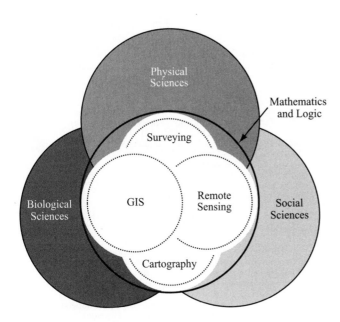

Figure 1-3   Interaction model depicting the relationship of the geographic information sciences (remote sensing, geographic information systems, cartography, and surveying) as they relate to mathematics and logic and the physical, biological, and social sciences.

between remote sensing, cartography, surveying, and GIS, where no subdiscipline dominates and all are recognized as having unique yet overlapping areas of knowledge and intellectual activity as they are used in physical, biological, and social science research (Figure 1-3).

The theory of science suggests that scientific disciplines go through four classic developmental stages. Wolter (1975) suggested that the growth of a scientific discipline, such as remote sensing, that has its own techniques, methodologies, and intellectual orientation seems to follow the sigmoid or logistic curve illustrated in Figure 1-4. The growth stages of a scientific field are: Stage 1 — a preliminary growth period with small increments of literature; Stage 2 — a period of exponential growth when the number of publications doubles at regular intervals; Stage 3 — a period when the rate of growth begins to decline but annual increments remain constant; and Stage 4 — a final period when the rate of growth approaches zero. The characteristics of a scholarly field during each of the stages may be briefly described as follows: Stage 1 — little or no social organization; Stage 2 — groups of collaborators and existence of invisible colleges, often in the form of ad hoc institutes, research units, etc.; Stage 3 — increasing specialization and increasing controversy; and Stage 4 — decline in membership in both collaborators and invisible colleges.

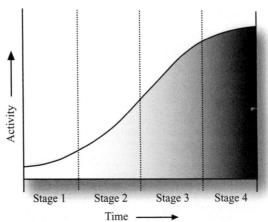

Figure 1-4   The developmental stages of a scientific discipline (Wolter, 1975; Jensen and Dahlberg, 1983).

Using this logic, it may be suggested that remote sensing is in Stage 2 of a scientific field, experiencing exponential growth since the mid-1960s with the number of publications doubling at regular intervals (Colwell, 1983; Cracknell and Hayes, 1993; Jensen, 2005). Empirical evidence is presented in Table 1-1, including: 1) the organization of many specialized institutes and centers of excellence associated with remote sensing, 2) the organization of numerous professional societies devoted to remote sensing research, 3) the publication of numerous new scholarly remote sensing journals, 4) significant technological advancement such as improved sensor systems and methods of image analysis, and 5) intense self-examination (e.g., Dehqanzada and Florini, 2000). We may be approaching Stage 3 with increasing specialization and theoretical controversy. However, the rate of growth of remote sensing has not begun to decline. In fact, there has been a tremendous surge in the numbers of persons specializing in remote sensing and commercial firms using remote sensing during the 1990s and early 2000s (Davis, 1999; ASPRS, 2004). Significant improvements in the spatial resolution of satellite remote sensing (e.g., more useful 1 × 1 m panchromatic data) has brought even more social science GIS practitioners into the fold. Hundreds of new peer-reviewed remote sensing research articles are published every month.

Art: The process of visual photo or image interpretation brings to bear not only scientific knowledge, but all of the background that a person has obtained through his or her lifetime. Such learning cannot be measured, programmed, or completely understood. The synergism of combining scientific knowledge with real-world analyst experience allows the interpreter to develop heuristic rules of thumb to extract

Table 1-1.   Major milestones in remote sensing.

1600 and 1700s
1687 - Sir Isaac Newton's *Principia* summarizes basic laws of mechanics

1800s
1826 - Joseph Nicephore Niepce takes first photograph
1839 - Louis M. Daguerre invents positive print daguerrotype photography
1839 - William Henry Fox Talbot invents Calotype negative/positive process
1855 - James Clerk Maxwell postulates additive color theory
1858 - Gaspard Felix Tournachon takes aerial photograph from a balloon
1860s - James Clerk Maxwell puts forth electromagnetic wave theory
1867 - The term *photogrammetry* is used in a published work
1873 - Herman Vogel extends sensitivity of emulsion dyes to longer wavelengths, paving the way for near-infrared photography

1900
1903 - Airplane invented by Wright Brothers (Dec 17)
1903 - Alfred Maul patents a camera to obtain photographs from a rocket

1910s
1910 - International Society for Photogrammetry (ISP) founded in Austria
1913 - First International Congress of ISP in Vienna
1914 to 1918 - World War I photo-reconnaissance

1920s
1920 to 1930 - Increase civilian photointerpretation and photogrammetry
1926 - Robert Goddard launches liquid-powered rocket (Mar 16)

1930s
1934 - American Society of Photogrammetry (ASP) founded
1934 - *Photogrammetric Engineering* (ASP)
1938 - *Photogrammetria* (ISP)
1939 to 1945 - World War II photo-reconnaissance advances

1940s
1940s - RADAR invented
1940s - Jet aircraft invented by Germany
1942 - Kodak patents first false-color infrared film
1942 - Launch of German V-2 rocket by Wernher VonBraun (Oct 3)

1950s
1950s - Thermal infrared remote sensing invented by military
1950 - 1953 Korean War aerial reconnaissance
1953 - *Photogrammetric Record* (Photogrammetric Society, U.K.)
1954 - Westinghouse, Inc. develops side-looking airborne radar system
1955 to 1956 - U.S. Genetrix balloon reconnaissance program
1956 to 1960 - Central Intelligence Agency U-2 aerial reconnaissance program
1957 - Soviet Union launched *Sputnik* satellite (Oct 4)
1958 - United States launched *Explorer 1* satellite (Jan 31)

1960s
1960s - Emphasis primarily on visual image processing
1960s - Michigan Willow Run Laboratory active — evolved into ERIM
1960s - First *International Symposium on Remote Sensing of Environment* at Ann Arbor, MI
1960s - Purdue Laboratory for Agricultural Remote Sensing (LARS) active
1960s - Forestry Remote Sensing Lab at U.C. Berkeley (Robert Colwell)
1960s - ITC– Delft initiates photogrammetric education for foreign students
1960s - Digital image processing initiated at LARS, Berkeley, Kansas, ERIM
1960s - Declassification of radar and thermal infrared sensor systems
1960 - 1972 United States COEONA spy satellite program
1960 - *Manual of Photo-interpretation* (ASP)
1960 - Term *remote sensing* introduced by Evelyn Pruitt and other U. S. Office of Naval Research personnel
1961 - Yuri Gagarin becomes first human to travel in space
1961 - 1963 Mercury space program
1962 - Cuban Missile Crisis — U-2 photo-reconnaissance shown to the public
1964 - SR-71 discussed in President Lyndon Johnson press briefing
1965 to 1966 - Gemini space program
1965 - *ISPRS Journal of Photogrammetry & Remote Sensing*
1969 - *Remote Sensing of Environment* (Elsevier)

Table 1-1. continued

1970s
1970s, 80s - Possible to specialize in remote sensing at universities
1970s - Digital image processing comes of age
1970s - Remote sensing integrated with geographic information systems
1972 - ERTS-1 (Earth Resource Technology Satellite) launched (NASA)
1973 - 1979 Skylab program (NASA)
1973 - *Canadian Journal of Remote Sensing* (Canadian RS Society)
1975 - ERTS-2 launched (renamed Landsat 2)
1975 - *Manual of Remote Sensing* (ASP)
1977 - METEOSAT-1 launched (European Space Agency)
1978 - Landsat 3 launched (NASA)
1978 - Nimbus 7 launched - Coastal Zone Color Scanner
1978 - TIROS-N launched with AVHRR sensor
1978 - SEASAT launched (NASA Jet Propulsion Laboratory)

1980s
1980s - AAG Remote Sensing Specialty Group > 500 members
1980s - Commercialization attempted and failed – EOSAT, Inc.
1980 - ISP becomes Intl. Soc. for Photogrammetry & Remote Sensing
1980 - *International Journal of Remote Sensing* (Remote Sensing Society)
1980 - European Space Agency (ESA) created (Oct 30)
1980 - *IEEE Transactions Geoscience and Remote Sensing* (GRSS Society)
1981 - First *International Geoscience and Remote Sensing Symposium*
1981 - NASA Space Shuttle program initiated (STS-1)
1981 - Space Shuttle Imaging Radar (SIR - A) launched (NASA)
1982 - Landsat 4 - Thematic Mapper and MSS launched (NASA)
1983 - *Manual of Remote Sensing*, 2nd Ed. (ASP)
1984 - Landsat 5 - Thematic Mapper launched (NASA)
1984 - Space Shuttle Imaging Radar (SIR-B) launched (NASA)
1986 - SPOT 1 launched (SPOT Image, Inc.)
1986 - *Geocarto International* initiated (Geocarto International Center)
1989 - *The Earth Observer* (NASA Goddard Space Flight Center)

1990s
1990s - Digital soft-copy photogrammetry comes of age
1990s - University degree programs in remote sensing available
1990s - Light Detection and Ranging (LIDAR) comes of age
1990s - NASA assists commercial remote sensing (Stennis Space Center)
1990s - Increased use of hyperspectral and LIDAR sensors
1990 - SPOT 2 launched (Spot Image, Inc.)
1991 - "Mission to Planet Earth" initiated (NASA)
1991 - ERS-1 launched (European Space Agency)
1992 - *U.S. Land Remote Sensing Policy Act* becomes law
1993 - Landsat 6 does not achieve orbit (EOSAT, Inc.)
1993 - SPOT 3 launched (SPOT Image, Inc).
1993 - Space Shuttle Imaging Radar (SIR-C) (NASA)
1995 - RADARSAT-1 launched (Canadian)
1995 - ERS-2 launched (European Space Agency)
1995 - IRS-1C launched (5×5 m) (Indian Remote Sensing Program)
1995 - CORONA imagery declassified, transferred to National Archives
1996 - *Manual of Photographic Interpretation*, 2nd Ed. (ASPRS)
1997 - Earlybird does not achieve orbit (EarthWatch, Inc.)
1998 - Mission to Planet Earth becomes Earth Science Enterprise (NASA)
1998 - *Manual of Remote Sensing - Radar* (ASPRS)
1998 - SPOT 4 launched (SPOT Image, Inc.)
1999 - *Manual of Remote Sensing - Geosciences* (ASPRS)
1999 - Landsat 7 ETM$^+$ launched (NASA; April 15)
1999 - IKONOS does not achieve orbit (Space Imaging, Inc.; Apr 27)
1999 - IKONOS 2 launched (Space Imaging, Inc.; Sept 24)
1999 - *Terra* Earth observing system launched (NASA)
1999 - ImageSat launched (Israel ImageSat International)

2000 – 2006
2000 - New Millennium program initiated (NASA)
2001 - QuickBird launched (DigitalGlobe, Inc.)
2002 - *Aqua* Earth observing system launched (NASA)
2002 - ENVISAT launched (European Space Agency)
2002 - Object-oriented image segmentation algorithms introduced
2003 - OrbView-3 launched (ORBIMAGE, Inc.)
2004 - *GIScience & Remote Sensing* (Bellwether Publishing, Inc.)
2004 - *Manual of Photogrammetry*, 5th Ed. (ASPRS)
2005 - *Google Earth* serves DigitalGlobe and Landsat TM data
2006 - ORBIMAGE purchases Space Imaging and changes name to GeoEye

valuable information from the imagery. It is a fact that some image analysts are superior to other image analysts because they: 1) understand the scientific principles better, 2) are more widely traveled and have seen many landscape objects and geographic areas, and/or 3) they can synthesize scientific principles and real-world knowledge to reach logical and correct conclusions. Thus, remote sensing image interpretation is both an art and a science.

Information About an Object or Area

Sensors can obtain very specific information about an object (e.g., the diameter of an oak tree crown) or the geographic extent of a phenomenon (e.g., the polygonal boundary of an entire oak forest). The electromagnetic energy emitted or reflected from an object or geographic area is used as a surrogate for the actual property under investigation. The electromagnetic energy measurements must be turned into information using visual and/or digital image processing techniques.

The Instrument (Sensor)

Remote sensing is performed using an instrument, often referred to as a *sensor.* The majority of remote sensing instruments record EMR that travels at a velocity of 3×10^8 m s^{-1} from the source, directly through the vacuum of space or indirectly by reflection or reradiation to the sensor. The EMR represents a very efficient high-speed communications link between the sensor and the remote phenomenon. In fact, we know of nothing that travels faster than the speed of light. Changes in the amount and properties of the EMR become, upon detection by the sensor, a valuable source of data for interpreting important properties of the phenomenon (e.g., temperature, color). Other types of force fields may be used in place of EMR, such as acoustic (sonar) waves (e.g., Dartnell and Gardner, 2004). However, the majority of remotely sensed data collected for Earth resource applications is the result of sensors that record electromagnetic energy.

Distance: How Far Is Remote?

Remote sensing occurs at a distance from the object or area of interest. Interestingly, there is no clear distinction about how great this distance should be. The intervening distance could be 1 cm, 1 m, 100 m, or more than 1 million m from the object or area of interest. Much of astronomy is based on remote sensing. In fact, many of the most innovative remote sensing systems and visual and digital image processing methods were originally developed for remote sensing extraterrestrial landscapes such as the moon, Mars, Io, Saturn, Jupiter, etc. This text, however, is concerned primarily with remote sensing of the terrestrial Earth, using sensors that are placed on suborbital air-breathing aircraft or orbital satellite platforms placed in the vacuum of space.

Remote sensing and digital image processing techniques can also be used to analyze inner space. For example, an electron microscope can be used to obtain photographs of extremely small objects on the skin, in the eye, etc. An x-ray instrument is a remote sensing system where the skin and muscle are like the atmosphere that must be penetrated, and the interior bone or other matter is the object of interest.

Remote Sensing Advantages and Limitations

Remote sensing has several unique advantages as well as some limitations.

Advantages

Remote sensing is *unobtrusive* if the sensor is passively recording the electromagnetic energy reflected from or emitted by the phenomenon of interest. This is a very important consideration, as *passive remote sensing* does not disturb the object or area of interest.

Remote sensing devices are programmed to collect data systematically, such as within a single 9×9 in. frame of vertical aerial photography or a matrix (raster) of Landsat 5 Thematic Mapper data. This systematic data collection can remove the sampling bias introduced in some *in situ* investigations (e.g., Karaska et al., 2004).

Remote sensing science is also different from cartography or GIS because these sciences rely on data obtained by others. Remote sensing science can provide fundamental, new scientific information. Under controlled conditions, remote sensing can provide fundamental biophysical information, including x,y location; z elevation or depth; biomass; temperature; and moisture content. In this sense, remote sensing science is much like surveying, providing fundamental information that other sciences can use when conducting scientific investigations. However, unlike much of surveying, the remotely sensed data can be obtained systematically over very large geographic areas rather than just single-point observations. In fact, remote sensing–derived information is now critical to the successful modeling of numerous natural (e.g., water-supply estimation; eutrophication studies; nonpoint source pollution) and cultural (e.g., land-use conversion at the urban fringe; water-demand estimation; population estimation) processes (Walsh et al., 1999; Stow et al., 2003; Nemani et al., 2003; Karaska et al., 2004). A good example is the digital elevation model that is so important in many spatially-distributed GIS models (Clarke, 2001). Digital elevation models are now produced mainly from stereoscopic imagery, light detection and ranging (LIDAR) (e.g., Maune, 2001; Hodgson et al., 2003b; 2005), radio detection and ranging (RADAR) measurements, or interferometric synthetic aperture radar (IFSAR) imagery.

Limitations

Remote sensing science has limitations. Perhaps the greatest limitation is that it is often oversold. *Remote sensing is not a panacea* that will provide all the information needed to conduct physical, biological, or social science research. It simply provides some spatial, spectral, and temporal information of value in a manner that we hope is efficient and economical.

Human beings select the most appropriate remote sensing system to collect the data, specify the various resolutions of the remote sensor data, calibrate the sensor, select the platform that will carry the sensor, determine when the data will be collected, and specify how the data are processed. Human method-produced error may be introduced as the remote sensing instrument and mission parameters are specified.

Powerful *active* remote sensor systems that emit their own electromagnetic radiation (e.g., LIDAR, RADAR, SONAR) can be intrusive and affect the phenomenon being investigated. Additional research is required to determine how intrusive these active sensors can be.

Remote sensing instruments may become uncalibrated, resulting in uncalibrated remote sensor data. Finally, remote sensor data may be expensive to collect and analyze. Hopefully, the information extracted from the remote sensor data justifies the expense. Interestingly, the greatest expense in a typical remote sensing investigation is for well-trained image analysts, not remote sensor data.

 ## The Remote Sensing Process

Scientists have been developing procedures for collecting and analyzing remotely sensed data for more than 150 years. The first photograph from an aerial platform (a tethered balloon) was obtained in 1858 by the Frenchman Gaspard Felix Tournachon (who called himself Nadar). Significant strides in aerial photography and other remote sensing data collection took place during World War I and II, the Korean Conflict, the Cuban Missile Crisis, the Vietnam War, the Gulf

War, the war in Bosnia, and the war on terrorism. Many of the accomplishments are summarized in Table 1-1 and in Chapter 3 (History of Aerial Photography and Aerial Platforms). Basically, military contracts to commercial companies resulted in the development of sophisticated electro-optical multispectral remote sensing systems and thermal infrared and microwave (radar) sensor systems whose characteristics are summarized in Chapters 7, 8, and 9, respectively. While the majority of the remote sensing systems may have been initially developed for military reconnaissance applications, the systems are also heavily used for monitoring the Earth's natural resources.

The remote sensing data-collection and analysis procedures used for Earth resource applications are often implemented in a systematic fashion that can be termed the *remote sensing process*. The procedures in the remote sensing process are summarized here and in Figure 1-5:

- The hypothesis to be tested is defined using a specific type of logic (e.g., inductive, deductive) and an appropriate processing model (e.g., deterministic, stochastic).

- *In situ* and collateral data necessary to calibrate the remote sensor data and/or judge its geometric, radiometric, and thematic characteristics are collected.

- Remote sensor data are collected passively or actively using analog or digital remote sensing instruments, ideally at the same time as the *in situ* data.

- *In situ* and remotely sensed data are processed using a) analog image processing, b) digital image processing, c) modeling, and d) *n*-dimensional visualization.

- Metadata, processing lineage, and the accuracy of the information are provided and the results communicated using images, graphs, statistical tables, GIS databases, Spatial Decision Support Systems (SDSS), etc.

It is useful to review the characteristics of these remote sensing process procedures.

Statement of the Problem

Sometimes the general public and even children look at aerial photography or other remote sensor data and extract useful information. They typically do this without a formal hypothesis in mind. More often than not, however, they interpret the imagery incorrectly because they do not understand the nature of the remote sensing system used to collect

The Remote Sensing Process

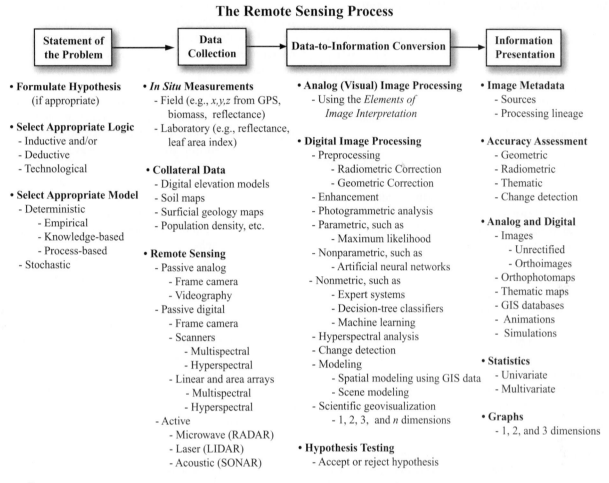

Figure 1-5   Scientists generally use the remote sensing process when extracting information from remotely sensed data.

the data or appreciate the vertical or oblique perspective of the terrain recorded in the imagery.

Scientists who use remote sensing, on the other hand, are usually trained in the *scientific method*—a way of thinking about problems and solving them. They use a formal plan that has at least five elements: 1) stating the problem, 2) forming the research hypothesis (i.e., a possible explanation), 3) observing and experimenting, 4) interpreting data, and 5) drawing conclusions. It is not necessary to follow this formal plan exactly.

The scientific method is normally used in conjunction with environmental models that are based on two primary types of logic:

• deductive logic

• inductive logic

Models based on deductive and/or inductive logic can be further subdivided according to whether they are processed *deterministically* or *stochastically* (Jensen, 2005). Some scientists extract new thematic information directly from remotely sensed imagery without ever explicitly using inductive or deductive logic. They are just interested in extracting information from the imagery using appropriate methods and technology. This *technological* approach is not as rigorous, but it is common in *applied remote sensing*. The approach can also generate new knowledge.

Remote sensing is used in both scientific (inductive and deductive) and technological approaches to obtain knowledge. There is debate as to how the different types of logic used in the remote sensing process yield new scientific knowledge (e.g., Fussell et al., 1986; Curran, 1987; Fisher and Lindenberg, 1989; Dobson, 1993; Skidmore, 2002).

Identification of In situ and Remote Sensing Data Requirements

If a hypothesis is formulated using inductive and/or deductive logic, a list of variables or observations are identified that will be used during the investigation. *In situ* observation and/or remote sensing may be used to collect information on the most important variables.

Scientists using remote sensing technology should be well trained in field and laboratory data-collection procedures. For example, if a scientist wants to map the surface temperature of a lake, it is usually necessary to collect some accurate empirical *in situ* lake-temperature measurements at the same time the remote sensor data are collected. The *in situ* observations may be used to 1) calibrate the remote sensor data, and/or 2) perform an unbiased accuracy assessment of the final results (Congalton and Green, 1998). Remote sensing textbooks provide some information on field and laboratory sampling techniques. The *in situ* sampling procedures, however, are learned best through formal courses in the sciences (e.g., chemistry, biology, forestry, soils, hydrology, meteorology). It is also important to know how to collect accurately socioeconomic and demographic information in urban environments based on training in human geography, sociology, etc.

Most *in situ* data are now collected in conjunction with global positioning system (GPS) *x, y, z* data (Jensen and Cowen, 1999). Scientists should know how to collect the GPS data at each *in situ* data-collection station and how to perform differential correction to obtain accurate *x, y, z* coordinates (Rizos, 2002).

Collateral Data Requirements

Many times *collateral* data (often called *ancillary* data), such as digital elevation models, soil maps, geology maps, political boundary files, and block population statistics, are of value in the remote sensing process. Ideally, the spatial collateral data reside in a GIS (Clarke, 2001).

Remote Sensing Data Requirements

Once we have a list of variables, it is useful to determine which of them can be remotely sensed. Remote sensing can provide information on two different classes of variables: *biophysical* and *hybrid*.

Biophysical Variables: Some biophysical variables can be measured directly by a remote sensing system. This means that the remotely sensed data can provide fundamental biological and/or physical (*biophysical*) information directly, generally without having to use other surrogate or ancillary data. For example, a thermal infrared remote sensing system can record the apparent temperature of a rock outcrop by measuring the radiant energy exiting its surface. Similarly, it is possible to conduct remote sensing in a very specific region of the spectrum and identify the amount of water vapor in the atmosphere. It is also possible to measure soil moisture content directly using microwave remote sensing techniques (Engman, 2000). NASA's Moderate Resolution Imaging Spectrometer (MODIS) can be used to measure absorbed photosynthetically active radiation (APAR) and leaf area index (LAI). The precise *x, y* location, and height (*z*) of an object can be extracted directly from stereoscopic aerial photography, overlapping satellite imagery (e.g., SPOT), light detection and ranging (LIDAR) data, or interferometric synthetic aperture radar (IFSAR) imagery.

Table 1-2 is a list of selected biophysical variables that can be remotely sensed and useful sensors to acquire the data. Characteristics of many of these remote sensing systems are discussed in Chapters 4, 7, 8, and 9. Great strides have been made in remotely sensing many of these biophysical variables. They are important to the national and international effort under way to model the global environment (Jensen et al., 2002; Asrar, 2004).

Hybrid Variables: The second general group of variables that can be remotely sensed include *hybrid* variables, created by systematically analyzing more than one biophysical variable. For example, by remotely sensing a plant's chlorophyll absorption characteristics, temperature, and moisture content, it might be possible to model these data to detect vegetation stress, a hybrid variable. The variety of hybrid variables is large; consequently, no attempt is made to identify them. It is important to point out, however, that nominal-scale land use and land cover are hybrid variables. For example, the land cover of a particular area on an image may be derived by evaluating several of the fundamental biophysical variables at one time [e.g., object location (*x, y*), height (*z*), tone and/or color, biomass, and perhaps temperature]. So much attention has been placed on remotely sensing this hybrid *nominal*-scale variable that the *interval*- or *ratio*-scaled biophysical variables were largely neglected until the mid-1980s. Nominal-scale land-use and land-cover mapping are important capabilities of remote sensing technology and should not be minimized. Many social and physical scientists routinely use such data in their research. However, there is now a dramatic increase in the extraction of interval- and ratio-scaled biophysical data that are incor-

Table 1-2.   Selected biophysical and hybrid variables and potential remote sensing systems used to obtain the information.

Biophysical Variables	Potential Remote Sensing Systems
***x, y, z* Geodetic Control** ***x, y* Location from Orthocorrected Imagery**	- Global Positioning Systems (GPS) - Analog and digital stereoscopic aerial photography, Space Imaging IKONOS, DigitalGlobe QuickBird, Orbimage OrbView-3, French SPOT HRV, Landsat (Thematic Mapper, Enhanced TM$^+$), Indian IRS-1CD, European ERS-1 and 2 microwave and ENVISAT MERIS, MODIS (Moderate Resolution Imaging Spectrometer), LIDAR, Canadian RADARSAT 1 and 2
***z* Topography/Bathymetry** - Digital Elevation Model (DEM) - Digital Bathymetric Model (DBM)	 - GPS, stereoscopic aerial photography, LIDAR, SPOT, RADARSAT, IKONOS, QuickBird, OrbView-3, Shuttle Radar Topography Mission (SRTM), Interferometric Synthetic Aperture Radar (IFSAR) - SONAR, bathymetric LIDAR, stereoscopic aerial photography
Vegetation - Pigments (e.g., chlorophyll *a* and *b*) - Canopy structure and height - Biomass derived from vegetation indices - Leaf area index (LAI) - Absorbed photosynthetically active radiation - Evapotranspiration	 - Color aerial photography, Landsat ETM$^+$, IKONOS, QuickBird, OrbView-3, Orbimage SeaWiFS, Advanced Spaceborne Thermal Emission and Reflection Radiometer (ASTER), Moderate Resolution Imaging Spectrometer (MODIS), ENVISAT, airborne hyperspectral (e.g., AVIRIS, HyMap, CASI) - Stereoscopic aerial photography, LIDAR, RADARSAT, IFSAR - Color-infrared (CIR) aerial photography, Landsat (TM, ETM$^+$), IKONOS, QuickBird, OrbView-3, Advanced Very High Resolution Radiometer (AVHRR), Multiangle Imaging Spectroradiometer (MISR), airborne hyperspectral systems (e.g., AVIRIS, HyMap, CASI)
Surface Temperature (land, water, atmosphere)	- ASTER, AVHRR, GOES, Hyperion, MISR, MODIS, SeaWiFS, airborne thermal infrared
Soil and Rocks - Moisture - Mineral composition - Taxonomy - Hydrothermal alteration	 - ASTER, passive microwave (SSM/1), RADARSAT, MISR, ALMAZ, Landsat (TM, ETM$^+$), ERS-1 and 2, Intermap Star 3*i* - ASTER, MODIS, hyperspectral systems (e.g., AVIRIS, HyMap, CASI) - High-resolution color and CIR aerial photography, airborne hyperspectral systems (e.g., AVIRIS, HyMap, CASI) - Landsat (TM, ETM$^+$), ASTER, MODIS, airborne hyperspectral (e.g., AVIRIS, HyMap, CASI)
Surface Roughness	- Aerial photography, ALMAZ, ERS-1 and 2, RADARSAT, Intermap Star 3*i*, IKONOS, QuickBird, ASTER, ENVISAT ASAR
Atmosphere - Aerosols (e.g., optical depth) - Clouds (e.g., fraction, optical thickness) - Precipitation - Water vapor (precipitable water) - Ozone	 - MISR, GOES, AVHRR, MODIS, CERES, MOPITT, MERIS - GOES, AVHRR, MODIS, MISR, CERES, MOPITT, UARS, MERIS - Tropical Rainfall Measurement Mission (TRMM), GOES, AVHRR, SSM/1, MERIS - GOES, MODIS, MERIS - MODIS
Water - Color - Surface hydrology - Suspended minerals - Chlorophyll/gelbstoffe - Dissolved organic matter	 - Color and CIR aerial photography, Landsat (TM, ETM$^+$), SPOT, IKONOS, QuickBird, OrbView-3, ASTER, SeaWiFS, MODIS, airborne hyperspectral systems (e.g., AVIRIS, HyMap, CASI), AVHRR, GOES, bathymetric LIDAR, MISR, CERES, Hyperion, TOPEX/POSEIDON, MERIS

Table 1-2.   Selected biophysical and hybrid variables and potential remote sensing systems used to obtain the information.

Biophysical Variables	Potential Remote Sensing Systems
Snow and Sea Ice - Extent and characteristics	- Color and CIR aerial photography, AVHRR, GOES, Landsat (TM, ETM+), SPOT, SeaWiFS, IKONOS, QuickBird, ASTER, MODIS, MERIS, ERS-1 and 2, RADARSAT
Volcanic Effects - Temperature, gases	- ASTER, MISR, Hyperion, MODIS, airborne hyperspectral systems
BRDF (bidirectional reflectance distribution function)	- MISR, MODIS, CERES
Selected Hybrid Variables	**Potential Remote Sensing Systems**
Land Use - Commercial, residential, transportation, etc. - Cadastral (property) - Tax mapping	- Very high spatial resolution panchromatic, color and /or CIR stereoscopic aerial photography, high spatial resolution satellite imagery (<1 x 1 m: IKONOS, QuickBird, OrbView-3), SPOT (2.5 m), LIDAR, high spatial resolution hyperspectral systems (e.g., AVIRIS, HyMap, CASI)
Land Cover - Agriculture, forest, urban, etc.	- Color and CIR aerial photography, Landsat (MSS, TM, ETM+), SPOT, ASTER, AVHRR, RADARSAT, IKONOS, QuickBird, OrbView-3, LIDAR, IFSAR, SeaWiFS, MODIS, MISR, MERIS, hyperspectral systems (e.g., AVIRIS, HyMap, CASI)
Vegetation - stress	- Color and CIR aerial photography, Landsat (TM, ETM+), IKONOS, QuickBird, OrbView-3, AVHRR, SeaWiFS, MISR, MODIS, ASTER, MERIS, airborne hyperspectral systems (e.g., AVIRIS, HyMap, CASI)

porated into quantitative models that can accept spatially distributed information.

Remote Sensing Data Collection

Remotely sensed data are collected using passive or active remote sensing systems. *Passive* sensors record electromagnetic radiation that is reflected or emitted from the terrain (Shippert, 2004). For example, cameras and video recorders can be used to record visible and near-infrared energy reflected from the terrain. A multispectral scanner can be used to record the amount of thermal radiant flux exiting the terrain. *Active* sensors such as microwave (RADAR), LIDAR, or SONAR bathe the terrain in machine-made electromagnetic energy and then record the amount of radiant flux scattered back toward the sensor system.

Remote sensing systems collect analog (e.g., hard-copy aerial photography or video data) and/or digital data [e.g., a matrix (raster) of brightness values obtained using a scanner, linear array, or area array]. A selected list of some of the most important remote sensing systems is presented in Table 1-3.

The amount of electromagnetic radiance, L (watts m^{-2} sr^{-1}; watts per meter squared per steradian), recorded within the IFOV of an optical remote sensing system (e.g., a picture element in a digital image), is a function of:

$$L = f(\lambda, s_{x,y,z}, t, \theta, P, \Omega) \qquad (1\text{-}1)$$

where

λ = wavelength (spectral response measured in various bands or at specific frequencies). Wavelength (λ) and frequency (υ) may be used interchangeably based on their relationship with the speed of light (c) where $c = \lambda \times \upsilon$.

$s_{x,y,z}$ = x, y, z location of the pixel and its size (x, y);

Table 1-3. Selected remote sensing systems and their characteristics.

Remote Sensing Systems	Blue	Green	Red	Near-infrared	Middle-infrared (SWIR)	Thermal infrared	Micro-wave	Spatial (m)	Temporal (days)
Suborbital Sensors									
Panchromatic film (black & white)		0.5 ——— 0.7 μm						Variable	Variable
Color film	0.4 ———— 0.7 μm							Variable	Variable
Color-infrared film		0.5 ———— 0.9 μm						Variable	Variable
Digital Frame Cameras (CCD)	1	1	1	1				0.25 – 5	Variable
CASI - 1500	0.40 ——288 bands——1.0 μm							variable	Variable
AVIRIS - Airborne Visible Infrared Imaging Spectrometer	0.40 ———224 bands———2.5 μm							2.5 or 20	Variable
Intermap Star-3*i* X-band radar							1	Variable	Variable
Satellite Sensors									
NOAA-9 AVHRR LAC			1	1		3		1100	14.5/day
NOAA- K, L, M			1	1	2	2		1100	14.5/day
Landsat Multispectral Scanner (MSS)		1	1	2				79	16 – 18
Landsat 4 and 5 Thematic Mappers (TM)	1	1	1	1	2	1		30 and 120	16
Landsat 7 Enhanced TM (ETM⁺) — Multispectral	1	1	1	1	2	1		30 and 60	16
— Panchromatic		0.52 ———— 0.9 μm						15	16
SPOT 4 HRV — Multispectral		1	1	1				20	Pointable
— Panchromatic		0.51 ———— 0.73 μm						10	Pointable
GOES Series (East and West)		0.52 ———— 0.72 μm				4		700	0.5/hr
European Remote Sensing Satellite (ERS-1 and 2)	VV polarization C-band (5.3 GHz)						1	26 – 28	—
Canadian RADARSAT (several modes)	HH polarization C-band (5.3 GHz)						1	9 – 100	1 – 6 days
Shuttle Imaging Radar (SIR-C)							3	30	Variable
Sea-Viewing Wide Field-of-View Sensor (SeaWiFS)	3	2	1	2				1130	1
MODIS - Moderate Resolution Imaging Spectrometer	0.405 ————— 36 bands ————— 14.385 μm							250, 500, 1000	1 – 2
ASTER - Advanced Spaceborne Thermal Emission and Reflection Radiometer	0.52 — 3 bands — 0.86 μm			1.6 — 6 bands — 2.43 μm / 8.12 — 5 bands — 11.6 μm				15 / 30 / 90	5 / 16 / 16
MISR - Multiangle Imaging SpectroRadiometer	Nine CCD cameras in four bands (440, 550, 670, 860 nm)							275 and 1100	1 – 2
NASA Topex/Poseidon — TOPEX radar altimeter — POSEIDON single-frequency radiometer	(18, 21, 37 GHz) (13.65 GHz)							315,000	10
Space Imaging IKONOS — Multispectral	1	1	1	1				4	Pointable
— Panchromatic	0.45 ———— 0.9 μm							1	
Digital Globe QuickBird — Multispectral	1	1	1	1				2.44	Pointable
— Panchromatic	0.45 ———— 0.9 μm							0.61	

t = temporal information, i.e., when, how long, and how often the data were acquired;

θ = set of angles that describe the geometric relationships between the radiation source (e.g., the Sun), the terrain target of interest (e.g., a corn field), and the remote sensing system;

P = polarization of back-scattered energy recorded by the sensor; and

Ω = radiometric resolution (precision) at which the data (e.g., reflected, emitted, or back-scattered radiation) are recorded by the remote sensing system.

It is useful to briefly review characteristics of the parameters associated with Equation 1-1 and how they influence the nature of the remote sensing data collected.

Spectral Information and Resolution

Most remote sensing investigations are based on developing a deterministic relationship (i.e., a model) between the amount of electromagnetic energy reflected, emitted, or back-scattered in specific bands or frequencies and the chemical, biological, and physical characteristics of the phenomena under investigation (e.g., a corn field canopy). *Spectral resolution* is the number and dimension (size) of specific wavelength intervals (referred to as *bands* or *channels*) in the electromagnetic spectrum to which a remote sensing instrument is sensitive.

Multispectral remote sensing systems record energy in multiple bands of the electromagnetic spectrum. For example, in the 1970s and early 1980s, the Landsat Multispectral Scanners (MSS) recorded remotely sensed data of much of the Earth that is still of significant value for change detection studies. The bandwidths of the four MSS bands are displayed in Figure 1-6a (band 1 = 500 – 600 nm; band 2 = 600 – 700 nm; band 3 = 700 – 800 nm; and band 4 = 800 – 1,100 nm). The nominal size of a band may be large (i.e., coarse), as with the Landsat MSS near-infrared band 4 (800 – 1,100 nm) or relatively smaller (i.e., finer), as with the Landsat MSS band 3 (700 – 800 nm). Thus, Landsat MSS band 4 detectors recorded a relatively large range of reflected near-infrared radiant flux (300 nm wide) while the MSS band 3 detectors recorded a much reduced range of near-infrared radiant flux (100 nm wide).

The four multispectral bandwidths associated with the Positive Systems ADAR 5500 digital frame camera are shown

for comparative purposes (Figure 1-6a, c, and d). The camera's bandwidths were refined to record information in more specific regions of the spectrum (band 1 = 450 – 515 nm; band 2 = 525 – 605 nm; band 3 = 640 – 690 nm; and band 4 = 750 – 900 nm). There are gaps between the spectral sensitivities of the detectors. Note that this digital camera system is also sensitive to reflected blue wavelength energy.

The aforementioned terminology is typically used to describe a sensor's *nominal spectral resolution*. However, it is difficult to create a detector that has extremely sharp band-pass boundaries such as those shown in Figure 1-6a. Rather, the more precise method of stating bandwidth is to look at the typical Gaussian shape of the detector sensitivity, such as the example shown in Figure 1-6b. The analyst then determines the Full Width at Half Maximum (FWHM). In this hypothetical example, the Landsat MSS near-infrared band 3 under investigation is sensitive to energy between 700 and 800 nm.

A *hyperspectral* remote sensing instrument typically acquires data in hundreds of spectral bands (Goetz, 2002). For example, the Airborne Visible and Infrared Imaging Spectrometer (AVIRIS) has 224 bands in the region from 400 to 2,500 nm spaced just 10 nm apart based on the FWHM criteria (Clark, 1999; NASA, 2006). An AVIRIS hyperspectral datacube of a portion of the Savannah River Site near Aiken, SC, is shown in Figure 1-7. *Ultraspectral* remote sensing involves data collection in many hundreds of bands.

Certain regions or spectral bands of the electromagnetic spectrum are optimal for obtaining information on biophysical parameters. The bands are normally selected to maximize the contrast between the object of interest and its background (i.e., object-to-background contrast). Careful selection of the spectral bands might improve the probability that the desired information will be extracted from the remote sensor data.

Spatial Information and Resolution

Most remote sensing studies record the spatial attributes of objects on the terrain. For example, each silver halide crystal in an analog aerial photograph and each picture element in a digital remote sensor image is located at a specific location in the image and associated with specific *x,y* coordinates on the ground. Once rectified to a standard map projection, the spatial information associated with each silver halide crystal or pixel is of significant value because it allows the remote sensing–derived information to be used with other spatial

Spectral Resolution

a. Nominal spectral resolution of the Landsat Multispectral Scanner and Positive Systems ADAR 5500 digital frame camera.

b. Precise bandpass measurement of a detector based on Full Width at Half Maximum (FWHM) criteria.

c. Single band of ADAR 5500 data.

d. Multispectral remote sensing.

Figure 1-6   a) The spectral bandwidths of the four Landsat Multispectral Scanner (MSS) bands (green, red, and two near-infrared) compared with the bandwidths of an ADAR 5500 digital frame camera. b) The true spectral bandwidth is the width of the Gaussian-shaped spectral profile at Full Width at Half Maximum (FWHM) intensity (Clark, 1999). This example has a spectral bandwidth of 0.1 µm (100 nm) between 700 and 800 nm. c) If desired, it is possible to collect reflected energy in a single band of the electromagnetic spectrum (e.g., 750 – 900 nm). This is a 1 × 1 ft spatial resolution ADAR 5500 near-infrared image. d) Multispectral sensors collect data in multiple bands of the spectrum (images courtesy of Positive Systems, Inc.).

data in a GIS or spatial decision support system (Jensen et al., 2002).

There is a general relationship between the size of an object or area to be identified and the spatial resolution of the remote sensing system. *Spatial resolution* is a measure of the smallest angular or linear separation between two objects that can be resolved by the remote sensing system. The spatial resolution of aerial photography may be measured by 1) placing calibrated, parallel black and white lines on tarps that are placed in the field, 2) obtaining aerial photography of the study area, and 3) computing the number of resolvable *line pairs per millimeter* in the photography. It is also possible to determine the spatial resolution of imagery by com-

Airborne Visible Infrared Imaging Spectrometer (AVIRIS) Datacube of the Savannah River Site near Aiken, SC

Near-infrared image on top of the datacube is just one of 224 bands at 10 nm nominal bandwidth acquired on July 26, 1999.

Figure 1-7   Hyperspectral imagery of an area on the Savannah River Site, SC obtained by NASA's Airborne Visible/Infrared Imaging Spectrometer (AVIRIS). The nominal spatial resolution is 3.4 × 3.4 m. The atmosphere absorbs most of the electromagnetic energy near 1,400 and 1,900 nm, causing the dark bands in the hyperspectral datacube.

puting its modulation transfer function, which is beyond the scope of this text (Joseph, 2000).

Many satellite remote sensing systems use optics that have a constant instantaneous-field-of-view (IFOV) (Figure 1-2). Therefore, a sensor system's *nominal spatial resolution* is defined as the dimension in meters (or feet) of the ground-projected IFOV where the diameter of the circle (D) on the ground is a function of the instantaneous-field-of-view (β) times the altitude (H) of the sensor above ground level (AGL) (Figure 1-2):

$$D = \beta \times H. \tag{1-2}$$

Pixels are normally represented on computer screens and in hard-copy images as rectangles with length and width. Therefore, we typically describe a sensor system's nominal

spatial resolution as being 10 × 10 m or 30 × 30 m. For example, DigitalGlobe's QuickBird has a nominal spatial resolution of 61 × 61 cm for its panchromatic band and 2.44 × 2.44 m for the four multispectral bands. The Landsat 7 Enhanced Thematic Mapper Plus (ETM[+]) has a nominal spatial resolution of 15 × 15 m for its panchromatic band and 30 × 30 m for six of its multispectral bands. Generally, the smaller the nominal spatial resolution, the greater the spatial resolving power of the remote sensing system.

Figure 1-8 depicts digital camera imagery of an area in Mechanicsville, N.Y., at resolutions ranging from 0.5 × 0.5 m to 80 × 80 m. Note that there is not a significant difference in the interpretability of 0.5 × 0.5 m data, 1 × 1 m data, and even 2 × 2 m data. However, the urban information content decreases rapidly when using 5 × 5 m imagery and is practically useless for urban analysis at spatial resolutions larger than 10 × 10 m. This is the reason historical Landsat MSS data (79 × 79 m) are of little value for most urban applications (Jensen and Cowen, 1999; Jensen et al., 2002).

A useful heuristic rule of thumb is that in order to detect a feature, the nominal spatial resolution of the remote sensing system should be less than one-half the size of the feature measured in its smallest dimension. For example, if we want to identify the location of all maple trees in a park, the minimum acceptable spatial resolution would be approximately one-half the diameter of the smallest maple tree's crown. Even this spatial resolution, however, will not guarantee success if there is no difference between the spectral response of the maple tree (the object) and the soil or grass surrounding it (i.e., its background).

Some sensor systems, such as LIDAR, do not completely "map" the terrain surface. Rather, the surface is "sampled" using laser pulses sent from the aircraft at some nominal time interval (Raber et al., 2002). The ground-projected laser pulse may be very small (e.g., 10 – 15 cm in diameter) with samples located approximately every 1 to 6 m on the ground. Spatial resolution would appropriately describe the ground-projected laser pulse (e.g., 15 cm) but *sampling density* (i.e., number of points per unit area) describes the frequency of ground observations (Hodgson et al., 2005).

Because we have spatial information about the location of each pixel (x,y) in the image matrix, it is also possible to examine the spatial relationship between a pixel and its neighbors. Therefore, the amount of spectral autocorrelation and other spatial geostatistical measurements can be determined based on the spatial information inherent in the imagery (Walsh et al., 1999; Jensen, 2005).

Spatial Resolution

Figure 1-8   Imagery of residential housing near Mechanicsville, N.Y., obtained on June 1, 1998, at a nominal spatial resolution of 0.3 × 0.3 m (approximately 1 × 1 ft) using a digital camera (courtesy of Litton Emerge, Inc.). The original data were resampled to derive the imagery with the simulated spatial resolutions shown.

Temporal Information and Resolution

One of the valuable things about remote sensing science is that it obtains a record of Earth landscapes at a unique moment in time. Multiple records of the same landscape obtained through time can be used to identify processes at work and to make predictions.

The *temporal resolution* of a remote sensing system generally refers to how often the sensor records imagery of a particular area. The temporal resolution of the sensor system shown in Figure 1-9 is every 16 days. Ideally, the sensor obtains data repetitively to capture unique discriminating characteristics of the object under investigation (Haack et

al., 1997). For example, agricultural crops have unique phenological cycles in each geographic region (discussed in Chapter 11). To measure specific agricultural variables, it is necessary to acquire remotely sensed data at critical dates in the phenological cycle (Johannsen et al., 2003). Analysis of multiple-date imagery provides information on how the variables are changing through time. Change information provides insight into processes influencing the development of the crop (Jensen et al., 2002). Fortunately, several satellite sensor systems such as SPOT, IKONOS, ImageSat and QuickBird are pointable, meaning that they can acquire imagery off-nadir. *Nadir* is the point directly below the spacecraft. This dramatically increases the probability that imagery will be obtained during a growing season or during

Temporal Resolution

Remote Sensor Data Acquisition

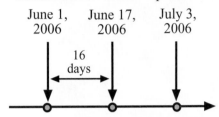

Figure 1-9   The temporal resolution of a remote sensing system refers to how often it records imagery of a particular area. This example depicts the systematic collection of data every 16 days, presumably at approximately the same time of day. Landsat Thematic Mappers 4 and 5 had 16-day revisit cycles.

an emergency. However, off-nadir oblique viewing also introduces bidirectional reflectance distribution function (BRDF) issues, discussed in the next section.

There are often trade-offs associated with the various resolutions that must be made when collecting remote sensing data (Figure 1-10; Color Plate 1-1). Generally, the higher the temporal resolution requirement (e.g., monitoring hurricanes every half-hour), the lower the spatial resolution requirement (e.g., the NOAA GOES weather satellite records images with 4×4 to 8×8 km pixels). Conversely, the higher the spatial resolution requirement (e.g., monitoring urban land-use with 1×1 m data), the lower the temporal resolution requirement (e.g., every 1 to 10 years). For example, Figure 1-11 documents significant residential and commercial land-use development for an area near Atlanta, GA, using high spatial resolution (1×1 m) aerial photography obtained in 1993 and 1999. Some applications such as crop type or yield estimation might require relatively high temporal resolution data (e.g., multiple images obtained during a growing season) and moderate spatial resolution data (e.g., 250×250 m pixels). Emergency response applications may require very high spatial and temporal resolution data collection that generates tremendous amounts of data.

Another aspect of temporal information is how many observations are recorded from a single pulse of energy that is directed at the Earth by an active sensor such as LIDAR. For example, most LIDAR sensors emit one pulse of laser energy and record multiple responses from this pulse. Measuring the time differences between multiple responses allows for the determination of object heights and terrain structure. Also, the length of time required to emit an energy

signal by an active sensor is referred to as the *pulse length*. Short pulse lengths allow precise distance (i.e., range) measurement.

Radiometric Information and Resolution

Some remote sensing systems record the reflected, emitted, or back-scattered electromagnetic radiation with more precision than other sensing systems. This is analogous to making a measurement with a ruler. If you want precisely to measure the length of an object, would you rather use a ruler with 16 or 1,024 subdivisions on it?

Radiometric resolution is defined as the sensitivity of a remote sensing detector to differences in signal strength as it records the radiant flux reflected, emitted, or back-scattered from the terrain. It defines the number of just discriminable signal levels. Therefore, radiometric resolution can have a significant impact on our ability to measure the properties of scene objects. The Landsat 1 Multispectral Scanner launched in 1972 recorded reflected energy with a precision of 6-bits (values ranging from 0 to 63). Landsat 4 and 5 Thematic Mapper sensors launched in 1982 and 1984, respectively, recorded data in 8 bits (values from 0 to 255) (Figure 1-12). Thus, the Landsat TM sensors had improved radiometric resolution (sensitivity) when compared with the original Landsat MSS. QuickBird and IKONOS sensors record information in 11 bits (values from 0 to 2,047). Several new sensor systems have 12-bit radiometric resolution (values ranging from 0 to 4,095). Radiometric resolution is sometimes referred to as the level of *quantization*. High radiometric resolution generally increases the probability that phenomena will be remotely sensed more accurately.

Polarization Information

The polarization characteristics of electromagnetic energy recorded by a remote sensing system are an important variable that can be used in many Earth resource investigations (Curran et al., 1998). Sunlight is polarized weakly. However, when sunlight strikes a nonmetal object (e.g., grass, forest, or concrete) it becomes depolarized and the incident energy is scattered differentially. Generally, the more smooth the surface, the greater the polarization. It is possible to use polarizing filters on passive remote sensing systems (e.g., aerial cameras) to record polarized light at various angles. It is also possible to selectively send and receive polarized energy using active remote sensing systems such as RADAR (e.g., horizontal send, vertical receive - HV; vertical send, horizontal receive - VH; vertical send, vertical receive - VV; horizontal send, horizontal receive - HH). Multiple-polarized

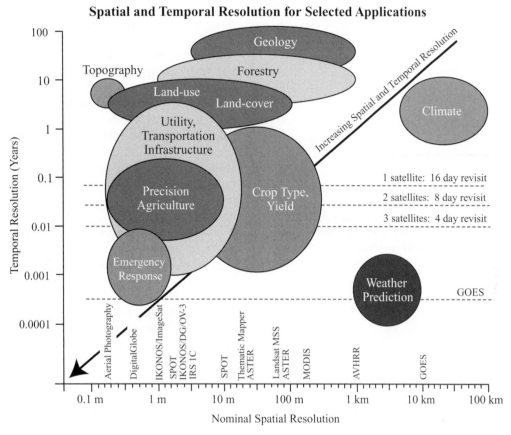

Figure 1-10   There are spatial and temporal resolution considerations that must be made for certain applications (Color Plate 1-1). A more detailed breakdown of the spatial and temporal requirements for urban applications is found in Chapter 13.

Digital Orthophotos of an Area near Atlanta, GA

a. 1993 orthophoto.   b. 1999 orthophoto.

Figure 1-11   Portions of digital-orthophoto-quarter-quads (DOQQ) of an area near Atlanta, GA. These data reside in the Georgia Spatial Data Infrastructure database and are useful for monitoring land-use change through time and the process of urbanization.

Radiometric Resolution

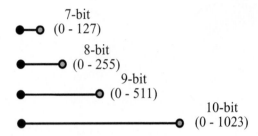

7-bit
(0 - 127)

8-bit
(0 - 255)

9-bit
(0 - 511)

10-bit
(0 - 1023)

Figure 1-12   The radiometric resolution of a remote sensing system is the sensitivity of its detectors to differences in signal strength as they record the radiant flux reflected, emitted, or back-scattered from the terrain. The energy is normally quantized during an analog-to-digital (A-to-D) conversion process to 8, 9, 10 bits or more.

RADAR imagery is an especially useful application of polarized energy.

Angular Information

Remote sensing systems record very specific angular characteristics associated with each exposed silver halide crystal or pixel (Barnsley, 1999). The angular characteristics are a function of (Figure 1-13a):

• the location in a three-dimensional sphere of the illumination source (e.g., the Sun for a passive system or the sensor itself in the case of RADAR, LIDAR, and SONAR) and its associated azimuth and zenith angles,

• the orientation of the terrain facet (pixel) or terrain cover (e.g., vegetation) under investigation, and

• the location of the suborbital or orbital remote sensing system and its associated azimuth and zenith angles.

There is always an angle of incidence associated with the incoming energy that illuminates the terrain and an angle of exitance from the terrain to the sensor system. This *bidirectional* nature of remote sensing data collection is known to influence the spectral and polarization characteristics of the at-sensor radiance, L, recorded by the remote sensing system.

A *goniometer* can be used to document the changes in at-sensor radiance, L, caused by changing the position of the sensor and/or the source of the illumination (e.g., the Sun) (Figure 1-13b). For example, Figure 1-13c presents three-

dimensional plots of smooth cordgrass (*Spartina alterniflora*) BRDF data collected at 8 a.m., 9 a.m., 12 p.m., and 4 p.m. on March 21, 2000, for band 624.20 nm. The only thing that changed between observations was the Sun's azimuth and zenith angles. The azimuth and zenith angles of the spectroradiometer were held constant while viewing the smooth cordgrass. Ideally, the BRDF plots would be identical, suggesting that it does not matter what time of day we collect the remote sensor data because the spectral reflectance characteristics from the smooth cordgrass remain constant. It is clear that this is not the case and that the time of day influences the spectral response. The Multiangle Imaging Spectrometer (MISR) onboard the *Terra* satellite was designed to investigate the BRDF phenomena. Research continues on how to incorporate the BRDF information into the digital image processing system to improve our understanding of what is recorded in the remotely sensed imagery (Sandmeier, 2000; Schill et al., 2004).

Angular information is central to the use of remote sensor data in photogrammetric applications. Stereoscopic image analysis is based on the assumption that an object on the terrain is remotely sensed from two angles. Viewing the same terrain from two vantage points introduces stereoscopic parallax, which is the foundation for all stereoscopic photogrammetric and radargrammetric analysis (Light and Jensen, 2002).

Suborbital (Airborne) Remote Sensing Systems

High-quality photogrammetric cameras mounted onboard aircraft continue to provide aerial photography for many Earth resource applications. For example, the U.S. Geological Survey's National Aerial Photography Program (NAPP) systematically collected 1:40,000-scale black-and-white or color-infrared aerial photography of much of the United States every 5 to 10 years. Some of these photogrammetric data are now being collected using digital frame cameras. In addition, sophisticated remote sensing systems are routinely mounted on aircraft to provide high spatial and spectral resolution remotely sensed data. Examples include hyperspectral sensors such as NASA's AVIRIS, the Canadian Airborne Imaging Spectrometer (CASI), and the Australian HyMap hyperspectral system. These sensors can collect data on demand when disaster strikes (e.g., oil spills or floods) if cloud-cover conditions permit. There are also numerous radars, such as Intermap's Star-3*i*, that can be flown on aircraft day and night and in inclement weather. Unfortunately, suborbital remote sensor data are usually expensive to acquire per km². Also, atmospheric turbulence can cause the data to have severe geometric distortions that can be difficult to correct.

Bidirectional Reflectance Distribution Function

a. Angular relationships.

b. Sandmeier Field Goniometer.

Band 624.20 nm   8:00 a.m. θ_i 20.4° φ_i 103.0°

Band 624.20 nm   9:00 a.m. θ_i 32.3° φ_i 113.0°

Band 624.20 nm   12:00 p.m. θ_i 57.1° φ_i 169.0°

Band 624.20 nm   4:00 p.m. θ_i 29.8° φ_i 248.7°

c. Comparison of hourly three-dimensional plots of BRDF for smooth cordgrass (*Spartina alterniflora*) data collected at 8 a.m., 9 a.m., 12 p.m., and 4 p.m. at the boardwalk site on March 21 – 22, 2000, for band 624.20 nm.

Figure 1-13   a) The concepts and parameters of the bidirectional reflectance distribution function (BRDF). A target is bathed in irradiance (dE_i) from a specific Sun zenith and azimuth angle, and the sensor records the radiance (dL_r) exiting the target of interest at a specific azimuth and zenith angle. b) The Sandmeier Field Goniometer collecting smooth cordgrass (*Spartina alterniflora*) BRDF measurements at North Inlet, SC. Spectral measurements are made at Sun zenith angle of θ_i and Sun azimuth angle of φ_i and a sensor zenith angle of view of θ_r and sensor azimuth angle of φ_r. A GER 3700 spectroradiometer, attached to the moving sled mounted on the zenith arc, records the amount of radiance leaving the target in 704 bands at 76 angles (Sandmeier, 2000; Schill et al., 2004). c) Hourly three-dimensional plots of BRDF data.

Current and Proposed Satellite Remote Sensing Systems

Remote sensing systems onboard satellites provide high-quality, relatively inexpensive data per km^2. For example, the European Remote Sensing satellites (ERS-1 and 2) collect 26×28 m spatial resolution C-band active microwave (RADAR) imagery of much of Earth, even through clouds. Similarly, the Canadian Space Agency RADARSAT obtains C-band active microwave imagery. The United States has progressed from multispectral scanning systems (Landsat MSS launched in 1972) to more advanced scanning systems (Landsat 7 Enhanced Thematic Mapper Plus in 1999). The Land Remote Sensing Policy Act of 1992 specified the future of satellite land remote sensing programs in the United States (Asker, 1992; Jensen, 1992). Unfortunately, Landsat 6 with its Enhanced Thematic Mapper did not achieve orbit when launched on October 5, 1993. Landsat 7 was launched on April 15, 1999, to relieve the United States' land remote sensing data gap. Unfortunately, it now has serious scan-line corrector problems. Chapter 7 reviews the plans for the proposed Landsat Continuity Mission. Meanwhile, the French have pioneered the development of linear array remote sensing technology with the launch of SPOT satellites 1 through 5 in 1986, 1990, 1993, 1998, and 2002.

The International Geosphere–Biosphere Program (IGBP) and the United States Global Change Research Program (USGCRP) call for scientific research to describe and understand the interactive physical, chemical, and biological processes that regulate the total Earth system. Space-based remote sensing is an integral part of these research programs because it provides the only means of observing global ecosystems consistently and synoptically. The National Aeronautics and Space Administration (NASA) Earth Science Enterprise (ESE) is the name given to the coordinated plan to provide the necessary satellite platforms and instruments and an Earth Observing System Data and Information System (EOSDIS), and related scientific research for IGBP. The Earth Observing System (EOS) is a series of Earth-orbiting satellites that will provide global observations for 15 years or more. The first satellites were launched in the late 1990s. EOS is complemented by missions and instruments from international partners. For example, the Tropical Rainfall Mapping Mission (TRMM) is a joint NASA/Japanese mission.

EOS Science Plan: Asrar and Dozier (1994) conceptualized the remote sensing science conducted as part of NASA's ESE. They suggested that the Earth consists of two subsystems, 1) the physical climate, and 2) biogeochemical cycles, linked by the global hydrologic cycle, as shown in Figure 1-14.

The *physical climate* subsystem is sensitive to fluctuations in the Earth's radiation balance. Human activities have caused changes to the planet's radiative heating mechanism that rival or exceed natural change. Increases in greenhouse gases between 1765 and 1990 have caused a radiative forcing of 2.5 W m^{-2}. If this rate is sustained, it could result in global mean temperatures increasing about 0.2 to 0.5 °C per decade during the next century. Volcanic eruptions and the ocean's ability to absorb heat may impact the projections. Nevertheless, the following questions are being addressed using remote sensing (Asrar and Dozier, 1994):

- How do clouds, water vapor, and aerosols in the Earth's radiation and heat budgets change with increased atmospheric greenhouse-gas concentrations?

- How do the oceans interact with the atmosphere in the transport and uptake of heat?

- How do land-surface properties such as snow and ice cover, evapotranspiration, urban/suburban land use, and vegetation influence circulation?

The Earth's *biogeochemical cycles* have also been changed by humans. Atmospheric carbon dioxide has increased by 30 percent since 1859, methane by more than 100 percent, and ozone concentrations in the stratosphere have decreased, causing increased levels of ultraviolet radiation to reach the Earth's surface. Global change research is addressing the following questions:

- What role do the oceanic and terrestrial components of the biosphere play in the changing global carbon budget?

- What are the effects on natural and managed ecosystems of increased carbon dioxide and acid deposition, shifting precipitation patterns, and changes in soil erosion, river chemistry, and atmospheric ozone concentrations?

The *hydrologic cycle* links the physical climate and biogeochemical cycles. The phase change of water between its gaseous, liquid, and solid states involves storage and release of latent heat, so it influences atmospheric circulation and globally redistributes both water and heat (Asrar and Dozier, 1994). The hydrologic cycle is the integrating process for the fluxes of water, energy, and chemical elements among components of the Earth system. Important questions to be addressed include these three:

- How will atmospheric variability, human activities, and climate change affect patterns of humidity, precipitation, evapotranspiration, and soil moisture?

Figure 1-14   The Earth system can be subdivided into two subsystems—the physical climate system and biogeochemical cycles—that are linked by the global hydrologic cycle. Significant changes in the external forcing functions and human activities have an impact on the physical climate system, biogeochemical cycles, and the global hydrologic cycle. Examination of these subsystems and their linkages defines the critical questions that the NASA Earth Observing System (EOS) is attempting to answer (adapted from Asrar and Dozier, 1994).

• How does soil moisture vary in time and space?

• Can we predict changes in the global hydrologic cycle using present and future observation systems and models?

These and other research questions are articulated in NASA's current Earth System Science focus areas (Asrar, 2004). The models that address these research questions require sophisticated remote sensing measurements. To this end, the EOS *Terra* satellite was launched on December 18, 1999. It contains five remote sensing instruments (MODIS,

ASTER, MISR, CERES, and MOPITT) designed to address many of the research topics (King, 2003). The EOS *Aqua* satellite was launched in May, 2002. The Moderate Resolution Imaging Spectrometer (MODIS) has 36 bands from 0.405 to 14.385 μm that collect data at 250×250 m, 500×500 m, and 1×1 km spatial resolutions. MODIS views the entire surface of the Earth every one to two days, making observations in 36 spectral bands of land- and ocean-surface temperature, primary productivity, land-surface cover, clouds, aerosols, water vapor, temperature profiles, and fires.

The Advanced Spaceborne Thermal Emission and Reflection Radiometer (ASTER) has five bands in the thermal infrared region between 8 and 12 μm with 90-m pixels. It also has three broad bands between 0.5 and 0.9 μm with 15-m pixels and stereo capability, and six bands in the short-wave infrared region (1.6 – 2.5 μm) with 30-m spatial resolution. ASTER is the highest spatial resolution sensor system on the EOS *Terra* platform and provides information on surface temperature that can be used to model evapotranspiration.

The Multiangle Imaging SpectroRadiometer (MISR) has nine separate charge-coupled-device (CCD) pushbroom cameras to observe the Earth in four spectral bands and at nine view angles. It provides data on clouds, atmospheric aerosols, and multiple-angle views of the Earth's deserts, vegetation, and ice cover. The Clouds and the Earth's Radiant Energy System (CERES) consists of two scanning radiometers that measure the Earth's radiation balance and provide cloud property estimates to assess their role in radiative fluxes from the surface of the Earth to the top of the atmosphere. Finally, the Measurements of Pollution in the Troposphere (MOPITT) scanning radiometer provides information on the distribution, transport, sources, and sinks of carbon monoxide and methane in the troposphere.

The National Polar-orbiting Operational Environmental Satellite System (NPOESS) Preparatory Project (NPP) to be launched will extend key EOS measurements in support of long-term monitoring of climate trends and global biological productivity until the NPOESS can be launched sometime in the future. The NPP will contain MODIS-like instruments such as the Visible Infrared Imaging Radiometer Suite (VIIRS). With a five-year design life NPP will provide data past the designed lifetimes of EOS *Terra* and *Aqua* satellites through the launch of NPOESS (NOAA NPOESS, 2006).

Commercial Vendors: Space Imaging, Inc., launched IKONOS-2 on September 24, 1999. The IKONOS-2 sensor system has a 1×1 m panchromatic band and four 4×4 m multispectral bands (Table 1-3). DigitalGlobe, Inc., launched QuickBird on October 18, 2001, with a 61×61 cm panchromatic band and four 2.44×2.44 m multispectral bands. Orbimage, Inc. launched OrbView-3 on June 26, 2003, with 1×1 m panchromatic and 4×4 m multispectral bands.

Remote Sensing Data Analysis

Remote sensor data are analyzed using a variety of image processing techniques (Figures 1-5 and 1-15), including:

- analog (visual) image processing, and

- digital image processing.

Analog and digital analysis of remotely sensed data seek to detect and identify important phenomena in the scene. Once identified, the phenomena are usually measured, and the information is used in solving problems (Estes et al., 1983; Haack et al., 1997). Thus, both manual and digital analysis have the same general goals. However, the attainment of these goals may follow different paths.

Human beings are adept at visually interpreting images produced by certain types of remote sensing devices, especially cameras. One could ask, "Why try to mimic or improve on this capability?" First, there are certain thresholds beyond which the human interpreter cannot detect "just noticeable differences" in the imagery. For example, it is commonly known that an analyst can discriminate only about nine shades of gray when interpreting continuous-tone, black-and-white photography. If the data were originally recorded with 256 shades of gray, there might be more subtle information present in the image than the interpreter can extract visually. Furthermore, the interpreter brings to the task all the pressures of the day, making the interpretation subjective and generally unrepeatable. Conversely, the results obtained by computer are repeatable (even when wrong!). Also, when it comes to keeping track of a great amount of detailed quantitative information, such as the spectral characteristics of a vegetated field throughout a growing season for crop identification purposes, the computer is very adept at storing and manipulating such tedious information and possibly making a more definitive conclusion as to what crop is being grown. This is not to say that digital image processing is superior to visual image analysis. Rather, there may be times when a digital approach is better suited to the problem at hand. Optimum results are often achieved using a synergistic combination of both visual and digital image processing.

Fundamental Image Analysis Tasks
• Detect, Identify, Measure
• Solve problems

Application of the *Multi* concept
- Multispectral - Multifrequency - Multipolarization
- Multitemporal - Multiscale - Multidisciplinary

Use of *Collateral Information*
- Literature - Laboratory spectra - Dichotomous keys - Prior probabilities
- Field training sites - Field test sites - Soil maps - Surficial geology maps

Analog (Visual) Image Processing	**Digital Image Processing**
Elements of Image Interpretation	*How the Elements of Image Interpretation Are Extracted or Used in Digital Image Processing*
• Grayscale tone (black to white)	• 8- to 12-bit brightness values or scaled to surface reflectance or emittance

• Color (RGB = red, green, blue)	• 24-bit color look-up table display
	  - Multiband RGB color composites
	  - Transforms (e.g., intensity, hue, saturation)

• Height (elevation) and depth	• Soft-copy photogrammetry, radargrammetry, RADAR interferometry, LIDAR, SONAR
• Size (length, area, perimeter, volume)	• Soft-copy photogrammetry, radargrammetry, RADAR interferometry
• Shape	• Soft-copy photogrammetry, radargrammetry, interferometry, landscape ecology metrics, object-oriented image segmentation
• Texture	• Texture transforms, geostatistical analysis, landscape ecology metrics, fractal analysis
• Pattern	• Autocorrelation, geostatistical analysis, landscape ecology metrics, fractal analysis
• Shadow	• Soft-copy photogrammetry, radargrammetry, measurement from rectified images
• Site	• Contextual, expert system, neural network analysis
• Association	• Contextual, expert system, neural network analysis
• Arrangement	• Contextual, expert system, neural network analysis

Figure 1-15   Analog (visual) and digital image processing of remotely sensed data use the elements of image interpretation.

Analog (Visual) Image Processing

Human beings use the fundamental elements of image interpretation summarized in Figure 1-15, including grayscale tone, color, height (depth), size, shape, shadow, texture, site, association, and arrangement. The human mind is amazingly good at recognizing and associating these complex elements in an image or photograph because we constantly process (a) profile views of Earth features every day and (b) images seen in books, magazines, the television and the Internet. Furthermore, we are adept at bringing to bear all the knowledge in our personal background and collateral information. We then converge all this evidence to identify phenomena in images and judge their significance. Precise measurement of objects (length, area, perimeter, volume, etc.) may be performed using photogrammetric techniques applied to either monoscopic (single-photo) or stereoscopic (overlapping) images. Numerous books have been written on how to perform visual image interpretation and photogrammetric measurement.

There is a resurgence in the art and science of visual image interpretation as the digital remote sensor systems provide increasingly higher spatial resolution imagery. Many people are displaying IKONOS 1 × 1 m and QuickBird 61 × 61 cm imagery on the computer screen and then visually interpreting the data. The data are also often used as a base map in GIS projects (Clarke, 2001).

Digital Image Processing

Scientists have made significant advances in digital image processing of remotely sensed data for scientific visualization and hypothesis testing (e.g., Estes and Jensen, 1998; Townshend and Justice, 2002; Kraak, 2003). The methods are summarized in Donnay et al. (2001), Bossler et al. (2002), Jensen (2005), and others. Digital image processing now makes use of many elements of image interpretation using the techniques summarized in Figure 1-15. The major types of digital image processing include image preprocessing (radiometric and geometric correction), image enhancement, pattern recognition using inferential statistics, photogrammetric image processing of stereoscopic imagery, expert system (decision- tree) and neural network image analysis, hyperspectral data analysis, and change detection (Figure 1-5).

Radiometric Correction of Remote Sensor Data: Analog and digital remotely sensed imagery may contain noise or error that was introduced by the sensor system (e.g., electronic noise) or the environment (e.g., atmospheric scattering of light into the sensor's field of view). Advances have been made in our ability to remove these deleterious effects through simple image normalization techniques and more advanced absolute radiometric calibration of the data to scaled surface reflectance (for optical data). Calibrated remote sensor data allows imagery and derivative products obtained on different dates to be compared (e.g., to measure the change in leaf area index between two dates). Fundamental digital image processing principles are discussed in Jensen (2005).

Geometric Correction of Remote Sensor Data: Most analog and digital remote sensor data are now processed so that individual picture elements are in their proper planimetric positions in a standard map projection. This facilitates the use of the imagery and derivative products in GIS or spatial decision support systems.

Image Enhancement: Images can be digitally enhanced to identify subtle information in the analog or digital imagery that might otherwise be missed. Significant improvements have been made in our ability to contrast stretch and filter data to enhance low and high frequency components, edges, and texture in the imagery (e.g., Emerson et al., 1999). In addition, the remote sensor data can be linearly and nonlinearly transformed into information that is more highly correlated with real-world phenomena through principal components analysis and various vegetation indices (Townshend and Justice, 2002).

Photogrammetry: Significant advances have been made in the analysis of stereoscopic remote sensor data obtained from airborne or satellite platforms using computer workstations and digital image processing photogrammetric algorithms (e.g., Adams and Chandler, 2002). Soft-copy photogrammetric workstations can be used to extract accurate digital elevation models (DEMs) and differentially corrected orthophotography from the triangulated aerial photography or imagery (Light and Jensen, 2002; Linder, 2003). The technology is revolutionizing the way DEMs and orthophotos are produced for rural and urban–suburban applications.

Parametric Information Extraction: Scientists attempting to extract land-cover information from remotely sensed data now routinely specify if the classification is to be:

- *hard*, with discrete mutually exclusive classes, or *fuzzy*, where the proportions of materials within pixels are extracted (Seong and Usery, 2001);

- based on individual pixels (referred to as a *per-pixel classification*) or if it will use object-oriented image segmentation algorithms that take into account not only the spectral characteristics of a pixel, but also the spectral characteristics of contextual surrounding pixels. Thus, the algorithms take into account spectral and spatial information (Herold et al., 2003; Hodgson et al., 2003a; Tullis and Jensen, 2003).

Once these issues are addressed, it is a matter of determining whether to use parametric, nonparametric, and/or nonmetric classification techniques. The maximum likelihood classification algorithm continues to be the most widely used parametric classification algorithm. Unfortunately, the algorithm requires normally distributed training data in n bands (rarely the case) for computing the class variance and covariance matrices. It is difficult to incorporate nonimage categorical data into a maximum likelihood classification. Fortunately, fuzzy maximum likelihood classification algorithms are now available (e.g., Foody, 1996).

Nonparametric Information Extraction: Nonparametric clustering algorithms, such as ISODATA, continue to be used extensively in digital image processing research. Unfortunately, such algorithms depend on how the seed training data are extracted and it is often difficult to label the clusters to turn them into information classes. For these reasons there has been a significant increase in the development and use of artificial neural networks (ANN) for remote sensing applications (e.g., Qiu and Jensen, 2005). The ANN does not require normally distributed training data. ANN may incorporate virtually any type of spatially distributed data in the classification. The only drawback is that sometimes it is difficult to determine exactly how the ANN came up with a certain conclusion because the information is locked within the weights in the hidden layer(s). Scientists are working on ways to extract hidden information so that the rules used can be more formally stated. The ability of an ANN to *learn* should not be underestimated.

Nonmetric Information Extraction: It is difficult to make a computer understand and use the heuristic rules of thumb and knowledge that a human expert uses when interpreting an image. Nevertheless, there has been progress in the use of artificial intelligence (AI) to try to make computers do things that, at the moment, people do better. One area of AI that has great potential for image analysis is the use of expert systems that place all the information contained within an image in its proper context with ancillary data and extract valuable information. Duda et al. (2001) describe various types of expert system decision-tree classifiers as *nonmetric*.

Parametric digital image classification techniques are based primarily on summary statistics such as the mean, variance, and covariance matrices. Decision-tree or rule-based classifiers are not based on inferential statistics, but instead "let the data speak for itself" (Gahegan, 2003). In other words, the data retains its precision and is not dumbed down by summarizing it through means, etc. Decision-tree classifiers can process virtually any type of spatially distributed data and can incorporate prior probabilities (McIver and Friedl, 2002). There are three approaches to rule creation: 1) explicitly extracting knowledge and creating rules from experts, 2) implicitly extracting variables and rules using cognitive methods (Lloyd et al., 2002), and 3) empirically generating rules from observed data and automatic induction methods (Tullis and Jensen, 2003). The development of a decision tree using human-specified rules is time-consuming and difficult. However, it rewards the user with detailed informa-

tion about how individual classification decisions were made (Zhang and Wang, 2003).

Ideally, computers can derive the rules from training data without human intervention. This is referred to as *machine-learning* (Huang and Jensen, 1997; Jensen, 2005). The analyst identifies representative training areas. The machine learns the patterns from these training data, creates the rules, and uses them to classify the remotely sensed data. The rules are available to document how decisions were made.

Hyperspectral: Special software is required to process hyperspectral data obtained by imaging spectroradiometers such as AVIRIS and MODIS. Kruse et al. (1992), Landgrebe and Biehl (2006), Digital Research Systems (2006) and others have pioneered the development of hyperspectral image analysis software. The software reduces the dimensionality of the data (number of bands) to a manageable degree, while retaining the essence of the data. Under certain conditions the software can be used to compare the remotely sensed spectral reflectance curves with a library of spectral reflectance curves. Analysts are also able to identify the type and proportion of different materials within an individual picture element (referred to as end-member spectral mixture analysis) (Lu and Weng, 2004; Platt and Goetz, 2004).

Modeling Remote Sensing Data Using a GIS Approach: Remotely sensed data should not be analyzed in a vacuum without the benefit of collateral information such as soil maps, hydrology, and topography (Ramsey et al., 1995). For example, land-cover mapping using remotely sensed data has been significantly improved by incorporating topographic information from digital terrain models and other GIS data (e.g., Stow et al., 2003). GIS studies require timely, accurate updating of the spatially distributed variables in the database that remote sensing can provide (Clarke, 2001). Remote sensing can benefit from access to accurate ancillary information to improve classification accuracy and other types of modeling. Such synergy is critical if successful expert system and neural network analyses are to be performed (Tullis and Jensen, 2003). A framework for modeling the uncertainty between remote sensing and geographic information systems was developed by Gahegan and Ehlers (2000).

Scene Modeling: Strahler et al. (1986) describe a framework for modeling in remote sensing. Basically, a remote sensing model has three components: 1) a scene model, which specifies the form and nature of the energy and matter within the scene and their spatial and temporal order; 2) an atmospheric model, which describes the interaction between the atmosphere and the energy entering and being emitted from the scene; and 3) a sensor model, which describes the behavior of the sensor in responding to the energy fluxes incident on it and in producing the measurements that constitute the image. They suggest that the problem of scene inference, then, becomes a problem of model inversion in which the order in the scene is reconstructed from the image and the remote sensing model. For example, Woodcock et al. (1997) inverted the Li-Strahler Canopy Reflectance Model for mapping forest structure.

Basically, successful remote sensing modeling predicts how much radiant flux in certain wavelengths should exit a particular object (e.g., a conifer canopy) even without actually sensing the object. When the model's prediction is the same as the sensor's measurement, the relationship has been modeled correctly. The scientist then has a greater appreciation for energy–matter interactions in the scene and may be able to extend the logic to other regions or applications with confidence. The remote sensor data can then be used more effectively in physical deterministic models (e.g., watershed runoff, net primary productivity, and evapotranspiration models) that are so important for large ecosystem modeling. Recent work allows one to model the utility of sensors with different spatial resolutions for particular applications, such as urban analysis (Collins and Woodcock, 1999).

Change Detection: Remotely sensed data obtained on multiple dates can be used to identify the type and spatial distribution of changes taking place in the landscape (Friedl et al., 2002; Zhan et al., 2002). The change information provides valuable insight into the *processes* at work (Alberti et al., 2004; Auch et al., 2004). Change detection algorithms can be used on per-pixel and object-oriented (polygon) classifications. Unfortunately, there is still no universally accepted method of detecting change or of assessing the accuracy of change detection map products. Digital image processing change detection principles are discussed in Jensen (2005).

Information Presentation

Information derived from remote sensor data are usually summarized as an enhanced image, image map, orthophotomap, thematic map, spatial database file, statistic, or graph (Figure 1-5). Thus, the final output products often require knowledge of remote sensing, cartography, GIS, and spatial statistics as well as the systematic science being investigated (e.g., soils, agriculture, urban studies). Scientists who understand the rules and synergistic relationships of the technologies can produce output products that communicate effectively. Those who violate fundamental rules (e.g., car-

tographic theory or database topology design) often produce poor output products that do not communicate effectively.

Image maps offer scientists an alternative to line maps for many cartographic applications. Thousands of satellite image maps have been produced from Landsat MSS (1:250,000 and 1:500,000 scale), TM (1:100,000 scale) and AVHRR, and MODIS data. Image maps at scales >1:24,000 are possible using imagery with a spatial resolution of $\leq 1 \times 1$ m (Light and Jensen, 2002). Because image map products can be produced for a fraction of the cost of conventional line maps, they provide the basis for a national map series oriented toward the exploration and economic development of the less-developed areas of the world, most of which have not been mapped at scales of 1:100,000 or larger.

Remote sensor data that have been geometrically rectified to a standard map projection are becoming indispensable in most sophisticated GIS databases. This is especially true of orthophotomaps, which have the metric qualities of a line map and the information content of an aerial photograph or other type of image.

Unfortunately, *error* is introduced in the remote sensing process and must be identified and reported. Innovations in error reduction include: 1) recording the lineage of the operations applied to the original remote sensor data, 2) documenting the geometric (spatial) error and thematic (attribute) error of the source materials, 3) improving legend design, especially for change detection map products derived from remote sensing, and 4) improved accuracy assessment. The remote sensing and GIS community should incorporate technologies that track all error in final map and image products. This will result in more accurate information being used in the decision-making process.

 Earth Observation Economics

The National Research Council recognized that there is an economic system at play when remote sensor data are used for earth resource management applications (Figure 1-16) (Miller et al., 2001). It consists of an information delivery system with three components: data collection, image processing, and information consumer (user).

The data collection system is composed of commercial vendors and public agencies that operate remote sensing systems. Private industry provides information at market value. Public agencies generally provide remote sensor data at the cost of fullfilling a user request (COFUR). Remote sensing

Remote Sensing Earth Observation Economics

Figure 1-16   Remote sensing Earth observation economics. The goal is to minimize the *knowledge gap* between the information delivery system, remote sensing experts, and the information consumer (user). The remote sensing–derived economic, social, strategic, environmental, and/or political information must be cost-effective, and easy to use to achieve equilibrium (adapted from Miller et al., 2003).

has been around since the 1960s. There is an increasing number of experts that can use analog and/or digital image processing techniques to extract information from the imagery. Finally, there is the information consumer (user) of the remote sensing–derived information. The user generally needs information of economic, social, strategic, environmental and/or political value (Liverman et al., 1998).

In order for the revenues generated by the information delivery system to be sufficient to support the capital and operating costs of the system, there must be a balance (equilibrium) between the value of the information, as perceived by the user (consumer), and the revenue necessary to support the system (Miller et al., 2001, 2003). The equilibrium has been achieved for airborne photogrammetric and LIDAR mapping applications for several decades. Time will tell if the balance between perceived value and cost can be maintained in the spaceborne case. Mergers are occurring. On January 12, 2006, ORBIMAGE acquired Space Imaging's assets and now functions as GeoEye, Inc., providing

IKONOS, OrbView-2 and Orbview-3 image products. GeoEye plans to launch a new sensor in 2007 with a spatial resolution of 0.41 × 0.41 m (GeoEye, 2006)..

The equilibrium can also be impacted by remote sensing technology experts that do not have a good understanding of the user information requirements. In fact, some remote sensing experts are often baffled as to why the consumers don't embrace the remote sensing-derived information. What they fail to consider is that the consumers generally have no motivation to switch to remote sensing-derived information on economic, social, environmental, strategic, or political attributes simply because it is based on new technology. Furthermore, the consumers on the right side of the diagram often have little knowledge of remote sensing technology or of how it is used to derive information.

Miller et al. (2001; 2003) suggest that this situation creates a *knowledge gap* between the remote sensing experts and the information consumers (user) (Figure 1-16). Bridging the

Organization of
Remote Sensing of the Environment

Chapter 1.
Remote Sensing of the Environment
- *In Situ* Data Collection
- Remote Sensing Data Collection
- The Remote Sensing Process

Chapter 2.
Electromagnetic Radiation Principles
- Conduction, Convection, and Radiation
- Energy-Matter Interactions in the Atmosphere
- Energy-Matter Interactions with the Terrain

Chapter 3.
History of Aerial Photography and
Aerial Platforms
- History of Photography
- Photography from Aerial Platforms
- Photo-Reconnaissance in WWI and WWII
- Cold War Photo-Reconnaissance
- Celestial Satellilte Sentinels
- Unmanned Aerial Vehicles

Chapter 4.
Aerial Photography
- Vertical and Oblique Vantage Points
- Aerial Cameras
- Filtration and Films
- Planning Aerial Photography Missions

Chapter 5.
Elements of Visual Image Interpretation
- Elements of Image Interpretation
- Methods of Search

Chapter 6.
Photogrammetry
- Flightlines of Aerial Photography
- Image Nomenclature
- Scale/Height Measurement on Single Photos
- Stereoscopic Measurement
- Orthophotos and Digital Elevation Models
- Area Measurement

Chapter 7.
Multispectral Remote Sensing
- Multispectral Data Collection
- Discrete Detectors and Sanning Mirrors
- Multispectral Imaging Using Linear Arrays
- Imaging Spectroscopy
- Digital Frame Cameras
- Satellite Photographic Systems

Chapter 8.
Thermal Infrared Remote Sensing
- History
- Thermal Infrared Radiation Properties
- Atmospheric Windows
- Thermal Radiation Laws
- Thermal Properties of Terrain
- Thermal Infrared Data Collection
- TIR Environmental Considerations

Chapter 9.
Active and Passive Microwave
- History
- Active Microwave System Components
- RADAR Environmental Considerations
- SAR Remote Sensing from Space
- RADAR Interferometry
- Passive Microwave Remote Sensing

Chapter 10.
LIDAR Remote Sensing
- Principles (returns, density, intensity)
- Processing to Create DEM, DSM, DTM
- Accuracy of LIDAR-derived products

Chapter 11.
Remote Sensing of Vegetation
- Photosynthesis Fundamentals
- Spectral Characteristics of Vegetation
- Temporal Characteristics of Vegetation
- Vegetation Indices
- Landscape Ecology Metrics
- Biodiversity and GAP Analysis
- Vegetation Change Detection

Figure 1-17   Book organization.

Organization of
***Remote Sensing of the Environment** - continued*

<table>
<tr>
<td valign="top">

Chapter 12.
Remote Sensing of Water
• Surface Water Biophysical Characteristics
• Precipitation
• Aerosols and Clouds
• Water Vapor and Snow
• Water-quality Modeling

</td>
<td valign="top">

Chapter 14.
Remote Sensing Soils, Minerals, and
Geomorphology
• Soil Characteristics and Taxonomy
• Remote Sensing Soil Properties
• Remote Sensing Rocks and Minerals
• Geology and Geomorphology

</td>
</tr>
<tr>
<td valign="top">

Chapter 13.
Remote Sensing the Urban Landscape
• Urban/Suburban Resolution Considerations
• Remote Sensing Land Use/Land Cover
• Residential
• Commercial and Services
• Industrial and Transportation
• Communications and Utilities
• Urban Meteorological Data
• Critical Environmental Area Assessment
• Disaster Emergency Response

</td>
<td valign="top">

Chapter 15.
***In Situ* Spectral Reflectance Measurement**
• Spectral Reflectance of a Material
• Illumination Considerations
• Radiometer Considerations

</td>
</tr>
</table>

Figure 1-17 Book organization (continued).

gap is mandatory if we are to use remote sensing to solve earth resource management problems. It is unlikely that the user community can devote the time to learn the physics of remote sensing and methods of analog and/or digital image processing and GIS modeling necessary to produce useful information. Conversely, there is considerable interest on the technology side of the problem to build a communication bridge. Therefore, one way to decrease the size of the knowledge gap is for the remote sensing technologists to work more closely with the user communities to understand their requirements. This will lead to more useful remote sensing-derived information of value to the user communities.

Advances in remote sensing image delivery systems by commercial firms such as Google, Inc., and their *Google Earth* application are having a tremendous impact on the public's use and appreciation of remote sensor data (Fallows, 2006).

 Earth Resource Analysis Perspective

Remote sensing is used for numerous applications such as medical image analysis (e.g., *x*-raying a broken arm), nondestructive evaluation of products on an assembly line, and analysis of Earth resources. *This book focuses on the art and science of applying remote sensing for the extraction of useful Earth resource information* (Figure 1-17). Earth resource information is defined as any information concerning terrestrial vegetation, soils, minerals, rocks, water, and urban infrastructure as well as certain atmospheric characteristics. Such information may be useful for modeling the global carbon cycle, the biology and biochemistry of ecosystems, aspects of the global water and energy cycle, climate variability and prediction, atmospheric chemistry, characteristics of the solid Earth, population estimation, and monitoring land-use change and natural hazards (Johannsen et al., 2003).

Book Organization

This chapter defined terms and provided a perspective on how remote sensing science can be useful for Earth resource investigations (Figure 1-17). Chapter 2 introduces principles of electromagnetic radiation and how it is used to perform remote sensing of the environment. Chapter 3 reviews the history of photography, and aerial and satellite platforms. Chapter 4 introduces aerial photography, filtration, and film. Chapter 5 presents the elements of image interpretation. Chapter 6 reviews principles of photogrammetry used to extract quantitative information from aerial photography. Chapter 7 reviews optical-mechanical remote sensing systems. Chapter 8 introduces thermal infrared remote sensing. Chapter 9 reviews active (RADAR) and passive microwave remote sensing. Chapter 10 introduces remote sensing using light detection and ranging (LIDAR) technology. Chapter 11 describes how remote sensing is used to extract biophysical characteristics of terrestrial and aquatic vegetation. Chapter 12 provides insight into remote sensing of water, ice, and snow as well as atmospheric water vapor and temperature. Chapter 13 demonstrates how remote sensing can provide unique urban/suburban infrastructure information. Chapter 14 describes how selected soil and mineral characteristics may be remotely sensed and how major geomorphic features on the surface of the Earth may be identified. Chapter 15 describes how *in situ* spectral reflectance measurements are obtained.

References

Adams, J. C. and J. H. Chandler, 2002, "Evaluation of Lidar and Medium Scale Photogrammetry for Detecting Soft-cliff Coastal Change," *Photogrammetric Record*, 17(99): 405–418.

Alberti, M., Weeks, R. and S. Coe, 2004, "Urban Land-Cover Change Analysis in Central Puget Sound," *Photogrammetric Engineering & Remote Sensing*, 70(9):1043–1052.

American Society for Photogrammetry and Remote Sensing, 1952, 1966, *Manual of Photogrammetry*, Bethesda: ASP&RS.

Asker, J. R., 1992, "Congress Considers Landsat 'Decommercialization' Move," *Aviation Week & Space Technology,* May 11, 18–19.

Asrar, G., 2004, *Earth Science Applications Plan*, Washington: NASA, 89 p.

Asrar, G. and J. Dozier, 1994, *EOS: Science Strategy for the Earth Observing System*, Woodbury, MA: American Institute of Physics, 342 p.

Auch, R., Taylor, J. and W. Acevedo, 2004, *Urban Growth in American Cities – Glimpses of U. S. Urbanization*, Circular #1252, Washington: U.S. Geological Survey, 52.

Barnsley, M., 1999, "Digital Remotely Sensed Data and Their Characteristics," in Longley, P. E., Goodchild, M. F., McGwire, D. J. and D. W. Rhind (Eds.), *Geographical Information Systems*, NY.: John Wiley, 451–466.

Bossler, J. D., Jensen, J. R., McMaster, R. B. and C. Rizos, 2002, *Manual of Geospatial Science & Technology*, London: Taylor & Francis, 623 p.

Clark, R. N., 1999, *Spectroscopy of Rocks and Minerals, and Principles of Spectroscopy*, Denver: U.S. Geological Survey, http://speclab.cr.usgs.gov, 58 p.

Clarke, K. C., 2001, *Getting Started with Geographic Information Systems*, Upper Saddle River: Prentice-Hall, 353 p.

Collins, J. B. and C. E. Woodcock, 1999, "Geostatistical Estimation of Resolution-Dependent Variance in Remotely Sensed Images," *Photogrammetric Engineering & Remote Sensing,* 65(1):41–50.

Colwell, R. N. (Ed.), 1960, *Manual of Photographic Interpretation*, Falls Church: ASP&RS.

Colwell, R. N. (Ed.), 1983, *Manual of Remote Sensing*, 2nd. Ed., Falls Church: ASP&RS.

Colwell, R. N., 1984, "From Photographic Interpretation to Remote Sensing," *Photogrammetric Engineering and Remote Sensing,* 50(9):1305.

Colwell, R. N., 1997, "History and Place of Photographic Interpretation," in *Manual of Photographic Interpretation*, 2nd Ed., W. R. Phillipson (Ed.), Bethesda: ASPRS, 33–48.

Cracknell, A. P. and L. W. B. Hayes, 1993, *Introduction to Remote Sensing*, London: Taylor & Francis, 293 p.

Curran, P. J., 1987, "Remote Sensing Methodologies and Geography," *Intl. Journal of Remote Sensing,* 8:1255–1275.

Curran, P. J., Milton, E. J., Atkinson, P. M. and G. M. Foody, 1998, "Remote Sensing: From Data to Understanding," in P.

E. Longley, et al. (Eds.), *Geocomputation: A Primer*, NY: John Wiley, 33–59.

Dahlberg, R. W. and J. R. Jensen, 1986, "Education for Cartography and Remote Sensing in the Service of an Information Society: The United States Case," *American Cartographer*, 13(1):51–71.

Dartnell, P. and J. V. Gardner, 2004, "Predicting Seafloor Facies from Multibeam Bathymetry and Backscatter Data," *Photogrammetric Engineering & Remote Sensing*, 70(9):1081–1091.

Davis, B. A., 1999, "Overview of NASA's Commercial Remote Sensing Program," *Earth Observation*, 8(3):58–60.

Digital Research, 2006, *Environment for Visualizing Images: ENVI*, www.digitalresearch.com

Dehqanzada, Y. A. and A. M. Florini, 2000, *Secrets for Sale: How Commercial Satellite Imagery Will Change the World*, Washington: Carneggie Endowment for Intl. Peace, 45 p.

Dobson, J. E., 1993, "Commentary: A Conceptual Framework for Integrating Remote Sensing, Geographic Information Systems, and Geography," *Photogrammetric Engineering & Remote Sensing,* 59(10):1491–1496.

Donnay, J., Barnsley, M. J. and P. A. Longley, 2001, *Remote Sensing and Urban Analysis*, NY: Taylor & Francis, 268 p.

Duda, R. O., Hart, P. E. and D. G. Stork, 2001, *Pattern Classification*, N.Y: John Wiley, 394–452.

Emerson, C. W., Lam, N. and D. A. Quattrochi, 1999, "Multiscale Fractal Analysis of Image Texture and Pattern," *Photogrammetric Engineering & Remote Sensing,* 65(1):51–61.

Engman, E. T., 2000, "Soil Moisture," in Schultz, G. A. and E. T. Engman (Eds.), *Remote Sensing in Hydrology and Water Management*, Berlin: Springer, 197–216.

Estes, J. E. and J. R. Jensen, 1998, "Development of Remote Sensing Digital Image Processing Systems and Raster GIS," *History of Geographic Information Systems*, T. Foresman (Ed.), NY: Longman, 163–180.

Estes, J. E., Hajic, E. J. and L. R. Tinney, 1983, "Fundamentals of Image Analysis: Visible and Thermal Infrared Data," *Manual of Remote Sensing,* R. N. Colwell, (Ed.), Bethesda: ASPRS, 987–1125.

Fallows, J., 2006, "Spy's-Eye View," *Atlantic Monthly*, (March):140–144.

Fisher, P. F. and R. E. Lindenberg, 1989, "On Distinctions among Cartography, Remote Sensing, and Geographic Information Systems," *Photogrammetric Engineering & Remote Sensing*, 55(10):1431–1434.

Foody, G. M., 1996, "Approaches for the Production and Evaluation of Fuzzy Land Cover Classifications from Remotely Sensed Data," *Intl. Journal of Remote Sensing*, 17(7):1317–1340.

Friedl, M. A., McIver, D. K., Hodges, J. C. F., Zhang, X. Y., Muchoney, D., Strahler, A. H., Woodcock, C. E., Gopal, S., Schneider, A., Cooper, A., Baccini, A., Gao, F. and C. Schaaf, 2002, "Global Land Cover Mapping from MODIS: Algorithms and Early Results," *Remote Sensing of Environment*, 83:287–302.

Fussell, J., Rundquist, D. and J. A. Harrington, 1986, "On Defining Remote Sensing," *Photogrammetric Engineering & Remote Sensing*, 52(9):1507–1511.

Gahegan, M., 2003, "Is Inductive Machine Learning Just Another Wild Goose (or Might It Lay the Golden Egg)?" *Intl. Journal of GIScience*, 17(1):69–92.

Gahegan, M. and M. Ehlers, 2000, "A Framework for the Modelling of Uncertainty Between Remote Sensing and Geographic Information Systems," *ISPRS Journal of Photogrammetry & Remote Sensing*, 55:176–188.

GeoEye, 2006, *ORBIMAGE Acquires Space Imaging and Changes Brand Name to GeoEye*, Dulles: GeoEye, Inc., http://www.orbimage-acquisition.com/news/geoEye.htm.

Goetz, S. J., 2002, "Recent Advances in Remote Sensing of Biophysical Variables: An Overview of the Special Issue," *Remote Sensing of Environment*, 79:145–146.

Haack, B., Guptill, S. C., Holz, R. K., Jampoler, S. M., Jensen, J. R. and R. A. Welch, 1997, "Urban Analysis and Planning," *Manual of Photographic Interpretation*, Bethesda: ASP&RS, 517–553.

Herold, M., Guenther, S. and K. Clarke, 2003, "Mapping Urban Areas in the Santa Barbara South Coast Using IKONOS Data and eCognition," *eCognition Application Note*, Munich: Definiens Imaging GmbH, 4(1):3.

Hodgson, M. E., Jensen, J. R., Tullis, J. A., Riordan, K. D. and C. M. Archer, 2003a, "Synergistic Use of LIDAR and Color Aerial Photography for Mapping Urban Parcel Imperviousness," *Photogrammetric Engineering & Remote Sensing*, 69(9):973–980.

Hodgson, M. E., Jensen, J. R., Schmidt, L., Schill, S. and B. A. Davis, 2003b, "An Evaluation of LIDAR- and IFSAR-derived Digital Elevation Models in Leaf-on Conditions with USGS Level 1 and Level 2 DEMS," *Remote Sensing of Environment,* 84(2003):295-308.

Hodgson, M. E., Jensen, J. R., Raber, G., Tullis, J., Davis, B., Thompson, G. and K. Schuckman, 2005, "An Evaluation of LIDAR derived Elevation and Terrain Slope in Leaf-off Conditions," *Photogrammetric Engineering & Remote Sensing*, 71(7):817–823.

Huang, X. and J. R. Jensen, 1997, "A Machine Learning Approach to Automated Construction of Knowledge Bases for Image Analysis Expert Systems That Incorporate Geographic Information System Data," *Photogrammetric Engineering & Remote Sensing*, 63(10):1185–1194.

Jensen, J. R., 1992, "Testimony on S. 2297, The Land Remote Sensing Policy Act of 1992," Senate Committee on Commerce, Science, and Transportation, *Congressional Record*, (May 6):55–69.

Jensen, J. R., 2005, *Introductory Digital Image Processing: A Remote Sensing Perspective*, Upper Saddle River: Prentice-Hall, 525 p.

Jensen, J. R. and D. C. Cowen, 1999, "Remote Sensing of Urban/Suburban Infrastructure and Socioeconomic Attributes," *Photogrammetric Engineering & Remote Sensing*, 65(5):611–622.

Jensen, J. R. and R. E. Dahlberg, 1983, "Status and Content of Remote Sensing Education in the United States," *Intl. Journal of Remote Sensing*, 4(2):235–245.

Jensen, J. R. and S. Schill, 2000, "Bi-directional Reflectance Distribution Function (BRDF) of Smooth Cordgrass (*Spartina alterniflora*)," *Geocarto International*, 15(2):21–28.

Jensen, J. R., Botchway, K., Brennan-Galvin, E., Johannsen, C., Juma, C., Mabogunje, A., Miller, R., Price, K., Reining, P., Skole, D., Stancioff, A. and D. R. F. Taylor, 2002, *Down to Earth: Geographic Information for Sustainable Development in Africc*, Washington: National Academy Press, 155 p.

Jensen, J. R., Qiu, F. and K. Patterson, 2001, "A Neural Network Image Interpretation System to Extract Rural and Urban Land Use and Land Cover Information fro Remote Sensor Data," *Geocarto International*, 16(1):19–28.

Jensen, J. R., Saalfeld, A., Broome, F., Cowen, D., Price, K., Ramsey, D., Lapine, L. and E. L. Usery, 2005, "Chapter 2: Spatial Data Acquisition and Integration," in R. B. McMaster and E. L. Usery (Eds.), *A Research Agenda for Geographic Information Science*, Boca Raton: CRC Press, 17–60.

Johannsen, C. J., Petersen, G. W., Carter, P. G. and M. T. Morgan, 2003, "Remote Sensing: Changing Natural Resource Management," *Journal of Soil & Water Conservation*, 58(2):42–45.

Joseph, G., 2000, "How Well Do We Understand Earth Observation Electro-optical Sensor Parameters?" *ISPRS Journal of Photogrammetry & Remote Sensing*, 55:9–12.

Karaska, M. A., Huguenin, R. L., Beacham, J. L., Wang, M., Jensen, J. R., and R. S. Kaufman, 2004, "AVIRIS Measurements of Chlorophyll, Suspended Minerals, Dissolved Organic Carbon, and Turbidity in the Neuse River, N.C.," *Photogrammetric Engineering & Remote Sensing*, 70(1):125–133.

Kraak, M., 2003, "Geovisualization Illustrated," *ISPRS Journal of Photogrammetry & Remote Sensing*, (2003):390–399.

Kruse, F. A., Lefkoff, A. B., Boardman, J. W., Heidebrecht, K. B., Shapiro, A. T., Barloon, P. J. and A. F. H. Goetz, 1992, "The Spectral Image Processing System," *Proceedings, Intl. Space Year Conference*, Pasadena, 10 p.

Landgrebe, D. and L. Biehl, 2006, *An Introduction to MULTISPEC*, W. Lafayette: Purdue University, 50 p.

Light, D. L. and J. R. Jensen, 2002, "Photogrammetric and Remote Sensing Considerations," in *Manual of Geospatial Science & Technology*, Bossler, J. D., Jensen, J. R., McMaster, R. B. and C. Rizos (Eds.), London: Taylor & Francis, 233–252.

Linder, W., 2003, *Digital Photogrammetry: Theory and Applications*, Berlin: Springer-Verlag, 189 p.

Liverman, D., Moran, E. F., Rindfuss, R. R. and P. C. Stern, 1998, People and Pixels: Linking Remote Sensing and Social Science, Washington: NRC, 244 p.

Lloyd, R., Hodgson, M. E. and A. Stokes, 2002, "Visual Categorization with Aerial Photographs," *Annals of the Association of American Geographers*. 92(2):241–266.

Lu, D. and Q. Weng, 2004, "Spectral Mixture Analysis of the Urban Landscape in Indianapolis with Landsat ETM$^+$ Imagery," *Photogrammetric Engineering & Remote Sensing*, 70(9):1053–1062.

Maune, D. F., (Ed.), 2001, *Digital Elevation Model Technologies and Applications*, Bethesda: ASP&RS, 538 p.

McIver, D. K. and M. A. Friedl, 2002, "Using Prior Probabilities in Decision-tree Classification of Remotely Sensed Data," *Remote Sensing of Environment*, 81:253–261.

Miller, R. B., Abbott, M. R., Harding, L. W., Jensen, J. R., Johannsen, C. J., Macauley, M., MacDonald, J. S. and J. S. Pearlman, 2001, *Transforming Remote Sensing Data into Information and Applications*, Washington: NRC, 75 p.

Miller, R. B., Abbott, M. R., Harding, L. W., Jensen, J. R., Johannsen, C. J., Macauley, M., MacDonald, J. S. and J. S. Pearlman, 2003, *Using Remote Sensing in State and Local Government: Information for Management and Decision Making*, Washington: NRC, 97 p.

NASA, 2006, *Airborne Visible/Infrared Imaging Spectrometer (AVIRIS)* home page, http://aviris.jpl.nasa.gov/.

NOAA NPOESS, 2006, *National Polar Orbiting Operational Environmental Satellite System*, http://www.ipo.noaa.gov/.

Nemani, R. R., Keeling, C. D., Hashimoto, H., Jolly, W. M., Piper, S. C., Tucker, C. J., Myneni, R. B. and S. W. Running, 2003, "Climate-Driven Increases in Global Terrestrial Net Primary Production from 1982 to 1999," *Science*, 300(6):1560–1563.

Platt, R. V. and A. F. H. Goetz, 2004, "A Comparison of AVIRIS and Landsat for Land Use Classification at the Urban Fringe," *Photogrammetric Engineering & Remote Sensing*, 70(7):813–819.

Pruitt, E. L., 1979, "The Office of Naval Research and Geography," *Annals,* Association of American Geographers, 69(1):106.

Qiu, F. and J. R. Jensen, 2005, "Opening the Neural Network Black Box and Breaking the Knowledge Acquisition Bottleneck of Fuzzy Systems for Remote Sensing Image Classification," *Intl. Journal of Remote Sensing*, 25(9):1749–1768.

Raber, G. T., Jensen, J. R., Schill, S. R. and K. Schuckman, 2002, "Creation of Digital Terrain Models using an Adaptive LI-

DAR Vegetation Point Removal Process," *Photogrammetric Engineering & Remote Sensing*, 68(12):1307–1315.

Ramsey, R. D., Falconer, A. and J. R. Jensen, 1995, "The Relationship Between NOAA-AVHRR Normalized Difference Vegetation Index and Ecoregions in Utah," *Remote Sensing of Environment*, 53:188–198.

Rizos, C., 2002, "Introducing the Global Positioning System," in *Manual of Geospatial Science and Technology*, Bossler, J. D., Jensen, J. R., McMaster, R. B. and C. Rizos (Eds.), London: Taylor & Francis, 77–94.

Robbins, J., 1999, "High-Tech Camera Sees What Eyes Cannot," *New York Times*, Science Section, September 14, D5.

Sandmeier, S. R., 2000, "Acquisition of Bidirectional Reflectance Factor Data with Field Goniometers," *Remote Sensing of Environment,* 73:257–269.

Schill, S., Jensen, J. R., Raber and D. E. Porter, 2004, "Temporal Modeling of Bidirectional Reflection Distribution Function (BRDF) in Coastal Vegetation," *GIScience & Remote Sensing*, 41(2):116-135.

Seong, J. C. and E. L. Usery, 2001, "Fuzzy Image Classification for Continental-Scale Multitemporal NDVI Images Using Invariant Pixels and an Image Stratification Method," *Photogrammetric Engineering & Remote Sensing*, 67(3):287–294.

Shippert, P., 2004, *Spotlight on Hyperspectral*, Boulder: Research Systems, www.geospatial_online.com/shippert, 5 p.

Skidmore, A. K., 2002, "Chapter 2: Taxonomy of Environmental Models in the Spatial Sciences," in *Environmental Modelling with GIS and Remote Sensing*, A. K. Skidmore (Ed.), London: Taylor & Francis, 8–25.

Stow, D, Coulter, L., Kaiser, J., Hope, A., Service, D., Schutte, K. and A. Walters, 2003, "Irrigated Vegetation Assessments for Urban Environments," *Photogrammetric Engineering & Remote Sensing*, 69(4):381–390.

Strahler, A. H., Woodcock, C. E. and J. A. Smith, 1986, "On the Nature of Models in Remote Sensing," *Remote Sensing of Environment*, 20:121–139.

Teillet, P. M., Gauthier, R. P., Chichagov, A. and G. Fedosejevs, 2002, "Towards Integrated Earth Sensing: Advanced Technologies for *in situ* Sensing in the Context of Earth Observation," *Canadian Journal of Remote Sensing*, 28(6):713–718.

Townshend, J. R. G. and C. O. Justice, 2002, "Towards Operational Monitoring of Terrestrial Systems by Moderate-resolution Remote Sensing," *Remote Sensing of Environment*, 83:351–359.

Tullis, J. A. and J. R. Jensen, 2003, "Expert System House Detection in High Spatial Resolution Imagery Using Size, Shape, and Context," *Geocarto International*, 18(1):5–15.

Walsh, S. J., Evans, T. P., Welsh, W. F., Entwisle, B. and R. R. Rindfuss, 1999, "Scale-dependent Relationships Between Population and Environment in N.E. Thailand," *Photogrammetric Engineering & Remote Sensing*, 65(1):97–105.

Wolter, J. A., 1975, *The Emerging Discipline of Cartography*, Minneapolis: University of Minnesota, Department of Geography, unpublished dissertation.

Woodcock, C. E., Collins, J. B., Jakabhazy, V., Li, X., Macomber, S. and Y. Wu, 1997, "Inversion of the Li-Strahler Canopy Reflectance Model for Mapping Forest Structure," *IEEE Transactions Geoscience & Remote Sensing*, 35(2):405–414.

Zhan, X., Sohlberg, R. A., Townshend, J. R. G., DiMiceli, C., Carrol, M. L., Eastman, J. C., Hansen, M. C. and R. S. DeFries, 2002, "Detection of Land Cover Changes Using MODIS 250 m Data," *Remote Sensing of Environment*, 83:336–350.

Zhang, Q. and J. Wang, 2003, "A Rule-based Urban Land Use Inferring Method for Fine-resolution Multispectral Imagery," *Canadian Journal of Remote Sensing*, 29(1):1–13.

Electromagnetic Radiation Principles 2

Energy recorded by a remote sensing system undergoes fundamental interactions that should be understood to properly interpret the remotely sensed data. For example, if the energy being remotely sensed comes from the Sun, the energy

- is radiated by atomic particles at the source (the Sun),

- travels through the vacuum of space at the speed of light,

- interacts with the Earth's atmosphere,

- interacts with the Earth's surface,

- interacts with the Earth's atmosphere once again, and

- finally reaches the remote sensor, where it interacts with various optics, filters, film emulsions, or detectors.

It is instructive to examine each of these fundamental interactions that electromagnetic energy undergoes as it progresses from its source to the remote sensing system detector(s). Understanding these interactions helps image analysts extract useful information from the remotely sensed data (e.g., Tuominen and Pekkarinen, 2004).

 Conduction, Convection, and Radiation

Energy is the ability to do work. In the process of doing work, energy is often transferred from one body to another or from one place to another. The three basic ways in which energy can be transferred include conduction, convection, and radiation (Figure 2-1). Most people are familiar with *conduction* that occurs when one body (molecule or atom) transfers its kinetic energy to another by colliding with it. This is how a metal pan is heated by a hot burner on a stove. In *convection*, the kinetic energy of bodies is transferred from one place to another by physically moving the bodies. A good example is the heating of the air near the ground in the morning hours. The warmer air near the surface rises, setting up convectional currents in the atmosphere, which may produce cumulus clouds. The transfer of energy by electromagnetic *radiation* is of primary interest to remote sensing science because it is the only form of energy transfer that can take place in a vacuum such as the region between the Sun and the Earth.

Energy Transfer

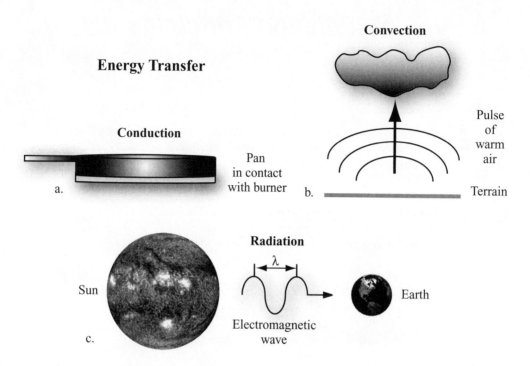

Figure 2-1 Energy may be transferred three ways: conduction, convection, and radiation. a) Energy may be conducted directly from one object to another as when a pan is in direct physical contact with a hot burner. b) The Sun bathes the Earth's surface with radiant energy causing the air near the ground to increase in temperature. The less dense air rises, creating convectional currents in the atmosphere. c) Electromagnetic energy in the form of electromagnetic waves may be transmitted through the vacuum of space from the Sun to the Earth.

Electromagnetic Radiation Models

To understand how electromagnetic radiation is created, how it propagates through space, and how it interacts with other matter, it is useful to describe the processes using two different models: the *wave* model and the *particle* model (Englert et al., 1994).

Wave Model of Electromagnetic Energy

In the 1860s, James Clerk Maxwell (1831–1879) conceptualized electromagnetic radiation (EMR) as an electromagnetic wave that travels through space at the speed of light. It took many years for scientists like Leon Foucault and Albert A. Michelson to determine the *speed of light*, c, as 299,792,458 meters per second (i.e., m s^{-1}), or 186,282.397 miles s^{-1}. These values are often generalized to 3×10^8 m s^{-1}, 300,000 km s^{-1} or 186,000 miles s^{-1}. A useful relation for quick calculations is that light travels about 1 ft per nanosecond (10^{-9} s) (Rinker, 1999). The *electromagnetic wave* consists of two fluctuating fields—one electric and the other

magnetic (Figure 2-2). The two vectors are at right angles (orthogonal) to one another, and both are perpendicular to the direction of travel.

How is an electromagnetic wave created? *Electromagnetic radiation* is generated whenever an electrical charge is accelerated. The wavelength (λ) of the electromagnetic radiation depends upon the length of time that the charged particle is accelerated. Its frequency (ν) depends on the number of accelerations per second. *Wavelength* is formally defined as the mean distance between consecutive maximums (or minimums) of a roughly periodic pattern (Figures 2-2 and 2-3) and is normally measured in micrometers (μm) or nanometers (nm). *Frequency* is the number of wavelengths that pass a point per unit time. A wave that sends one crest by every second (completing one cycle) is said to have a frequency of one cycle per second, or one *hertz*, abbreviated 1 Hz. Frequently used measures of wavelength and frequency are found in Table 2-1.

The relationship between the wavelength (λ) and frequency (ν) of electromagnetic radiation is based on the following formula, where c is the speed of light (Rott, 2000):

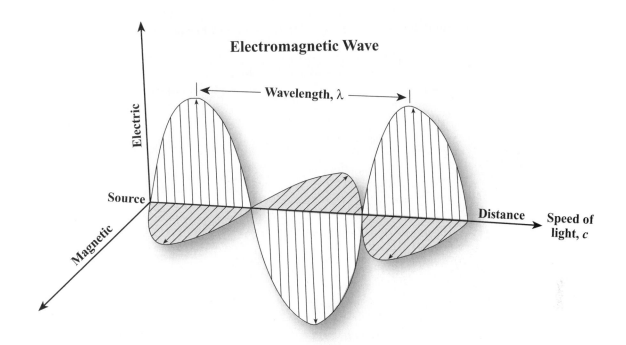

Figure 2-2 An electromagnetic wave is composed of both electric and magnetic vectors that are orthogonal (at 90° angles) to one another. The waves travel from the source at the speed of light (3×10^8 m s^{-1}).

$$c = \lambda v, \qquad\qquad (2\text{-}1)$$

$$v = \frac{c}{\lambda}, \qquad\qquad (2\text{-}2)$$

and

$$\lambda = \frac{c}{v}. \qquad\qquad (2\text{-}3)$$

Note that frequency is *inversely* proportional to wavelength. This relationship is shown diagrammatically in Figure 2-3, where the longer the wavelength, the lower the frequency; the shorter the wavelength, the higher the frequency. When electromagnetic radiation passes from one substance to another, the speed of light and wavelength change while the frequency remains the same.

All objects above absolute zero (–273°C or 0 K) emit electromagnetic energy, including water, soil, rock, vegetation, and the surface of the Sun. The Sun represents the initial source of most of the electromagnetic energy recorded by remote sensing systems (except RADAR, LIDAR, and SONAR) (Figure 2-4; Color Plate 2-1). We may think of the Sun as a 5770 – 6000 K *blackbody* (a theoretical construct that absorbs and radiates energy at the maximum possible rate per unit area at each wavelength (λ) for a given temper-

ature). The total emitted radiation from a blackbody (M_λ) measured in watts per m^{-2} is proportional to the fourth power of its absolute temperature (T) measured in kelvin (K). This is known as the *Stefan-Boltzmann law* and is expressed as (Rott, 2000):

$$M_\lambda = \sigma T^4 \qquad\qquad (2\text{-}4)$$

where σ is the Stefan-Boltzmann constant, 5.6697×10^{-8} W m^{-2} K^{-4}. The important thing to remember is that the amount of energy emitted by an object such as the Sun or the Earth is a function of its temperature. The greater the temperature, the greater the amount of radiant energy exiting the object. The amount of energy emitted by an object is computed by summing (integrating) the area under its curve (Figure 2-5). It is clear from this illustration that the total emitted radiation from the 6000 K Sun is far greater than that emitted by the 300 K Earth.

In addition to computing the total amount of energy exiting a theoretical blackbody such as the Sun, we can determine its dominant wavelength (λ_{max}) based on *Wien's displacement law*:

$$\lambda_{max} = \frac{k}{T} \qquad\qquad (2\text{-}5)$$

Inverse Relationship Between Wavelength and Frequency

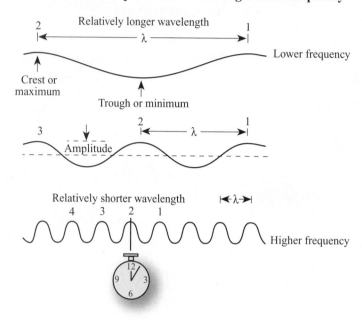

Figure 2-3 This cross-section of several electromagnetic waves illustrates the inverse relationship between wavelength (λ) and frequency (ν). The longer the wavelength, the lower the frequency; the shorter the wavelength, the higher the frequency. The amplitude of an electromagnetic wave is the height of the wave crest above the undisturbed position. Successive wave crests are numbered 1, 2, 3, and 4. An observer at the position of the clock records the number of crests that pass by in a second. This frequency is measured in cycles per second, or *hertz*.

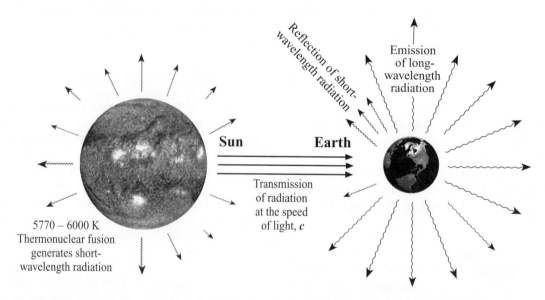

Figure 2-4 The thermonuclear fusion taking place on the surface of the Sun yields a continuous spectrum of electromagnetic energy. The 5770 – 6000 kelvin (K) temperature of this process produces a large amount of relatively short wavelength energy that travels through the vacuum of space at the speed of light. Some of this energy is intercepted by the Earth, where it interacts with the atmosphere and surface materials. The Earth reflects some of the energy directly back out to space or it may absorb the short wavelength energy and then re-emit it at a longer wavelength (after Strahler and Strahler, 1989).

Table 2-1. Wavelength and frequency standard units of measurement.

Wavelength (λ)	
kilometer (km)	1,000 m
meter (m)	1.0 m
centimeter (cm)	$0.01\ m = 10^{-2}\ m$
millimeter (mm)	$0.001\ m = 10^{-3}\ m$
micrometer (μm)	$0.000001 = 10^{-6}\ m$
nanometer (nm)	$0.000000001 = 10^{-9}\ m$
Angstrom (A)	$0.0000000001 = 10^{-10}\ m$

Frequency (cycles per second)	
hertz (Hz)	1
kilohertz (kHz)	$1,000 = 10^3$
megahertz (MHz)	$1,000,000 = 10^6$
gigahertz (GHz)	$1,000,000,000 = 10^9$

where k is a constant equaling 2898 μm K, and T is the absolute temperature in kelvin. Therefore, as the Sun approximates a 6000 K blackbody, its dominant wavelength (λ_{max}) is 0.483 μm:

$$0.483\mu m = \frac{2898\ \mu m\ K}{6000\ K}.$$

Electromagnetic energy from the Sun travels in 8 minutes across the intervening 93 million miles (150 million km) of space to the Earth. As shown in Figure 2-5, the Earth approximates a 300 K (27°C) blackbody and has a dominant wavelength at approximately 9.66 μm:

$$9.66\ \mu m = \frac{2898\ \mu m\ K}{300\ K}.$$

Although the Sun has a dominant wavelength at 0.48 μm, it produces a continuous spectrum of electromagnetic radiation ranging from very short, extremely high-frequency gamma and cosmic waves to long, very-low-frequency radio waves (Figures 2-6 and 2-7). The Earth intercepts only a

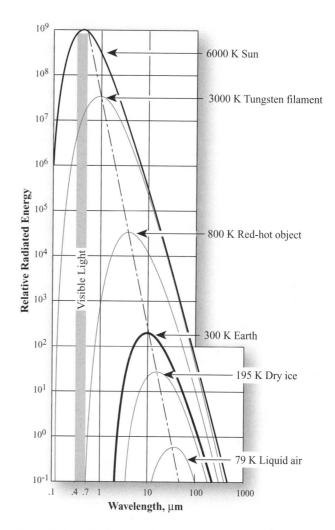

Figure 2-5   Blackbody radiation curves for several objects, including the Sun and the Earth, which approximate 6000 K and 300 K blackbodies, respectively. The area under each curve may be summed to compute the total radiant energy (M_λ) exiting each object (Equation 2-4). Thus, the Sun produces more radiant exitance than the Earth because its temperature is greater. As the temperature of an object increases, its dominant wavelength (λ_{max}) shifts toward the shorter wavelengths of the spectrum.

very small portion of the electromagnetic energy produced by the Sun.

As mentioned in Chapter 1, in remote sensing research we often specify a particular region of the electromagnetic spectrum (e.g., red light) by identifying a beginning and ending wavelength (or frequency) and then attaching a description. This wavelength (or frequency) interval in the electromag-

Table 2-2.   Methods of describing the color spectrum (after Nassau, 1983).

Color[a]	Wavelength Descriptions				Energy Descriptions	
	Angstrom (A)	Nanometer (nm)	Micrometer (μm)	Frequency (Hz x 10^{14})	Wave Number[c] (ψ cm^{-1})	Electron Volt (eV) = 1240/λ
Ultraviolet, sw	2,537	254	0.254	11.82	39,400	4.89
Ultraviolet, lw	3,660	366	0.366	8.19	27,300	3.39
Violet (limit)[b]	4,000	400	0.40	7.50	25,000	3.10
Blue	4,500	450	0.45	6.66	22,200	2.75
Green	5,000	500	0.50	6.00	20,000	2.48
Green	5,500	550	0.55	5.45	18,200	2.25
Yellow	5,800	580	0.58	5.17	17,240	2.14
Orange	6,000	600	0.60	5.00	16,700	2.06
Red	6,500	650	0.65	4.62	15,400	1.91
Red (limit)[b]	7,000	700	0.70	4.29	14,300	1.77
Infrared, near	10,000	1,000	1.0	3.00	10,000	1.24
Infrared, far	300,000	30,000	30.00	0.10	333	0.041

[a]Typical values only; lw = long wavelength; sw = short wavelength. [b]Exact limit depends on the observer, light intensity, eye adaptation, and other factors. [c]The wave number (ψ) is the number of waves in a unit length (usually per cm). Therefore, $\psi = 1 / \lambda$ (cm) = 10,000 / λ (μm) =100,000,000 / λ (A) in cm^{-1}.

netic spectrum is commonly referred to as a *band, channel,* or *region*. The major subdivisions of visible light are summarized in Table 2-2 and presented diagrammatically in Figure 2-7 and Color Plate 2-2. We generally think of visible light as being composed of energy in the blue (0.4 – 0.5 μm), green (0.5 – 0.6 μm), and red (0.6 – 0.7 μm) bands of the electromagnetic spectrum. Reflected near-infrared energy in the region from 0.7 to 1.3 μm is commonly used to expose black-and-white and color-infrared-sensitive film.

The middle-infrared region (often referred to as the short wavelength infrared, SWIR) includes energy with a wavelength of 1.3 to 3 μm. The thermal infrared region has two very useful bands at 3 to 5 μm and 8 to 14 μm. The microwave portion of the spectrum consists of much longer wavelengths (1 mm – 1 m). The radio-wave portion of the spectrum may be subdivided into UHF, VHF, radio (HF), LF, and ULF frequencies.

The spectral resolution of most remote sensing systems is described in terms of bands of the electromagnetic spectrum. For example, the spectral dimensions of the four bands of the Landsat Multispectral Scanner (MSS) and SPOT High Resolution Visible (HRV) sensors are shown in Figure 2-8, along with the spatial resolution of each band for comparison. The exact Landsat MSS and SPOT band specifications are provided in Chapter 7

Electromagnetic energy may be described not only in terms of wavelength and frequency but also in photon energy units such as joules (J) and electron volts (eV), as shown in Figure 2-7. Several of the more important mass, energy, and power conversions are summarized in Table 2-3.

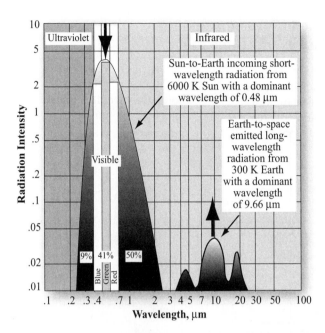

Figure 2-6   The Sun approximates a 6000 K blackbody with a dominant wavelength of about 0.48 μm. The Earth approximates a 300 K blackbody with a dominant wavelength of about 9.66 μm. The 6000 K Sun produces approximately 41% of its energy in the visible region from 0.4 to 0.7 μm (blue, green, and red light). The other 59% of the energy is in wavelengths shorter than blue light (<0.4 μm) and longer than red light (>0.7 μm). Our eyes are only sensitive to light from 0.4 to 0.7 μm (after Strahler and Strahler, 1989). Fortunately, it is possible to make remote sensor detectors sensitive to energy in these nonvisible regions of the spectrum.

Table 2-3.   Mass, energy, and power conversions.

Conversion from English to SI Units

To get:	Multiply:	By:
newtons[a]	pounds	4.448
joules[b]	BTUs[c]	1055
joules	calories[d]	4.184
joules	kilowatt-hours[e]	3.6 x 10^6
joules	foot-pounds[f]	1.356
joules	horsepower[g]	745.7

Conversion from SI to English Units

To get:	Multiply:	By:
BTUs	joules	0.00095
calories	joules	0.2390
kilowatt-hours	joules	2.78 x 10^{-7}
foot-pounds	joules	0.7375
horsepower	watts	0.00134

[a]newton: force needed to accelerate a mass of 1 kg by 1 m s^{-2}
[b]joule: a force of 1 newton acting through 1 meter.
[c]British thermal unit, or BTU: energy required to raise the temperature of 1 pound of water by 1 degree Fahrenheit.
[d]calorie: energy required to raise the temperature of 1 kilogram of water by 1 degree Celsius.
[e]kilowatt-hour: 1000 joules per second for 1 hour.
[f]foot-pound: a force of 1 pound acting through 1 foot.
[g]horsepower: 550 foot-pounds per second.

The Particle Model: Radiation from Atomic Structures

In *Opticks* (1704), Sir Isaac Newton stated that light was a stream of particles, or corpuscles, traveling in straight lines. He also knew that light had wavelike characteristics based on his work with glass plates. Nevertheless, during the two hundred years before 1905, light was thought of primarily as a smooth and continuous wave. Then, Albert Einstein (1879–1955) found that when light interacts with electrons, it has a different character (Figure 2-9). He concluded that when light interacts with matter, it behaves as though it is composed of many individual bodies called *photons*, which carry such particle-like properties as energy and momentum (Meadows, 1992). As a result, most physicists today would answer the question, "What is light?" by saying that light is a *particular* kind of matter (Feinberg, 1985). Thus, we sometimes describe electromagnetic energy in terms of its wavelike properties. But when the energy interacts with matter, it is useful to describe it as discrete packets of energy, or *quanta*. It is practical to review how electromagnetic energy is generated at the atomic level; this provides insight as to how light interacts with matter.

Electrons are the tiny negatively charged particles that move around the positively charged nucleus of an atom (Figure 2-10). Atoms of different substances are made up of varying numbers of electrons arranged in different ways. The interaction between the positively charged nucleus and the nega-

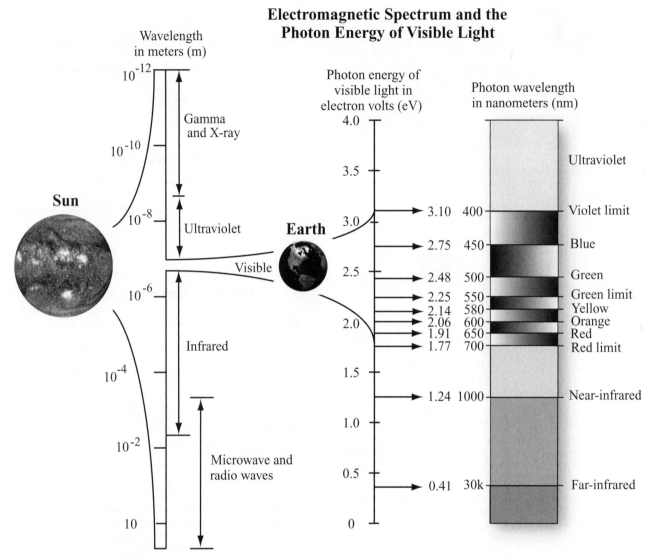

Figure 2-7   The electromagnetic spectrum and the photon energy of visible light. The Sun produces a continuous spectrum of energy from gamma rays to radio waves that continually bathe the Earth in energy. The visible portion of the spectrum may be measured using wavelength (measured in micrometers or nanometers, i.e., μm or nm) or electron volts (eV).

tively charged electron keeps the electron in orbit. While its orbit is not explicitly fixed, each electron's motion is restricted to a definite range from the nucleus. The allowable orbital paths of electrons moving around an atom might be thought of as energy classes or levels (Figure 2-10a). In order for an electron to climb to a higher class, work must be performed. However, unless an amount of energy is available to move the electron up at least one energy level, it will accept no work. If a sufficient amount of energy is received, the electron will jump to a new level and the atom is said to be *excited* (Figure 2-10b). Once an electron is in a higher orbit, it possesses potential energy. After about 10^{-8} seconds, the electron falls back to the atom's lowest empty energy

level or orbit and gives off radiation (Figure 2-10c). The wavelength of radiation given off is a function of the amount of work done on the atom, i.e., the quantum of energy it absorbed to cause the electron to become excited and move to a higher orbit.

Electron orbits are like the rungs of a ladder. Adding energy moves the electron up the energy ladder; emitting energy moves it down. However, the energy ladder differs from an ordinary ladder in that its rungs are unevenly spaced. This means that the energy an electron needs to absorb, or to give up, in order to jump from one orbit to the next may not be the

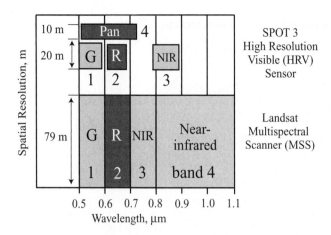

Figure 2-8   The nominal spectral bandwidths of the individual bands of the Landsat MSS and SPOT HRV sensor systems are summarized along the abscissa. The spatial resolution of each individual band is shown on the ordinate axis.

Figure 2-9   Albert Einstein (1879–1955) (courtesy Library of Congress).

same as the energy change needed for some other step. Furthermore, an electron does not necessarily use consecutive rungs. Instead, it follows what physicists call *selection rules*. In many cases, an electron uses one sequence of rungs as it climbs the ladder and another sequence as it descends (Nassau, 1983; 1984). The energy that is left over when the electrically charged electron moves from an excited state (Figure 2-10b) to a de-excited state (Figure 2-10c) is emitted by the atom as a single packet of electromagnetic radiation, a particle-like unit of light called a *photon*. Every time an electron jumps from a higher to a lower energy level, a photon moves away at the speed of light.

Somehow an electron must disappear from its original orbit and reappear in its destination orbit without ever having to traverse any of the positions in between. This process is called a *quantum leap* or *quantum jump*. If the electron leaps from its highest excited state to the ground state in a single leap, it will emit a single photon of energy. It is also possible for the electron to leap from an excited orbit to the ground state in a series of jumps (e.g., from 4 to 2 to 1). If it takes two leaps to get to the ground state, then each of these jumps will emit photons of somewhat less energy. The energies emitted in the two jumps must sum to the total of the single large jump (Trefil and Hazen, 1995).

Niels Bohr (1885–1962) and Max Planck recognized the discrete nature of exchanges of radiant energy and proposed the *quantum theory* of electromagnetic radiation. This theory states that energy is transferred in discrete packets called quanta or photons, as discussed. The relationship between the frequency of radiation expressed by wave theory and the quantum is:

$$Q = h\nu \tag{2-6}$$

where Q is the energy of a quantum measured in joules, h is the Planck constant (6.626×10^{-34} J s), and ν is the frequency of the radiation (Figure 2-11). Referring to Equation 2-3, we can multiply the equation by h/h, or 1, without changing its value:

$$\lambda = \frac{hc}{h\nu}. \tag{2-7}$$

By substituting Q for $h\nu$ (from Equation 2-6), we can express the wavelength associated with a quantum of energy as:

$$\lambda = \frac{hc}{Q} \tag{2-8}$$

or

$$Q = \frac{hc}{\lambda}. \tag{2-9}$$

Thus, we see that the energy of a quantum is inversely proportional to its wavelength, i.e., the longer the wavelength involved, the lower its energy content. This inverse relationship is important to remote sensing because it suggests that it is more difficult to detect longer-wavelength energy being emitted at thermal infrared wavelengths than those at shorter

Creation of Light from Atomic Particles and the Photoelectric Effect

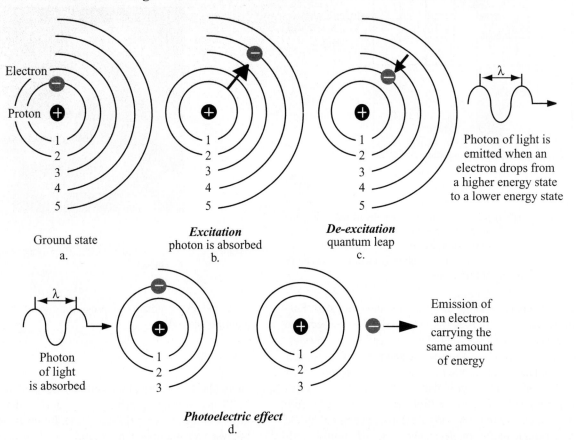

Figure 2-10   a–c) A photon of electromagnetic energy is emitted when an electron in an atom or molecule drops from a higher-energy state to a lower-energy state. The light emitted (i.e., its wavelength) is a function of the changes in the energy levels of the outer, valence electron. For example, yellow light is produced from a sodium vapor lamp in Figure 2-12. d) Matter can also be subjected to such high temperatures that electrons, which normally move in captured, nonradiating orbits, are broken free. When this happens, the atom remains with a positive charge equal to the negatively charged electron that escaped. The electron becomes a free electron, and the atom is called an ion. If another free electron fills the vacant energy level created by the free electron, then radiation from all wavelengths is produced, i.e., a continuous spectrum of energy. The intense heat at the surface of the Sun produces a continuous spectrum in this manner.

visible wavelengths. In fact, it might be necessary to have the sensor look at or dwell longer on the parcel of ground if we are trying to measure the longer wavelength energy.

Substances have color because of differences in their energy levels and the selection rules. For example, consider energized sodium vapor that produces a bright yellow light that is used in some street lamps. When a sodium vapor lamp is turned on, several thousand volts of electricity energize the vapor. The outermost electron in each energized atom of sodium vapor climbs to a higher rung on the energy ladder and then returns down the ladder in a certain sequence of rungs, the last two of which are 2.1 eV apart (Figure 2-12).

The energy released in this last leap appears as a photon of yellow light with a wavelength of 0.58 μm with 2.1 eV of energy (Nassau, 1983).

Matter can be heated to such high temperatures that electrons that normally move in captured, nonradiating orbits break free. This is called the *photoelectric effect* (Figure 2-10d). When this happens, the atom remains with a positive charge equal to the negatively charged electron that escaped. The electron becomes a free electron and the atom is called an *ion*. In the ultraviolet and visible (blue, green, and red) parts of the electromagnetic spectrum, radiation is produced by changes in the energy levels of the outer valence elec-

Figure 2-11 The energy of quanta (photons) ranging from gamma rays to radio waves in the electromagnetic spectrum.

trons. The wavelengths of energy produced are a function of the particular orbital levels of the electrons involved in the excitation process. If the atoms absorb enough energy to become ionized and if a free electron drops in to fill the vacant energy level, then the radiation given off is unquantized and a *continuous spectrum* is produced rather than a band or a series of bands. Every encounter of one of the free electrons with a positively charged nucleus causes rapidly changing electric and magnetic fields, so that radiation at all wavelengths is produced. The hot surface of the Sun is largely a *plasma* in which radiation of all wavelengths is produced. As previously shown in Figure 2-7, the spectrum of a plasma like the Sun is continuous.

In atoms and molecules, electron orbital changes produce the shortest-wavelength radiation, molecule vibrational motion changes produce near- and/or middle-infrared energy, and rotational motion changes produce long-wave-

length infrared or microwave radiation. More will be said about how thermal infrared radiation is produced and recorded by remote sensing systems in Chapter 8 (Thermal Infrared Remote Sensing).

 Atmospheric Energy–Matter Interactions

Radiant energy is the capacity of radiation within a spectral band to do work. Once electromagnetic radiation is generated, it is propagated through the Earth's atmosphere almost at the speed of light in a vacuum. Unlike a vacuum in which nothing happens, however, the atmosphere may affect not only the speed of radiation but also its wavelength, its intensity, and its spectral distribution. The electromagnetic radiation may also be diverted from its original direction due to refraction.

Creation of Light from Atomic Particles in a Sodium Vapor Lamp

Figure 2-12 Creation of light from atomic particles in a sodium vapor lamp. After being energized by several thousand volts of electricity, the outermost electron in each energized atom of sodium vapor climbs to a high rung on the energy ladder and then returns down the ladder in a predictable fashion. The last two rungs in the descent are 2.1 eV apart. This produces a photon of yellow light, which has 2.1 eV of energy (see Table 2-2).

Refraction

The speed of light in a vacuum is 3×10^8 m s^{-1}. When electromagnetic radiation encounters substances of different density, like air and water, refraction may take place. *Refraction* refers to the bending of light when it passes from one medium to another of different density. Refraction occurs because the media are of differing densities and the speed of EMR is different in each. The *index of refraction (n)* is a measure of the optical density of a substance. This index is the ratio of the speed of light in a vacuum, c, to the speed of light in a substance such as the atmosphere or water, c_n:

$$n = \frac{c}{c_n}. \tag{2-10}$$

The speed of light in a substance can never reach the speed of light in a vacuum. Therefore, its index of refraction must always be greater than 1. For example, the index of refraction for the atmosphere is 1.0002926 and 1.33 for water. Light travels more slowly through water because of water's higher density.

Refraction can be described by Snell's law, which states that for a given frequency of light (we must use frequency since, unlike wavelength, it does not change when the speed of light changes), the product of the index of refraction and the sine of the angle between the ray and a line normal to the interface is constant:

$$n_1 \sin \theta_1 = n_2 \sin \theta_2. \tag{2-11}$$

From Figure 2-13 we can see that a nonturbulent atmosphere can be thought of as a series of layers of gases, each with a slightly different density. Anytime energy is propagated through the atmosphere for any appreciable distance at any angle other than vertical, refraction occurs.

The amount of refraction is a function of the angle made with the vertical (θ), the distance involved (in the atmosphere the greater the distance, the more changes in density), and the density of the air involved (air is usually more dense near sea level). Serious errors in location due to refraction can occur in images formed from energy detected at high altitudes or at acute angles. However, these location errors are predictable by Snell's law and thus can be removed. Notice that

$$\sin \theta_2 = \frac{n_1 \sin \theta_1}{n_2}. \tag{2-12}$$

Therefore, if one knows the index of refraction of medium n_1 and n_2 and the angle of incidence of the energy to medium n_1, it is possible to predict the amount of refraction that will take place ($\sin \theta_2$) in medium n_2 using trigonometric relationships. It is useful to note, however, that most image analysts never concern themselves with computing the index of refraction.

Scattering

One very serious effect of the atmosphere is the scattering of radiation by atmospheric particles. *Scattering* differs from

Atmospheric Refraction

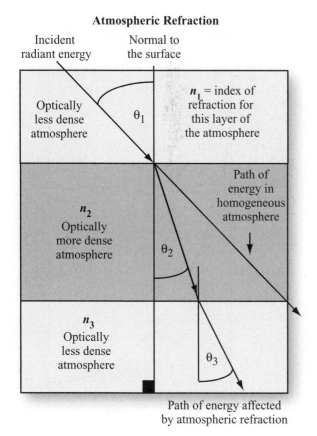

Path of energy affected
by atmospheric refraction

Figure 2-13   Refraction in three nonturbulent atmospheric layers. The incident energy is bent from its normal trajectory as it travels from one atmospheric layer to
another. Snell's law can be used to predict how
much bending will take place, based on a knowledge of the angle of incidence (θ) and the index of
refraction of each atmospheric level, n_1, n_2, and n_3.

Atmospheric Layers and Constituents

Figure 2-14   Major subdivisions of the atmosphere and the types
of molecules and aerosols found in each layer
(adapted from Miller and Vermote, 2002).

wavelength of the incident electromagnetic radiation (Figure
2-15a). Rayleigh scattering is named after the English physicist Lord Rayleigh, who offered the first coherent explanation for it. All scattering is accomplished through absorption
and re-emission of radiation by atoms or molecules in the
manner previously described in the section on radiation from
atomic structures. It is impossible to predict the direction in
which a specific atom or molecule will emit a photon, hence
scattering. The energy required to excite an atom is associated with powerful short-wavelength, high-frequency radiation.

The approximate amount of Rayleigh scattering in the atmosphere in optical wavelengths (0.4 – 0.7 μm) may be computed using the Rayleigh scattering cross-section (τ_m)
algorithm (Cracknell and Hayes, 1993):

$$\tau_m = \frac{8\pi^3(n^2-1)^2}{(3N^2\lambda^4)} \qquad (2\text{-}13)$$

where n = refractive index, N = number of air molecules per
unit volume, and λ = wavelength. The amount of scattering
is inversely related to the fourth power of the radiation's
wavelength. For example, ultraviolet light at 0.3 μm is scattered approximately 16 times more than red light at 0.6 μm,
i.e., $(0.6/0.3)^4 = 16$. Blue light at 0.4 μm is scattered about 5
times more than red light at 0.6 μm, i.e., $(0.6/0.4)^4 = 5.06$.
The amount of Rayleigh scattering expected throughout the
visible spectrum (0.4 – 0.7 μm) is shown in Figure 2-16.

reflection in that the direction associated with scattering is
unpredictable, whereas the direction of reflection (to be
defined shortly) is predictable. There are essentially three
types of scattering: Rayleigh, Mie, and nonselective scattering. Major subdivisions of the atmosphere and the types of
molecules and aerosols found in each layer are shown in Figure 2-14 (Miller and Vermote, 2002). The relative size of the
wavelength of the incident electromagnetic radiation, the
diameter of the gases, water vapor, and/or dust with which
the energy interacts, and the type of scattering that should
occur are summarized in Figure 2-15.

Rayleigh scattering (often referred to as *molecular scattering*) occurs when the effective diameter of the matter (usually air molecules such as oxygen and nitrogen in the
atmosphere) are many times smaller (usually < 0.1) than the

Atmospheric Scattering

Rayleigh Scattering

a. ◯ Gas molecule

Mie Scattering

b. diameter   Smoke, dust

Photon of electromagnetic
energy modeled as a wave

Nonselective Scattering

c. Water
vapor

Figure 2-15   Type of scattering is a function of 1) the wavelength
of the incident radiant energy and 2) the size of the
gas molecule, dust particle, and/or water vapor
droplet encountered.

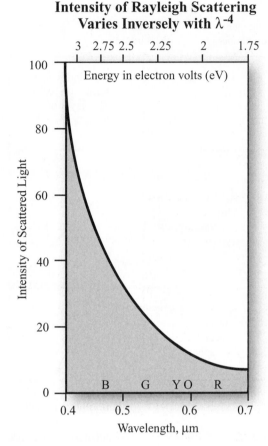

**Intensity of Rayleigh Scattering
Varies Inversely with λ^{-4}**

Figure 2-16   The intensity of Rayleigh scattering varies inversely
with the fourth power of the wavelength (λ^{-4}).

Most Rayleigh scattering by gas molecules takes place in the
atmosphere 2 to 8 km above the ground (Figure 2-14). Rayleigh scattering is responsible for the blue appearance of the
sky. The shorter violet and blue wavelengths are more efficiently scattered than the longer orange and red wavelengths. When we look up on a cloudless day and admire the
blue sky, we are witnessing the preferential scattering of the
short-wavelength sunlight. Rayleigh scattering is also
responsible for red sunsets. Since the atmosphere is a thin
shell of gravitationally bound gas surrounding the solid
Earth, sunlight must pass through a longer slant path of air at
sunset (or sunrise) than at noon. Since the violet and blue
wavelengths are scattered even more during their now
longer path through the air than when the Sun is overhead,
what we see when we look toward the sunset is the residue—
the wavelengths of sunlight that are hardly scattered away at
all, especially the oranges and reds.

Mie scattering (sometimes referred to as non-molecular or
aerosol particle scattering) takes place in the lower 4.5 km of
the atmosphere, where there may be many essentially spherical particles present with diameters approximately equal to
the size of the wavelength of the incident energy (Figure 2-
15b). The actual size of the particles may range from 0.1 to

10 times the wavelength of the incident energy. For visible
light, the main scattering agents are dust and other particles
ranging from a few tenths of a micrometer to several
micrometers in diameter. The amount of scatter is greater
than Rayleigh scatter, and the wavelengths scattered are
longer. The greater the amount of smoke and dust particles
in the atmospheric column, the more that violet and blue
light will be scattered away and only the longer orange and
red wavelength light will reach our eyes. Pollution also contributes to beautiful sunsets and sunrises.

Nonselective scattering takes place in the lowest portions of
the atmosphere where there are particles >10 times the
wavelength of the incident electromagnetic radiation (Figure
2-15c). This type of scattering is nonselective, i.e., all wavelengths of light are scattered, not just blue, green, or red.
Thus, the water droplets and ice crystals that make up clouds
and fogbanks scatter all wavelengths of visible light equally
well, causing the cloud to appear white. Nonselective scattering of approximately equal proportions of blue, green,

and red light always appears as white light to the casual observer. This is the reason why putting our automobile high beams on in fog only makes the problem worse as we nonselectively scatter even more light into our visual field of view.

Scattering is a very important consideration in remote sensing investigations. It can severely reduce the information content of remotely sensed data to the point that the imagery loses contrast and it becomes difficult to differentiate one object from another.

Absorption

Absorption is the process by which radiant energy is absorbed and converted into other forms of energy. The absorption of the incident radiant energy may take place in the atmosphere or on the terrain. An *absorption band* is a range of wavelengths (or frequencies) in the electromagnetic spectrum within which radiant energy is absorbed by a substance. The effects of water (H_2O), carbon dioxide (CO_2), oxygen (O_2), ozone (O_3), and nitrous oxide (N_2O) on the transmission of light through the atmosphere are summarized in Figure 2-17a. The cumulative effect of the absorption by the various constituents can cause the atmosphere to "close down" completely in certain regions of the spectrum. This is bad for remote sensing because no energy is available to be sensed. Conversely, in the visible portion of the spectrum ($0.4 - 0.7$ μm), the atmosphere does not absorb all of the incident energy but transmits it rather effectively. Portions of the spectrum that transmit radiant energy effectively are called *atmospheric windows*.

Absorption occurs when incident energy of the same frequency as the resonant frequency of an atom or molecule is absorbed, producing an excited state. If instead of reradiating a photon of the same wavelength, the energy is transformed into heat motion and is subsequently reradiated at a longer wavelength, absorption occurs. When dealing with a medium like air, absorption and scattering are frequently combined into an *extinction coefficient* (Konecny, 2003). Transmission is inversely related to the extinction coefficient times the thickness of the layer. Certain wavelengths of radiation are affected far more by absorption than by scattering. This is particularly true of infrared and wavelengths shorter than visible light. The combined effects of atmospheric absorption, scattering, and reflectance (from cloud tops) can dramatically reduce the amount of solar radiation reaching the Earth's surface at sea level, as shown in Figure 2-17b.

Chlorophyll in vegetation absorbs much of the incident blue and red light for photosynthetic purposes. Chapter 10 describes the importance of these chlorophyll absorption bands and their role when remotely sensing vegetation. Similarly, water is an excellent absorber of energy (Chapter 11). Many minerals have unique absorption characteristics in very specific portions of the electromagnetic spectrum that allow us to use remote sensing to identify them (Chapter 13), assuming there is no overlying vegetation (Clark, 1999).

Energy reaching the remote sensing system must pass through the atmosphere twice. Therefore, it is common to identify two *atmospheric transmission coefficients*: one for energy coming into the atmosphere (T_{θ_o}) at an incidence angle related to the source of the energy (e.g., the Sun), and one for the atmosphere that the Earth surface–reflected or emitted energy must pass through to reach the remote sensor system (T_{θ_v}). We will see how important these two parameters are in the discussion on atmospheric correction later in this chapter.

Reflectance

Reflectance is the process whereby radiation "bounces off" an object like the top of a cloud, a water body, or bare soil. Actually, the process is more complicated, involving reradiation of photons in unison by atoms or molecules in a layer approximately one-half wavelength deep. Reflection exhibits fundamental characteristics that are important in remote sensing. First, the incident radiation, the reflected radiation, and a vertical to the surface from which the angles of incidence and reflection are measured all lie in the same plane. Second, the angle of incidence and the angle of reflection (exitance) are approximately equal, as shown in Figure 2-18.

There are various types of reflecting surfaces. *Specular reflection* occurs when the surface from which the radiation is reflected is essentially smooth (i.e., the average surface-profile height is several times smaller than the wavelength of radiation striking the surface). Several features, such as calm water bodies, act like *near-perfect specular reflectors* (Figure 2-18a,b). If there are very few ripples on the surface, the incident energy will leave the water body at an angle equal and opposite to the incident energy. We know this occurs from our personal experience. If we shine a flashlight at night on a tranquil pool of water, the light will bounce off the surface and into the trees across the way at an angle equal to and opposite from the incident radiation.

If the surface has a large surface height relative to the size of the wavelength of the incident energy, the reflected rays go

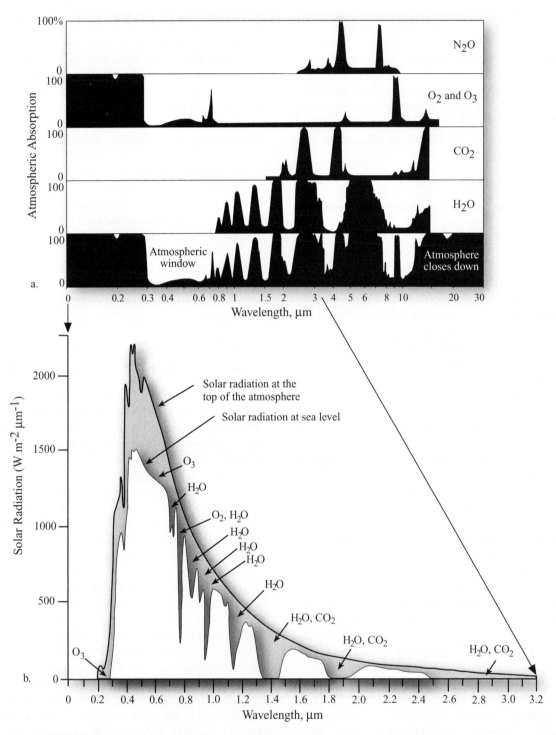

Figure 2-17   a) The absorption of the Sun's incident electromagnetic energy in the region from 0.1 to 30 μm by various atmospheric gases. The first four graphs depict the absorption characteristics of N_2O, O_2 and O_3, CO_2, and H_2O, while the final graphic depicts the cumulative result of all these constituents being in the atmosphere at one time. The atmosphere essentially "closes down" in certain portions of the spectrum while "atmospheric windows" exist in other regions that transmit incident energy effectively to the ground. It is within these windows that remote sensing systems must function. b) The combined effects of atmospheric absorption, scattering, and reflectance reduce the amount of solar irradiance reaching the Earth's surface at sea level.

Specular versus Diffuse Reflectance

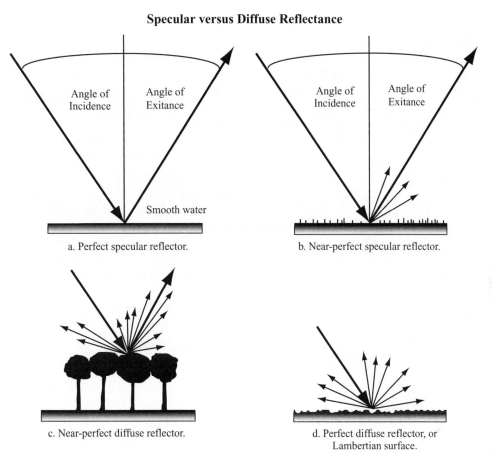

a. Perfect specular reflector.

b. Near-perfect specular reflector.

c. Near-perfect diffuse reflector.

d. Perfect diffuse reflector, or Lambertian surface.

Figure 2-18   The nature of specular and diffuse reflectance.

in many directions, depending on the orientation of the smaller reflecting surfaces. This *diffuse reflection* does not yield a mirror image, but instead produces diffused radiation (Figure 2-18c). White paper, white powders, and other materials reflect visible light in this diffuse manner. If the surface is very rough, then unpredictable scattering may occur. Lambert defined a perfectly diffuse surface; hence, the commonly designated *Lambertian surface* is one for which the radiant flux leaving the surface is constant for any angle of reflectance (Figure 2-18d).

Much of the incident radiant flux from the Sun is reflected from the top of clouds and other materials in the atmosphere. A substantial amount of this energy is reradiated back to space. As we shall see, the specular and diffuse reflection principles that apply to clouds also apply to the terrain.

Spectral reflectance may be measured from great distances using multispectral remote sensing systems. Optical remote sensing systems that record spectral reflectance are summa-

rized in Chapters 4 and 7. It is also possible to obtain spectral reflectance measurements of the terrain (e.g., crops, soil, and water) by holding a spectroradiometer just a few centimeters above the ground as shown in Figure 1-1. Methods of obtaining *in situ* spectral reflectance measurements are presented in Chapter 15.

Terrain Energy–Matter Interactions

The time rate of flow of energy onto, off of, or through a surface is called *radiant flux* (Φ) and is measured in watts (W) (Table 2-4). The characteristics of the radiant flux and what happens to it as it interacts with the Earth's surface is of critical importance in remote sensing. In fact, this is the fundamental focus of much remote sensing research. By carefully monitoring the exact nature of the incoming (incident) radiant flux in selective wavelengths and how it interacts with

Table 2-4.   Radiometric concepts (Colwell, 1983).

Name	Symbol	Units	Concept
Radiant energy	Q_λ	joules, J	Capacity of radiation within a specified spectral band to do work.
Radiant flux	Φ_λ	watts, W	Time rate of flow of energy onto, off of, or through a surface.
Radiant flux density at the surface			
Irradiance	E_λ	watts per square meter, W m^{-2}	Radiant flux incident upon a surface per unit area of that surface.
Radiant exitance	M_λ	watts per square meter, W m^{-2}	Radiant flux leaving a surface per unit area of that surface.
Radiance	L_λ	watts per square meter, per steradian, W m^{-2} sr^{-1}	Radiant intensity per unit of projected source area in a specified direction.
Hemispherical reflectance	ρ_λ	dimensionless	$\dfrac{\Phi_{\text{reflected}_\lambda}}{\Phi_{i_\lambda}}$
Hemispherical transmittance	τ_λ	dimensionless	$\dfrac{\Phi_{\text{transmitted}_\lambda}}{\Phi_{i_\lambda}}$
Hemispherical absorptance	α_λ	dimensionless	$\dfrac{\Phi_{\text{absorbed}_\lambda}}{\Phi_{i_\lambda}}$

the terrain, it is possible to learn important information about the terrain.

Radiometric quantities have been identified that allow analysts to keep a careful record of the incident and exiting radiant flux (Table 2-4). We begin with the simple *radiation budget equation,* which states that the total amount of radiant flux in specific wavelengths (λ) incident to the terrain (Φ_{i_λ}) must be accounted for by evaluating the amount of radiant flux reflected from the surface ($\Phi_{\text{reflected}_\lambda}$), the amount of radiant flux absorbed by the surface ($\Phi_{\text{absorbed}_\lambda}$), and the amount of radiant flux transmitted through the surface ($\Phi_{\text{transmitted}_\lambda}$):

$$\Phi_{i_\lambda} = \Phi_{\text{reflected}_\lambda} + \Phi_{\text{absorbed}_\lambda} + \Phi_{\text{transmitted}_\lambda}. \qquad (2\text{-}14)$$

It is important to note that these radiometric quantities are based on the amount of radiant energy incident to a surface from any angle in a hemisphere (i.e., a half of a sphere).

Hemispherical Reflectance, Absorptance, and Transmittance

Hemispherical reflectance (ρ_λ) is defined as the dimensionless ratio of the radiant flux reflected from a surface to the radiant flux incident to it (Table 2-4):

$$\rho_\lambda = \frac{\Phi_{\text{reflected}_\lambda}}{\Phi_{i_\lambda}}. \qquad (2\text{-}15)$$

Hemispherical transmittance (τ_λ) is defined as the dimensionless ratio of the radiant flux transmitted through a surface to the radiant flux incident to it:

$$\tau_\lambda = \frac{\Phi_{\text{transmitted}_\lambda}}{\Phi_{i_\lambda}}. \qquad (2\text{-}16)$$

Hemispherical absorptance (α_λ) is defined by the dimensionless relationship:

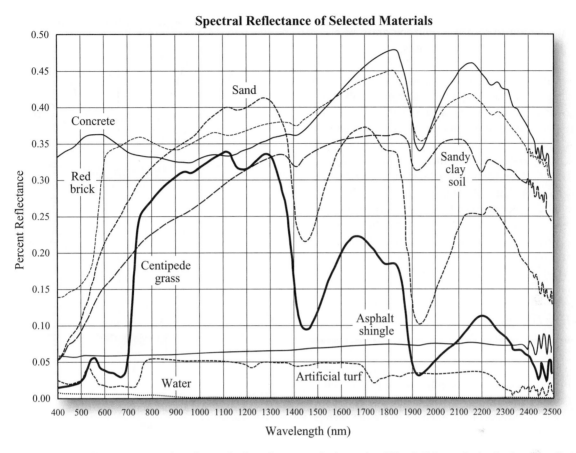

Figure 2-19 Spectral reflectance curves for urban–suburban phenomena in the region 400 – 2,500 nm obtained using a handheld spectro-radiometer (Chapter 15 describes how *in situ* spectral reflectance measurements are collected).

$$\alpha_\lambda = \frac{\Phi_{absorbed_\lambda}}{\Phi_{i_\lambda}} \qquad (2\text{-}17)$$

or

$$\alpha_\lambda = 1 - (\rho_\lambda + \tau_\lambda). \qquad (2\text{-}18)$$

These definitions imply that radiant energy must be conserved, i.e., it is either returned back by reflection, transmitted through a material, or absorbed and transformed into some other form of energy inside the terrain. The net effect of absorption of radiation by most substances is that the energy is converted into heat, causing a subsequent rise in the substance's temperature.

These radiometric quantities are useful for producing general statements about the spectral reflectance, absorptance, and transmittance characteristics of terrain features. In fact, if we take the simple hemispherical reflectance equation and multiply it by 100, we obtain an expression for percent reflectance ($\rho_{\lambda_\%}$):

$$\rho_{\lambda_\%} = \frac{\Phi_{reflected_\lambda}}{\Phi_{i_\lambda}} \times 100, \qquad (2\text{-}19)$$

which is often used in remote sensing research to describe the spectral reflectance characteristics of various phenomena. Examples of spectral percent reflectance curves for selected urban–suburban phenomena are shown in Figure 2-19. Spectral reflectance curves provide no information about the absorption and transmittance of the radiant energy. But because many of the sensor systems such as cameras and some multispectral scanners record only reflected energy, this information is still quite valuable and can form the basis for object identification and assessment. For example, it is clear from Figure 2-19 that the Centipede grass reflects only approximately 3 to 4 percent of the incident red radiant energy (0.6 – 0.7 μm) while reflecting up to approximately 35 percent of the incident near-infrared radiant flux (0.7 – 0.9 μm). If we wanted to discriminate between grass and artificial turf, the ideal portion of the spectrum to remotely sense in would be the near-infrared region because artificial turf reflects only about 5 percent of the incident near-infrared energy. This would cause a black-and-white infrared image of the terrain to display Centipede grass in bright tones and the artificial turf in darker tones.

Hemispherical reflectance, transmittance, and absorptance radiometric quantities do not provide information about the exact amount of energy reaching a specific area on the ground from a specific direction or about the exact amount of radiant flux exiting the ground in a certain direction. Remote sensing systems can be located in space only at a single point in time, and they usually look only at a relatively small portion of the Earth at a single instant. Therefore, it is important to refine our radiometric measurement techniques so that more precise radiometric information can be extracted from the remotely sensed data (Du et al., 2002). This requires the introduction of several radiometric quantities that provide progressively more precise radiometric information.

Radiant Flux Density

A flat area (e.g., 1×1 m in dimension) being bathed in radiant flux (Φ) in specific wavelengths (λ) from the Sun is shown in Figure 2-20. The amount of radiant flux intercepted divided by the area of the plane surface is the average *radiant flux density*.

Irradiance and Exitance

The amount of radiant flux incident upon a surface per unit area of that surface is called *irradiance (E_λ)*:

$$E_\lambda = \frac{\Phi_\lambda}{A}. \tag{2-20}$$

The amount of radiant flux leaving a surface per unit area of that surface is called *exitance (M_λ)*:

$$M_\lambda = \frac{\Phi_\lambda}{A}. \tag{2-21}$$

Both quantities are usually measured in Watts per meter squared (W m^{-2}). Although we do not have information on the direction of either the incoming or outgoing radiant energy (i.e., the energy can come and go at any angle throughout the entire hemisphere), we have now refined the measurement to include information about the size of the study area of interest on the ground in m^2. Next we need to

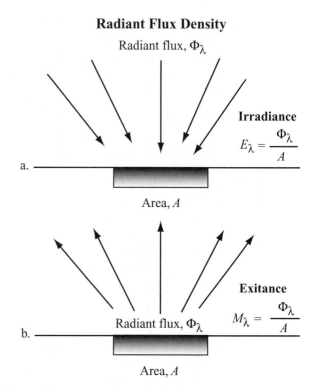

Figure 2-20   The concept of radiant flux density for an area on the surface of the Earth. a) *Irradiance* is a measure of the amount of radiant flux incident upon a surface per unit area of the surface measured in Watts m^{-2}. b) *Exitance* is a measure of the amount of radiant flux leaving a surface per unit area of the surface measured in Watts m^{-2}.

refine our radiometric measurement techniques to include information on what direction the radiant flux is leaving the study area.

Radiance

Radiance is the most precise remote sensing radiometric measurement. *Radiance (L_λ)* is the radiant intensity per unit of projected source area in a specified direction. It is measured in Watts per meter squared per steradian (W m^{-2} sr^{-1}). The concept of radiance is best understood by evaluating Figure 2-21. First, the radiant flux leaves the projected source area in a specific direction toward the remote sensor. We are not concerned with any other radiant flux that might be leaving the source area in any other direction. We are interested only in the radiant flux in certain wavelengths (Φ_λ) leaving the projected source area (A) within a certain direction (cos θ) and solid angle (Ω) (Milman, 1999):

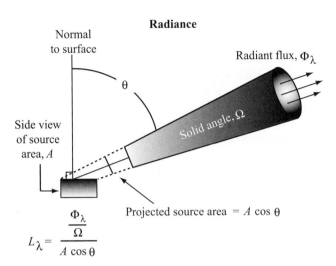

Radiance

Normal to surface

θ

Radiant flux, Φ_λ

Solid angle, Ω

Side view of source area, A

Projected source area $= A \cos \theta$

$$L_\lambda = \frac{\frac{\Phi_\lambda}{\Omega}}{A \cos \theta}$$

Figure 2-21   The concept of *radiance* leaving a specific projected source area on the ground, in a specific direction, and within a specific solid angle.

$$L_\lambda = \frac{\frac{\Phi_\lambda}{\Omega}}{A \cos \theta}. \qquad (2-22)$$

One way of visualizing the solid angle is to consider what you would see if you were in an airplane looking through a telescope at the ground. Only the energy that exited the terrain and came up to and through the telescope in a specific solid angle (measured in steradians) would be intercepted by the telescope and viewed by your eye. Therefore, the solid angle is like a three-dimensional cone (or tube) that funnels radiant flux from a specific point source on the terrain toward the sensor system. Hopefully, energy from the atmosphere or other terrain features does not become scattered into the solid angle field of view and contaminate the radiant flux from the area of interest on the ground. Unfortunately, this is not often the case because scattering in the atmosphere and from other nearby areas on the ground can contribute spurious spectral energy, which enters into the solid angle field of view.

 Atmospheric Energy–Matter Interactions

The radiant flux reflected or emitted from the Earth's surface once again enters the atmosphere, where it interacts with the various gases, water vapor, and particulates. Thus, atmospheric scattering, absorption, reflection, and refraction influence the radiant flux once again before the energy is recorded by the remote sensing system.

 Energy–Matter Interactions at the Sensor

Finally, energy–matter interactions take place when the energy reaches the remote sensor. If an aerial camera is being used, then the radiance will interact with the camera filter, the optical glass lens, and finally the film emulsion with its light-sensitive silver halide crystals. The emulsion must then be developed and printed before an analog copy is available for analysis. Rather than storing a latent image on film, an optical–mechanical detector will digitally record the number of photons in very specific wavelength regions reaching the sensor.

Target and Path Radiance

Ideally, the radiance (L) recorded by the camera or detector is a true function of the amount of radiance leaving the target terrain within the instantaneous-field-of-view (IFOV) at a specific solid angle, as previously discussed. Unfortunately, other radiant energy may enter into the field of view from various other paths and introduce confounding noise into the remote sensing process. Therefore, additional radiometric variable definitions are needed to identify the major sources and paths of this energy (Cracknell and Hayes, 1993; Green, 2003). The variables are summarized in Table 2-5. The various paths and factors that determine the radiance reaching the remote sensor are summarized in Figure 2-22, including:

- *Path 1* contains spectral solar irradiance (E_{o_λ}) that was attenuated very little before illuminating the terrain within the IFOV. Notice in this case that we are interested in the solar irradiance from a specific solar zenith angle (θ_o) and that the amount of irradiance reaching the terrain is a function of the atmospheric transmittance at this angle (T_{θ_o}). If all of the irradiance makes it to the ground, then the atmospheric transmittance (T_{θ_o}) equals one. If none of the irradiance makes it to the ground, then the atmospheric transmittance is zero.

- *Path 2* contains spectral diffuse sky irradiance (E_{d_λ}) that never even reaches the Earth's surface (the target study area) because of scattering in the atmosphere. Unfortunately, such energy is often scattered directly into the IFOV of the sensor system. As previously discussed, Rayleigh scattering of blue light contributes much to this diffuse sky irradiance. That is why the blue band image

Table 2-5.　Radiometric variables used in remote sensing.

Radiometric Variables

E_o = solar irradiance at the top of the atmosphere (W m^{-2})

E_{o_λ} = spectral solar irradiance at the top of the atmosphere (W m^{-2} μm^{-1})

E_d = diffuse sky irradiance (W m^{-2})

E_{d_λ} = spectral diffuse sky irradiance (W m^{-2} μm^{-1})

E_{du_λ} = the *upward* reflectance of the atmosphere

E_{dd_λ} = the *downward* reflectance of the atmosphere

E_g = global irradiance incident on the surface (W m^{-2})

E_{g_λ} = spectral global irradiance on the surface (W m^{-2} μm^{-1})

τ = normal atmospheric optical thickness

T_θ = atmospheric transmittance at an angle θ to the zenith

θ_o = solar zenith angle

θ_v = view angle of the satellite sensor (or scan angle)

μ = cos θ

ρ_λ = surface target reflectance at a specific wavelength

ρ_{λ_n} = reflectance from a neighboring area

L_s = total radiance at the sensor (W m^{-2} sr^{-1})

L_T = total radiance from the target of interest toward the sensor (W m^{-2} sr^{-1})

L_i = intrinsic radiance of the target (W m^{-2} sr^{-1}) (i.e., what a handheld radiometer would record on the ground without intervening atmosphere)

L_p = path radiance from multiple scattering (W m^{-2} sr^{-1})

produced by a remote sensor system is often much brighter than any of the other bands. It contains much unwanted diffuse sky irradiance that was inadvertently scattered into the IFOV of the sensor system. Therefore, if possible, we want to minimize its effects. Green (2003) refers to the quantity as the upward reflectance of the atmosphere (E_{du_λ}).

- *Path 3* contains energy from the Sun that has undergone some Rayleigh, Mie, and/or nonselective scattering and perhaps some absorption and reemission before illuminating the study area. Thus, its spectral composition and polarization may be somewhat different from the energy that reaches the ground from path 1. Green (2003)

refers to this quantity as the downward reflectance of the atmosphere (E_{dd_λ}).

- *Path 4* contains radiation that was reflected or scattered by nearby terrain (ρ_{λ_n}) covered by snow, concrete, soil, water, and/or vegetation into the IFOV of the sensor system. The energy does not actually illuminate the study area of interest. Therefore, if possible, we would like to minimize its effects.

- *Path 5* is energy that was also reflected from nearby terrain into the atmosphere, but then scattered or reflected onto the study area.

Therefore, for a given spectral interval in the electromagnetic spectrum (e.g., λ_1 to λ_2 could be 0.6 − 0.7 μm or red light), the total solar irradiance reaching the *Earth's surface*, E_{g_λ}, is an integration of several components:

$$E_{g_\lambda} = \int_{\lambda_1}^{\lambda_2} (E_{o_\lambda} T_{\theta_o} \cos\theta_o + E_{d_\lambda}) d\lambda \quad \text{(W m}^{-2}\text{ μm}^{-1}\text{). (2-23)}$$

It is a function of the spectral solar irradiance at the top of the atmosphere (E_{o_λ}) multiplied by the atmospheric transmittance (T_{θ_o}) at a certain solar zenith angle (θ_o) plus the contribution of spectral diffuse sky irradiance (E_{d_λ}).

Only a small amount of this irradiance is actually reflected by the terrain in the direction of the satellite sensor system. If we assume the surface of Earth is a diffuse reflector (a Lambertian surface), the total amount of radiance exiting the target study area (L_T) toward the sensor is:

$$L_T = \frac{1}{\pi} \int_{\lambda_1}^{\lambda_2} \rho_\lambda\, T_{\theta_v} (E_{o_\lambda} T_{\theta_o} \cos\theta_o + E_{d_\lambda}) d\lambda. \quad \text{(2-24)}$$

The average surface target reflectance (ρ_λ) is included because the vegetation, soil, and water within the IFOV selectively absorb some of the incident energy. Therefore, not all of the energy incident to the IFOV (E_{g_λ}) leaves the IFOV. In effect, the terrain acts like a filter, selectively absorbing certain wavelengths of light while reflecting others. Note that the energy exiting the terrain is at an angle (θ_v), requiring the use of an atmospheric transmittance factor T_{θ_v} once again.

It would be wonderful if the total radiance recorded by the sensor, L_S, equaled the radiance returned from the target study area of interest, L_T. Unfortunately, $L_S \neq L_T$ because there is some additional radiance from different *paths* that

Various Paths of Radiance Received by a Remote Sensing System

Figure 2-22 *Radiance* (L_T) *from paths 1, 3, and 5 contains intrinsic valuable spectral information about the target of interest. Conversely, the* path radiance *(L_p) from paths 2 and 4 includes diffuse sky irradiance or radiance from neighboring areas on the ground. This path radiance generally introduces unwanted radiometric noise in the remotely sensed data and complicates the image interpretation process.*

may fall within the IFOV of the sensor system detector (Figure 2-22). This is often called *path radiance*, L_P. Thus, the total radiance recorded by the sensor becomes:

$$L_S = L_T + L_P \quad (\text{W m}^{-2} \text{ sr}^{-1}). \qquad (2\text{-}25)$$

We see from Equation 2-25 and Figure 2-22 that the path radiance (L_P) is an intrusive (bad) component of the total amount of radiance recorded by the sensor system (L_S). It is composed of radiant energy primarily from the diffuse sky irradiance (E_d) from path 2 as well as the reflectance from nearby ground areas ρ_{λ_n} from path 4. Path radiance introduces error to the remote sensing data-collection process. It

can impede our ability to obtain accurate spectral measurements.

A great deal of research has gone into developing methods to remove the contribution of path radiance (L_P). Methods for computing path radiance are summarized in Richards and Jia (1999). Radiative transfer model programs such as MODTRAN, Second Simulation of the Satellite Signal in the Solar Spectrum (6S), and others may be used to predict path radiance on a particular day for a particular study area (e.g., Matthew et al., 2000; InSpec, 2002; Vermote et al., 2002). Such information can then be used to remove the path radiance (L_P) contribution to the remote sensing signal (L_S) (e.g., Research Systems, 2003; Green, 2003). Song et al. (2001) and Jensen

(2005) provide information about when and how to remove the deleterious effects of path radiance. The process is commonly referred to as radiometric correction.

Methods of collecting *in situ* spectral reflectance measurements that can be used to atmospherically correct remote sensor data are presented in Chapter 15.

 ### References

Colwell, R. N., (Ed.), 1983, *Manual of Remote Sensing*, 2nd ed., Bethesda: ASP&RS, 2440 p.

Cracknell, A. P. and L. W. Hayes, 1993, "Atmospheric Corrections to Passive Satellite Remote Sensing Data," Chapter 8 in *Introduction to Remote Sensing*, NY.: Taylor & Francis, 116–158.

Du, Y., Teillet, P. M. and J. Cihlar, 2002, "Radiometric Normalization of Multitemporal High-resolution Satellite Images with Quality Control for Land Cover Change Detection," *Remote Sensing of Environment*, 82:123.

Englert, B., Scully, M. O. and H. Walther, 1994, "The Duality in Matter and Light," *Scientific American*, 271(6):86–92.

Feinberg, G., 1985, "Light," in R. K. Holz (Ed.), *The Surveillant Science: Remote Sensing of the Environment*, 2nd Ed., NY.: John Wiley, 2–11.

Green, R. O., 2003, "Introduction to Atmospheric Correction," Chapter 2 in *ACORN Tutorial*, Boulder: Analytical Imaging and Geophysics, LLC, 12–18.

InSpec, 2002, *ACORN 4.0 User's Guide*, Boulder: Analytical Imaging and Geophysics, LLC, 76 p.

Jensen, J. R., 2005, *Introductory Digital Image Processing: A Remote Sensing Perspective*, Upper Saddle River: Prentice-Hall, 525 p.

Konecny, G., 2003, *Geoinformation: Remote Sensing, Photogrammetry and GIS*, N.Y.: Taylor & Francis, 248 p.

Matthew, M. W. and 13 co-authors, 2000, "Status of Atmospheric Correction Using a MODTRAN4-based Algorithm," *SPIE*

Proc. Algorithms for Multispectral, Hyperspectral, and Ultraspectral Imagery VI, 4049:199–207.

Miller, S. W. and E. Vermote, 2002, *NPOESS Visible/Infrared Imager/Radiometer Suite: Algorithm Theoretical Basis Document*, Version 5, Lanham: Raytheon, 83 p.

Milman, A. S., 1999, *Mathematical Principles of Remote Sensing: Making Inferences from Noisy Data*, Ann Arbor: Ann Arbor Press, 37 p.

Nassau, K., 1983, *The Physics and Chemistry of Color: The Fifteen Causes of Color*, NY.: John Wiley.

Nassau, K., 1984, "The Physics of Color," in *Science Year 1984*, Chicago: World Book, 126–139.

Research Systems, 2003, *FLAASH—Fast Line-of-sight Atmospheric Analysis of Spectral Hypercubes*, Boulder: Research Systems, http://www.rsinc.com/envi/flaash.asp.

Richards, J. A. and X. Jia, 1999, *Remote Sensing Digital Image Analysis*, N.Y.: Springer-Verlag, 363 p.

Rinker, J. N., 1999, *Introduction to Spectral Remote Sensing*, Alexandria: U.S. Army Topographic Engineering Center, http://www.tec.army.mil/terrain/desert/tutorial.

Rott, H., 2000, "Physical Principles and Technical Aspects of Remote Sensing," in Schultz, G. A. and E. T. Engman (Eds.), *Remote Sensing in Hydrology and Water Management*, Berlin: Springer, 16–39.

Song, C., Woodcock, C. E., Soto, K. C., Lenney, M. P. and S. A. Macomber, 2001, "Classification and Change Detection Using Landsat TM Data: When and How to Correct Atmospheric Effects?" *Remote Sensing of Environment*, 75:230–244.

Trefil, J. and R. M. Hazen, 1995, *The Sciences: An Integrated Approach*, NY.: John Wiley, 634 p.

Tuominen, S. and A. Pekkarinen, 2004, "Local Radiometric Correction of Digital Aerial Photographs for Multi Source Forest Inventory," *Remote Sensing of Environment*, 89:72–82.

Vermote, E. F., El Saleous, N. Z. and C. O. Justice, 2002, "Atmospheric Correction of MODIS Data in the Visible to Middle Infrared: First Results," *Remote Sensing of Environment*, 83:97–111.

History of Aerial Photography and Aerial Platforms

3

Electromagnetic energy reflected from the Earth's surface may be recorded by a variety of remote sensing systems. The *camera* is still one of the most reliable and useful remote sensing instruments. This chapter first reviews the history of photography. It then documents the development of suborbital platforms used to obtain aerial photography. Finally, a brief history of orbital aerial photography data collection is presented.

History of Photography

Photography was an invention waiting to happen from 1833 to 1839. Two of the technological components had been around for centuries, waiting for the right inventor to put everything together, including: 1) a correct theory of light and color, and 2) a recording instrument (the camera obscura). All that was lacking was the invention of a light-sensitive emulsion that could be made permanent (fixed).

Light and Color

For more than a thousand years India's astrologers have taught that the Sun's white light is composed of all colors. Unfortunately, Aristotle's belief that all colors are created by mixing black and white prevailed well into the 17th century (Wolinsky, 1999). Even Leonardo da Vinci could not make up his mind, declaring on different occasions that there were six primary colors — or eight. It took the genius of Sir Isaac Newton to put forth the correct concept of light and color. In 1672, when making public his *New Theory about Light and Colours*, Newton wrote,

> In the begining of the yeare 1666... I procured me a Triangular glasse Prisme, to try therewith the Celebrated Phaenemena of Colours.

Newton found that in using the prism, he could disperse white light into a spectrum of colors — red, orange, yellow, green, blue, indigo, and violet. Utilizing a second prism, he found that he could recombine the colors into white light. He was the first to scientifically document the spectral nature of light. His ideas helped launch the era of modern optics.

Figure 3-1   A portable camera obscura focused reflected light from people or landscapes through a lens (A), onto a mirror (B), and subsequently onto a clear plate of glass (C). The person then sketched the relative proportions and shape of objects onto transparent paper. The camera obscura was an important technological stepping stone in the development of cameras and photography (after Ganot, 1855).

The Camera Obscura

Fundamental camera principles were known as early as the 4th century BC in China and in Greece. A device known as the *camera obscura* (a dark chamber) had long been utilized by artists as a drawing aid (Figure 3-1). The camera obscura uses a simple lens inserted into a box to focus images of the outside world onto a mirror and subsequently onto a plate of glass. The person then sketches the desired information onto relatively translucent paper (e.g., thin onionskin paper). In this way, the relative proportions and shape of objects can be recorded. The problem facing those who would invent photography was how to capture and preserve the image on the glass (Fanton, 1989).

Invention of Light-Sensitive Emulsions and Methods of Permanently Fixing the Image

Joseph Nicephore Niepce (1763–1833) lived in France (Figure 3-2). He produced the world's first photographic image of nature, circa 1826 (Gernsheim and Gernsheim, 1952). It was obtained from an upper window at his estate "Le Gras" in the village of Saint Loup de Varenne, near Charlon-sur-

Figure 3-2   Joseph Nicephore Niepce (1763– 1833).

Saone, France, and records the buildings and vegetation in the courtyard (Figure 3-3). On the left of the photograph is the upper loft of the Niepce family house; to the right is a pear tree with a patch of sky showing through an opening in the branches; in the middle is the slanting roof of the barn; to the far right is another wing of the house.

This is the first successful *permanent* photograph (i.e., remotely sensed image). But how was the first photograph produced? Niepce dissolved bitumen of Judea (a kind of asphalt) in lavender oil (a solvent used in varnishes), then coated a sheet of pewter (metal) with the mixture. He placed the sheet in a camera obscura and aimed the camera through an open window at his courtyard. He exposed the sheet for eight hours. The shading of the buildings is odd because of the diverse lighting caused by the Sun traversing across the sky from east to west during the eight-hour exposure. Incident light from the Sun hardened the bitumen of Judea in bright areas and left it soft in dark areas. Niepce then washed the plate with a mixture of lavender oil and white petroleum. This time it removed the soft bitumen that had not been struck by light. "The result was a permanent direct positive picture, in which the lights were represented by bitumen and the shades by bare metal" (Gernsheim and Gernsheim, 1952). Niepce had found a way to remove the unexposed and still light-sensitive material so that the image stayed visible. Niepce called it a *heliograph,* from the Greek *helios* for "Sun" and *graphos* for "drawing" (London and Upton, 1994). Niepce died in 1833 and was never able to appreciate the impact that photography would have on society.

First Photograph

Figure 3-3   Joseph Nicephore Niepce took the world's first pho-tographic image in 1826 of his estate courtyard. He used a camera obscura and a metal plate of pewter coated with an emulsion of bitumen of Judea (a kind of asphalt). The exposure lasted eight hours. The parts of the plate exposed to light hardened. Areas that did not receive light remained soft. He removed the unexposed soft bitumen by washing the plate in lavender oil and white petroleum. This resulted in the first permanent photographic image (Gernsheim and Gernsheim, 1952; © Gernsheim Collection, Harry Ransom Humanities Research Center, University of Texas, Austin; used with permission).

Louis Jacques Mande Daguerre (1787–1851) was a scenic artist who created stage sets for operas and theater (Figure 3-4). He and Niepce corresponded often and became business partners in 1829. After Niepce's death in 1833, Daguerre continued experimenting with various emulsions and image "fixers." On January 7, 1839, he presented the *daguerreo-type* process to the French Academy of Sciences. The daguerreotype process involved the following steps (Quack-enbush, 1960; Newhall, 1993; London and Upton, 1994):

- A polished surface of silver was plated onto a sheet of copper.

- The plate was made light-sensitive by exposing it to the vapors from iodine crystals in a box. The vapor reacted with the silver, producing the light-sensitive compound silver iodide.

- The photo-sensitive plate was placed in a camera obscura, and an exposure was made. The exposure of the plate often lasted several minutes, during which time the participants

Figure 3-4   Photograph of Louis Jacques Mande Daguerre (1787–1851).

(or landscape) had to remain very still. During exposure in the camera, the plate recorded a *latent image* of the scene, which was chemically present but undetectable to the human eye.

- The latent image was developed by putting the exposed plate in a box that had a dish of heated mercury at the bottom. Where many photons of light had impacted the plate, the mercury vapor formed a frosty amalgam, or alloy, with the silver. This amalgam made up the bright areas of the image. Where few photons of light impacted the plate, no amalgam was formed. The plate was then placed in a bath of common salt (sodium chloride), which caused any unexposed silver iodide to become insensitive to further light action. The plate was then washed in water and dried. The result was a one-of-a-kind positive image. A few years later sodium thiosulphate (hypo), which Herschel had discovered in 1819, was used as the fixative agent instead of sodium chloride.

This direct positive process yielded a single positive print called a "Sun-drawn miniature," as the word "photograph" had not been coined. Daguerre made it clear that he consid-ered the invention to be his own, but agreed to transfer it to the partnership with Isidore Niepce (Joseph's son) "on con-dition that this new process shall bear the name of Daguerre alone; it may, however, only be published simultaneously with the first process, in order that the name of M. Joseph

Daguerreotypes

a. United States Capitol, 1846.   b. President Abraham Lincoln, 1864.

Figure 3-5   Daguerreotypes of a) the United States Capitol and b) President Abraham Lincoln (courtesy of Library of Congress).

Nicephore Niepce may always figure, as it should in this invention" (Fouque, 1867).

The secretary of the French Academy of Sciences marveled at Daguerre's invention and stated that "It upsets all scientific theories of light and optics, and will revolutionize the art of drawing." Other imminent members of the French Academy of Sciences, such as Alexander Von Humboldt (the geographer who conceived of the "isoline"), verified the importance of the invention. The daguerreotype process was purchased by the French government and both Daguerre and Isidore Niepce were granted lifetime pensions. Daguerre wrote a 79 page booklet describing the daguerreotype process, which was published in more than 30 editions and distributed throughout the world. A serious problem for early photographers making daguerreotypes, however, was that some of them went insane; a photographer exposed to the mercury fumes over a prolonged period of time could obtain mercury poisoning.

In the United States, thousands of daguerreotypes were made of landscapes and people by such famous photographers as Matthew Brady. The daguerreotypes that remain today in the Library of Congress are important historical documents. Daguerreotypes of the United States Capitol and President Abraham Lincoln are shown in Figure 3-5.

Daguerreotypes recorded very fine detail, and the public loved them. Unfortunately, there was no way of producing multiple copies except by rephotographing the original. What was needed was a process where any number of positive images could be made from a single exposure. In 1839, an Englishman named William Henry Fox Talbot (1800–1877), who was a member of the Royal Society (equivalent to the French Academy of Sciences), discovered that an image formed on paper coated with silver iodide, though barely visible, could be developed or strengthened with gallic acid and silver nitrate. He used waxed, transparent paper negatives, fixed in hypo, to make positive copies on silver chloride paper. Photographs made in this way did not equal daguerreotypes in brilliancy or sharpness, but the *calotype* (Greek *kalos* for "beautiful" and *typos* for "impression") process made it possible to produce a number of positive copies from the negative. Talbot's *Pencil of Nature* (1844) was the first book to be illustrated with photographs.

Talbot's negative-positive process is basically the same one we use today in terrestrial and aerial photography. Talbot contended that his process predated Daguerre's process and presented his case to the Royal Society. Interestingly, when the famed astronomer Sir John F. W. Herschel of England was asked to come to France and compare Daguerre's process with that of Talbot's, he replied, "I must tell you that compared to these masterpieces of Daguerre, Monsieur Talbot produces nothing but vague, foggy things. There is as much difference between these two products as there is between the moon and the sun" (Newhall, 1993). Herschel also introduced the word *photography* instead of Talbot's term "photogenic drawing" and coined the terms *positive* and *negative*.

Photography was considered miraculous at the time because prior to 1839 it was impossible to know exactly what any person or place looked like without actually knowing the person or visiting the location. Portraiture was available only to the wealthy and was subject to embellishment by the artist. Even the best descriptions of faraway wonders of the world were still only words. Photography allowed the common man and woman to reach into the flow of time, stop it for an instant, and then preserve that moment for posterity.

In 1851, the Englishman Frederick Scott Archer found that a glass plate coated with *collodion* (nitrocellulose dissolved in ether and alcohol) could be used as an emulsion. He developed a way to coat a glass plate with the substance (Newhall, 1969). The plate was made sensitive by dipping it in silver nitrate. The silver ions combined with the iodine ions to form light-sensitive iodide within the collodion. The plate was exposed while still moist and then developed in pyrogallic acid, fixed in hypo, and washed. The *collodion wet-plate* process had to be performed on location. This required a complete mobile darkroom if photography was acquired in the field. From 1851 until 1888 this was the most popular form of photography, completely replacing the daguerreotype and calotype processes.

In the *Proceedings* of the Royal Society in 1855, the noted Scottish physicist James Clerk Maxwell (1831–1879) (Figure 3-6) made what is believed to be the earliest suggestion that objects could be reproduced in color by photography. His paper consists of a basic discussion on the theory of color vision. Referring to early works by Thomas Young, whom he says seems to have been the first to understand Newton's suggestion on the mixing of colors, Maxwell makes the following statement:

> This theory of colour may be illustrated by a supposed case taken from the art of photography. Let it be required to ascertain the colours of a landscape, by means of Impressions taken on a preparation equally sensitive to rays of every colour.

> Let a plate of red be placed before the camera and an impression taken. The positive of this will be transparent wherever the red light has been abundant in the landscape, and opaque where it has been wanting. Let

Figure 3-6   Photograph of Scottish physicist James Clerk Maxwell (1831–1879).

it now be put in a magic lantern (i.e., a projector) along with the red glass, and a red picture will be thrown on the screen.

> Let this operation be repeated with a green and violet glass and by means of three magic lanterns let the three images be superimposed on the screen. The colour of any point on the screen will then depend on that of the corresponding point of the landscape; and, by properly adjusting the intensities of the light, etc., a complete copy of the landscape as far as visible colour is concerned, will be thrown on the screen (Niven, 1890).

On May 17, 1861, with the help of photographer Thomas Sutton, Maxwell demonstrated this *additive color combining* technique. He photographed a bow of multicolored ribbon four times using black-and-white film. Each exposure was made through a different filter: red (sulfo-cyanide of iron), green (copper chloride), blue (ammoniacal sulfate of copper), and yellow (lemon-colored glass). He then projected light through the red, green, and blue filtered black-and-white images and was able to recreate an image of the bow of multicolored ribbon. He found that the yellow filter was unnecessary. Chapter 4 provides additional information on additive color-combining principles.

Great improvements in photography were realized after Richard L. Maddox, a London physician and photo-micrographer, invented the dry-plate process in 1871, which used gelatin as the medium for suspending light-sensitive silver salts (Maddox, 1871). This was greatly superior to the collodion wet-plate process in that the emulsion was 1) much more sensitive (approximately 60 times faster than collodion), and 2) could be developed when the emulsion was dry. It was no longer necessary to have a wet laboratory nearby. The more sensitive emulsion was also important because it stopped action more rapidly, greatly improving the quality of detail in terrestrial and aerial photographs. Building on this logic, roll film was invented by Leon Warnerke in 1875.

In 1873, Herman Vogel discovered that by soaking silver halide emulsions in various dyes, he could extend their sensitivity to longer wavelengths. Subsequent investigations revealed that dyes could be rendered that extend emulsion sensitivity into the infrared portion of the spectrum (Thompson, 1966).

Prior to 1888, the general public could buy a camera and film, but they had to know how to develop the film and print the photographs. In 1888, George Eastman revolutionized photography. Building upon the work by Warnerke, he used a gelatin emulsion in which light-sensitive silver salts were suspended. This material was then coated on paper and produced in a roll format. This clear, flexible film support did much to advance and increase the usefulness of photography (Thompson, 1966). George Eastman founded the Eastman Kodak Company and developed a photofinishing system that lived up to the slogan, "You press the button, we do the rest." The general public bought the inexpensive box cameras, exposed 100 negatives, and sent the exposed film still inside the camera to Eastman's company, where it was processed. The company then sent the 100 prints back to the customer, along with the camera and a new roll of film inside. This sounds very much like the recyclable cardboard or plastic cameras sold today!

Finally, in 1924, Mannes and Godousky patented the first part of their work on multilayer film, which led to the marketing of the three-layer color film Kodachrome in 1935 (McCamy, 1960).

 Photography from Aerial Platforms

Mankind has had the desire to fly like a bird and have an aerial perspective or bird's-eye view of the Earth for a long time. Even our earliest mythology is consumed with flight. For example, in Greek mythology Daedalus built wings of

Figure 3-7   A man-powered ornithopter. Flapping the arms was supposed to generate sufficient aerodynamic lift to sustain flight. Experimentation with ornithopters often resulted in serious accidents.

feathers and wax for himself and his son, Icarus. They escaped from prison using the wings. Unfortunately, Icarus became caught up in the joy of flying and flew too close to the Sun. The wings melted, and he fell to Earth. The Greek hero Perseus used winged sandals when he flew to rescue Andromeda (Lopez, 1995). After dreaming about flying for centuries, mankind's creativity eventually resulted in a series of inventions that allowed us to escape gravity and obtain an aerial perspective. Some of the aerial platforms invented were more useful and safe than others.

Ornithopters

Not surprisingly, the first real attempts at flight involved people trying to imitate birds. They built flapping devices called *ornithopters* (Figure 3-7). They jumped from high places strapped to these devices and often died. For example, in 1010 a monk named Eilmer, equipped with an ornithopter, took off from an abbey in England and broke his legs. Denis Bolori of France in 1536 tried to fly using wings flapped by a spring. He died when the spring broke. Leonardo da Vinci's notebooks written between 1488 and 1514 contain designs of ornithopters and simple helicopters. We do not know if da Vinci actually built the designs. He did suggest that the devices should be tested over water for safety.

Lighter-than-Air Flight Using Balloons

The hot-air balloon was invented by Joseph and Etienne Montgolfier in 1783 in France (Lopez, 1995). They burned straw and wool to produce the less dense air that lifted the balloon and basket off the ground. The first balloon passengers were animals. The first humans to ride in a Montgolfier

Figure 3-8   Photograph of Gaspard Felix Tournachon (1820–1910), the famous Parisian photographer. He called himself Nadar. Here he is seen kneeling in a fragile balloon gondola. He obtained the first aerial photograph from a balloon in 1858 near Paris, France, and patented the aerial survey as we know it today. Unfortunately, the first aerial photograph did not survive (© Roger-Viollet, Paris, France; used with permission).

balloon were J. F. Pilatre de Rozier and the Marquis d'Arlandes, who flew over Paris. People who ventured into the sky in balloons often referred to themselves as *aeronauts*.

The first-known aerial photograph was obtained in 1858 by the Parisian portrait photographer and passionate aeronaut Gaspard Felix Tournachon (he called himself *Nadar*) (Figure 3-8). Tournachon had a vision of what aerial photography might contribute in the future, and on October 23, 1858, he applied for a patent for what we now call an aerial survey

— the mapping of the land from a series of overlapping aerial photographs (Newhall, 1969). Later in the month, he ascended in a captive balloon over Paris. He suffered many failures attempting to obtain an aerial photograph, because the gas escaping from the mouth of the balloon desensitized the collodion-coated glass plates he was using. He finally achieved success over the Val de Bievre on the outskirts of Paris when he went aloft in a tethered balloon only 80 m (264 ft) above the ground, exposed a photographic plate, and then was rapidly hauled back down to Earth, where he dashed into an inn and developed the picture. The world's first aerial photograph no longer exists but Nadar described its contents:

> I develop my picture... Good luck! There is something!... It cannot be denied: here right under me are all of the three houses in the little village: the farm, the inn and the police station... You can distinguish perfectly a delivery van on the road whose driver has stopped short before the balloon, and on the roof-tiles two white pigeons who have just landed there. Thus I was right! (*Le Monde Illustre*, October 30, 1858).

Nadar himself referred to the photograph as "a simple positive upon glass," made with "detestable materials." Nevertheless, aerial photography was born. The first successful aerial photographs that we have a record of were recorded two years later on October 13, 1860, from the tethered balloon *Queen of the Air* in the United States by James W. Black and Samuel A. King from a height of 1,200 ft over Boston, Massachusetts (Figure 3-9). The photographs were obtained using wet collodion plates (Tennant, 1903). Black was a professional photographer from the firm of Black & Bathelder. King had his own photography business — King & Allen. At the conclusion of the days' aerial photography, King (1860) commented,

> This is only the precursor, no doubt, of numerous other experiments; for no one can look upon these pictures, obtained by aid of the balloon, without being convinced that the time has come when what has been used only for public amusement can be made to serve some practical end.

In July, 1863, Sir Oliver Wendell Holmes (a photographic student of Samuel King) photointerpreted the contents of this photograph for *The Atlantic Monthly* (Newhall, 1969):

> Boston, as the eagle and the wild goose see it, is a very different object from the same place as the solid citizen looks up at its eaves and chimneys. The Old South and

**First Aerial Photograph from A
Captive Balloon in the United States: Boston, MA**

Figure 3-9   A portion of an aerial photograph of downtown Boston, MA, obtained by aeronauts James W. Black and Samuel A. King from a tethered balloon at an altitude of 1,200 ft on October 13, 1860. It is believed to be the first aerial photograph taken from a captive balloon in the United States and the earliest aerial photograph still in existence. It was obtained using a wet collodion plate (used with permission of the Smithsonian Institution, Washington, DC; #3B-15472).

Trinity Church are two landmarks not to be mistaken. Washington Street slants across the picture as a narrow cleft. Milk Street winds as if the cowpath which gave it a name had been followed by the builders of its commercial palaces. Windows, chimneys, and skylights attract the eye in the central parts of the view, exquisitely defined, bewildering in numbers.... As a first attempt it is on the whole a remarkable success; but its greatest interest is in showing what we may hope to see accomplished in the same direction.

During the Civil War in America, Union General McClellan used balloons to observe Confederate Army positions and movement. For example, in June, 1862, the Union Army used balloons tethered at an altitude of 1,400 ft to gather intelligence and draw maps of the defenses of Richmond,

Figure 3-10   The *Intrepid* balloon being tethered by Union troops at the battle of Fair Oaks on June 1, 1862 (used with permission of the Smithsonian Institution, Washington, DC; #2A-03710).

Virginia. The balloon *Intrepid* is shown tethered by Union troops on the ground at the battle of Fair Oaks on June 1, 1862, in Figure 3-10 (Quackenbush, 1960; Fischer, 1975). The *Intrepid* was inflated using a special hydrogen-producing system (Figure 3-11). It is believed that some aerial photographs were obtained by observers using tethered balloons. However, not a single Civil War (1861–1865) aerial photograph has survived (Colwell, 1997).

Meanwhile, in Europe, Gaspard Felix Tournachon (Nadar) was still actively involved in balloon aerial photography. In fact, he designed and built a tremendous balloon called *Le*

Figure 3-11   The balloon *Intrepid* being inflated by using Thaddeus Lowe's portable hydrogen generating system during the Civil War battle of Fair Oaks on June 1, 1862 (used with permission of the Smithsonian Institution, Washington, DC; #2A-03675).

Geant (The Giant) in 1863. The balloon held 210,000 ft^3 of gas, and its gondola consisted literally of a two-story house fitted with three-decker beds, a bathroom, and even a printing press. The gondola could comfortably house 12 passengers (Newhall, 1969). Nadar wrote:

> The balloon *Le Geant* will be employed in various aerostatic photography... the results of which will be so valuable for all planispheric, cadastral, strategical, and other surveys.

Unfortunately, in October 1863, the gigantic balloon descended too rapidly. Nine passengers aboard, including Ms. Nadar, were almost killed when the balloon was dragged 25 miles across the French countryside. No aerial photographs obtained from *Le Geant* have survived. However, Nadar did successfully obtain oblique aerial photographs using perhaps the first aerial multiple-lens camera from the *Hippodrome Balloon* tethered 1,700 ft above Paris in 1868 (Figure 3-12).

The dry-plate process pioneered by Richard Maddox in 1871 was a real boon to obtaining quality aerial photographs from balloons because no wet laboratory was required in the balloon gondola or nearby on the ground, and the emulsion was much faster, resulting in less image blur. This resulted in a

Figure 3-12   Oblique photograph of Paris obtained by Gaspard Felix Tournachon (Nadar) from the *Hippodrome Balloon* tethered 1,700 ft above the ground in 1868 (after Newhall, 1969).

Figure 3-13   Aerial photograph of "San Francisco in Ruins" obtained by George R. Lawrence after the April 18, 1906 earthquake using a 49-lb camera attached to a *captive airship* consisting of 17 kites tethered from a ship in San Francisco Bay. The kites achieved an altitude of 2,000 ft above sea level (courtesy Library of Congress; #LC-USZ 62-16440).

great deal of balloon aerial photography taking place during the last few decades of the nineteenth century. Gaston Tissandier published the first manual on air photography from balloon platforms, *La Photographie en Balloon* (Newhall, 1969).

Aerial photography from tethered and free-flying balloons continues today. Stabilizing the balloon during the instant of exposure continues to be a major problem. Also, untethered balloons are at the mercy of the prevailing wind, making it difficult to navigate the balloon over the intended landscape. Nevertheless, many scientists and groups continue to obtain aerial photography using balloons. A few of the most active organizations are identified in Appendix A.

Lighter-than-Air Flight Using Kites

Cameras carried aloft by large kites have also been used to obtain aerial photography. Frenchman Arthur Batut pioneered the development of kite aerial photography. In 1890 he published *La Photographie Aerienne par Cerf-volant,* in which he outlined the uses of kite aerial photography for the explorer (to view the land beyond), the archaeologist (to locate ruins), the military (to reconnoiter), and the agronomist (to locate vine diseases). Batut felt that the kite would bring aerial photography within the grasp of the common man (Batut, 1890; Tennant, 1903).

George R. Lawrence of Chicago was almost killed twice trying to obtain aerial photographs from captive balloons. On one occasion he fell 228 ft in a gondola from a balloon sus-

pended over the Chicago stockyards and was only saved because the free-falling gondola landed on telegraph wires. Therefore, it is not surprising that he switched to kite aerial photography. He often used as many as 17 kites in a connected "train" (although five to 10 usually sufficed) to carry aloft cameras weighing as much as 2,000 lbs. He called this collection of tethered kites a *Captive Airship* (Baker, 1994). A mount hanging below the lowest kite in the series allowed him to fix the camera in any direction before sending it up. A system of booms, lines, and lead weights prevented the camera from turning horizontally while at the same time decreased the camera's tendency to swing. Lawrence tripped the shutter by incorporating an insulated wire into the steel kite line, which was used to carry an electric current to the camera (Newhall, 1969; Baker, 1994). Some of Lawrence's most celebrated aerial photographs were acquired six weeks after the April 18, 1906, San Francisco earthquake using 17 of his kites at one time (Figure 3-13). At first, rain soaked the line and shorted out the command to trip the shutter. Then, the Sun came out and dried the line, allowing the shutter to function. Panoramic negatives were acquired using a 49-lb panoramic camera that held a celluloid-film plate 18.75 by 48 in. Contact prints created from the negatives at the time were the largest ever produced from an airborne platform. The camera covered 160° in a single sweep of the lens, producing extremely detailed wide-angle photographs of the ruined city (Baker, 1994).

The major problem with kites and balloons is that they are not navigable in the strict sense of the word. Still, many excellent aerial photographs have been obtained using them.

Relationship of Wing Design to Aerodynamic Lift

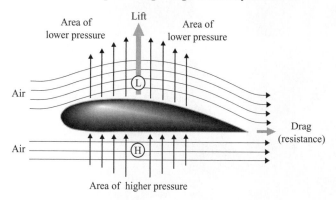

Figure 3-14   A wing that is curved on the top and relatively flat on the bottom creates *aerodynamic lift* (after Lopez, 1995).

There is an Internet user group devoted to obtaining aerial photography using kites (Appendix A).

Heavier-than-Air Flight Using Rockets

Even before airplanes were invented, aerial photographs were obtained by rocketing a camera into the sky and then retrieving the camera and film. In 1888, *La Nature* described a "photo rocket" invented by Amadee Denisse. In 1891, Ludwig Rahrmann received a German patent for a photographic system that was shot into the air using a rocket or large-caliber gun. The camera returned to the ground via parachute (Newhall, 1969). In 1903, Alfred Maul patented a rocket camera. Aerial photographs from a rocket that achieved an altitude of 2,600 ft were published in the *Illustrated London News* on December 7, 1912.

Heavier-than-Air Flight Using Pigeons, Gliders, and Aircraft

All heavier-than-air birds, gliders, and airplanes achieve sustained flight based on the principle of *aerodynamic lift*. Wind blowing above and below a wing will cause the wing to achieve aerodynamic lift if the wing has the proper shape. A flat wing shape fights airflow, causing drag (resistance), while a curved wing shape (Figure 3-14) allows air to flow smoothly around it. A wing that is curved on the top and almost flat on the bottom creates aerodynamic lift. The molecules of air passing over the top of the wing surface have a longer distance to travel and therefore must move more rap-

idly, creating less pressure than the slower air flowing below the wing. The higher pressure air below the wing exerts pressure upward, causing the wing to lift. Tilting the wing upward will increase the aerodynamic lift even more. However, if it is tilted too much, the aerodynamic lift will be lost and the wing (and aircraft!) will stall and fall (Lopez, 1995).

Pigeons

In 1903, Julius Neubronner patented a breast-mounted aerial camera for carrier pigeons that weighed only 2.5 oz (Figure 3-15a,b). An article of the day described the pigeon data collection process:

As a carrier pigeon, after starting, at first describes a spiral line, it is quite easy to take a number of views of a given portion of the ground from different points of view. After once determining the position of its cote (which it recognizes from a distance upward of 20 miles) the pigeon flies towards its goal in a straight line and at the uniform speed of an express train, so that the route to be recorded photographically can be readily determined in advance (*Scientific American*, 1909).

Exposures were made automatically at 30-second intervals (Quackenbush, 1960; Fischer, 1975). For obvious reasons, pigeons are not an ideal remote sensing platform.

Gliders

George Cayley and John Stringfellow dramatically improved the aerodynamics and control mechanisms of heavier-than-air gliders. Cayley's coachman was the first person to successfully fly in a glider in 1853. From 1891 to 1896, a German civil engineer named Otto Lilienthal completed nearly 2,000 glider flights. Sometimes he would glide for more than 1,000 ft. He died in a glider flight in 1896 (Lopez, 1995). An American engineer named Octave Chanute built on Lilienthal's designs and dramatically improved glider control. He was also a great supporter of the aviation work of Orville and Wilbur Wright.

Orville and Wilbur Wright (Figure 3-16) were fascinated with the concept of flight. Initially they experimented with gliders at Kitty Hawk, NC, where a consistent wind could be found and the only obstructions were sand dunes and the surf. They built on the glider designs of Lilienthal and Chanute and eventually designed their own gliders in 1900. A photograph of one of the Wright Brothers lying prone in a glider in 1902 is shown in Figure 3-17a.

a. Pigeons as remote sensing platforms.

b. Aerial photograph obtained by a pigeon.

Figure 3-15   a) A squadron of pigeons equipped with lightweight (approximately 2.5 oz) 70-mm aerial cameras. b) Portion of an oblique aerial photograph obtained from a camera carried by a pigeon. The pigeon's wings are visible (© Deutsches Museum, Munich, Germany; used with permission).

Aircraft

In 1903, the Wright Brothers built a 12-horsepower engine and the first operational aircraft propeller. By the flip of a coin, Wilbur Wright was given the chance to fly the first heavier-than-air motor-driven machine on December 14, 1903. He stalled and crashed without injury. On December 17, 1903, at Kill Devil Hills at Kitty Hawk, NC. Orville Wright flew for 12 seconds and 120 ft (Figure 3-17b). Both brothers flew several times that day, with Wilbur flying the longest distance of 852 ft. Manned heavier-than-air powered flight had begun.

The original Wright Flyer was controlled by the pilot lying prone on the lower wing (Figure 3-17b). Turns were made by swinging the body from one side to another. This caused the rudder to move and the wings to become warped. The pilot's left hand controlled the up and down motion (ascent and descent) of the aircraft using an "elevator." The pilot held on firmly to the aircraft with the right hand. Not surprisingly, no aerial photographs were obtained using the original Wright Flyer because all hands were required just to maintain control and keep from crashing. The Wright Brothers built a much improved airplane in 1907 that allowed the pilot and a passenger to sit upright on the lower wing.

a. Orville Wright.   b. Wilbur Wright.

Figure 3-16   Orville and Wilbur Wright were instrumental in the development of heavier-than-air gliders and motor-driven aircraft (courtesy of Wikipedia).

a. Glider.   b. The Wright Flyer.

Figure 3-17   a) One of the Wright Brothers in an unpowered glider at Kitty Hawk, NC in 1902. b) The Wright Flyer had a 12-horsepower engine and specially designed propellers. Orville took the first successful heavier-than-air engine powered flight on December 17, 1903. It lasted 12 seconds and covered 120 ft (used with permission of the Smithsonian Institution, Washington DC).

The first time that an airplane was used as a platform to obtain aerial photography was in 1908, when Pathe motion-picture photographer L. P. Bonvillain accompanied Wilbur Wright. He took motion pictures over Camp d'Auvours, near Le Mans, France. The original motion-picture film has not been found (Newhall, 1969), but an enlargement of one of the frames of the motion picture was published in a French magazine in 1908.

a. Aerial photographer and pilot.

b. Aerial photography in WWI.

Figure 3-18   a) An aerial photographer and pilot in a Curtiss AH-13 airplane with Graflex camera in 1915. b) Close-up view of a World War I Curtiss JN-4 Jenny with the camera mounted on the right side (used with permission of the Smithsonian Institution, Washington, DC; #2B-41742 and 2B-41668, respectively).

Photo-Reconnaissance in WWI and WWII

Numerous new aircraft companies came into existence from 1907 to 1930. Many of the founder's names are still associated with United States' aircraft companies today (Lopez, 1995):

- Glenn Curtiss in 1907 — Curtiss

- Wright Brothers in 1909 — eventually Curtiss-Wright

- Glenn Martin in 1912 — now Martin Marietta

- William Boeing in 1916 — Boeing

- Donald Douglas in 1920 — now McDonnell Douglas

- Alan Lockheed in 1926 — now Lockheed-Martin

- John Northrop in 1929 — Northrop

- LeRoy Grumman in 1929 — now Northrop/Grumman

Aircraft produced by these companies and others in Germany, France, and Britain were used for military photo-reconnaissance and other purposes in World War I and World War II.

Aerial Photography in World War I

In order to take an aerial photograph in the early years of World War I, it was necessary for the airman to point the handheld camera out the side of the plane, make an exposure, and then change the plate before taking another photograph. Cameras were initially strapped to the photographer's chest or attached to the side of the plane (Figure 3-18ab.) Aerial photography was very dangerous duty, especially in the early formative years.

The wartime use of aerial photography was not appreciated in the beginning of the war. In fact, World War I trench maps derived from photographs taken by the pilots on their own initiative were at first dismissed as being "a most disgraceful thing to have attempted" (Newhall, 1969). Eventually, however, commanders learned the value of aerial reconnaissance and it completely changed the military tactics used in World War I. First, aerial photographs were used to make relatively accurate maps for planning military strategy over poorly mapped terrain. A knowledge of existing and destroyed roads, barriers, and constriction points was very valuable for planning the movement of troops and materials. Second, a vast amount of military material was simply impossible to conceal from the aerial camera lens (Quackenbush, 1960). Troop movements and the stockpiling of arms and supplies could be documented. For example, on August 22, 1914, World War I British reconnaissance aerial photography revealed a major change in direction of the German forces

Figure 3-19 Vertical photography of World War I trenches in Europe. Examination of stereoscopic photography revealed the location of men, gun emplacements, and ammunition bunkers. Millions of men died in trench warfare during World War I (used with permission of the Smithsonian Institution, Washington, DC; #2B-41711).

Figure 3-20 Vertical aerial photograph of a V-2 rocket launching facility at Peenemunde in World War II. Note the large circular earth revetment around the launch facilities designed to deflect the blast during liftoff or to minimize the destruction during an accident. These facilities were eventually destroyed (courtesy of Federation of American Scientists, Intelligence Research Program).

advancing on Paris. This timely information allowed the Allied army to fortify its position on the Marne River and hold off the German advance to Paris (Lopez, 1995). World War I often consisted of trench warfare. Battlefield trenches dug in World War I for an unknown area are displayed in Figure 3-19. Trained photo-interpreters routinely analyzed stereoscopic photography of trenches to locate gun emplacements and ammunition dumps. By 1918, French aerial units were developing and printing as many as 10,000 photographs each night during periods of intense activity. During the Meuse-Argonne offensive, 56,000 aerial photographs were made and delivered to American Expeditionary Forces in four days. At the end of World War I, Lieutenant Edward Steichen of the Photographic Section of the American Expeditionary Forces stated,

The consensus of expert opinion, as expressed at the various inter-Allied conferences on aerial photography, is that at least two-thirds of all military information is either obtained or verified by aerial photography.

Aerial Photography in World War II

Aerial photo-reconnaissance played a significant role in World War II. In 1938, General Werner von Fritsch, Chief of the German General Staff, made a prophetic statement: "The nation with the best photo-reconnaissance will win the next war." By 1940, Germany led the world in military photo-reconnaissance. Fortunately, German photo-interpretation stagnated as the war progressed. The United States had almost no capability in military photo-interpretation when it entered World War II (Fischer, 1975). However, British and American aerial photography data-collection and photo-interpretation capabilities improved quickly. After the retreat from the mainland at Dunkirk in 1940, the British were almost completely cut off from their normal sources of military intelligence and had to rely on aerial photography as their chief source of information. The photo-identification of German invasion barges in canals near the coast of France in the summer of 1940 constituted the major evidence that an invasion of England would take place in less than 48 hours. The British launched such an effective air attack on the invasion forces that Germany was forced to postpone the invasion and finally to abandon it (Quackenbush, 1960). Aerial photography also assisted in the destruction of German V-2 rocket facilities late in the war (Figure 3-20).

Most reconnaissance World War II aircraft had little defense capability. Therefore, they were often escorted by fighter aircraft. Millions of reconnaissance aerial photographs were obtained during World War II by modified aircraft such as

a.   b.

Figure 3-21   a) A B-17 Flying Fortress of the U.S. 8th Air Force in a lower group flew under an upper group just as bombs were released.
b) The next exposure documents that the bomb crashed through the port horizontal stabilizer of the lower aircraft (used with
permission of the Smithsonian Institution, Washington, DC; #3A-20850 and 20851, respectively).

the P-38. In addition, bombers often obtained aerial photographs during bombing raids that were used in subsequent bomb-damage assessment. For example, consider the accident documented during a bombing and aerial reconnaissance mission over Berlin, Germany (Figure 3-21ab). A B-17 Flying Fortress of the U.S. 8th Air Force in a lower group flew under an upper group just as bombs were released. The first aerial photograph reveals details about the B-17 Flying Fortress, such as several bombs being released, clouds, and the Berlin countryside below. The next exposure documents that a bomb crashed through the port horizontal stabilizer of the lower aircraft, clearing the way for subsequent bombs. It is unlikely that the damaged B-17 could have returned safely. Germany introduced jet aircraft platforms near the end of World War II (Boyne, 1993).

 Cold War Photo-Reconnaissance

After World War II, the arms race between the United States and the Soviet Union caused the U.S. government to think that there was a tremendous bomber and/or missile gap. Great emphasis was placed at the highest levels on developing technology to collect information. Similarly, the Soviet Union desired information about U.S. activities. Any aerial reconnaissance overflight of another state without authorization was considered to be an illegal and hostile act unless national leaders agreed to it beforehand. But, such overflights were absolutely essential if the United States was to know what was taking place in the Soviet Union and other countries. United States President Dwight Eisenhower approved the U-2 aerial reconnaissance program in December 1954 (Brugioni and Doyle, 1997). Subsequently, President Eisenhower attended the Four-Power Summit

Conference in Geneva, Switzerland, on July 21, 1955, and presented what the press eventually called the historic Open Skies Doctrine. He suggested that the absence of trust and the presence of "terrible weapons" among states provoked in the world "fears and dangers of surprise attack." To eliminate those fears, he urged that the Soviet Union and the United States provide "facilities for aerial photography to the other country" and conduct mutually supervised reconnaissance overflights. First Secretary of the Communist Party, Khruschev, privately rejected the Open Skies Doctrine the same day (Hall, 1996). Nevertheless, the concept had been presented.

Just a few days later on July 29, 1955, President Eisenhower publicly announced plans for launching "small unmanned, Earth-circling satellites as part of the U.S. participation in the 1957–1958 International Geophysical Year." The real purpose, however, was to establish the principle in international law of "freedom of space" with all that it implied for strategic reconnaissance conducted at altitudes above the "airspace" to which the states beneath claimed exclusive sovereignty (Hall, 1996). In effect, the President was laying the groundwork for the already conceived Genetrix Reconnaissance Balloon Project, a U-2 suborbital aircraft project, and subsequent satellite orbital reconnaissance.

Genetrix Reconnaissance Balloons

Considerable research was conducted by the U.S. Air Force and the Central Intelligence Agency (CIA) during 1954 and 1955 on the utility of using unmanned balloons operating at high altitudes with cameras in their gondolas to obtain reconnaissance information over vast geographic areas. This was called the Grayback Program. A trial balloon was launched by the CIA from Scotland, which drifted across the Soviet Union and was retrieved near South Korea. Similar balloons were also launched and recovered in the United States (Peebles, 1997).

Based on these successes, the Genetrix Reconnaissance Balloon Project was initiated in October 1955. The goal of the program was to obtain complete photographic coverage of the Soviet landmass based on the proposed launch and recovery of 2,500 unmanned high-altitude balloons. President Eisenhower gave approval for the launches to begin on December 27, 1955. From January 10, 1956, through February 6, 1956, approximately 448 balloons were launched from Scotland, Norway, West Germany, and Turkey (Peebles, 1997). The gondolas were to be retrieved by aircraft stationed in a large arc from Okinawa through Japan to Alaska. Many of the early balloons were recovered. How-

ever, Eastern European, Soviet, and Communist Chinese air defenses responded quickly and were able to stop many of the balloons.

On February 4, 1956, Soviet Deputy Foreign Minister Andrei A. Gromyko protested to the U.S. Ambassador. On February 7, 1956, anticipating the Soviet response, President Eisenhower suggested to Secretary of State John Foster Dulles that the operation be suspended and "we should handle it so it would not look as though we had been caught with jam on our fingers" (Day et al., 1998). On February 9, 1956, the Soviets held a press conference and displayed about 50 balloons and instrument containers. The United States said they were weather balloons. It was a major embarrassment to the United States (Day et al., 1998). Only 44 of the 448 balloons were recovered. However, the cameras on these balloons obtained 13,813 aerial photographs covering 1,116,449 mi^2 of Soviet and Chinese terrain, including the very significant nuclear refining facility at Dononovo in Siberia (Peebles, 1997). This marked the end of this early U.S. balloon reconnaissance program. The U-2 aircraft reconnaissance program was initiated just a few months later, in July 1956.

U-2 Aircraft Reconnaissance Program

In December 1954, President Eisenhower instructed the CIA to contract with Lockheed (in cooperation with the U.S. Air Force) to develop a photo-reconnaissance jet aircraft that could fly above the Soviet Union at will to document their military capability. This led to the creation of the secret Lockheed Skunk Works (named after a smelly location in the Lil' Abner cartoon strip) in Burbank, and later Palmdale, CA, where Kelly Johnson and his engineers developed the aluminum U-2 aircraft (Rich and Janos, 1994).

The original U-2s were painted dull black to prevent them from glinting in the Sun (Figure 3-22a). It could fly at 70,000 ft above sea level and expose 3,600 ft of high-resolution film on a single mission. The camera was developed by Dr. Edwin Land, inventor of the Polaroid camera. The original U-2 payload consisted of two high-resolution cameras — one a special long-focal-length spotting camera able to resolve objects two to three feet across from a height of 70,000 ft, and the other a tracking camera that could produce a continuous strip of film of the whole flight path (Rich and Janos, 1994). The first Russian overflight occurred on July 4, 1956. It took off from Weisbaden, West Germany to survey Soviet naval shipyards and submarine construction. It overflew Poland, Belorussia, Moscow, Leningrad, and the Soviet Baltic states. Much to the consternation of the Amer-

Lockheed U-2

a. U-2 aircraft.

b.

c. ER-2 aircraft.

Figure 3-22　a) The Lockheed U-2 high-altitude reconnaissance aircraft was developed at
the Skunk Works in Burbank and
Palmdale, CA, for President Eisenhower. The program was supervised
by the Central Intelligence Agency. b)
Francis Gary Powers in front of a U-2
aircraft (courtesy Lockheed Martin,
Inc.). c) Many U-2s are still in service
as Earth resource observation aircraft.
In this example there is a U-2 in the
foreground and an ER-2 in the background in flight near San Francisco,
CA (courtesy NASA and Lockheed
Martin, Inc.).

icans, the Soviet radar detected and tracked the first U-2 at
its design altitude of 70,000 ft (Hall, 1996).

Soviet surface-to-air missiles and aircraft could not touch
the U-2 for four years. Information provided from the U-2
overflights had a tremendous impact on our knowledge
about the Soviet Union during the 1950s and early 1960s.
However, missile technology continued to improve, and on
May 1, 1960, a cluster of 14 SA-2 air-to-air missiles shot
down U-2 pilot Francis Gary Powers (Figure 3-22b). He was
interrogated for 10 to 16 hours a day for 61 days. In August
1960, Soviet authorities staged a widely publicized, open
trial that was designed to embarrass the United States. They
sentenced Francis Gary Powers to 10 years in a Soviet
prison; however, he was exchanged after 21 months for
Soviet spy Rudolph Abel, who was being held in the United
States (Powers, 1997).

After the Francis Gary Powers incident, the U-2 still provided important strategic reconnaissance information
throughout the world. For example, in July 1962, the director of the CIA sent a memo to President John F. Kennedy
saying that he believed the Soviets would deploy medium-
range ballistic missiles (MRBM) just 90 miles from the U.S.
mainland on Cuba; exactly when was unclear (Walter, 1992;
Goldberg, 1993). On October 9, 1962, President Kennedy
approved reconnaissance flights over western Cuba using
high-altitude U-2s. Hurricane Ella kept the planes on the
ground until October 14, 1962. The photographs were interpreted on October 15 and presented to President Kennedy on
October 16, 1962. A portion of one of the U-2 photographs
shown to President Kennedy is found in Figure 3-23a. An
oblique aerial photograph obtained by an RF-101 aircraft
recorded MRBM facilities near San Cristobal, Cuba, on
October 25, 1962 (Figure 3-23b).

Cuban Missile Crisis, October, 1962

a. Vertical aerial photograph of San Cristobal, Cuba.

b. Oblique aerial photograph of San Cristobal, Cuba.

Figure 3-23 a) A portion of the U-2 aerial photograph of San Cristobal, Cuba, shown to President John F. Kennedy by the U.S.'s top photointerpreter Arthur C. Lundahl on October 16, 1962. The photograph was obtained on October 14, 1962, at an altitude of 21,300 m. It depicts missile trailers, a few launchers, and tents that were used to prepare missiles and warheads for launch. President Kennedy asked, "Are you sure?" "Mr. President," Lundahl replied, "I am as sure of this as a photointerpreter can be sure of anything. And I think, sir, you might agree that we have not misled you on anything we have reported to you. Yes, I am convinced they are missiles" (Brugioni, 1996) (U.S. Air Force photograph). b) Low-oblique aerial photograph of Medium Range Ballistic Missile (MRBM) Launch Site 1 near San Cristobal, Cuba, obtained on October 25, 1962, by low-flying RF-101 aircraft. Long missile shelter tents and missile transporters are visible (U.S. Air Force photograph; courtesy Federation of American Scientists).

On October 25, 1962, United Nations U.S. Ambassador Adlai Stevenson challenged the Soviet Ambassador, Valerian A. Zorin, to deny the U.S. charge that the Russians had installed offensive missile bases on Cuba. "In due course, sir, you will have your reply. Do not worry," replied Zorin. Stevenson responded, "I am prepared to wait until hell freezes over, if that's your decision. I am also prepared to present the evidence in this room." Aerial photographs on easels were then brought in, and the peaceful Cuban countryside in August was shown to contain missile facilities by mid-October. Twenty-four hours later there were unmistakable signs of missile installations. President Kennedy then

Lockheed SR-71

Figure 3-24   Two views of the Lockheed SR-71 reconnaissance aircraft. It can fly at greater than 70,000 ft above sea level and achieve airspeeds greater than 2,000 m.p.h. (courtesy Lockheed Martin, Inc.).

initiated a naval blockade of Cuba. After much political brinkmanship, Soviet Premier Khruschev relented on October 28, 1962, and the world returned from the brink of nuclear war.

Interestingly, the detail in the photographs was largely lost on the untrained presidential staff. Attorney General Robert F. Kennedy wrote about the CIA's explanation of the photographs at the first emergency meeting: "I, for one, had to take their word for it. I examined the pictures carefully, and what I saw appeared to be no more than the clearing of a field for a farm or the basement of a house. I was relieved to hear later that this was the same reaction of virtually everyone at the meeting, including President Kennedy. Even a few days later, when more work had taken place on the site, he remarked that it looked like a football field" (Kennedy, 1969).

Other Presidents have used the suborbital U-2 platform as an intelligence asset. For example, President Reagan ordered U-2 overflights of Nicaragua to identify Soviet arms buildup in support of rebel forces. President Bush used U-2 assets in the Gulf War in 1991. President Clinton used them in the war in Bosnia in 1998–1999. Specially modified U-2 aircraft are also used to support various NASA Earth resource (designated ER) remote sensing projects (Brugioni, 1985).

The SR-71

"We knew in 1958, two years before it happened, that the Russians were going to shoot down a U-2. The U-2 could fly higher than any other plane, but it was slow. We needed a replacement that could fly higher, farther, and faster," said Kelly Johnson. The Skunk Works subsequently developed the SR-71 (Figure 3-24). It was delivered to the U.S. Air Force in 1965. In 1976, it set the world speed and altitude records of 2,193 m.p.h. in level flight at 85,126 ft, which exceeds the velocity of a 30.06 rifle bullet. It is still the fastest, highest-flying aircraft in the world as far as we know. At Mach-3 the glass of the cockpit is blistering hot, over 6,000°F. The SR-71's first mission was in 1965 over Hanoi, the capital of North Vietnam. It flew daily reconnaissance missions over North Vietnam and was never touched. The SR-71 detected the Soviet combat brigade in Cuba in 1979. The SR-71 carries only remote sensing and signal intelligence equipment. Several are still in use although they have been officially retired numerous times.

Airborne Stealth Technology

From 1975 to the present, the Lockheed Skunk Works and other aerospace companies throughout the world have developed *stealth* technology. A stealth aircraft is constructed of computer-designed facets of metal that have special shapes and orientations designed to deflect incoming radar waves away from the source of transmission on the ground or from another plane in the air. In addition, special composite material is usually applied to the aircraft, which absorbs incident radar energy (Rich and Janos, 1994). These conditions make the aircraft relatively invisible to radar detection. The radar cross-section (signature) of a non-stealth airplane looks as big as an 18-wheel tractor trailer. Conversely, the radar

F-117 Nighthawk

Figure 3-25   Various views of the Lockheed F-117 Nighthawk stealth aircraft. The unique fuselage geometry and energy-absorbent materials on the plane make it difficult to detect on conventional radars.

cross-section of an F-117 is equivalent to that of a small steel marble. This stealth capability was dramatically unveiled to the world in the 1990 Gulf War when previously secret F-117 aircraft (Figure 3-25) were used to fly night reconnaissance and tactical missions over Iraq and Kuwait. Stealth technology was also used extensively in the conflict in Bosnia in 1998–1999.

Satellite Remote Sensing

The first man-made satellite to orbit the Earth was *Sputnik I,* launched by the Soviet Union on October 24, 1957. In a single moment it effectively confirmed a worldwide "open skies" policy for objects launched into orbit. The United States initiated its *Corona* orbital satellite reconnaissance program in the late 1950s and began launch operations on February 28, 1959. It was managed jointly by the CIA and the U.S. Air Force. The first eight reconnaissance missions did not produce any imagery (McDonald, 1997b). However, on August 18, 1960, Mission 9009 was successful. This ninth mission was recovered in midair and became not only the first Corona to return from space with reconnaissance

film, but also the first object to return from space and be recovered in midair (McDonald, 1997a). The age of space reconnaissance had begun. In just one mission it provided more photographic coverage of the Soviet Union than all previous U-2 missions.

Corona

Keyhole (KH) was the codeword assigned to the U.S. space reconnaissance activities for the Corona, Argon, and Lanyard programs. The KH abbreviation referred to the camera systems associated with these reconnaissance programs, e.g., KH-1, KH-2, KH-3, and KH-4 were Corona sensors; Argon's camera was KH-5, and Lanyard's camera was KH-6. The initial Corona spatial resolution was 40 ft for Corona KH-1 in 1960 and eventually 4.5 to 6 ft for KH-4B in 1972 (Ruffner, 1995; McDonald, 1997b).

The first Corona photograph was of the Mys Shmidta Air Field (Figure 3-26a). All of the KH cameras had a focal length of 24 inches. All KH-4 satellites (1962–1972) contained twin panoramic cameras that could obtain stereo-

First Corona Satellite Reconnaissance Photograph
Mys Shmidta Air Field, U.S.S.R. on August 18, 1960

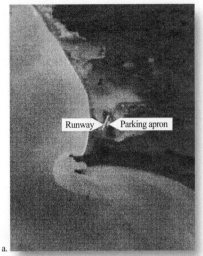

a.

Artist's Rendition of the Corona KH-4B Camera in Flight

b.

Figure 3-26   a) The first photograph collected by Corona KH-1. The spatial resolution was approximately 40 ft. b) Artist's rendition of the internal components of the Corona KH-4B orbital platform (courtesy National Reconnaissance Office).

scopic photographs. An artist's rendition of the KH-4B platform and internal components is shown in Figure 3-26b. The details of each Corona launch from June 25, 1959, to the final mission on May 25, 1972, are found in McDonald (1997b). A Corona KH-4 panchromatic photograph obtained on October 20, 1964, four days after China's first above-ground nuclear test at Lop Nor is shown in Figure 3-27a. Corona played a major role monitoring nuclear proliferation.

Despite their great value, the early reconnaissance satellites were far from perfect. They returned their images by parachuting the film back to Earth in a capsule, sometimes days or even weeks after they were taken (Figure 3-27b). That delay could be crippling. Both the 1967 Six-Day War in Israel and the Soviet invasion of Czechoslovakia in 1968 ended before the United States could obtain satellite imagery of the trouble spots (Richelson, 1992).

It is important to point out that both the Soviet Union and the United States initiated their orbital space reconnaissance systems in 1956. However, due to its technological superiority, the United States was able to attempt launching Corona beginning in 1959 while the Soviet's Zenit-2 program was first launched in 1961. Both the Zenit and Corona platforms met their design objectives as reliable reconnaissance systems operating in an entirely new orbital dimension (Gorgin, 1997). The last Corona image was obtained on May 31, 1972. Over 800,000 images were acquired consisting of approximately 2.1 million feet of film in 39,000 cans (Clarke, 1999).

In 1967, President Lyndon Johnson addressed a group of educators,

> I wouldn't want to be quoted on this, but we've spent thirty-five or forty billion dollars on the space program. And if nothing else had come out of it except the knowledge we've gained from space photography, it would be worth ten times what the whole program has cost. Without satellites, I'd be operating by guess. But tonight we know how many missiles the enemy has, and it turned out our guesses were way off. We were doing things we didn't need to do. We were building things we didn't need to build. We were harboring fears we didn't need to harbor (Richelson, 1992; Walter, 1992).

President Johnson was responding to critics that said he had spent too much on the space program and not enough on poverty.

Only recently has it been made known that imagery from the Corona program were also used for domestic map compilation and updating. For example, the U.S. Geological Survey has used the data to update the 1:250,000 and 1:24,000-scale map series (McDonald, 1997a; Clarke, 1999). A domestic Corona image of the Pentagon in Washington, DC is shown in Figure 3-27c.

**Corona Photography of
Lop Nor, China, on October 20, 1964**

Ground zero

a.

b. Data capture.

c. Pentagon

Figure 3-27   a) Corona KH-4 vertical panchromatic photograph of Lop Nor, China, above-ground nuclear test site on October 20, 1964. b) Photograph of the parachute of a Corona reconnaissance satellite capsule being captured. The film in the capsule was then developed and photointerpreted (courtesy Federation of American Scientists). c) Domestic image of the Pentagon in Washington, DC, recorded by Corona Mission 1101 on September 25, 1967 (courtesy USGS).

Executive Order Number 12951, issued by President William Clinton on February 22, 1995, changed the world of photo-satellite reconnaissance. The order directed:

> Imagery acquired by the space-based national intelligence reconnaissance systems known as the Corona, Argon, and Lanyard Missions shall, within 18 months of the date of this order be declassified.

The Web site for browsing and obtaining duplicates of Corona film is found in Appendix A. An excellent overview of the Corona program and the potential utility of the data for Earth resource analysis is found in Clarke (1999).

Ongoing Satellite Sentinels

Bailey and Kearney wrote in *Defense News* (1991):

> Satellite data and airborne radars have replaced the cavalry scout and the foot patrol as the commander's eyes.... Although the fog of war was not eliminated, General Schwarzkopf's view of the battlefield exceeded anything

Gulf War Bomb Damage Assessment

Figure 3-28   A warehouse south of Kuwait City suspected of housing Iraqi aircraft (U. S. Navy TARPS image released under the Freedom of Information Act to William M. Arkin).

Keyhole 12 (KH-12) Imagery

a. Zhawar Kili suspected terrorist training camp in Afghanistan.
b. Shifa pharmaceutical plant in the Sudan suspected of producing chemical weapons.

Figure 3-29   Examples of Keyhole 12 (KH-12) imagery. a) Zhawar Kili training camp in Afghanistan (spatial resolution was approximately 10 x 10 cm). b) Shifa pharmaceutical plant in the Sudan. These satellite images were used by Secretary of State William S. Cohen in his debriefing after cruise missile attacks on August 20, 1998 (courtesy of Federation of American Scientists, Intelligence Research Program).

before possible. There was far less uncertainty regarding the enemy's vulnerabilities.

They called the Gulf War a *hyperwar* because so many decisions were based on satellite and aircraft remote sensing and signal intelligence (Walter, 1992). An example of Gulf War image intelligence is found in Figure 3-28.

In the United States, at least four types of satellites now gather intelligence information. LaCrosse bathes the Earth in microwaves (radar) and can resolve through cloud-cover objects less than three feet across, augmenting the work of the optical remote sensing satellites when bad weather blocks their view. Optical satellite remote sensing systems are far more powerful than the earlier Corona systems. It is generally believed that some optical sensors can be used to resolve objects as small as 6 in. across using linear or area array charge-coupled-device (CCD) technology discussed in Chapters 4 and 7.

Ferrets (signal intelligence satellites — SIGINT) listen with very sensitive receivers to radio and microwave transmissions. Even in the early 1970s, a ferret named Rhyolite could record signals that aimed Soviet and Chinese missiles and at the same time monitor as many as 11,000 conversations going on between telephones and walkie-talkies. Tracking and Data Relay Satellite Systems (TDRS) orbit the globe at

22,300 miles above the equator in geostationary orbit just like Arthur C. Clarke, the noted author, wrote about in *Wireless World* in 1945 (Walter, 1992). They may be used to transmit remote sensor data from satellites to the ground.

Treaty compliance, arms-control agreements, and border disputes continue to be monitored by nations using high-resolution imagery obtained from satellite platforms (Richelson, 1990). For example, consider the KH-12 panchromatic images of Zhawar Kili, a suspected terrorist training camp support complex in Afghanistan (Figure 3-29a), and the Shifa pharmaceutical plant in the Sudan, suspected of producing chemical weapons (Figure 3-29b). These satellite images were used by Secretary of State William S. Cohen in his debriefing after cruise missile attacks by the United States on August 20, 1998.

Fortunately, humankind has also devoted some of its resources to placing remote sensing platforms in space for peaceful, Earth-resource applications. The most noteworthy include *Skylab*, the Landsat Multispectral Scanner (MSS) (1972-1984), the Landsat Thematic Mapper series of satellites (1984 to present), the Landsat Enhanced Thematic Mapper+ (1999-present), the Space Shuttle Photography Program (ongoing), and numerous meteorological satellites (e.g., U.S. GOES and AVHRR, European METEOSAT and ENVISAT). NASA's Earth Observing System launched its

QuickBird Image of Boston, MA

Figure 3-30   Panchromatic 61 × 61 cm image of Boston, MA, collected by QuickBird on December 27, 2001. Circular high-rises overlook Boston Inner Harbor. The Wharf at Rowe can be seen in the lower-right hand corner (courtesy DigitalGlobe, Inc.).

first Earth-resource oriented (*Terra*) satellite in 1999 and the second (*Aqua*) in 2002. Commercial firms launched IKONOS 2 in 1999, ImageSat in 1999; QuickBird in 2001, and OrbView-3 in 2003 all with nominal panchromatic spatial resolutions of $\leq 2 \times 2$ m. Chapter 7 provides details about many of these remote sensing platforms and sensor systems. A QuickBird 61 × 61 cm panchromatic image of Boston, MA is shown in Figure 3-30. High spatial resolution digital imagery such as this contains so much spatial detail that it can be photo-interpreted as if it were a typical analog aerial photograph.

Unmanned Aerial Vehicles

It is sad that many of the improvements in analog and digital aerial photography data collection and in the art and science of photo-interpretation have taken place during World War I, World War II, the Korean War, the Vietnam War, the Gulf War, the conflict in Bosnia, and Operation Iraqi Freedom. One of the most important recent advancements has been the use of *unmanned aerial vehicles* (UAV) (e.g., McDaid and Oliver, 1997; Lockheed Martin, 2006). More expensive UAVs are controlled from a home base by an analyst who knows exactly where the UAV is located at all times through on-board GPS and forward-looking videography and/or infrared (FLIR).

Tactical and Endurance UAVs

The Department of Defense (DOD) has invested billions in the development of unmanned aerial vehicles since the 1980s. The DOD wants UAVs to satisfy mission unique tactical or endurance reconnaissance and surveillance (designated RQ-1) requirements. Tactical UAVs function within 200 km of their lift off point and usually stay aloft for relatively brief periods of time. Conversely, endurance UAVs can travel > 200 km and can stay aloft for many hours. Some UAVs can stay aloft virtually undetectable due to their size for hours on end, constantly monitoring the same geographic area.

Several of the more important UAVs and their characteristics are presented in Table 3-1. UAVs range from small, light weight handheld units (e.g., Desert Hawk is 32 in. long and 6.5 lbs) to runway-operated behemoths that weigh thousands of pounds (e.g., Global Hawk is 44.4 ft. long and 25,600 lbs). Military UAVs are costly but not as expensive as losing human life or a more expensive aircraft. Each Global Hawk costs > $15 million. A single predator costs $4.5 million. A Shadow 200 costs $300,000. A package of six Desert Hawks cost $300,000. These figures are subject to change.

The United States' Predator UAV was used in surveillance missions over sensitive areas of Bosnia in 1995 (Figure 3-31a,b) and in Afghanistan and Operation Iraqi Freedom. The smaller Shadow 200 is another UAV used extensively in Operation Iraqi Freedom. Table 3-1 depicts a soldier with the U.S. Army 101st Military Intelligence Battalion getting a Shadow 200 UAV ready for a mission over Iraq on September 22, 2004.

Desert Hawk was developed by the Lockheed Martin Skunk Works. It is made of mold-injected polypropylene foam and is powered by an electric motor driving a pusher propeller. It is launched by two persons, who attach a bungee cord to it, extend the cord to about 100 m and then simply let the UAV go. Desert Hawk's operational altitude is 500 ft. (150 m). It can cruise for about one hour at a speed of up to 57 mph (92 km/h), and its operational radius is about 6 nautical miles (11 km). Desert Hawk lands on its Kevlar belly. Its payload consisting of color cameras for day or FLIR for night-time surveillance. Desert Hawk was used in Operation Iraqi Freedom (Lockheed Martin, 2006).

Predator Unmanned Aerial Vehicle (UAV) Imagery of Vogosca Ammunition Plant

a.   b.

Figure 3-31   Predator unmanned aerial vehicle (UAV) pre-strike (a) and post-strike (b) imagery of the Vogosca Ammunition Plant in Bosnia
(courtesy John Pike, Federation of American Scientists, Intelligence Research Program).

UAV Payload

The UAVs summarized in Table 3-1 can carry payloads ranging from a few pounds to 2,000 pounds. UAVs may carry digital cameras, forward-looking infrared (FLIR), hyperspectral sensors, and/or synthetic aperture radar sensor systems for all-weather capability. UAVs typically use wireless technology to transmit images in realtime to the home base, or they store the imagery on-board using data recorders.

UAV Considerations

UAV technology represents a double-edged sword. On the one hand, nonmilitary UAV platforms can be especially useful for low-cost remote sensing data collection for law enforcement, Earth-resource analysis, urban planning, etc. For example, scientists have for some time equipped model airplanes and helicopters with lightweight cameras to obtain quality, high spatial resolution aerial photography (e.g., Hinckley and Walker, 1993). Jackson and Hardin (2005) documented the use of relatively inexpensive model airplanes for collecting large-scale aerial photography using

digital camera technology. An example of their imagery is shown in Figure 3-32. Details about the platform are provided in Chapter 4.

UAVs are also extremely useful for military reconnaissance. "The soldier of the future will have his own mechanical bird which will provide him with pictures, sound and even smells of what is inside a building or bunker," McDaid and Oliver, (1997) predicted. This is true today. Unfortunately, the same technology can be used to spy on the general public or used to conduct illegal activities such as monitoring the conditions around a bank preparatory to a robbery or for industrial espionage, where one company spies on another. Our generation of remote sensing scientists must be careful how it utilizes this new remote sensing technology (Slonecker et al., 1998).

 Commercial Remote Sensing Platforms

The majority of nonmilitary commercial aerial photography is collected using single-wing aircraft. More affluent photogrammetric engineering and remote sensing companies and nonmilitary government agencies have access to jet plat-

Table 3-1. Characteristics of several unmanned aerial vehicles (UAV). The UAV images are not to scale relative to one another (images courtesy of DOD, 2005 and NASA UAV, 2006).

Vendor	Unmanned Aerial Vehicle	Endurance (hours)	Weight (pounds)	Altitude (ft) Range (nm)	Height Length Wingspan (feet)
Aerosonde (Aerosonde Robotic Aircraft, LLC)		40	30.9	20,000 1,800	n.a. 5.4 9.5
Cypher (United Technologies/ Sikorsky, Inc.)		2.5	300	5,000 16	2 n.a. n.a.
Desert Hawk (Skunk Works: Lockheed Martin)		1	6.5	500 ft 6.7	n.a. 32 in. 52 in.
Global Hawk (Teledyne Ryan, Inc.)		42	25,600	65,000 14,000	n.a. 44.4 116.2
Predator (General Atomics, Inc.)		29	6,500	>40,000 n.a.	7.4 36 64
Shadow 200 (AAI, Inc.) (DOD, 2005)		4	215	15,000 32	n.a. 9 12.75

**Digital Aerial Photography Obtained Using
An Inexpensive Unmanned Aerial Vehicle**

Figure 3-32 A large-scale aerial photograph of a portion of South Padre Island, TX obtained in 2005 using an inexpensive unmanned aerial vehicle. Note the detail present in the photograph sufficient to identify jet skis, jet ski docks, small boats, and automobiles. The UAV system used to collect this image is discussed in Chapter 4 (courtesy of Perry Hardin and Mark Jackson, Department of Geography, Brigham Young University).

forms. These types of platforms normally cannot fly as high or as fast as military platforms, but they are sufficient for acquiring quality metric aerial photography and other types of remote sensor data for managing our cities and natural resources. Most of the aircraft now have onboard GPS that keep track of the aircraft's exact location during remote sensing data collection.

 ### References

Bailey, R. and T. Kearney, 1991, "Combat Enters Hyperwar Era," *Defense News*, July 22, 1991.

Baker, S., 1994, "San Francisco in Ruins: The 1906 Aerial Photographs of George R. Lawrence," *Landscape*, 10(10):9–14.

Batut, A., 1890, *La Photographie Aerienne par Cerf-volant*, Paris: Gauthier-Villars.

Boyne, W. J., 1993, *Silver Wings: A History of the United States Air Force*, NY: Simon & Schuster, 366 p.

Brugioni, D., 1985, "New Roles for Recce," *Air Force Magazine*, October, 94–101.

Brugioni, D. A., 1996, "The Art and Science of Photoreconnaissance," *Scientific American*, 274(3):78–85.

Brugioni, D. A. and F. J. Doyle, 1997, "Arthur C. Lundahl: Founder of the Image Exploitation Discipline," *Corona: Between the Sun & the Earth*, R. A. McDonald (Ed.), Bethesda: ASP&RS, 159–166.

Clarke, A. C., 1945, "Extra-terrestrial Relays: Can Rocket Stations Give World-wide Radio Coverage?" *Wireless World*, October, 305–309.

Clarke, K., 1999, *Project Corona*, sponsored by the National Science Foundation, Santa Barbara: UCSB http://www.geog.ucsb.edu/~kclarke/Corona/Corona.html.

Colwell, R. N., 1997, "History and Place of Photographic Interpretation," *Manual of Photographic Interpretation*, W. Philipson (Ed.), 2nd Ed., Bethesda: ASP&RS, 3–48.

Day, D. A., J. M. Logsdon and B. Latell, Eds., 1998, *Eye in the Sky: The Story of the Corona Spy Satellites*, History of Aviation Series, Washington: Smithsonian Institution, p. 128.

DOD, 2005, *September 22, 2004 image of Shadow 200 UAV*, DOD photograph #040922-A-69575-035, Washington: Department of Defense, U.S. Army, on-line Oct. 29.

Eder, J. M., 1945, *History of Photography*, translated by Edward Epstean, NY: Columbia University Press, 318–324.

Fanton, B., 1989, "Photography: 150th Anniversary," *Sky*, 84–91.

Fischer, W. A., 1975, "History of Remote Sensing," in R. G. Reeves, (Ed.), *Manual of Remote Sensing*, Bethesda: ASP&RS, 27–50.

Fouque, V., 1867, *La Verite sur l'invention de la photographie: Nicephore Niepce, sa vie, ses essais, ses travaux*, Paris: Libraire des Auteurs et de l'Academie des Bibliophiles, p. 61.

Ganot, A., 1855, *Traite elementaire de physique*, Paris.

Gernsheim, H. and A. Gernsheim, 1952, "Re-discovery of the World's First Photograph," *The Photographic Journal*, Section A (May): 118–121.

Goldberg, V., 1993, *The Power of Photography*, NY: Abbeville, 287.

Gorgin, P. A., 1997, "Zenit: Corona's Soviet Counterpart," *Corona: Between the Sun & the Earth*, R. A. McDonald (Ed.), Bethesda: ASP&RS, 85–107.

Hall, R. C, 1996, "Post War Strategic Reconnaissance and the Genesis of Project Corona," *Corona: Between the Sun & the Earth*, R. A. McDonald (Ed.), Bethesda: ASP&RS, 25–58.

Hardin, P. J. and M.W. Jackson, 2005, "An Unmanned Aerial Vehicle for Rangeland Photography," *Rangeland Ecology and Management*, 58:439–442.

Hinckley, T. K., and J. W. Walker, 1993, "Obtaining and Using Low-altitude/Large-scale Imagery," *Photogrammetric Engineering & Remote Sensing*. 59:310–318.

Kennedy, R. F., 1969, *Thirteen Days: A Memoir of the Cuban Missile Crisis*, New York: Mentor Co., 192 p.

King, S. A., 1860, "The Late Balloon Photographic Experiment," *Boston Herald*, October 16, 1860.

Lockheed Martin, 2006, *Desert Hawk specifications*, www.lockheedmartin.com.

London, B. and J. Upton, 1994, *Photography*, 5th Ed., NY: Harper Collins, 422 p.

Lopez, D. S., 1995, *Aviation*, NY: Macmillan, 256 p.

Maddox, R. L., 1871, *British Journal of Photography*, 18:422–423.

McCamy, C. S., 1960, "A Demonstration of Color Perception with Abridged Color Projections Systems," *Photographic Science & Engineering*, 4(3):156.

McDaid, H. and D. Oliver, 1997, *Smart Weapons: Top Secret History of Remote Controlled Airborne Weapons*, NY: Barnes & Noble, 208 p.

McDonald, R. A. (Ed.), 1997a, *CORONA: Between the Sun and the Earth: The First NRO Reconnaissance Eye in Space*, Bethesda: ASP&RS, 400 p.

McDonald, R. A., 1997b, "Corona, Argon, and Lanyard: A Revolution for U.S. Overhead Reconnaissance," *Corona: Between the Sun & the Earth*, Bethesda: ASP&RS, 61–74.

NASA UAV, 2006, *Unmanned Aerial Vehicles*, Washington: NASA Wallops Space Flight Center; NASA Goddard Space Flight Center, http://uav.wff.nasa.gov/.

Newhall, B., 1969, *Airborne Camera: The World from the Air and Outer Space*, NY: Hastings House, 144 p.

Newhall, B., 1993, *The History of Photography*, NY: Museum of Modern Art, 319 p.

Niven, W. D., Ed., 1890, *The Scientific Papers of James Clerk Maxwell*, Cambridge: Cambridge Univ. Press, 1:136–137.

Peebles, C., 1997, *The Corona Project: America's First Spy Satellites*, ISBN 1-55750-688-4, 31–33.

Pike, J., 1998-1999, *Intelligence Research Program*, Washington: Federation of American Scientists, www.FAS.org.

Powers, F. G., Jr., 1997, "Foreward: From the U-2 to Corona," *CORONA: Between the Sun and the Earth: The First NRO Reconnaissance Eye in Space*, R. A. McDonald, Ed., Bethesda: ASP&RS, vii–ix.

Quackenbush, R. S., 1960, "Development of Photo Interpretation," *Manual of Photographic Interpretation*, Bethesda: ASP&RS, 1–18.

Rich, B. R. and L. Janos, 1994, *Skunk Works — A Personal Memoir of My Years at Lockheed*, NY: Little Brown, 370 p.

Richelson, J. T., 1990, "From CORONA to LACROSSE: A Short History of Satellites," *The Washington Post*, February 25, 1990, B1–B4.

Richelson, J. T., 1992, "Spies in Space," *Air & Space*, 6(5):75–80.

Ruffner, K. C., 1995, *CORONA: America's First Satellite Program,* Washington: CIA, 360 p.

Scientific American, 1909, "Pigeon Cameras," *Scientific American*, 100(4):27.

Slonecker, E. T., D. M. Shaw and T. M. Lillesand, 1998, "Emerging Legal and Ethical Issues in Advanced Remote Sensing Technology," *Photogrammetric Engineering & Remote Sensing*, 64(6):589–595.

Tennant, J. A., 1903, "Aerial Photography," *The Photo-Miniature,* 5(52):144–173.

Thompson, M. M., 1966, *Manual of Photogrammetry*, Bethesda: ASP&RS, 3rd Ed., 1:244–245.

Walter, W. J., 1992, *Space Age*, NY: Random House, 335 p.

Wolinsky, C., 1999, "The Quest for Color," *National Geographic*, 196(1):72–93.

Aerial Photography — Vantage Point, Cameras, Filters, and Film

4

At first glance it might appear that obtaining and interpreting aerial photography is a routine task because the technology has been available for more than a century. This is not the case. Important decisions must be made to ensure that high-quality aerial photography is collected and interpreted properly (Narumalani et al., 2002). Many of these decisions are based on principles of optics and how energy interacts with light-sensitive materials. Some of the most important issues to be addressed include:

- vantage point (e.g., vertical, low-oblique, high-oblique),

- camera (e.g., analog single- and multiple-lens, digital),

- filtration (e.g., haze, minus-blue, band-pass filters), and

- film emulsion (black-and-white, black-and-white infrared, color, and color-infrared).

The more that an image analyst knows about how an aerial photograph was collected, the better image interpreter he or she will be. This chapter also describes how imagery is digitized.

 Vertical and Oblique Vantage Points

Aerial photography may be obtained from vertical or oblique vantage points depending upon project requirements.

Vertical Aerial Photography

A *vertical photograph* is obtained when the camera's optical axis is within ± 3° of being vertical (perpendicular) to the Earth's level surface (Figure 4-1a). A portion of a vertical black-and-white photograph of the Goosenecks of the San Juan River in Utah is shown in Figure 4-1b. This aerial photograph can be visually interpreted to extract detailed qualitative information about the surface geology, geomorphology, and hydrology within the vertical field of view. This aerial photograph in conjunction with another overlapping vertical aerial photograph obtained from a slightly different viewing position may be analyzed using quantitative photogrammetric principles (discussed in Chapter 6) to derive the following types of information:

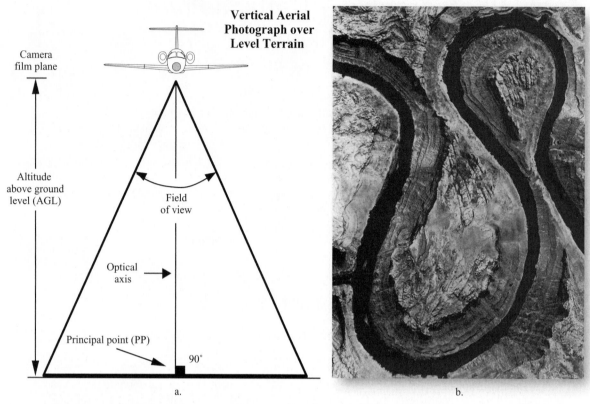

Figure 4-1   a) A vertical aerial photograph has ≤ 3° of tilt from a ray perpendicular to the terrain. b) A vertical aerial photograph of the Goosenecks of the San Juan River in Utah.

- *planimetric* (*x,y* location) *maps* of natural (e.g., stream network, rock outcrops) and man-made features (e.g., trails, roads, buildings);

- *topographic* (*z*-elevation above sea level) *maps*;

- *digital elevation models* (DEMs),

- *orthophotographs* (aerial photographs that are geometrically accurate in *x,y*).

Resource managers sometimes forget that the planimetric, topographic, and orthophotomaps used in almost all geographic information systems (GIS) are created using fundamental photogrammetric principles applied to near-vertical aerial photography (Jensen et al., 2002; 2005). For example, the U.S. Geological Survey's 7.5-minute 1:24,000 map series and many digital elevation models are derived photogrammetrically (Linder, 2003).

Vertical aerial photography is collected for a wide variety of earth resource management and planning applications (Miller et al., 2003; Paine and Kiser, 2003). For example,

consider the vertical aerial photograph of Gulfport, Mississippi shown in Figure 4-2, which documents wind and storm surge damage to the Grand Casino. A portion of the casino was broken off and moved by the storm surge onto Highway 90. Ship cargo containers from a nearby wharf were strewn across the terrain. A color version of this vertical aerial photograph is found in Color Plate 4-1.

Unfortunately, the general public is not used to viewing the tops of objects such as buildings, trees, roads, etc. It takes considerable practice and experience to efficiently and accurately interpret a vertical aerial photograph of the terrain.

Oblique Aerial Photography

An *oblique* or *tilted aerial photograph* is obtained if the camera's optical axis deviates more than a few degrees from vertical (Wolf and DeWitt, 2000). It is called a *low-oblique aerial photograph* if the horizon is not visible (Figure 4-3a). A low-oblique photograph of a bridge over the Congaree River in South Carolina is shown in Figure 4-3b. It was acquired at an altitude of 600 ft above ground level. Visual

Vertical Aerial Photograph of Hurricane Katrina Damage at Gulfport, MS, on August 30, 2005

Figure 4-2   A vertical aerial photograph of wind and storm surge damage of Gulfport, Mississippi caused by Hurricane Katrina recorded on August 30, 2005 (courtesy of NOAA Remote Sensing Division).

interpretation of the photograph yields detailed information about the bridge superstructure.

A *high-oblique aerial photograph* is obtained if the horizon is visible (Figure 4-4a). A high-oblique photograph of the Grand Coolee Dam in Washington is shown in Figure 4-4b. This impressive 1940 oblique photograph represents a valuable historical record of the dam and the cultural landscape that was created adjacent to the dam during its construction to house the workers.

People are generally better able to photo-interpret oblique aerial photography than vertical aerial photography because they have a lifetime of experience looking at the façade (side) of objects as they navigate in their daily environment. Oblique aerial photographs record a familiar façade or side view of the terrain. For example, it is much easier for most

image analysts to identify features in the low-oblique aerial photograph of Gulfport, Mississippi (Figure 4-5) than the vertical aerial photograph found in Figure 4-2. Note the detailed information found in the section of the Grand Casino sitting on Highway 90 and the façade of the Grand Hotel in the background. A color version of this low-oblique aerial photograph is found in Color Plate 4-1.

During image interpretation, analysts should orient oblique photographs so that the features of interest appear in natural perspective. Usually this will mean that objects with height appear to lean away from the person viewing the image.

Scientists and/or resource managers may live their entire lives and never extract quantitative information from oblique aerial photography. Therefore, oblique aerial photography photogrammetric principles will not be presented here. If it is necessary to extract quantitative information

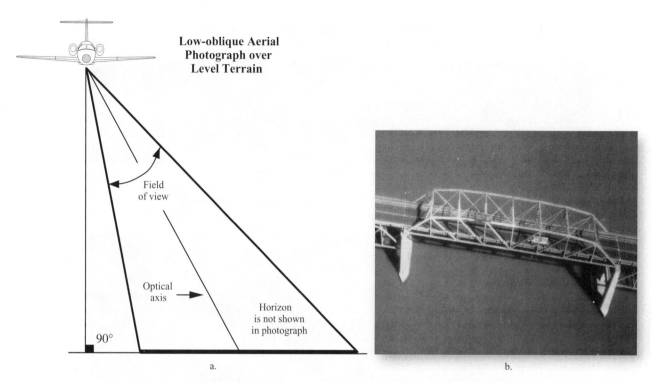

Figure 4-3   a) A low-oblique aerial photograph is tilted > 3° from vertical, and the horizon is not visible. b) A low-oblique aerial photograph of a bridge on the Congaree River near Columbia, SC. The photograph was obtained at an elevation of approximately 600 ft above ground level. One should always view an oblique aerial photograph so that the important features such as the bridge fall away from the viewer.

Figure 4-4   a) A high-oblique aerial photograph is tilted > 3° from vertical, and the horizon is visible. b) A high-oblique aerial photograph of the Grand Coulee Dam in Washington taken in 1940. The homes were built for workers during construction of the dam.

Low Oblique Aerial Photograph of Hurricane Katrina Damage at Gulfport, MS, on August 30, 2005

Figure 4-5   A low-oblique aerial photograph of damage caused by Hurricane Katrina obtained on August 30, 2005 (courtesy of USGS Coastal and Marine Geology Program).

Figure 4-6   One of the first commercially available box cameras was created for Louis Daguerre by Samuel F. B. Morse, inventor of the Morse code.

from oblique photography, consult the algorithms in the *Manual of Photogrammetry* (McGlone, 2004). This discussion will focus on obtaining near-vertical aerial photography and extracting both qualitative and quantitative information.

Aerial Cameras

One of the very first box cameras made for commercial purchase was developed for Louis Daguerre in France by Samuel F. B. Morse, who invented the Morse code (Figure 4-6). While modern cameras are much more sophisticated than this simple box camera, they nevertheless share certain fundamental characteristics. A good way to understand how a modern camera functions is to compare its components with those of the human eye (Figure 4-7).

Aerial Camera Components Compared with the Eye

The light-sensitive *retina* in the human eye is analogous to the light-sensitive *film* located at the *film plane* at the back of the camera. Both the eye and the camera use a *lens* to focus reflected light from the real world onto the retina or film. In the eye, the amount of light allowed to illuminate the retina is controlled by the iris, which can expand or contract in dark or light conditions and by the eyelid, which acts as a shutter. In the camera, the amount of light reaching the film plane is controlled by 1) the *size of the lens aperture opening*, and 2) a *shutter*, which controls the length of time – *exposure* – that the lens aperture remains open and allows light to pass through. The shutter can be placed in the camera lens assembly or just in front of the film at the back of the camera, in which case it is referred to as a *focal plane shutter*.

Focal Plane and Focal Length

The area in which the film is held flat during an exposure is called the *focal plane* or *film plane*. When a camera is focused at infinity, the distance from the rear nodal point of the lens to the film plane is known as the *focal length*. Some focal-length lenses used for aerial mapping include 88 mm (3.5 in.), 152 mm (6 in.), 210 mm (8.25 in.), and 305 mm (12 in.). Zoom lenses have a constantly changing focal length. Conversely, most metric aerial cameras have a fixed focal length, such as 152 mm (6 in.). Longer focal-length lenses, such as the 305 mm (12 in.) lens, are especially useful for high-altitude aerial photography data collection. Military photo-reconnaissance operations commonly employ lenses of 3 to 6 ft to obtain detailed photographs from extremely high altitudes or to acquire oblique photographs across borders.

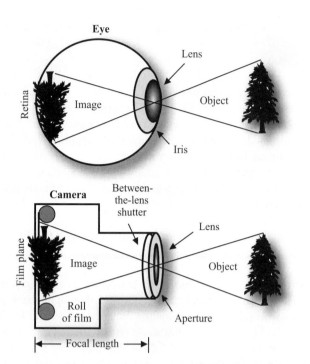

Figure 4-7   A comparison of the optical components of the simple camera with those of the human eye.

f/stop

The ratio of the camera lens focal length (f) to the diameter of the lens opening (d) is known as the *f/stop*:

$$f/\text{stop} = \frac{\text{lens focal length}}{\text{diameter of lens opening}} = \frac{f}{d} \qquad (4\text{-}1)$$

The f/stop ratio is often used to designate the relative aperture setting or *speed* of the camera lens system. For example, a camera with a focal length of 100 mm and a maximum lens diameter opening of 10 mm (called *full aperture*) would have an f/10 speed lens:

$$f/10 = \frac{100 \text{ mm}}{10 \text{ mm}}.$$

If the lens diameter at full aperture were twice as large, say 20 mm instead of 10 mm, the lens rating would be f/5. Basically, the *smaller* the f-rating, the faster the lens, i.e., the more light admitted through the lens opening per unit of time. In the previous example, the f/5 lens is said to be faster than the f/10 lens. The aircraft is always moving during aerial photography data collection (except when using a helicopter or tethered balloon or kite). Therefore, it is important to be able to acquire the photograph very rapidly. We do

Table 4-1.   International series of f/stops and shutter speeds.

Relative Aperture of f/stop
Larger lens openings --------------- Smaller lens openings
f/1---1.4---2---2.8---4---5.6---8---11---16---22---32---etc.
Shutter Speeds
Slower -- Faster
1-1/2-1/4-1/8-1/16-1/30-1/60-1/125-1/250-1/500-1/1000-1/2000-etc.

not want to have the image of a building smeared across the camera film plane as the aircraft moves forward. Therefore, aerial photography cameras generally have very fast lenses that can in effect "stop action." The fast lenses are often coupled with very sensitive film (to be discussed).

A complete system of apertures beginning at f/1 can be produced by multiplying any aperture by 1.4142136 (the square root of 2), which yields the succeeding smaller aperture. For example, the sequence of full-stop increments is f/1, f/1.4142, f/2, f/2.828, f/4, f/5.656, f/8, f/11, f/16, f/22, f/32, etc., as shown in Table 4-1. Each lens opening in the series transmits one-half as much light as the preceding lens opening (i.e., f/8 transmits one-half as much light as f/5.6). For example, if we "stop down" from f/11 to f/16, we decrease both the aperture of the lens opening and the amount of light reaching the film plane by two. Conversely, if we "stop up" from f/16 to f/11, we would increase the diameter of the lens diaphragm and the amount of light illuminating the film plane by a factor of two. This is shown diagrammatically in Figure 4-8. The *lens speed* is the f-number of the maximum effective diameter of the lens when the diaphragm is wide open (full aperture). Thus, we have an f/2.8 lens in Figure 4-8 (London and Upton, 1994).

Shutter Speed

To ensure that the film emulsion receives the correct amount of light during an exposure, the aerial photographer must select the correct relationship between the size of the lens aperture opening (i.e., the f/stop) and how long the light is allowed to illuminate the film plane. The length of time the shutter is open is called *exposure time*. It is controlled by the shutter mechanism. Shutter speeds on a camera usually range from "bulb," which lasts as long as the photographer pushes the exposure button, to 1 second, 0.75, 0.5, 0.25, 1/100, 1/200, 1/400, 1/500, 1/1000, and 1/2,000 seconds. For example, perhaps an ideal aerial photography exposure at 1:00 p.m. in the afternoon in clear Sunlight is f/16 at 1/200 second. If the sky above the aircraft suddenly becomes com-at 1/200-sec, but for some reason the pilot said that the plane's speed would need to be increased? The photographer would then want to increase the shutter speed to perhaps 1/400-sec so that blurring at the film plane would not take place. This would effectively cut the exposure time in half. The photographer would then have to compensate for this by doubling the aperture of the lens by going from f/16 to f/11. Hopefully, the new 1/400-sec exposure at f/11 would produce the desired photographic results.

Why is it important to know about f/stops and shutter speeds, when most of the aerial photography you will ever analyze will be acquired by engineers working for photogrammetric engineering firms who pay meticulous attention to these parameters? First, high-stratus clouds, cloud shadows, and Sun elevation differences throughout the day can cause light to fall off or increase, resulting in underexposed or overexposed aerial photography, respectively. Proper adjustment of the f/stop and shutter speed can maintain proper exposure throughout the day. Sometimes it is necessary to reject a batch of aerial photography if it has been improperly exposed. Second, most persons who use aerial photography will at some time desire to acquire their own photography even if it is obtained by pointing a 35-mm camera out the window of a light plane. At this time you will need to select an optimum combination of f/stop and shutter speed to obtain a good exposure of the terrain on film. It is also likely that relatively inexpensive unmanned aerial vehicles (UAVs) specifically designed for obtaining aerial photography will soon be available. For example, Hinckley and Walker (1993) and Hardin and Jackson (2005) provide information on how to acquire aerial photography using low-altitude UAVs.

Types of Aerial Cameras

Several types of cameras are often used to obtain aerial photography, including:

- single-lens mapping (metric) cameras;

- multiple-lens (multiple-band) cameras;

- digital cameras;

- miscellaneous cameras.

Single-Lens Mapping (Metric) Cameras

Single-lens metric cameras obtain most of the aerial photography used to map the planimetric (x,y) location of features and to derive topographic (contour) maps. These *carto-*

Figure 4-8   The f/stops for a camera lens and the size of their aperture openings. In this example, the f/stops range from f/2.8 to f/22, with the size of the apertures shown by the white circles. The lowest f/stop has the largest opening and lets in the greatest amount of light. The greater the f/stop, the less light admitted. Each f/stop lets in half (or double) the light of the next setting. For example, it takes four circles the size of an f/5.6 aperture to equal the size of an f/2.8 aperture (after London and Upton, 1994).

pletely overcast and allows much less light to illuminate the terrain below, the aerial photographer might have to increase the amount of light illuminating the film plane. He could do this in one of two ways: 1) by increasing the size of the f/stop, e.g., from f/16 to f/11, which would double the area of the aperture opening, or 2) by allowing a longer exposure, e.g., a 1/100-sec exposure would double the length of time the aperture was open. Hopefully, either of these adjustments would result in an acceptable aerial photograph. But what if the photographer already had a good exposure at f/16

Figure 4-9   a) System components of a metric camera consist of 1) vibration isolation elements, 2) film platen pressure plate, 3) drive unit, 4) control unit, 5) film feed rollers, 6) film take-up reel, 7) film magazine, 8) unexposed film-supply reel, 9) drive unit, 10) forward-motion compensation device, 11) mount, 12) exchangeable universal shutter, 13) lens cone assembly, and 14) lens. b) Profile view of a metric camera and system components. c) Example of a metric camera installed in the floor of an aircraft complete with operator console and a terrain bore-sighting instrument. d) A plane with two camera ports can expose two types of emulsions at exactly the same time, e.g., color and color-infrared film. Note that in this photograph two camera ports and two retracted wheels are present (courtesy E. Coyote Enterprises, Inc., Marilyn O'Cuilinn; Z/I Imaging).

graphic cameras are calibrated to provide the highest geometric and radiometric quality aerial photography. They usually consist of a camera body, lens cone assembly, shutter, film feed and uptake motorized transport assembly at the film plane, and an aircraft mounting platform. Filter(s) placed in front of the lens determine the wavelengths of light that are allowed to illuminate the film plane. An artist's rendition of the internal components of a metric camera is shown in Figure 4-9a. A profile view of a typical metric camera is shown in Figure 4-9b.

In the United States, Federal Aviation Administration (FAA) approval is required to cut a hole in an airplane's fuselage to accommodate an aerial camera. An example of a single camera mounted in the floor of an aircraft fuselage is shown in Figure 4-9c. If two aerial cameras are mounted in the aircraft

**Camera Lens
Angle of View**

a.

b.

Figure 4-10   a) The greater the camera lens angle of view, the greater the amount of terrain photographed at a constant altitude above ground level. Narrow (40°), normal (70°), wide-angle (90°), and super-wide-angle (110°) lenses are portrayed. b) Using the same narrow (40°) angle-of-view camera lens at two different altitudes results in dramatically different amounts of terrain being photographed.

(Figure 4-9d), it is possible to expose two types of emulsions at the same time (e.g., color and color-infrared) by synchronizing the camera shutter release mechanisms.

The lens cone assembly is the most important part of the camera. It usually consists of a single, expensive *multiple-element lens* that projects undistorted images of the real world onto the film plane. The multiple-element lens is focused at infinity because the aircraft typically flies at thousands of meters above ground level during data collection. Metric mapping cameras use various lenses with different angular fields of view, depending on the mission requirements. *Narrow* camera lenses have an angular field of view of < 60°, *normal* 60° – 75°, *wide-angle* 75° – 100°, and *super-wide-angle* > 100°. The wider the angular field of view, the greater the amount of Earth recorded on the film at a given altitude above ground level. The higher the altitude, the greater the amount of Earth recorded on the film by each lens. These relationships are summarized in Figure 4-10.

An *intervalometer* is used to expose the photographic film at specific intervals of time (dependent upon the aircraft altitude above ground level and speed) that will result in the proper amount of endlap to be obtained for overlapping (stereoscopic) coverage.

Aerial cameras usually expose film that is 24 cm (9.5 in.) wide in rolls ≥ 100 to 500 ft in length, depending upon the thickness of the film. Individual exposures are typically 9 × 9 in. (23 × 23 cm). At the instant of exposure, the film is held in place against a flat platen surface located at the focal plane. Vacuum pressure is applied to the film via the platen just prior to the instant of exposure to remove any bubbles, bumps, or irregularities in the unexposed film. After exposure, the vacuum is released and the drive mechanism moves the exposed film onto the take-up reel in preparation for the next exposure.

Depending on the velocity of the aircraft (v) and the aircraft altitude above ground level (h), the film might be advanced slightly during exposure to compensate for image motion. Special-purpose *image motion compensation* (IMC) magazines move the film across the focal plane anywhere from 0 to 64 mm per second. Correction is achieved by shifting the platen pressure plate with the film attached via vacuum in the flight direction in accordance with a velocity-to-height ratio (v/h) and the focal length of the lens. This greatly

Importance of Forward Image Motion Compensation

a. Without compensation.

b. With compensation.

Figure 4-11   a) Vertical aerial photograph of a backyard swimming pool obtained without forward image motion compensation. Note the resolution target lying on the ground. b) Photograph obtained with forward image motion compensation. More elements of the resolution target can be discerned (courtesy E. Coyote Enterprises, Inc., Marilyn O'Cuilinn; Z/I Imaging).

Annotation

Figure 4-12   Annotation around the perimeter of an aerial photograph is very important. This particular metric camera provides the following information: 1) a grayscale step wedge, 2) notepad, 3) altimeter, 4) fiducial marks, 5) clock, 6) lens cone serial number, 7) focal length in mm, 8) frame number, 9) mission name and date, and 10) navigation data (not visible). Only one-half of this aerial photograph of Mineral Wells, TX, is visible (courtesy E. Coyote Enterprises, Inc., Marilyn O'Cuilinn; Z/I Imaging).

increases the quality of the aerial photography. An example of the effectiveness of utilizing image motion compensation and stabilization is shown in Figure 4-11. Notice the improved detail in the aerial photograph that was obtained using image motion compensation

Most modern metric cameras provide detailed *image annotation* around the 9 × 9 in. image area of the film (Paine and Kiser, 2003). For example, numerous types of ancillary information are displayed around the perimeter of the vertical aerial photograph of Mineral Wells, TX, shown in Figure 4-12. A programmable light-emitting diode inside the camera exposed text information onto the film. Important information present includes: 1) a grayscale step wedge used to determine if a proper exposure has been obtained, 2) a notepad where the aerial photographer can enter mission critical notes in pencil if necessary, 3) altimeter, 4) white cross-hair fiducial marks, 5) clock, 6) lens cone serial number, 7) focal length in mm, 8) project frame number, 9) mission name and date, and 10) navigation data (not visible). Fiducial marks are discussed in Chapter 6.

Sometimes we analyze aerial photographs that are many years old. Having detailed image annotation information is critical to successful information extraction, especially if sophisticated photogrammetric instruments have to be calibrated or if the photography will be used in computer softcopy photogrammetry applications discussed in Chapter 6.

Multiple-Lens (Multiple-Band) Cameras

More information can usually be obtained about the environment from a study of photographs taken simultaneously in several regions (bands) of the electromagnetic spectrum than from photographs taken in any single band. When conducting multiband spectral reconnaissance (Colwell, 1997), each of the cameras simultaneously records photographs of the same geographic area, but using different film and/or filter combinations. For example, a four-camera Hasselblad 70-mm configuration is shown in Figure 4-13a. By carefully selecting the film and filter combinations, specific wavelengths of light reflected from the scene can be recorded by each of the cameras.

Multiple-band aerial photography of Century City, Los Angeles, CA, is shown in Figure 4-13b. Notice the different information recorded in the individual blue, green, red, and near-infrared photographs. A natural-looking color composite of these photographs can be produced by simultaneously projecting blue light through the blue photograph onto a screen, green light through the green photograph, and red light through the red photograph. A color-infrared color composite could be created by simultaneously projecting blue light through the green photograph, green light through the red photograph, and red light through the near-infrared photograph. More will be said about color composites later in this chapter.

Airborne Digital Cameras

Digital cameras are revolutionizing airborne photogrammetric data collection. The imaging sensor of a digital camera is usually a charge-coupled device (CCD) or Complimentary Metal Oxide Semiconductor (CMOS) computer chip. The sensor detectors convert light into electrons that can be measured and converted into a radiometric intensity value (McGlone, 2004). Digital cameras may be based on *linear* or *area* arrays of detectors (Figure 4-14). Table 4-2 summarizes the number of detectors found in some of the commercial area array chips currently available.

Digital cameras utilize a lens with its associated diaphragm to control the *f*/stop, a shutter to control the length of exposure, and a focusing device. However, the major difference is that instead of using film, CCD linear or area arrays are located at the film plane. The lens focuses the light from the outside world onto the linear- or area-array bank of detectors. The photons of light illuminating each of the detectors cause an electrical charge to be produced that is directly related to the amount of incident radiant energy. This analog signal is then sampled electronically and converted into a digital brightness value ranging from 8-bit (values from 0 − 255) to 12-bit (values from 0 to 4095). The brightness values obtained from the analog-to-digital (A-to-D) conversion may be stored and read by computer systems. *The CCDs are actually more sensitive to spectral reflectance changes in the scene than the silver halide crystals used in conventional analog aerial photography.*

Small-Format Digital Cameras: These cameras typically have < 15 megapixels (MP). For example, the digital camera shown in Figure 4-15a is based on an area array with 3,456 columns and 2304 rows yielding almost 8 MP (Table 4-2). At the instant of exposure, the camera rapidly records three versions of the scene using internal filtration. The result is one image based solely on blue light reflected from the terrain, another based on only green light reflected from the terrain, and a final image produced only from reflected red light. The three individual black-and-white images are recorded in the camera's random access memory (RAM) and can be color-composited using additive color theory to produce a natural-looking color photograph. It is also possible to make the detectors sensitive to near-infrared light.

This inexpensive UAV shown in Figure 4-15b contains an intervalometer to obtain photographs at specific intervals to obtain the necessary end-lap. The images are rapidly stored on a data recorder. The operator on the ground knows where the plane is based on the use of an onboard GPS antenna (not shown). The 8 MP digital camera obtains very high resolu-

Multiband Aerial Photography

a. Multiband cameras.

b. Blue (0.4 - 0.5 μm)   c. Green (0.5 - 0.6 μm)

d. Red (0.6 - 0.7 μm)   e. Near-infrared (0.7 - 1.0 μm)

Figure 4-13   a) Four 70-mm Hasselblad cameras arranged in a mount that may be installed in a specially designed hole in the bottom of an aircraft. All four cameras are exposed electronically at exactly the same instant. Different film and filter combinations may be used in each camera. b-e) Multiple-band aerial photography of Century City, Los Angeles, CA. The large panels in the parking lot were used to calibrate the photography. Notice how bright the vegetation is around the curvilinear hotel in the center of the black-and-white near-infrared photograph and how dark it appears in the blue, green, and red photographs.

tion aerial photography (Figures 4-15cd). As with any aircraft, great care must be exercised when using UAVs to ensure that they do not crash causing injury to life and property. This is no trivial matter.

Medium-format Digital Cameras: These cameras are usually based on area arrays with approximately $4,000 \times 4,000$ detectors (yielding 16 MP) using Fairchild CCD486 or Kodak KAF-1680 chips (Table 4-2). Due to the rather limited number of pixels, the ground coverage is not as good as

traditional large-format analog frame cameras. Nevertheless, medium-format digital cameras have proven useful for many applications. The Emerge Digital Sensor System (DSS) is a good example of a medium-format digital camera (discussed in Chapter 7).

Large-Format Digital Cameras: These cameras surpass the performance of large-format analog frame cameras. Some cameras such as the Leica Geosystems Airborne Digital Sensor 40 (ADS-40) are based on multiple linear arrays.

Figure 4-14   a) Enlarged view of a 2048 element charge-coupled-device (CCD) linear array.
b) Enlarged view of a 3456 × 2304 area array CCD.

Table 4-2.   Characteristics of several commercially available charge-coupled-device (CCD) arrays (McGlone, 2004; Nikon, 2005).

Vendor	Sensor	Size (horizontal × vertical)	Pixels
Nikon	D70	3008 × 2000	6,016,000
Kodak	DCS-460	3060 × 2036	6,230,160
Canon	APS-C	3456 × 2304	7,962,624
DALSA	DALSA 4027	4008 × 2672	10,709,376
Kodak	KA1-11000	4008 × 2672	10,709,376
Nikon	D2X	4288 × 2848	12,212,224
Fairchild	CCD486	4096 × 4096	16,777,216
Kodak	KAF-1680	4096 × 4096	16,777,216
DALSA	FTF4052M	5344 × 4008	21,418,752
Phillips	DALSA-DMC	7168 × 4096	29,360,128
BAE	CCD595	9216 × 9216	84,346,656

The ADS-40 uses several forward, nadir, and aft-looking linear arrays consisting of 12,000 detector elements each (Figure 4-15b). Other cameras, such as the Z/1 Imaging Digital Modular Camera (DMC) and the Vexcel large-format digital camera, are based on the use of multiple area arrays. The Z/1 uses four 7000 × 4000 arrays to obtain panchromatic data and four 3000 × 2000 arrays to obtain multispectral data. The Vexcel digital camera uses 13 CCD arrays with 4000 × 2700 elements. Characteristics of the Emerge DSS, Leica ADS40, Z/1 DMC, and the Vexcel large-format digital camera are presented in Chapter 7 (Multispectral Remote Sensing Systems).

Because of the large number of large-format analog cameras already in existence, digital cameras will be used along with the traditional analog cameras for some time. Eventually, however, most photogrammetric aerial photography will be obtained using large-format digital metric cameras (McGlone, 2004).

Miscellaneous Cameras

There are a variety of relatively simple, inexpensive cameras used both commercially and in a research mode that provide high-quality aerial photography. For example, extensive use is made of 35-mm cameras mounted inside a plane or hand-

Inexpensive Unmanned Aerial Vehicle Characteristics

a. Unmanned aerial vehicle fuselage.

Data recorder and
intervalometer   Digital camera

b. System components.

c. Citrus grove near Weslaco, TX.

d. Water treatment plant on South Padre Island, TX.

Figure 4-15   a) An unmanned aerial vehicle modified to carry a digital camera. b) System components include a data recorder, intervalometer, and GPS antenna (not shown). c and d) Examples of aerial photography obtained using the UAV [courtesy of Perry Hardin and Mark Jackson, Department of Geography, Brigham Young University; Hardin and Jackson (2005)].

held by the scientist to obtain aerial photography of small research areas (Warner et al., 1996). Such systems can provide excellent, inexpensive aerial photography if properly mounted, exposed, and processed.

Aerial Photography Filtration

In 1666, Sir Isaac Newton (1642-1727) (Figure 4-17a), while experimenting with a prism, found that he could disperse white light into a *spectrum* of colors (red, orange, yellow, green, blue, indigo, and violet — ROYGBIV) (Figure 4-18). A color example of what takes place when white light is directed through a prism is provided in Color Plate 4-2a. Utilizing a second prism, Newton found that he could recombine the colors into white light (Figure 4-19). He published his research on the properties of light in *Opticks* in 1704 (Figure 4-17b).

Additive color theory is based on what happens when light is mixed. Color Plate 4-2b reveals that white light consists of all the colors of the visible spectrum. Black is the absence of all these colors. Blue, green, and red are the *primary colors*. Additively combining green and red light creates yellow light. Additively combining blue and green light creates cyan. Additively combining blue and red light creates magenta. Yellow, magenta, and cyan are referred to as *complementary colors* because, when paired, they produce white light, e.g., yellow plus blue, magenta plus green, and cyan plus red. Color relationships are summarized in Table 4-3.

Large-format Leica Geosystems Airborne Digital Sensor 40 (ADS-40)

Figure 4-16   The Leica Geosystems ADS-40 digital camera uses multiple linear arrays consisting of 12,000 detectors (courtesy of Leica Geosystems, Inc.).

Table 4-3.   The color of an object as seen in white light.

Color as Seen in White Light	Colors of Light Absorbed
Blue	Green and red
Green	Blue and red
Red	Blue and green
Yellow (green and red)	Blue
Magenta (blue and red)	Green
Cyan (blue and green)	Red
Black	Blue, green, and red
White	None
Gray	Equal proportions of blue, green, and red

a. Sir Isaac Newton (1642-1727)

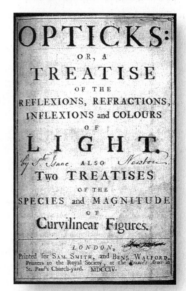

b. *Opticks*, Newton's published research on the properties of light.

Figure 4-17   Sir Isaac Newton published *Opticks* in 1704. This work eventually led to the invention of the reflecting telescope.

Additive color theory is used to display images on television screens and on computer monitors. Each picture element (pixel) on a monitor screen (there are usually 480 rows by 525 lines of pixels) consists of three color guns — blue, green, and red. Each color gun's intensity in each pixel is modulated according to the amount of primary color present in the scene being transmitted from the TV station or from the computer central processing unit (Jensen, 2005). The result is a color visual display of the phenomena based on additive color theory.

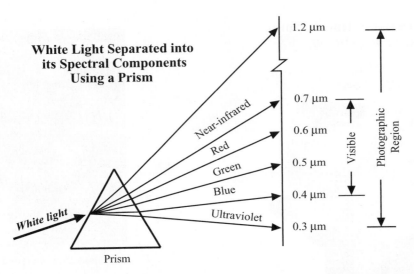

Figure 4-18   White light can be separated into its spectral components using a prism, as originally discovered by Sir Isaac Newton in 1666 and published in *Opticks* in 1704. Only blue, green, and red light are visible to the human eye.

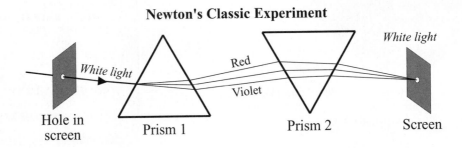

Figure 4-19   A diagrammatic representation of Newton's classic experiment in which white light is decomposed into its spectral components using one prism, and then recombined back into white light using a second prism, before being projected onto a screen.

Subtractive color theory is based on the use of pigments or dyes and not light. We use subtractive color theory when we paint or work with filters. For example, we know that if we mix equal proportions of blue, green, and red paint, we do not get white but obtain some dark gray color. Subtractive color theory is based on the use of the complementary color dyes — yellow, magenta, and cyan. An example of subtractive color theory is found in Color Plate 4-2c. If we projected white light onto a translucent filter made of yellow dye, the filter would subtract the blue light and allow the green and red light to pass through. Similarly, a magenta dye filter subtracts the green light and allows the blue and red to be transmitted. A cyan dye filter subtracts the red light and allows the blue and green light to be transmitted. But what if we superimposed two of these dye filters? If we superimposed the magenta dye filter and the cyan dye filter, everything but

blue light would be subtracted. This is because the magenta filter subtracts the green light while the cyan filter subtracts the red light. If we superimposed the yellow and cyan filters only green light would be perceived. Finally, if we superimposed the yellow and magenta dye filters, only red light would be perceived. If all three of the subtractive dyes (yellow, magenta, and cyan) were used, they would subtract (filter out) all of the white light and we would not perceive any color; it would appear black. It will be demonstrated how subtractive color dyes are used to create the colors we see in 1) developed color negatives, and 2) color positive prints.

A scientist can selectively record very specific wavelengths of light reflected from the scene onto a photographic emulsion. This is done using specific filter and film combinations. We will first investigate various filter options.

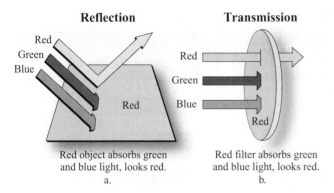

Figure 4-20   a) A red object absorbs most of the blue and green incident light and reflects most of the red light toward our eyes or a remote sensing instrument. b) A red filter looks red because it selectively absorbs blue and green light while allowing red light to be transmitted through it.

Types, Sizes, and Colors of Filters

A filter placed in front of the camera lens does exactly what the name implies, i.e., it filters out certain wavelengths of light before they can reach the film plane and expose the film. To understand how filters work, consider why a red piece of paper looks red (Figure 4-20). White light, consisting of equal proportions of blue, green, and red wavelength energy, illuminates a piece of red paper. The paper absorbs the blue and green light but reflects the red light. Therefore, anything that absorbs (subtracts) both blue and green light and reflects red light will look red. Refer to Table 4-3 for a summary of these relationships. Conversely, if we hold a red filter up to a white light, it will appear red to us because it lets through (i.e., transmits) only the red light while it absorbs both the blue and green light. This is the key to an understanding of photographic filtration. *Filters* subtract some of the light reflected from a scene before the light reaches the film in the camera. Get into the habit of thinking of a red filter not so much as one that looks red, but as one that absorbs (subtracts) blue and green light.

Using this logic, a yellow filter absorbs blue light and allows green and red light to be transmitted (Table 4-3). Our eyes perceive a mixture of red and green light as yellow (i.e., the lack of blue). Due to Rayleigh scattering (Chapter 2), blue light is scattered in the atmosphere to a much greater degree than either green or red light and can therefore cause aerial photography to record considerable unwanted, scattered blue light. Therefore, it is common to use a yellow filter to selectively remove some of the scattered path radiance

(especially ultraviolet and some blue light) before it ever reaches the emulsion. This *minus-blue filter* will be shown to be particularly important when collecting near-infrared aerial photography.

Most aerial photography is acquired using at least one standard filter. The spectral-transmittance characteristics of selected Kodak Wratten filters over the wavelength interval 200 to 1,100 nm (0.2 – 1.1 μm) are shown in Figure 4-21. In addition, a transmittance curve for Kodak filters HF3 and Wratten 12 are shown in Figures 4-22a and 4-22b, respectively (Kodak, 1999). These filters are important to aerial photography. When collecting natural color aerial photography, it is desirable to eliminate much of the scattering of ultraviolet radiation caused by atmospheric haze. For this purpose and to obtain a more satisfactory color balance, *haze filters* (HF) were developed that absorb light shorter than 400 nm. Similarly, when collecting color-infrared aerial photography, a yellow filter is used, which subtracts almost all of the blue light (wavelengths shorter than 500 nm). This minus-blue filter reduces the effects of atmospheric scattering and allows the proper energy to interact with each of the film's layers, to be discussed shortly. If desired, it is possible to configure a camera film/filter combination so that it selectively records a very specific band of reflected electromagnetic energy on the film. This is called spectral *band-pass filtering*. For example, if one wanted to photograph only reflected green light for a specific aerial photography project, a Kodak Wratten 58 filter could be used (refer to Figure 4-21).

In addition to the normal filters, *polarizing filters* may also be used with aerial photography. To understand how polarizing filters work, it is necessary to briefly review the nature of light. Remember that an electromagnetic quanta of light discussed in Chapter 2 vibrates in all directions perpendicular to its direction of travel as shown in Figure 4-23. When a quanta of light hits a nonmetallic surface, the vibration in only one direction, or plane, is reflected completely. Conversely, all vibrations are reflected by a bare metallic surface. Now, depending upon the angle at which the camera or our eyes are viewing the object, vibrations in other planes are reduced or eliminated completely. This reflected light — vibrating in only one plane — is called *polarized light*. The light from a blue sky is polarized because it is reflected from nonmetallic particles in the atmosphere. Likewise, light reflected from a wheat field or a body of water into the field of view of a camera is polarized since wheat and water are nonmetallic.

Sometimes it is useful to only view a certain "angle" of polarized light leaving the scene. For example, Figure 4-24

Spectral-Transmittance Bar Charts for Selected Wratten Filters

Figure 4-21   Transmission characteristics of selected Wratten filters (courtesy Eastman Kodak Co.).

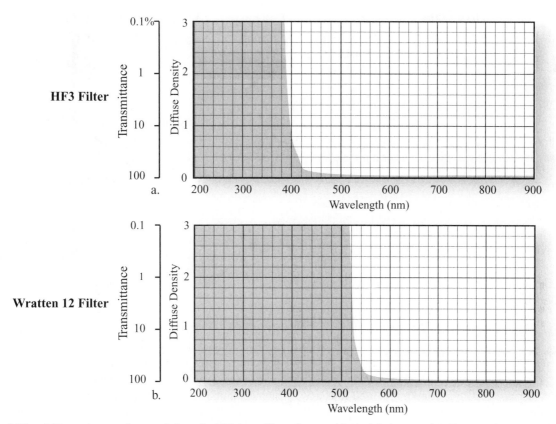

Figure 4-22 a) Transmittance characteristics of a HF3 haze filter often used in aerial photography. b) Transmittance characteristics of a Wratten 12 (minus-blue) filter used when acquiring color-infrared photography (courtesy Eastman Kodak Co.).

demonstrates how a polarizing filter placed in front of a camera lens will pass the vibration of a light ray in just one plane at a time. Some filters even have handles with which to adjust the filters so that they pass the light vibration in a plane parallel to the handle.

One of the most widely used applications of polarizing filters is when photographing waterbodies. By manipulating the polarizing filter, it is possible to filter out the unwanted reflections of some types of polarized light reflected from the water surface. This can improve our ability to see further into the water column. Polaroid glasses worn by people are simply polarized filters that perform much the same function. Interestingly, no one ever asks us whether we want vertically or horizontally polarized glasses, which provide different views of the world to our eyes. Polarized energy will also be important in active microwave (radar) remote sensing discussed in Chapter 9.

Thus, filters allow us to selectively filter out certain types of unwanted light while allowing very specific wavelengths of light to pass through the filter. The light that the filter trans-

mits through it is then allowed to pass through the lens system and illuminate the light-sensitive silver halide crystals in the film emulsion.

 Aerial Photography Films

Film is usually exposed in a camera mounted in an aircraft that is moving relatively fast (e.g., > 150 m.p.h.). Fortunately, films made especially for aerial photography can stop action and yield high-quality aerial photography if 1) the proper film/filter combination is used, 2) the aperture opening (*f*/stop) is appropriate, and 3) the length of the exposure is correct. Unfortunately, analysts sometimes are forced to interpret underexposed (too dark; not enough light was allowed to create an exposure) or overexposed (too bright; an excessive amount of light was allowed to create an exposure) aerial photography. Therefore, it is important for photo-interpreters to understand the fundamental nature of film emulsions and their speed in order to properly interpret

Polarized Light

Figure 4-23   Light entering the upper portion of the atmosphere is unpolarized. It is polarized when it is reflected from nonmetallic surfaces such as atmospheric particles, water, trees, concrete, etc.

aerial photography or to understand why problems occur in some aerial photography.

General Characteristics of Photographic Emulsions

Sensitometry is the science of measuring the sensitivity of photographic materials (McGlone, 2004). It is used to control the operations of exposing and processing photographic materials. Cross-sections through four typical films (Figure 4-25) reveal that they consist of:

- *emulsion* layer(s) containing light-sensitive silver halide crystals (AgX where X is the halogen symbol) embedded in a gelatin;

- a *base* or *support* material that may be transparent (acetate film or glass) or opaque (paper);

- an *anti-halation* layer that absorbs light that passes through the emulsion and the base to prevent reflection back to the emulsion.

The sensitivity of a photographic emulsion is a function of the size, shape, and number of *silver halide crystals* in the emulsion per unit area and the wavelengths of light to which the grains are sensitive. This is real silver, and although it is present in relatively small quantities in a roll of film, it does account for the relatively high cost of film.

Silver Halide Crystal Grain Size, Density, and Shape

To understand the relationship between grain size and density, we will depict a piece of photographic negative film as

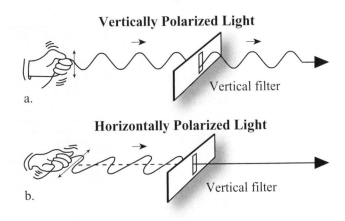

Figure 4-24   a) Vertically polarized light is allowed to pass through a vertically polarized filter. b) Horizontally polarized light is not allowed to pass through the vertically polarized filter.

being composed of simply silver halide crystals and support backing (Figure 4-26a). Film A has 10 silver halide crystals per unit distance while Film B has only seven crystals. As grain size increases, the total number of grains in the emulsion per unit distance decreases. If an emulsion is composed of grains one-half as large as those in another emulsion, it will require approximately twice as much light to expose it. Therefore, Film B is said to be faster than Film A because it requires less light for proper exposure. Under the scanning electron microscope, conventional silver halide crystals appear as cubes, octahedra (eight-sided solids), or irregularly shaped pebbles (Figure 4-26b). Also, note that the crystals are not all exactly the same size.

Faster films can be used advantageously when photographing objects that are moving rapidly across the film plane, as in aerial photography. Unfortunately, as sensitivity and grain size increase, the resulting image becomes more coarse, and resolution (sharpness or crispness of the image) may be reduced. Film resolution can be tested by photographing a standard test pattern that consists of groups of line pairs (parallel lines of varying thickness separated by spaces equal to the line thickness). An example of a resolution test pattern target was shown previously in Figure 4-11. The number of lines per millimeter in the smallest line pattern that can clearly be discerned on the developed film is the resolution of the film. It is common to be able to resolve 25 to 100 line pairs per millimeter on 1:10,000 scale aerial photography positive prints. This type of spatial resolution results in aerial photography with very high spatial detail.

Generalized Cross-Sections of Black-and-White Panchromatic, Black-and-White Infrared, Color, and Color-Infrared Film

Black-and-White Film

Panchromatic — blue, green and red sensitive emulsion of silver halide crystals

Base

Anti-halation layer

a.

Black-and-White Infrared Film

Near-infrared sensitive layer

Base

Anti-halation layer

b.

Normal Color Film

Blue sensitive layer [yellow dye-forming layer]

Yellow internal filter blocks blue light
Green (and blue) sensitive layer [magenta dye-forming layer]

Red (and blue) sensitive layer [cyan dye-forming layer]

Base

Anti-halation layer

c.

Color-infrared Film

Near-infrared (and blue) sensitive layer [cyan dye-forming layer]

Green (and blue sensitive layer [yellow dye-forming layer]

Red (and blue) sensitive layer [magenta dye-forming layer]

Base

Anti-halation layer

d.

Figure 4-25   Generalized cross-sections of a) panchromatic black-and-white film, b) black-and-white infrared film, c) normal or natural color film, and d) color-infrared film.

Black-and-White Photographic Emulsions

Just as the retina of the human eye is sensitive to different wavelengths of light ranging from blue through red (0.4 – 0.7 μm), it is possible for film manufacturers to create black-and-white photographic emulsions that have the following sensitivities (McGlone, 2004):

• *orthochromatic* emulsions are sensitive to blue and green light to approximately 0.6 μm;

• *panchromatic* emulsions are sensitive to ultraviolet, blue, green, and red light to approximately 0.7 μm;

• *near-infrared* emulsions are sensitive to blue, green, red, and near-infrared light to approximately 0.9 μm.

The lower wavelength-limit of light sensitivity is approximately 0.35 μm and results from the light absorption characteristics of the gelatin associated with the emulsion.

The spectral sensitivity of standard black-and-white printing paper, panchromatic film, and near-infrared film are summarized in Figure 4-27. The diagram reveals why it is possible to use a red "safe light" in a photographic darkroom. The printing paper is simply not sensitive to red light. Conversely, black-and-white panchromatic film records ultraviolet, blue, green, and red reflected light. Therefore, a haze filter (such as the Kodak HF3 previously discussed) is often used to keep ultraviolet and blue light from exposing the film. Panchromatic film is the most widely used black-and-white aerial film, as it produces gray tones that are expected and recognized by human beings, i.e., water is dark, sand is white, concrete is gray, clouds are white. Much of the aerial

Two Films with Different Sizes and Densities of Silver Halide Crystals

Electron Microscope Photograph of Silver Halide Crystals

Figure 4-26   a) Films A and B represent hypothetical profiles of two films with different sizes and densities of silver halide crystals. b) An electron microscope photograph of actual silver halide crystals. Note that the crystals are not perfectly uniform in size, shape, or spacing (courtesy Eastman Kodak Co.).

Spectral Sensitivity of Selected Black-and-White Films and Printing Paper

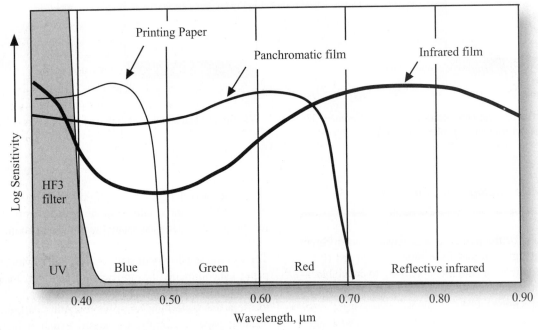

Figure 4-27   Spectral sensitivity of black-and-white film and paper emulsions over the wavelength interval 0.35 – 0.9 μm (courtesy Eastman Kodak Co.).

photography acquired for photogrammetric purposes to make planimetric and topographic maps is panchromatic aerial photography. Typical panchromatic aerial photogra-

phy films marketed by Eastman Kodak Company are summarized in Table 4-4.

Table 4-4.   Eastman Kodak aerial photography films (courtesy Eastman Kodak Co.; all film names are trademarked).

Black-and-White Aerial Films	Characteristics
Plus-X AEROGRAPHIC 2402	Medium speed, high-dimensional stability for aerial mapping and reconnaissance.
Plus-X AERECON II 3404	Medium speed, fine-grain, medium- to high-altitude reconnaissance film.
Tri-X AEROGRAPHIC 2403	High speed, high-dimensional stability for mapping under low illumination.
Tri-X AERECON SO-50	Similar to 2403; thin base for increased spool capacity.
Double-X AEROGRAPHIC 2405	Medium- to high-speed film for mapping and charting, dimensional stability.
Aero LX 2408	Intermediate speed, very fine-grain, medium- to high-altitude mapping and reconnaissance.
Panatomic-X AEROGRAPHIC II 2412	Intermediate speed, very fine-grain, medium- to high-altitude mapping and reconnaissance.
Panatomic-X AEROGRAPHIC II 3412	Similar to 2412; thin base; greater spool capacity for medium- to high-altitude mapping.
AERECON High Altitude 3409	Thin base, extremely fine-grain, high definition film for high-altitude, stabilized platform aerial cameras with high-quality optical systems.
Infrared AEROGRAPHIC 2424	Reduces haze effects in vegetation surveys, highlights water, multispectral photography.
Color Aerial Films	
AEROCOLOR II Negative 2445	Medium speed for mapping and reconnaissance.
AEROCHROME II MS 2448	Color reversal film for low- to medium-altitude aerial mapping and reconnaissance.
AEROCHROME HS SO-358	High-speed color negative film for low-altitude aerial photography.
AEROCHROME HS SO-359	High-speed color reversal film for low- to medium-altitude mapping and reconnaissance.
AEROCHROME II Infrared 2443	False-color reversal film, high dimensional stability for vegetation surveys, camouflage detection, and Earth resource investigations.
AEROCHROME II Infrared NP SO-134	Similar to 2443; greater infrared response, suitable for altitudes above 15,000 ft.

A black-and-white panchromatic aerial photograph is shown in Figure 4-28a. A black-and-white near-infrared photograph of the same area is shown in Figure 4-28b. Care must be exercised when interpreting near-infrared photography because the camera has recorded energy that is beyond our human perception. Consequently, certain tones might not represent what we would anticipate in the image. For example, healthy green vegetation is dark on panchromatic photography because it absorbs almost 80–90 percent of the incident blue, green, and red radiant energy for photosynthetic purposes. Conversely, healthy vegetation is bright on the black-and-white near-infrared photography because it reflects 40–70 percent of the incident near-infrared energy. Human beings have very little idea how much near-infrared radiant energy is reflected from commonplace materials such as vegetation, water, and concrete. Therefore, we should constantly refer to spectral reflectance curves of these phenomena that summarize how much green, red, and near-infrared energy these objects typically reflect (refer to Chapter 2). In this manner, we can understand and even predict how certain objects in the real world will look on infrared photography. It is instructive to review the process whereby simple black-and-white aerial photography negatives and positive prints are produced.

Creating a Black-and-White Aerial Photography Negative

Consider the simple binary terrain depicted in Figure 4-29 that consists of calm, non-turbid ocean water and a sandy beach with no vegetation on it. Assume that the silver halide crystals in the film in the aerial camera are sensitive to blue, green, and red light from $0.4 - 0.7$ μm. The non-turbid ocean water would absorb much of the incident blue, green, and red radiant flux from the Sun. Conversely, the sandy beach would reflect much of the incident radiant flux. Some portion of this energy would be collected by the optics of the camera lens and focused onto the silver halide crystals at the film plane. When the required amount of light exposes a silver halide crystal suspended in the emulsion, the entire crystal becomes exposed, regardless of its size. Basically, the bond between the silver and the halide is weakened when silver halide crystals are exposed to light. Notice in our example that there was sufficient energy to expose the silver halide crystals on the left side of the film but insufficient energy to expose the crystals on the right side of the film. An emulsion that has been exposed to light contains an invisible image of the object called the *latent image*. To turn the latent image on the emulsion into a negative, it must be *developed*.

Panchromatic

Black-and-White Infrared

a.

b.

Figure 4-28   a) Black-and-white panchromatic photograph of Tivoli North Bay on the Hudson River, NY. Vegetation is recorded in relatively dark tones because it absorbs much of the incident green and red energy. Some green and red energy passes through the water column and provides information on suspended sediment patterns and submerged aquatic vegetation (V = wild celery, *Vallisneria americana*). b) A black-and-white near-infrared photograph depicts the land-water interface well, because water absorbs most of the incident near-infrared energy causing it to appear dark. Conversely, deciduous trees on the island and tidal wetland vegetation (T = cattail, *Typha angustifolia*; P = common reed, *Phragmites aus.*) are bright because healthy vegetation reflects much of the incident near-infrared energy (Berglund, 1999) (please refer to Color Plate 4-4ab).

**Creation of an Aerial Photographic Negative by
Exposing and then Developing Silver Halide Crystals**

Figure 4-29   A diagrammatic representation of how a black-and-white negative of an ocean-beach scene is exposed and developed.

Creation of a Positive Print from an Aerial Photography Negative

White light from the enlarger

Dense (the beach) Clear (the ocean)

Developed silver → halide crystals

Negative after development

Little radiant flux passing through the negative

Lots of radiant flux passing through the negative

Enlarger lens

Undeveloped latent image on photographic paper

Paper after development

Ocean Sandy beach

Positive print of the landscape with dark water and bright sandy beach

Figure 4-30   Diagrammatic representation of how a positive black-and-white print of the ocean-beach scene is produced from a developed negative.

When the latent image is developed with the proper chemicals, areas of the emulsion that were exposed to intense light turn to free silver and become black (dense or opaque), as shown. Areas that received no light become clear if the support is the typical transparent plastic film. The degree of darkness of the developed negative is a function of the total exposure (product of illuminance and exposure time), which caused the emulsion to form the latent image.

Creating a Positive Aerial Photographic Print from a Black-and-White Negative

What good is a negative? Most people do not photo-interpret *negatives,* because they are a reversal of both the tone and geometry of the real world, as shown in Figure 4-29. Therefore, it is customary to produce a positive print of the scene from the negative (Figure 4-30). A *positive print* is produced by placing the developed negative in an enlarger with the emulsion side of the negative facing the light source. White light is transmitted through the negative, passed through the enlarger lens, and onto photographic paper that has its own silver halide crystal sensitivity and film speed. In this case, the dense (dark) beach area on the negative allows very little radiant flux to pass through the negative while the clear (ocean) area on the negative allows a lot of radiant flux to pass through it. The energy is focused through the lens onto the undeveloped photographic paper, and once again the individual silver halide crystals are either completely exposed or unexposed, forming a latent image on the photographic paper. After development, the exposed ocean area becomes dark on the paper print and the sand beach area becomes light. Thus, we now have the correct tone and geometry of the terrain in the form of a positive paper print that can be visually interpreted. If we want a positive film transparency to view, then positive transparency material is substituted for the photographic paper. This typically creates an airphoto with improved fine detail and more gray tones.

Advanced Radiometric Characteristics of Black-and-White Films

Transmittance: The ability of a portion of a developed film to pass light is called its *transmittance* ($T_{i,j}$). A black portion of the film may transmit no light, while a clear portion of the film may transmit almost 100 percent of the incident light. Therefore, the transmittance at location i,j in the photograph is (Mikhail et al., 2001):

$$T_{i,j} = \frac{\text{light passing through the film}}{\text{total incident light}}. \qquad (4\text{-}2)$$

Opacity: There is an inverse relationship between transmittance and how opaque an area on the film is. Something that is very opaque does not transmit light well. *Opacity ($O_{i,j}$)* is the reciprocal of transmittance:

$$O_{i,j} = \frac{1}{T_{i,j}}. \qquad (4\text{-}3)$$

Density: Transmittance and opacity are two good measures of the darkness of any portion of a developed negative. However, psychologists have found that the human visual system does not respond linearly to light stimulation, but rather we respond logarithmically. Therefore, it is common to use *density ($D_{i,j}$)*, which is the common logarithm of opacity, as our measure of choice:

$$D_{i,j} = \log_{10} O_{i,j} = \log_{10}\left(\frac{1}{T_{i,j}}\right). \qquad (4\text{-}4)$$

If 10 percent of the light can be transmitted through a film at a certain i,j location, transmittance is 1/10, opacity is 1/0.10 or 10, and density is the common logarithm of 10 or 1.0. Table 4-5 summarizes the general relationship between transmittance, opacity, and density values.

Characteristic Curve: Several variables, such as the length of the exposure, length of film development, type of film emulsion, and film speed can cause differences in how a given object is recorded on the film and its density. Therefore, it is important to understand the nature of the characteristic curve of films and how this impacts the creation of properly exposed areas on the photographic negative.

The *characteristic curves* of two hypothetical black-and-white negative films, X and Y, are shown in Figure 4-31. These are called *D log E* curves where the density of the portion of the film under investigation is plotted on the y-axis (already in logarithmic form, as previously discussed) and

Table 4-5.   Relationship between Transmittance, Opacity, and Density.

Percent Transmittance	Transmittance $T_{i,j}$	Opacity $O_{i,j}$	Density $D_{i,j}$
100	1	1	0.00
50	0.50	2	0.30
25	0.25	4	0.60
10	0.10	10	1.00
1.0	0.01	100	2.00
0.1	0.001	1000	3.00

the total exposure of the film is plotted in logarithmic units on the x-axis (McGlone, 2004). We will use relative log Exposure units instead of ergs/cm^2 or meter-candle-second used in photometric science. Different films have different D log E curves, but the shapes are generally similar, consisting of three parts — the *toe*, a *straight-line* or *linear section*, and the *shoulder*. The curves provide important information about the film emulsion.

Unfortunately, even if no exposure has been made (e.g., an aerial photograph has not even been taken yet), there is some density already present in the film. The density of the unexposed emulsion is called *fog*. There is also some density from the film base material. *Gross fog* is the sum of the unexposed density from the film base and the emulsion *fog*. The *gross fog* level is located in the lower position of the toe and labeled D_{min} in Figure 4-31. In order to begin to create an image of value in the photograph, we must have an exposure greater than D_{min}.

At this point it is useful to give a formal definition of *exposure (E)*, which is a function of several factors previously discussed, including *f*/stop (i.e., *f/d*), the amount of radiant energy coming from the scene (*s*) measured in Joules mm^{-2} sec^{-1}, and the exposure time (*t*) in seconds:

$$E = \frac{st}{4 \cdot \left(\frac{f}{d}\right)^2}. \qquad (4\text{-}5)$$

As exposure begins, the density increases curvilinearly from point *a* to point *b*. This is called the *toe* of the curve. If objects are recorded with just this length of exposure, they may be underexposed. As the length of the exposure increases, there is a portion of the curve from *b* to *c* where the increase in density is nearly linear with changes in the

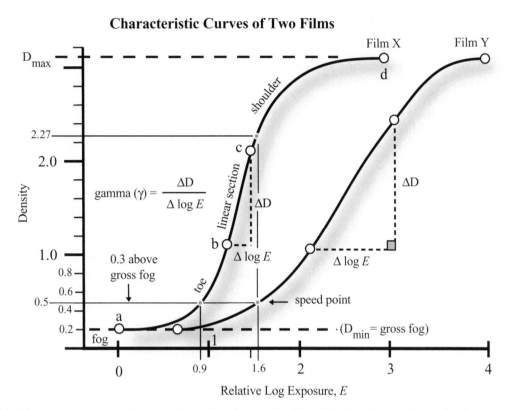

Characteristic Curves of Two Films

Figure 4-31 Characteristic curves of two hypothetical black-and-white films. To obtain densities in the developed negative of 0.5, Film X requires a relative log Exposure of 0.9 units while Film Y requires 1.6 units. Thus, Film X is *faster* than Film Y. This means that we could obtain the same densities in Film X using a shorter exposure time. This could be very important if we want to stop action when collecting aerial photography to obtain quality aerial photographs. Film X also has a much steeper gamma (γ), meaning that it produces a much greater range of density (i.e., contrast) than Film Y in the linear (straight-line) section of the D log E curve.

log of exposure. This is called the *linear* or straight-line part of the characteristic curve. High-quality aerial photography may be obtained if the exposure is made in the linear portion of the curve and perhaps a small part of the toe. As the length of the exposure time increases from c to d, density increases at a decreasing curvilinear rate. This is the *shoulder* of the curve. The maximum density of the film is D_{max}. Objects recorded in the shoulder of the curve are usually overexposed.

The slope of the linear section of the D log E curve is called *gamma* (γ) and provides valuable information about the *contrast* of the film. The greater the slope of the line, the higher the gamma and the higher the contrast of the film. Notice that films X and Y in Figure 4-31 have significantly different gammas. If the same length of exposure (e.g., 1.6 relative log Exposure units) were applied to both films, Film X would have a greater range of density (approximately 2.2) while Film Y would only yield a density of approximately 0.5 at this same exposure. Hopefully, Film X would provide

a full continuum of grays from white to black, with plenty of gray tones in between at this exposure. Conversely, Film Y might not have such a full range of gray tones at this exposure. We often want to retain the subtle gray tones in aerial photography because most natural landscapes (soils, rocks, vegetation, water) are relatively low in contrast.

This brings us to *film speed* once again. Some films are more sensitive to light than others. In our example, it appears that Film X is the faster film. It requires a shorter exposure time to produce a wider range of density (and therefore contrast). This can be of value for aerial photography because we may want to "stop" the action as the aircraft flies above the terrain at hundreds of miles per hour. Unfortunately, fast films require larger silver halide crystals that can also reduce the spatial resolution of the film and impede our ability to resolve objects that are spectrally different yet spatially near to one another. Conversely, Film Y would require a longer exposure to produce the same range of density. It would probably be composed of smaller silver halide crystals. If

Film Y were properly exposed, it might provide high spatial resolution and subtle grayscale information. But if Film Y was not fast enough to stop action, the result might be blurred aerial photography. Obviously there must be some quantitative way of comparing one film speed with another.

Every film speed system requires the selection of a *speed point* to compare the speed of two films. For aerial photography films, the speed point is "density = 0.3 + density above gross fog" (Kodak, 1988). Therefore, since gross fog is 0.2 for both hypothetical aerial films, the speed point for both films is 0.3 + 0.2 = 0.5, as shown in Figure 4-31. The 0.5 density line intersecting the two films can be used to determine how much faster one film is than another. For example, to produce a density of 0.5 on Film X, approximately 0.9 units of relative log Exposure are required. To produce the same density (0.5) on Film Y, approximately 1.6 units of relative log Exposure are required. Thus, Film X is faster than Film Y by approximately 0.7 relative log units of Exposure. For every 0.3 log E, there is one *f*/stop difference in exposure (Kodak, 1988). Thus, these two films differ in speed by greater than two *f*/stops (i.e., 0.7/0.3 = 2.33). If we required a great range of image density in our photographs such as a value of 2.0 (i.e., well-exposed blacks, whites, and intermediate gray tones) but we had to have short exposure times to stop action, as is common in aerial photography, then we might want to select Film X because it would provide the density range we desire with a much shorter relative log Exposure than Film Y. It is possible to compute the exact film speed (*FS*) of an aerial film using the formula (Kodak, 1988):

$$FS = \frac{3}{2E} \qquad (4\text{-}6)$$

where E is the exposure measured in meter-candle-seconds (MCS) at the point on the characteristic curve where the density is 0.3 units above D_{min} (i.e., gross fog), as discussed.

Sometimes we acquire handheld aerial photography using non-aerial photography films. The speed of these films is determined by the American Standards Association (ASA) film speed system that assigns a number to a film that is approximately equal to the inverse of the shutter speed (in seconds) required for proper exposure in good Sunlight for a lens opening of *f*/16. Therefore, if a film is properly exposed in good Sunlight at *f*/16 and 1/200 second, it is classified as an ASA 200 film (Wolf and DeWitt, 2000). The original Kodacolor Film introduced in 1942 had an ASA speed of just 25. That speed jumped to ASA 32 in 1955 and to ASA 64 in 1963 with the introduction of Kodacolor-X Film. In 1977, Kodak improved Kodacolor II Film to an ASA of 100

and introduced Kodacolor 400 Film. In 1983, the Kodacolor VR 1000 film was introduced with an ASA of 1000, the most significant single advance in silver halide technology in more than 50 years.

If one must plan an aerial photography mission, the *Kodak Aerial Exposure Computer* can be used to determine the most appropriate lens aperture (*f*/stop) and shutter speed once information is provided on the following variables: effective aerial film speed, haze factor, speed of aircraft, altitude above ground level, latitude, and time of day.

Digitizing Black-and-White (and Color) Film

Densitometry is the measurement of transmission or reflection characteristics of objects or features in photographic images. The density (*D*) characteristics of a negative or positive transparency film can be measured using a *densitometer*. There are several types of densitometers, including flatbed (Figure 4-32) and drum microdensitometers, video densitometers, and linear or area array charge-coupled-device densitometers.

Microdensitometer Digitization: The characteristics of a typical *flatbed microdensitometer* are shown in Figure 4-32. This instrument can measure the density characteristics of very small portions of a negative or positive transparency, down to just a few micrometers in size, hence the term *microdensitometer*. Basically, a known quantity of light is sent from the light source toward the receiver. If the light encounters a very dense portion of the film, very little light will be transmitted to the receiver. If the light encounters a very clear portion of the film, then much of the light will be transmitted to the receiver. The densitometer can output the characteristics at each *i,j* location in the photograph in terms of transmittance, opacity, or density, as previously discussed. The amount of light recorded by the receiver is more commonly converted into a digital brightness value designated, $BV_{i,j,k}$ which refers to the location in the photograph at row *i* and column *j* and band *k*. At the end of each scan line, the light source steps in the *y* direction some Δy to scan along a line contiguous and parallel to the previous one. As the light source is scanned across the image, the continuous output from the receiver is converted to a series of discrete numerical values on a pixel-by-pixel basis. This analog-to-digital (A-to-D) conversion process results in a matrix of values that are usually recorded in 8-bit bytes (values ranging from 0 to 255) or more (e.g., 12-bit data). These data are then stored on disk or tape for future analysis.

Flatbed Densitometer

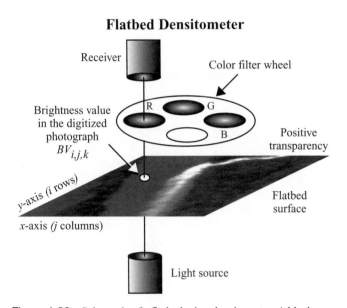

Figure 4-32 Schematic of a flatbed microdensitometer. A black-and-white negative or positive transparency is converted from an analog photographic image into a single matrix of digital brightness values, $BV_{i,j,k}$. A color negative or positive transparency is separated into three registered matrices based on the density of the three dyes (yellow, magenta, and cyan) found at each location in the photography. The spot size that is analyzed during the digitization process may be as small as a few micrometers; hence the term *microdensitometer*.

Scanning imagery at spot sizes <12 μm may result in noisy digitized data, because the spot size approaches the dimension of the film's silver halide crystals (McGlone, 2004). Table 4-6 summarizes the relationship between digitizer scanning spot size (IFOV) measured in dots per inch or micrometers and the pixel ground resolution at various scales of aerial photography or imagery. For simple photogrammetric investigations, a value of 600 DPI may be used because a scanning resolution of 600 DPI (42 μm) is near to the geometric accuracy of most flatbed scanners (about 50 μm) (Linder, 2003). The algorithms for converting from DPI to μm and vice versa are presented in Table 4-6.

A simple black-and-white photograph has only a single band, $k=1$. However, we may need to digitize color photography. In such circumstances, we use three specially designed filters that determine the amount of light transmitted by each of the dye layers in the film (Figure 4-32). The negative or positive transparency is scanned three times ($k = 1, 2$ and 3), each time with a different filter. This extracts spectral information from the respective dye layers found in color and color-infrared aerial photography and results in a registered three-band digital data set for subsequent image processing.

Video Digitization: It is possible to digitize hard-copy imagery by sensing it through a video camera and then performing an analog-to-digital conversion on the 525 lines by 512 rows of data that are within the standard field of view (as established by the National Television System Committee). Video digitizing involves freezing and then digitizing a frame of analog video camera input. A full frame of video input can be read as rapidly as 1/30-sec. A high-speed analog-to-digital converter, known as a *frame grabber*, digitizes the data and stores them in a buffer memory. The memory is then read by the host computer and the digital information stored on disk or tape.

Video digitization of hard-copy imagery is performed very rapidly, but the results are not always useful for digital image processing purposes. For example, there are dramatic differences in the radiometric sensitivity and repeatability of various video cameras. A serious problem is vignetting (light fall-off) away from the center of the image being digitized. This can affect the spectral signatures extracted from the scene. Also, any distortion in the vidicon optical system will be transferred to the digital remote sensor data, making it difficult to edge-match between adjacent images that have been digitized in this manner.

Linear and Area Array Charge-Coupled-Device (CCD) Digitization: Advances in the personal computer industry have spurred the development of flatbed, desktop linear array digitizers based on linear array charge-coupled-devices that can be used to digitize hard-copy negatives, paper prints, or transparencies at 300 to 3,000 pixels per inch (Figure 4-33ab). The hard-copy photograph is placed on the glass. The digitizer optical system illuminates an entire line of the hard-copy photograph at one time with a known amount of light. A linear array of detectors records the amount of light reflected from or transmitted through the photograph along the array and performs an A-to-D conversion. There are linear CCDs with up to 20,000 elements and trilinear CCDs with up to 14,000 elements (McGlone, 2004). The linear array is stepped in the y direction, and another line of data is digitized. It is possible to purchase useful desktop color scanners for less than $300. Many digital image processing laboratories use these inexpensive desktop digitizers to convert hard-copy remotely sensed data into a digital format. Desktop scanners provide surprising good spatial precision and a reasonable characteristic curve when scanning black-and-white images. An optional "transilluminater" can be purchased for the backlighting of film (or 35-mm slide) to be scanned. Unfortunately, most desktop

Table 4-6.   Relationship between digitizer instantaneous field of view (IFOV) measured in dots per inch or micrometers, and the pixel ground resolution at various scales of photography.

Digitizer Detector IFOV		Pixel Ground Resolution at Various Scales of Photography (meters)					
Dots per inch	Micrometers	1:40,000	1:20,000	1:9,600	1:4,800	1:2,400	1:1,200
100	254.00	10.16	5.08	2.44	1.22	0.61	0.30
200	127.00	5.08	2.54	1.22	0.61	0.30	0.15
300	84.67	3.39	1.69	0.81	0.41	0.20	0.10
400	63.50	2.54	1.27	0.61	0.30	0.15	0.08
500	50.80	2.03	1.02	0.49	0.24	0.12	0.06
600	42.34	1.69	0.85	0.41	0.20	0.10	0.05
700	36.29	1.45	0.73	0.35	0.17	0.09	0.04
800	31.75	1.27	0.64	0.30	0.15	0.08	0.04
900	28.23	1.13	0.56	0.27	0.14	0.07	0.03
1000	25.40	1.02	0.51	0.24	0.12	0.06	0.03
1200	21.17	0.85	0.42	0.20	0.10	0.05	0.03
1500	16.94	0.67	0.34	0.16	0.08	0.04	0.02
2000	12.70	0.51	0.25	0.12	0.06	0.03	0.02
3000	8.47	0.33	0.17	0.08	0.04	0.02	0.01
4000	6.35	0.25	0.13	0.06	0.03	0.02	0.008

Useful Scanning Conversions:
DPI = dots per inch;   μm = micrometers;   I = inches;   M = meters
From DPI to micrometers: μm = (2.54 / DPI)10,000
From micrometers to DPI: DPI = (2.54 / μm)10,000
From inches to meters: M = I × 0.0254
From meters to inches: I = M × 39.37

Computation of Pixel Ground Resolution:
PM = pixel size in meters;   PF = pixel size in feet;   S = photo scale
Using DPI: PM = (S/DPI)/39.37   PF = (S/DPI)/12
Using micrometers: PM = (S × μm) 0.000001   PF = (S × μm) 0.00000328
For example, if a 1:6,000 scale aerial photograph is scanned at 500 DPI, the pixel size will be (6000/500)/39.37 = 0.3048 meters per pixel or (6000/500)/12 = 1.00 foot per pixel. If a 1:9,600 scale aerial photograph is scanned at 50.8 μm, the pixel size will be (9,600 × 50.8)(0.000001) = 0.49 meters or (9,600 × 50.8)(0.00000328) = 1.6 feet per pixel.

scanners are designed for 8.5 × 14 in. originals, and most aerial photographs are 9 × 9 in. Under such conditions, the analyst must digitize the 9 × 9 in. photograph in two sections (e.g., 8.5 × 9 in. and 0.5 × 9 in.) and then digitally *mosaic* the two pieces together. The mosaicking process can introduce both geometric and radiometric error.

Linear array CCDs are used in many inexpensive scanning digitizers, such as the one shown in Figure 4-31a. Area array

Area Array CCD Image Digitizer

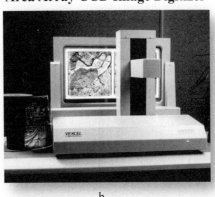

Linear Array CCD Flatbed Digitizer

a.

b.

Figure 4-33 a) Inexpensive desktop linear array CCD digitizer. b) An image digitizer based on area array CCD technology (courtesy Vexcel, Inc.).

CCDs are used in certain digitizers like the one shown in Figure 4-33b, which digitizes from 160 dpi to 3,000 dpi (approximately 160 μm to 8.5 μm) over a 10 × 20 in. image area (254 mm × 508 mm). The system scans the film (the original negative or positive transparency) as a series of rectangular image segments, or tiles. It then illuminates and scans a *reseau grid*, which is an array of precisely located crosshatches etched into the glass of the film carrier. The reseau grid coordinate data are used to locate the exact orientation of the CCD camera during scanning and to geometrically correct each digitized tile of the image relative to all others. Radiometric calibration algorithms are then used to compensate for uneven illumination encountered in any of the tile regions. When scanning a color image, the scanner stops on a rectangular image section and captures that information sequentially with each of four color filters (blue, green, red, and neutral) before it moves to another section. Most other scanners digitize an entire image with one color filter and then repeat the process with the other color filters. This can result in color misregistration and loss of image quality. Area array digitizing technology has obtained geometric accuracy of ≤ 2 μm over 23 × 23 cm images when scanned at 25 μm per pixel, and repeatability of ≤ 3 μm.

Hopefully there is a relationship between the brightness value $(BV_{i,j,k})$ or density $(D_{i,j})$ at any particular location in the film and the energy reflected from the real-world object space $(O_{x,y})$ at the exact location. Scientists take advantage of this relationship by 1) making careful *in situ* observations in the field, such as the amount of biomass for a 1 × 1 m spot on the Earth located at $O_{x,y}$ and then 2) measuring the brightness value $(BV_{i,j,k})$ or density of the object at that exact location in the photograph using a *densitometer*. If enough samples are located in the field and in the photography, it

may be possible to develop a correlation between the real-world object space and the image space. This is an important use of digitized aerial photography.

Digitized National Aerial Photography Program (NAPP) Data: The NAPP was initiated in 1987 as a replacement for the National High Altitude Aerial Photography (NHAP) Program. The objective of the NAPP was to acquire and archive photographic coverage of the conterminous United States at 1:40,000 scale using either color-infrared or black-and-white film. The photography was acquired at an altitude of 20,000 ft above ground level (AGL) with a 6-in. focal-length metric camera. The photography was acquired ideally on a five-year cycle, resulting in a nationwide photographic database that is readily available through the EROS Data Center in Sioux Falls, SD, or the Aerial Photography Field Office in Salt Lake City, UT (USGS, 1994).

This high spatial resolution NAPP photography represents a wealth of information for on-screen photo-interpretation and can become a high-resolution base map upon which other GIS information (e.g., parcel boundaries, utility lines, and tax data) may be overlaid after it is digitized and rectified to a standard map projection. Light (1993) summarized the optimum methods for converting the NAPP data into a national database of digitized photography that meets National Map Accuracy Standards. Microdensitometer scanning of the photography, using a spot size of 15 μm, preserves the 27 resolvable line pair per millimeter (lp/mm) spatial resolution in the original NAPP photography. This process generally yields a digital dataset that has a ground spatial resolution of 1 × 1 m, depending on original scene contrast. This meets most land-cover and land-use mapping user requirements.

The digitized information can be color separated into separate bands of information if desired. The 15 μm scanning spot size will support most digital soft-copy photogrammetry for which coordinate measurements are made using a computer and the monitor screen (Light, 1993). Because the digitized NAPP data are so useful as a high spatial resolution GIS base map, many states are entering into cost-sharing relationships with the U.S. Geological Survey and having their NAPP coverage digitized and output as digital orthophotomaps. A large amount of NAPP data have been digitized and converted into digital orthophotoquads. Much of the digital data is available and can be browsed and ordered at the USGS orthophotoquad Website in Appendix A.

Color Photographic Emulsions

Normal color and color-infrared photographic emulsions are heavily used in photogrammetry and photo-interpretation.

Normal Color Aerial Photography

Normal color photography records energy in the region from 0.4 – 0.7 μm (blue, green, and red light) and depicts the terrain in the same hues (colors) as our eyes perceive the landscape. A haze filter is normally used to prevent ultraviolet light from exposing the film (e.g., Kodak HF3). Color film emulsions generally consist of three layers of silver halide crystals arranged as shown in Figure 4 -25c. The top layer is sensitive to blue light, the second layer is sensitive to blue and green light, and the bottom layer is sensitive to red and blue light. To prevent blue light from exposing the bottom two layers, a yellow blue-blocking filter is placed in the emulsion between the top two layers (McGlone, 2004). The spectral sensitivity of the blue, green, and red emulsion layers is shown in Figure 4-34a. Color aerial photography films marketed by Eastman Kodak Company are summarized in Table 4-4.

Blue, green, and red light reflected from the various objects in the scene interact with and activate the blue, green, and red-sensitive silver halide crystals in the film, forming a latent image. A color rendition of the process is shown in Color Plate 4-3a.

Color film development is exactly like the first step of black-and-white film development, i.e., the exposed silver halide crystals in each layer are turned into black crystals of silver. The remainder of the process, however, depends on whether the film is color negative or color reversal film. With *color negative film*, a negative is produced and color prints are

Table 4-7. Primary and complementary colors.

Primary Colors	Complementary Colors
blue	yellow
green	magenta
red	cyan

made from the negative. *Color reversal film* produces a true color transparency directly on the film. This process is used for making slides or positive transparencies.

The second step in the creation of a positive normal color photograph from a color negative involves the use of special dyes. Silver halides in each layer that turned black in the first step are replaced with *dyes* of the *complementary colors* of the layer (Color Plate 4-3a; Table 4-7). Dense black silver grains in the blue-sensitive layer are replaced with yellow dye (yellow is the complementary color of blue and is composed of green and red light). *However, the amount of dye placed at each location in the film layer is inversely proportional to the intensity of light that illuminated that location.* Therefore, the more blue light that illuminated the area, the less yellow dye that will be present. Black silver crystals in the green-sensitive layer are replaced with magenta dye. Black silver crystals in the red-sensitive layer are replaced with cyan dye. This process results in a negative that is composed of yellow, magenta, and cyan dyes. This is the reason that when we look at a color negative we see strange colors (yellow, magenta, and cyan) instead of the traditional blue, green, and red found in the landscape. We are not used to seeing nature in hues of yellow, magenta, and cyan; therefore, we rarely photo-interpret color negatives.

To produce a color positive print from the negative that we can photo-interpret, white light is projected through the color negative to expose a three-layered color printing paper emulsion. The color negative acts as a filter, exposing the three layers on the photographic paper to the three colors yellow, magenta, and cyan. When developing the positive print paper, however, the complementary colors of yellow, magenta, and cyan (i.e., blue, green, and red) are produced. This second color reversal yields the original colors of the scene on the positive print that are then suitable for photo-interpretation (Color Plate 4-3a). A normal color aerial photograph of Tivoli North Bay, NY is shown in Color Plate 4-4a.

Spectral Sensitivity of the Three Layers of Normal Color Film

a.

Spectral Sensitivity of the Three Layers of Color-Infrared Film

b.

Figure 4-34   a) The spectral sensitivity of the three dye layers of normal color film. A haze filter (e.g., HF3) may be used that filters out much of the light less than 0.4 μm. b) The spectral sensitivity of the three dye layers of color-infrared film. A Wratten 12 (yellow) filter is often used, which filters out much of the light shorter than 0.5 μm (courtesy Eastman Kodak Co.).

Color-Infrared Aerial Photography

Color-infrared film was originally developed during World War II, when there was great interest in increasing the sensitivity of films in the infrared region of the spectrum to detect camouflage. In 1941, S. Q. Duntley of the Massachusetts Institute of Technology headed up a study on optical and visual camouflage detection. The research concluded that the vegetation chlorophyll absorption band located in the red spectral region (0.60 – 0.68 μm) and the high amount of near-infrared reflectance from vegetation in the region > 0.7 μm would be difficult to be replicated in a camouflage paint. Therefore, in 1942, Duntley requested that Eastman Kodak Company, produce a false-color camouflage-detection film. Dr. Mannes of the Eastman Kodak Company demonstrated the test results on October 19, 1942. It was during this time that the decision was made for foliage to appear red in the finished color-infrared aerial photography. It was made available to the military under the name "Aero Kodacolor Reversal Film, Camouflage Detection" (Colwell, 1997). The film was successful in the war for locating targets covered with camouflage netting or paint. It has also proven to be of significant value for monitoring the condition and location of vegetation in Earth resource surveys. Color-infrared aerial photography films marketed by Eastman Kodak Co. are summarized in Table 4-4. The most popular continues to be Kodak AEROCHROME II Infrared Film 2443.

Color-infrared film records reflected energy in the region from approximately 0.35 μm to just above 0.9 μm (ultraviolet, blue, green, red, and near-infrared) and portrays the energy on the film in false colors (Rundquist and Sampson, 1988; Kodak, 1999). Like normal color film, color-infrared film also has three emulsion layers, each sensitive to a different part of the spectrum (Figure 4-25d; 4-34b). The top layer has a sensitivity from ultraviolet through near-infrared. The middle layer is most sensitive to green and blue light. The bottom layer is most sensitive to blue and red light.

The color-infrared film is exposed through a dark yellow filter (usually a Kodak Wratten 12 minus-blue filter) that allows the green, red, and near-infrared light from the terrain to expose the film while preventing light shorter than 0.5 μm from reaching the film (McGlone, 2004). Color Plate 4-3b depicts how the dyes introduced into each of the film layers are different when compared to the development of a standard color photo. Basically, they are "offset by one," meaning that the green-sensitive layer receives yellow dye, the red-sensitive layer receives magenta dye, and the infrared-sensitive layer receives cyan dye. Thus, when white light is eventually projected through this negative and a positive print is produced, anything that reflected *only* green light in the scene shows up as bright blue, anything that reflected *only* red light in the scene will appear in bright shades of green, and anything in the scene that reflected *only* near-infrared energy will be recorded in bright shades of red. If something reflected equal proportions of red and near-infrared light but no green light, then it would show up in shades of yellow on the color-infrared film. Similarly, equal amounts of green, red, and near-infrared reflected from the scene, such as from a concrete surface, will yield a medium gray tone.

Thus, color-infrared film exhibits a *color balance shift* when compared to traditional color aerial photography. What can be confusing to interpreters is the fact that as human beings we have no idea how much near-infrared energy is reflected by certain phenomena since we cannot see or appreciate energy in this region. For example, a color-infrared aerial photograph is shown in Color Plate 4-4b. This photograph was obtained using Kodak AEROCHROME II Infrared 2443 film with a 6 in. focal-length lens. Deep, clear water

absorbs almost all of the near-infrared radiant flux incident to it while reflecting somewhat more green and red light. Therefore, if the water is deep and free of suspended sediment or any organic matter, it will appear dark or almost black on color-infrared film. However, if there is substantial suspended sediment in the water, it may appear in relatively dark shades of blue and green on the near-infrared photography. Conversely, vegetation absorbs approximately 80 percent of the green and red radiant flux incident to it while it reflects approximately 40 – 70 percent of the incident near-infrared radiant flux. This condition causes healthy green vegetation to be recorded in shades of red because the near-infrared energy is dominant. Concrete streets or gravel railroad beds reflect approximately equal proportions of green, red, and near-infrared energy and are recorded as a "steel gray" signature on the color-infrared photography.

Similar color reversals are evident in the example of a building at the University of Nebraska photographed using color and color-infrared film (Color Plate 4-5ab). Objects that reflect blue light exclusively are recorded in black on the color-infrared photograph (e.g., the car on the left). The red carpet on the steps of the building is recorded in shades of green. The tile roof is recorded in shades of yellow, meaning that the tile absorbs much of the green light while reflecting heavily in the red and near-infrared portions of the spectrum.

Color-infrared aerial photography will continue to be one of the most important remote sensing tools. In fact, it was the emulsion of choice for the USGS National Aerial Photography Program. New high spatial resolution ($\leq 1 \times 1$ m panchromatic and $\leq 4 \times 4$ m multispectral) satellite remote sensor digital data from commercial vendors such as Space Imaging and DigitalGlobe, Inc. are reducing our dependency on large-scale analog color-infrared photography.

 Planning Aerial Photography Missions

Time of Day — Sun Angle

The ideal time of day to obtain aerial photography is when the Sun is between 30° and 52° above the horizon, i.e., within two hours of solar noon. A Sun angle less than 30° may not allow proper illumination or adequate reflectance from the scene. This can result in underexposed aerial photography that is difficult to interpret. Also, photography taken at low Sun angles is usually unacceptable because of the extremely long shadows cast by trees and buildings that can obscure valuable information. However, some scientists (especially geologists) prefer low Sun angle photography to enhance terrain representation (Lyon and Falkner, 1995).

A Sun angle greater than 52° may yield *hotspots* on the photography. These are unusually bright areas on the photograph, which are a function of Sun and camera orientation. They become especially bothersome if there is a lot of water in the study area.

Photographic flightlines can be oriented in any direction as long as terrain and safety factors are considered. The angle of the Sun at a particular latitude and specific time of day may be obtained from solar ephemeris tables.

Weather

Aerial photography is ideally collected a few days after the passage of a frontal system, which removes water vapor and particulates (especially smog) from the air and allows the wind associated with the frontal system to decrease. Aerial photography obtained in very humid conditions is degraded because of atmospheric scattering of unwanted light into the field of view of the camera. Also, the water vapor may absorb some of the signal we are trying to record on film.

Strong winds can cause the plane to drift off course, resulting in poor flightline coverage, loss of stereoscopic overlap, and sometimes oblique rather than vertical photography. Clouds in aerial photography, of course, are unacceptable. Not only is a portion of the terrain obscured by a cloud, but the cloud shadow causes tonal variations that might be interpreted incorrectly. Aerial photography obtained prior to the passage of a frontal system may be ideal if the relative humidity is low and the winds are not strong.

Flightline Layout

To obtain the desired aerial photography, it is necessary to know the desired photo scale (e.g., 1:10,000), the scale of the base map on which all information is to be plotted (e.g., 1:12,000), the x,y coordinates of the four corner points of the study area (latitude, longitude or UTM), the size of the geographic area to be photographed (e.g., 100 km²), the average forward overlap of each frame of photography (e.g., 60 percent), the average sidelap of each frame (e.g., 20 percent), the film format (e.g., 9 × 9 in.), and camera focal-length. Once this information is known, it is possible to compute 1) the necessary flight altitude above ground level (AGL), 2) the number of flightlines required, 3) map distance between

flightlines, 4) ground distance between exposures, 5) map distance between exposures, and 6) total number of exposures required. Many of these variables are described in Chapter 6 (Photogrammetry).

References

Berglund, J. A., 1999, *Evaluating the Use of RADARSAT SAR and JERS-1 SAR for Estimating LAI and Delineating Upland Land Cover and Tidal Freshwater Wetland Vegetation Species in the Tivoli North Bay, Area, NY,* Columbia, SC: Department of Geography, unpublished masters thesis, 310 p.

Bossler, J. D., Jensen, J. R., McMaster, R. B. and C. Rizos, 2002, *Manual of Geospatial Science and Technology*, London: Taylor & Francis, 623 p.

Colwell, R. N., 1997, "History and Place of Photographic Interpretation," Philipson, W., (Ed.), *Manual of Photographic Interpretation*, Bethesda: ASP&RS, 2nd Ed., 3–47.

USGS, 1994, *Digital Orthophotos Fact Sheet*, Washington, DC: U.S. Geological Survey, 2 p.

Hinckley, T. K. and J. W. Walker, 1993, "Obtaining and Using Low Altitude, Large-Scale Imagery," *Photogrammetric Engineering & Remote Sensing*, 59(3):310–318.

Hooper, N. J. and G. C. Gustafson, 1983, "Automation and Recording of the Image Interpreter's Mensuration Tasks for Man-made Objects," *SPIE Proceedings*, 424:56–67.

Jensen, J. R., 1995, "Issues Involving the Creation of Digital Elevation Models and Terrain Corrected Orthoimagery Using Soft-copy Photogrammetry," *Geocarto International*, 10(1):5–21.

Jensen, J. R., 2005, *Introductory Digital Image Processing: A Remote Sensing Perspective*, 3rd Ed., Upper Saddle River: Prentice-Hall, Inc., 525 p.

Jensen, J. R., Botchway, K., Brennan-Galvin, E., Johannsen, C., Juma, C., Mabogunje, A., Miller, R., Price, K., Reining, P., Skole, D., Stancioff, A. and D. R. F. Taylor, 2002, *Down to Earth: Geographic Information for Sustainable Development in Africa*, Washington: National Research Council, 155 p.

Jensen, J. R., Saalfeld, A., Broome, F., Cowen, D., Price, K., Ramsey, D., Lapine, L. and E. L. Usery, 2005, "Chapter 2: Spatial Data Acquisition and Integration," in R. B. McMaster and E. L. Usery (Eds.), *A Research Agenda for Geographic Information Science*, Boca Raton: CRC Press, 17–60.

Kodak, 1988, "Kodak Technical Information (Part III): The Characteristic Curve," *Kodak Technical Bits*, Summer:12–16.

Kodak, 2006, Kodak Home Page, Rochester, NY: Eastman Kodak, Co., http://www.kodak.com.

Liang, S., 2004, *Quantitative Remote Sensing of Land Surfaces*, New York: John Wiley & Sons, 534 p.

Light, D. L., 1993, "The National Aerial Photography Program as a Geographic Information System Resource," *Photogrammetric Engineering & Remote Sensing*, 48(1):61–65.

Light, D. L., 1996, "Film Cameras or Digital Sensors? The Challenge Ahead for Aerial Imaging," *Photogrammetric Engineering & Remote Sensing*, 62(3):285–291.

Linder, W., 2003, *Digital Photogrammetry: Theory and Applications*, Berlin: Springer-Verlag, 189 p.

London, B. and J. Upton, 1994, *Photography*, 5th Ed., NY: Harper Collins, 422 p.

Lunetta, R. S. and C. D. Elvidge, 1998, *Remote Sensing Change Detection*, Ann Arbor: Ann Arbor Press, 318 p.

Lyon, J. G. and E. Falkner, 1995, "Estimating Cost for Photogrammetric Mapping and Aerial Photography," *Journal of Surveying Engineering*, 121(2):63–86.

McGlone, J. C. (Ed.), 2004, *Manual of Photogrammetry*, 5th Ed., Bethesda: ASP&RS, 1151 p.

Meadows, J., 1992, *The Great Scientists*, NY: Oxford University Press, 250 p.

Mikhail, E. M., Bethel, J. S. and J. C. McGlone, 2001, *Introduction to Modern Photogrammetry*, N.Y.: John Wiley, 479 p.

Miller, R. B., Abbott, M. R., Harding, L. W., Jensen, J. R., Johannsen, C. J., Macauley, M., MacDonald, J. S. and J. S. Pearlman, 2003, *Using Remote Sensing in State and Local Government: Information for Management and Decision Making*, Washington: National Research Council, 97 p.

Narumalani, S., Hlady, J. and J. R. Jensen, 2002, "Information Extraction from Remotely Sensed Data," in Bossler, J. D., Jensen, J. R., McMaster, R. B. and C. Rizos, *Manual of Geospatial Science and Technology*, London: Taylor & Francis, 298–324 p.

Nikon, 2005, *Nikon home page*, www.Nikon.com.

Paine, D. P. and J. D. Kiser, 2003, *Aerial Photography and Image Interpretation*, NY: John Wiley, 632 p.

Philipson, W., 1997, *Manual of Photographic Interpretation*, Bethesda, MD: ASP&RS, 2nd Ed., 689 p.

Rasher, M. E. and W. Weaver, 1990, *Basic Photo Interpretation*, Washington: USDA Soil Conservation Service, 320 p.

Rundquist, D. C. and S. A. Sampson, 1988, *A Guide to the Practical Use of Aerial Color-Infrared Photography in Agriculture*, Lincoln, NB: Conservation and Survey Division, 27 p.

Warner, W. S., R. W. Graham and R. E. Read, 1996, *Small Format Aerial Photography*, Scotland: Whittles, 348 p.

White, L., 1995, *Infrared Photography Handbook*, Amherst: Amherst Media, 108 p.

Wolf, P. R. and B. A. DeWitt, 2000, *Elements of Photogrammetry with Applications in GIS*, 3rd Ed., NY: John Wiley, 608 p.

Elements of Visual Image Interpretation 5

Humans are adept at interpreting images of objects. After all, they have been doing this all their lives. With some instruction they can become excellent image analysts. *Photo* or *image interpretation* is defined as

the examination of images for the purpose of identifying objects and judging their significance (Philipson, 1997; McGlone, 2004).

This chapter introduces the fundamental concepts associated with the visual interpretation of images of objects recorded primarily by remote sensing systems operating in the optical blue, green, red, and reflective near-infrared portions of the electromagnetic spectrum. The imagery that is interpreted may be acquired using a variety of sensors, including traditional analog cameras (e.g., Leica RC 30), digital cameras (e.g., Leica ADS 40), multispectral scanners (e.g., Landsat Thematic Mapper), and linear or area-array sensor systems (e.g., SPOT, IRS-1C, MODIS, IKONOS, QuickBird, OrbView-3, ImageSat). Methods of extracting quantitative information from remotely sensed images obtained in the optical portion of the spectrum are presented in Chapter 6 (Photogrammetry). Distinctive elements of image interpretation associated with thermal infrared (e.g., surface temperature) and active microwave (RADAR) imagery are presented in Chapters 8 and 9, respectively.

 Introduction

There are a number of important reasons why photo or image interpretation is such a powerful scientific tool, including:

- the aerial/regional perspective;

- three-dimensional depth perception;

- the ability to obtain knowledge beyond our human visual perception;

- the ability to obtain a historical image record to document change.

This chapter discusses these considerations and then introduces the fundamental elements of image interpretation used by image analysts to implement them. Various methods of search are also presented, including the use of collateral (ancillary) information, convergence of evidence, and application of the multi-concept in image analysis.

The Aerial/Regional Perspective

A vertical or oblique aerial photograph or other type of visible/near-infrared image records a detailed but much reduced version of reality. A single image usually encompasses much more geographic area than human beings could possibly traverse or really appreciate in a given day. For example, consider the photograph obtained by the astronauts through a porthole on *Apollo 17* that captures one-half of the entire Earth (a hemisphere) at one time (Figure 5-1). Much of Africa is visible, from the arid Sahara to the dark vegetation of the Congo to the Cape of Good Hope shrouded in clouds. Conversely, a single 9×9 in. 1:63,360-scale (1 in. = 1 mi) vertical aerial photograph records only 81 mi^2 of geography at one time.

Examination of the Earth from an aerial perspective allows scientists and the general public to identify objects, patterns, and human-land interrelationships that may never be completely understood if we were constrained to a terrestrial, Earth-bound vantage point. It does not matter whether the aerial perspective is from the top of a tall building, an elevated hillside, a light plane, a high-altitude jet, or a satellite platform. The resultant remotely sensed image provides spatial terrain information that we would not be able to acquire and appreciate in any other manner. This is why remote sensing image interpretation is so important for military reconnaissance and civilian Earth resource investigation.

Care must be exercised, however, when interpreting vertical and oblique imagery. Human beings are accustomed to looking at the façade (side) of objects from a terrestrial vantage point and do not normally have an appreciation for what objects look like when they are recorded from a vertical or oblique perspective (Haack et al., 1997). In addition, we are not used to looking at and interpreting the significance of many square kilometers of terrain at one time. Our line of sight on the ground is usually less than a kilometer. Therefore, the regional analysis of vertical and oblique remote sensor data requires training and practice.

Three-Dimensional Depth Perception

We can view a single aerial photograph or image with our eyes and obtain an appreciation for the geographic distribution of features in the landscape. However, it is also possible to obtain a *three-dimensional view* of the terrain as if we were actually in an airborne balloon or aircraft looking out the window. One way to obtain this three-dimensional effect

Earth as Seen from Apollo 17

Figure 5-1   A photograph of the Earth obtained by the astronauts onboard *Apollo 17*, shooting through a porthole of the spacecraft. Almost the entire continent of Africa is visible as well as Saudi Arabia and part of Iraq and India. Note the arid Sahara and the dark, vegetated terrain of the rain forest along the equator in central Africa. Antarctica is especially apparent at the South Pole. Photographs like this helped mankind to realize how vulnerable and precious the Earth is as it rests like a multicolored jewel in the blackness of space (courtesy of NASA).

is to obtain two photographs or images of the terrain from two slightly different vantage points. We can train our eyes to view the two images of the terrain at the same time. Our mind fuses this stereoscopic information into a three-dimensional model of the landscape that we perceive in our minds as being real (Linder, 2003). For example, the stereopair of downtown St. Louis, Missouri, shown in Figure 5-2 provides detailed three-dimensional information when viewed using a stereoscope.

We know from life that it is important to not only know the size and shape of an object but that its height, depth, and volume are also very diagnostic characteristics. Analysis of stereoscopic imagery in three dimensions allows us to appreciate the three-dimensional nature of the undulating terrain and the slope and aspect of the land. In addition, the stereoscopic analysis process usually exaggerates the height or depth of the terrain, allowing us to appreciate very subtle differences in object height and terrain slope and aspect that we might never appreciate from a terrestrial vantage point. Chapter 6 (Photogrammetry) introduces the principles of stereoscopy and how they are used to extract three-dimensional information from remote sensor data. Three-dimen-

Three-dimensional Perspective:
Stereoscopic Image of St. Louis, Missouri

Figure 5-2   This stereopair of St. Louis, Missouri, consists of two views of the terrain obtained at two different exposure stations along a single flightline. Three-dimensional information about the terrain can be obtained by viewing the model using a stereoscope. The human mind uses the parallax information inherent in the images to produce a three-dimensional model that can yield detailed terrain information.

sional information can also be obtained by analyzing RADAR, LIDAR, and SONAR remote sensor data.

Obtaining Knowledge Beyond our Human Visual Perception

Our eyes are sensitive primarily to blue, green, and red light. Therefore, we sample a very limited portion of the electromagnetic energy that is actually moving about in the environment and interacting with soil, rock, water, vegetation, the atmosphere, and urban structure. Fortunately, ingenious sensors have been invented that can measure the activity of X-rays, ultraviolet, near-infrared, middle-infrared, thermal

infrared, microwave, and radiowave energy. Carefully calibrated remote sensor data provides new information about an object that humans might never be able to appreciate in any other manner (Robbins, 1999). For example, consider the imagery of an agricultural area in Saudi Arabia shown in Figure 5-3. Healthy vegetation absorbs much of the green and red light from the Sun for photosynthesis. Therefore, agricultural fields show up in dark shades of gray in green and red multispectral imagery. Conversely, the greater the amount of biomass present in an agricultural field, the greater the amount of near-infrared energy reflected causing heavily-vegetated fields to appear bright in near-infrared imagery. The green and red images suggest that vegetation is present in almost all of the dark center-pivot fields. The

Remote Sensing Can Provide Knowledge by Measuring Energy Characteristics in Spectral Regions Beyond Our Human Visual Perception

a. Green reflectance.   b. Red reflectance.   c. Near-infrared reflectance.

Figure 5-3   Indian IRS-1C LISS III imagery (23 × 23 m) of an agricultural area in Saudi Arabia. a,b) Vegetation absorbs most of the green and red incident energy causing vegetated fields to appear dark. c) Conversely, vegetation reflects a substantial amount of incident near-infrared energy causing it to appear bright. In this example, several fields appear dark in the green and red images (possibly due to recent irrigation, stubble from a previous crop, or plowing) suggesting that vegetation is present. Careful examination of the same fields in the near-infrared image reveals that very little vegetation is present. The near-infrared image also provides detailed information about the spatial distribution of the biomass present in each field. A color composite of this imagery is found in Color Plate 5-1 (images courtesy of Indian National Remote Sensing Agency).

near-infrared imagery provides more definitive information about the spatial distribution and amount of vegetation (biomass) found within the fields. A color composite of the bands is found in Color Plate 5-1.

Historical Image Record and Change Detection Documentation

A single aerial photograph or image captures the Earth's surface and atmosphere at a unique moment in space and time, not to be repeated again. These photographs or images are valuable historical records of the spatial distribution of natural and human-made phenomena. When we acquire multiple images of the Earth, we can compare the historic imagery with the new imagery to determine if there are any subtle, dramatic, or particularly significant changes (Jensen and Cowen, 1999; McCoy, 2005). The study of change usually increases our understanding about the natural and human-induced *processes* at work in the landscape. Knowledge about the spatial and temporal dynamics of phenomena allows us to develop predictive models about what has happened in the past and what may happen in the future (Lunetta and Elvidge, 1998). Predictive modeling is one of the major goals of science. Remote sensing image interpretation is playing an increasingly important role in predictive modeling and simulation (Goetz, 2002; Miller et al., 2003; Jensen et al., 2005).

Remote sensing is especially useful for monitoring human activity through time, which can hopefully lead to sustainable development and good governance. For example, consider Figure 5-4, which documents the effects of President Robert G. Mugabe's order to demolish rural poor informal settlements in Mozambique in 2005. President Mugabe said his urban clean-up campaign *Operation Murambatsvina* was needed to "restore sanity" to Zimbabwe's cities, which he said had become overrun with criminals. The opposing political party has its support base among the urban poor, and says *Operation Murambatsvina* was aimed at forcing them to rural areas where the Mugabe government could more easily control them. It is estimated that more than 200,000 people became homeless. Remote sensing provided a detailed historic record of the tragic events (BBC, 2005).

 ### Elements of Image Interpretation

To perform regional analysis, view the terrain in three dimensions, interpret images obtained from multiple regions of the electromagnetic spectrum, and perform change detection, it is customary to use principles of image interpretation that have been developed through empirical experience for more than 150 years (Estes et al., 1983; Kelly et al., 1999; McGlone, 2004). The most fundamental of these principles are the *elements of image interpretation* that are routinely

Remote Sensing Imagery as A Historical Record:
Informal City Demolition in Harare, Zimbabwe in 2005

a. QuickBird 61-cm image obtained on April 16, 2005.　　　b. QuickBird 61-cm image obtained on June 4, 2005.

Figure 5-4　　High spatial resolution (61 × 61 cm) panchromatic satellite imagery captured the destruction (razing) of informal housing in Harare, Zimbabwe, that began May 25, 2005. President Mugabe ordered police with bulldozers and sledgehammers to demolish more than 20,000 informal housing structures, causing more than 200,000 people to become homeless (images courtesy of DigitalGlobe, Inc.).

used when visually photo-interpreting an image (Bossler et al., 2002). The elements of image interpretation include location, tone and color, size, shape, texture, pattern, shadow, height and depth, volume, slope, aspect, site, situation, and association (Figure 5-5). Some of the adjectives associated with each of these elements of image interpretation are summarized in Table 5-1. Each image is composed of individual silver halide crystals or pixels that have a unique color or tone at a certain geographic location. This is the fundamental building block upon which all other elements are based. Therefore, we may consider these to be the primary or first-order elements of image interpretation (Konecny, 2003). The secondary and tertiary elements are basically spatial arrangements of tone and color. The higher order elements of site, situation, and association are often based on different methods of search, including: the use of collateral information, convergence of evidence, and the use of the multi-concept.

A well-trained image interpreter uses many of the elements of image interpretation during analysis without really thinking about them (Lloyd et al., 2002). However, a novice interpreter may have to systematically force himself or herself to consciously evaluate an unknown object with respect to these elements to finally identify it and judge its significance in relationship to all the other phenomena in the scene.

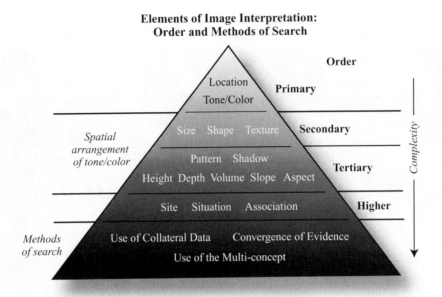

Figure 5-5   Photo or image interpretation is usually based on the use of the elements of image interpretation. The location of individual silver halide crystals in a photograph or pixels in an image represent the primary (first order) elements of image interpretation. Secondary and tertiary elements are spatial arrangements of tone and color. The higher-order elements of site, situation, and association often make use of various search methods to perform accurate image interpretation.

x,y Location

There are two primary methods of obtaining precise x,y coordinate information about an object: 1) survey it in the field using traditional surveying techniques or global positioning system (GPS) instruments, or 2) collect remote sensor data of the object, register (rectify) the image to a basemap, and then extract the x,y coordinate information directly from the rectified image.

If option one is selected, most scientists now use relatively inexpensive GPS instruments in the field (Figure 5-6) to obtain a precise measurement of an object's location in degrees of longitude and latitude on the Earth's graticule or in meters easting and northing in a map projection (e.g., Universal Transverse Mercator) (McCoy, 2005). Scientists must then transfer the coordinates of the point (e.g., a specific tree location) or polygon (e.g., the perimeter of a small lake) onto an accurate planimetric map. Most scientists in the United states use U.S. Geological Survey 7.5-minute quadrangle maps. In Britain they use Ordinance Survey maps.

Most aircraft or spacecraft used to collect remote sensor data now have a GPS receiver. This allows the remote sensing instrument onboard to obtain accurate x,y,z GPS coordinates at each photograph exposure station or the center of each line scan. In the case of aerial photography, this means that we can obtain information about the exact location of the center of each aerial photograph (i.e., the *principal point*) at the instant of exposure. We can use the GPS information collected by the sensor (and perhaps some collected on the ground) to register (rectify) the uncontrolled photo or image to UTM or another map projection. If we also correct for the relief displacement of the topography, then the photo or image becomes an *orthophoto* or *orthoimage* with all the metric qualities of a line map (McGlone, 2004). Geographic coordinates (x,y) of points and polygons can then be extracted directly from the rectified image. Chapter 6 (Photogrammetry) discusses how orthophotos are created. Jensen (2005) describes methods used to digitally rectify (i.e., register) remote sensor data to a standard map projection.

Tone and Color

Real-world surface materials such as vegetation, water, and bare soil often reflect different proportions of energy in the blue, green, red, and near-infrared portions of the electromagnetic spectrum. We can plot the amount of energy reflected from each of these materials at specific wavelengths and create a spectral reflectance curve, sometimes called a *spectral signature* (Jensen et al., 2005). For exam-

Table 5-1.   Elements of image interpretation.

Element	Common Adjectives (quantitative and qualitative)
x,y location	• *x,y* image coordinates: column (*x*) and row (*y*) coordinates in an unrectified image • *x,y* image map coordinates: silver halide crystals or pixels in photograph or image are rectified to a map projection (e.g., UTM)
Tone/Color	• gray tone: light (bright), intermediate (gray), dark (black) • color: IHS = intensity, hue (color), saturation; RGB = red, green, and blue; Munsell
Size	• length, width, perimeter, area (m^2) • small, medium (intermediate), large
Shape	• an object's geometric characteristics: linear, curvilinear, circular, elliptical, radial, square, rectangular, triangular, hexagonal, pentagonal, star, amorphous, etc.
Texture	• characteristic placement and arrangement of repetitions of tone or color • smooth, intermediate (medium), rough (coarse), mottled, stippled
Pattern	• spatial arrangement of objects on the ground: systematic, unsystematic or random, linear, curvilinear, rectangular, circular, elliptical, parallel, centripetal, serrated, striated, braided
Shadow	• a silhouette caused by solar illumination from the side
Height/Depth Volume/Slope/ Aspect	• *z*-elevation (height), *z*-bathymetry (depth), volume (m^3), slope °, aspect °
Site Situation Association	• *Site*: elevation, slope, aspect, exposure, adjacency to water, transportation, utilities • *Situation*: objects are placed in a particular order or orientation relative to one another • *Association*: related phenomena are usually present

ple, generalized spectral reflectance curves for objects found in a south Florida mangrove ecosystem are shown in Figure 5-7a (Davis and Jensen, 1998). Spectral reflectance curves of selected materials provide insight as to why they appear as they do on black-and-white or color imagery. We will first consider why objects appear in certain grayscale tones on black-and-white images.

GPS Data Collection

Figure 5-6   Scientist collecting global positioning system (GPS) *x, y*, location data in smooth cordgrass (*Spartina Alterniflora*) in Murrells Inlet, SC.

Tone

A band of electromagnetic energy (e.g., green light from 0.5 – 0.6 µm) recorded by a remote sensing system may be displayed in shades of gray ranging from black to white. These shades of gray are usually referred to as *tone*. We often say, "This part of an image has a 'bright' tone, this area has a 'dark' tone, and this feature has an intermediate 'gray' tone." Of course, the degree of darkness or brightness is a function of the amount of light reflected from the scene within the specific wavelength interval (band). For example, consider three black-and-white images of a south Florida mangrove ecosystem (Figure 5-7 b–d). The three images record the amount of green, red, and near-infrared energy reflected from the scene, respectively.

Incident green light (0.5 – 0.6 µm) penetrates the water column farther than red and near-infrared energy and is reflected off the sandy bottom or the coral reef (Figure 5-7b). Therefore, the green band provides subsurface detail about the reef structure surrounding the mangrove islands. Mangrove vegetation absorbs approximately 86 percent of the incident green light for photosynthetic purposes and reflects approximately 14 percent. This causes mangroves to appear relatively dark in a single-band green image. Sand reflects high, equal proportions of blue, green, red, and near-infrared incident energy, so it appears bright in all images.

Mangroves reflect approximately 9 percent of the incident red energy (0.6 – 0.7 µm) while absorbing approximately 91

Tone and Color

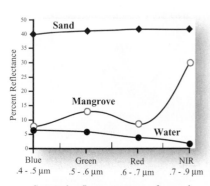

a. Spectral reflectance curves for sand, mangrove, and water in Florida.

b. Black-and-white photograph of green reflected energy from Florida mangroves.

c. Black-and-white photograph of red reflected energy.

d. Black-and-white photograph of near-infrared reflected energy.

e. Stand of pine (evergreen) surrounded by hardwoods (courtesy Emerge, Inc).

f. Vegetation is dark, fallow is bright, and turbid water is gray (GeoEye, Inc.).

g. U-2 photograph of a Russian Sputnik launch site (courtesy John Pike, FAS).

h. High-contrast terrestrial photograph of a Dalmation.

i. High-contrast terrestrial photograph of a cow.

Figure 5-7   Elements of Image Interpretation — Tone and Color.

percent of the incident energy for photosynthetic purposes. This causes the mangroves to appear very dark in the red photograph (Figure 5-7c). Red light does not penetrate as well into the water column, so the water has a slightly darker tone, especially in the deeper channels. As expected, sandy areas have bright tones.

In the black-and-white image recording only near-infrared energy (0.7 – 0.92 μm), vegetation is displayed in bright tones (Figure 5-7d). Healthy vegetation reflects much of the incident near-infrared energy (approximately 28 percent). Generally, the brighter the tone from a vegetated surface, the greater the amount of biological matter (biomass) present (Jensen et al., 1999). Conversely, water absorbs most of the incident near-infrared energy, causing the water to appear dark. There is a great contrast between the bright upland consisting of vegetation and sand, and the dark water. Therefore, it is not surprising that the near-infrared region is considered to be the best for discriminating between the upland and water interface.

A black-and-white infrared image of an evergreen pine stand surrounded by deciduous hardwood forest is shown in Figure 5-7e. The tonal contrast makes it easy to discriminate between the two major species.

One must be careful, however, when interpreting individual band black-and-white images. For example, consider the Landsat Thematic Mapper band 3 (red) image of a Colorado agricultural area (Figure 5-7f). As expected, the greater the amount of vegetation, the greater the absorption of the incident red light by the vegetation, and the darker the vegetated area within the center-pivot irrigation system. Conversely, fallow fields and areas not in agricultural production show up in much brighter tones. Unfortunately, the lake also shows up as an intermediate shade of gray. This is because the lake has received a substantial amount of suspended sediment, causing it to reflect more red radiant energy than it normally would if it were deep, nonturbid water. If we had additional blue, green, and perhaps near-infrared bands to analyze, it would probably be clear that this is a water body. However, when viewing only a single band of imagery displayed in black-and-white tones, we could come to the conclusion that the lake is simply a large vegetated field perhaps in the early stages of development when some of the sandy bare soil is still visible through the canopy. In fact, it has approximately the same gray tone as several of the adjacent fields within the center-pivot irrigation systems.

Human beings can differentiate between approximately 40-50 individual shades of gray in a black-and-white photograph or remote sensor image. However, it takes practice and skill to extract useful information from broad-band panchromatic black-and-white images or black-and-white images of individual bands. For example, consider the U-2 photograph of a Russian *Sputnik* launching site shown in Figure 5-7g. Careful examination of the gray tones and the shadows by a trained analyst reveals that the excavated earth from the blast area depression was deposited in a large mound nearby. Human beings simply are not used to viewing the tops of objects in shades of gray. They must be trained. Furthermore, humans often have a very difficult time identifying features if the scene is composed of very high contrast information. This is exemplified by viewing terrestrial photographs of two very well known objects: a Dalmatian and a cow in Figure 5-7h,i, respectively. Many novice analysts simply cannot find the Dalmatian or the cow in the photographs. This suggests that extremely high contrast aerial photographs or images are difficult to interpret and that it is best to acquire and interpret remotely sensed imagery that has a continuum of grayscale tones from black to gray to white, if possible.

Color

We may use additive color-combining techniques to create color composite images from the individual bands of remote sensor data as previously discussed in Chapter 4. This introduces *hue* (color) and *saturation* in addition to grayscale tone (*intensity*). A color composite of the green, red, and near-infrared bands of the mangrove study area is found in Color Plate 5-2. Notice how much more visual information is present in the color composite. Humans can discriminate among thousands of subtle colors. In this false-color image, all vegetation is depicted in shades of red (magenta), sand is bright white, and the water is in various shades of blue. Most scientists prefer to acquire some form of multispectral data so that color composites can be made. This may include the collection of natural color aerial photography, color-infrared aerial photography, or multispectral data, where perhaps many individual bands are collected and a select few are additively color-combined to produce color images.

Unfortunately, some people's color perception is impaired. This means that they do not experience the same mental impression of a color (e.g., green) as does the vast majority of the population. While this may be somewhat of a disadvantage when selecting a shirt or tie to wear, many excellent image analysts have some color perception disorder. There are special tests such as those shown in Color Plate 5-3 that can be viewed to determine if color blindness is present.

Size — Length, Width, Perimeter, and Area

The size of an object is one of its most distinguishing characteristics and one of the most important elements of image interpretation. The most commonly measured parameters are length (m), width (m), perimeter (m), area (m^2), and occasionally volume (m^3). The analyst should routinely measure the size of unknown objects. To do this it is neces-

Size

a. Automobiles: diverse, but approximately 15 ft in length and 6 ft wide.

b. Railroad: 4.71 ft between rails and 8-in. between the railroad ties.

c. A single trailer on a tractor-trailer rig is approximately 45 to 50 ft in length.

d. Baseball: 90 ft between bases; 60 ft from home plate to the pitcher's mound.

e. Diving board: approximately 12 ft in length.

f. Cars and trucks can be used to scale the size of the air-conditioning units.

Figure 5-8   Elements of Image Interpretation — Size.

sary to know the scale of the photography (e.g., 1:24,000) and its general unit equivalent or verbal scale (i.e., 1 in = 2,000 ft). In the case of digital imagery it is necessary to know the nominal ground spatial resolution of the sensor system (e.g., 1×1 m).

Measuring the size of an unknown object allows the interpreter to rule out many possible alternatives. One must be careful, however, because all of the objects in remote sensor data are at a scale less than 1:1, and we are not used to looking at a miniature version of an object that may measure only a few centimeters in length and width on the image. Measuring the size of a few well-known objects in an image such as car length, road and railroad width, size of a typical single-family house, etc., allows us to understand the size of unknown features in the image and eventually to identify them. There are several subjective relative size adjectives, including small, medium, and large. These adjectives should be used sparingly.

Objects that have relatively unique sizes can be used to judge the size of other objects in the scene. For example, midsize cars are approximately 15 ft long and 6 ft wide in the United States (Figure 5-8a). They may be two-thirds that size in Europe, Asia, etc. Notice that it is possible to differentiate between automobiles and pickup trucks. Also note that the 6-in. white line separating parking spaces is quite visible, giving some indication of the high spatial resolution of this aerial photography. The distance between regular-gauge railroad tracks is 4.71 ft in the United States (Figure 5-8b). This provides diagnostic information about the length of the individual railroad cars. The average length of a trailer on a tractor-trailer rig (Figure 5-8c) is 45 to 50 ft, allowing us to appreciate the size of the adjacent warehouse.

Field dimensions of major sports such as soccer, baseball (Figure 5-8d), football, and tennis are standardized worldwide. The distance between the bases on a baseball diamond is 90 ft, while the distance from the pitcher's mound to home plate is 60 ft. Most swimming pool diving boards (Figure 5-8e) are 12 ft long. Additional examples are found in Chapter 13 (Remote Sensing the Urban Landscape).

If these objects are visible within an image, it is possible to determine the size of other objects in the scene by comparing their dimensions with those of the known object's dimension. For example, the diameter of the two rooftop air-conditioning units shown in Figure 5-8f is at least the length of the car and truck also visible in the image.

It is risky to measure the precise length, perimeter, and area of objects in unrectified aerial photography or other types of unrectified remote sensor data. The terrain is rarely completely flat within the instantaneous-field-of-view of an aerial photograph or other type of image. This causes points that are higher than the average elevation to be closer to the sensor and points that are lower than the average elevation to be farther away from the sensor system. Thus, different parts of the image have different scales. Tall buildings, hills, and depressions may have significantly different scales than those at the average elevation within the photograph.

Therefore, the optimum situation is where the aerial photography or other image data have been geometrically rectified and terrain-corrected to become, in effect, an orthophotograph or orthoimage where all objects are in their proper planimetric x,y location. It is then possible to measure the length, perimeter, and area of features using several methods, including polar planimeter, tablet digitization, dot-grid analysis, or digital image analysis. These size measurement methods are discussed in Chapter 6 (Photogrammetry).

Shape

It would be wonderful if everything had a unique shape that could be easily discerned from a vertical or oblique perspective. Unfortunately, novice interpreters sometimes have difficulty even identifying the shape of the building they are in, much less appreciating the planimetric x,y shape of natural and man-made objects recorded in aerial photography or other imagery. Nevertheless, many features do have very unique shapes. There are numerous shape adjectives such as (Table 5-1) linear, curvilinear, circular, elliptical, radial, square, rectangular, triangular, hexagonal, star, elongated, and amorphous (no unique shape).

There are an infinite variety of uniquely shaped natural and man-made objects in the real world. Unfortunately, we can only provide a few examples (Figure 5-9). Modern jet aircraft (Figure 5-9a) typically have a triangular (delta) shape

and distinctively shaped shadows. Humankind's residential housing and public/commercial buildings may range from very simple rectangular mobile homes for sale (Figure 5-9b) to complex geometric patterns such as the Pentagon in Washington, DC (Figure 5-9c). The 0.5×0.5 m black-and-white infrared image of the Pentagon was obtained using a digital camera. Human transportation systems (Figure 5-9d) in developed countries usually have a curvilinear shape and exhibit extensive engineering.

Humankind modifies nature in a tremendous variety of ways, some of them very interesting. For example, Figure 5-9e depicts the curvilinear shape of carefully engineered levees (rising just 2 ft above the ground) that direct water continuously through a rice field in Louisiana. An adjacent field has been systematically plowed. But nature designs the most beautiful shapes, patterns, and textures, including the radial frond pattern of palm trees shown in Figure 5-9f.

The best image interpreters spend a great amount of time in the field viewing and appreciating natural and man-made objects and their shapes. He or she is then in a good position to understand how these shapes appear when recorded on vertical or oblique imagery.

Texture

Texture is the characteristic placement and arrangement of repetitions of tone or color in an image. In an aerial photograph, it is created by tonal repetitions of groups of objects that may be too small to be discerned individually. Sometimes two features that have very similar spectral characteristics (e.g., similar black-and-white tones or colors) exhibit different texture characteristics that allow a trained interpreter to distinguish between them. We often use the textural adjectives smooth (uniform, homogeneous), intermediate, and rough (coarse, heterogeneous).

It is important to understand that the texture in a certain portion of a photograph is strictly a function of scale. For example, in a very large-scale aerial photograph (e.g., 1:500) we might be able to actually see the leaves and branches in the canopy of a stand of trees and describe the area as having a coarse texture. However, as the scale of the imagery becomes smaller (e.g., 1:5,000), the individual leaves and branches and even the tree crowns might coalesce, giving us the impression that the stand now has an intermediate texture, i.e., it is not smooth but definitely not rough. When the same stand of trees is viewed at a very small scale (e.g., 1:50,000), it might appear to be a uniform forest stand with smooth texture. Thus, texture is a function of the scale of the

Shape

a. Triangular (delta) shape of
a typical passenger jet.

b. Rectangular single- and
double-wide mobile homes for sale.

c. Black-and-white infrared image of the
Pentagon (courtesy Positive Systems, Inc.).

d. A curvilinear cloverleaf highway
intersection in the United States.

e. The curvilinear shape of carefully engineered rice field levees in Louisiana.

f. Radial palm tree fronds
in San Diego, CA.

Figure 5-9   Elements of Image Interpretation — Shape.

imagery and the ability of the interpreter to perceive and describe it.

Several other texture adjectives are often used, including mottled, stippled, etc. It is difficult to define exactly what is meant by each of these textures. It is simply better to present a few examples, as shown in Figure 5-10. Both the avocado orchard and the trees in the courtyard have a coarse texture on this large-scale photograph (Figure 5-10a). Conversely, the concrete road and much of the grass yard have a smooth texture. Just behind the pool, the soil exhibits varying degrees of moisture content, causing a mottled texture.

In Figure 5-10b, the pine forest on the left has a relatively coarse texture as the individual tree crowns are visible. The bright sandy beach has a smooth texture. Both cattails near the shore and waterlilies farther out into L-Lake on the Savannah River Site exhibit intermediate to rough textures.

Finally, the waterlilies give way to dark, smooth-textured water.

Two piles of 50-ft pine logs at a sawmill in Georgia are shown in Figure 5-10c. The logs exhibit a coarse, heterogeneous texture with a linear pattern. The shadow between the stacks has a smooth texture.

Figure 5-10d is an interesting photograph of systematically placed circular marijuana plants interspersed in a field of corn. The physiology (structure) of the two types of plants, their spacing, and orientation combine to produce a coarse-textured agricultural field. The shadows produced by the marijuana plants contribute substantially to the texture of the area. Interestingly, the goal of the farmers appears to be working. Few novice interpreters appreciate the subtle differences in texture visible in the field.

Texture

a. Relatively coarse-texture avocado field. The grass and road have a smooth texture.

b. Pine has coarse texture; cattails and waterlilies have intermediate texture.

c. Coarse texture of freshly cut pine logs at a sawmill in Georgia.

d. Coarse-texture South Carolina corn field interspersed with circular marijuana plants.

e. Mottled texture on fallow soil in a Georgia center-pivot irrigation system.

f. A variety of textures along a tributary of the Mississippi.

Figure 5-10   Elements of Image Interpretation — Texture.

Part of the agricultural field in Figure 5-10e is being cultivated while the remainder is in fallow. The vegetated southwest portion of the center-pivot irrigation system has a relatively smooth texture. However, the remaining fallow portion of the field appears to have areas with varying amounts of soil moisture or different soil types. This causes this area to have a mottled texture. One part of the mottled texture region still bears the circular scars of six wheels of the center-pivot irrigation system.

Various vegetation and sand bar textures are present in the large scale photograph of a tributary to the Mississippi River in Figure 5-10f. A dense stand of willows parallels the lower shoreline, creating a relatively fine texture when compared with the hardwood behind it with its coarse texture. The sand bars interspersed with water create a unique, sinuous texture as well as a serrated pattern. Some of the individual tree crowns in the upper portion of the image are spaced well apart, creating a more coarse texture.

Pattern

Pattern is the spatial arrangement of objects in the landscape (Figure 5-11). The objects may be arranged randomly or systematically. They may be natural, as with a drainage pattern, or human-made, as with the Township and Range land tenure system present in the western United States. Pattern is a very diagnostic characteristic of many features. Typical pattern adjectives include random, systematic, circular, centripetal, oval, curvilinear, linear, radiating, rectangular, hexagonal, pentagonal, octagonal, etc.

Examples of typical patterns captured on remote sensor data are shown in Figure 5-11. The first example depicts the systematic, triangular pattern of B-52s being dismantled at Montham Air Force Base (Figure 5-11a). A large metal blade cuts the fuselage into a specific number of parts. The parts must remain visible for a certain number of days so that

Pattern

a. Systematic, triangular pattern of B-52s being dismantled (courtesy USGS).

b. Seven circular grain silos exhibit a curvilinear pattern on this southeastern farm.

c. Random, sinuous braided stream pattern on a sandy soil at Pen Branch, SC.

d. Township & Range survey pattern on mottled soil in Texas.

e. Systematic, linear rows of potatoes with some damaged (dark) by late blight.

f. Radiating road pattern in Paris (Sovinformsputnik and Aerial Images, Inc.).

Figure 5-11   Elements of Image Interpretation — Pattern.

foreign countries can use their own aerial reconnaissance technology to verify that the specified number of B-52s have been removed from service as part of the strategic arms limitation process. Heavy equipment moving between the aircraft creates a unique curvilinear transportation pattern.

Seven large silos used to store agricultural grain are seen in Figure 5-11b. The individual silos are circular but they are situated in a curvilinear pattern on the landscape. Numerous rectangular farm buildings oriented north-south are arranged in a random fashion.

A random, sinuous braided stream pattern is present at the mouth of Pen Branch, SC, in Figure 5-11c. This particular pattern resembles braided hair, hence the terminology. The major drainage patterns visible in remotely sensed data are discussed in Chapter 13.

Figure 5-11d depicts the systematically surveyed Township & Range cadastral system superimposed on an agricultural region in Texas. The NAPP photograph reveals small farmsteads separated by large tracts of agricultural land. The soil moisture and soil type differences in the fields combine to create an unsystematic, mottled soil texture.

Potatoes arranged in systematically spaced linear rows are shown in Figure 5-11e. Various rows are arranged in a rectangular pattern. This near-infrared photograph reveals that some of the fields that appear dark have experienced late blight damage.

A KVR-1000 Russian satellite photograph reveals the systematic, radiating road pattern centered on the Arch de Triumph in Paris (Figure 5-11f).

Shadow

a. People and benches recorded in kite photography (courtesy Cris Benton).

b. Shadows cast from La Gloriette - Arch of Glory - in Vienna, Austria.

c. Bridge and sign shadows provide valuable information.

d. Pyramids of Giza (courtesy of Sovin-forsputnik and Aerial Images, Inc.).

e. Shadows provide information about object heights (Emerge, Inc).

f. Orient images so that shadows fall toward the viewer during image analysis.

Figure 5-12   Elements of Image Interpretation — Shadow.

Shadow

Most remote sensor data is collected within ± 2 hours of solar noon to avoid extensive shadows in the imagery. This is because shadows from objects can obscure other objects that might otherwise be detected and identified. On the other hand, the shadow or *silhouette* cast by an object may be the only real clue to an object's identity. For example, consider the shadows cast by two people standing on a pier and the shadows cast by benches in Figure 5-12a. The shadows in the image actually provide more information than the objects themselves. La Gloriette — Arch of Glory — in Vienna, Austria, has unique statues on top of it (Figure 5-12b). Through careful evaluation of the shadows in the vertical photograph, it is possible to determine the location of the statues on top of the building. Similarly, shadows cast by signs or bridges (Figure 5-12c) are often more informative than the objects themselves in vertical aerial photography.

Very small-scale photography or imagery usually does not contain shadows of objects unless they protrude a great distance above surrounding terrain such as mountains, extremely tall buildings, etc. For example, consider the shadows cast by the great pyramids of Giza in Egypt (Figure 5-12d). The distinctive shadows are very diagnostic during image interpretation.

In certain instances, shadows can provide clues about the height of an object when the image interpreter does not have access to stereoscopic imagery. For example, the building shadows in Figure 5-12e provide valuable information about the relative height of the building above the ground, i.e., that it is a one-story single-family residence. Chapter 6 (Photogrammetry) describes how shadow information can be used to measure the height of objects.

When interpreting imagery with substantial shadows, it is a good practice to orient the imagery so that the shadows fall

Height and Depth

a. Relief displacement is an important monoscopic cue about object height.

b. Shadows and masking in downtown San Francisco (courtesy GeoEye, Inc.).

c. Bathymetry of Monteray Bay, CA (courtesy HJW, Inc.; SPOT Image, Inc.).

Figure 5-13   Elements of Image Interpretation — Height and Depth.

toward the image interpreter such as those shown in Figure 5-12f. This keeps the analyst from experiencing *pseudoscopic illusion* where low points appear high and vice versa. For example, it is difficult to interpret the photograph of the forest and wetland shown in Figure 5-12f when it is viewed with the shadows falling away from the viewer. Please turn the page around 180° and see how difficult it is to interpret correctly. Unfortunately, most aerial photography of the northern hemisphere is obtained during the leaf-off spring months when the Sun casts shadows northward. This can be quite disconcerting. The solution is to reorient the photographs so that south is at the top. Unfortunately, if we have to make a photomap or orthophotomap of the study area, it is cartographic convention to orient the map with north at the top. This can then cause some problems when laypersons interpret the photomap because they do not know about pseudoscopic illusion.

Shadows on radar imagery are completely black and contain no information. Fortunately, this is not the case with shadows on aerial photography. While it may be relatively dark in the shadow area, there may still be sufficient light scattered into the area by surrounding objects to illuminate the terrain to some degree and enable careful image interpretation to take place.

Height and Depth

The ability to visually appreciate and measure the height (elevation) or depth (bathymetry) of an object or landform is one of the most diagnostic elements of image interpretation (Figure 5-13). Stereoscopic parallax is introduced to remotely sensed data when the same object is viewed from two different vantage points along a flightline. Viewing these overlapping photographs or images using *stereoscopic* instruments is the optimum method for visually appreciating the three-dimensionality of the terrain and for extracting accurate *x,y,* and *z* topographic and/or bathymetric information. Techniques of stereoscopic analysis are presented in Chapter 6 (Photogrammetry).

However, there are also *monoscopic* cues that we can use to appreciate the height or depth of an object. For example, any object such as a building or utility pole that protrudes above the local datum will exhibit radial relief displacement outward from the principal point of a typical vertical aerial photograph. In effect, we are able to see the side of the feature, as demonstrated in Figure 5-13a. Also, all objects protruding above the local datum cast a shadow that provides diagnostic height or elevation information such as the various buildings in San Francisco shown in Figure 5-13b. Also, masking takes place in some images where tall objects obscure objects behind them, making it clear that one object has greater elevation than another. For example, the building at the top of Figure 5-13b is masking the buildings behind it, suggesting that it has greater height.

The optimum method of obtaining bathymetric measurements is to use a sonar remote sensing device which sends out a pulse of sound and measures how long it takes for the sound to pass through the water column, bounce off the bottom, and be recorded by the sensor. The image of Monterey Bay, CA, in Figure 5-13c was obtained using SONAR and merged with a SPOT image of the terrestrial landscape.

Site, Situation, and Association

a. Thermal electric Haynes Steam
Plant in Long Beach, CA.

b. A sawmill with its associated piles
of raw and finished lumber.

c. Vogtle Nuclear Power Plant
near Augusta, GA.

Figure 5-14   Elements of Image Interpretation — Site, Situation, and Association.

Site, Situation, and Association

Site, situation, and *association* characteristics are very important when trying to identify an object or activity. A *site* has unique physical and/or socioeconomic characteristics. The physical characteristics might include elevation, slope, aspect, and type of surface cover (e.g., bare soil, grass, shrub/scrub, rangeland, forest, water, asphalt, concrete, housing, etc.). Socioeconomic site characteristics might include the value of the land, the land-tenure system at the site (metes and bounds versus Township and Range), adjacency to water, and/or adjacency to a certain type of population (professional, blue-collar, retired, etc.).

Situation refers to how certain objects in the scene are organized and oriented relative to one another. Often, certain raw materials, buildings, pipelines, and finished products are situated in a logical, predictable manner.

Association refers to the fact that when you find a certain phenomena or activity, you almost invariably encounter related or associated features or activities. Site, situation, and association elements of image interpretation are rarely used independently when analyzing an image. Rather, they are used synergistically to arrive at a logical conclusion. For example, sewage disposal plants are almost always located on flat sites situated near a water source so they can dispose of the treated water, and they exist relatively close to the producing community. Large commercial shopping malls typically have multiple large buildings on level sites, massive parking lots, and are ideally situated near major transportation arteries and population centers.

Thermal electric power plants such as the Haynes Steam Plant in Long Beach, CA, shown in Figure 5-14a are usually located on flat, well-engineered sites with large tanks of petroleum (or other type of natural resource) nearby that is burned to create steam to propel the electric generators. Man-made levies, called "revetments," often encompass the tank farm to contain the petroleum in the event of an accident. Thermal electric power plants are often associated with some type of recirculating cooling ponds. The water is used to cool critical steam-generating components.

Sawmills such as the one shown in Figure 5-14b are usually sited on flat terrain within 20 km of many stands of trees and associated with large piles of raw timber, well-organized piles of finished lumber, a furnace to dispose of wood waste products, and an extensive processing facility. Railroad spurs are often used to transport the finished lumber or wood-chip products to market.

Nuclear power plants exist on extremely well-engineered level sites. They have large concrete reactor containment building(s). The site may contain large recirculating cooling water ponds or enormous cooling towers such as those under construction at the Vogtle Nuclear Power Plant near Augusta, GA, in Figure 5-14c. Power-generating plants do not need to be adjacent to the consuming population as electricity can be transported economically great distances.

Expert image analysts bring to bear site, situation, and association knowledge to an image interpretation problem. Such knowledge is obtained by observing phenomena in the real world. The best image analysts have seen and appreciate a diverse array of natural and man-made environments. It is difficult to identify an object in an image if one has never seen the object in the real world and does not appreciate its site, situation, and association characteristics.

Methods of Search

Photo-interpretation has been taking place since Gaspard Felix Tournachon (Nadar) took the first successful aerial photograph in France in 1858. Over the years, scientists have developed some valuable approaches to interpreting remotely sensed data, including: 1) utilizing collateral (ancillary) information, 2) converging the evidence, and 3) applying the multi-concept in image analysis.

Using Collateral Information

Trained image interpreters rarely interpret aerial photography or other remote sensor data in a vacuum. Instead, they collect as much *collateral* (often called *ancillary*) information about the subject and the study area as possible. Some of the major types of collateral information are summarized in Table 5-2, including the use of a variety of maps for orientation, political boundary information, property line cadastral data, geodetic control (*x,y,z*), forest stand data, geologic data, hazard information, surface and subsurface hydrologic data, socioeconomic data, soil taxonomy, topographic and bathymetric data, transportation features, and wetland information. Ideally, these data are stored in a geographic information system (GIS) for easy retrieval and overlay with the remote sensor data.

It is useful to contact the local National Weather Service to obtain quantitative information on the meteorological conditions that occurred on the day the remote sensor data were collected (cloud cover, visibility, and precipitation) and for the days preceding data collection. USGS water-supply and stream-gauge reports are also useful.

Scientists also obtain local street maps, terrestrial photographs, local and regional geography books, and journal and popular magazine articles about the locale or subject matter. They talk with local experts. Well-trained image analysts get into the field to appreciate firsthand the lay of the land, its subtle soil and vegetation characteristics, the drainage and geomorphic conditions, and human cultural impact.

Often much of this collateral spatial information is stored in a GIS. This is particularly useful since the remote sensor

Table 5-2. Collateral information often used in the interpretation of aerial photography and other remotely sensed data in the United States.

Topic	Collateral Information
General orientation	International Map of the World 1:1,000,000 National Geospatial-Intelligence Agency (NGA) 1:100,000; 1:250,000 USGS 7.5-min 1:24,000 USGS 15-min 1:63,360 Image browsing systems: Google Earth, Space Imaging, DigitalGlobe, SPOT
Boundaries or districts	USGS 7.5-min 1:24,000 USGS 15-min 1:63,360 Boards – state, county, city, school, fire, voting, water/sewer
Cadastral	City and county tax maps
Geodetic control	USGS digital line graph – elevation NGS – nautical and bathymetric charts
Forestry	USFS – forest stand information
Geology	USGS – surface and subsurface
Hazards	FEMA – flood insurance maps USCG – environmental sensitivity index
Hydrology	USGS digital line graph – surface hydrology NGS – nautical and bathymetric charts USGS – water-supply reports USGS – stream gauge reports
Socio-economic	Bureau of the Census – demographic data – TIGER block data – census tracts
Soils	SCS, NRCS – soil taxonomy maps
Topography/ bathymetry	USGS – National elevation dataset (NED) NGA - digital terrain elevation data (DTED) USCG – nautical and bathymetric charts
Trans-portation	USGS digital line graph – transportation County and state transportation maps
Weather/ atmosphere	National Weather Service – NEXRAD
Wetland	USGS – National Wetland Inventory maps NOAA – Coastal change analysis program

data can be geometrically registered to the spatial information in the GIS database and important interrelationships evaluated.

Convergence of Evidence

It is a good idea to work from the known to the unknown. For example, perhaps we are having difficulty identifying a particular type of industry in an aerial photograph. Careful examination of what we do know about things surrounding and influencing the object of interest can provide valuable clues that could allow us to make the identification. This might include a careful interpretation of the building characteristics (length, width, height, number of stories, type of construction), the surrounding transportation pattern (e.g., parking facilities, railroad spur to the building, adjacent to an interstate), site slope and aspect, site drainage characteristics, unique utilities coming into or out of the facility (pipelines, water intake or output), unusual raw materials or finished products in view outside the building, and methods of transporting the raw and finished goods (tractor trailers, loading docks, ramps, etc.). We bring all the knowledge we have to the image interpretation problem and *converge our evidence* to identify the object or process at work.

Let us consider another example of convergence of evidence. Suppose we were asked to describe the type of airport facility shown in Figure 5-15. At first glance we might conclude that this is a civilian airport with commercial jets. However, upon closer inspection we see that jets 2, 3, and 6 appear normal (e.g., large delta-shaped Boeing 707s), but jets 4 and 5 exhibit substantially different shadow patterns on the fuselage and on the ground. Jet number 1 also exhibits some unusual shadow characteristics. Furthermore, we note that jets 4 and 5 have an unusual dark circular object with a white bar on it that appears to lie on top of the jets.

An image analyst that has seen a Boeing E-3 Airborne Warning and Control System (AWACS) aircraft on the ground would probably identify the AWACS aircraft quickly. Non-military image analysts would need to

- note the absence of commercial airport passenger boarding/unloading ramp facilities suggesting that this is a military airport;

- examine the unusual shadow patterns;

- consult manuals containing examples of various candidate aircraft that could cast such shadows (e.g., Figure 5-11b);

- converge the evidence to arrive at the correct conclusion.

The first step to camouflage an E-3 sitting on the tarmac would be to align the white bar with the fuselage (please refer to jet 1).

The Multi-concept

Robert Colwell of the Forestry Department at the University of California at Berkeley put forth the *multi-concept* in image interpretation in the 1960s (Colwell, 1997). He suggested that the most useful and accurate method of scientific image interpretation consisted of performing the following types of analysis: *multispectral, multidisciplinary, multiscale,* and *multitemporal.* The multi-concept was further elaborated upon by Estes et al. (1983) and Teng (1997).

Colwell pioneered the use of *multiband* aerial photography and *multispectral* remote sensor data. He documented that in agriculture and forest environments, measurements made in multiple discrete wavelength regions (bands) of the electromagnetic spectrum were usually more valuable than acquiring single broadband panchromatic-type imagery. For example, Figure 5-7 documented the significant difference in information content found in green, red, and near-infrared multispectral images of mangrove.

Colwell also suggested that *multiscale* (often called multistage) photography or imagery of an area was very important. Smaller-scale imagery (e.g.,1:80,000) was useful for placing intermediate scale imagery (e.g., 1:40,000) in its proper regional context. Then, very large-scale imagery (e.g., > 1:10,000) could be used to provide detailed information about local phenomena. *In situ* field investigation is the largest scale utilized and is very important. Each scale of imagery provides unique information that can be used to calibrate the others.

Professor Colwell was a great believer in bringing many *multidisciplinary* experts together to focus on a remote sensing image analysis or information extraction problem. The real world consists of soils, surface and subsurface geology, vegetation, water, atmosphere, and man-made urban structure. In this age of increasing scientific specialization, it is difficult for any one person to be able to understand and extract all the pertinent and valuable information present within a remote sensor image. Therefore, Colwell strongly suggested that image analysts embrace the input of other multidisciplinary scientists in the image interpretation process. This philosophy and process often yields synergistic, novel, and unexpected results as multidisciplinary scientists each bring their expertise to bear on the landscape appreciation problem (Donnay et al., 2001). Table 5-3 lists the disci-

Convergence of Evidence

a. Aircraft on the tarmac in Riyadh, Saudi Arabia.

b. Royal Saudi Air Force Boeing E-3 Airborne
Warning and Control System (AWACS).

Figure 5-15   a) Panchromatic image (61 × 61 cm) of an airport in Riyadh, Saudi Arabia, obtained on December 9, 2001 (courtesy of DigitalGlobe, Inc.). b) Terrestrial view of the Royal Saudi Air Force version of the E-3 Airborne Warning and Control System (AWACS) aircraft. The E-3 is equipped with a "look-down" radar that can separate airborne targets from the ground and sea clutter returns. Its radar "eye" has a 360° view of the horizon and can "see" more than 320 km (200 mi). The U.S. Air Force, NATO, United Kingdom, France and Saudi Arabia operate AWACS. [Speed > 800 km/hr (500 mi/hr); Ceiling > 10,670 m (35,000 ft); Endurance > 11 hours without refueling; Range > 9,250 km (5,000 nautical mi)] (Boeing, 2005).

plines of colleagues that often collaborate when systematically studying a certain topic.

While single-date remote sensing investigations can yield important information, they do not always provide informa-

tion about the processes at work. Conversely, a *multitemporal* remote sensing investigation obtains more than one image of an object. Monitoring the phenomena through time allows us to understand the processes at work and to develop predictive models (Lunetta and Elvidge, 1998; Schill et al.,

Table 5-3. Multidisciplinary scientists bring their unique training to the image interpretation process.

Topic	Disciplines
Agriculture	Agronomy, agricultural engineering, biology, biogeography, geology, landscape ecology, soil science
Biodiversity, habitat	Biology, zoology, biogeography, landscape ecology, marine science, soil science
Database and algorithm preparation	Cartography, GIS, computer science, photogrammetry, programming, analytical modeling
Forestry, rangeland	Forestry, agronomy, rangeland ecology, landscape ecology, biogeography, soil science
Geodetic control	Geodesy, surveying, photogrammetry
Geology soils	Geology, geomorphology, agronomy, geography
Hazards	Geology, hydrology, urban and physical geography, human geography, sociology
Hydrology	Hydrology, chemistry, geology, geography
Topography/ bathymetry	Geodesy, surveying, photogrammetry
Transportation	Transportation engineering, city planning, urban geography
Urban studies	Urban, economic, and political geography, city planning, transportation engineering, civil engineering, landscape ecology
Weather/ atmosphere	Meteorology, climatology, physics, chemistry, atmospheric science
Wetland	Biology, landscape ecology, biogeography

1999). Colwell pioneered the concept of developing crop phenological calendars in order to monitor the spectral changes that take place as plants progress through the growing season. Once crop calendars are available, they may be used to select the optimum dates during the growing season to acquire remote sensor data. Many other phenomena, such as residential urban development, have been found to undergo predictable cycles that can be monitored using remote sensor data. A trained image analyst understands the phenological cycle of the phenomena he or she is interpreting and uses this information to acquire the optimum type of remote sensor data on the optimum days of the year.

Conclusion

We now have an understanding of the fundamental elements of image interpretation. We can utilize the elements of image interpretation to carefully analyze aerial photography or other types of optical (blue, green, red, and near-infrared wavelength) remote sensor data. Based on this foundation, we are prepared to progress to more sophisticated image analysis techniques, including the extraction of quantitative information from remote sensor data using principles of photogrammetry.

References

BBC, 2005, "What Lies Behind the Zimbabwe Demolitions?" *BBC News World Edition*, London: BBC NEWS, July 26.

Boeing, 2005, *Airborne Warning and Control (AWACS) Aircraft*, Seattle: Boeing Aircraft Corporation, www.boeing.com/defense-space/ic/awacs/e3svcww/e3overview.html.

Bossler, J. D., Jensen, J. R., McMaster, R. B. and C. Rizos, 2002, *Manual of Geospatial Science and Technology*, London: Taylor & Francis, 623 p.

Colwell, R. N., 1997, "History and Place of Photographic Interpretation," *Manual of Photographic Interpretation*, W. Philipson (Ed.), 2nd Ed., Bethesda: ASP&RS, 3–48.

Davis, B. A. and J. R. Jensen, 1998, "Remote Sensing of Mangrove Biophysical Characteristics," *Geocarto International*, 13(4):55–64.

Donnay, J., Barnsley, M. J. and P. A. Longley, 2001, *Remote Sensing and Urban Analysis*, NY.: Taylor & Francis, 268 p.

Estes, J. E., Hajic, E. J. and L. R. Tinney, 1983, "Fundamentals of Image Analysis: Analysis of Visible and Thermal Infrared Data," *Manual of Remote Sensing*, R. N. Colwell, (Ed.), Bethesda: ASP&RS, 1:1039–1040.

Goetz, S. J., 2002, "Recent Advances in Remote Sensing of Biophysical Variables: An Overview of the Special Issue," *Remote Sensing of Environment*, 79:145–146.

Haack, B., Guptill, S., Holz, R., Jampoler, S., Jensen, J. R. and R. A. Welch, 1997, "Urban Analysis and Planning," *Manual of Photographic Interpretation*, W. Philipson (Ed.), 2nd Ed., Bethesda: ASP&RS, 517–547.

Jensen, J. R., 2005, *Introductory Digital Image Processing: A Remote Sensing Perspective*, 3rd Ed., Upper Saddle River: Prentice-Hall, Inc., 525 p.

Jensen, J. R. and D. C. Cowen, 1999, "Remote Sensing of Urban/Suburban Infrastructure and Socioeconomic Attributes," *Photogrammetric Engineering & Remote Sensing*, 65(5):611–622.

Jensen, J. R., Coombs, C., Porter, D., Jones, B., Schill, S. and D. White, 1999, "Extraction of Smooth Cordgrass (*Spartina alterniflora*) Biomass and LAI Parameters from High Resolution Imagery," *Geocarto International*, 13(4):25–34.

Jensen, J. R., Saalfeld, A., Broome, F., Cowen, D., Price, K., Ramsey, D., Lapine, L. and E. L. Usery, 2005, "Chapter 2: Spatial Data Acquisition and Integration," in R. B. McMaster and E. L. Usery (Eds.), *A Research Agenda for Geographic Information Science*, Boca Raton: CRC Press, 17–60.

Jensen, R. R., Gatrell, J. D. and D. D. McLean (Eds.), 2005, *Geo-Spatial Technologies in Urban Environments*, N.Y.: Springer, 176 p.

Kelly, M., Estes, J. E. and K. A. Knight, 1999, "Image Interpretation Keys for Validation of Global Land-Cover Data Sets," *Photogrammetric Engineering & Remote Sensing*, 65:1041–1049.

Konecny, G., 2003, *Geoinformation: Remote Sensing, Photogrammetry and GIS*, London: Taylor & Francis, 248 p.

Linder, W., 2003, *Digital Photogrammetry: Theory and Applications*, Berlin: Springer-Verlag, 189 p.

Lloyd, R., Hodgson, M. E. and A. Stokes, 2002, "Visual Categorization with Aerial Photographs," *Annals of the Association of American Geographers*. 92(2):241–266.

Lunetta, R. S. and C. D. Elvidge, 1998, *Remote Sensing Change Detection: Environmental Monitoring Methods and Applications*, Ann Arbor: Ann Arbor Press, 318 p.

McCoy, R. M., 2005, *Field Methods in Remote Sensing*, N.Y.: Guilford Press, 159 p.

McGlone, J. C., 2004, *Manual of Photogrammetry*, 5th Ed., Bethesda: ASP&RS, 1151 p.

Miller, R. B., Abbott, M. R., Harding, L. W., Jensen, J. R., Johannsen, C. J., Macauley, M., MacDonald, J. S. and J. S. Pearlman, 2003, *Using Remote Sensing in State and Local Government: Information for Management and Decision Making*, Washington: National Academy Press, 97 p.

Philipson, W., 1997, *Manual of Photographic Interpretation*, 2nd Ed., Bethesda: ASP&RS, 555 p.

Robbins, J., 1999, "High-Tech Camera Sees What Eyes Cannot," *New York Times*, Science Section, September 14, D5.

Schill, S., Jensen, J. R. and D. C. Cowen, 1999, "Bridging the Gap Between Government and Industry: the NASA Affiliate Research Center Program," *Geo Info Systems*, 9(9):26–33.

Teng, W. L., 1997, "Fundamentals of Photographic Interpretation," *Manual of Photographic Interpretation*, W. Philipson (Ed.), 2nd Ed., Bethesda: ASP&RS, 49–113.

Wolf, P. R. and B. A Dewitt, 2000, *Elements of Photogrammetry with Applications in GIS*, NY.: McGraw Hill, 608 p.

Photogrammetry

6

*P*hotogrammetry is the art and science of making accurate measurements by means of aerial photography. *Analog photogrammetry* is performed using data that are in a hard-copy format such as a 9×9 in. aerial photographic print or positive transparency. *Digital photogrammetry* is performed using a computer and digital or digitized aerial photography. Neither method is superior to the other. Each approach has an appropriate role depending upon the nature of the remote sensor data and the image analysis goals. This chapter focuses on methods to extract *quantitative* information from aerial photography using both analog and digital photogrammetric techniques.

Below are important measurements that can be obtained from a *single vertical aerial photograph* using analog or digital photogrammetric techniques:

- scale of the photography;

- object height;

- object length;

- area of an object or polygon;

- perimeter of an object or polygon;

- the grayscale tone or color of an object.

The following quantitative measurements may be made using *multiple (overlapping) stereoscopic aerial photographs* and analog or digital measurement of stereoscopic parallax (to be defined):

- precise planimetric (x,y) object location of building footprints, streets, hydrology, and shorelines in a standard map projection;

- precise object height (z).

In addition, digital photogrammetric techniques applied to stereoscopic aerial photography can yield:

- digital elevation models (DEM);

- bathymetric models;

Flightline of Aerial Photography

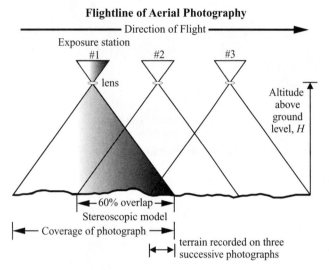

Figure 6-1   A single flightline of vertical aerial photography with 60 percent overlap (endlap) obtained at three exposure stations H meters above ground level.

Block of Aerial Photography

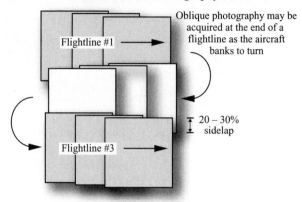

Figure 6-2   A block of aerial photography is produced by photographing multiple flightlines that have 20–30 percent sidelap. Some oblique aerial photography may be acquired at the end of a flightline if the frame camera is not turned off during the turn.

• slope and aspect information derived from the digital elevation or bathymetric models;

• planimetrically accurate (x,y) orthophotographs.

Flightlines of Vertical Aerial Photography

It is necessary to understand how individual aerial photographs are acquired at various *exposure stations* along a flightline before we can correctly apply photogrammetric techniques to extract quantitative measurements.

A *flightline* of vertical aerial photography is obtained by mounting a camera in an aircraft, helicopter, or spacecraft and obtaining photographs of the terrain directly beneath the platform at specific exposure stations. The geometry of three hypothetical vertical aerial photographs taken in succession over relatively level terrain are depicted in Figure 6-1. The time between individual exposures along a flightline is determined by setting the camera *intervalometer*. The aerial photographer takes into account the speed of the aircraft and the scale of the desired photography and sets the intervalometer so that each vertical aerial photograph overlaps the next photograph in the flightline by approximately 60 percent (referred to as *stereoscopic overlap* or *endlap*) (Linder, 2003). This *overlap* is very important because it provides at least two and sometimes three photographic views of each object in the real world along a flightline (Figure 6-1).

Sometimes aerial photographs are acquired with >80 percent overlap in mountainous terrain (Konecny, 2003).

Most aerial photography projects require multiple flightlines of photography to cover the geographic area of interest. When this occurs, the flightlines are normally overlapped by 20–30 percent, referred to as *sidelap* (Figures 6-2 and 6-3). To acquire multiple flightlines, a pilot must make a 180° turn at the end of a flightline and then fly in the opposite direction. Care should be exercised when analyzing vertical aerial photographs acquired at the end of a flightline because some photography obtained while the plane was banking during the turn may be oblique. The algorithms presented in this chapter work best with near-vertical aerial photographs ($\leq 3°$ from nadir). Sometimes wind or pilot error will cause the plane to drift to the left or right of the desired line of flight. When this occurs, the footprint of each photograph or the entire flightline may be offset as shown in Figure 6-3b. This is not a serious condition as long as it does not become too severe and the 60 percent overlap and 20–30 percent sidelap are maintained.

Multiple flightlines with 20–30 percent sidelap are referred to as a *block* of aerial photography (Figure 6-2). For example, consider a portion of a block of vertical aerial photography of downtown Columbia, SC, shown in Figure 6-3a. The block consists of three photographs in each of two flightlines (#3 and #4). The photography were acquired on March 30, 1993, at approximately 3,000 ft above ground level with a 6-in. (152.82 mm) focal-length lens and a Wild RC 10 camera yielding approximately 1:6,000-scale photography. There is

Block of Aerial Photography Compiled to Create an Uncontrolled Photomosaic

Figure 6-3   a) Two flightlines (#3 and #4) of 1:6,000-scale vertical aerial photography of Columbia, SC, obtained at 3,000 ft. above local ground level using a 6-in. (152.82 mm) focal-length lens. b) An uncontrolled photomosaic of the block of aerial photography.

60 percent overlap along each flightline and 20 percent side-lap between the two flightlines.

It is possible to combine several vertical photographs in the block of photography to create an *uncontrolled photomosaic*

(Figure 6-3b). This example depicts only six 1:6,000-scale photographs. Acquiring photography of a county, state or country sometimes requires thousands of photographs, depending upon the size of the country and the scale of the photography. Table 6-1 provides several metric and English

Figure 6-4   Flightline #4, Photo #5 of vertical panchromatic aerial photography of Columbia, SC, obtained on March 30, 1993, at approximately 3,000 ft above ground level using a Wild RC 10 camera with a 6-in. focal-length lens. The photography depicts the state capitol in the lower right corner and the University of South Carolina campus in the center of the image. Eight fiducial marks are present. The "titling" ancillary data contains very valuable information, including a verbal scale (1 in. = 500 ft) but unfortunately does not show the exact altitude, time of day, amount of tilt, etc., that many metric cameras now routinely provide. This is a reduction of the original 9 × 9 in. 1:6,000-scale photograph.

unit equivalents that can be of significant value when measuring the length and width of flightlines.

Most of the aerial photography acquired each year is obtained in the early spring leaf-off period when snow is not present and deciduous trees are still dormant. This makes it easier for the image analyst to see through the branches of the canopy to the ground surface. This is especially important if one is trying to extract topographic information from stereoscopic aerial photography (to be discussed).

 Fiducial Marks and Principal Points

A typical 9 × 9 in. panchromatic vertical aerial photograph is shown in Figure 6-4 (reduced for printing purposes). Note the eight *fiducial marks* located in the four corners and in the centers of the four sides. Some metric cameras have only four fiducial marks. Drawing a line between opposite fiducial marks locates the *principal point (PP)* of the photograph, which is the exact point on the Earth where the optical

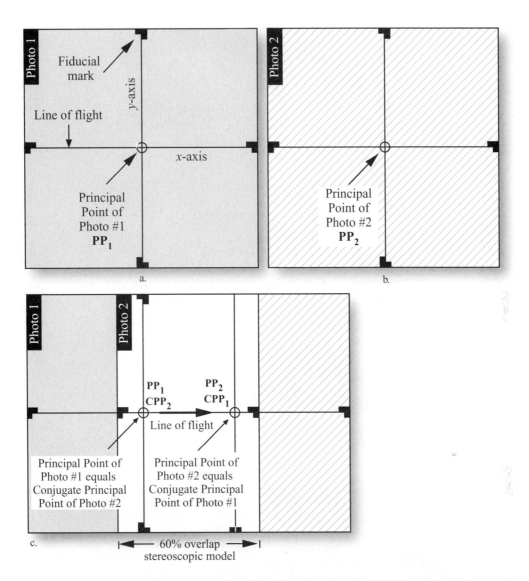

Figure 6-5   a) Geometry of a single aerial photograph (Photo #1). The principal point (PP) is located at the intersection of lines drawn through opposing fiducial marks. b) Geometry of Photo #2. c) Superposition of Photo #1 and Photo #2. The 60 percent overlap area can be viewed stereoscopically. Note the location of the principal point (PP) and conjugate principal points (CPP) on each photograph. A line drawn between the principal points identifies the line of flight.

axis of the camera was pointing during the instant of exposure (Figure 6-5). By carefully examining the vicinity of the principal point on an airphoto, its location can be visually transferred to each of the two adjacent photos in the flight-line. The transferred principal point is called a *conjugate principal point (CPP)*. The actual line of flight (as opposed to the desired line of flight) can be determined by laying out the photography and drawing a line through the principal points and conjugate principal point (Figure 6-5).

Geometry of Vertical Aerial Photography

A diagram of the geometry of a single vertical aerial photograph of downtown Columbia, SC, is shown in Figure 6-6. In this example, a 9×9-in. negative was obtained at exposure station L at 3,200 ft above sea level (H) with a local elevation of 200 ft (h). The developed negative image space (a', b', c', and d') is a reversal in both *tone* and *geometry* of the Earth object space (A, B, C, and D) and is situated a distance

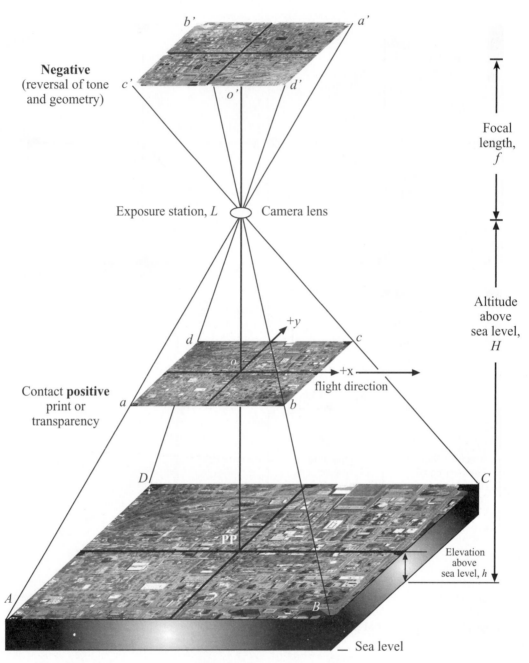

Negative
(reversal of tone
and geometry)

Focal
length,
f

Exposure station, *L*   Camera lens

Altitude
above
sea level,
H

+*y*

o

+x
flight direction

Contact **positive**
print or
transparency

PP

Elevation
above
sea level, *h*

Sea level

Figure 6-6   Geometry of a vertical aerial photograph obtained over flat terrain (Columbia, SC).

equal to the focal length *f* (distance *o'L*) behind the rear nodal point of the camera lens. A positive version of the scene can be produced by contact-printing the negative with either photographic paper or positive transparency material. The result is a positive print or positive transparency with tone and geometry the same as that of the original object space. The reversal in geometry from negative to positive space is seen by comparing the locations of negative image

points *a'*, *b'*, *c'*, and *d'* with their corresponding positive image points *a, b, c,* and *d* in Figure 6-6.

The photographic coordinate axes *x* and *y* radiate from the principal point in the positive contact print. These axes are important for photogrammetric measurements to be discussed. Because photointerpreters often work with positive prints or positive transparencies, they are primarily interested in the positions of images in the positive plane. Conse-

Table 6-1.   Selected units and their equivalents.

Metric	English	Conversion
Linear Measurement		
centimeter (cm)	inch (in.)	1 cm = 0.3937 in. 1 in. = 2.54 cm
meter (m)	foot (ft)	1 m = 3.28 ft 1 ft. = 0.305 m
kilometer (km)	mile (mi)	1 km = 0.621 mi 1 mi = 1.61 km
Area Measurement		
square centimeter (cm^2)	square inch (sq in.)	1 cm^2 = 0.155 sq in. 1 sq in. = 6.4516 cm^2
square meter (m^2)	square foot (sq ft)	1 m^2 = 10.764 sq ft 1 sq ft. = 0.0929 m^2
square kilometer (km^2)	square mile (sq mi)	1 km^2 = 0.3861 sq mi 1 sq mi = 2.59 km^2
hectare (ha)	acre	1 ha = 10,000 m^2 1 ha = 2.471 acres 1 acre = 43,560 sq ft 1 acre = 0.4047 ha
Volume Measurement		
cubic centimeter (cm^3)	cubic inch (cu in.)	1 cm^3 = 0.061 cu in. 1 cu in. = 16.387 cm^3
cubic meter (m^3)	cubic foot (cu ft.)	1 m^3 = 35.315 cu ft 1 cu ft. = 0.02832 m^3
liter (li)	gallon (gal)	1 li = 1,000 cm^3 1 li = 0.264 gal 1 gal = 231 cu in. 1 gal = 3.7853 li

quently, most of the line drawings in this chapter depict the positive rather than the negative plane.

Scale and Height Measurement on Single Vertical Aerial Photographs

The scale of an aerial photograph may be expressed as a *verbal scale* or as a *representative fraction* (dimensionless). For

example, if 1 in. on an aerial photograph represents 2,000 ft (24,000 in.) on the ground, the scale may be expressed as a:

- *Verbal scale:* 1 in. = 2,000 ft

- *Representative fraction:* $\dfrac{1}{24,000}$ or 1:24,000.

When comparing two scales, remember that the larger the number in the scale expression, the *smaller* the scale. For example, compare the scale expressions 1/11,000 and 1/12,000. The image of a given object on the aerial photograph will actually be larger on the larger scale 1/11,000-scale photograph, i.e., a 100-yard football field will be larger on the 1/11,000-scale photo than on the 1/12,000-scale photograph. Table 6-2 is a useful scale conversion chart.

A verbal scale is sometimes found in the titling ancillary information on the border of an aerial photograph. While this scale may be correct, it is much better to calculate the exact scale of every photograph that will be used to extract photogrammetric measurements. Several methods are available for computing the scale of aerial photographs obtained over level or variable relief terrain.

Scale of a Vertical Aerial Photograph over Level Terrain

There are two main methods of determining the scale of single aerial photographs obtained over level terrain. One involves comparing the size of objects measured in the real world or from a map (e.g., the length of a section of road) with the same object measured on the aerial photograph. The second method involves computing the relationship between camera lens focal length and the altitude of the aircraft above ground level.

Computing Scale by Comparing Real-World Object Size versus Photographic Image Size

The scale, *s*, of a vertical aerial photograph obtained over nearly level terrain is the ratio of the size of the object measured on the aerial photograph, *ab*, compared to the actual length of the object in the real world, *AB*:

$$s = \frac{ab}{AB}.\qquad(6\text{-}1)$$

This relationship is based on the geometry of similar triangles *Lab* and *LAB* in Figures 6-6 and 6-7. To compute scale using Equation 6-1, the image analyst first identifies an object in the vertical aerial photograph whose length (*AB*) was measured on the ground or, more commonly, was mea-

Table 6-2.   Scale conversion chart.

Scale	ft per in. (a)	in. per mi	mi per in.	acres per sq. in. (b)	ha per sq. in.	m per cm	cm per km	km per cm
1:500	41.67	126.72	0.008	0.03986	0.0161	5.00	200	0.005
1:600	50.00	105.60	0.009	0.05739	0.0232	6.00	166.66	0.006
1:1,000	83.33	63.36	0.016	0.15940	0.0645	10.00	100	0.010
1:2,000	166.67	31.68	0.032	0.63771	0.2581	20.00	50.00	0.020
1:3,000	250.00	21.12	0.047	1.4348	0.5807	30.00	33.33	0.030
1:4,000	333.33	15.84	0.063	2.5507	1.0323	40.00	25.00	0.040
1:5,000	416.67	12.67	0.079	3.9856	1.6129	50.00	20.00	0.050
1:6,000	500.00	10.56	0.095	5.7392	2.3226	60.00	16.66	0.060
1:10,000	833.33	6.336	0.158	15.9421	6.4517	100	10.00	0.100
1:12,000	1000.00	5.280	0.189	22.9568	9.29058	120.00	8.33	0.120
1:15,000	1250.00	4.224	0.237	35.8700	14.5164	150.00	6.66	0.150
1:20,000	1666.67	3.168	0.316	63.7692	25.8070	200.00	5.0	0.200
1:24,000	2000.00	2.640	0.379	91.8273	37.1620	240.00	4.166	0.240
1:25,000	2083.33	2.534	0.395	99.6387	40.3234	250.00	4.00	0.250
1:48,000	4000.00	1.320	0.758	367.309	148.6481	480.00	2.08	0.480
1:50,000	4166.67	1.267	0.789	398.556	161.2935	500.00	2.00	0.500
1:63,360	5280.00	1	1	640.000	259.0056	633.60	1.58	0.634
1:100,000	8333.33	0.634	1.578	1594.00	645.174	1,000.00	1	1
1:250,000	20833.33	0.2534	3.946	9963.90	4032.338	2,500.00	0.40	2.50
1:500,000	41666.67	0.1267	7.891	39855.63	16129.35	5,000.00	0.2	5.00
1:1,000,000	83333.33	0.0634	15.783	159422.49	64517.41	10,000.00	0.100	10.00
	$\dfrac{scale}{12}$	$\dfrac{63360}{scale}$	$\dfrac{scale}{63360}$	$\dfrac{(a)^2}{43560}$	$\dfrac{(b)}{2.471}$	$\dfrac{scale}{100}$	$\dfrac{100000}{scale}$	$\dfrac{scale}{100000}$

sured from a reference map. The analyst then measures the corresponding distance on the photograph (ab) and solves for s.

For example, consider the vertical aerial photograph of downtown Columbia, SC, previously shown in Figure 6-4. The titling information says the original photograph was obtained at a nominal scale of 1 in. = 500 ft (1:6,000). But is this correct? Many subsequent computations will make use of the scale parameter, so it is important to make sure that we know the actual scale of the photograph and not just a nominal scale.

In this example, we will first determine the scale of the photograph using the road shown in Figure 6-8. The road width from curb to curb was 56.1 ft (AB) as measured in the field using a surveyor's tape. The width of the same road was

Figure 6-7 The geometry of a vertical aerial photograph collected over relatively flat terrain.

Figure 6-8 An enlargement of a portion of the aerial photograph shown in Figure 6-4. The graphic shows the width of the road from curb to curb (56.1') and the sidewalk (6') as measured using a surveyor's tape and measured on the aerial photograph (0.113" and 0.012", respectively). These measurements may be used to compute the scale of the photograph in this portion of the photograph.

0.113 in. (*ab*) on the vertical aerial photograph (Figure 6-8). Using Equation 6-1, the scale of the photograph is:

$$s = \frac{ab}{AB} = \frac{0.113''}{56.1'} = \frac{0.113''}{673.2''} = \frac{1''}{5957.52''}$$

Representative fraction: $\frac{1}{5957}$

Verbal scale: 1 in. = 496.46 ft .

Just to be sure, the width of a nearby sidewalk was measured. The sidewalk was 6 ft in the real world and 0.012 in. on the aerial photograph. If we use the width of the sidewalk to compute the scale of the photography, we get 1:6,000:

$$s = \frac{ab}{AB} = \frac{0.012''}{6'} = \frac{0.012''}{72''} = \frac{1''}{6000''}$$

Representative fraction: $\frac{1}{6000}$

Verbal scale: 1 in. = 500 ft.

When computing scale, the goal is to modify the input values so that the numerator has a value of one and both the numerator and denominator are in the same units, e.g., in inches, feet, or meters. Sometimes this requires multiplying both the numerator and denominator by the reciprocal of the numerator. The best way to conceptualize scale is as a representative fraction. Therefore, in this last example 1 in. on the photograph equals 6,000 in. in the real world; 1 ft on the photograph equals 6,000 ft; and 1 m on the photograph equals 6,000 m. With a representative fraction, we can work in any units we desire.

To offset the effects of aircraft tilt at the instant of exposure, the measurement of scale using this method should be based on the average of several scale checkpoints in the photograph. When possible, some analysts like to compute the scale by selecting objects along lines such as roads that intersect approximately at right angles and are centrally located on the photograph. They then compute an average scale using several scale measurements (Mikhail et al., 2001).

Vertical aerial photography scale can also be determined if objects whose lengths are already known appear on the photograph. A baseball diamond, football field, or soccer field may be measured on the photograph and the photo scale determined as the ratio of the photo distance to the known ground distance. For example, what is the scale of a vertical aerial photograph on which the distance between home plate and first base equals 0.5 in? The distance from home plate to first base is 90 ft. Therefore,

$$s = \frac{ab}{AB} = \frac{0.5''}{90'}$$

$$s = \frac{0.5''}{90' \times 12''}$$

$$s = \frac{0.5''}{1080''}$$

$$s = \frac{1''}{2160''}$$

Representative fraction: $\frac{1}{2160}$.

Examples of sports field dimensions found throughout the world are summarized in Chapter 12 (Remote Sensing the Urban Landscape).

Computing Scale by Relating Focal Length to Altitude Above Ground Level (AGL)

Scale may also be expressed in terms of camera focal length, f, and flying height above the ground, H, by equating the geometrically similar triangles Loa and LPA in Figure 6-7:

$$s = \frac{f}{H}. \tag{6-2}$$

From Equation 6-2 it is evident that the scale of a vertical aerial photograph is directly proportional to camera focal length (image distance) and inversely proportional to flying height above ground level (object distance). This means that if the altitude above ground level is held constant, increasing the size of the focal length will result in larger images of objects at the film plane. Conversely, if focal length is held constant, the images of objects will be smaller as we gain additional altitude above ground level. If the focal length of the camera and the altitude of the aircraft above ground level are unknown, the image analyst will have to consult the ancillary information appearing on the edges of the photograph, the flight roll, or the aerial photographer's mission summary.

For example, a vertical aerial photograph is obtained over flat terrain with a 12-in. focal-length camera lens from an altitude of 60,000 ft AGL. Using Equation 6-2, the scale is:

$$s = \frac{f}{H} = \frac{12''}{60000'}$$

$$s = \frac{1'}{60000'}$$

Representative fraction: $= \frac{1}{60000}$

Verbal scale: 1 in. = 5000 ft.

By transposing Equation 6-2 to read,

$$H = \frac{f}{s} \tag{6-3}$$

or

$$f = H \times s, \tag{6-4}$$

it is possible to determine the altitude above ground level, H, of the photograph at the instant of exposure, or the focal length, f, of the camera if the scale of the photograph, s, is known along with one of the other variables. For example, in the case of the Columbia, SC, photograph (Figure 6-4) we know that the scale of the photograph based on the measurement of road width is 1:5,957 or 1" = 5,957", and the focal length is 6 in. Using Equation 6-3, the altitude of the aircraft above local datum at the instant of exposure is:

$$H = \frac{6''}{\left(\frac{1''}{5957''}\right)} = 35742'' = 2978.5'.$$

Similarly, if we only knew the altitude above datum (2,978.5') and the scale (1:5,957), we could compute the focal length of the camera system using Equation 6-4:

$$f = 2978.5' \times \frac{1'}{5957'} = 0.5' = 6''.$$

Scale of a Vertical Aerial Photograph over Variable Terrain

One of the principal differences between a near-vertical aerial photograph and a planimetric map is that for photographs taken over variable terrain, there are an infinite number of different scales present in the aerial photograph. If topographic elevation decreases within a certain portion of the aerial photograph relative to other areas, then that portion of the photograph will have a smaller scale than the rest of the photograph because the land will have "moved away" from the aerial camera that is flown at a constant altitude. Conversely, if a topographic feature such as a mountain or a

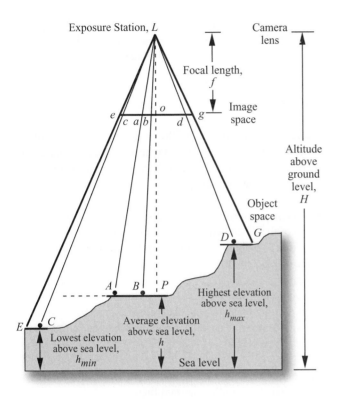

Figure 6-9   Geometry of a vertical aerial photograph obtained over terrain with variable relief.

$$\frac{Lb}{LB} = \frac{Lo}{LP} = \frac{f}{H-h}. \tag{6-6}$$

Substituting Equation 6-6 into Equation 6-5,

$$s_{ab} = \frac{ab}{AB} = \frac{f}{H-h}. \tag{6-7}$$

If the line between points a and b in the vertical aerial photograph is considered to be infinitely small, Equation 6-7 reduces to an expression for photo scale at a specific point. Dropping subscripts, the scale at any point whose elevation above sea level is h and whose camera altitude above sea level is H, may be expressed as:

$$s = \frac{f}{H-h}. \tag{6-8}$$

In our example in Figure 6-9, different scale values would be computed at locations c and d in the aerial photograph. One scale value would be a function of the minimum elevation above sea level within the photograph (s_{min}) while the other would be a function of the maximum elevation above sea level within the photograph (s_{max}):

$$s_{min} = \frac{f}{H-h_{min}} \tag{6-9}$$

$$s_{max} = \frac{f}{H-h_{max}}. \tag{6-10}$$

Usually an average or *nominal* scale is computed to define the overall scale of a vertical aerial photograph taken over variable terrain:

$$s_{avg} = \frac{f}{H-h_{avg}}. \tag{6-11}$$

It should be remembered that the average scale is only at those points that lie at average elevation, and it is only an approximate scale for all other locations on the photograph.

To demonstrate these relationships, let us consider the maximum elevation, h_{max}, average elevation, h_{avg}, and minimum elevation, h_{min}, of the terrain in Figure 6-9 to be 10,000, 8,000, and 6,000 ft above sea level, respectively. The flying height of the aircraft above sea level is 20,000 ft, and the camera focal length is 6 in. (152.82 mm). The maximum, minimum, and average scale of the vertical aerial photograph would be:

building protrudes above the average elevation of the local terrain, then the scale in that area of the photograph will be larger because the land will have "moved closer" to the aerial camera.

The geometry of a single vertical aerial photograph taken over terrain with variable local relief from exposure station L is shown in Figure 6-9. Points A and B in the real-world object space are located on level terrain and recorded on the positive print image space at a and b, respectively. Points A and B are located at an elevation of h units above sea level. As previously discussed, the scale of the aerial photograph at location ab in the positive print is equal to the ratio of the photo distance ab to ground distance AB. By similar triangles Lab and LAB, the photo scale at location ab in the vertical aerial photograph may be computed using several relationships, including:

$$s_{ab} = \frac{ab}{AB} = \frac{Lb}{LB}. \tag{6-5}$$

Also, by similar triangles Lob and LPB,

$$s_{max} = \frac{f}{H - h_{max}}$$

$$s_{max} = \frac{6''}{20000' - 10000'}$$

$$s_{max} = \frac{6''}{10000'}$$

$$s_{max} = 1 : 20000$$

$$s_{min} = \frac{f}{H - h_{min}}$$

$$s_{min} = \frac{6''}{20000' - 6000'}$$

$$s_{min} = \frac{6''}{14000'}$$

$$s_{min} = 1 : 28000$$

$$s_{avg} = \frac{f}{H - h_{avg}}$$

$$s_{avg} = \frac{6''}{20000' - 8000'}$$

$$s_{avg} = \frac{6''}{12000'}$$

$$s_{avg} = 1 : 24000 \, .$$

Height Measurement from Single Aerial Photographs

There are two primary methods of computing the heights of objects on single vertical aerial photographs. The first involves the measurement of image relief displacement, and the second is based on the measurement of shadow length.

Height Measurement Based on Relief Displacement

The image of any object lying above or below the horizontal plane passing through the elevation of the principal point is displaced on a truly vertical aerial photograph from its true planimetric (x,y) location. The *relief displacement* is outward from the principal point for objects whose elevations

are above the local datum, and inward or toward the principal point for objects whose elevations are below the local datum. The direction of relief displacement is radial from the principal point of the photograph.

The amount of the relief displacement, *d*, is:

- *directly proportional* to the difference in elevation, *h*, between the top of the object whose image is displaced and the local datum, i.e., the greater the height of the object above the local datum, the greater its displacement.

- *directly proportional* to the radial distance, *r*, between the top of the displaced image and the principal point, i.e., the farther the object is from the principal point, the greater the displacement.

- *inversely proportional* to the altitude, *H*, of the camera above the local datum. Therefore, a reduction in relief displacement of an object can be achieved by increasing the flying height.

These relationships are depicted in Figure 6-10. Notice from similar triangles in the diagram that:

$$\frac{h}{H} = \frac{d}{r} \, . \tag{6-12}$$

Rearranging the relationship, we can see that the amount of displacement, *d*, is directly proportional to the height of the object, *h*, and its distance from principal point, *r*, and inversely proportional to the altitude above local datum, *H*:

$$d = \frac{h \times r}{H} \, . \tag{6-13}$$

If we solve for the height of the object, *h*, the equation becomes:

$$h = \frac{d \times H}{r} \, . \tag{6-14}$$

Therefore, we can compute the height of an object from its relief displacement characteristics on a single vertical aerial photograph. It is important that both the top and the bottom of the object being measured are clearly visible and that the base is on level terrain.

For example, let us compute the height of the Senate Condominium shown in Figure 6-10a. The photograph has been rotated 90° counterclockwise to facilitate viewing (shadows now fall toward the viewer, and the displaced façade of the

Figure 6-10   Measurement of object height (in this case, the Senate Condominium in Columbia, SC) from a single vertical aerial photograph based on relief displacement principles. a) The radial distance from the principal point (PP) and the top of the building is r. The distance from the base of the building to the top of the building is d. b) The height of the building (h) above local datum is computed based on the relationship between similar triangles.

building leans comfortably away from the viewer). Based on previous measurements, the altitude of the camera above local datum, H, is known to be 2,978.5 ft, the radial distance from the principal point to the top of the building, r, is measured as 2.23 in., and building relief displacement, d, is 0.129 in. The photogrammetrically computed height, h, of the condominium is:

The actual height of the building measured with a surveyor's tape is 172.75 ft. To obtain accurate object height measurements using this technique, it is imperative that the altitude of the aircraft above the local datum be as precise as possible. Also, great care should be exercised when measuring r and d on the photograph. Keep in mind that r is measured from the principal point to the top of the object.

$$h = \frac{0.129'' \times 2978.5'}{2.23''} = 172.3'.$$

Height Measurement Based on Shadow Length

The height of an object, h, may be computed by measuring the length of the shadow cast, L, on vertical aerial photography. Because the rays of the Sun are essentially parallel throughout the area shown on vertical aerial photographs, the length of an object's shadow on a horizontal surface is proportional to its height. Figure 6-11 illustrates the trigonometric relationship involved in determining object heights from shadow measurements. Notice that the tangent of angle a would be equal to the opposite side, h, over the adjacent side, which is the shadow length, L, i.e.,

$$\tan a = \frac{h}{L}. \tag{6-15}$$

Solving for height yields:

$$h = L \times \tan a. \tag{6-16}$$

The Sun's elevation angle, a, above the local horizon can be predicted using a solar ephemeris table. This requires a knowledge of the geographic coordinates of the site (longitude and latitude), the acquisition date, and time of day. Alternatively, the solar altitude may be empirically computed if sharply defined shadows of known height are formed on the photograph. For example, we know from previous discussion that the height of the Senate Condominium in Figure 6-12 is 172.75 ft. It casts a shadow onto level ground that is 0.241" in length on the photograph. The scale of the photography is 1:5,957 or 1" = 496.46'. Therefore, the shadow length on the photograph is 119.65 ft. The tangent of angle a can be found using Equation 6-15:

$$\tan a = \frac{h}{L} = \frac{172.75'}{119.65'} = 1.44.$$

Other shadow lengths on the same aerial photograph can be measured and their lengths multiplied by 1.44 to determine their heights. For example, the tower on the nearby parking structure in Figure 6-12 casts a shadow that is 0.119", or 59.1' long onto level ground in the photograph. Therefore, the height of the tower is:

$$h = L \times \tan a = 59.1' \times 1.44 = 85.10'.$$

The actual height of the tower is 86 ft measured with a surveyor's tape.

Care must be exercised when computing the height of objects based on shadow length in aerial photography. Important factors to be considered include shadows falling

Measurement of the Height of Objects Based on Shadow Length

$$\tan a = \frac{\text{opposite}}{\text{adjacent}}$$

$$= \frac{\text{height, } h}{\text{shadow, } L}$$

$$h = L \times \tan a$$

Figure 6-11   The height of objects can be measured from vertical aerial photography based on the length of shadows cast on level terrain.

on unlevel terrain, shadows produced from leaning objects, shadows not cast from the true top of the object, and snow or other types of groundcover obscuring the true ground level.

 Stereoscopic Measurement of Object Height or Terrain Elevation

A single aerial photograph captures a precise record of the positions of objects in the scene at the instant of exposure. If we acquire multiple photographs along a flightline, we record images of the landscape from different vantage points. For example, the top of a tall building might be on the left side of photo #1 and in the middle of overlapping photo #2 because the aircraft has moved hundreds of meters between exposures. If we opened up the back of the aerial camera, held the shutter open, and looked at the groundglass at the focal plane while the aircraft was flying along a flightline, we would literally see the tall building first enter the groundglass field of view and then traverse across the groundglass until it eventually leaves the camera's field of view.

The change in position of an object with height, from one photograph to the next relative to its background, caused by the aircraft's motion, is called *stereoscopic parallax. Parallax* is the apparent displacement in the position of an object, with respect to a frame of reference, caused by a shift in the position of observation. Parallax is a normal characteristic of aerial photography and is the basis for three-dimensional stereoscopic viewing. Differences in the parallax of various

Figure 6-12   The height of the parking garage tower may be computed by determining the relationship between its shadow length and the length of the shadow cast by a building of known height such as the nearby Senate Condominium. It is important that the shadow falls on level ground and that both the beginning and end of the shadow are visible.

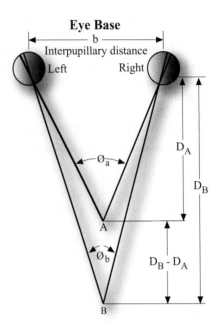

Figure 6-13   Parallactic angles are formed when our eyes focus on objects in the real world. The mind associates differences in parallactic angles with differences in distance to the various objects. This allows us to have very sensitive depth perception.

objects of interest (called *differential parallax*) can be used to measure the heights of objects and to extract topographic information such as contour lines from aerial photographs by means of stereoscopic instruments. The measurement of stereoscopic parallax is the basis for the creation of almost all topographic maps (x, y, and z) and planimetric maps (x,y) and is a very important element of photogrammetry and remote sensing.

Fundamentals of Human Stereoscopy

Stereoscopy is the science of perceiving depth using two eyes. When a human being's two eyes (*binocular vision*) are focused on a certain point, the optical axes of the eyes converge on that point, forming a *parallactic angle* (ϕ). The nearer the object, the greater the parallactic angle. For example, in Figure 6-13 the optical axes of the left and right eyes, L and R, are separated by the *eye base* or *interpupillary distance* (Konecny, 2003). The eye base of the average adult is

between 63 and 69 mm (approximately 2.5 to 2.7 in.). When the eyes are focused on point A, the optical axes converge, forming parallactic angle ϕ_a. Similarly, when looking at point B, the optical axes converge, forming parallactic angle ϕ_b. The brain has learned to associate distances D_A and D_B with corresponding parallactic angles ϕ_a and ϕ_b and gives the viewer the visual and mental impression that object A is closer than object B. This is the basis of *depth perception*. If both objects were exactly the same distance from the viewer, then $\phi_a = \phi_b$ and the viewer would perceive them as being the same distance away.

When we walk outside, the maximum distance at which distinct stereoscopic depth perception is possible is approximately 1,000 m for the average adult. Beyond that distance, parallactic angles are extremely small, and changes in parallactic angle necessary for depth perception may not be discerned. This is why humans have trouble determining whether one house is behind another house, or one car is behind another car when these objects are thousands of meters away from us. Conversely, if we could somehow stretch our eyes to be a meter or even hundreds of meters apart, then we would be able to resolve much more subtle differences in parallactic angles and determine which objects are closer to us over much greater distances. Such *hyperstereoscopy* depth perception would be ideal for hunting and sports activities, but it would require a substantial modification of the human head. Fortunately, there is a simple method that we can use to obtain a hyperstereoscopy condition when collecting and interpreting stereoscopic aerial photography.

Stereoscopy Applied to Aerial Photography

Overlapping aerial photography (usually 60 percent endlap) obtained at exposure stations along a flightline contain stereoscopic parallax. The exposure stations are separated by large distances. Nevertheless, it is possible to let our eyes view the photographs as if our eyes were present at the two exposure stations at the instant of exposure (Figure 6-14a). This results in a *hyperstereoscopy* condition that allows us to view the terrain in three dimensions. We normally view the stereoscopic aerial photography using a lens or mirror stereoscope with magnifying lenses. These instruments enhance the three-dimensional nature of the stereoscopic model. There are other stereoscopic viewing alternatives.

Methods of Stereoscopic Viewing

The photo-interpreter can view the vertically exaggerated stereoscopic model of the 60 percent endlap area of two successive aerial photographs using one of four methods: 1) keeping the lines of sight parallel with the aid of a stereoscope, 2) keeping the lines of sight parallel without the aid of a stereoscope, 3) crossing the eyes and reversing the order of the stereoscopic images, or 4) using anaglyph or polarized glasses.

The vast majority of image analysts prefer to use a simple lens pocket stereoscope or mirror stereoscope that assists the eyes in keeping parallel lines of sight and in addition, usually magnifies the photographs (Figure 6-14b,e). This produces some eye strain. It is suggested that novice interpreters only view photographs in stereo for 10 to 15 minutes at a time in the beginning while the eyes become accustomed to using a stereoscope.

Some people are adept at forcing their eyes to remain parallel and thus do not need to use a stereoscope (Figure 6-14c,f). They simply situate the overlapping portion of two stereo photographs adjacent to one another, position their head approximately 8 in. from the photographs, and then let their eyes relax as if they were looking at infinity. Gradually, the mind will fuse the two stereoscopic images into a third image directly in the middle of the two stereo photos. This is a good skill to acquire since one can then easily view stereo-

scopic photographs without a stereoscope whenever the need arises. It is particularly useful when conducting field work. However, this is unnatural for the eyes and may be uncomfortable and cause eye strain.

Some image analysts are able to reverse the order of the stereo photographs (Figure 6-14d,g) and then fixate on a point directly in front of the photos. This causes the eyes to cross. This produces a true stereoscopic impression, but it is very strenuous on the eyes and is not recommended.

Another way of making sure that the left and right eyes view distinct images is to use *anaglyph* or *polarized* glasses in conjunction with specially prepared image materials. It is possible to produce aerial photography where the left image is depicted in shades of blue and the right image is projected in shades of red. The analyst then wears anaglyphic glasses with red (left) and blue (right) lenses. The mind fuses the two views and creates a three-dimensional scene. Similarly, it is possible to view the left and right photographs through specially prepared polarized glasses that accomplish the same goal (Konecny, 2003).

Lens and Mirror Stereoscopes and Stereo Cameras

The *stereoscope* is a binocular viewing system specially developed to analyze terrestrial stereoscopic photographs (*not* aerial photographs). It was invented by the English physicist Charles Wheatstone in 1833, although he did not describe it publicly until 1838. Wheatstone used a pair of mirrors, one before each eye, oriented at 45° to allow pictures placed at either side to be fused by the eyes, as shown in Figure 6-15a. David Brewster invented an alternative stereoscopic system using lenses in 1849 (Figure 6-15b). The pictures were smaller, but they were enlarged by the lenses. This became the parlor stereoscope so popular during the 1800s and early 1900s for viewing specially prepared stereoscopic photographs.

Throughout most of the 1800s, stereo photographs were obtained by taking one exposure and then picking up the camera on a tripod and moving it a certain distance to the left or right and taking another picture. This introduced stereoscopic parallax between the two photographs. For example, consider the 1899 stereogram of the Salt Lake Temple, in Salt Lake City, Utah, in Figure 6-16a. A wealth of historical information about the temple is available when the stereopair is viewed using a stereoscope.

Stereoscopic photographs may also be acquired using special stereoscopic cameras that contain two identical lenses that are situated 60 to 70 mm apart (2.36–2.75 in.) that take

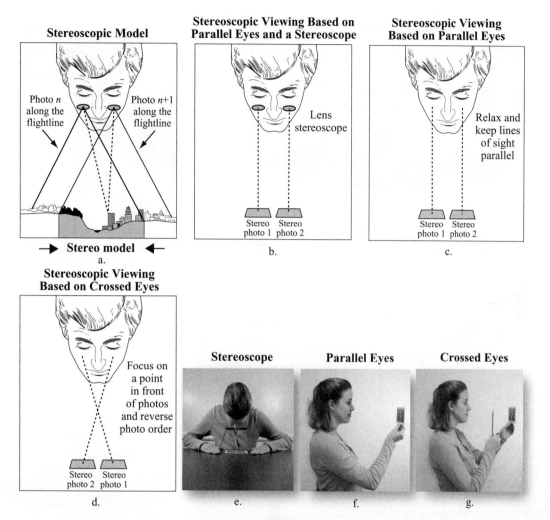

Figure 6-14 a) A hyperstereoscopy condition can be achieved by taking overlapping vertical aerial photographs along a flightline. The analyst then views one photo of the stereopair with the left eye and the adjacent photo with the right eye. When the viewer focuses both eyes on a single feature within the stereoscopic model, such as the top or base of a building, the mind subconsciously associates differences in the parallactic angle between the base and the top of the building as depth-perception information. We perceive a realistic three-dimensional model of the terrain in our minds. This means that we can literally view any landscape in the world from the comfort of our office in very realistic three dimensions if we have stereoscopic aerial photography and the proper viewing equipment. b,e) A lens or mirror stereoscope assists our eyes' lines of sight to remain parallel rather than converging, which is our natural tendency. This analyst is looking at a specially prepared stereopair of photographs. c,f) It is possible to obtain a stereoscopic view by training the eyes' lines of sight to remain parallel without using a stereoscope. This is very useful in the field. d,g) Some people can see stereo by focusing on a point in front of the stereoscopic photographs and letting their lines of sight cross. This is not a natural thing to do and can cause eye strain.

two photographs of the scene at exactly the same time. The cameras obtain photographs of objects from slightly different vantage points, introducing stereoscopic parallax. Note the two separate lenses on the vintage stereoscopic camera in Figure 6-16b. Companies continue to produce stereoscopic cameras.

The development of the concept of stereoscopic photographs and the invention of stereoscopic cameras allowed people to view marvelous stereoscopic views of local and foreign landscapes in three dimensions as if they were really at the location. There continues to be significant interest in the collection of stereoscopic photography by the public. The National Stereoscopic Association promotes the study and

Wheatstone's Mirror Stereoscope

a.   b.

Figure 6-15   a) Wheatstone's mirror stereoscope used two 45° angled mirrors (a and b) to reflect the left and right images of photographs (A and B) toward the viewer. A screw mechanism adjusted the views to accommodate various interpupillary distances. b) An example of Brewster's lens stereoscope.

Stereogram

Stereo Camera

a.   b.

Figure 6-16   a) A terrestrial stereogram of the temple in Salt Lake City, UT, obtained in 1899. b) A vintage stereo camera.

collection of stereographs, stereo cameras, and related materials and publishes *Stereo World*.

Viewing Stereoscopic Aerial Photographs

The same stereoscopic principles used in the original stereoscopes are used in our current photogrammetric stereoscopes. The simple *pocket lens stereoscope* consists of two convex lenses mounted on a rigid metal or plastic frame (Figure 6-17a). The distance between the lenses can be varied to accommodate various eye bases (interpupillary distances). The special lenses help keep the viewer's lines of sight parallel and also magnify the photography. The proper method of arranging stereoscopic photographs for analysis using a pocket lens stereoscope is demonstrated in Figure 6-

18ab. First, the principal point and conjugate principal points are located (PP and CPP, respectively) on each photograph. A line is then drawn through them on each photograph. This is the line of flight previously discussed. The flightlines on each of the photographs are oriented so that they form a continuous line (i.e., they become *colinear*). The analyst then slides one of the photographs left or right so that a portion of the stereoscopic overlap area is visible. Then, the stereoscope is placed above the overlap area and stereoscopic viewing takes place. The common overlap area that can be viewed in stereo of a pair of 9 × 9 in. aerial photographs taken with 60 percent overlap is about 5.4 in. Unfortunately, when the photographs are aligned for stereovision using the pocket stereoscope, not all of the 5.4 in. of the stereo model can be seen at one time. When this occurs, the

Stereoscopes

a. Lens stereoscope.

b. Mirror stereoscope.

c. Zoom stereoscope.

d. Analog stereoscopic zoom-transfer-scope.

e. Digital stereoscopic zoom-transfer-scope.

Figure 6-17   a) Lens stereoscope with attached parallax bar, situated over stereoscopic aerial photographs having 60 percent overlap. Only a portion of the entire stereoscopic model can be viewed at one time using a lens stereoscope. b) An analyst using a mirror stereoscope can view the entire stereoscopic model at one time. c) A zoom stereoscope functions like a mirror stereoscope. In this example, positive transparency photographs are being viewed on a light table. d) An analog stereoscopic zoom-transfer-scope can be used to view stereoscopic aerial photographs and a map at the same time (courtesy Image Interpretation Systems, Inc.). e) A digital stereoscopic zoom-transfer-scope (courtesy Optem, Inc.).

interpreter can gently lift up the edge of the top photograph to see what is underneath.

A *mirror stereoscope* (Figure 6-17b) permits the entire stereoscopic model of the two overlapping aerial photographs to be viewed. Mirror stereoscopes often have magnification options (e.g., 2×, 3×, 6×) available that greatly increase the interpreter's ability to magnify and interpret fine detail in the stereo model. For example, a more sophisticated and expen-

sive mirror stereoscope with zoom magnification is shown attached to a light table in Figure 6-17c. This configuration allows the image analyst to view stereoscopically successive overlapping aerial photographs on a roll of film without having to cut the roll of film. Still more sophisticated instruments, based on the camera-lucida principle, allow the analyst to view stereoscopic imagery while simultaneously viewing a map in superposition (Figure 6-17d). This is commonly called a *zoom-transfer-scope* and is often used to

Stereopair Preparation

Figure 6-18   Preparation of a stereopair for viewing using a lens stereoscope. a) The principal point and conjugate principal point are identified on each photograph, and a line is drawn through them with a straightedge. This locates the line of flight. b) A portion of
the common overlap area that is of interest to the interpreter is moved parallel with the line of flight. The interpreter then views
the overlap area using the lens stereoscope.

transfer information interpreted from aerial photography
onto a map base. Digital stereoscopic zoom-transfer-scopes
are now available (Figure 6-17e).

Stereoscopic Aerial Photography – How Does It Work?

A hypothetical example will demonstrate how the image
analyst perceives the third dimension in stereoscopic aerial
photography using a stereoscope. First, consider the profile
view of two stereoscopic aerial photographs taken from two
exposure stations, L_1 and L_2 in Figure 6-19. Let us evaluate
the characteristics of the top of a very tall building (which
we will designate as object A in the real-world object space)
and the top of a smaller building (object B in the real-world
object space) lying on extremely flat ground just above sea
level. The distance between the two exposure stations, L_1
and L_2, is called the *air base* (*A-base*). The air base in effect
becomes a stretched interpupillary distance. This condition
is responsible for the exaggerated third dimension when analyzing the photographs stereoscopically.

In our example, objects A and B are recorded on the right
photograph (L_1) at locations a and b, respectively (Figure 6-
19b). The aircraft then continues along its designated line of

flight. The forward progress of the aircraft causes the *image*
of objects a and b to move across the camera's focal plane
parallel to the line of flight. When exposure L_2 is made, it
captures the image of objects A and B at locations a' and b',
on the left photograph (Figure 6-19a). This change in position of an image of an object from one photograph to the
next caused by the aircraft's motion is referred to as stereoscopic *x-parallax* (Wolf and Dewitt, 2000).

There are some very interesting properties associated with *x-
parallax* that allow us to obtain precise photogrammetric
measurements from vertical aerial photography. To understand these relationships, consider Figure 6-19c, which
depicts the two photographs taken at exposure stations L_1
and L_2 in *superposition*. Superposition means that we adjust
the profile views of photos L_1 and L_2 so that the vertical lines
running through each of the photos' principal points (PP_1
and PP_2) are superimposed on top of one another. This
allows us to determine how much the objects have moved
across the film plane from one exposure to the next.

Notice that the image of object A moved from a to a' on the
two successive aerial photographs. The parallax of point a is
$p_a = x_a - x_{a'}$. Similarly, the image of object B moved from b
to b'. The parallax of point b is $p_b = x_b - x_{b'}$. The important

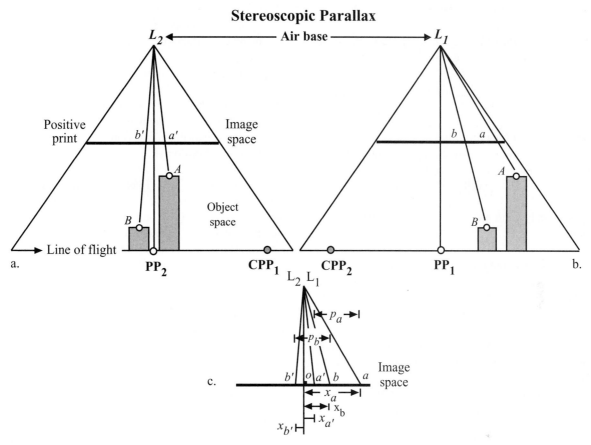

Figure 6-19 Stereoscopic parallax is introduced when an object is viewed from two different vantage points. a,b) Objects *A* and *B* in the real world are recorded as points *a* and *b* in exposure L_1 and at *a'* and *b'* in exposure L_2. c) When the locations of *a* and *b* and *a'* and *b'* are placed in superposition, the image of object *A* has greater *x*-parallax (p_a) than the image of object *B* (p_b). This difference in parallax (called *differential parallax, dp*) can be related to the actual height of the objects.

thing to recognize is that *a* (the taller building) moved a greater distance across the film plane than *b* (the shorter building) because *a* had greater stereoscopic parallax than *b*.

It is also important to understand that *all objects in the scene at exactly the same elevation will have an identical amount of x-parallax*. For example, any point located on the top of building *A* will have exactly p_a *x*-parallax when the stereoscopic photographs are analyzed. Similarly, any other building in the entire stereoscopic model that had the same height as building *A* would have the same *x*-parallax.

Another way of thinking about *x*-parallax is to place a beach ball (which we will call object *A)* anywhere on a very flat plane, such as the Bonneville Salt Flats in Utah or on a lake, and then acquire stereoscopic photographs. No matter where the beach ball is in the stereoscopic model, its image, *a,* on the first photograph would travel exactly the same distance along the image coordinate system to *a'* on the adjacent

overlapping photograph because the ball would have exactly the same amount of *x*-parallax everywhere in the stereo model! Therefore, the *x*-parallax of any point is:

- directly related to the elevation of the point above mean terrain;

- greater for high points (e.g., those closer to the aerial camera) than for low points.

The ability to measure small differences in the parallax (i.e., the *differential parallax*) between any two points using two overlapping photographs allows us to determine elevation differences using stereoscopic parallax equations. This is how almost all topographic maps are made and how very precise tree and building height measurements can be obtained using stereo photographs. In order to do this, we use the fundamental stereoscopic parallax equation.

Measurement of Absolute and Differential Parallax

To compute the height of an object, h_o, using stereoscopic photogrammetric techniques, we use the *parallax equation*:

$$h_o = (H - h) \times \frac{dp}{(P + dp)} \qquad (6\text{-}17)$$

where $H - h$ is the altitude of the aircraft above ground level (AGL); P is the *absolute stereoscopic parallax* at the base of the object being measured (we usually use the air base for this measurement), and dp is the differential parallax.

This equation may yield incorrect results unless the following conditions are met:

• the vertical aerial photographs have $\leq 3°$ tilt;

• the adjacent photographs are exposed from almost exactly the same altitude above ground level;

• the principal points (PPs) of both photographs lie at approximately the same elevation above ground level;

• the base of the objects of interest are at approximately the same elevation as that of the principal points.

If these conditions are met, then we only need to obtain three measurements on the stereoscopic photographs to compute the absolute height of an object found within the stereoscopic overlap portion of a stereopair. For example, consider the stereoscopic characteristics of Photo$_{4-4}$ and Photo$_{4-5}$ of Columbia, SC, shown previously in Figure 6-3 and shown diagrammatically in Figure 6-20. Remember that the aerial photography was acquired at an altitude of 2,978.5 ft above ground level with a 6-in. focal-length lens yielding a nominal scale of approximately 1:5,957. We will compute the height of the Senate Condominium. The computation of the height of the condominium is accomplished by:

1) determining the altitude of the aircraft above ground level *(H – h)* which we know to be 2,978.5 ft. When using the parallax equation, it is necessary to express the aircraft height above ground level *(H – h)* in the units desired. In this case we will use feet.

2) locating the *principal point* (PP) on each of the photographs by drawing lines through the opposing four fiducial marks on each photograph.

3) finding the *conjugate principal point* (CPP) on both photos by locating the position of each photo's principal point on the other photograph.

4) positioning the photographs along the flightline by aligning the PP and CPP of each photograph so that they are in a straight line. This represents the *line of flight*. Note that the line of flight is from right to left in Figure 6-20. Novice interpreters often think that flightlines must progress from left to right. They do not.

5) determining the average photo air base (absolute stereoscopic parallax, P). First, measure the distance between the principal point (PP$_{4-4}$) and the conjugate principal point (CPP$_{4-5}$) on Photo$_{4-4}$, which we will call A-base$_{4-4}$ (Figure 6-20b,d). This was 3.41 in. Do the same thing for Photo$_{4-5}$. In this case A-base$_{4-5}$ was 3.39 in. (Figure 6-20a,c). The mean of these two measurements is the average photo air base (P) between the two exposure stations (i.e., absolute stereoscopic parallax, $P = 3.4$ in.). Absolute stereoscopic parallax (P) and differential parallax (dp) must be measured in the same units, e.g., either millimeters or thousandths of an inch.

6) measuring the differential parallax (dp) between the base of the building and the top of the building. This measurement is best obtained using a *parallax bar* (often called a *stereometer*), which will be discussed shortly. However, it may also be possible to compute the difference in parallax between the objects using a quality ruler. In this example, the top corner of the building is designated as being *Object A* and the base of the building is *Object B*. Notice in Figure 6-20ab that the building is just visible in the lower left-hand portion of Photo$_{4-4}$ but that it has moved dramatically across the film plane and is located near the vertical line through the principal point in Photo$_{4-5}$. The location of the top of the building is labeled a in Photo$_{4-4}$ and a' in Photo$_{4-5}$. The location of the base of the building is labeled b in Photo$_{4-4}$ and b' in Photo$_{4-5}$. The x-parallax of the top (from a to a') and the x-parallax of the base of the building (from b to b') can be measured using a ruler in one of three ways: 1) measurement using fiducial lines, 2) measurement based on superposition, or 3) measurement using a parallax bar (stereometer).

X-Parallax Measurement Using Fiducial Lines

Figure 6-20c,d depict a diagrammatic plan view of the information content of Photo$_{4-4}$ and Photo$_{4-5}$. Note that the location of the principal points (PP), conjugate principal points (CPP), base of the building, and top of the building are located on each photograph. Also note that the diagram depicts the building leaning away from the principal point due to relief displacement, as previously discussed.

Figure 6-20 Computing the height of the Senate Condominium in Columbia, SC, by measuring the stereoscopic x-parallax of the top and base of the building from Photo$_{4-4}$ and Photo$_{4-5}$. When placed in superposition (e), the differential parallax (dp) of the building is determined and can be used to compute the height of the building. Refer to the next illustration to view the actual aerial photographs used in the computation.

In conjunction with the planimetric map, it is useful to actually view the image of the Senate Condominium as recorded on Photo$_{4-4}$ and Photo$_{4-5}$. There is a much enlarged view in Figure 6-21. Note that the principal point line (fiducial line) is visible near the building in Photo$_{4-5}$ (Figure 6-21a), but that the principal point line is far away from the building in

Photo$_{4-4}$, necessitating a break in the line (Figure 6-21b). Using the real aerial photography, it is possible to measure the x-parallax of the top of the building on Photo$_{4-4}$ (designated $x_a = -3.82$ in.) and x-parallax of the base of the building (designated $x_b = -3.606$ in.) from the fiducial line. Similarly, it is possible to measure the x-parallax of the top of the build-

**Methods of Measuring
Stereoscopic *x*-parallax from
Overlapping Aerial Photographs:**
- Measurement using fiducial lines (a,b)
- Measurement based on superposition (c)

Figure 6-21   a,b) The measurement of stereoscopic *x*-parallax using the fiducial lines for Photo$_{4-5}$ and Photo$_{4-4}$, respectively. c) Measurement of stereoscopic *x*-parallax using superposition.

ing on Photo$_{4-5}$ (designated $x_{a'}$ = -0.270 in.) and the base of the building (designated $x_{b'}$ = -0.267 in.) from the fiducial line. The absolute value of the *x*-parallax of the top of the building is p_a = 3.55 in. and the *x*-parallax of the base of the building is p_b = 3.339 in. The differential parallax (*dp*) is the difference between the two values, *dp* = 0.211 in.

Substituting the measured values into Equation 6-17

$$h_o = (H - h) \times \frac{dp}{(P + dp)}$$

yields

$$h_o = 2978.5' \times \frac{0.211''}{(3.4'' + 0.211'')}$$

$$h_o = 174' \quad .$$

The actual height of the building is 172 ft.

X-Parallax Measurement Based on Superposition

It is possible to orient the two overlapping photographs as shown in Figures 6-20e and 6-21c where the base and top of the building are aligned parallel with the line of flight. It does not matter how much distance separates the two photographs. The x-parallax of the top of the building is measured as the distance from the top of the building on $Photo_{4-5}$ to the top of the same corner of the building on $Photo_{4-4}$ ($p_a = 0.30$ in.). The x-parallax of the base of the building is the distance from the base of the building on $Photo_{4-5}$ to the base on $Photo_{4-4}$ ($p_b = 0.511$ in.). The differential parallax between the top and base of the building is identical to what was computed using the fiducial line method, $dp = 0.211$ in., and would yield the same building height estimate. Some image analysts prefer this simple parallax measurement technique.

Now that we have the altitude of the aircraft above ground level $(H - h)$ and the absolute stereoscopic parallax (P) computed for these two photographs, it is possible to measure the bottom and top of other features within the stereoscopic model and compute their differential parallax measurements (dp). While it is simple and useful to be able to measure the amount of differential parallax using a ruler as just described, it is also possible to use a parallax bar that was designed specifically for measuring stereoscopic parallax.

X-Parallax Measurement Using A Parallax Bar

The parallax of specific objects in a stereopair can be computed accurately and quickly using a *parallax bar* (sometimes referred to as a *stereometer*). The parallax bar consists of a bar with two clear plastic or glass plates attached to it (Figure 6-22). One of the plates is fixed, while the other plate can be moved back and forth along the length of the bar by adjusting a vernier dial. There is usually a bright red dot etched into each of the plastic (glass) plates. These are called *measuring marks*.

When using a simple lens stereoscope, the image analyst arranges the photographs so that the object of interest (such as the corner of the top of a building) in the left photograph is viewed by the left eye and the same object in the right-hand photograph is viewed by the right eye. Next, the analyst adjusts the stereometer so that 1) the left measuring mark is placed exactly on the corner of the top of the building in the left-hand photograph, and 2) the right measuring mark is adjusted so that it is placed on the same corner of the top of the building in the right-hand photograph. The measuring marks are shifted in position on the stereometer using the vernier screw device until they visually fuse together

Lens Stereoscope and Parallax Bar

Figure 6-22   Close-up view of a parallax bar with the floating marks and vernier measuring mechanism. It is attached to the lens stereoscope and placed over the stereopair to make parallax measurements.

into a single mark, which appears to exist in the stereo model as a three-dimensional red ball that lies at the particular elevation of the corner of the top of the building. This produces the same effect as if a red ball had actually existed at the corner of the top of the building at the time the two photographs were exposed. If the measuring marks are successfully fused and hover exactly at the corner of the top of the building, then the stereoscopic x-parallax of this point may be easily read from the vernier scale on the parallax bar. We have just used the principle of the *floating mark*.

If we wish to measure the parallax of the base of the corner of the same building, we simply focus our stereoscopic vision on the base and adjust the two (left and right) measuring marks until they once again fuse as a three-dimensional ball that just touches the ground at the base of the building. We can then read the vernier scale to obtain the x-parallax measurement of the base of the building.

The significance of the *floating mark* for photogrammetric measurement is that it can be used to accurately measure the parallax at any two points over the stereo model. It can be placed at the top of an object such as a building and the parallax of the point read, e.g., p_a. Then, the floating mark can be placed at the bottom of the object and the parallax read, e.g., p_b. The difference $(p_a - p_b)$ results in the *differential parallax* (dp) measurement required as input to Equation 6-17. It makes no difference if the object is man-made with vertical sides or if it is a terrain feature, i.e., it can be the top and base of a building or tower, the peak of a mountain or a river floodplain. From the *differential parallax* between the two measurement points, the elevation difference may be derived.

The principle of the floating mark can be used to map lines of constant elevation in the terrain. For example, if the floating mark (the fused, red three-dimensional ball) were moved around the stereo model so that it maintained contact with the terrain, i.e., it was not allowed to float above the terrain or go down below the terrain (which is possible when viewing in stereo!), then a line of constant stereoscopic x-parallax would be identified. If this were done along the side of a mountain, the line of constant x-parallax could be determined, which could then be converted directly into elevation above sea level. This line of constant parallax is related to actual elevation through the use of horizontal (x,y) and vertical (z) ground-control markers that were surveyed in the field. This is how USGS topographic maps with contour lines are produced. The stereoplotter operator places the floating mark on the terrain (representing a specific elevation) and then moves the floating mark about the terrain in the stereo model while keeping the mark firmly on the ground. One can convert the x-parallax measurement to actual elevation by locating horizontal-vertical ground-control points (e.g., 100 ft. above sea level at location x,y) in the stereoscopic model and determining its x-parallax. Any other point within the stereoscopic model with the same x-parallax must then lie at 100 ft above sea level. It is also possible to map building perimeters by placing the three-dimensional floating mark so that it just rests on the rooftop, and then tracing the mark around the edge of the rooftop. This results in a map of the planimetric (x,y) location of the building.

The simple parallax bar (stereometer) is the least expensive and one of the least accurate of all instruments that are based on the concept of the floating mark. Very expensive analog and digital stereoscopic plotting instruments are available that allow precise parallax measurements to be made. For example, a Zeiss analytical stereoplotter is shown in Figure 6-23. When used in conjunction with ground-control information collected in the field, the system analytically correlates and computes the amount of x-parallax for each picture element (pixel) within the two stereoscopic aerial photographs inside the system. A stereoplotter operator interacts with the analytical instrument to produce planimetric and topographic maps. In addition, they can be used to produce raster digital elevation models. An example of typical x,y planimetry and contours (z) derived from a stereoplotter is shown in Figure 6-24. For a detailed discussion of how analytical stereoplotters function, refer to the *Manual of Photo-*

Lens Stereoscope and Parallax Bar

Figure 6-23   A Zeiss P3 Planicomp analytical stereoplotter. The analyst views the stereo model through the binocular lens system and can adjust the floating mark using the cursor on the table (courtesy Carl Zeiss).

grammetry published by the American Society for Photogrammetry and Remote Sensing (McGlone, 2004).

Stereoscopic photography may also be acquired on the ground. The stereoscopic photographs can be analyzed using the principle of the floating mark to produce a detailed three-dimensional representation of the façade of buildings if desired (of course, field X, Y, Z ground-control measurements are required to scale the stereo model). This is called *close-range photogrammetry* (Warner et al., 1996; Wolf and Dewitt, 2000). Stereoscopic photogrammetric techniques can be applied to terrestrial photography to restore historical buildings or reconstruct a human's arm, leg, or face.

 Digital Elevation Models, Orthophotos and Planimetric Features using Soft-copy Photogrammetry

One of the most important events in the history of remote sensing and photogrammetry has been the development of *soft-copy photogrammetry*. The term *soft-copy* means that a digital version of the stereoscopic aerial photography is analyzed rather than a hardcopy. James Case developed the first photogrammetric soft-copy system in the early 1980s. Current soft-copy photogrammetric systems (e.g., the Leica GeoSystems Photogrammetry Suite) allow the analyst to process digital imagery on a personal computer, view the stereo model in three dimensions on a computer screen,

Figure 6-25　An image analyst viewing a stereo model in three-dimensions using anaglyph glasses and the Leica Geosystems Photogrammetry Suite.

Figure 6-24　a) One aerial photograph of a stereopair of Covina, CA. b) A line drawing of the planimetric (x,y) location of the roads, building footprints, fences, and landscape elements (trees, shrubs), as well as 1-ft contours derived from the stereo model.

• extraction of spot elevations, contours, and digital elevation models;

• creation of digital orthoimages;

• extraction of thematic features of interest.

Collection of Accurate Horizontal and Vertical Ground Control

derive digital elevation models (DEMs), extract contours, produce orthophotographs, and extract planimetric feature detail. An analyst using soft-copy photogrammetry software and wearing anaglyph glasses to analyze a stereo model in three dimensions is shown in Figure 6-25.

The stages of development required to produce DEMs and terrain-corrected orthoimagery, and extract thematic features of interest using soft-copy photogrammetry are outlined in Figure 6-26. Important considerations include:

• collection of ground-control point data;

• collection of stereoscopic aerial photography or other remote sensor data;

• image digitization technology;

• soft-copy photogrammetry software;

For many years the ground-control point (GCP) data were only obtained by registered land surveyors using traditional ground surveying equipment. This situation has changed in two very important ways. First, the use of GPS for GCP data collection is now routine (Figure 6-26a). The availability of survey-grade GPS technology makes it possible to collect accurate GCP information with X,Y,Z root-mean-squared-error (RMSE) of <3 cm when the data are differentially corrected (McCoy, 2005). Remember, however, that GPS is most often used to compute horizontal position. GPS-derived vertical position (*Z*-elevation) is related to the ellipsoid, not the geoid or mean sea level. Therefore, to relate the GPS-derived elevation (ellipsoid height) to conventional elevation (orthometric height), a geoid model is necessary (Wolf and Dewitt, 2000). Second, the GPS-derived GCP data necessary to prepare a block of aerial photography for information extraction can be obtained by non-surveyors (i.e., geographers, foresters, geologists).

Extraction of Digital Elevation Models, Orthoimagery, and Planimetric Features Using Soft-copy Photogrammetry

a. Plan Flightline, Ground Control, and Remote Sensor Data Collection
- collect ground-control data using GPS
- acquire remotely sensed data and digitize if necessary
- assess quality of GCP and remote sensor data

b. Photogrammetric Triangulation and Block Adjustment
- register frames of imagery to ground control using interior and exterior orientation
- weight-constrained least squares bundle adjustment

c. Create and View Registered Stereopairs
- view stereo model using anaglyph or polarized glasses
- extract spot elevations using the *floating mark* if desired

GPS ground-control point data collection

Interior

Exterior

Remote sensor data

Anaglyph stereopair
(shown in black and white)

d. Extract Digital Elevation Model (DEM)
- automated collection of DEM in a user-specified grid
- edit elevation "postings" while draped over 3D view of the terrain
- write DEM to output file

e. Extract Digital Orthoimage
- use DEM to create orthoimage
- create orthoimage mosaic from individual frames
- overlay orthoimage with thematic data to produce orthoimage map

f. Extract Planimetric Features
- view anaglyphic or polarized triangulated stereomodel
- place *floating mark* on feature of interest and outline it
- create contours by keeping the floating mark on the ground at a constant elevation (i.e. constant amount of parallax)

Uncorrected DEM

Corrected DEM

Orthoimage created using corrected DEM

road centerline

building footprint

Figure 6-26 Typical stages in the creation of digital elevation models, orthoimagery, and the extraction of planimetric features using soft-copy photogrammetric techniques.

Collection of Stereoscopic Aerial Photography or Other Remote Sensor Data

High-quality stereoscopic aerial photography are required to produce accurate DEMs, orthophotography and extract planimetric detail (Figure 6-26a). Considerable medium-scale (1:20,000 to 1:40,000) metric aerial photography has been acquired by subcontractors for government agencies (e.g., U.S. Geological Survey National Aerial Photography Program, U.S. Coast and Geodetic Survey, Bureau of Land Management). Commercial photogrammetric engineering firms continue to collect most of the large-scale (>1:20,000) metric aerial photography. They use analog and digital sensor systems, which provide improved geometric and radiometric resolution (e.g., Leica ADS 40). Most remote sensing data collection platforms are now equipped with GPS and inertial navigation systems. Very high spatial resolution DEMs, orthophotographs, and planimetric feature detail can be derived from such imagery. Unfortunately, the photography and derived products are still expensive.

Some satellite remotely sensed data suitable for medium-scale photogrammetric applications are available. For example, it is possible to obtain stereo panchromatic remote sensor data with a nominal spatial resolution of 5 × 5 m from recent SPOT and Indian IRS-1C sensors (refer to Chapter 7). The panchromatic data may be used to derive medium-resolution DEMs and orthoimages, despite the fact that individual houses and small buildings cannot be resolved.

Imagery with a spatial resolution of 0.25 to 2.5 m is required to resolve trailers, houses, small buildings, narrow roads, and drainage networks, so important in many urban-suburban applications (Jensen and Cowen, 1999). Fortunately, several commercial firms now provide high spatial resolution satellite-derived remote sensor data that can be used to provide DEMs, orthoimagery, and some planimetric feature extraction, (refer to Chapter 7):

- GeoEye IKONOS 1 × 1 m panchromatic data;

- OrbImage OrbView-3 1 × 1 m panchromatic data;

- DigitalGlobe QuickBird 61 × 61 cm panchromatic data.

While such digital satellite remote sensor data may never replace the demand for high-quality large-scale aerial photography, there will be many applications where the DEMs and ortho-rectified satellite data are sufficient. Thus, another major stumbling block is being overcome as relatively high spatial resolution satellite remote sensor data becomes available for use in the creation of DEMs and orthoimages.

Image Digitization Technology

Linear and area-array digitization technology based on charge coupled devices (CCDs) or photo-diodes have revolutionized image digitization. Scientists now have access to desktop systems that will digitize black-and-white imagery to 12 bits (values from 0 to 4095) and color imagery to 30 bits (>1 billion colors) at repeatable spatial resolutions approaching <10 μm. Scientists can inexpensively scan historical images at high spatial resolution for photogrammetric projects (Konency, 2003). The importance of image digitization will become less important as more of the data analyzed in soft-copy photogrammetric systems are collected by digital remote sensing systems.

Soft-Copy Photogrammetry

Soft-copy photogrammetric software has made it possible for scientists and laypersons to create DEMs, prepare orthophotographs, extract contours, and map thematic features of interest. This capability is largely due to improvements in the photogrammetric software that performs a) interior orientation, b) exterior orientation, and c) aero-triangulation.

Interior Orientation

Interior orientation is the procedure whereby the geometric characteristics of an aerial photograph are mathematically related to the geometric characteristics (including deformities) of the camera system that took the photograph. This means establishing the relationship between 1) the camera internal coordinate system and 2) the image pixel coordinate system (Linder, 2003). This step requires information about the camera system, which is typically found in the camera calibration report. Most frame cameras have a camera calibration report that was created at the time the camera was produced or recalibrated. Typical information required for interior orientation that is available in the camera calibration report includes:

- x,y location of the principal point (e.g., $x,y = 0,0$; refer to Figure 6-6);

- x,y location of all fiducial marks (mm);

- lens focal length (cm);

• deformation characteristics of the lens.

The analyst obtains this information from the camera calibration report and imports it into the interior orientation program. The image analyst then identifies the *x,y* location of the fiducial marks associated with each photograph produced by the camera. For example, consider Figure 6-27 where an analyst is in the process of performing interior orientation for one of the photographs of Columbia, SC. The analyst is collecting information about fiducial mark #2, located in a corner of the photograph. The coordinates of all four fiducial marks from the camera calibration report are labeled Film X and Film Y in the display. The image coordinates of all four fiducial marks measured by the analyst are labeled Image X and Image Y in the display. The fiducial mark calibration report coordinates are then related to the coordinates of the fiducial marks measured by the image analyst and residuals computed. The relationship between this particular image and the camera calibration information has a root-mean-squared-error (RMSE) of 1.12 pixels (56.76 μm) which is very good. Interior orientation is performed for each photograph in the block of photography. It relates the geometric characteristics of the aerial photograph to the *internal* geometric characteristics of the camera that produced it.

Exterior Orientation

All aerial photographs are tilted somewhat. We need to know how to model this tilt if we are going to extract useful measurements from aerial photography. There are six elements of exterior orientation that express the spatial location and angular orientation of a tilted aerial photograph at the moment of exposure $(X_L, Y_L, Z_L, \omega, \phi, \kappa)$. The three-dimensional coordinates of the aircraft at the moment of exposure are X_L, Y_L, Z_L where Z_L is the altitude of the camera above the local datum. At the instant of exposure the camera might be rolling, pitching, or yawing. These three angles of orientation are omega, phi, and kappa $(\omega\text{-}\phi\text{-}\kappa)$. All the methods developed to determine these six parameters for each aerial photograph require photographic images of at least three ground-control points whose X, Y, Z coordinates are known (Wolf and Dewitt, 2000). If we can determine these parameters for each aerial photograph, we can use the information to relate image coordinates to real-world (*exterior*) map coordinates.

Exterior orientation determines the mathematical relationship between image coordinates (x,y,z) and real-world map coordinates (X, Y, Z) for selected ground-control points. A *ground control point* (*GCP*) is defined as any object in an image for which real-world X, Y, Z ground coordinates are known (Linder, 2003). We need to locate at least three well-distributed GCPs in each image. This means that they should be distributed so that they form a triangle in the image. They should not be located in a straight line.

High-quality *artificial* (or *panel*) GCPs are usually marked in the field using a white cross with each of the four legs of the cross being 5D long and the width of the bars being D wide (e.g., 0.6 m) (Wolf and Dewitt, 2001). The color of an artificial GCP should contrast with the background material (e.g., a white cross on dark asphalt). Most analysts have seen artificial GCPs at selected street intersections in the city. Sometimes GCPs are located at strategic locations in the countryside. The horizontal location (X, Y) and/or Z elevation of the GCPs are obtained using differentially corrected (static or kinematic) GPS measurements and the proper geoid model. The more horizontal/vertical ground control points that can be located throughout a block of aerial photography, the better. The most accurate exterior orientation will take place in areas surrounded by artificial GCPs. Horizontal control points are usually symbolized with a triangle. Vertical control points are symbolized using a circle. Horizontal and vertical control at a point is symbolized using a triangle within a circle.

In addition to artificial horizontal/vertical GCPs, the image analyst can select *pass points* (sometimes referred to as natural points) within the overlap area of a stereopair. Pass points should be clearly visible in each photograph of the stereopair (e.g., the corner of a street, base of a telephone pole, intersection of two fences). The analyst obtains the *x,y* location of these pass points in each image of the stereopair. Pass points can be used to pass control from one photograph to the next in a strip of aerial photography. *Tie points* are pass points located in the 20 percent sidelap area (refer to Figure 6-2) that can be used to pass control from one flight-line strip of aerial photography to another.

Unlike interior orientation, which only works on one photograph at a time, *exterior orientation* requires that the analyst locate the image coordinates of the control points in both images in a stereopair. For example, consider the exterior orientation taking place in Figure 6-28. The analyst is in the process of collecting image coordinates for ground-control point #1, located on the grounds of the South Carolina State capitol. The left image is Photo 4-4, and the right image is Photo 4-5. The analyst has already input the X, Y, Z ground reference coordinates (X Reference, Y Reference, Z Reference).

Aero-triangulation

Figure 6-27 Interior orientation being performed on one of the aerial photographs of Columbia, SC. The analyst is in the process of collecting information about fiducial mark #2. The coordinates of all four fiducial marks from the camera calibration report are labeled Film X and Film Y. The image coordinates of all four fiducial marks are located in the imagery and labeled Image X and Image Y.

Once the interior and exterior orientation are completed, the information can be used to perform *aero-triangulation*. Aero-triangulation is the process whereby the real-world X, Y, and Z ground coordinates can be determined for any pixel in a stereopair based on photo coordinate measurements (Wolf and Dewitt, 2000). This is a powerful capability. It means that the geometric characteristics obtained from the relatively small number of GCPs used during exterior orientation can be extended to every pixel within a digital stereo model and that we can obtain the X, Y location and Z elevation at each of these pixels without having to go out in the field and do formal GPS surveying. Aero-triangulation

allows the relatively small amount of ground control to be passed (or *bridged*) from one photograph to another in a block of aerial photography using image-derived pass and tie points.

The aero-triangulation process creates stereopairs that can be viewed and used to extract digital elevation models, orthophotography, and thematic features of interest (Figure 6-25 and 6-26c). The analyst typically views the triangulated imagery on the screen using anaglyph classes. This produces a three-dimensional model in the observer's mind as previously discussed. Color Plate 6-1 presents an anaglyph ste-

Exterior Orientation:
Stereoscopic Collection of Image Coordinates of Ground Control Points

Figure 6-28   Exterior orientation is being performed on part of the Columbia, SC, stereopair. The analyst is in the process of collecting image coordinates (X File, Y File) for ground-control point #1, located on the grounds of the South Carolina state capitol. The analyst has already input the *X,Y,Z* ground reference coordinates [X Reference, Y Reference, Z Reference (masked from view)]. The image coordinates of all ground-control points must be determined.

reopair of Columbia, SC, produced from the aforementioned interior and exterior orientation and aero-triangulation.

Aero-triangulation uses a weight-constrained least-squares bundle adjustment to properly orient all of the images in a block of aerial photography. This adjustment is based on the use of collinearity equations. *Collinearity* is the condition where the camera exposure station (*L*), an object on the ground (*A*), and its photo image (*a*) all lie on a straight line in three-dimensional space (e.g., Figure 6-29) (Wolf and DeWitt, 2000). The collinearity condition for any point (*a*)

in an aerial photograph can be expressed by two equations: one for the *x* photo coordinate (x_a) and one for the *y* photo coordinate (y_a):

$$x_a = x_o - f\left[\frac{m_{11}(X_A - X_L) + m_{12}(Y_A - Y_L) + m_{13}(Z_A - Z_L)}{m_{31}(X_A - X_L) + m_{32}(Y_A - Y_L) + m_{33}(Z_A - Z_L)}\right] \quad (6\text{-}18)$$

$$y_a = y_o - f\left[\frac{m_{21}(X_A - X_L) + m_{22}(Y_A - Y_L) + m_{23}(Z_A - Z_L)}{m_{31}(X_A - X_L) + m_{32}(Y_A - Y_L) + m_{33}(Z_A - Z_L)}\right] \quad (6\text{-}19)$$

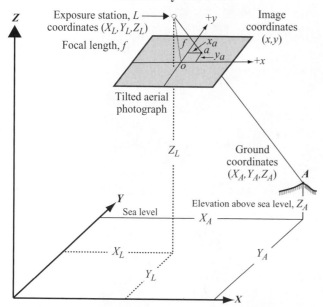

Collinearity Condition

Figure 6-29 Collinearity condition associated with a tilted aerial photograph.

• along lines of equal elevation (creating a contour);

• in a systematic grid creating a digital elevation model.

Elevations extracted by the analyst at very specific locations (e.g., at road intersections, corners of buildings, manhole covers) in the stereo model are commonly referred to as *spot elevations* or *spot heights*. The analyst obtains the X,Y,Z coordinates of a point by placing the *floating mark* (previously discussed) firmly on the ground (not below the surface and not hovering above the surface) and taking a reading.

As previously mentioned, contours are obtained by the analyst specifying the desired elevation, e.g., 100 m ASL. The analyst then places the floating mark on the ground at a location known to be 100 m ASL (based on its parallax properties). She then moves the floating mark along the ground at the same elevation creating a contour. When the contour line is complete, the analyst increments to the next contour interval (e.g., 110 m ASL), places the cursor on the ground at 110 m ASL, and proceeds as before. This is an accurate yet physically demanding method to extract contours.

Using soft-copy photogrammetric systems, ground control, and camera calibration information, it is possible to perform aero-triangulation of the photography (or imagery) and utilize information obtained during the process to generate a lattice (grid) of elevation values within each stereo model of a flightline (Figure 6-26d). The software automatically performs stereo-correlation on orthorectified patches of imagery and computes the parallax (and related elevation) associated with each new point in the stereo model. The stereo-correlation is performed for all points in a user-specified grid. Typical systems can process thousands of points in the stereo model per second (Konecny, 2003). An example of a digital elevation model produced from the Columbia, SC, stereopair is shown in Figure 6-30a. The DEM was processed to appear as a plastic shaded-relief model.

Scientists can resample and mosaic the final DEM to whatever spatial resolution is desired (e.g., 5×5 m, 20×20 m). In this manner the DEM can be brought into geometric congruence with data in a GIS. *Thus, scientists and laypersons have the capability of creating very accurate DEMs on demand for site-specific projects using desktop soft-copy photogrammetry.*

The accuracy of the digital elevation model that can be extracted using photogrammetric methods is a function of the scale and resolution of the remote sensor data, the flying height at which the imagery were acquired, the base/height ratio (i.e., geometry) of the stereoscopic imagery, accuracy

where x_o and y_o are the coordinates of the principle point (PP) from the camera calibration report; X_L, Y_L, and Z_L are the ground coordinates of exposure station L; the ground coordinates of point A are X_A, Y_A, and Z_A; camera focal length is f, and the m's are functions of rotation angles omega, phi, and kappa (ω-ϕ-κ) (Wolf and Dewitt, 2000). The collinearity equations are used in *space resection* to determine the six elements of exterior orientation for both photographs of a stereopair. *Space intersection* can then take place to identify the real-world ground coordinates of any point in the stereomodel. The intricacies of space resection and space intersection based on collinearity are beyond the scope of this introductory book. Please refer to Wolf and DeWitt (2000), Mikhail et al. (2001), or the *Manual of Photogrammetry* (McGlone, 2004) for additional information.

Extracting Spot Elevations, Contours, and DEMs using Soft-Copy Photogrammetry

The ground control incorporated into the exterior orientation and aero-triangulation makes it possible to assign an elevation (actually a certain amount of parallax) to each conjugate (matching) pair of pixels found within a stereopair. Elevation information can be extracted from the triangulated stereo model:

• at very specific locations identified by the operator;

a. Uncorrected DEM.   b. Corrected DEM.

c. DEM with buildings and trees removed.   d. Slope map.

Figure 6-30   a) Uncorrected digital elevation model derived using soft-copy photogrammetric techniques. b) Corrected DEM. c) DEM with buildings and trees removed. d) Slope map produced from DEM with buildings and trees removed.

of the ground control, and the quality of the soft-copy photo-grammetry software used. Users sometimes feel that a DEM derived using soft-copy photogrammetry will be error free. This is not the case.

Tall Structures and Trees Impact the Creation of Photogrammetrically Derived Digital Elevation Models

The automatic stereo correlation used to create DEMs works well when the terrain is devoid of trees, buildings, over-passes, bridges, etc., which extend above the nominal ter-rain. When such objects are present, however, the algorithms assume these objects are terrain and computes the differen-tial parallax and resultant height of such surfaces. The heights are then placed in the DEM. For example, Figure 6-30a depicts a DEM of a four-block region of the University of South Carolina campus derived from 1:6,000-scale pho-tography. The blocky appearance of the DEM is due to

detailed elevation information for each building and tree in the study area.

Methods Used to Edit a DEM Impact Its Accuracy

Soft-copy photogrammetric systems allow users to view the grid of DEM elevation "postings" superimposed on a ste-reopair in three-dimensions. The analyst can edit individual elevation postings by moving them so that they come in con-tact with the ground using the "principle of the floating mark," whereby each posting becomes a floating mark. The analyst may 1) correct individual postings, 2) select a poly-gon of postings and change all of them to the same elevation, or 3) select a polygon of postings along a slope and have them scaled to lie between the highest and lowest points encountered within the polygon. When carefully used, the analyst can correct most problems encountered in the DEM. For example, the DEM in Figure 6-30b was edited so that the top of each building was at the correct elevation. DEMs that include elevation information about buildings and trees may be of use if the analyst desires to drape an orthophoto on top of the DEM and perhaps do a "fly-by" through the city. However, if the analyst wanted a DTM of just the nominal ground terrain in the four-block region, this is certainly not present in Figures 6-30ab.

To create a DEM of the region that does not have building and tree information in it, the analyst must usually manually edit the elevation "postings" in the DEM that correspond with the buildings and trees and effectively drive or push them to the nominal terrain height in the area. This can be difficult if a building or stand of trees is large. However, if the buildings and trees are not too large it is possible to identify the general trend of the terrain between buildings and large trees such that the "postings" of buildings and trees can be moved to the nominal terrain elevation. Careful editing of the original DEM in this manner can produce a revised DEM that depicts just the local relief of the area, without buildings and trees, as shown in Figure 6-30c. A percent slope database of the region (important in many environmental and hydrologic studies) cannot accurately be computed from the DEM with buildings and trees in it. It can be produced from the DEM with buildings and trees removed, as shown in Figure 6-30d.

Urbanized areas with buildings and trees may have to be manually edited to obtain a DEM of just the nominal terrain. This can be a laborious process and is subject to error being introduced by the analyst. The larger the scale of aerial photography and the greater the height of the buildings and structures in the study area, the greater the difficulty of obtaining an accurate DEM of the terrain. DEMs are most accurate and require the least editing when produced for rural areas that do not have significant "leaf-on" tree cover or man-made structures (buildings, bridges, etc.).

For example, consider the high-quality DEM extracted from large-scale aerial photography with few trees or buildings shown in Figure 6-31. These four test pads on the Savannah River Site near Aiken, SC are used to evaluate the effectiveness of various hazardous waste site clay cap materials. Large-scale stereoscopic aerial photography of the test pads takes place routinely so that any subsidence (i.e., depression) in the clay caps can be documented immediately. Figure 6-31a documents the selection of ground-control point #5 during the exterior orientation phase of the project. Control point #5 is a 15-in. plastic pipe extending 30 cm above the ground, which was surveyed to within ±3 cm using GPS. The digital elevation model extracted after aero-triangulated stereo model is shown in Figure 6-31b. This view was created by draping one of the photographs in the stereopair over the DEM and viewing it from an oblique perspective. A color version of this example is found in Color Plate 6-2.

Creation of Digital Orthoimages

More than 30 years ago, Thrower and Jensen (1976) reviewed how analog (hard-copy) orthophotography was created, and identified numerous cartographic applications. They stated:

> Orthophoto-mapping represents a technique by which spatially arrayed data might be both more accurately measured and communicated because of the special attributes of the orthophoto map, namely, the image of an aerial photograph and the metric qualities of a controlled line map.

As predicted, there has been a tremendous increase in the use of digital orthoimagery and digital elevation models in many applications. Lay persons, planners, GIS practitioners, cartographers, and the media (e.g., newspapers, magazines, television) use orthoimages as cartographic backdrops upon which thematic information is overlaid (e.g., property lines, utility lines, drainage networks, contours, troop deployment) (Rumsey and Punt, 2004). *Many image analysts are now using desktop soft-copy photogrammetric systems to generate their own personal DEMs and orthophotographs for input to GIS.* This section describes how orthophotographs are created using soft-copy photogrammetry and important issues that should be considered when creating and using them.

Orthoimages are created from remotely sensed images. The geometry of an unrectified digital photograph is changed from that of a conical bundle of rays to parallel rays that are orthogonal to the ground and to the image plane. Therefore, instead of having a perspective center, the viewing perspective is modeled as being an infinite distance from the ground. Figure 6-32 demonstrates the change in image geometry and the removal of terrain-induced displacement (Δs). The point P that lies at a specific elevation above mean sea level on the ground is found at p in the image plane coordinate system when it should be located at c. The correction of the shift from p to c (i.e., Δs) is the goal of orthorectification. Each part of the terrain is independently corrected during the rectification process.

The effects of topographic relief displacement and camera altitude variations are removed in orthoimagery. The result is a planimetrically correct orthoimage. For example, Figure

**Exterior Orientation: Westinghouse Savannah River Site
Stereoscopic Collection of Image Coordinates of Ground-Control Points**

Point #	Point ID	> Description	Type	Usage	Active	X Reference	Y Reference	Z Reference
1	5	> laymax_pip	Full	Control	X	438912.700	3682631.232	91.489
2	6	laymax_pip	Full	Control	X	438907.990	3682623.710	91.483
3	7	laymax_pip	Full	Control	X	438912.870	3682623.110	91.461
4	8	entonite_to	Full	Check	X	438931.951	3682650.304	88.752
5	9	entonite_pip	Full	Control	X	438945.249	3682631.695	91.193
6	10	entonite_pip	Full	Control	X	438945.250	3682623.192	91.056

Image #	Image Name	Active	X File	Y File
1	srs95_frl684	X	10254.924	6110.706
2	srs95_frl685	X	9202.335	11632.155

a.

Aerial Photography Draped over Photogrammetrically-derived Digital Elevation Model

b.

Figure 6-31   a) Exterior orientation taking place on a stereopair of the test pads on the Westinghouse Savannah River Site near Aiken, SC. The photography was obtained in 1995. b) Display of the photogrammetrically-derived DEM draped with one of the stereopair photographs.

6-33 depicts a power transmission line in rugged terrain. The powerline was surveyed to be very straight in the real world, but the substantial local relief causes relief displacement to take place in the photograph. This causes the powerline to not be in its proper planimetric position in the unrectified aerial photograph. The process of differential bit-by-bit orthographic rectification removes distortions caused by tilt of the camera at the instant of exposure and the effects of relief displacement. The powerline is now in its proper planimetric position, and its shape and geometry are correct.

Figure 6-32   a) A perspective projection aerial photograph displaces the image of objects away from their true planimetric position. For example, P is located at *p* in the photograph when it should be located at *c*. b) An orthographic projection places all images of objects in the correct planimetric position. The goal when creating an orthophotograph is to move objects recorded on an aerial photograph specific distances and directions, $\acute{y}s$, to their proper planimetric position.

Figure 6-33   a) An uncorrected aerial photograph of a power transmission line in rugged terrain. b) Planimetrically accurate orthophotograph after correction to remove roll, pitch, and yaw errors introduced by the aircraft at the instant of exposure and the effects of relief displacement.

The photograph is now an orthophotograph with the accuracy of a traditional line map.

This *planimetric* accuracy allows analysts to use orthoimages like maps for making measurements of geographic location, distances, angles, and area. On unrectified imagery such measurements can only be approximated because of image displacement and scale change caused by variations in local relief (i.e., relief displacement) and airphoto tilt.

Soft-copy photogrammetric instruments can produce terrain-corrected orthoimages on demand. The general procedure was summarized in Figure 6-26a-e.   The user selects any triangulated image in the block of aerial photography (or SPOT data) and its associated DEM to generate a digital orthoimage. During the orthorectification process, the effects of elevation upon the image perspective are removed to produce a geocoded data set with an even pixel spacing in map space. For each orthoimage pixel of known latitude and longitude, the algorithm first uses the DEM to determine the height of the point. The rational functions for the image are then used to determine the pixel in the triangulated image that corresponds to the point in ground space. The intensity

of this point (resampled based on its neighbors using a bilinear interpolation scheme) is then assigned to the output orthoimage pixel. This process typically runs at a rate >200,000 pixels/sec until the rectified output orthoimage is completely filled. The result is a terrain-corrected orthoimage that can be used as a cartographic map.

Problems and Solutions Associated with Orthoimagery Derived Using Soft-copy Photogrammetry

The accuracy of a digital orthoimage is a function of the quality of the imagery, the ground control, the photogrammetric triangulation, and the DEM used to create it (Figure 6-26). An orthoimage may be produced from the original DEM, a DEM with building rooftops cleaned up, or even a DEM with buildings and trees removed. A DEM produced from the collection of field surveying or even digitized contours may also be used to create the orthoimage. Therefore, the analyst should always have access to the metadata (history) of how the DEM was created. In this way only the most appropriate DEM data will be used in the creation of the orthoimagery.

Large-scale (e.g., 1:6,000) urban orthoimages derived from uncorrected DEMs often exhibit severe distortion of building edges (Nale, 1994). For example, the orthoimage in Figure 6-34a was derived using an uncorrected DEM while Figure 6-34b was produced using a DEM with buildings and

a. Orthophoto derived from uncorrected DEM.　　b. Orthophoto derived from corrected DEM.

Figure 6-34　The quality of the DEM influences the quality of the orthophoto.

a. Orthophoto draped over uncorrected DEM.　　b. Orthophoto draped over corrected DEM.

Figure 6-35　The quality of the DEM influences the quality of the orthophoto. The impact of trees is minimized and buildings are shaped correctly, when the orthophoto is draped over a corrected DEM.

tree elevations corrected. The roof edges at "a" and "b" are correct in Figure 6-34b

The importance of the quality of the DEM can be appreciated even more when the orthoimage is draped back onto the DEM used in its creation. For example, consider the two orthophotos in Figure 6-35ab created and draped over uncorrected and corrected DEMs, respectively. The edges of buildings are "smeared" due to the effects of adjacent trees in Figure 6-35a. Conversely, when the rooftop and tree elevations are corrected in the DEM, the resultant orthoimage depicts the same buildings with sharp, distinct edges (Figure 6-35b). This is the type of orthoimagery that should be used in simulated fly-bys through urban environments. More sophisticated soft-copy photogrammetric systems can produce extremely accurate digital elevation models and very

Figure 6-36 A quality orthophotograph of Rosslyn, VA, draped over a carefully edited digital elevation model (courtesy L-H Systems, Inc., and Marconi, Inc.; © SOCET set).

clean orthophotography, such as the three-dimensional view of Rosslyn, VA, in Figure 6-36.

Traditional Ortho-rectification Does Not Eliminate Radial and Relief Displacement of Tall Structures

Most users have the mistaken impression that the tops of tall buildings or extremely tall trees in an orthoimage are in their proper planimetric location, i.e., that the roofs of tall buildings are over their foundations. This is only the case when the sensor acquires the data from a great altitude and the relief displacement of the buildings or other structures caused by radial distortion is minimal. Unfortunately, when using large-scale aerial photography with significant building and tree relief displacement, only the base of the buildings or trees are in their proper planimetric location in an orthoimage. A traditional digital orthophoto does not correct for "building lean" caused by radial distortion. Indeed, the "building lean" of tall buildings or other structures (bridges, overpasses, etc.) may completely obscure the image of the Earth's surface for several hundred feet. If we attempted to digitize buildings from a traditional digital orthoimage, we would discover that the building locations are misplaced on the Earth's surface relative to the building heights (Nale, 1994).

Fortunately, an accurate method of creating *true orthophotos* has been developed. In order to understand how it is derived, it is first useful to review how a traditional orthophoto is produced. The brightness value of pixel *a* is obtained by starting

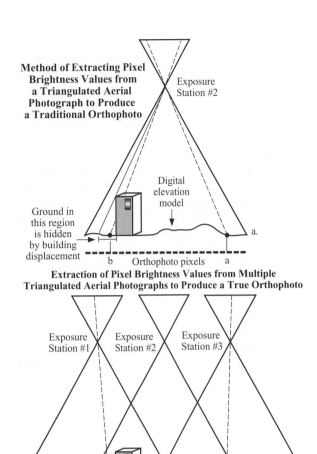

Figure 6-37 a) The brightness value of pixel *b* in a traditional orthophoto would be extracted from the top of the building due to relief displacement. Pixel *a* would have the correct gray shade because it is not obscured from view. Note that both pixels are extracted from a single photo obtained at a single exposure station. b) Pixels *a* and *b* in a true orthophoto are derived from the triangulated photo in the flightline that is most appropriate. For example, pixel *a* is derived from the photo obtained at Exposure Station #3. The tone of *b* is extracted from Exposure Station #1, where it is not obscured by building displacement (Southard, 1994; courtesy of L-H Systems, Inc., and Marconi, Inc.; © SOCET set).

at *a*'s ground x,y position, interpolating the elevation from the DEM, tracing up through the math model to the image, interpolating the proper shade of gray from the image, and assigning the resulting gray shade to pixel *a* in the new orthophoto (Figure 6-37a). The process is acceptable for pixel *a* because there is no obstruction (e.g., building) between the ground x,y,z location at *a* that obscures the view of the ground from the exposure station. The problem that arises is illustrated by pixel *b*. In the original image, the ground "behind" the building at *b* is not depicted because of the height of the building. Pixel *b* will have the tone or color of the roof of the building. The final traditional orthophoto will show the building leaning away from the center of the photograph as in Figure 6-37a. The roof will be shown where there should be ground at the back of the building, and the side of the building will be shown where the roof should be. The ground behind the building will not be shown at all. These displacements are related to the height of the building and the position of the building in the original photo. The greater the height of the building and the closer it is to the side of the original photograph, the worse the displacement will be.

An elegant solution to the above orthophoto problems was developed (Walker, 1994; BAE Systems, 2005). In Figure 6-37b we see three triangulated aerial photographs and a DEM covering the entire footprint of the project area. Using traditional three-dimensional stereoscopic feature extraction tools, the outlines of buildings, bridges, and other obstructions are identified. However, the brightness value or gray tone for pixel *a* is interpolated from the most nadir (directly overhead) Exposure Station (#3 in Figure 6-37b) that has the best view of the ground at location *a*. The algorithm then examines the DEM and feature data and determines that the view of the ground for pixel *b* is obscured by the building at Exposure Station #2 and automatically selects imagery from Exposure Station #1 to obtain the proper pixel color for pixel *b*. The application of these algorithms results in a *true orthophoto* where:

- building rooftops are shown in their correct planimetric *x,y* location;

- the ground on all sides of all buildings is shown in its proper location;

- tops and bottoms of overpasses are shown in their proper locations;

- orthophotos and map sheets can be made that are larger than any of the input images.

A comparison between a traditional orthophotograph and a true orthophoto of the U.S. Capitol is shown in Figure 6-38. Building lean (displacement) has been removed, and all building rooftops are in their proper planimetric location.

a. Conventional orthophoto.

b. True orthophoto.

Figure 6-38   The difference between a conventional orthophotograph (a) and a true orthophoto (b) of the U.S. Capitol. Building lean has been removed, and all building rooftops are in their proper planimetric position (Southard, 1994; courtesy of L-H Systems, Inc., and Marconi, Inc.; © SOCET set).

The ability to create true orthoimages is accelerating the application of orthoimagery in GIS (BAE Systems, 2005).

Extraction of Thematic Features of Interest Using Soft-copy Photogrammetry

Almost all of the detail found in traditional planimetric and/or topographic maps such as the U.S. geological Survey's 7.5-minute (1:24,000-scale) products is extracted using photogrammetric techniques. With the proper ground control, high-quality stereoscopic imagery, and soft-copy photogrammetric software it is now possible for laypersons to extract many types of thematic information. This is a significant event in the history of cartography and remote sensing data analysis as scientists and other researchers are no longer dependent on any agency to provide the information they require. They can generate their own thematic products according to their specifications.

Extraction of Planimetric Features Using Soft-copy Photogrammetry

building footprint

road centerline

building footprint

building footprint

reflecting pond footprint

Figure 6-39   Building footprints, reflecting pond footprint, and a street centerline extracted from the anaglyph stereopair shown in Color Plate 6-3. Please view this stereopair with a stereoscope.

The thematic information extraction procedure is exactly like the one previously described for creating contour lines except that in this case the analyst extracts various thematic features of interest. For example, Figure 6-39 depicts three types of features being extracted from the aero-triangulated stereopair of Columbia, SC, shown in Color Plates 6-1 and 6-3. To extract building footprint information, the analyst systematically places the three-dimensional *floating mark* on the several corners of a building. The software automatically creates a rubber-band line from one building corner to the next. The results is a polygon that contains the planimetric location of the building footprint as if the X, Y coordinates of the *foundation* were measured on the ground. The planimetric location of the reflecting pond is produced in a similar manner.

Emergency response agencies (e.g., 911, police, fire) require detailed street centerline information. This is obtained by keeping the floating mark firmly on the ground as the analyst traces the center of the road while viewing the three-dimensional anaglyph stereopair. Similar procedures are used to map telephone transmission lines and other utilities visible on the surface of the earth. A drainage network map can be created by keeping the floating mark on the ground while tracing the centerline of major streams and their tributaries.

Area Measurement

It is possible to obtain area measurements directly from unrectified vertical aerial photography if the terrain is very level. However, if the local topographic elevation within the field of view of the photograph varies by more than about 5 percent of the flying height above ground level (e.g., 200 m if $H = 4,000$ m), then serious area measurement error could occur. If the local relief varies by more than 5 percent, the image interpreter could carefully stratify the photograph into geographic areas that have approximately the same scale. For example, if a single aerial photograph consisted primarily of the image of a flat plateau at one elevation (e.g., 400 m above sea level) and a gently sloping river valley at another elevation (e.g., 150 m above sea level), the interpreter could stratify the photograph into two parts. He could then compute the exact scale of the plateau portion of the photograph and the exact scale of the river valley portion of the photograph. These two different scales would be used when measuring the area of selected features found in these two different landscapes. The optimum solution is to convert the aerial photography into orthoimagery as previously discussed. The area measurements will then be accurate because the orthophotograph has a uniform scale.

Area Measurement of Well-Known Geometric Shapes

It is a straightforward task to determine the area of well-known geometric shapes such as rectangles, squares, and circles in aerial photography. The analyst first measures the length, width, side, or diameter of the feature on the image. He or she then converts this measurement to a ground distance based on the scale of the photography. The ground distance is then used with the appropriate mathematical relationship to compute the area:

$$\text{Area of a circle} = \pi r^2$$

$$\text{Area of a square} = s^2$$

$$\text{Area of a rectangle} = l \times w$$

where s is the side of a square; l is the length of a rectangle; and w is the width of a rectangle. For example, if the diameter of an agricultural center-pivot irrigation system is 0.5 in. when measured on 1:24,000-scale (1 in. = 2,000 ft) aerial photography, then the ground distance diameter (d) is 1,000 ft and the radius (r) is 500 ft. The center-pivot irrigation system would contain 3.1416 x 500^2 or 785,400 ft^2. An acre contains 43,560 ft^2. A hectare (ha) contains 10,000 m^2 (Table 6-1). One acre = 0.4047 ha; 1 ha = 2.471 acres. Therefore, the center-pivot irrigation system contains 18.03 acres (7.3 ha).

Area Measurement of Irregularly Shaped Polygons

There are several methods for obtaining accurate area measurements from irregularly shaped polygonal features such as lakes, forest stands, and urbanized areas found in vertical aerial photography. The most popular methods include the use of dot grids and on-screen digitization using digital image processing or soft-copy photogrammetric techniques.

Dot Grids

A dot grid is simply a piece of clear acetate that has a user-specified number of dots scribed onto it per sq. in. (or cm^2), e.g., 25 dots per sq. in. (Figure 6-40). The goal is to use a dot grid that has a dot density sufficient to capture the essence of the subject matter you want to measure (e.g., forest cover acreage) yet not so dense that a tremendous number of dots must be counted. It is useful to create a grid that is most suitable for the scale of the photography or map under investigation.   For example, if an analyst was going to use a dot grid

Dot Grid

Figure 6-40   A dot grid consisting of 25 dots per sq. in. overlaid on 1:600-scale (1in. = 50 ft) aerial photography. In this example, the goal is to determine the acres (hectares) of terrain covered by forest.

on 1:24,000-scale aerial photography, it would be wise to make a dot grid with nine dots per in.^2 because at 1:24,000 scale, 1 in.^2 equals approximately 9 acres. If more precision were desired, 18 dots could be systematically placed in each square inch of the grid. The analyst then randomly drops the grid over a polygon of interest and counts the number of dots falling within it. An alternative method of increasing the accuracy of the measurement is, after the first count, to lift up the clear plastic dot grid and randomly drop it again and recount. When the results from the two counts are averaged, the answer will generally be closer to the true value. This can be repeated any number of times. Again, the best results are obtained when the photointerpreted polygons have been transferred onto a base map and the dot grid is overlaid on the map.

The 6×6 in. orthophoto in Figure 6-40 is 1:600 scale (1" = 50 ft). Therefore, it contains 90,000 ft^2 of terrain (2.066 acres or 0.836 ha). It is overlaid with a dot grid consisting of 25 dots per in^2. At this scale, each square inch contains 2,500 ft^2. Each dot therefore represents 100 ft^2. The interpreter analyzes the image and highlights those dots that appear to coincide with the location of forest (they are exaggerated in this example to improve visibility). The analyst simply

On-screen Digitization

Digitized Polygons

a.

b.

Figure 6-41   a) Example of on-screen digitization using a digital image processing system. b) Polygon attribute tables for each of the individual polygons contain information about the perimeter and area.

counts the number of dots representing forest, 336, which equates to 33,600 ft^2 or 0.771 acre (0.316 ha). Forest occupies 37.3 percent of the study area (0.771 / 2.066 acres).

It is obvious from the example that there are some small, dark vegetated forest features that did not fall neatly below a dot and were not counted. Another interpreter might have included the dots closest to these features. Thus, the use of a dot grid involves "method-induced error," which is primarily a function of the methodology (logic) used by the image analyst as he or she allocates the dots. The important thing to remember when using a dot grid is to be consistent. Dot grids are available from forestry-supply companies.

On-screen Digitization

Perhaps the most widely used method of obtaining polygon area estimates from aerial photography or other remote sensor data is to analyze the image using a digital image processing system or GIS (Figure 6-41). While it is easy to scan an unrectified 9 × 9 in. aerial photograph and perform area calculations using the system, it is better if the aerial photography is scanned and then geometrically rectified to a map base. All major digital image processing systems (ERDAS, PCI, ER Mapper, ENVI, etc.) and GIS (ESRI, Autodesk,

etc.) have this fundamental capability. Details on how the image rectification is performed are found in Jensen (2005).

With the aerial photograph rectified to a map base, it is a straightforward task to utilize the system software and the cursor to place a "rubber band" polygon around the area of interest. For example, Figure 6-41a depicts hand-drawn polygons around the forest land cover. The vertices of the polygon can be easily edited. The beauty of this on-screen measurement approach is that each of the individual polygons has a record in a "polygon attribute table" associated with it that summarizes the attribute type (e.g., forest or nonforest), perimeter, and area (in hectares or acres). For example, there are only 14 forest polygons in Figure 6-41b. The attributes for all 14 polygons (e.g., hectares) can be summed by the image processing or GIS software to generate totals.

In summary, it is not wise to digitize polygons directly from aerial photography that have not been geometrically corrected, especially when the terrain has substantial local relief. For example, one should never extract area information by digitizing polygons drawn on unrectified soil photomaps. If the local relief is minimal, some relatively accurate area measurements might be made if the imagery has been geometrically rectified to a standard map projection using techniques described in Jensen (2005). However, it should be remembered that any relief displacement present in the imagery will introduce error into these area measurements. The preferred method to extract area information is to aero-triangulate the aerial photography and then use soft-copy photogrammetry thematic information extraction techniques previously discussed.

References

BAE Systems, 2005, *SOCET Set: Ortho Module*, San Diego: BAE Systems, Inc., www.tailsin.com.

Bossler, J. D., Jensen, J. R., McMaster, R. B. and C. Rizos, 2002, *Manual of Geospatial Science and Technology*, London: Taylor & Francis, 623 p.

Jensen, J. R., 2005, *Introductory Digital Image Processing: A Remote Sensing Perspective*, 3rd Ed., Upper Saddle River: Prentice-Hall, 525 p.

Jensen, J. R. and D. J. Cowen, 1999, "Remote Sensing of Urban/Suburban Infrastructure and Socio-Economic Attributes," *Photogrammetric Engineering & Remote Sensing*, 65(5):611–622.

Konecny, G., 2003, *Geoinformation: Remote Sensing, Photogrammetry and Geographic Information Systems*, London: Taylor & Francis, 248 p.

Linder, W., 2003, *Digital Photogrammetry: Theory and Applications*, Berlin: Springer-Verlag, 189 p.

McCoy, R. M., 2005, *Field Methods in Remote Sensing*, N.Y.: Guilford Press, 159 p.

McGlone, J. C., 2004, *Manual of Photogrammetry*, 5th Ed., Bethesda: ASP&RS, 1151 p.

Mikhail, E. M., Bethel, J. S. and J. C. McGlone, 2001, *Introduction to Modern Photogrammetry*, NY: John Wiley, 479 p.

Nale, D. K., 1994, "Digital Orthophotography: "What It Is and Isn't," *GIS World*, 7(6):22.

Rumsey, D. and E. M. Punt, 2004, *Cartographica Extraordinaire: The Historical Map Transformed*, Redlands: ESRI Press, 147 p.

Southard, G. W., 1994, *True Orthophoto Capabilities of SOCET SET*, Englewood, CO: Leica Helava, Inc., 8 p.

Thrower, N. J. W. and J. R. Jensen, 1976, "The Orthophoto and Orthophotomap: Characteristics, Development and Application," *The American Cartographer*, 3(1):39–56.

Walker, A. S., 1994, Correspondence concerning the SOCET SET, San Diego: Leica AG Photogrammetry and Metrology, Inc., December 20.

Warner, W. S., R. W. Graham and R. E. Read, 1996, *Small Format Aerial Photography*, Scotland: Whittles, 348 p.

Wolf, P. R. and B. A Dewitt, 2000, *Elements of Photogrammetry with Applications in Geographic Information Systems*, NY: McGraw Hill, 608 p.

Multispectral Remote Sensing Systems 7

ultispectral remote sensing systems record reflected or emitted energy from an object or area of interest in multiple bands (regions) of the electromagnetic spectrum. *Hyperspectral remote sensing* systems record data in hundreds of bands. *Ultraspectral remote sensing* systems collect data in thousands of bands. Most multispectral and hyperspectral remote sensing systems collect data in a digital format. This chapter first introduces the fundamental structure of a digital image. The characteristics of historical, current, and proposed multispectral and hyperspectral remote sensing systems are then presented.

 Digital Multispectral Data Collection

Chapter 4 discussed the digitization of hard-copy aerial photography. Digitized color and color-infrared aerial photography may be considered three-band multispectral data. Digitized natural color aerial photography can be converted to blue, green, and red bands of registered digital data. Digitized color-infrared aerial photography can be converted to green, red, and near-infrared bands of digital data. Although these three-band multispectral datasets are sufficient for many applications, there are times when additional spectral bands located at optimum locations in the electromagnetic spectrum can be useful for a specific application. Fortunately, engineers have developed detectors that are sensitive to hundreds of bands in the electromagnetic spectrum.

An overview of how digital remote sensing data are turned into useful information is shown in Figure 7-1. The remote sensing system first detects electromagnetic energy that exits from the phenomena of interest and passes through the atmosphere. The energy detected is recorded as an analog electrical signal, which is usually converted into a digital value through an analog-to-digital (A-to-D) conversion. If an aircraft platform is used, the digital data are simply returned to Earth. If a spacecraft platform is used, the digital data are telemetered to Earth receiving stations directly or indirectly via tracking and data relay satellites (TDRS). In either case, it may be necessary to perform some radiometric and/or geometric preprocessing of the digital remotely sensed data to improve its interpretability. The data may then be enhanced for subsequent human visual analysis or processed further using digital image processing algorithms. Biophysical and/or land-cover information extracted using visual or digital image processing is distributed and used to make decisions. Hopefully, the decisions result in sustainable development (Jensen et al., 2002).

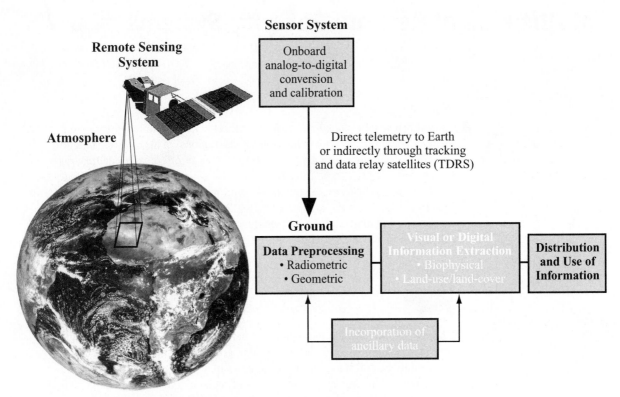

Figure 7-1   An overview of the way digital remotely sensed data are transformed into useful information. The data recorded by the detectors are often converted from an analog electrical signal to a digital value and calibrated. Ground preprocessing removes geometric and radiometric distortions. This may involve the use of ephemeris or ancillary (collateral) data such as map *x,y* coordinates, a digital elevation model, etc. The data are then ready for visual or digital analysis to extract biophysical or land-use/land-cover information. Future sensor systems may conduct preprocessing and information extraction onboard the remote sensing system.

Digital Image Terminology

Digital remote sensor data are usually stored as a matrix (array) of numbers. Each digital value is located at a specific row (i) and column (j) in the matrix (Figure 7-2). A *pixel* is defined as a two-dimensional picture element that is the smallest nondivisible element of a digital image. Each pixel at row (i) and column (j) in the image has an original *brightness value* (*BV*) associated with it. Some scientists call it a *digital number* (*DN*) value. The dataset may consist of n individual bands (k) of multispectral or hyperspectral imagery. Thus, it is possible to identify the brightness value (*BV*) of a particular pixel in the dataset by specifying its row (i), column (j), and band (k) coordinate, i.e., $BV_{i,j,k}$. It is important to understand that the n bands are all geometrically registered to one another. Therefore, a road intersection in band 1 at row 4, column 4 (i.e., $BV_{4,4,1}$) should be located at the same row and column coordinate in the fourth band (i.e., $BV_{4,4,4}$). Hopefully, the brightness values at the two locations

are different; otherwise, the information content of the two images is redundant.

The analog-to-digital conversion that takes place onboard the sensor system usually creates pixels with a brightness value range of 8 to 12 bits. This is called the *quantization* level of the remote sensor data. Remote sensor data quantized to 8 bits have brightness values that range from 0 to 255. Data quantized to 12 bits range from 0 to 1023, etc. The greater the range of possible brightness values, the more precise we may be able to measure the amount of radiance recorded by the detector. One can think of quantization as if it were a ruler. We can obtain more accurate measurements of an object using a ruler that has 1,024 subdivisions than with a ruler that only has 256 subdivisions.

Remote Sensing Systems to be Examined

There are a variety of digital multispectral and hyperspectral remote sensing systems. It is beyond the scope of this book

Figure 7-2 Digital remote sensor data are stored in a matrix format. Picture element (pixel) brightness values (*BV*) are located at row *i*, column *j*, and band *k* in the multispectral or hyperspectral dataset. The digital remote sensor brightness values are normally stored as 8-bit bytes with values ranging from 0 to 255. However, several image digitization systems and some remote sensing systems now routinely collect 10-, 11-, or 12-bit data.

to provide information on all of them. However, it is possible to review selected remote sensing systems that are or will be of significant value for Earth resource investigations. They are organized according to the type of remote sensing technology used, as summarized in Figure 7-3, including:

Multispectral Imaging Using Discrete Detectors and Scanning Mirrors

- Landsat Multispectral Scanner (MSS)

- Landsat Thematic Mapper (TM)

- Landsat 7 Enhanced Thematic Mapper Plus (ETM$^+$)

- NOAA Geostationary Operational Environmental Satellite (GOES)

- NOAA Advanced Very High Resolution Radiometer (AVHRR)

- NASA and ORBIMAGE, Inc., Sea-viewing Wide Field-of-view Sensor (SeaWiFS)

- Daedalus, Inc., Aircraft Multispectral Scanner (AMS)

- NASA Airborne Terrestrial Applications Sensor (ATLAS)

Multispectral Imaging Using Linear Arrays

- SPOT 1, 2, and 3 High Resolution Visible (HRV) sensors and SPOT 4 and 5 High Resolution Visible Infrared (HRVIR) and *Vegetation* sensor

- Indian Remote Sensing System (IRS) Linear Imaging Self-scanning Sensor (LISS-III and LISS-IV)

- NASA *Terra* Advanced Spaceborne Thermal Emission and Reflection Radiometer (ASTER)

- NASA *Terra* Multiangle Imaging Spectroradiometer (MISR)

- DigitalGlobe, Inc. (QuickBird)

- Space Imaging, Inc. (IKONOS)

- ImageSat International, Inc. (EROS A1)

- ORBIMAGE, Inc. (OrbView-3 and OrbView-5)

- Leica Geosystems, Inc. Airborne Digital Sensor System (ADS-40)

Remote Sensing Systems Used to Collect Multispectral and Hyperspectral Imagery

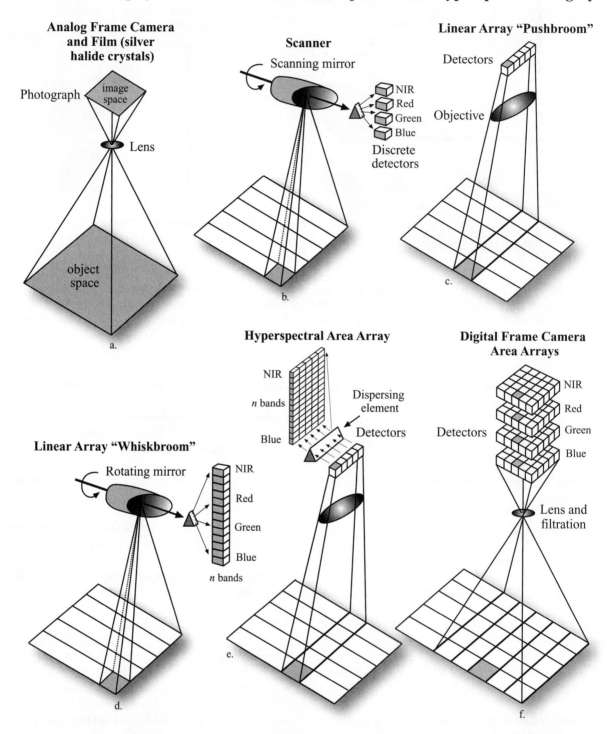

Figure 7-3   Six types of remote sensing systems used for multispectral and hyperspectral data collection: a) traditional aerial photography, b) multispectral imaging using a scanning mirror and discrete detectors, c) multispectral imaging with linear arrays (often referred to as "pushbroom" technology), d) imaging with a scanning mirror and linear arrays (often referred to as "whiskbroom" technology), e) imaging spectrometry using linear and area arrays, and f) digital frame camera aerial photography based on area arrays.

Imaging Spectrometry Using Linear and Area Arrays

- NASA Jet Propulsion Laboratory Airborne Visible/Infrared Imaging Spectrometer (AVIRIS)

- Compact Airborne Spectrographic Imager 1500 (CASI 1500) and hyperspectral SWIR sensor (SASI 600)

- NASA *Terra* Moderate Resolution Imaging Spectrometer (MODIS)

- NASA Earth Observer (EO-1) Advanced Land Imager (ALI), Hyperion, and LEISA Atmospheric Corrector (LAC)

Digital Frame Cameras Based on Area Arrays

- Leica Geosystems Emerge Digital Sensor System

- Vexcel UltraCam Large Format Camera

- Z/1 Digital Modular Camera

Astronaut Photographic Systems

- NASA Space Shuttle and International Space Station Imagery

The following discussion identifies the spatial, spectral, temporal, and radiometric characteristics of the remote sensing systems.

 Multispectral Imaging Using Discrete Detectors and Scanning Mirrors

The collection of multispectral remote sensor data using discrete detectors and scanning mirrors began in the mid-1960s. Despite the technology's age, several new remote sensing systems still use it.

Earth Resource Technology Satellites and Landsat Sensor Systems

In 1967, the National Aeronautics & Space Administration (NASA), encouraged by the U.S. Department of the Interior, initiated the Earth Resource Technology Satellite (ERTS) program. This program resulted in the launch of seven satellites carrying a variety of remote sensing systems designed primarily to acquire Earth resource information. The most

noteworthy sensors have been several Landsat Multispectral Scanners and Landsat Thematic Mappers (Figure 7-4; Tables 7-1 and 7-2). The Landsat program is the United States' oldest land-surface observation satellite system, having obtained data since 1972. It has had a tumultuous history of management and funding sources.

The chronological launch and retirement history of the satellites is shown in Figure 7-4. The ERTS-1 satellite, launched on July 23, 1972, was an experimental system designed to test the feasibility of collecting Earth resource data by unmanned satellites. Prior to the launch of ERTS-B on January 22, 1975, NASA renamed the ERTS program *Landsat*, distinguishing it from the *Seasat* active-microwave satellite launched on June 26, 1978. At this time, ERTS-1 was retroactively named Landsat-1 and ERTS-B became Landsat-2 at launch. Landsat-3 was launched March 5, 1978; Landsat-4 on July 16, 1982; and Landsat-5 on March 1, 1984.

The Earth Observation Satellite Company (EOSAT) obtained control of the Landsat satellites in September, 1985. Unfortunately, Landsat 6, with its Enhanced Thematic Mapper (ETM) (a 15×15 m panchromatic band was added), failed to achieve orbit on October 5, 1993. Landsat 7, with its Enhanced Thematic Mapper Plus (ETM$^+$) sensor system, was launched on April 15, 1999. Please refer to the NASA Landsat 7 home page for a detailed history of the Landsat program (NASA Landsat 7, 2006).

Landsats 1 through 3 were launched into Sun-synchronous polar orbits at a nominal altitude of 919 km (570 mi). The platform is shown in Figure 7-5a. The satellites had an orbital inclination of 99°, which made them nearly polar (Figure 7-5b) and caused them to cross the equator at an angle of approximately 9° from normal. The satellites orbited Earth once every 103 minutes, resulting in 14 orbits per day (Figure 7-5c). During each polar orbit, the satellites crossed the equator at approximately the same local time (9:30 to 10:00 a.m.) on the illuminated side of Earth.

Figures 7-5c and 7-6 illustrate how repeat coverage of a geographic area was obtained. From one orbit to the next, a position directly below the spacecraft moved 2,875 km (1,785 mi) at the Equator as the Earth rotated beneath it. The next day, 14 orbits later, it was back to its original location, with orbit 15 displaced westward from orbit 1 by 159 km (99 mi) at the equator. This continued for 18 days, after which orbit 252 fell directly over orbit 1 once again. Thus, the Landsat sensor systems had the capability of observing the entire globe (except poleward of 81°) once every 18 days, or about 20 times a year. There were approximately 26 km (16 mi) of sidelap between successive orbits. This sidelap was a

Chronological Launch and Retirement History of the Landsat Satellites

Figure 7-4   Chronological launch and retirement history of the Landsat series of satellites (1 through 7) from 1972 to 2004.

maximum at 81° north and south latitudes (about 85%) and a minimum at the equator (about 14%). The sidelap has proven useful for stereoscopic analysis applications.

The nature of the orbiting Landsat system has given rise to a Path and Row *Worldwide Reference System* (*WRS*) for locating and obtaining Landsat imagery for any area on Earth. The WRS has catalogued the world's landmass into 57,784 scenes. Each scene is approximately 185 km wide by 170 km long.

An elegant method of determining if remote sensor data (e.g., Landsat MSS, Thematic Mapper, Enhanced Thematic Mapper Plus) are available for a specific location is to use the U.S. Geological Survey's Global Visualization Viewer (USGS GloVis, 2006). For example, suppose we are interested in locating a Landsat Thematic Mapper image of Charleston, SC. We can enter the Global Visualization

Viewer and specify WRS Path 16 and Row 37 and search the database as shown in Figure 7-7. If we do not know the path and row designation, we could a) move the cursor on the regional map and place it on Charleston, SC, or b) input the latitude and longitude coordinates of Charleston, SC (33.2° N, -81° W). We can specify the month and year (e.g., May, 2005) and the amount of acceptable cloud cover. We can also specify whether the search should be conducted regionally as in Figure 7-7a (pixels resampled to 1000 m) or locally as in Figure 7-7b (pixels resampled to 240 m). A color version of the Global Visualization Viewer interface is shown in Color Plate 7-1.

In this section, we are interested in the type of sensors carried aloft by the Landsat satellites and the nature and quality of remote sensor data provided for Earth resource investigations. The most important sensors are the Multispectral Scanner and several Thematic Mappers.

Table 7-1.   Landsat Multispectral Scanner (MSS) and Landsat Thematic Mapper (TM) sensor system characteristics.

	Landsat Multispectral Scanner (MSS)			Landsat 4 and 5 Thematic Mapper (TM)		
Band	Spectral Resolution (μm)	Radiometric Sensitivity (NEΔP)[a]		Band	Spectral Resolution (μm)	Radiometric Sensitivity (NEΔP)
4[b]	0.5 – 0.6	0.57		1	0.45 – 0.52	0.8
5	0.6 – 0.7	0.57		2	0.52 – 0.60	0.5
6	0.7 – 0.8	0.65		3	0.63 – 0.69	0.5
7	0.8 – 1.1	0.70		4	0.76 – 0.90	0.5
8[c]	10.4 – 12.6	1.4K (NEΔT)		5	1.55 – 1.75	1.0
				6	10.40–12.5	0.5 (NEΔT)
				7	2.08–2.35	2.4
IFOV at nadir	79 × 79 m for bands 4 through 7 240 × 240 m for band 8			30 × 30 m for bands 1 through 5, 7 120 × 120 m for band 6		
Data rate	15 Mb/s			85 Mb/s		
Quantization levels	6 bit (values from 0 to 63)			8 bit (values from 0 to 255)		
Earth coverage	18 days Landsat 1, 2, 3 16 days Landsat 4, 5			16 days Landsat 4, 5		
Altitude	919 km			705 km		
Swath width	185 km			185 km		
Inclination	99°			98.2°		

[a] The radiometric sensitivities are the noise-equivalent reflectance differences for the reflective channels expressed as percentages (NEΔP) and temperature differences for the thermal infrared bands (NEΔT).

[b] MSS bands 4, 5, 6, and 7 were renumbered bands 1, 2, 3, and 4 on Landsats 4 and 5.

[c] MSS band 8 was present only on Landsat 3.

Landsat Multispectral Scanner

The Landsat Multispectral Scanner (MSS) was placed on Landsat satellites 1 through 5. The MSS multiple-detector array and the scanning system are shown diagrammatically in Figure 7-8a. Sensors such as the Landsat MSS (and Thematic Mapper to be discussed) are optical-mechanical systems in which a mirror scans the terrain perpendicular to the flight direction. While it scans, it focuses energy reflected or emitted from the terrain onto discrete detector elements. The detectors convert the radiant flux measured within each instantaneous field of view (IFOV) in the scene into an electronic signal (Figures 7-3b and 7-8a). The detector elements

are placed behind filters that pass broad portions of the spectrum. The MSS had four sets of filters and detectors, whereas the TM had seven. The primary limitation of this approach is the short viewing residence time of the detector in each IFOV. To achieve adequate signal-to-noise ratio without sacrificing spatial resolution, such a sensor must operate in broad spectral bands of ≥ 100 nm or must use optics with unrealistically small ratios of focal length to aperture (f / stop).

The MSS scanning mirror oscillates through an angular displacement of ±5.78° off-nadir. This 11.56° field-of-view

Figure 7-5   a) Nimbus-style platform used for Landsats 1, 2, and 3 and associated sensor and telecommunication systems. b) Inclination of the Landsat orbit to maintain a Sun-synchronous orbit. c) From one orbit to the next, the position directly below the satellite moved 2,875 km (1,785 mi) at the equator as Earth rotated beneath it. The next day, 14 orbits later, it was approximately back to its original location, with orbit 15 displaced westward from orbit 1 by 159 km (99 mi). This is how repeat coverage of the same geographic area was obtained.

resulted in a swath width of 185 km (115 mi) for each orbit. Six parallel detectors sensitive to four spectral bands (channels) in the electromagnetic spectrum viewed the ground simultaneously: 0.5 to 0.6 µm (green), 0.6 to 0.7 µm (red), 0.7 to 0.8 µm (reflective near-infrared), and 0.8 to 1.1 µm (reflective near-infrared). These bands were originally num-

bered 4, 5, 6, and 7, respectively, because a Return-Beam-Vidicon (RBV) sensor system also onboard the satellite recorded energy in three bands labeled 1, 2, and 3.

When not viewing the Earth, the MSS detectors were exposed to internal light and Sun calibration sources. The

Landsat Orbital Tracks

Figure 7-6   Orbital tracks of Landsat 1, 2, or 3 during a single day of coverage. The satellite crossed the equator every 103 minutes, during which time the Earth rotated a distance of 2,875 km under the satellite at the equator. Every 14 orbits, 24 hours elapsed.

spectral sensitivity of the bands is summarized in Table 7-1 and shown diagrammatically in Figure 7-8b. Note that there is spectral overlap between the bands.

Prior to the launch of the satellite, the engineering model of the ERTS MSS was tested by viewing the scene behind the Santa Barbara Research Center in Goleta, CA. Bands 4 and 6 (green and near-infrared, respectively) of the area are shown in Figure 7-9. Note the spatial detail present when the sensor is located only 1 to 2 km from the mountains. The spatial resolution is much lower when the sensor is placed 919 km above Earth in orbit.

The IFOV of each detector was square and resulted in a ground resolution element of approximately 79 × 79 m (67,143 ft^2). The analog voltage signal from each detector was converted to a digital value using an onboard A-to-D converter. The data were quantized to 6-bits with a range of values from 0 to 63. These data were then rescaled to 7-bits (0 to 127) for three of the four bands in subsequent ground processing (i.e., bands 4, 5, and 6 were decompressed to a range of 0 to 127). It is important to remember that the early 1970s Landsat MSS data were quantized to 6-bits when comparing MSS data collected in the late 1970s and 1980s, which were collected at 8-bits.

During each scan, the voltage produced by each detector was sampled every 9.95 μs. For one detector, approximately

3300 samples were taken along a 185-km line. Thus, the IFOV of 79 m × 79 m became about 56 m on the ground between each sample (Figure 7-10). The 56 × 79 m area is called a Landsat MSS picture element. Thus, although the measurement of landscape brightness was made from a 6,241 m^2 area, each pixel was reformatted as if the measurement were made from a 4,424 m^2 area (Figure 7-10). Note the overlap of the areas from which brightness measurements were made for adjacent pixels.

The MSS scanned each line across-track from west to east as the southward orbit of the spacecraft provided the along-track progression. Each MSS scene represents a 185 × 170 km parallelogram extracted from the continuous swath of an orbit and contains approximately 10 percent overlap. A typical scene contains approximately 2340 scan lines with about 3240 pixels per line, or about 7,581,600 pixels per channel. All four bands represent a data set of more than 30 million brightness values. Landsat MSS images provided an unprecedented ability to observe large geographic areas while viewing a single image. For example, approximately 5000 vertical aerial photographs obtained at a scale of 1:15,000 are required to equal the geographic coverage of a single Landsat MSS image. This allows regional terrain analysis to be performed using one data source rather than a multitude of aerial photographs.

Use of the USGS *Global Visualization Viewer* to Locate Landsat Remote Sensor Data

a. Search for images of Charleston, SC, using the criteria: Path 16 and Row 37, Landsat 4-5 TM, May 2005, 100% maximum cloud cover; resampled to 1,000 m pixels.

b. Search for Charleston, SC using pixels resampled to 240 m.

Figure 7-7   Use of the USGS *Global Visualization Viewer* to locate Landsat Thematic Mapper imagery of Charleston, SC.

Figure 7-8 a) Major components of the Landsat Multispectral Scanner system on Landsats 1 through 5 (Landsat 3 also had a thermal infrared band). A bank of 24 detectors (six for each of the four bands) measured reflected energy from Earth from an instantaneous field of view of 79 m × 79 m. b) Landsat Multispectral Scanner bandwidths. Notice that the bands do not end abruptly, as suggested by the usual nomenclature, e.g., band 4 (0.5 – 0.6 µm).

Landsat Thematic Mapper

Landsat Thematic Mapper (TM) sensor systems were launched on July 16, 1982 (Landsat 4), and on March 1, 1984 (Landsat 5). The TM is an optical-mechanical whisk-broom sensor that records energy in the visible, reflective-infrared, middle-infrared, and thermal infrared regions of the electromagnetic spectrum. It collects multispectral imagery that has higher spatial, spectral, temporal, and radiometric resolution than the Landsat MSS.

The Landsat 4 and 5 platform is shown in Figure 7-11. The Thematic Mapper sensor system configuration is shown in Figure 7-12. A telescope directs the incoming radiant flux obtained along a scan line through a scan line corrector to 1) the visible and near-infrared primary focal plane, or 2) the

middle-infrared and thermal infrared cooled focal plane. The detectors for the visible and near-infrared bands (1 – 4) are four staggered linear arrays, each containing 16 silicon detectors. The two middle-infrared detectors are 16 indium-antimonide cells in a staggered linear array, and the thermal infrared detector is a four-element array of mercury-cadmium-telluride cells.

Landsat TM data have a ground-projected IFOV of 30 × 30 m for bands 1 through 5 and 7. The thermal infrared band 6 has a spatial resolution of 120 × 120 m. The TM spectral bands represent important departures from the bands found on the traditional MSS, also carried onboard Landsats 4 and 5. The original MSS bandwidths were selected based on their utility for vegetation inventories and geologic studies. Conversely, the TM bands were chosen after years of analy-

Terrestrial Images of Goleta, CA, obtained using the Landsat Multispectral Scanner

a. Landsat MSS band 4 (0.5 - 0.6 μm).

b. Landsat MSS band 6 (0.7 - 0.8 μm).

Figure 7-9   Two terrestrial images acquired by the engineering model of the Landsat MSS on March 4, 1972, at the Santa Barbara Research Center of Hughes Aircraft, Inc. The top image (a) was acquired using the MSS band 4 detectors (0.5 – 0.6 μm), and the bottom image (b) was acquired using band 6 detectors (0.7 – 0.8 μm). Note the high spatial fidelity of the images, which is possible when the terrain is close and not 919 km away.

sis for their value in water penetration, discrimination of vegetation type and vigor, plant and soil moisture measurement, differentiation of clouds, snow, and ice, and identification of hydrothermal alteration in certain rock types (Table 7-2). The refined bandwidths and improved spatial resolution of the Landsat TM versus the Landsat MSS and several other sensor systems (Landsat 7 and SPOTs 1 – 5) are shown graphically in Figure 7-13. Examples of individual bands of Landsat Thematic Mapper imagery of Charleston, SC, obtained in 1994 are shown in Figure 7-14.

The Landsat TM bands were selected to make maximum use of the dominant factors controlling leaf reflectance, such as leaf pigmentation, leaf and canopy structure, and moisture content, as demonstrated in Figure 7-15. Vegetation absorbs much of the incident blue, green, and red radiant flux for

Ground-projected IFOV

6241 m²

4424 m²

Sampling interval of MSS data

Figure 7-10   Relationship between the original 79 × 79 m ground-projected IFOV of the Landsat MSS and the rate at which it was resampled (i.e., every 9.95 μs). This resulted in pixels that were 56 × 79 m on data obtained from the EROS Data Center at Sioux Falls, SD.

Landsat 4 and 5 Thematic Mappers

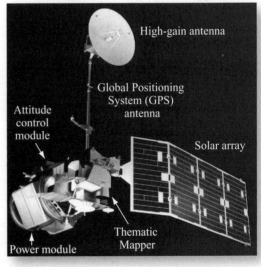

High-gain antenna

Global Positioning System (GPS) antenna

Attitude control module

Solar array

Power module

Thematic Mapper

Figure 7-11   Landsat 4 and 5 platform and associated sensor and telecommunication systems.

photosynthetic purposes; therefore, vegetated areas appear dark in TM band 1 (blue), 2 (green), and 3 (red) images, as seen in the Charleston, SC, Landsat TM data (see Figure 7-14). Vegetation reflects approximately half of the incident near-infrared radiant flux, causing it to appear bright in the band 4 (near-infrared) image. Bands 5 and 7 both provide more detail in the wetland because they are sensitive to soil

Table 7-2.   Characteristics of the Landsat 4 and 5 Thematic
              Mapper spectral bands.

Band 1: 0.45 – 0.52 μm (blue). This band provides increased penetration of waterbodies, as well as supporting analyses of land-use, soil, and vegetation characteristics. The shorter wavelength cutoff is just below the peak transmittance of clear water, and the upper-wavelength cutoff is the limit of blue chlorophyll absorption for healthy green vegetation. Wavelengths < 0.45 μm are substantially influenced by atmospheric scattering and absorption.

Band 2: 0.52 – 0.60 μm (green). This band spans the region between the blue and red chlorophyll absorption bands and reacts to the green reflectance of healthy vegetation.

Band 3: 0.63 – 0.69 μm (red). This is the red chlorophyll absorption band of healthy green vegetation and is useful for vegetation discrimination. It is also useful for soil-boundary and geological-boundary delineations. This band may exhibit more contrast than bands 1 and 2 because of the reduced effect of atmospheric attenuation. The 0.69-μm cutoff is significant because it represents the beginning of a spectral region from 0.68 to 0.75 μm, where vegetation reflectance crossovers take place that can reduce the accuracy of vegetation investigations.

Band 4: 0.76 – 0.90 μm (near-infrared). For reasons discussed, the lower cutoff for this band was placed above 0.75 μm. This band is very responsive to the amount of vegetation biomass and/or leaf area present. It is useful for crop identification and emphasizes soil/crop and land/water contrasts.

Band 5: 1.55 – 1.75 μm (mid-infrared). This band is sensitive to the turgidity or amount of water in plants. Such information is useful in crop drought studies and in plant vigor investigations. This is one of the few bands that can be used to discriminate among clouds, snow, and ice.

Band 6: 10.4 – 12.5 μm (thermal infrared). This band measures the amount of infrared radiant energy emitted from surfaces. The apparent temperature is a function of the emissivities and the true (kinetic) temperature of the surface. It is useful for locating geothermal activity, thermal inertia mapping for geologic investigations, vegetation classification, vegetation stress analysis, and soil moisture studies. The band often captures unique information on differences in topographic aspect in mountainous areas.

Band 7: 2.08 – 2.35 μm (mid-infrared). This is an important band for the discrimination of geologic rock formations. It has been shown to be effective for identifying zones of hydrothermal alteration in rocks.

and plant moisture conditions (Table 7-2). The band 6 (thermal) image provides limited information of value.

The equatorial crossing time was 9:45 a.m. for Landsats 4 and 5 with an orbital inclination of 98.2°. Landsats 4 and 5 were placed in a 705 km orbit. The lower orbit also increased the amount of relief displacement introduced into the imagery obtained over mountainous terrain. The new orbit also caused the period between repetitive coverage to change from 18 to 16 days for both the MSS and TM data collected by Landsats 4 and 5.

There was a substantial improvement in the level of quantization from 6 to 8 bits per pixel (Table 7-1). This, in addition to a greater number of bands and a higher spatial resolution, increased the data rate from 15 to 85 Mb/s. Ground receiving stations were modified to process the increased data flow. Based on the improvements in spectral, spatial, and radiometric resolution, Solomonson (1984) suggested that "it appears that the TM can be described as being twice as effective in providing information as the Landsat MSS. This is based on its ability to provide twice as many separable classes over a given area as the MSS, numerically provide two more independent vectors in the data or demonstrate through classical information theory that twice as much information exists in the TM data."

Efforts to move the Landsat Program into the commercial sector began under the Carter Administration in 1979 and resulted in legislation passed in 1984 that charged the National Oceanic and Atmospheric Administration to transfer the program to the private sector. The Earth Observing Satellite Company (EOSAT) took over operation in 1985 and was given the rights to market Landsat TM data. This was a very difficult time for satellite-based remote sensing in the United States with each Landsat TM scene costing approximately $4,500.

It is important to point out the amazing longevity of the Landsat 5 Thematic Mapper system launched March 1, 1984. Even after Landsat 6 failed to achieve orbit on October 5, 1993, Landsat 5 TM continued to provide high-quality remote sensor data. Data collection was suspended for a time on November 26, 2005, due to satellite platform backup solar array problems (USGS Landsat 5, 2006).

Landsat 7 Enhanced Thematic Mapper Plus

On October 28, 1992, President Clinton signed the Land Remote Sensing Policy Act of 1992 (Public Law 102-555). The law authorized the procurement of Landsat 7 and called for its launch within 5 years of the launch of Landsat 6. In

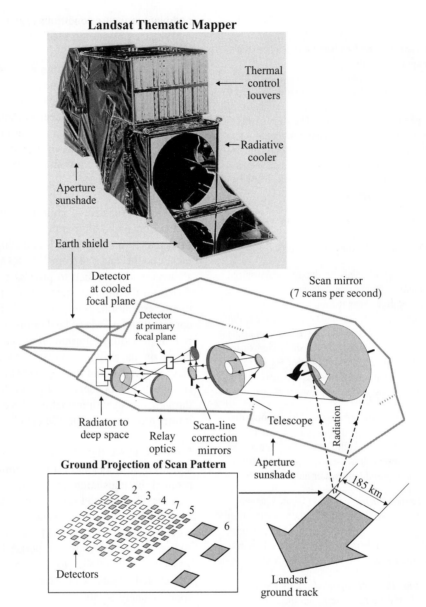

Figure 7-12   Major components of the Landsats 4 and 5 Thematic Mapper sensor system. The sensor is sensitive to the seven bands of the electromagnetic spectrum summarized in Table 7-1. Six of the seven bands have a spatial resolution of 30 × 30 m; the thermal infrared band has a spatial resolution of 120 × 120 m. The lower diagram depicts the sensor in its operational position.

parallel actions, Congress funded Landsat 7 procurement and stipulated that data from publicly funded remote sensing satellite systems like Landsat must be sold to United States government agencies and their affiliated users at the cost of fulfilling user requests.

With the passage of the Land Remote Sensing Policy Act of 1992, oversight of the Landsat program began to be shifted from the commercial sector back to the federal government. NASA was responsible for the design, development, launch and on-orbit checkout of Landsat 7 (Figure 7-16), and the

installation and operation of the ground system. The U.S. Geological Survey (USGS) was responsible for data capture, processing, and distribution of the Landsat 7 data, mission management, and maintaining the Landsat 7 data archive.

Landsat 7 was launched on April 15, 1999, from Vandenburg Air Force Base, CA, into a Sun-synchronous orbit (Figure 7-16). Landsat 7 was designed to work in harmony with NASA's EOS *Terra* and *Aqua* satellites. It was designed to achieve three main objectives (NASA Landsat 7, 2006):

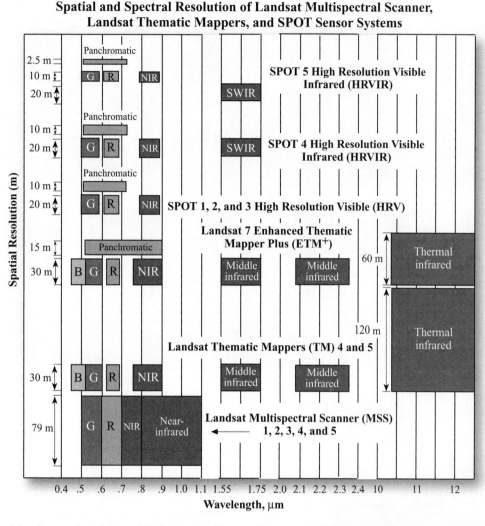

Figure 7-13 Spatial and spectral resolution of the Landsat Multispectral Scanner (MSS), Landsats 4 and 5 Thematic Mapper (TM), Landsat 7 Enhanced Thematic Mapper Plus (ETM+), SPOTs 1, 2, and 3 High Resolution Visible (HRV), and SPOTs 4 and 5 High Resolution Visible Infrared (HRVIR) sensor systems. The SPOTs 4 and 5 *Vegetation* sensor characteristics are not shown (it consists of four 1.15×1.15 km bands).

- maintain data continuity by providing data that are consistent in terms of geometry, spatial resolution, calibration, coverage characteristics, and spectral characteristics with previous Landsat data;

- generate and periodically refresh a global archive of substantially cloud-free, sunlit landmass imagery;

- continue to make Landsat-type data available to U.S. and international users at the cost of fulfilling user requests and to expand the use of such data for global-change research and commercial purposes.

Landsat 7 is a three-axis stabilized platform carrying a single nadir-pointing instrument, the ETM+ (Figure 7-16). The ETM+ instrument is a derivative of the Landsat 4 and 5 Thematic Mapper sensors. Therefore, it is possible to refer to Figure 7-12 for a review of its mirror and detector design. The ETM+ is based on scanning technology (Figure 7-3b) despite the fact that linear array "pushbroom" technology has been commercially available since the launch of the French SPOT 1 satellite in 1986. Nevertheless, the ETM+ instrument was an excellent sensor with several notable improvements over its predecessors Landsat 4 and 5.

The characteristics of the Landsat 7 ETM+ are found in Table 7-3. The ETM+ bands 1 through 5 and 7 are identical

Landsat 5 Thematic Mapper Data of Charleston, SC

a. Band 1.

b. Band 2.

c. Band 3.

d. Band 4.

e. Band 5.

f. Band 6 (thermal infrared).

g. Band 7.

Figure 7-14   Landsat Thematic Mapper data of Charleston, SC, obtained on February 3, 1994. Bands 1 through 5 and 7 are 30×30 m spatial resolution. Band 6 is 120×120 m.

to those found on Landsats 4 and 5 and have the same 30×30 m spatial resolution. The thermal infrared band 6 has 60×60 m spatial resolution (instead of 120×120 m). Perhaps most notable is the new 15×15 m panchromatic band ($0.52 - 0.90$ μm). Landsat 7 ETM+ images of San Diego, CA, are shown in Figure 7-17. An ETM+ color composite of San Diego, CA, is displayed in Color Plate 7-2.

Landsat 7 is in orbit 705 km above the Earth, collects data in a swath 185 km wide, and cannot view off-nadir. Its revisit interval is 16 days. It has a 378-gigabit solid-state recorder that can hold 42 minutes of sensor data and 29 hours of housekeeping telemetry data. The ETM+ records 150 megabits of data each second. Landsat 7 can transmit data directly to ground receiving stations at the EROS Data Center in Sioux Falls, SD, or to Fairbanks, AK. Landsat 7 international data may by acquired by transmission using TDRS satellites or by international receiving stations.

The Landsat 7 ETM+ has excellent radiometric calibration, which is accomplished using partial and full aperture solar calibration. Ground look calibration is performed by acquiring images of certain Earth landmass calibration targets. Biophysical and atmospheric characteristics of these targets are well instrumented on the ground.

At one time approximately 250 ETM+ images were processed by the EROS Data Center each day. Unfortunately, the ETM+ Scan Line Corrector (SLC) failed on May 31, 2003, resulting in imagery with significant missing data. The

Figure 7-15   Progressive changes in percent reflectance for a sycamore leaf at varying oven-dry-weight moisture content. The dominant factors controlling leaf reflectance and the location of six of the Landsat Thematic Mapper bands are superimposed.

Figure 7-16   Landsat 7 satellite with its Enhanced Thematic Mapper Plus (ETM⁺) sensor system (courtesy of Lockheed Martin, Inc.).

SLC compensates for the forward motion of the satellite. Efforts to recover the SLC were not successful.

A USGS/NASA Landsat team has refined ETM⁺ gap-filling techniques that merge data from multiple ETM⁺ acquisitions. They also developed modifications to the Landsat-7 acquisition scheme to acquire two or more clear scenes as near in time as possible to facilitate this gap-filling process.

These merged images resolve some of the missing data problems (USGS Landsat 7, 2006).

Landsat Data Continuity

A memorandum from the President's Office of Science and Technology Policy (OSTP) signed on December 23, 2005, summarizes the future of Landsat (OSTP, 2005). The

Landsat 7 Enhanced Thematic Mapper Plus Imagery of San Diego, CA

a. Band 1 (0.45 - 0.515 µm).   b. Band 2 (0.525 - 0.605 µm).   c. Band 3 (0.63 - 0.69 µm).

d. Band 4 (0.75 - 0.90 µm).   e. Band 5 (1.55 - 1.75 µm).   f. Band 6 (10.4 - 12.5 µm) thermal.

g. Band 7 (2.08 - 2.35 µm).   h. Band 8 (0.52 - 0.90 µm) pan.   i. Band 8 enlarged.

Figure 7-17   Landsat 7 ETM⁺ imagery of San Diego, CA, obtained on April 24, 2000. Bands 1 through 5 and 7 are 30 × 30 m. Thermal infrared band 6 is 60 × 60 m. Panchromatic band 8 is 15 × 15 m (courtesy of NASA).

Departments of Commerce, Defense, the Interior and NASA are to take the following near-term actions:

• Proceed with the NPOESS program without incorporating a Landsat-type instrument;

• NASA will acquire a single Landsat data continuity mission in the form of a free-flyer spacecraft to collect the required land surface data and deliver its data to the Department of the Interior U.S. Geological Survey (USGS);

• DOI, through the USGS, will be responsible for the operations of the Landsat data continuity mission and for the col-

lection, archiving, processing, and distribution of the land surface data to U.S. Government and other users;

• The agencies will seek to implement an approach for this mission in a manner that does not preclude a long-term solution for continuity of Landsat-type data.

It remains the goal of the U.S. Government to transition the Landsat program from a series of independently planned missions to a sustained operational program funded and managed by a U.S. Government operational agency or agencies, international consortium, and/or commercial partnership. Concurrent with the actions cited above, the National

Table 7-3.   Landsat Enhanced Thematic Mapper Plus (ETM$^+$) compared with the Earth Observer (EO-1) sensors.

Landsat 7 Enhanced Thematic Mapper Plus (ETM+)			**EO-1 Advanced Land Imager (ALI)**		
Band	**Spectral Resolution (μm)**	**Spatial Resolution (m) at Nadir**	**Band**	**Spectral Resolution (μm)**	**Spatial Resolution (m) at Nadir**
1	0.450 – 0.515	30 × 30	MS-1	0.433 – 0.453	30 × 30
2	0.525 – 0.605	30 × 30	MS-1	0.450 – 0.510	30 × 30
3	0.630 – 0.690	30 × 30	MS-2	0.525 – 0.605	30 × 30
4	0.750 – 0.900	30 × 30	MS-3	0.630 – 0.690	30 × 30
5	1.55 – 1.75	30 × 30	MS-4	0.775 – 0.805	30 × 30
6	10.40 – 12.50	60 × 60	MS-4'	0.845 – 0.890	30 × 30
7	2.08 – 2.35	30 × 30	MS-5'	1.20 – 1.30	30 × 30
8 (panchromatic)	0.52 – 0.90	15 × 15	MS-5	1.55 – 1.75	30 × 30
			MS-7	2.08 – 2.35	30 × 30
			Panchromatic	0.480 – 0.690	10 × 10
			EO-1 Hyperion Hyperspectral Sensor 220 bands from 0.4 to 2.4 μm at 30 × 30 m		
			EO-1 LEISA Atmospheric Corrector (LAC) 256 bands from 0.9 to 1.6 μm at 250 × 250 m		
Sensor Technology	Scanning mirror spectrometer		Advanced Land Imager is a pushbroom radiometer. Hyperion is a pushbroom spectroradiometer. LAC uses area arrays.		
Swath Width	185 km		ALI = 37 km; Hyperion = 7.5 km; LAC = 185 km		
Data Rate	250 images per day @ 31,450 km^2		—		
Revisit	16 days		16 days		
Orbit and Inclination	705 km, Sun-synchronous Inclination = 98.2° Equatorial crossing 10:00 a.m. ±15 min.		705 km, Sun-synchronous Inclination = 98.2° Equatorial crossing = Landsat 7 + 1 min		
Launch	April 15, 1999		November 21, 2000		

Science and Technology Council, in coordination with NASA, DOI/USGS, and other agencies and EOP offices as appropriate, will lead an effort to develop a long-term plan to achieve technical, financial, and managerial stability for operational land imaging in accord with the goals and objectives of the U.S. Integrated Earth Observation System (OSTP, 2005).

Earth Observer (EO-1)

The Earth Observer (EO-1) mission was launched on November 21, 2000 (NASA EO-1, 2006). It was placed in a 705-km Sun-synchronous orbit at a 98.7° inclination such that it was flying in formation 1 minute behind Landsat 7 in the same ground track. This close separation enabled EO-1 to observe the same ground location (scene) through the same atmospheric region so that scene comparisons between the two satellites could be made.

Table 7-4.   NOAA Geostationary Operational Environmental Satellite (GOES I-M) Imager sensor system characteristics.

GOES-8, 9 10, 12 Band	Spectral Resolution (μm)	Spatial Resolution (km)	Band Utility
1	0.52 – 0.72	1 × 1	Clouds, pollution, haze detection, and identification of severe storms
2	3.78 – 4.03	4 × 4	Fog detection, discriminates between water, clouds, snow or ice clouds during daytime, detects fires and volcanoes, nighttime sea-surface temperature (SST)
3	6.47 – 7.02	8 × 8	Estimation of mid- and upper-level water vapor, detects advection, and tracks mid-level atmospheric motion
4	10.2 – 11.2	4 × 4	Cloud-drift winds, severe storms, cloud-top heights, heavy rainfall
5	11.5 – 12.5	4 × 4	Identification of low-level water vapor, SST, and dust and volcanic ash

EO-1 specifications are summarized in Table 7-3. It contains a linear array Advanced Land Imager (ALI) with 10 bands from 0.4 to 2.35 μm at 30 × 30 m spatial resolution. The Hyperion hyperspectral sensor records data in 220 bands from 0.4 to 2.4 μm at 30 × 30 m spatial resolution. The Linear Etalon Imaging Spectrometer Array (LEISA) Atmospheric Corrector is a 256-band hyperspectral instrument sensitive to the region from 0.9 to 1.6 μm at 250 × 250 m spatial resolution. It is designed to correct for water-vapor variations in the atmosphere. All three of the EO-1 land imaging instruments view all or subsegments of the Landsat 7 swath.

NPOESS Preparatory Project (NPP)

NASA's Goddard Space Flight Center selected Ball Aerospace to build a spacecraft for the NPOESS Preparatory Project (NPP), the precursor to the NPOESS mission. NPP is designed to function as a bridge between the NASA EOS program and NPOESS for the development of the following sensors: Visible/Infrared Imager Radiometer Suite (VIIRS), Cross-track Infrared Sounder, Advanced Technology Microwave Sounder, ATMS Photos (NOAA NPOESS, 2006).

National Atmospheric and Oceanic Administration Multispectral Scanner Sensors

NOAA operates two series of remote sensing satellites: the Geostationary Operational Environmental Satellites (GOES) and the Polar-orbiting Operational Environmental Satellites (POES). Both are currently based on multispectral scanner technology. The U.S. National Weather Service uses data from these sensors to forecast the weather. We often see GOES images of North and South America weather patterns on the daily news. The Advanced Very High Resolution

Radiometer was developed for meteorological purposes. However, global climate change research has focused attention on the use of AVHRR data to map vegetation and sea-surface characteristics.

Geostationary Operational Environmental Satellite (GOES)

GOES is operated by the NOAA National Environmental Satellite Data and Information Service (NESDIS). NOAA's most recent generation of geostationary satellites began with GOES-8, which was launched in April, 1994. GOES-9 was launched on May 23, 1995. GOES-10 West was launched April 25, 1997, and became operational July 7, 1998. GOES-12 East was launched July 23, 2001, and became operational on April 1, 2003 (NOAA GOES, 2006).

The GOES system consists of several observing subsystems:

• GOES Imager (provides multispectral image data);

• GOES Sounder (provides hourly 19-channel soundings);

• a data-collection system (DCS) that relays data from *in situ* sites at or near the Earth's surface to other locations.

The GOES spacecraft is a three-axis (x,y,z) stabilized design capable of continuously pointing the optical line of sight of the imaging and sounding radiometers toward the Earth. GOES are placed in geostationary orbits approximately 35,790 km (22,240 statute miles) above the equator. The satellites remain at a stationary point above the equator and rotate at the same speed and direction as Earth. This enables the sensors to stare at a portion of the Earth from the geosynchronous orbit and thus more frequently obtain images of clouds, monitor the Earth's surface temperature and water

Figure 7-18   a) Geographic coverage of GOES East (75° W) and GOES West (135° W). b) Radiant flux from the terrain is reflected off a scanning mirror (not shown) onto the primary and secondary mirrors. A dichroic beamsplitter separates the visible light from the thermal infrared energy. Subsequent beamsplitters separate the thermal energy into specific bands (after Loral Space Systems, 1996).

vapor characteristics, and sound the Earth's atmosphere for its vertical thermal and water vapor structures.

GOES-12 East is normally situated at 75° W longitude and GOES-10 West is at 135° W longitude. The geographic coverage of GOES East and GOES West is summarized in Figure 7-18a. These sensors view most of the Earth from approximately 20° W to 165° E longitude. Poleward coverage is between approximately 77° N and S latitude. GOES East and West view the contiguous 48 states, South America, and major portions of the central and eastern Pacific Ocean and the central and western Atlantic Ocean. Pacific coverage includes the Hawaiian Islands and Gulf of Alaska, the latter known to weather forecasters as "the birthplace of North American weather systems" (NOAA GOES, 2006).

GOES Imager: The Imager is a five-channel multispectral scanner. The bandwidths and spatial resolution are summarized in Table 7-4. By means of a two-axis gimballed mirror in conjunction with a 31.1-cm (12.2-in.) diameter telescope, the Imager's multispectral channels can simultaneously sweep an 8-km (5-statute mile) north-to-south swath along an east-to-west/west-to-east path, at a rate of 20° (optical) per second. The telescope concentrates both the visible and thermal radiant flux from the terrain onto a 5.3-cm secondary mirror (Figure 7-18b). Dichroic beamsplitters separate the incoming scene radiance and focus it onto 22 detectors (8 visible and 14 thermal). The visible energy passes through the initial beamsplitter and is focused onto 8 silicon visible

detector elements. Each of the 8 visible detectors has an IFOV of approximately 1×1 km at the satellite's suborbital point on the Earth.

Thermal infrared energy is routed to the specialized detectors in the radiative cooler. The thermal infrared energy is further separated into the 3.9-, 6.75-, 10.7-, and 12-μm channels. Each of the four infrared channels has a separate set of detectors: four-element indium-antimonide (InSb) detectors for band 2; two-element mercury-cadmium-telluride (Hg:Cd:Te) detectors for band 3; and four-element mercury-cadmium-telluride (Hg:Cd:Te) detectors for both bands 4 and 5.

The GOES channels have 10-bit radiometric precision. The primary utility of the visible band 1 (1×1 km) is in the daytime monitoring of thunderstorms, frontal systems, and tropical cyclones. Band 2 (4×4 km) responds to both emitted terrestrial radiation and reflected solar radiation. It is useful for identifying fog and discriminating between water and ice clouds, and between snow and clouds, and for identifying large or very intense fires. It can be used at night to track low-level clouds and monitor near-surface wind circulation. Band 3 (8×8 km) responds to mid- and upper-level water vapor and clouds. It is useful for identifying the jet stream, upper-level wind fields, and thunderstorms. Energy recorded by band 4 (4×4 km) is not absorbed to any significant degree by atmospheric gases. It is ideal for measuring cloud-top heights, identifying cloud-top features, assessing the

Geostationary Operational Environmental Satellite Imagery

a. GOES-8 East visible August 25, 1998.

b. GOES-8 East thermal infrared August 25, 1998.

c. GOES-8 East visible August 25, 1998.

d. GOES-8 East thermal infrared August 25, 1998.

e. GOES-8 East water vapor August 25, 1998.

f. GOES-8 satellite.

Figure 7-19   a–e) Examples of GOES-8 imagery obtained on August 25, 1998. f) The GOES-8 satellite (images courtesy of NOAA).

severity of some thunderstorms, and tracking clouds and frontal systems at night. Thermal band 5 (4 × 4 km) is similar to band 4 except that this wavelength region has a unique sensitivity to low-level water vapor. GOES-8 East visible, thermal infrared, and water vapor images of Hurricane Bonnie on August 25, 1998, are shown in Figure 7-19a-e.

The Imager scans the continental United States every 15 min; scans most of the hemisphere from near the North Pole to approximately 20° S latitude every 30 min; and scans the entire hemisphere once every 3 hr in "routine" scheduling mode. Optionally, special imaging schedules are available, which allow data collection at more rapid time intervals over

Table 7-5.   NOAA Advanced Very High Resolution Radiometer (AVHRR) sensor system characteristics.

Band	NOAA-6, 8, 10 Spectral Resolution (μm)[a]	NOAA-7, 9, 11, 12, 13, 14 Spectral Resolution (μm)[a]	NOAA-15 to 18 AVHRR/3 Spectral Resolution (μm)[a]	Band Utility
1	0.580 – 0.68	0.580 – 0.68	0.580 – 0.68	Daytime cloud, snow, ice, and vegetation mapping; used to compute NDVI
2	0.725 – 1.10	0.725 – 1.10	0.725 – 1.10	Land/water interface, snow, ice, and vegetation mapping; used to compute NDVI
3	3.55 – 3.93	3.55 – 3.93	3A: 1.58 - 1.64 3B: 3.55 – 3.93	Snow/ice discrimination Day/night cloud and surface-temperature mapping
4	10.50 – 11.50	10.30 – 11.30	10.30 – 11.30	Day/night cloud and surface-temperature mapping
5	None	11.50 – 12.50	11.50 – 12.50	Cloud and surface temperature, day-and-night cloud mapping; removal of atmospheric water vapor path radiance

IFOV at nadir	1.1×1.1 km
Swath width	2700 km at nadir

[a] TIROS-N was launched on October 13, 1978; NOAA-6 on June 27, 1979; NOAA-7 on June 23, 1981; NOAA-8 on March 28, 1983; NOAA-9 on December 12, 1984; NOAA-10 on September 17, 1986; NOAA-11 on September 24, 1988; NOAA-12 on May 14, 1991; NOAA-13 on August 9, 1993; NOAA-14 on December 30, 1994; NOAA (K)-15 on May 13, 1998; NOAA (L)-16 on September 21, 2000; NOAA (M)-17 on June 24, 2002; NOAA (N)-18 on May 20, 2005.

reduced geographic areas. During Rapid Scan Operations (RSO) and Super Rapid Scan Operations (SRSO), images are collected over increasingly reduced-area sectors at 7.5-min intervals (RSO) and at either 1-min or 30-sec intervals (SRSO). A typical image acquired at 30-second intervals covers a rectangle of about 10° of latitude and 15° of longitude. The 1-min SRSO collects 22 images per hour with 2 segments of 1-min interval images, allowing for regularly scheduled 15-min interval operational scans.

GOES Sounder: The GOES Sounder uses 1 visible and 18 infrared sounding channels to record data in a north-to-south swath across an east-to-west path. The Sounder and Imager both provide full Earth imagery, sector imagery, and area scans of local regions. The 19 bands yield the prime sounding products of vertical atmospheric temperature profiles, vertical moisture profiles, atmospheric layer mean temperature, layer mean moisture, total precipitable water, and the lifted index (a measure of stability). These products are used to augment data from the Imager to provide information on atmospheric temperature and moisture profiles, surface and cloud-top temperatures, and the distribution of atmospheric ozone (Loral Space Systems, 1996).

Advanced Very High Resolution Radiometer (AVHRR)

The Satellite Services Branch of the National Climatic Data Center, under the auspices of NESDIS, has established a digital archive of data collected from the NOAA Polar-orbiting Operational Environmental Satellites (POES) (Kidwell, 1998). This series of satellites commenced with TIROS-N (launched in October 1978) and continued with NOAA-A (launched in March, 1983 and renamed NOAA-8). NOAA-18 was launched May 20, 2005. These Sun-synchronous polar-orbiting satellites carry the Advanced Very High Resolution Radiometer (AVHRR) (Table 7-5). Substantial progress has been made in using AVHRR data for land-cover characterization and the mapping of daytime and nighttime clouds, snow, ice, and surface temperature. Unlike the Landsat TM and Landsat 7 ETM$^+$ sensor systems with nadir revisit cycles of 16 days, the AVHRR sensors acquire images of the entire Earth two times each day (NOAA AVHRR, 2006). This high frequency of coverage enhances the likelihood that cloud-free observations can be obtained for specific temporal windows and makes it possible to monitor change in land-cover conditions over short periods, such as a growing season. Moreover, the moderate resolution (1.1 × 1.1 km) of the AVHRR data makes it feasible to collect,

NOAA 17 Overpasses on October 17, 2003

Figure 7-20   Overpasses of the NOAA-17 satellite on October 17, 2003. The AVHRR is onboard NOAA-17 (courtesy of NOAA).

store, and process continental or global datasets. For these reasons, NASA and NOAA initiated the AVHRR Pathfinder Program to create universally available global long-term remotely sensed datasets that can be used to study global climate change.

The AVHRR satellites orbit at approximately 833 km above Earth at an inclination of 98.9° and continuously record data in a swath 2700 km wide at 1.1×1.1 km spatial resolution at nadir. Normally, two NOAA-series satellites are operational at one time (one odd, one even). The odd-numbered satellite typically crosses the equator at approximately 2:30 p.m. and 2:30 a.m., and the even-numbered satellite crosses the equator at 7:30 p.m. and 7:30 a.m. local time. Each satellite orbits Earth 14.1 times daily (every 102 min) and acquires complete global coverage every 24 hr. Orbital tracks of the NOAA-17 satellite on October 17, 2003, are shown in Figure 7-20. Because the number of orbits per day is not an integer, the suborbital tracks do not repeat on a daily basis, although the local solar time of the satellite's passage is essentially unchanged for any latitude. However, the satellite's orbital drift over time causes a systematic change of illumination conditions and local time of observation, which is a source of nonuniformity when analyzing multidate AVHRR data.

The AVHRR is a cross-track scanning system. The scanning rate of the AVHRR is 360 scans per minute. A total of 2,048 samples (pixels) are obtained per channel per Earth scan, which spans an angle of $\pm 55.4°$ off-nadir. The IFOV of each band is approximately 1.4 milliradians leading to a resolu-

tion at the satellite subpoint of 1.1×1.1 km (Figure 7-21a). The more recent AVHRR systems have five channels (Table 7-5; Figure 7-21b).

Full resolution AVHRR data obtained at 1.1×1.1 km is called *local area coverage* (*LAC*) data. It may be resampled to 4×4 km *global area coverage* (*GAC*) data. The GAC data contains only one out of three original AVHRR lines and the data volume and resolution are further reduced by starting with the third sample along the scan line, averaging the next four samples, and skipping the next sample. The sequence of average four, skip one is continued to the end of the scan line. Some studies use GAC data while others use the full-resolution LAC data.

The AVHRR provides regional information on vegetation condition and sea-surface temperature. For example, a portion of an AVHRR image of the South Carolina coast obtained on May 13, 1993, at 3:00 p.m. is shown in Figure 7-22. Band 1 is approximately equivalent to Landsat TM band 3. Vegetated land appears in dark tones due to chlorophyll absorption of red light. Band 2 is approximately equivalent to TM band 4. Vegetation reflects much of the near-infrared radiant flux, yielding bright tones, while water absorbs much of the incident energy. The land–water interface is usually quite distinct. The three thermal bands provide information about Earth's surface and water temperature. The gray scale is inverted for the thermal infrared data with cold, high clouds in black and warm land and water in lighter tones. This particular image captured a large lobe of warm Gulf Stream water.

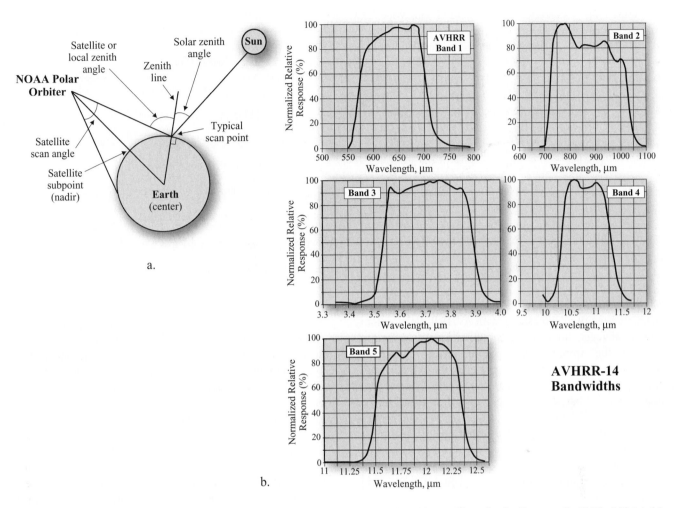

Figure 7-21   a) Relationships among the Earth, Sun, and the NOAA Polar Orbiter. The satellite subpoint lies at nadir. b) The NOAA-14 AVHRR bandwidths for bands 1 through 5.

AVHRR data are also routinely used to inventory sea surface temperature (SST). Color Plate 7-3 is a sea-surface temperature map derived from NOAA-16 AVHRR imagery obtained on October 16, 2003 (Gasparovic, 2003).

Scientists often compute a normalized difference vegetation index (NDVI) from the AVHRR data using the visible (AVHRR$_1$) and near-infrared (AVHRR$_2$) bands to map the condition of vegetation on a regional and national level. It is a simple transformation based on the following ratio:

$$NDVI = \frac{\rho_{nir} - \rho_{red}}{\rho_{nir} + \rho_{red}} = \frac{AVHRR_2 - AVHRR_1}{AVHRR_2 + AVHRR_1}. \quad (7\text{-}1)$$

The NDVI equation produces values in the range of –1.0 to 1.0, where increasing positive values indicate increasing green vegetation, and negative values indicate nonvegetated surfaces such as water, barren land, ice, and snow or clouds. To obtain the most precision, the NDVI is derived from calibrated, atmospherically corrected AVHRR channel 1 and 2 data in 16-bit precision, prior to geometric registration and sampling. The final NDVI results from –1 to 1 are normally scaled from 0 to 200. Vegetation indices are discussed in detail in Chapter 10.

NDVI data obtained from multiple dates of AVHRR data can be composited to provide summary seasonal information. The *n*-day NDVI composite is produced by examining each NDVI value pixel by pixel for each observation during the compositing period to determine the maximum value. The retention of the highest NDVI value reduces the number of cloud-contaminated pixels.

Table 7-6. Characteristics of the Sea-viewing Wide Field-of-view Sensor (SeaWiFS).

SeaWiFS Band	Band Center (nm)	Bandwidth (nm)	Band Utility
1	412	402 – 422	Identify yellow substances
2	443	433 – 453	Chlorophyll concentration
3	490	480 – 500	Increased sensitivity to chlorophyll concentration
4	510	500 – 520	Chlorophyll concentration
5	555	545 – 565	Gelbstoffe (yellow substance)
6	670	660 – 680	Chlorophyll concentration
7	765	745 – 785	Surface vegetation, land–water interface, atmospheric correction
8	865	845 – 885	Surface vegetation, land–water interface, atmospheric correction

The NDVI and other vegetation indexes (refer to Chapter 10) have been used extensively with AVHRR data to monitor natural vegetation and crop condition, identify deforestation in the tropics, and monitor areas undergoing desertification and drought. For example, the U.S. Geological Survey developed the Global Land Cover Characterization dataset based primarily on the unsupervised classification of 1-km AVHRR 10-day NDVI composites. The AVHRR source imagery dates from April 1992 through March 1993. Ancillary data sources include digital elevation data, ecoregions interpretation, and country- or regional-level vegetation and land-cover maps (USGS Global Land-cover, 2006).

NOAA Global Vegetation Index products based on AVHRR data are summarized in NOAA GVI (2006): 1st-generation (May 1982–April, 1985); 2nd-generation (April 1985–present), and 3rd-generation new products (April 1985–present). A color-composite AVHRR image of the conterminous United States is shown in Color Plate 7-4a. The average Normalized Difference Vegetation Index (NDVI) for much of Canada and the United States in August derived from AVHRR data from 1981 through 2000 is displayed in Color Plate 7-4b.

ORBIMAGE, Inc., and NASA Sea-viewing Wide Field-of-view Sensor

Oceans cover more than two-thirds of the Earth's surface and play an important role in the global climate system. The SeaWiFS is an advanced scanning system designed specifically for ocean monitoring. The *SeaStar* satellite (OrbView-2), developed by ORBIMAGE, Inc., in conjunction with NASA, carried the SeaWiFS into orbit using a Pegasus rocket on August 1, 1997 (NASA/Orbimage SeaWiFS, 2006). The Pegasus rocket was flown aloft by a Lockheed L-1011 and released at an altitude of about 39,000 ft, whereupon the rocket was ignited and the spacecraft was lifted into orbit. The final orbit was 705 km above the Earth. The equatorial crossing time is 12 p.m.

SeaWiFS builds on what was learned about ocean remote sensing using the Nimbus-7 satellite Coastal Zone Color Scanner (CZCS) launched in 1978. CZCS ceased operation in 1986. The SeaWiFS instrument consists of an optical scanner with a 58.3° total field of view. Incoming scene radiation is collected by a telescope and reflected onto the rotating half-angle mirror. The radiation is then relayed to dichroic beamsplitters that separate the radiation into eight wavelength intervals (Table 7-6). SeaWiFS has a spatial resolution of 1.13×1.13 km (at nadir) over a swath of 2800 km. It has a revisit time of 1 day.

SeaWiFS records energy in eight spectral bands with very narrow wavelength ranges (Table 7-6) tailored for the detection and monitoring of very specific ocean phenomena, including ocean primary production and phytoplankton processes, ocean influences on climate processes (heat storage and aerosol formation), and the cycles of carbon, sulfur, and nitrogen. In particular, SeaWiFS has specially designed bands centered at 412 nm (to identify yellow substances through their blue wavelength absorption), at 490 nm (sensitive to chlorophyll concentration), and in the 765 and 865 nm near-infrared (to assist in the removal of atmospheric attenuation).

a. AVHRR band 1.

b. AVHRR band 2.

c. AVHRR band 3.

d. AVHRR band 4.

**NOAA-11 AVHRR Imagery
of the South Carolina Coast
Obtained on May 13, 1993**

e. AVHRR band 5.

Figure 7-22   Portion of a NOAA-11 AVHRR image of the South Carolina coast obtained on May 13, 1993, at 3:00 p.m. (refer to Table 7-5 for band specifications). Vegetation appears dark in band 1 due to chlorophyll absorption of red light. Vegetation appears bright in band 2 because it reflects much of the incident near-infrared radiant flux. Water absorbs much of the incident energy; therefore, the land–water interface is usually distinct. Three thermal bands (3, 4, and 5) provide surface-temperature information. The grayscale is inverted with cold, high clouds in black and warm land and water in lighter tones. A large lobe of warm Gulf Stream water is easily identified (images courtesy of NOAA).

SeaWiFS observations help scientists understand the dynamics of ocean and coastal currents, the physics of mixing, and the relationships between ocean physics and large-scale patterns of productivity. The data fill the gaps in ocean biological observations between those of the test-bed CZCS and MODIS. SeaWiFS also provides laser water penetration depth imagery for naval operations.

Figure 7-23 Characteristics of the Daedalus airborne multispectral scanner (AMS) and associated electronics that are carried onboard the aircraft during data collection. The diameter of the circular ground area viewed by the sensor, D, is a function of the instantaneous field of view, β, of the scanner and the altitude above ground level of the aircraft, H, at the time of data collection. Radiant flux from Earth's surface is passed from the optical system onto a dichroic grate, which sends the various wavelengths of light to detectors that are continuously monitored by the sensor system electronics.

Aircraft Multispectral Scanners

Orbital sensors such as the Landsat MSS, TM, and ETM⁺ collect data on a repetitive cycle and at set spatial and spectral resolutions. Often it is necessary to acquire remotely sensed data at times that do not coincide with the scheduled satellite overpasses and at perhaps different spatial and spectral resolutions. Rapid collection and analysis of high-resolution remotely sensed data may be required for specific studies and locations. When such conditions occur or when a sensor configuration different from the Landsat or SPOT sensors is needed, agencies and companies often use a multispectral scanner placed onboard an aircraft to acquire remotely sensed data. There are several commercial and

publicly available MSS that can be flown onboard aircraft, including the Daedalus Airborne Multispectral Scanner (AMS) and the NASA Airborne Terrestrial Applications Sensor (ATLAS).

Daedalus DS-1260, DS-1268, and Airborne Multispectral Scanner

Numerous remote sensing laboratories and government agencies in many countries purchased Daedalus DS-1260, DS-1268, or the Airborne Multispectral Scanner over the last 40 years (e.g., SenSyTech, 2004). These expensive sensor systems have provided much of the useful high spatial and spectral resolution multispectral scanner data (including thermal infrared) for monitoring the environment. For exam-

Table 7-7.   System characteristics of the Daedalus Airborne Multispectral Scanner (AMS) and NASA Airborne Terrestrial Applications Sensor (ATLAS).

		Airborne Multispectral Scanner (AMS)		NASA Airborne Terrestrial Applications Sensor (ATLAS)		
	Band	Spectral Resolution (μm)	Spatial Resolution (m)	Band	Spectral Resolution (μm)	Spatial Resolution (m)
	1	0.42 – 0.45	Variable,	1	0.45 – 0.52	2.5 to 25 m
	2	0.45 – 0.52	depending	2	0.52 – 0.60	depending
	3	0.52 – 0.60	upon altitude	3	0.60 – 0.63	upon altitude
	4	0.60 – 0.63	above ground	4	0.63 – 0.69	above ground
	5	0.63 – 0.69	level	5	0.69 – 0.76	level
	6	0.69 – 0.75		6	0.76 – 0.90	
	7	0.76 – 0.90		7	1.55 – 1.75	
	8	0.91 – 1.05		8	2.08 – 2.35	
	9	3.00 – 5.50		9	Removed	
	10	8.50 – 12.5		10	8.20 – 8.60	
				11	8.60 – 9.00	
				12	9.00 – 9.40	
				13	9.60 – 10.20	
				14	10.20 – 11.20	
				15	11.20 – 12.20	
IFOV	2.5 mrad			2.0 mrad		
Quantization	8 – 12 bits			8 bits		
Altitude	variable			variable		
Swath width	714 pixels			800 pixels		

ple, the AMS has detectors that span the region from the ultraviolet through near-infrared (0.42 – 1.05 μm), a hot-target thermal infrared detector (3.0 – 5.5 μm), and a standard thermal infrared detector (8.5 – 12.5 μm). Table 7-7 summarizes the characteristics of the AMS sensor system.

The basic principles of operation and components of the Airborne Multispectral Scanner (AMS) are shown in Figure 7-23. Radiant flux reflected or emitted from the terrain is collected by the scanning optical system and projected onto a dichroic grating. The grating separates the reflected radiant flux from the emitted radiant flux. Energy in the reflective part of the spectrum is directed from the grating to a prism (or refraction grating) that further separates the energy into specific bands. At the same time, the emitted thermal incident energy is separated from the reflective incident energy. The electromagnetic energy is then focused onto a bank of discrete detectors situated behind the grating and the prism. The detectors that record the emitted energy are usually cooled by a dewar of liquid nitrogen or some other sub-

stance. The signals recorded by the detectors are amplified by the system electronics and recorded.

The flight altitudes for aircraft MSS surveys are determined by evaluating the size of the desired ground-resolution element (pixel) and the size of the study area. The diameter of the circular ground area viewed by the sensor, D, is a function of the instantaneous field of view, β, of the scanner and the altitude above ground level, H, where

$$D = \beta \times H. \qquad (7\text{-}2)$$

For example, if the IFOV of the scanner is 2.5 mrad, the ground size of the pixel in meters is a product of the IFOV (0.0025) and the altitude AGL in meters. Table 7-8 presents flight altitudes and corresponding pixel sizes at nadir for an IFOV of 2.5 mrad.

The following factors should be considered when collecting aircraft MSS data:

Table 7-8.   Aircraft multispectral scanner flight altitude AGL and pixel size assuming an IFOV of 2.5 milliradians (mrad).

Flight Altitude AGL (m)	Pixel Size (m)
1,000	2.5
2,000	5.0
4,000	10.0
16,000	40.0
50,000	125.0

- The IFOV of the MSS optical system and the altitude AGL dictate the width of a single flight line of coverage. At lower altitudes, the high spatial resolution may be outweighed by the fact that more flight lines are required to cover the area compared to more efficient coverage at higher altitudes with larger pixels. The pixel size and the geographic extent of the area to be surveyed are considered, objectives are weighed, and a compromise is reached.

- Even single flightlines of aircraft MSS data are difficult to rectify to a standard map series because of aircraft roll, pitch, and yaw during data collection. Notches in the edge of a flightline of data indicate aircraft roll. Such data require significant human and machine resources to make the data planimetrically accurate. Several agencies place GPS units on the aircraft to obtain precise flightline coordinates, which are useful when rectifying the aircraft MSS data.

Daedalus multispectral scanning systems have been modified on many occasions to simulate satellite remote sensor data collection. For example, the NASA Ames Airborne Sensor Facility at Moffett Field, CA, continues to operate the following systems: Thematic Mapper Simulator, MODIS Airborne Simulator, Airborne Ocean Color Simulator, and the Multispectral Atmospheric Mapping Sensor (NASA Airborne, 2006).

A Daedalus DS-1260 near-infrared band 10 image of the Four Mile Creek delta on the Westinghouse Savannah River Site in South Carolina is shown in Color Plate 7-5a. A color composite of bands 10, 6, and 4 (near-infrared, red, and green) is shown in Color Plate 7-5b. Near-infrared band 10 imagery of the same region collected on April 23, 1992, is shown in Color Plate 7-5c. Color-Plate 7-5d is a color composite of bands 10, 6, and 4. Thermal effluent was not allowed to enter Four Mile Creek after 1985. Examination of the imagery reveals that revegetation has taken place in many of the wetland sloughs.

NASA Airborne Terrestrial Applications Sensor

NASA's ATLAS multispectral scanner is operated by the Stennis Space Center, MS. ATLAS has 14 channels with a spectral range from 0.45 to 12.2 μm. There are six visible and near-infrared bands, two short-wavelength infrared bands (identical to Thematic Mapper bands 5 and 7) and six thermal infrared bands. The bandwidths are summarized in Table 7-7. The sensor has a total field of view of 72° and an IFOV of 2.0 mrad. ATLAS is flown on a Learjet 23 aircraft from 6000 to 41,000 ft. above ground level, yielding pixels with a ground resolution of approximately 2.5 × 2.5 m to 25 × 25 m, depending upon user specifications. There are normally 800 pixels per line plus three calibration source pixels. The data are quantized to 8-bits.

Calibration of the thermal data is performed using two internal blackbodies. Visible and near-infrared calibration is accomplished on the ground between missions using an integrating sphere. Onboard GPS documents the location in space of each line of data collected.

The ATLAS sensor is ideal for collecting high spatial resolution data in the visible, near-infrared, middle-infrared, and thermal infrared regions so important for many commercial remote sensing applications. The Learjet is an ideal suborbital platform because of its stability in flight and its ability to travel to the study area quickly. It is particularly useful for obtaining data immediately after a disaster such as Hurricanes Hugo and Andrew. An example of 2.5 × 2.5 m ATLAS band 6 (near-infrared) data obtained in 1998 for an area adjacent to Sullivans Island, SC, is shown in Figure 7-24.

 Multispectral Imaging Using Linear Arrays

Linear array sensor systems use diodes or charge-coupled-devices to record the reflected or emitted radiance from the terrain. Linear array sensors are often called *pushbroom* sensors because, like a single line of bristles in a broom, the linear array stares constantly at the ground while the aerial platform moves forward (see Figure 7-3c). The result is a more accurate measurement of the reflected radiant flux because 1) there is no moving mirror, and 2) the linear array detectors are able to dwell longer on a specific portion of the terrain, resulting in a more accurate measurement.

Airborne Terrestrial Applications Sensor (ATLAS) Image of Sullivans Island, SC

Figure 7-24   Near-infrared band 6 (0.76 – 0.90 μm) Airborne Terrestrial Applications Sensor (ATLAS) image of a portion of the smooth cordgrass (*Spartina alterniflora*) marsh behind Sullivans Island, SC. The 2.5 × 2.5 m data were obtained on October 15, 1998.

Chronological Launch History of the SPOT Satellites

| 1986 | 1987 | 1988 | 1989 | 1990 | 1991 | 1992 | 1993 | 1994 | 1995 | 1996 | 1997 | 1998 | 1999 | 2000 | 2001 | 2002 | 2003 |

Launch Dates
SPOT 1 - February 21, 1986
SPOT 2 - January 22, 1990
SPOT 3 - September 25, 1993
SPOT 4 - March 24, 1998
SPOT 5 - May 3, 2002

Figure 7-25   Chronological launch history of the SPOT satellites (SPOT, 2006).

SPOT Sensor Systems

The first SPOT satellite was launched on February 21, 1986 (Figure 7-25). It was developed by the French Centre National d'Etudes Spatiales (CNES) in cooperation with Belgium and Sweden. It had a spatial resolution of 10 × 10 m (panchromatic mode) and 20 × 20 m (multispectral mode) and provided several other innovations in remote sensor system design (Table 7-9). SPOT satellites 2 and 3 with identical payloads were launched on January 22, 1990, and September 25, 1993, respectively. SPOT 4 was launched on March 24, 1998, with a *Vegetation* sensor (1 × 1 km) and a short-wavelength infrared band. SPOT 5 was launched on

Table 7-9.   SPOTs 1, 2, and 3 High Resolution Visible (HRV), SPOTs 4 and 5 High Resolution Visible and Infrared (HRVIR), and SPOTs 4 and 5 *Vegetation* sensor system characteristics.

SPOTs 1, 2, 3 HRV and 4 HRVIR			SPOT 5 HRVIR			SPOT 4 and 5 *Vegetation*		
Band	Spectral Resolution (μm)	Spatial Resolution (m) at Nadir	Band	Spectral Resolution (μm)	Spatial Resolution (m) at Nadir	Band	Spectral Resolution (μm)	Spatial Resolution (km) at Nadir
1	0.50 – 0.59	20 × 20	1	0.50 – 0.59	10 × 10	0	0.43 – 0.47	1.15 × 1.15
2	0.61 – 0.68	20 × 20	2	0.61 – 0.68	10 × 10	2	0.61 – 0.68	1.15 × 1.15
3	0.79 – 0.89	20 × 20	3	0.79 – 0.89	10 × 10	3	0.78 – 0.89	1.15 × 1.15
Pan Pan (4)	0.51 – 0.73 0.61 – 0.68	10 × 10 10 × 10	Pan	0.48 – 0.71	2.5 × 2.5			
SWIR (4)	1.58 – 1.75	20 × 20	SWIR	1.58 – 1.75	20 × 20	SWIR	1.58 – 1.75	1.15 × 1.15
Sensor	Linear array pushbroom		Linear array pushbroom			Linear array pushbroom		
Swath	60 km ± 50.5°		60 km ± 27°			2250 km ± 50.5°		
Rate	25 Mb/s		50 Mb/s			50 Mb/s		
Revisit	26 days		26 days			1 day		
Orbit	822 km, Sun-synchronous Inclination = 98.7° Equatorial crossing 10:30 a.m.		822 km, Sun-synchronous Inclination = 98.7° Equatorial crossing 10:30 a.m.			822 km, Sun-synchronous Inclination = 98.7° Equatorial crossing 10:30 a.m.		

May 3, 2002, with visible, near-infrared, and shortwave infrared (SWIR) bands at 10 × 10 m and a panchromatic band at 2.5 × 2.5 m.

Since 1986, SPOT Earth observation satellites have been a consistent, dependable source of high-resolution Earth resource information. While many countries have seen their Earth resource monitoring sensor systems come and go depending primarily upon politics, one could always count on SPOT Image, Inc., to provide quality imagery. Unfortunately, the imagery has always been relatively expensive, usually more than $2,000 per panchromatic *or* multispectral scene, although it has been reduced in recent years. The cost can exceed $4,000 if both panchromatic and multispectral imagery of the study area are required.

SPOTs 1, 2, and 3

These satellites are all identical and consist of two parts: 1) the SPOT bus, which is a standard multipurpose platform, and 2) the sensor system instruments (Figure 7-26a,b) consisting of two identical high-resolution visible (HRV) sensors and a package comprising two data recorders and a telemetry transmitter. The satellites operate in a Sun-syn-

chronous, near-polar orbit (inclination of 98.7°) at an altitude of 822 km. The satellites pass overhead at the same solar time; the local clock time varies with latitude.

The HRV sensors operate in two modes in the visible and reflective-infrared portions of the spectrum. The first is a *panchromatic* mode corresponding to observation over a broad spectral band (similar to a typical black-and-white photograph). The second is a *multispectral* (color) mode corresponding to observation in three relatively narrow spectral bands (Table 7-9). Thus, the spectral resolution of SPOTs 1 through 3 is not as good as that of the Landsat Thematic Mapper. The ground spatial resolution, however, is 10 × 10 m for the panchromatic band and 20 × 20 m for the three multispectral bands when the instruments are viewing at nadir, directly below the satellite.

Reflected energy from the terrain enters the HRV via a plane mirror and is then projected onto two CCD arrays. Each CCD array consists of 6,000 detectors arranged linearly. An electron microscope view of some of the individual detectors in the linear array is shown in Figure 7-27a,b. This linear array *pushbroom* sensor images a complete line of the ground scene in the cross-track direction in one look as the

Figure 7-26   The SPOT satellites consist of the SPOT bus, which is a multipurpose platform, and the sensor system payload. Two identical high-resolution visible (HRV) sensors are found on SPOTs 1, 2, and 3 and two identical high-resolution visible infrared (HRVIR) sensors on SPOTs 4 and 5. Radiant energy from the terrain enters the HRV or HRVIR via a plane mirror and is then projected onto two CCD arrays. Each CCD array consists of 6,000 detectors arranged linearly. This results in a spatial resolution of 10×10 or 20×20 m, depending on the mode in which the sensor is being used. The swath width at nadir is 60 km. The SPOT HRV and HRVIR sensors may also be pointed off-nadir to collect data. SPOTs 4 and 5 carry a *Vegetation* sensor with 1.15×1.15 km spatial resolution and 2,250-km swath width (adapted from SPOT Image, Inc.).

sensor system progresses downtrack (refer to Figure 7-26c). This capability breaks tradition with the Landsat MSS, Landsat TM, and Landsat 7 ETM$^+$ sensors because no mechanical scanning takes place. A linear array sensor is superior because there is no mirror that must scan back and forth to collect data (mirror-scan velocity is a serious issue) and this allows the detector to literally 'stare' at the ground for a longer time, obtaining a more accurate record of the spectral radiant flux exiting the terrain. The SPOT satellites

pioneered this linear array pushbroom technology in commercial Earth resource remote sensing as early as 1986.

When looking directly at the terrain beneath the sensor system, the two HRV instruments can be pointed to cover adjacent fields, each with a 60-km swath width (Figure 7-26c). In this configuration the total swath width is 117 km and the two fields overlap by 3 km. It is also possible to selectively point the mirror to off-nadir viewing angles through com-

SPOT Linear Array Enlargements

100 µm   a.   10 µm   b.

Figure 7-27   a) Scanning electron microscope images of the surface of a CCD linear array like that used in the SPOT HRV sensor systems. Approximately 58 CCD detectors are visible, with rows of readout registers on both sides. b) Seven detectors of a CCD linear array are shown at higher magnification (© CNES 2004, Spot Image Corporation).

mands from the ground station. In this configuration it is possible to observe any region of interest within a 950-km-wide strip centered on the satellite ground track (i.e., the observed region may not be centered on the ground track) (Figure 7-28a). The width of the swath actually observed varies between 60 km for nadir viewing and 80 km for extreme off-nadir viewing.

If the HRV instruments were only capable of nadir viewing, the revisit frequency for any given region of the world would be 26 days. This interval is often unacceptable for the observation of phenomena evolving on time scales ranging from several days to a few weeks, especially where cloud cover hinders the acquisition of usable data. During the 26-day period separating two successive SPOT satellite passes over a given point on Earth and taking into account the steering capability of the instruments, the point in question could be observed on seven different passes if it were on the equator and on 11 occasions if at a latitude of 45° (Figure 7-28b). A given region can be revisited on dates separated alternatively by 1 to 4 (or occasionally 5) days.

The SPOT sensors can also acquire cross-track stereoscopic pairs of images for a geographic area (Figure 7-28c). Two observations can be made on successive days such that the two images are acquired at angles on either side of the vertical. In such cases, the ratio between the observation base (distance between the two satellite positions) and the height (satellite altitude) is approximately 0.75 at the equator and 0.50 at a latitude of 45°. SPOT data with these base-to-height ratios may be used for topographic mapping.

SPOT 10×10 m panchromatic data are of such high geometric fidelity that they can be photointerpreted like a typical aerial photograph in many instances. For this reason, SPOT panchromatic data are often registered to topographic base maps and used as orthophotomaps. Such image maps are useful in GIS databases because they contain more accurate planimetric information (e.g., new roads, subdivisions) than out-of-date 7.5-min topographic maps. The improved spatial resolution available is demonstrated in Figure 7-29, which presents a TM band 3 image and a SPOT panchromatic image of Charleston, SC.

SPOT sensors collect data over a relatively small 60×60 km ($3,600$ km^2) area compared with Landsat MSS and TM image areas of 170×185 km ($31,450$ km^2) (Figure 7-30). It takes 8.74 SPOT images to cover the same area as a single Landsat TM or MSS scene. This may be a limiting factor for regional studies. However, SPOT does allow imagery to be purchased by the km^2 (e.g., for a watershed or school district) or by the linear km (e.g., along a highway).

SPOTs 4 and 5

SPOT Image, Inc., launched SPOT 4 on March 24, 1998, and SPOT 5 on May 3, 2002. Their characteristics are summarized in Table 7-9. The viewing angle can be adjusted to $\pm 27°$ off-nadir. SPOTs 4 and 5 have several notable features of significant value for Earth resource remote sensing, including:

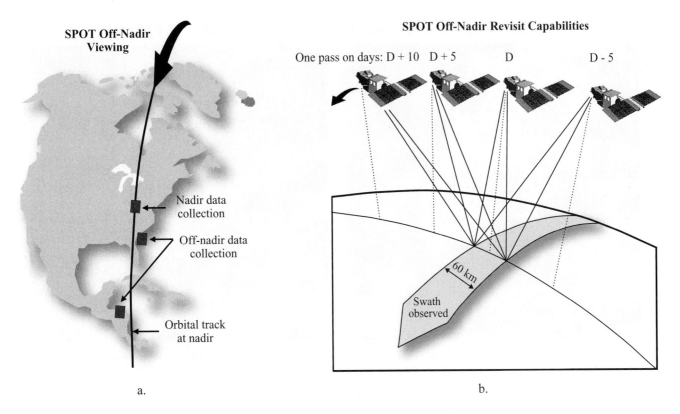

SPOT Off-Nadir Viewing

Nadir data collection

Off-nadir data collection

Orbital track at nadir

a.

SPOT Off-Nadir Revisit Capabilities

One pass on days: D + 10   D + 5   D   D - 5

60 km

Swath observed

b.

SPOT Stereoscopic Viewing Capabilities

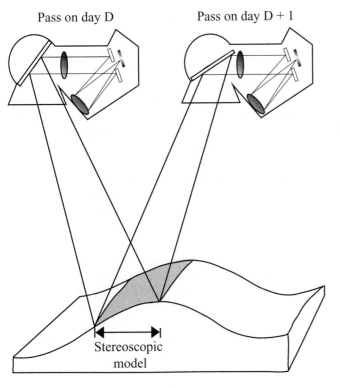

Pass on day D   Pass on day D + 1

Stereoscopic model

c.

Figure 7-28   a) The SPOT HRV and HRVIR instruments are pointable and can be used to view areas that are not directly below the satellite (i.e., off-nadir). This is useful for collecting information in the event of a disaster, when the satellite track is not optimum for collecting stereoscopic imagery. b) During the 26-day period separating two successive SPOT satellite overpasses, a point on the Earth could be observed on 7 different passes if it is at the Equator and on 11 occasions if at a latitude of 45°. A given region can be revisited on dates separated alternatively by 1, 4, and occasionally 5 days. c) Two observations can be made on successive days such that the two images are acquired at angles on either side of the vertical, resulting in stereoscopic imagery. Such imagery can be used to produce topographic and planimetric maps (adapted from SPOT Image, Inc.).

Comparison of Landsat TM (30 x 30 m) and SPOT HRV (10 x 10 m)

a. Landsat Thematic Mapper
Band 3 (30 x 30 m) on February 3, 1994.

b. SPOT HRV Panchromatic
Band (10 x 10 m) on January 10, 1996.

Figure 7-29   Comparison of the detail in 30 × 30 m Landsat TM band 3 data and SPOT 10 × 10 m panchromatic data of Charleston, SC (© Spot Image Corporation, Inc.).

- the addition of a short-wavelength infrared (SWIR) band (1.58 – 1.75 μm) for vegetation and soil moisture applications at 20 × 20 m;

- an independent sensor called *Vegetation* for small-scale vegetation, global change, and oceanographic studies;

- SPOT 4 has onboard registration of the spectral bands, achieved by replacing the original HRV panchromatic sensor (0.51 – 0.73 μm) with band 2 (0.61 – 0.68 μm) operating in both 10- *and* 20-m resolution mode;

- SPOT 5 panchromatic bands (0.48 – 0.7 μm) can collect 2.5 × 2.5 m imagery. A color composite of Los Angeles, CA, incorporating 2.5 × 2.5 m panchromatic data is found in Color Plate 7-6.

Because the SPOT 4 and 5 HRV sensors are sensitive to SWIR energy, they are called HRVIR 1 and HRVIR 2.

The SPOT 4 and 5 *Vegetation* sensor is independent of the HRVIR sensors. It is a multispectral electronic scanning radiometer operating at optical wavelengths with a separate objective lens and sensor for each of the four spectral bands; (blue = 0.43 – 0.47 μm used primarily for atmospheric cor-

rection; red = 0.61 – 0.68 μm; near-infrared = 0.78 – 0.89 μm; SWIR = 1.58 – 1.75 μm). Each sensor is a 1,728 CCD linear array located in the focal plane of the corresponding objective lens. The spectral resolution of the individual bands are summarized in Table 7-9. The *Vegetation* sensor has a spatial resolution of 1.15 × 1.15 km. The objective lenses offer a field-of-view of ±50.5°, which translates into a 2,250-km swath width. The *Vegetation* sensor has several important characteristics:

- multidate radiometric calibration accuracy better than 3% and absolute calibration accuracy better than 5% is superior to the AVHRR, making it more useful for repeatable global and regional vegetation surveys;

- pixel size is uniform across the entire swath width, with geometric precision better than 0.3 pixels and interband multidate registration better than 0.3 km because of the pushbroom technology;

- 10:30 a.m. equatorial crossing time versus AVHRR's 2:30 p.m. crossing time;

Table 7-10. Selected Indian National Remote Sensing Agency (NRSA) Indian Remote Sensing (IRS) satellite characteristics.

IRS-1A and 1B			IRS-1C and 1D		
LISS-I and LISS-II Bands	Spectral Resolution (μm)	Spatial Resolution (m) at Nadir	LISS-III, Pan, and WiFS Bands	Spectral Resolution (μm)	Spatial Resolution (m) at Nadir
1	0.45 – 0.52	LISS-I @72.5 m LISS-II @36.25 m	1	-	-
2	0.52 – 0.59	LISS-I @72.5 m LISS-II @36.25 m	2	0.52 – 0.59	23.5 × 23.5
3	0.62 – 0.68	LISS-I @72.5 m LISS-II @36.25 m	3	0.62 – 0.68	23.5 × 23.5
4	0.77 – 0.86	LISS-I @72.5 m LISS-II @36.25 m	4	0.77 – 0.86	23.5 × 23.5
			5	1.55 – 1.70	70.5 × 70.5
			Pan	0.50 – 0.75	5.2 × 5.2
			WiFS 1	0.62 – 0.68	188 × 188
			WiFS 2	0.77 – 0.86	188 × 188
Sensor	Linear array pushbroom		Linear array pushbroom		
Swath width	LISS-I = 148 km; LISS-II = 146 km		LISS-III = 141 km for bands 2, 3, and 4; band 5 = 148 km Pan = 70 km; WiFS = 692 km		
Revisit	22 days at equator		LISS-III is 24 days at equator; Pan is 5 days ± 26° off-nadir viewing; WiFS is 5 days at equator		
Orbit	904 km, Sun-synchronous Inclination = 99.5° Equatorial crossing 10:26 a.m.		817 km, Sun-synchronous Inclination = 98.69° Equatorial crossing 10:30 a.m. ± 5 min		
Launch	IRS-1A on March 17, 1988 IRS-1B on August 29, 1991		IRS-1C in 1995 IRS-1D in September 1997		

- a short-wavelength infrared band for improved vegetation mapping;

- it is straightforward to relate the HRVIR 10 × 10 m or 20 × 20 m data nested within the *Vegetation* 2,250 × 2,250 km swath width data;

- individual images can be obtained or data can be summarized over a 24-hr period (called a daily synthesis). Daily synthesis data can be compiled into *n*-day syntheses. NDVI datasets can be computed based on daily or 10-day synthesis images. A portion of a Ten-day vegetation index synthesis centered on Europe, the Middle East, and Africa is shown in Color Plate 7-7.

The dependable constellation of SPOT satellites offers unrivaled ability to acquire imagery of almost any point on the globe daily.

Indian Remote Sensing Systems

The Indian National Remote Sensing Agency (NRSA) has launched several Indian Remote Sensing (IRS) satellites: IRS-1A on March 17, 1988, IRS-1B on August 29, 1991, IRS-1C in 1995, and IRS-1D in September, 1997 (Table 7-10). IRS-P3 and IRS-P4 were launched on March 21, 1996, and May 26, 1999, respectively. IRS-P5 (CARTOSAT-1) was launched on May 5, 2005. IRS-P6 (RESOURCESAT-1)

Figure 7-30   Geographic coverage of the SPOT HRV and Landsat Multispectral Scanner and Thematic Mapper remote sensing systems.

was launched on October 17, 2003. The sensors onboard the satellites use linear array sensor technology (NRSA, 2006).

IRS-1A, -1B, -1C, and -1D

The IRS-1A and IRS-1B satellites acquired data with Linear Imaging Self-scanning Sensors (LISS-I and LISS-II) at spatial resolutions of 72.5×72.5 m and 36.25×36.25 m, respectively (Table 7-10). The data were collected in four spectral bands that were almost identical to the Landsat TM visible and near-infrared bands. The satellite altitude was 904 km, the orbit was Sun-synchronous, repeat coverage was every 22 days at the Equator (11-day repeat coverage with two satellites), and orbital inclination was 99.5°. The swath width was 146 to 148 km.

The IRS-1C and IRS-1D satellites carry three sensors (Table 7-10): the LISS-III multispectral sensor, a panchromatic sensor, and a Wide Field Sensor (WiFS). The LISS-III has four bands with the green, red, and near-infrared bands at 23.5×23.5 m spatial resolution and the short-wavelength infrared (SWIR) band at 70.5×70.5 m spatial resolution. The swath width is 141 km for bands 2, 3, and 4 and 148 km for the SWIR band. Repeat coverage is every 24 days at the equator.

The panchromatic sensor has a spatial resolution of approximately 5.2×5.2 m and stereoscopic imaging capability. The panchromatic band has a 70-km swath width with repeat coverage every 24 days at the Equator and a revisit time of 5 days with $\pm 26°$ off-nadir viewing. An example of the 5.2×5.2 m panchromatic data of downtown San Diego, CA (resampled to 5×5 m), is shown in Figure 7-31.

Figure 7-31   Indian Remote Sensing Satellite (IRS-1D) panchromatic image of downtown San Diego, CA. The 5.2×5.2 m image was resampled to 5×5 m (courtesy of Indian National Remote Sensing Agency).

The Wide Field Sensor (WiFS) has 188×188 m spatial resolution. The WiFS has two bands comparable to NOAA's AVHRR satellite ($0.62 - 0.68$ μm and $0.77 - 0.86$ μm) with a swath width of 692 km. Repeat coverage is five days at the Equator.

IRS-P3 and IRS-P4

The IRS-P3 WiFS is similar to the IRS-1D WiFS except for the inclusion of an additional band in the middle-infrared region ($1.55 - 1.70$ μm). IRS-P3 also has a Modular Optoelectronics Scanner (MOS), which collects data at three spatial resolutions (MOS A, B, C = 1569×1395 m; 523×523 m; and 523×644 m, respectively) in three bands (MOS A, B, C = $0.755 - 0.768$ μm; $0.408 - 1.01$ μm; $1.5 - 1.7$ μm).

The IRS-P4 satellite is devoted to oceanographic applications based primarily on its Ocean Color Monitor (OCM) sensor, which collects data in 8 bands from 402 to 885 nm at a spatial resolution of 360×236 m at 12-bit radiometric resolution. The swath width is 1,420 km. The IRS-P4 also carries a Multifrequency Scanning Microwave Radiometer (MSMR) (NRSA, 2006).

IRS-P5 (CARTOSAT-1)

CARTOSAT-1 was designed to provide imagery for large-scale cartographic applications. The satellite is in a Sun-syn-

Table 7-11.   NASA Advanced Spaceborne Thermal Emission and Reflection Radiometer (ASTER) characteristics.

Advanced Spaceborne Thermal Emission and Reflection Radiometer (ASTER)

Band	VNIR Spectral Resolution (μm)	Band	SWIR Spectral Resolution (μm)	Band	TIR Spectral Resolution (μm)
1 (nadir)	0.52 – 0.60	4	1.600 – 1.700	10	8.125 – 8.475
2 (nadir)	0.63 – 0.69	5	2.145 – 2.185	11	8.475 – 8.825
3 (nadir)	0.76 – 0.86	6	2.185 – 2.225	12	8.925 – 9.275
3 (backward)	0.76 – 0.86	7	2.235 – 2.285	13	10.25 – 10.95
		8	2.295 – 2.365	14	10.95 – 11.65
		9	2.360 – 2.430		
Technology (detector)	Pushbroom Si		Pushbroom PtSi:Si		Whiskbroom Hg:Cd:Te
Spatial resolution (m)	15 × 15		30 × 30		90 × 90
Swath width	60 km		60 km		60 km
Quantization	8-bits		8-bits		12-bits

chronous polar orbit (618 km) that records images of the entire globe in 1,867 orbits with a repeat cycle of 126 days. CARTOSAT-1 has two panchromatic cameras (0.5 - 0.85 μm) with 2.5 × 2.5 m spatial resolution that acquire two images simultaneously, one looking forward 26° and one looking aft -5° in stereo-mode. Each camera uses a linear array containing 12,000 detectors (Krishnaswamy and Kalyanaraman, 2006). The swath width is 26 km. All data are collected at 10-bit quantization. The platform can be maneuvered to obtain two side-by-side panchromatic images covering 55 km. Users can specify standard radiometrically and geometrically corrected data or precision ortho-rectified data (NRSA, 2006).

IRS-P6 (RESOURCESAT-1)

RESOURCESAT-1 is in an 817 km Sun-synchronous polar orbit (NRSA, 2006). It carries three sensor systems: LISS-III, LISS-IV, and AWiFS. The LISS-III sensor is identical to the LISS-III sensor flown on IRS-1C/1D except for a few significant changes. LISS-IV has exactly the same multispectral bands as LISS III. However, the multispectral data are recorded at 5.8 × 5.8 m spatial resolution. The user can select one of the bands (e.g., the red band) and obtain panchromatic data. In multispectral mode the images are 23 kms

wide while in panchromatic mode they are 70 kms wide. The spatial resolution of band 5 (1.55 – 1.70 μm) has been improved to 23.5 × 23.5 m. The data are recorded using 10,000 element linear arrays with quantization to 10-bits. The LISS IV sensor is pointable allowing a revisit period of 5 days.

The AWiFS sensor is an improved version of the WiFS sensor flown on IRS-1C/1D. AWiFS operates in four spectral bands identical to LISS-III, with a spatial resolution of 56 × 56 m and covers a swath of 740 kms.

Advanced Spaceborne Thermal Emission and Reflection Radiometer

The ASTER is a cooperative effort between NASA and Japan's Ministry of International Trade and Industry. ASTER obtains detailed information on surface temperature, emissivity, reflectance, and elevation (NASA ASTER, 2006). It is the only relatively high-spatial-resolution instrument on the *Terra* satellite (Table 7-11). It is used in conjunction with MODIS, MISR, and CERES sensors that monitor the Earth at moderate to coarse spatial resolutions. ASTER serves as a zoom lens for the other *Terra* instru-

ASTER Near-infrared Image of Pearl Harbor, Hawaii

Figure 7-32   *Terra* ASTER 15 × 15 m near-infrared band 3 (0.76 – 0.86 µm) image of Pearl Harbor, Hawaii, obtained on June 3, 2000 (courtesy NASA/GSFC/MITI/ERSADC/JAROS and U.S./Japan ASTER Science Team and the California Institute of Technology).

ments and is important for change detection and calibration/validation studies (King, 2003).

ASTER obtains data in 14 channels from the visible through the thermal infrared regions of the electromagnetic spectrum. It consists of three separate instrument subsystems. Individual bandwidths and subsystem characteristics are summarized in Table 7-11.

The VNIR detector subsystem operates in three spectral bands in the visible and near-infrared wavelength region with a spatial resolution of 15 × 15 m. It consists of two telescopes—one nadir-looking with a three-spectral-band CCD detector and another backward-looking with a single-band CCD detector. The backward-looking telescope provides a second view of the study area in band 3 for stereoscopic observations. Across-track pointing to 24° off-nadir is accomplished by rotating the entire telescope assembly. A 15 × 15 m near-infrared band 3 (0.76 – 0.86 µm) image of Pearl Harbor, HI, obtained on June 3, 2000, is shown in Figure 7-32. ASTER color composites of the island of Oahu and Pearl Harbor obtained on June 3, 2000, are found in Color Plate 7-8.

The SWIR subsystem operates six spectral bands in the 1.6 to 2.43 µm region through a single nadir-pointing telescope

that provides 30 × 30 m spatial resolution. Cross-track pointing (± 8.55°) is accomplished by a pointing mirror.

The TIR subsystem operates in five bands in the thermal infrared region using a single, fixed-position, nadir-looking telescope with a spatial resolution of 90 × 90 m. Unlike the other subsystems, it has a whiskbroom scanning system instead of a pushbroom system (Figure 7-3d). Each band uses 10 detectors in a staggered array with optical bandpass filters over each detector element. The scanning mirror functions both for scanning and cross-track pointing (±8.55°). During scanning, the mirror rotates 90° from nadir to view an internal blackbody. Multiple-date nighttime 90 × 90 m thermal infrared band 14 (10.95 – 11.65 µm) images of Pu'u O'o lava flows entering the sea at Kamokuna on the southeast side of the Island of Hawaii are shown in Figure 7-33.

Multiangle Imaging Spectroradiometer

The Multiangle Imaging Spectroradiometer (MISR) was built by NASA's Jet Propulsion Laboratory and is one of the five *Terra* satellite instruments. The MISR instrument measures the Earth's brightness in four spectral bands, at each of nine look angles spread out in the forward and aft directions along the flight line. Spatial samples are acquired every 275 m. Over a period of 7 min, a 360-km wide swath of Earth

ASTER Thermal-infrared Images of a Lava Flow on Hawaii

a.   b.   c.

Figure 7-33   A sequence of *Terra* ASTER nighttime 90×90 m thermal infrared band 14 ($10.95 - 11.65$ μm) images of Pu'u O'o lava flows entering the sea at Kamokuna on the southeast side of the Island of Hawaii. The images were obtained on a) May 22, 2000, b) June 30, 2000, and c) August 1, 2000. The brighter the pixel, the greater the lava temperature (courtesy of NASA/GSFC/MITI/ ERSADC/JAROS and U.S./Japan ASTER Science Team and the California Institute of Technology).

comes into view at all nine angles (King, 2003; NASA MISR, 2006).

An illustration of the nine look angles is shown in Figure 7-34. The digital pushbroom sensors image the Earth at 26.1°, 45.6°, 60°, and 70.5° forward and aft of the local vertical (nadir 0°). Note that the fore and aft camera angles are the same—the cameras are arranged symmetrically about nadir. In general, large viewing angles provide enhanced sensitivity to atmospheric aerosol effects and to cloud reflectance effects, whereas more modest angles are required for landsurface viewing.

Each MISR camera sees instantaneously a single row of pixels at right angles to the ground track in a pushbroom format. It records data in four bands: blue, green, red, and near-infrared. The individual band wavelengths are identified in Figure 7-34. Each camera has four independent linear CCD arrays (one per filter), with 1,504 active pixels per linear array.

The nadir-viewing camera (labeled An in Figure 7-34) provides imagery that is less distorted by surface topographic effects than that of any other MISR camera. It also is the least affected by atmospheric scattering. It provides 1) useful reference for navigating within all the MISR imagery, and 2) a base image to compare with images acquired at different angles of view. Such comparisons provide important "bidirectional reflectance distribution function, BRDF" information introduced in Chapter 1. The nadir-viewing camera also

offers an opportunity to compare observations with other nadir-viewing sensors such as Landsat TM and ETM[+]. The nadir-viewing camera also facilitates calibration.

The fore and aft 26.1° view angle cameras (Af and Aa) provide useful stereoscopic information that can be of benefit for measuring topographic elevation and cloud heights. The fore and aft 45.6° view angle cameras (Bf and Ba) are positioned to be especially sensitive to atmospheric aerosol properties. The fore and aft 60° view angle cameras (Cf and Ca) provide observations looking through the atmosphere with twice the amount of air than the vertical view. This provides unique information about the hemispherical albedo of land surfaces. The fore and aft 70.5° view angle cameras (Df and Da) provide the maximum sensitivity to off-nadir effects. The scientific community is interested in obtaining quantitative information about clouds and the Earth's surface from as many angles as possible.

Very-High-Resolution Linear Array Remote Sensing Systems

In 1994, the U.S. government allowed civil commercial companies to market high spatial resolution remote sensor data (approximately 1×1 to 4×4 m). This resulted in the creation of a number of commercial consortia that have the capital necessary to create, launch, and market high spatial resolution digital remote sensor data. The most notable companies are Space Imaging, Inc., ORBIMAGE, Inc., Digital-

Multiangle Imaging Spectroradiometer

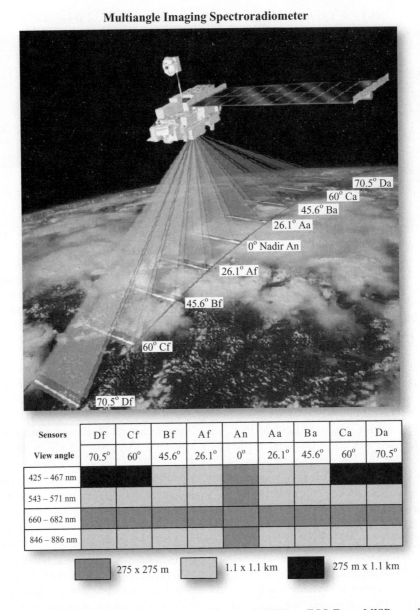

Sensors	Df	Cf	Bf	Af	An	Aa	Ba	Ca	Da
View angle	70.5°	60°	45.6°	26.1°	0°	26.1°	45.6°	60°	70.5°
425 – 467 nm									
543 – 571 nm									
660 – 682 nm									
846 – 886 nm									

275 x 275 m　　　1.1 x 1.1 km　　　275 m x 1.1 km

Figure 7-34 Artist's rendition of the Multiangle Imaging Spectroradiometer (MISR) on EOS *Terra*. MISR uses linear array technology to acquire imagery of the terrain in four bands at nine angles: at Nadir (0°) and at 26.1°, 45.6°, 60°, and 70.5° forward and aft of nadir (adapted from NASA Jet Propulsion Laboratory).

Globe, Inc., and ImageSat International, Inc. These companies targeted the geographic information system (GIS) and cartographic mapping markets traditionally serviced by the aerial photogrammetric industries. Some estimate the growing Earth observation industry to be 5 to 15 billion dollars a year. The commercial remote sensing firms hope to have an impact in markets as diverse as agriculture, natural resource management, local and regional government, transportation, emergency response, mapping, and eventually an array of average consumer applications as well

(Space Imaging, 2006). War during the early 21st century has created a significant military demand for commercial satellite-derived high-spatial-resolution imagery.

All commercial vendors offer an Internet on-line ordering service. All vendors offer a suite of standard and nonstandard products that can be tailored to user requirements, including the creation of digital elevation models from the remote sensor data. The commercial remote sensing companies typically price the imagery according to the type of

Table 7-12.   Sensor characteristics of Space Imaging/GeoEye, Inc., *IKONOS* satellite; ORBIMAGE, Inc., *OrbView-3* satellite; and Digital-Globe, Inc., *QuickBird* satellite.

| | Space Imaging/GeoEye, Inc. | | ORBIMAGE, Inc. | | | DigitalGlobe, Inc. | | |
| | IKONOS | | OrbView-3 | | | QuickBird | | |
Band	Spectral Resolution (μm)	Spatial Resolution (m) at Nadir	Band	Spectral Resolution (μm)	Spatial Resolution (m) at Nadir	Band	Spectral Resolution (μm)	Spatial Resolution (m) at Nadir
1	$0.45 - 0.52$	4×4	1	$0.45 - 0.52$	4×4	1	$0.45 - 0.52$	2.44×2.44
2	$0.52 - 0.60$	4×4	2	$0.52 - 0.60$	4×4	2	$0.52 - 0.60$	2.44×2.44
3	$0.63 - 0.69$	4×4	3	$0.625 - 0.695$	4×4	3	$0.63 - 0.69$	2.44×2.44
4	$0.76 - 0.90$	4×4	4	$0.76 - 0.90$	4×4	4	$0.76 - 0.89$	2.44×2.44
Pan	$0.45 - 0.90$	1×1	Pan	$0.45 - 0.90$	1×1	Pan	$0.45 - 0.90$	0.61×0.61
Sensor	Linear array pushbroom		Linear array pushbroom			Linear array pushbroom		
Swath	11 km		8 km			20 to 40 km		
Rate	25 Mb/s		50 Mb/s			50 Mb/s		
Revisit	< 3 days		< 3 days			1 to 5 days depending on latitude		
Orbit	681 km, Sun-synchronous Equatorial crossing 10–11 a.m.		470 km, Sun-synchronous Equatorial crossing 10:30 a.m.			600 km, not Sun-synchronous Equatorial crossing variable		
Launch	April 27, 1999 (failed) September 24, 1999		June 26, 2003			October 18, 2001		

product ordered and the amount of geographic coverage desired (km^2). The sensors used by these companies are based primarily on linear array CCD technology (Figure 7-3c). The sensor system characteristics are summarized in Table 7-12.

DigitalGlobe, Inc., *EarlyBird*, and *QuickBird*

EarthWatch, Inc., launched *EarlyBird* in 1996 with a 3×3 m panchromatic band and three visible to near-infrared (VNIR) bands at 15×15 m spatial resolution. Unfortunately, Earth-Watch lost contact with the satellite. EarthWatch (now Digi-talGlobe, Inc.) launched *QuickBird* on October 18, 2001, into a 600-km orbit. Interestingly, it is in a 66° orbit which is not Sun-synchronous. Revisit times range from 1 to 5 days, depending on latitude. It has a swath width of 20 to 40 km. QuickBird has a 0.61×0.61 m panchromatic band and four visible/near-infrared bands at 2.44×2.44 m spatial resolu-tion (Table 7-12). The data are quantized to 11 bits (bright-ness values from 0 to 2047). The sensor can be pointed fore and aft and across-track to obtain stereoscopic data (Digital-Globe, 2006).

DigitalGlobe plans to launch a new generation of high reso-lution remote sensing systems called *WorldView I* and *II* (Digital Globe, 2005). *WorldView I* will provide 0.5×0.5 m spatial resolution panchromatic imagery with an average revisit time of 1.7 days and a swath width of 16 km. It will be capable of collecting up to 500,000 km^2 (200,000 sq. mi.) per day of half-meter imagery. *WorldView I* will be able to downlink imagery to customer locations.

WorldView II will collect 0.5×0.5 m panchromatic data and 1.8×1.8 m multispectral data. It will incorporate the indus-try standard four multispectral bands plus four new bands (coastal, yellow, red edge, and near-infrared 2).

Space Imaging/GeoEye, Inc., *IKONOS*

Space Imaging, Inc., launched *IKONOS* on April 27, 1999. Unfortunately, the satellite never achieved orbit. Space Imaging successfully launched a second IKONOS on Sep-tember 24, 1999. The IKONOS satellite sensor has a 1×1 m panchromatic band and four multispectral visible and near-infrared bands at 4×4 m spatial resolution (Space Imaging,

IKONOS Panchromatic Imagery of Washington, DC

a. IKONOS panchromatic 1 x 1 m image of Washington, DC.

b. Enlarged view of the Washington Monument.

Figure 7-35   a) IKONOS 1 × 1 m panchromatic image of Washington, DC, obtained on September 30, 1999. The Washington Monument and White House are visible. b) An enlargement of the Washington Monument (courtesy Space Imaging/GeoEye, Inc.).

2006). Sensor characteristics are summarized in Table 7-12. IKONOS is in a Sun-synchronous 681-km orbit, with a descending equatorial crossing time of between 10 and 11 a.m. It has both cross-track and along-track viewing instruments, which enables flexible data acquisition and frequent revisit capability: < 3 days at 1 × 1 m spatial resolution (for look angles < 26°) and 1.5 days at 4 × 4 m spatial resolution. The nominal swath width is 11 km. Data are quantized to 11 bits. The first IKONOS 1 × 1 m panchromatic image was obtained on September 30, 1999, of downtown Washington, DC (Figure 7-35). The image contains high spatial resolution detail sufficient for many city planning and earth resource investigations. IKONOS imagery of Columbia, SC, is shown in Color Plate 7-9.

On January 12, 2006, ORBIMAGE Holdings Inc. acquired Space Imaging, Inc. (GeoEye, 2006). The combined company now does business under the brand name GeoEye, Inc. GeoEye operates three remote-sensing satellites including IKONOS, OrbView-2 and OrbView-3. GeoEye plans to launch another satellite in 2007 with a spatial resolution of 0.4 × 0.4 m (GeoEye, 2006).

ImageSat International, Inc., *EROS A1*

ImageSat International, Inc., successfully launched the *EROS A1* satellite using a Start-1 rocket from the Russian Cosmodrome in Svobodni, Siberia, on December 5, 2000

(ImageSat, 2006). The *EROS A1* was built by Israel Aircraft Industry's MBT Division. It weighs only 250 kg. The satellite is in a 480-km Sun-synchronous orbit with an equatorial crossing time of 9:45 a.m. EROS A1 is a single-band panchromatic remote sensing system. The spatial resolution is 1.8 × 1.8 m (standard) or 1 × 1 m in "over-sampled" mode. An EROS A1 image of the Kamchatka Submarine Base in Russia is shown in Figure 7-36. ImageSat plans to launch EROS B and other satellites to create a constellation of commercial high-resolution satellites.

ORBIMAGE, Inc., *OrbView-3*

ORBIMAGE, Inc., launched *OrbView-3* on June 26, 2003, with 1 × 1 m panchromatic data and four visible and near-infrared multispectral bands at 4 × 4 m spatial resolution (Orbimage, 2006). OrbView-3 has a 470 km Sun-synchronous orbit with a 10:30 a.m. equatorial crossing time and an 8-km swath width. The sensor revisits each location on Earth in less than three days with an ability to turn from side to side 45°. *OrbView-3* sensor specifications are summarized in Table 7-12.

ORBIMAGE, Inc. plans to launch OrbView-5 (Orbimage, 2006). It will have the same bandwidths as OrbView-3, but the panchromatic band will be 0.41 × 0.41 m and the multispectral bands will be 1.64 × 1.64 m. It will be pointable

EROS A1 Panchromatic Image of Kamchatka Submarine Base, Russia

Figure 7-36 Panchromatic 1.8×1.8 m image of the Kamchatka Submarine Base on the far-eastern frontier of Russia, obtained by the EROS A1 remote sensing system on December 25, 2001 (courtesy ImageSat, International, Inc.).

with a swath width of 15.2 km. OrbView-5 data will be useful for photogrammetric engineering applications.

Leica Geosystems *Airborne Digital Sensor System-40*

The Leica Geosystems ADS-40 digital camera makes use of linear array technology. It is based on the use of 12,000 element linear array CCDs. The linear arrays function in harmony to obtain stereoscopic imagery for photogrammetric applications. The ASD-40 collects data according to the specifications summarized in Table 7-13 and the logic shown in Figure 7-37. Three of the linear arrays collect panchromatic data: one looks forward 28.4°, one looks at nadir, and one looks aft -14.2° [there are actually two 12,000 element linear arrays at each of these three locations staggered by half a pixel (3.25 mm), which are analyzed together to yield panchromatic data] (McClone, 2004).

The blue, green, and red sensitive linear arrays collect data 14.2° off-nadir. The blue, green, and red measurements are obtained with a trichroid optical system that splits incoming light into red, green, and blue components using cascaded dichrotic filters. This results in perfectly registered blue, green, and red information. A near-infrared sensitive linear array looks 2° off-nadir adjacent to the panchromatic linear array at nadir.

Spectrometry Using Linear & Area Arrays

This section describes a major advance in remote sensing, *imaging spectrometry*, defined as the simultaneous acquisition of images in many relatively narrow, contiguous and/or non-contiguous spectral bands throughout the ultraviolet, visible, and infrared portions of the spectrum.

In the past, most remotely sensed data were acquired in 4 to 12 spectral bands. Imaging spectrometry makes possible the acquisition of data in hundreds of spectral bands simultaneously. Because of the very precise nature of the data acquired by imaging spectrometry, some Earth resource problems can be investigated in greater detail.

The value of an imaging spectrometer lies in its ability to provide a high-resolution reflectance spectrum for each picture element in the image. The reflectance spectrum in the region from 0.4 to 2.5 μm can be used to identify a large range of surface cover materials that cannot be identified with broadband, low-spectral-resolution imaging systems such as the Landsat MSS, TM, and SPOT. Many, although not all, surface materials have diagnostic absorption features that are only 20 to 40 nm wide. Therefore, spectral imaging systems that acquire data in contiguous 10-nm bands may

Figure 7-37   Characteristics of the Leica Geosystems Airborne Digital Sensor System ADS-40. a) Digital camera housing. b) Trichroid optical system. c) Layout of forward, aft, and nadir-looking 12,000-element linear arrays (adapted from Leica Geosystems, Inc.).

Table 7-13. Leica GeoSystems Airborne Digital Sensor 40 (ADS-40) characteristics.

Band	Spectral Resolution (nm)	Spatial Resolution (cm) at Nadir
Pan	465 – 680	variable 15 cm @ 1,450 m AGL
Blue	430 – 490	variable
Green	535 – 585	variable
Red	610 – 660	variable
Near-infrared	835 – 885	variable
Sensor	7 linear array CCDs	
Swath	12,000 pixels	
Detector Size	6.5 µm	
Detectors per Linear Array	*Panchromatic* = 2 × 12,000 *Multispectral* = 12,000	
Radiometric Resolution	12-bit	
Focal Length	62.7 mm	
Largest Application Scale	1:500 - 1:5,000 (depending upon aircraft altitude AGL)	
Geo-referencing	Applanix POS IMU with GPS and INS	

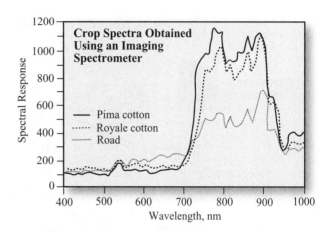

Figure 7-38 Imaging spectrometer crop spectra for Pima cotton, Royale cotton, and road surface extracted from 2 × 2 m data obtained near Bakersfield, CA.

produce data with sufficient spectral resolution for the direct identification of those materials with diagnostic spectral absorption features. For example, Figure 7-38 depicts high spectral resolution crop spectra over the interval 400 to 1000 nm obtained using an imaging spectrometer for an agricultural area near Bakersfield, CA. The absorption spectra for the Pima and Royale cotton differ from one another from about 725 nm, where the "red edge" is located, to about 900 nm, leading to the possibility that species within the same crop type might be distinguishable (SBRC, 1994). The Landsat scanners and SPOT HRV sensors, which have relatively large bandwidths, may not be able to resolve these spectral differences.

Simultaneous imaging in many contiguous spectral bands requires a new approach to remote sensor system design. One approach is to increase the residence time of a detector in each IFOV using a linear array of detector elements (Figure 7-3c). In this configuration, there is a dedicated detector element for each cross-track pixel, which increases the residence time to the interval required to move one IFOV along the flight direction.

Two more practical approaches to imaging spectrometry are shown in Figures 7-3d and 3e. The whiskbroom scanner linear array approach (Figure 7-3d) is analogous to the scanner approach used for Landsat MSS and ETM[+], except that radiant flux from within the IFOV is passed to a spectrometer, where it is dispersed and focused onto a linear array of detectors. Thus, each pixel is simultaneously sensed in as many spectral bands as there are detector elements in the linear array. For high spatial resolution imaging, this approach is suited only to an airborne sensor that flies slowly and when the readout time of the detector array is a small fraction of the integration time. Because of high spacecraft velocities, orbital imaging spectrometry might require the use of two-dimensional area arrays (Figure 7-3e). This eliminates the need for the optical scanning mechanism. In this situation, there is a dedicated column of spectral detector elements for each linear array cross-track pixel in the scene.

Thus, traditional broadband remote sensing systems such as Landsat MSS and SPOT HRV *undersample* the information available from a reflectance spectrum by making only a few measurements in spectral bands up to several hundred nanometers wide. Conversely, imaging spectrometers sample at close intervals (bands on the order of tens of nanome-

NASA Airborne Visible Infrared Imaging Spectrometer

Figure 7-39   Conceptual representation of imaging spectroscopy as implemented by the NASA JPL Airborne Visible-Infrared Imaging Spectrometer (AVIRIS). The scanner mirror focuses radiant flux onto linear arrays that contain 224 detector elements with a spectral sensitivity ranging from 400 to 2500 nm. A spectra of radiance (*L*) or percent reflectance can be obtained for each picture element. The AVIRIS scene was acquired over the Kennedy Space Center, FL. Only band 30 (655.56 nm) is displayed (adapted from Filippi, 1999).

ters wide) and have a sufficient number of spectral bands to allow construction of spectra that closely resemble those measured by laboratory instruments. Analysis of imaging spectrometer data allows extraction of a detailed spectrum for each picture element in the image (Figure 7-39). Such spectra often allow direct identification of specific materials within the IFOV of the sensor based upon their reflectance characteristics, including minerals, atmospheric gases, vegetation, snow and ice, and dissolved matter in water bodies.

Analysis of hyperspectral data often requires the use of sophisticated digital image processing software (e.g.,

ENVI). This is because it is usually necessary to calibrate (convert) the raw hyperspectral radiance data to scaled reflectance before it can be properly interpreted. This means removing the effects of atmospheric attenuation, topographic effects (slope, aspect), and any sensor anomalies. Similarly, to get the most out of the hyperspectral data it is usually necessary to use algorithms that 1) analyze spectra to determine its constituent materials, and 2) compare the spectra with a library of spectra obtained using handheld spectroradiometers such as that provided by the U.S. Geological Survey (Jensen, 2005).

Table 7-14.   Characteristics of the NASA Airborne Visible/Infrared Imaging Spectrometer (AVIRIS) and the ITRES Research, Ltd., Compact Spectrographic Imager 1500 (CASI 1500) hyperspectral remote sensing systems.

Sensor	Technology	Spectral Resolution (nm)	Spectral Interval (nm)	Number of Bands	Quantization (bits)	IFOV (mrad)	Total field of view (°)
AVIRIS	Whiskbroom linear array	400 – 2500	10	224	12	1.0	30°
CASI 1500	Linear (1480) and area array CCD (1480 × 288)	400 – 1050	2.2	288 possible at 2.2 nm; the number of bands and the number of pixels in the across-track are programmable.	14	0.49	40.5°

Government agencies and commercial firms have designed hundreds of imaging spectrometers capable of acquiring hyperspectral data. It is beyond the scope of this book to list all of them. Only three systems are summarized: NASA JPL's Airborne Visible/Infrared Imaging Spectrometer, the commercially available Compact Airborne Spectrographic Imager 1500, and NASA's Moderate Resolution Imaging Spectrometer onboard the *Terra* satellite.

Airborne Visible/Infrared Imaging Spectrometer

The first airborne imaging spectrometer (AIS) was built to test the imaging spectrometer concept with infrared area arrays (Vane and Goetz, 1993). The spectral coverage of the instrument was 1.9 to 2.1 μm in the *tree mode* and 1.2 to 2.4 μm in *rock mode* in contiguous bands that were 9.3 nm wide.

To acquire data with greater spectral and spatial coverage, AVIRIS was developed at NASA's Jet Propulsion Laboratory in Pasadena, CA (Table 7-14). Using a whiskbroom scanning mirror and linear arrays of silicon (Si) and indium-antimonide (InSb) configured as in Figure 7-3d, AVIRIS acquires images in 224 bands, each 10 nm wide in the 400 to 2500 nm region (NASA AVIRIS, 2006). The sensor is typically flown onboard the NASA/ARC ER-2 aircraft at 20 km above ground level and has a 30° total field of view and an instantaneous field of view of 1.0 mrad, which yields 20 × 20 m pixels. The data are recorded in 12 bits (values from 0 to 4095)

AVIRIS characteristics are summarized in Figure 7-39. The diagram depicts a single band of AVIRIS imagery (band 30; 655.56 nm) obtained over the Kennedy Space Center, FL, and radiance data extracted for a single pixel of saw palmetto vegetation. Color Plate 7-10 depicts a portion of an AVIRIS dataset acquired over Moffett Field, CA, at the southern end of San Francisco Bay. Three of the 224 spectral bands of data were used to produce the color composite on top of the *hyperspectral datacube*. The black areas in the datacube represent atmospheric absorption bands at 1.4 and 1.9 μm.

Compact Airborne Spectrographic Imager 1500

ITRES Research, Ltd., of Canada markets the CASI remote sensing system. The CASI 1500 is a pushbroom imaging spectrometer based on the use of a 1480 element across-track linear array and a 1480 × 288 area array CCD (Figure 7-3e). The instrument operates over a 650-nm spectral range (400 to 1050 nm) and has a 40.5° total field of view across 1480 pixels (ITRES, 2006).

A single line of terrain 1,480 pixels wide perpendicular to the flight path is sensed by the spectrometer optics (Figure 7-40). The radiance from within the IFOV is spectrally dispersed along the axis of the area array CCD so that a spectrum of energy (from blue through near-infrared) is obtained for each pixel across the swath. By repetitively reading the contents of the area array CCD as the aircraft moves along the flight path, a two-dimensional image at high spectral resolution is acquired. Since the radiant flux for all pixels in a particular swath are recorded simultaneously, spatial and spectral co-registration is assured. The across-track spatial resolution is determined by the altitude of the CASI above ground level and the IFOV, while the along-track resolution depends upon the velocity of the aircraft and the rate at which the CCD is read.

CASI 1500 may be programmed to collect 14-bit data in several modes (Table 7-14) (ITRES, 2006):

Hyperspectral Data Collection Using Linear and Area Arrays

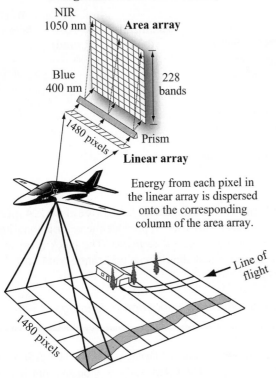

Figure 7-40   Hyperspectral data collection using a linear array pushbroom sensor that distributes the energy onto a 1480 × 228 element area array sensitive to the wavelength region from 400 to 1,050 nm.

- *Spatial mode*—where the full across-track resolution of 1,480 pixels are obtained for up to 19 nonoverlapping spectral bands with programmable center wavelengths and bandwidths.

- *Hyperspectral mode*—up to 1,480 programmable adjacent spatial pixels and up to 288 spectral bands over the full spectral range.

- *Full frame*—1,480 spatial pixels (across-track) and 288 spectral bands.

The specific bandwidths are selected according to the application (e.g., bathymetric mapping, inventorying chlorophyll *a* concentration). The result is a programmable area array remote sensing system that may be the precursor of future satellite hyperspectral sensor systems.

ITRES also markets the SASI 600. This is the first hyperspectral SWIR system to collect infrared data in 100 bands

from 950 to 2450 nm at 15 nm intervals with a swath width of 600 pixels (ITRES, 2006).

Moderate Resolution Imaging Spectrometer

The Moderate Resolution Imaging Spectrometer is flown on NASA's EOS *Terra* (a.m. equatorial crossing time) and *Aqua* (p.m. equatorial crossing time) satellites. MODIS provides long-term observations to derive an enhanced knowledge of global dynamics and processes occurring on the surface of the Earth and in the lower atmosphere (King, 2003; NASA MODIS, 2006). It yields simultaneous observations of high-atmospheric (cloud cover and associated properties), oceanic (sea-surface temperature and chlorophyll), and land-surface (land-cover changes, land-surface temperature, and vegetation properties) features.

MODIS is in a 705-km Sun-synchronous orbit. It views the entire surface of the Earth every one to two days. It has a field of view of ±55° off-nadir, which yields a swath width of 2,330 km. MODIS obtains high radiometric resolution images (12-bit) of daylight-reflected solar radiation and day/night thermal emission over all regions of the globe. MODIS is a whiskbroom scanning imaging radiometer consisting of a cross-track scan mirror, collecting optics, and a set of linear detector arrays with spectral interference filters located in four focal planes (Figure 7-3d). It collects data in 36 coregistered spectral bands: 20 bands from 0.4 to 3 μm and 16 bands from 3 to 15 μm. The bandwidths and their primary uses are summarized in (Table 7-15).

MODIS' coarse spatial resolution ranges from 250 × 250 m (bands 1 and 2) to 500 × 500 m (bands 3 through 7) and 1 × 1 km (bands 8 through 36). A MODIS band 4 (green: 0.545 – 0.565 μm) image of the Nile Delta, the Nile River, and much of the area surrounding the Red Sea is shown in Figure 7-41a. Healthy vegetation absorbs much of the incident green radiant flux, therefore, the vegetated areas are recorded in dark tones in this MODIS image. MODIS provides daylight reflection and day/night emission spectral imaging of any point on Earth at least every two days, with a continuous duty cycle. The swath width is 2330 km across-track. A composite image of global MODIS coverage on April 19, 2000, is shown in Figure 7-41b.

MODIS has one of the most comprehensive calibration subsystems ever flown on a remote sensing instrument. The calibration hardware includes a solar diffuser, a solar diffuser stability monitor, a spectroradiometric calibration instrument, a blackbody for thermal calibration, and a space viewport. The calibration allows the optical data to be converted

Table 7-15.   Characteristics of the *Terra* satellite Moderate Resolution Imaging Spectrometer (MODIS).

Band	Spectral Resolution (μm)	Spatial Resolution	Band Utility
1	0.620 – 0.670	250 × 250 m	Land-cover classification, chlorophyll absorption, leaf-area-index mapping
2	0.841 – 0.876	250 × 250 m	
3	0.459 – 0.479	500 × 500 m	Land, cloud, and aerosol properties
4	0.545 – 0.565	500 × 500 m	
5	1.230 – 1.250	500 × 500 m	
6	1.628 – 1.652	500 × 500 m	
7	2.105 – 2.155	500 × 500 m	
8	0.405 – 0.420	1 × 1 km	Ocean color, phytoplankton, biogeochemistry
9	0.438 – 0.448	1 × 1 km	
10	0.483 – 0.493	1 × 1 km	
11	0.526 – 0.536	1 × 1 km	
12	0.546 – 0.556	1 × 1 km	
13	0.662 – 0.672	1 × 1 km	
14	0.673 – 0.683	1 × 1 km	
15	0.743 – 0.753	1 × 1 km	
16	0.862 – 0.877	1 × 1 km	
17	0.890 – 0.920	1 × 1 km	Atmospheric water vapor
18	0.931 – 0.941	1 × 1 km	
19	0.915 – 0.965	1 × 1 km	
20	3.600 – 3.840	1 × 1 km	Surface–cloud temperature
21	3.929 – 3.989	1 × 1 km	
22	3.929 – 3.989	1 × 1 km	
23	4.020 – 4.080	1 × 1 km	
24	4.433 – 4.498	1 × 1 km	Atmospheric temperature
25	4.482 – 4.549	1 × 1 km	
26	1.360 – 1.390	1 × 1 km	Cirrus clouds
27	6.535 – 6.895	1 × 1 km	Water vapor
28	7.175 – 7.475	1 × 1 km	
29	8.400 – 8.700	1 × 1 km	
30	9.580 – 9.880	1 × 1 km	Ozone
31	10.780 – 11.280	1 × 1 km	Surface–cloud temperature
32	11.770 – 12.270	1 × 1 km	
33	13.185 – 13.485	1 × 1 km	Cloud-top altitude
34	13.485 – 13.785	1 × 1 km	
35	13.785 – 14.085	1 × 1 km	
36	14.085 – 14.385	1 × 1 km	

into scaled reflectance. MODIS data are being processed to create numerous global datasets, including (NASA MODIS, 2006):

• land ecosystem variables (e.g., vegetation indices, leaf area index, fraction of photosynthetically active radiation, vegetation net primary production);

Moderate Resolution Imaging Spectrometer (MODIS) Imagery

a. MODIS band 4 image of the Nile Delta
obtained on February 28, 2000.

b. A composite band 1 image of global *Terra* MODIS
coverage on April 19, 2000.

Figure 7-41   *Terra* MODIS band 4 (green: 0.545 – 0.565 μm) image of the heavily vegetated Nile Delta, the Nile River, and much of the area surrounding the Red Sea. Band 4 has a spatial resolution of 500 × 500 m. b) A composite band 1 (red: 0.62 – 0.67 μm) image of the global *Terra* MODIS coverage on April 19, 2000. The MODIS swath width is 2,330 km across track (courtesy NASA Goddard Space Flight Center).

- atmospheric variables (e.g., cloud fraction, cloud optical thickness, aerosol optical depth, etc.);

- ocean variables (e.g., sea-surface temperature and chlorophyll).

Digital Frame Cameras Based on Area Arrays

The charge-coupled-device (CCD) was invented in the late 1960s by scientists at the Bell Labs. It was originally conceived as a new type of computer memory circuit, but it soon became apparent that it had many other applications, including image data collection, because of the sensitivity of silicon to light.

The general characteristics of small-, medium-, and large-format digital cameras were introduced in Chapter 4. This section summarizes the characteristics and provides examples of imagery associated with three representative examples of modern digital cameras. We begin with the medium format Leica GeoSystems Emerge Digital Sensor System

(DSS) based on area array technology (Figure 7-3f) and then progress to more sophisticated large-format digital cameras based on *area* array (frame) technology (Z/1 DMC and Vexcel UltraCam).

Small-format Digital Data Collection: Leica Geosystems Emerge Digital Sensor System

The Digital Sensor System (DSS) uses a digital camera area array that acquires imagery containing 4,092 × 4,079 pixels (Table 7-16). Each pixel in the area array is 9 × 9 microns. Users can specify color (blue, green, and red) or color-infrared (green, red, and near-infrared) multiband imagery in the spectral region from 0.4 to 0.9 μm. In near-infrared mode, the sensor has a spectral response similar to that of Kodak Aerochrome 2443 color-infrared film, but with higher dynamic range. The data may be recorded at 8 or 16 bits per pixel. Emerge collects real-time differentially corrected GPS data about each digital frame of imagery. These data are used to mosaic and orthorectify the imagery using photogrammetric techniques introduced in Chapter 6. The pixel placement accuracy meets national map accuracy standards (Light, 2001; Leica Emerge, 2006).

Table 7-16.   Leica Geosystems Emerge Digital Sensor System (DSS) characteristics.

Band	Spectral Resolution Natural Color Mode (nm)	Spectral Resolution Near-infrared Mode (nm)	Spatial Resolution (m)
1	400 – 500	–	Variable, but usually 0.15 to 3 m
2	500 – 600	510 – 600	Variable, but usually 0.15 to 3 m
3	600 – 680	600 – 700	Variable, but usually 0.15 to 3 m
4	–	800 – 900	Variable, but usually 0.15 to 3 m

Sensor	Area arrays (4092 × 4079)
Swath	4092 pixels
Detector size	0.009 mm
Radiometric Resolution	8- and 16-bit

A variety of flying heights and different focal-length Nikon lenses can be used to obtain imagery with spatial resolutions ranging from 0.15 to 3 m. Figure 7-42 depicts multispectral bands of Emerge 1 × 1 m imagery obtained over Dunkirk, N.Y. Such high-resolution data can be collected on demand in good weather. A color-infrared composite of the same scene is found in Color Plate 7-11a. A natural color Emerge scene of an area adjacent to a New York Power Authority lake is found in Color Plate 7-11b.

Large-format Digital Data Collection: Z/1 Digital Modular Camera

The Z/1 Digital Modular Camera (DCM) uses four 7000 × 4000 CCD area arrays to obtain one complete frame of imagery. All CCDs have their own optics and function as stand-alone digital cameras that are synchronized to collect data at the exact same instant. Four additional CCDs with 3000 × 2000 detectors are used to obtain blue, green, red, and near-infrared multispectral data at the same instant. Therefore, the multispectral bands have reduced ground res-

olution compared to the panchromatic data. Individual frames of imagery are obtained just like a traditional frame camera with user-specified end-lap between frames (Hinz et al., 2001).

Large-format Digital Data Collection: Vexcel UltraCam based on Area Arrays

The Vexcel large-format digital camera uses eight optical cones (i.e., viewing ports) with 13 area array CCDs strategically located behind the eight ports. Nine of the 4000 × 2700 CCD arrays are used to collect panchromatic data. Four of the CCD arrays are used to collect blue, green, red, and near-infrared multispectral data. Each frame of imagery is 11,500 × 7500 pixels in dimension. Individual frames of imagery are obtained with user-specified end-lap between frames (Vexcel, 2006).

U.S. Space Shuttle Digital Photography

NASA astronauts document Earth processes during Space Shuttle Space Transportation System (STS) missions using both analog and digital camera systems (Lulla and Dessinov, 2000). Photography are cataloged into a database according to the specific mission (e.g., STS-74) or by thematic topic (NASA Shuttle Photography, 2006).

Space Shuttle Analog Cameras

Hasselblad and Linhof analog cameras are used on Space Shuttle missions. The four windows in the aft part of the Space Shuttle are used to obtain photography of the Earth. The windows allow 0.4 to 0.8 µm light to pass through. Very low-Sun angle photography obtained from the Space Shuttle often provides unique topographic views of remote mountainous areas otherwise poorly mapped. Seventy-five percent of the photographs in the archive cover the regions between 28° N and 28° S latitude, providing coverage for many little-known tropical areas. The Space Shuttle Earth Observations Project (SSEOP) at NASA Johnson Space Center contains the records of more than 400,000 photographs (http://images.jsc.nasa.gov).

Space Shuttle and Space Station Digital Photography

The International Space Station (ISS) was launched November 2, 2000. The United States ISS Laboratory Module has a specially designed optical window with a clear aperture 50.8 cm in diameter that is perpendicular to the Earth's surface most of the time. In 2001, astronauts began acquiring digital

Digital Frame Camera Imagery of Dunkirk, NY

a. Green band.

b. Red band.

c. Near-infrared band.

Figure 7-42   Leica Geosystems Emerge Digital Sensor System (DSS) multispectral imagery of Dunkirk, NY. The data were collected on December 12, 1998, at a spatial resolution of 1 × 1 m (courtesy Littton Emerge, Inc.).

images using a Kodak DCS 760 camera with Nikon lenses. Both color and monochrome digital images may be obtained. The camera uses a 3032 × 2008 CCD. Digital images are transmitted directly to the ground. A digital photograph of the Toquepala Copper Mine in Southern Peru taken from the International Space Station on September 22, 2003, is shown in Figure 7-43.

 ## References

DigitalGlobe, 2005, *DigitalGlobe Unveils Plans for WorldView I and II*, Oct., 4; www.digitalglobe.com.

DigitalGlobe, 2006, *DigitalGlobe*, www.digitalglobe.com.

Filippi, A., 1999, *Hyperspectral Image Classification Using A Batch Descending Fuzzy Learning Vector Quantization Artificial Neural Network: Vegetation Mapping at the John F. Kennedy Space Center*, unpublished Master's thesis, Columbia: University of South Carolina, 276 p.

Gasparovic, R. F., 2003, *Ocean Remote Sensing Program*, John Hopkins University Applied Physics Laboratory, http://fermi.jhuapl.edu/avhrr/index.html.

GeoEye, 2006, *ORBIMAGE Completes Acquisition of Space Imaging; Changes Brand Name to GeoEye*, Dulles, VA: GeoEye, Inc. Press release January 12; www.geoeye.com.

Hinz, A., C. Doerstel, and H. Meier, 2001, *DMC – The Digital Sensor Technology of Z/I Imaging*, in D. Fritsch and R. Spiller (Eds.), *Photogrammetric Week 2001* Heidelberg: Wichmann, 93-103.

ImageSat, 2006, *EROS-1A*, www.imagesatintl.com/.

ITRES, 2006, *Calibrated Airborne Spectrographic Imager 1500*, Canada: ITRES Research Ltd., www.itres.com.

Jensen, J. R., 2005, *Introductory Digital Image Processing: A Remote Sensing Perspective*, Upper Saddle River: Prentice-Hall, 525 p.

Jensen, J. R., Botchway, K., Brennan-Galvin, E., Johannsen, C., Juma, C., Mabogunje, A., Miller, R., Price, K., Reining, P., Skole, D., Stancioff, A. and D. R. F. Taylor, 2002, *Down to Earth: Geographic Information for Sustainable Development in Africa*, Washington: National Academy Press, 155 p.

Kidwell, K. B., 1998, *NOAA Polar Orbiter Data User's Guide: TIROS-N through NOAA-14*, Washington: NOAA U.S. Dept. of Commerce, 50 p.

Digital Frame Camera Image of the Toquepala Copper Mine in Southern Peru Taken from the International Space Station

Figure 7-43   Astronaut photograph of the Toquepala Copper Mine in southern Peru taken from the International Space Station on September 22, 2003, with a Kodak DCS 760 digital camera (photo #ISS007-E-15222). The open pit mine is 6.5 km across and descends more than 3,000 m into the earth. A dark line on the wall of the pit is the main access road to the bottom. Spoil dumps of material mined from the pit are arranged in tiers along the northwest lip of the pit (courtesy of NASA Earth Observatory and Dr. Kamlesh Lulla, NASA Johnson Space Center).

King, M., 2003, *EOS Data Product Handbook - Volume 1*, Washington: NASA, 258 p.

Krishnaswamy, M. and S. Kalyanaraman, 2006, "Indian Remote Sensing Satellite Cartosat-1: Technical Features and Data Products," *GIS Development: The Geospatial Portal*, http://www.gisdevelopment.net/technology/rs/techrs023.htm.

Leica Emerge, 2006, *Leica Emerge*, http://gis.leica-geosystems.com/products/emerge_digital_sensor/emerge_dss.asp.

Light, D. L., 2001, "An Airborne Direct Digital Imaging System," *Photogrammetric Engineering & Remote Sensing*, 67(11):1299–1305.

Loral Space Systems, 1996, *GOES I-M DataBook*, Loral Space Systems, #DRL 101-08, Revision 1; 200 p.

Lulla, K. and L. Dessinov, 2000, *Dynamic Earth Environments: Remote Sensing Observations from Shuttle-Mir Missions*, N.Y.: John Wiley, 268 p.

NASA Airborne, 2006, *NASA Airborne Sensor Facility*, Moffett Field, CA: NASA Ames Airborne Sensor Facility, http://asapdata.arc.nasa.gov/.

NASA ASTER, 2006, *Advanced Spaceborne Thermal Emission and Reflection Radiometer*, http://asterweb.jpl.nasa.gov/.

NASA AVIRIS, 2006, *Airborne Visible Infrared Imaging Spectrometer*, http://aviris.jpl.nasa.gov/.

NASA EO-1, 2006, *Earth Observer-1*, http://eo1.gsfc.nasa.gov/.

NASA ESE, 2006, *Earth Science Enterprise*, http://science.hq.nasa.gov/strategy/.

NASA Landsat 7, 2006, *Landsat 7*, http://landsat.gsfc.nasa.gov/.

NASA LDCM, 2006, *Landsat Data Continuity Mission*, http://ldcm.nasa.gov.

NASA MISR, 2006, *Multiangle Imaging Spectrometer*, http://www-misr.jpl.nasa.gov/index.html.

NASA MODIS, 2006, *Moderate Resolution Imaging Spectrometer*, http://modis.gsfc.nasa.gov/.

NASA/Orbimage SeaWiFS, 2006, *Sea-viewing Wide Field-of-view Sensor*, http://oceancolor.gsfc.nasa.gov/SeaWiFS/.

NASA Shuttle Photography, 2006, *Space Shuttle Earth Observation Photography*, http://earth.jsc.nasa.gov/sseop/efs/.

NOAA AVHRR, 2006, *Advanced Very High Resolution Radiometer*, http://www.oso.noaa.gov/poes/.

NOAA DMSP, 2006, *Defense Meteorological Satellite Program*, http://www.oso.noaa.gov/dmsp/.

NOAA GOES, 2006, *Geostationary Operational Environmental Satellite*, http://www.nesdis.noaa.gov/satellites.html.

NOAA GVI, 2006, *Global Vegetation Index Products*, http://www2.ncdc.noaa.gov/docs/gviug/.

NOAA/NASA POES, 2006, *NOAA/NASA Polar Operational Environmental Satellites (NPOES)*, http://goespoes.gsfc.nasa.gov/.

NOAA NPOESS, 2006, *National Polar Orbiting Operational Environmental Satellite System*, http://www.ipo.noaa.gov/.

NOAA NPP, 2006, *National Polar Orbiting Operational Environmental Satellite System Preparatory Project*, http://www.ipo.noaa.gov/Projects/npp.html.

NRSA, 2006, *Indian National Remote Sensing Agency*, http://www.nrsa.gov.in/engnrsa/sitemap.html.

Orbimage, 2006, *OrbView-3 and OrbView-5*, www.orbimage.com.

OSTP, 2005, *Memorandum: Landsat Data Continuity Strategy Adjustment*, Washington: Office of Science, Technology, and Policy, http://ldcm.gsfc.nasa.gov/docs/OSTP_ Landsat_memo_12-1AF496.pdf

SBRC, 1994, *Space Sensors*, Goleta: Santa Barbara Research Center, 33 p.

SenSyTech, 2004, *Airborne Multispectral Scanner*, www.sensystech.com.

Solomonson, V., 1984, "Landsat 4 and 5 Status and Results from Thematic Mapper Data Analyses," *Proceedings*, *Machine Processing of Remotely Sensed Data*, W. Lafayette: Lab for the Applications of Remote Sensing, 13–18.

Space Imaging, 2006, *Space Imaging*, www.spaceimaging.com.

SPOT Image, 2006, *SPOT*, www.spotimage.com/home/.

USGS Global Landcover, 2006, *Global Landcover Characterization*, http://edcsns17.cr.usgs.gov/glcc/.

USGS GloVis, 2006, *Global Visualization Viewer*, http://glovis.usgs.gov.

USGS Landsat 5, 2006, *Landsat 5 Update*, Sioux Falls: USGS EROS Data Center, http://landsat.usgs.gov.

Vane, G. and A. F. H. Goetz, 1993, "Terrestrial Imaging Spectrometry: Current Status, Future Trends," *Remote Sensing of Environment*, 44:117–126.

Vexcel, 2006, *UltraCam Large Format Digital Camera*, http://www.vexcel.com/.

Thermal Infrared Remote Sensing 8

Objects that have a temperature above absolute zero (0 K) emit electromagnetic energy. Therefore, all the features we encounter in the landscape everyday, such as vegetation, soil, rock, water, and people, emit thermal infrared electromagnetic radiation in the 3.0 – 14 μm portion of the spectrum. Humans experience this thermal energy primarily through our sense of touch. For example, we feel the thermal radiant energy from the Sun or the radiant energy from a fire on our face. However, our eyes cannot detect subtle differences in thermal infrared energy emitted from real-world objects because our eyes are primarily sensitive to short-wavelength visible light from 0.4 – 0.7 μm. Our eyes are not sensitive to the reflective infrared (0.7 – 3.0 μm) or thermal infrared energy (3.0 – 14 μm). Fortunately, engineers have developed detectors that are sensitive to thermal infrared radiation. These thermal infrared sensors allow humans to sense a previously invisible world of information as they monitor the thermal characteristics of the landscape.

When we are ill, one of the first things we do is take our temperature to determine if it is different from the normal 98.6 °F (37 °C). If our temperature is elevated, it usually means something is wrong. Similarly, the various components of the landscape such as vegetation, soil, rock, water, concrete, and asphalt shingles have predictable thermal characteristics based on how they selectively absorb solar short-wavelength energy and radiate thermal infrared energy. Thermal infrared remote sensing systems record thermal infrared radiation that can be used to determine 1) the type of material based on its thermal emission characteristics and/or 2) if significant changes have taken place in the thermal characteristics of these phenomena through time (Quattrochi and Luvall, 2004). In this manner, it is possible to identify the surface physical manifestation of certain diseases in humans (e.g., perhaps the existence of a tumor), stress in plants, thermal pollution in water bodies, or the loss of heat from buildings due to faulty insulation.

Aerial thermal infrared imagery has not been widely available to the general public due to 1) the relatively high cost of the thermal sensor, 2) the cost of mobilizing an aircraft to acquire the remotely sensed data, and 3) the difficulty of calibrating and correctly interpreting the imagery. The public knows about thermal infrared remote sensing largely from close-range applications such as those shown in Figure 8-1, including: residential and commercial heat-loss insulation studies, the use of handheld thermal imaging units to locate hot spots in homes and to find other human beings in a fire or at night, thermal images of humans to detect a variety of medical problems (sometimes referred to as *thermography)*, and nondestructive testing and evaluation of electronic components. The public also sees thermal infrared imagery being used to allocate firefighting resources during a forest fire or to target enemy facilities as in the Gulf War in 1991 or the Iraq War in 2006. The public in general does not realize that the night-time GOES images of weather fronts displayed on the nightly news are thermal infrared images.

Thermal infrared remote sensing will become more important in the future as additional orbital sensors obtain thermal data and costs decrease. Also, government agencies (e.g., police, drug enforcement, immigration border-patrol officers) now routinely use handheld thermal infrared sensors and forward-looking infrared sensors (FLIR) mounted in aircraft to look for missing persons and criminal activity.

 History of Thermal Infrared Remote Sensing

Sir Frederick William Herschel (1738–1822; Figure 8-2), probably the most famous astronomer of the 18th century, discovered the infrared portion of the electromagnetic spectrum and described it in his famous paper "Investigations of the Powers of the Prismatic Colours to Heat and Illuminate Objects: with Remarks." In 1800, Herschel created a spectrum of colors using a prism and measured different parts of the spectrum with a thermometer to see if some colors were warmer than others. As he moved the thermometer toward the red portion of the spectrum, the temperature increased. It seemed reasonable to move the thermometer just past the red end in order to watch the heating effect disappear. It did not. Instead, the temperature continued to increase beyond the red end of the spectrum (Figure 8-3). Herschel called this region of the electromagnetic spectrum *infrared*, which means "below the red" (Omega, 2005). Herschel was also the first person to correctly describe the form of our galaxy, The Milky Way.

In 1879, S. P. Langley began a research program to find a superior radiation detector. A year later he invented the bolometer that was able to measure temperature variations of 1/10,000 °C. In World War I, S. O. Hoffman was able to detect men at 120 m and eventually aircraft. In the 1930s, Germany developed the Kiel system for discriminating between bombers and night fighters. The British and the United States also developed infrared surveillance techniques in World War II. In fact, the single most important development in infrared technology was the invention of the detector element by warring nations during World War II.

Early infrared detectors consisted of lead salt photodetectors (Fischer, 1983). Now we have very fast detectors consisting of mercury-doped germanium (Ge:Hg), indium antimonide (In:Sb), and other substances that are very sensitive to infrared radiation. We also have computers to rapidly process the energy recorded by the sensors and to display the thermal characteristics of the scene.

Thus, it took about a century for governments to understand that remote sensing in the thermal infrared region could provide valuable tactical reconnaissance information, especially since the images could be recorded both in the daytime and at night. In the 1950s, the government contracted with civilian firms to improve thermal infrared technology (e.g., Texas Instruments). In the 1960s, some of these contractors received permission from the government to use the classified sensors to produce thermal infrared images for a few select civilian clients (Estes, 1966). In 1968, the government declassified the production of thermal infrared remote sensing systems that did not exceed a certain spatial resolution and temperature sensitivity. Thermal infrared remote sensing systems developed by Texas Instruments, Inc., Daedalus Enterprises, Inc., Rockwell International, Inc., etc., were first carried aloft by aircraft. Oil companies conducting geological exploration requested much of the early thermal infrared data and continue to be major consumers. Thermal infrared remote sensing systems mounted onboard aircraft continue to collect much of the on-demand thermal infrared data for public agencies (e.g., Environmental Protection Agency, Department of Energy, state departments of natural resources) and foreign governments.

The first declassified satellite remote sensor data were collected by the U.S. Television IR Operational Satellite (TIROS) launched in 1960. The coarse resolution thermal infrared data were ideal for monitoring regional cloud patterns and frontal movement. NASA launched the Heat Capacity Mapping Mission (HCMM) on April 26, 1978. It obtained 600×600 m spatial resolution thermal infrared data (10.5 – 12.6 μm) both day (1:30 p.m.) and night (2:30 a.m.). This was one of the first scientifically oriented (geology) thermal infrared systems. NASA's *Nimbus 7,* launched on October 23, 1978, had a Coastal Zone Color Scanner (CZCS) that included a thermal infrared sensor for monitoring sea-surface temperature. In 1980, NASA and the Jet Propulsion Laboratory developed the six-channel Thermal Infrared Multispectral Scanner (TIMS) that acquired thermal infrared energy in six bands at wavelength intervals of ≤1.0 μm (Quattrochi and Ridd, 1994). Successful studies using TIMS resulted in the development of the 15-channel Airborne Terrestrial Applications Sensor (ATLAS) (Lo et al., 1997).

The NOAA Geostationary Operational Environmental Satellite (GOES) collects thermal infrared data at a spatial reso-

Thermal Infrared Imagery

Figure 8-1   a) Thermal infrared image of an F-116 jet in flight. Note the hot afterburners. b) Thermal infrared image of radiant energy leaving a residential house, especially through the windows. Note the cool, insulated roof and metal water downspout. c) This house on blocks is losing much of its energy through the floor. d) Thermal image of a crime in progress at night. e) Thermal infrared image of an Apache helicopter with hot jet engines. f) Thermal infrared image of a F-117 jet. g) Thermal image of a power transformer revealing several very hot wires. h) Nondestructive thermal testing of a printed circuit board. i) Thermal image of a paddlewheel boat and passengers (courtesy FLIR Systems, Inc. and others).

Frederick William Herschel

Figure 8-2   Sir Frederick William Herschel (1738–1822) discovered the infrared portion of the electromagnetic spectrum in 1800 (from Herschel Space Observatory, NASA Jet Propulsion Lab).

lution of 8 × 8 km for weather prediction. Full-disk images of the Earth are obtained every 30 minutes both day and night by the thermal infrared sensor. Also, the NOAA Advanced Very High Resolution Radiometer (AVHRR) collects thermal infrared local area coverage (LAC) data at 1.1 × 1.1 km and global area coverage (GAC) at 4 × 4 km.

Landsat Thematic Mapper 4 and 5 sensors were launched on July 16, 1982, and March 1, 1984, respectively, and collected 120 × 120 m thermal infrared data (10.4 – 12.5 μm) along with two bands of middle-infrared data (1.55 – 1.75 and 2.08 – 2.35 μm). Landsat 7 was launched on April 15, 1999 with a 60 × 60 m well-calibrated thermal infrared sensor (10.4 – 12.5 μm). The Advanced Spaceborne Thermal Emission and Reflection Radiometer (ASTER) onboard *Terra* has six channels from 1.60 – 2.43 μm and five channels from 8.125 – 11.65 μm. Chapter 7 provides detailed information about the spatial, spectral, temporal, and radiometric characteristics of these and other thermal infrared sensor systems.

 ### Thermal Infrared Radiation Properties

An image analyst should not interpret a thermal infrared image as if it were an aerial photograph or a typical image produced by an optical-mechanical multispectral scanner or

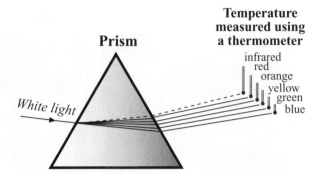

Figure 8-3   Herschel used a prism and a thermometer to discover that temperature increased beyond the red portion of the spectrum.

charge-coupled-device (CCD) sensor system. Rather, the analyst must think *thermally*. He or she must understand 1) how the short-wavelength energy radiated from the Sun interacts with the atmosphere, 2) how it interacts with Earth surface materials (i.e., some of the energy is transformed into longer-wavelength energy), 3) how the energy emitted by the terrain interacts with the atmosphere once again, and finally, 4) how a remote sensing detector records the *emitted* thermal infrared electromagnetic radiation. The analyst should also understand how both the sensor system itself and the terrain can introduce noise into the thermal infrared image that might make the data less useful or lead to incorrect image interpretation.

Kinetic Heat, Temperature, Radiant Energy, and Radiant Flux

All objects in the real world having a temperature above absolute zero (0 K; -273.16 °C; -459.69 °F) exhibit random motion. The energy of particles of molecular matter in random motion is called *kinetic heat* (also referred to as internal, real, or true heat). When these particles collide, they change their energy state and emit electromagnetic radiation as discussed in Chapter 2. The amount of heat can be measured in *calories*. We can measure the true kinetic temperature (T_{kin}) or concentration of this heat using a thermometer. We perform *in situ* (in-place) temperature measurement by placing the thermometer in direct physical contact with a plant, soil, rock or water body.

Fortunately, an object's internal kinetic heat is also converted to *radiant energy* (often called external or apparent energy), which allows us to utilize remote sensing technology. The electromagnetic radiation exiting an object is called *radiant flux* (Φ) and is measured in Watts as discussed in

Figure 8-4   Linear regression of remote sensing–derived water temperature (T_{rad}) versus true kinetic water temperature (T_{kin}) obtained by *in situ* measurement for four survey years in five different streams (Torgersen et al., 2001; reprinted with permission from Elsevier).

Chapter 2. The concentration of the amount of radiant flux exiting (emitted from) an object is its *radiant temperature* (T_{rad}). For most real world objects (except those composed of glass and metal) there is usually a high positive correlation between the true kinetic temperature of the object (T_{kin}) and the amount of radiant flux radiated from the object (T_{rad}). For example, Figure 8-4 correlates remote sensing-derived radiant water temperature with *in situ* water temperature measurement in streams (Torgersen et al., 2001). The remote sensing-derived temperatures were consistently within ± 0.5 °C of the ground reference measurements over four survey years in five different streams. Recent studies confirm these results (Oregon DEC, 2003). Examples of water surface temperature derived from forward-looking infrared (FLIR) data for two streams in Oregon are shown in Color Plate 8-1.

Relationships such as these suggest that we can utilize radiometers placed some distance from an object to measure its radiant temperature, which hopefully correlates well with the object's true kinetic temperature. *This is the basis of thermal infrared remote sensing temperature measurement.*

Unfortunately, the relationship is not perfect, with the remote measurement of the radiant temperature always being somewhat less than the true kinetic temperature of the object, especially for land surfaces. This is due to a thermal property called *emissivity,* to be discussed shortly (Schmugge et al., 2002).

Methods of Transferring Heat

The heat generated by the random motion of particles may be transferred from one location to another by conduction, convection, and radiation, as discussed in Chapter 2. Thermonuclear fusion taking place on the Sun produces a plasma of radiant flux consisting primarily of short-wavelength visible light that travels 93 million miles through the vacuum of space at the speed of light (3×10^8 m sec^{-1}). Some of this short-wavelength energy passes through the atmosphere and is absorbed by the Earth's surface materials and reradiated (emitted) at longer wavelengths. Some of this emitted longer-wavelength electromagnetic radiation passes through the atmosphere once again and can be recorded using airborne thermal infrared detectors. Hopefully, the longer-wavelength radiation recorded by the detectors provides valuable information about the temperature characteristics of the Earth's surface.

 Thermal Infrared Atmospheric Windows

Beyond the *visible* region of the electromagnetic spectrum, we encounter the *reflective infrared* region from 0.7 – 3 μm and the *thermal infrared* region from 3 – 14 μm (Figure 8-5). The only reason we can use remote sensing instruments to detect infrared energy in these regions is because the atmosphere allows a portion of the infrared energy to be transmitted from the terrain to the detectors (Schmugge et al., 2002). We call regions that pass energy *atmospheric windows*. Conversely, the black areas in Figure 8-5 denote regions of the electromagnetic spectrum where the atmosphere absorbs most of the infrared energy present. We call these regions *absorption bands*. Water vapor (H_2O), carbon dioxide (CO_2), and ozone (O_3) are responsible for most of the absorption. The atmosphere "closes down," making it almost impossible to perform remote sensing of the environment in these regions. For example, atmospheric water vapor (H_2O) absorbs most of the energy exiting the terrain in the region from 5 – 7 μm, making it almost useless for thermal infrared remote sensing.

Remote sensing instruments can be engineered to be sensitive to the infrared energy present within just the atmospheric windows. For example, film emulsions can be made

Atmospheric Windows in the Electromagnetic Spectrum

Figure 8-5   Atmospheric windows in the electromagnetic spectrum shown here are of significant value for remote sensing reflective and thermal infrared energy. Photographic films can be made sensitive to reflective energy from $0.7 - 1.3$ μm. Electro-optical sensing systems can record infrared energy from $0.7 - 14$ μm. The $3 - 5$ μm region is especially useful for monitoring hot targets such as forest fires and geothermal activity. Vegetation, soil, and rock are best monitored using the $8 - 14$ μm region for suborbital data collection. The $10.5 - 12.5$ μm region is used when thermal imagery is acquired from orbital sensors above the Earth's ozone layer.

sensitive to reflected infrared energy in the window from $0.7 - 1.3$ μm. Eastman Kodak's 2443 color-infrared film works within this *photographic infrared* region and is ideal for monitoring vegetation and water (Chapter 4). Electro-optical detectors on Landsat Thematic Mapper 4 and 5 are sensitive to the reflective middle-infrared windows from $1.55 - 1.75$ μm (TM band 5) and $2.08 - 2.35$ μm (TM band 7).

Electronic detectors can also be made sensitive to photons of thermal infrared radiant energy exiting the terrain in the two primary thermal infrared windows: $3 - 5$ μm and $8 - 14$ μm. Suborbital thermal infrared remote sensing systems utilize these spectral bands. However, the Earth's ozone (O_3) layer absorbs much of the thermal energy exiting the terrain in an absorption band from approximately $9.2 - 10.2$ μm. Therefore, satellite thermal infrared remote sensing systems often only record data in the region from $10.5 - 12.5$ μm (Figure 8-5) to avoid this absorption band. For example, ASTER band 12 is $8.925 - 9.275$ μm and band 13 is $10.25 - 10.95$ μm. The region from $9.276 - 10.24$ μm is not sensed due to atmospheric absorption.

Thermal Radiation Laws

A *blackbody* is a theoretical construct that absorbs all the radiation that falls on it and radiates energy at the maximum possible rate per unit area at each wavelength for any given temperature. No objects in nature are true blackbodies; how-

ever, we may think of the Sun as approximating a 6,000 K blackbody and the Earth as a 300 K blackbody. If we pointed a sensor at a blackbody, we would be able to record quantitative information about the total amount of radiant energy in specific wavelengths exiting the object and the dominant wavelength of the object. In order to do this, we utilize two important physical laws: the *Stefan-Boltzmann law* and *Wien's displacement law*.

Stefan-Boltzmann Law

The total spectral radiant exitance (M_b) measured in Watts m^{-2} leaving a blackbody (refer to Table 2-4) is proportional to the fourth power of its temperature (T). This is known as the *Stefan-Boltzmann law* and is expressed as:

$$M_b = \sigma T^4 \tag{8-1}$$

where σ is the Stefan-Boltzmann constant equaling 5.6697×10^{-8} W m^{-2} K^{-4}, and T is temperature in degrees Kelvin. The total radiant exitance is the integration of all the area under the blackbody radiation curve (Figure 8-6). Notice how the Sun produces more spectral radiant exitance (M_b) at 6,000 K than the Earth at 300 K. As the temperature increases, the total amount of radiant energy measured in Watts per m^2 (the area under the curve) increases and the radiant energy peak shifts to shorter wavelengths. To determine the dominant wavelength for a blackbody at a specific temperature, we use Wien's displacement law.

Wien's Displacement Law

The relationship between the true temperature of a blackbody (T) in degrees Kelvin and its peak spectral exitance or dominant wavelength (λ_{max}) is described by *Wien's displacement law*:

$$\lambda_{max} = \frac{k}{T} \qquad (8\text{-}2)$$

where k is a constant equaling 2898 µm K. We can determine the dominant wavelength of any object by substituting its temperature into Equation 8-2. Remember from Chapter 2 and Figure 8-6 that the dominant wavelength of the 6000 K Sun is 0.48 µm. The dominant wavelength for an 800 K red-hot object (Figure 8-6) is:

$$\lambda_{max} = \frac{2898\,\mu m\ k}{800\ K}$$

$$\lambda_{max} = 3.62\,\mu m\,.$$

We see a shift from longer to shorter wavelengths as the temperature of the blackbody increases. We can observe Wien's displacement law in real life. For example, when a poker is placed in the fire the tip progresses from dark red through orange and then to yellow. It never shifts into the green or blue portion of the spectrum because it is not that hot. Conversely, an acetylene torch has a hot flame and appears blue.

Why is knowing an object's dominant wavelength important to thermal infrared remote sensing? The dominant wavelength provides valuable information regarding the part of the thermal infrared spectrum in which we might want to sense the object. For example, if we are looking for 800 K forest fires that have a dominant wavelength of approximately 3.62 µm, then the most appropriate remote sensing system might be a 3 – 5 µm thermal infrared detector. Conversely, if we are interested in soil, water, and rock ambient temperatures on the Earth's surface (300 K) with a dominant wavelength of 9.67 µm, then a thermal infrared detector operating in the 8 – 14 µm region might be most appropriate.

Emissivity

The world is not composed of radiating blackbodies. Rather, it is composed of *selectively radiating bodies*, such as rock, soil, vegetation, and water, that emit a certain proportion of the energy emitted from a blackbody at the same tempera-

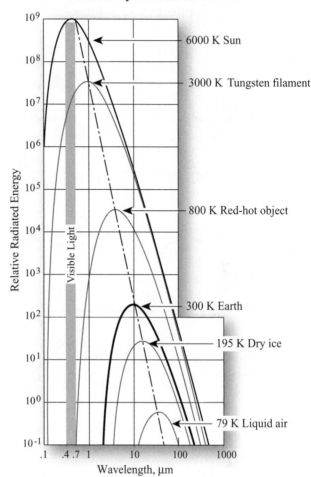

Blackbody Radiation Curves

Relative Radiated Energy

Visible Light

6000 K Sun

3000 K Tungsten filament

800 K Red-hot object

300 K Earth

195 K Dry ice

79 K Liquid air

Wavelength, µm

Figure 8-6   Blackbody radiation curves for several objects, including the Sun and the Earth, which approximate 6,000 K and 300 K blackbodies, respectively. The dominant wavelength, λ_{max}, shifts toward the short wavelength portion of the spectrum as the temperature of the object increases.

ture. *Emissivity*, (ε) is the ratio between the actual radiance emitted by a real world selective radiating body (M_r) and a blackbody at the same thermodynamic (kinetic) temperature (M_b) (Jacob et al., 2004):

$$\varepsilon = \frac{M_r}{M_b}\,. \qquad (8\text{-}3)$$

All selectively radiating bodies have emissivities ranging from 0 to ≤ 1 that fluctuate depending upon the wavelengths of energy being considered. A *graybody* outputs a constant emissivity that is less than one at all wavelengths. Figure 8-7a depicts the emissivity of a blackbody, a graybody, and a a.

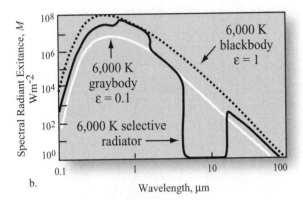

b.

Figure 8-7   a) Spectral emissivity of a blackbody, a graybody, and a hypothetical selective radiator. b) Spectral radiant exitance distribution of the blackbody, graybody, and hypothetical selective radiator (after Slater, 1980).

hypothetical selective radiator over the wavelength interval $0.1 - 100$ μm (Slater, 1980). Notice how the spectral emissivity of the hypothetical selective radiator fluctuates among levels of 0, 0.1, 0.3, and 1.0. The spectral radiant exitance of each of these 6,000 K bodies is shown in Figure 8-7b. Notice how the different emissivities give rise to dramatic changes in the spectral radiant exitance distribution of the selective radiator. Where the selective radiator's emissivity is 1.0, it outputs the same amount of radiant energy as the blackbody. Where the selective radiator's emissivity is 0, it emits no spectral radiant exitance.

The spectral radiant exitance for several real world radiating bodies is shown in Figure 8-8. Note that the radiant energy exiting the substance is approximately the same as a blackbody at the same temperature for much of the spectral range, but that the curves depart in certain areas. If the area beneath each of the curves was summed (integrated) over the spectral wavelength interval of interest on the x-axis, we would

find that the real-world spectral radiant exitance was always less than the blackbody radiant exitance at the same temperature. Thus, the emissivity of the real-world material would lie somewhere between 0 and 1 but would never be equal to 1. Some materials like distilled water have emissivities close to one (0.99) over the wavelength interval from $8 - 14$ μm, as summarized in Table 8-1. Others such as polished aluminum (0.08) and stainless steel (0.16) have low emissivities.

Why is it important to know about emissivity when conducting a thermal infrared remote sensing investigation? The reason is that two objects lying next to one another on the ground could have the same true kinetic temperature but have different apparent temperatures when sensed by a thermal radiometer simply because their emissivities are different. The emissivity of an object may be influenced by a number of factors, including (Schmugge et al., 2002; Weng et al., 2004):

- color—darker-colored objects are better absorbers and emitters (i.e., they have a higher emissivity) than lighter colored objects, which tend to reflect more of the incident energy;

- chemical composition;

- surface roughness—the greater the surface roughness of an object relative to the size of the incident wavelength, the greater the surface area of the object and potential for absorption and reemission of energy;

- moisture content—the more moisture an object contains, the greater its ability to absorb energy and become a good emitter. Wet soil particles have a high emissivity similar to water;

- compaction—the degree of soil compaction can affect emissivity;

- field of view—the emissivity of a single leaf measured with a very-high-resolution thermal radiometer will have a different emissivity than an entire tree crown viewed using a more coarse spatial resolution radiometer;

- wavelength—the emissivity of an object is generally considered to be wavelength-dependent. For example, while the emissivity of an object is often considered to be constant throughout the $8 - 14$ μm region, its emissivity in the $3 - 5$ μm region may be different; and

- viewing angle—the emissivity of an object can vary with sensor viewing angle.

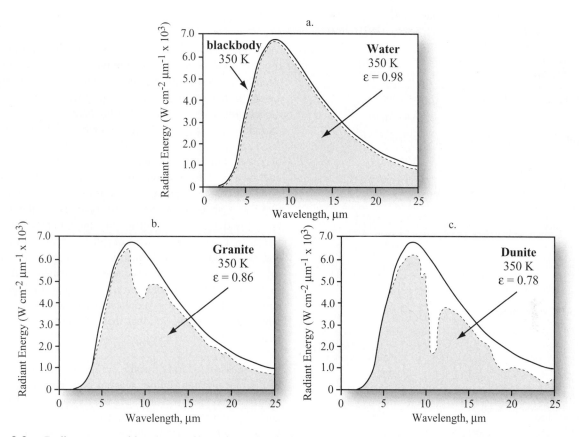

Figure 8-8   Radiant energy exiting a) water, b) granite, and c) dunite heated to 350 K compared with a blackbody at the same temperature.

Salisbury and D'Aria (1992) computed the emissivity for a variety of terrestrial materials in the 8 – 14 µm region, including igneous, metamorphic, and sedimentary rocks, desert varnish, soil, vegetation, water, and ice. Snyder et al. (1997) provided additional information on the bidirectional reflectance measurements of sands and soils in the 3 – 14 µm region. We must take into account an object's emissivity when we attempt to use thermal infrared remote sensing to measure an object's true kinetic temperature (Peres and DaCamara, 2004). This is done by applying Kirchoff's radiation law.

Kirchoff's Radiation Law

In Chapter 2 we learned that the terrain intercepts incident (incoming) radiant flux (Φ_i). This incident energy interacts with terrain materials. The amount of radiant flux reflected from the surface (Φ_r), the amount of radiant flux absorbed by the surface (Φ_α), and the amount of radiant flux transmitted through the surface (Φ_τ) can be carefully measured as we apply the principle of conservation of energy and attempt to keep track of what happens to all the incident energy. The

general equation for the interaction of *spectral (λ)* radiant flux with the terrain is:

$$\Phi_{i_\lambda} = \Phi_{r_\lambda} + \Phi_{\alpha_\lambda} + \Phi_{\tau_\lambda}. \tag{8-4}$$

Dividing each of the variables by the original incident radiant flux, Φ_{i_λ}:

$$\frac{\Phi_{i_\lambda}}{\Phi_{i_\lambda}} = \frac{\Phi_{r_\lambda}}{\Phi_{i_\lambda}} + \frac{\Phi_{\alpha_\lambda}}{\Phi_{i_\lambda}} + \frac{\Phi_{\tau_\lambda}}{\Phi_{i_\lambda}} \tag{8-5}$$

allows us to rewrite Equation 8-4 as:

$$1 = r_\lambda + \alpha_\lambda + \tau_\lambda \tag{8-6}$$

where r_λ is spectral hemispherical reflectance by the terrain, α_λ is spectral hemispherical absorptance, and τ_λ is spectral hemispherical transmittance (Slater, 1980).

The Russian physicist *Kirchoff* found that in the infrared portion of the spectrum the spectral emissivity of an object generally equals its spectral absorptance, i.e., $\alpha_\lambda = \varepsilon_\lambda$. This is often phrased as *"good absorbers are good emitters and*

Table 8-1.   Emissivity of selected materials from $8 - 14 \, \mu m$ (McDonnell Douglas, 1982; Sabins, 1997; Lillesand and Kiefer, 2003).

Material	Emissivity, ε
water, distilled	0.99
water	0.92 – 0.98
water with petroleum film	0.972
concrete	0.71 – 0.90
asphalt	0.95
tar/stone	0.97
loamy soil, dry	0.92
loamy soil, wet	0.95
soil, sandy	0.90
brick, red and rough	0.93
vegetation, closed canopy	0.98
vegetation, open canopy	0.96
grass	0.97
wood, planed oak	0.90
deciduous forest	0.97 – 0.98
coniferous forest	0.97 – 0.99
stainless steel	0.16
aluminum, foil	0.05
aluminum, polished	0.08
aluminum, paint	0.55
polished metals	0.16 – 0.21
oxidized steel	0.70
granite	0.86
dunite	0.78
basalt, rough	0.95
snow	0.83 – 0.85
paint	0.90 – 0.96
human skin	0.98

good reflectors are poor emitters". Also, most real-world materials are usually opaque to thermal infrared radiation, meaning that little radiant flux exits from the other side of the terrain element. Therefore, we may assume transmittance, $\tau_\lambda = 0$. Substituting emissivity for absorptance and removing transmittance from the equation yields (Vaughn et al., 2003):

$$1 = r_\lambda + \varepsilon_\lambda . \qquad (8\text{-}7)$$

This relationship is important because it describes why objects appear as they do on thermal infrared imagery. *Because the terrain theoretically does not lose any incident energy to transmittance, all energy leaving the object must be accounted for by the relationship between reflectance (r_λ) and emissivity (ε_λ).* If reflectivity increases, then emissivity must decrease. If emissivity increases, then reflectivity must decrease. For example, water absorbs almost all incident energy and reflects very little. Therefore, water is a very good emitter and has a high emissivity close to 1. Conversely, a sheet-metal roof reflects most of the incident energy and absorbs very little, yielding an emissivity much less than 1. Therefore, metal objects such as cars, aircraft, and metal roofs usually appear very cold (dark) on thermal infrared imagery. For example, the metal hangar and aircraft in the nighttime thermal infrared imagery in Figure 8-9 appear cold. Note that several aircraft have their engines on, which appear bright, and that their jet blast is warming the tarmac.

When measuring land or water surface temperature, the goal of thermal infrared remote sensing is to be able to point a radiometer at an object and have the recorded apparent radiant temperature (T_{rad}) equal the true kinetic temperature of the object (T_{kin}). Unfortunately, the radiant flux from a real-world object at a given temperature is not the same as the radiant flux from a blackbody at the same temperature, largely due to the effects of *emissivity*. Knowing the emissivity of an object makes it possible to modify the Stefan-Boltzmann law originally applicable to blackbodies ($M_b = \sigma T^4$) so that it pertains to the total spectral radiant flux of real world materials (M_r):

$$M_r = \varepsilon \, \sigma \, T_{kin}^{\; 4} . \qquad (8\text{-}8)$$

The equation takes into account the temperature of the object and its emissivity to create a more accurate estimate of the radiant flux exiting an object and recorded by the thermal infrared sensor.

Thermal infrared remote sensing systems generally record the apparent radiant temperature, T_{rad} of the terrain rather than the true kinetic temperature, T_{kin}. If we assume that the incorporation of emissivity in Equation 8-1 has improved our measurement to the point that

$$M_r = \varepsilon \, \sigma \, T_{kin}^{\; 4} \quad \text{and we assume that}$$

Table 8-2. Emissivity, true kinetic, and radiant temperatures of selected materials at 300 K and 27°C.

Material	Emissivity	True Kinetic Temperature T_{kin}		Radiant Temperature T_{rad} $T_{rad} = \varepsilon^{1/4} T_{kin}$	
	ε	K	°C	K	°C
blackbody	1.00	300	27	300.0	27.0
distilled water	0.99	300	27	299.2	26.2
rough basalt	0.95	300	27	296.2	23.2
vegetation	0.98	300	27	298.5	25.5
dry loam soil	0.92	300	27	293.8	20.8

Figure 8-9 Nighttime thermal infrared imagery of an airport with several different types of jet aircraft. The metal hangar and the aircraft appear cool (dark) because of metal's low emissivity. The concrete tarmac has a relatively high emissivity and appears in lighter shades of gray. Seven of the aircraft jet engines are active, as evidenced by the bright bloom along their fuselages and their jet-wash heating the tarmac. One or two aircraft engines were recently turned off.

$$M_b = \sigma T_{rad}^{\;4} \text{ and}$$

$$M_r = M_b \text{ then,}$$

$$\sigma T_{rad}^{\;4} = \varepsilon \, \sigma T_{kin}^{\;4}. \tag{8-9}$$

Therefore, the radiant temperature of an object recorded by a remote sensor is related to its true kinetic temperature and emissivity by the following relationship (Sabins, 1997):

$$T_{rad} = \varepsilon^{1/4} T_{kin}. \tag{8-10}$$

The relationship between true kinetic and radiant temperature for several different types of material are summarized in Table 8-2. It is clear from this table that, except for water, if the effect of emissivity is not accounted for when analyzing remotely sensed apparent radiant temperature (T_{rad}), then the true temperature (T_{kin}) of the object will be underestimated (Torgersen et al., 2001; Peres and DaCamara, 2004). What if we wanted to determine the emissivity of the various objects within our study area? This can be done using a thermometer and a handheld thermal infrared radiometer. If we measure an object's true temperature with a thermometer (T_{kin}) and its

apparent radiant temperature (T_{rad}) with a thermal radiometer, we can estimate its emissivity, ε, using the equation:

$$\varepsilon = \left(\frac{T_{rad}}{T_{kin}}\right)^4. \tag{8-11}$$

Ideally we would collect both the *in situ* temperature measurement and the remote measurement simultaneously. Also, to get the best emissivity approximation, it would be good to collect the remote measurement from a helicopter to try and simulate 1) the size of the remote sensor IFOV of interest (e.g., 20 × 20 m), and 2) the atmospheric effects (Peres and DaCamara, 2004) that might be encountered.

Many geologic studies are not interested in the temperature information that can be extracted from thermal infrared imagery. Instead, the apparent surface radiance exiting the terrain is analyzed so that the effects of temperature are removed and all that is left is terrain emissivity information. The emissivity information is then used to map minerals on the surface of the earth based on their documented emissivity characteristics (e.g., Kirkland et al., 2002; Vaughn et al., 2003; Jacob et al., 2004). Petitcolin and Vermote (2002) describe how MODIS middle-thermal infrared surface reflectance and derived emissivity information can be used in conjunction with thermal infrared radiance to solve the emissivity/temperature separation problem. Sobrino et al (2004) used visible and near infrared bands and a Normalized Difference Vegetation Index (NDVI) to extract land surface emissivity values from Landsat 5 Thematic Mapper band 6 thermal imagery.

Thermal Properties of Terrain

Water, rock, soil, vegetation, the atmosphere, and human tissue all have the ability to conduct heat directly through them (thermal conductivity) onto another surface and to store heat (thermal capacity). Some materials respond to changes in temperature more rapidly or slowly than others (thermal inertia). It is useful to review these thermal properties, as they have an impact on our ability to remotely sense thermal information about various types of materials.

Heat or *thermal capacity* (c) is a measure of the ability of a material to absorb heat energy. It is the quantity of heat required to raise the temperature of one gram of that material by 1 °C (cal g^{-1} °C^{-1}). Table 8-3 summarizes the thermal capacity of several materials at 20 °C. Water has the largest heat capacity of any common substance (1.00). The temperature of a lake usually varies very little between night and day. Conversely, rocks do not store heat well and exhibit significantly different temperatures in the night and day.

Thermal conductivity (K) is a measure of the rate that a substance transfers heat through it. It is measured as the number of calories that will pass through 1 cm^3 of material in 1 sec when two opposite faces are maintained at 1°C difference in temperature (cal cm^{-1} sec^{-1} °C^{-1}). The conductivity of a material can be variable due to the amount of moisture present. The thermal conductivity of a variety of materials is summarized in Table 8-3. Notice that many rocks and soils are poor conductors of heat.

Thermal inertia (P) is a measurement of the thermal response of a material to temperature changes and is measured in calories per cm^2 per second square root per 1°C (cal cm^{-2} sec$^{-1/2}$ °C^{-1}). Thermal inertia is computed using the equation

$$P = \sqrt{(K \cdot p \cdot c)} \qquad (8\text{-}12)$$

where K is thermal conductivity, p is density (g cm^{-3}), and c is thermal capacity. Density is a very important biophysical variable in this equation because thermal inertia generally increases linearly with increasing material density. Table 8-3 summarizes the thermal inertia of a variety of materials.

It would be wonderful if we could remotely sense each of the aforementioned variables and then simply compute

Table 8-3.   Thermal properties of common materials at 20°C (Janza, 1975).

	Thermal conductivity K	Thermal density p	Thermal capacity c	Thermal inertia P
Materials				
glass	0.0021	2.6	0.16	0.029
water	0.0013	1.0	1.0	0.036
wood	0.0050	0.5	0.327	0.009
Geologic Materials				
basalt	0.0050	2.8	0.20	0.053
dolomite	0.0120	2.6	0.18	0.075
granite	0.0075	2.6	0.16	0.056
gravel, sandy	0.0060	2.1	0.20	0.050
limestone	0.0048	2.5	0.17	0.045
obsidian	0.0030	2.4	0.17	0.035
sandstone	0.0120	2.5	0.19	0.075
shale	0.0042	2.3	0.17	0.041
slate	0.0050	2.8	0.17	0.049
soil, sandy	0.0014	1.8	0.24	0.024
soil, clay moist	0.0030	1.7	0.35	0.042

thermal inertia. Unfortunately, this is not the case, because conductivity, density, and thermal capacity must all be measured *in situ*. Nevertheless, it is possible to remotely sense and compute an *apparent thermal inertia* measurement per pixel in the following manner. A thermal infrared image is acquired over the same terrain in the nighttime and in the early daytime. The two images are geometrically and radiometrically registered to one another, and the change in temperature, ΔT, for a specific pixel is determined by subtracting the nighttime apparent temperature from the daytime apparent temperature. The apparent thermal inertia (ATI) per pixel is

$$ATI = \frac{1 - A}{\Delta T} \qquad (8\text{-}13)$$

with A being the albedo (reflectance) measured in the visible spectrum during the daytime for the pixel of interest (Kahle et al., 1981; Sabins, 1997).

The best way to think about thermal inertia is to associate it with an inverse relationship with the measured temperature change, ΔT. Basically, a high ΔT value is usually associated with terrain materials that have a low thermal inertia value.

Conversely, a low ΔT is usually associated with terrain materials that have a high thermal inertia value.

Geologists and other remote sensing scientists label areas in the image that have heterogeneous or homogeneous apparent thermal inertia characteristics to distinguish boundaries between bedrock and alluvial material, discriminate among rock units with similar spectral properties, and identify zones of hydrothermal alteration (Kahle et al., 1984). Price (1985) cautioned that apparent thermal inertia images should not be used in regions having variability in surface moisture (evaporation) like agricultural areas.

One of the first thermal infrared satellite remote sensing systems to collect both day (1:30 p.m.) and nighttime (2:30 a.m.) thermal infrared images of significant value for apparent thermal inertia mapping was the short-lived 1978 Heat Capacity Mapping Mission (HCMM) that acquired 600 × 600 m data in the region from 10.5 – 12.6 μm. The ASTER sensor onboard the *Terra* satellite collects five bands of day and nighttime thermal infrared data with 90 × 90 m spatial resolution and a 60 km swath width (Jacob et al., 2004). The five thermal ASTER bands are centered at 8.28, 8.63, 9.07, 10.65, and 11.28 μm. The MODIS sensor onboard *Terra* and *Aqua* has three middle infrared bands centered around 3.75, 3.95 and 4.05 μm, and three thermal infrared channels centered around 8.55, 11.03 and 12.02 μm. Jacob et al. (2004) compared the land surface emissivity and radiometric temperature derived from MODIS and ASTER sensors. Day and nighttime thermal infrared imagery can also be collected from suborbital aircraft using a variety of sensors, e.g., NASA (TIMS and ATLAS) and Itres TABI (Quattrochi and Ridd, 1998; Schmugge et al., 2002; Itres TABI, 2006).

Thermal Infrared Data Collection

Thermal infrared remote sensor data may be collected by:

- across-track thermal scanners; and

- pushbroom linear and area-array charge-coupled-device (CCD) detectors.

It is useful to review the nature of thermal infrared sensor systems and their components and how various system parameters influence the type and quality of the thermal infrared data collected.

Thermal Infrared Remote Sensing Based on Multispectral Scanners

Chapter 7 introduced how multispectral scanners function. This section provides additional information about thermal infrared scanners.

Daedalus DS-1260, DS-1268, Airborne Multispectral Scanner (AMS), NASA TIMS, and NASA ATLAS

These scanners have provided much of the useful high spatial and spectral resolution thermal infrared data for monitoring the environment. The DS-1260 records data in 10 bands including a thermal infrared channel (8.5 – 13.5 μm). The DS-1268 incorporates the Landsat Thematic Mapper middle-infrared bands (1.55 – 1.75 μm and 2.08 – 2.35 μm). The AMS contains a hot-target, thermal infrared detector (3.0 – 5.5 μm) in addition to the standard thermal infrared detector (8.5 – 12.5 μm).

Scientists continue to use thermal infrared imagery acquired by the NASA Thermal Infrared Multispectral Scanner (TIMS) which has six bands ranging from 8.2 – 12.2 μm (e.g., Schmugge et al., 2002; Coll et al., 2003). Many scientists also use the NASA Airborne Terrestrial Applications Sensor (ATLAS), which has six visible and near-infrared bands, two Thematic Mapper middle-infrared bands, and six thermal infrared bands from 8.2 – 12.2 μm. Specific bandwidths are summarized in Chapter 7. Both the TIMS and ATLAS sensors have a 2.0 or 2.5 milliradian (mrad) IFOV (Wallace, 1999; Coll et al., 2003).

The basic principles of operation and components of the AMS, TIMS, and ATLAS are shown in Figure 8-10. The diameter of the circular ground area viewed by the sensor, D, is a function of the instantaneous field of view, β, of the scanner measured in milliradians and the altitude of the scanner above ground level, H, where

$$D = H \times \beta. \qquad (8\text{-}14)$$

For example, if the IFOV of the scanner is 2.5 mrad, the ground size of the pixel in meters is a product of the IFOV (0.0025) and the altitude AGL in meters. Table 8-4 presents flight altitudes and corresponding pixel sizes at nadir (the point directly below the aircraft) for an IFOV of 2.5 mrad. IFOVs typically range from 0.5 – 5 mrad.

When performing *across-track* scanning, an electric motor is oriented parallel with the aircraft fuselage and direction of

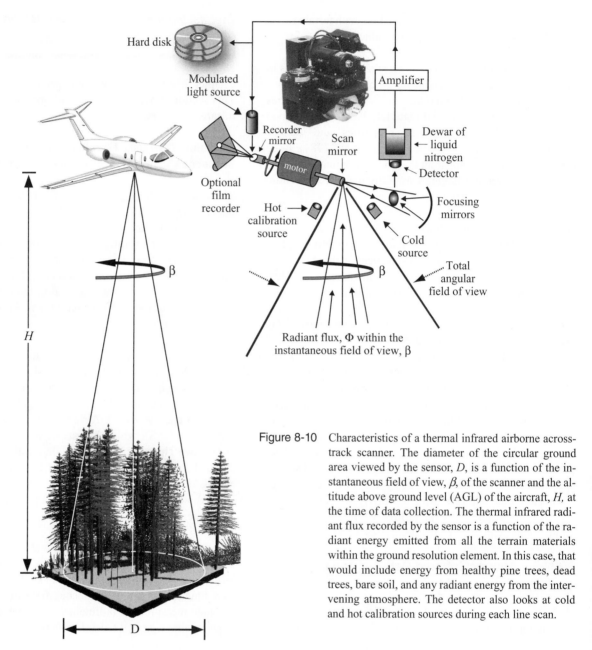

Figure 8-10   Characteristics of a thermal infrared airborne across-track scanner. The diameter of the circular ground area viewed by the sensor, *D*, is a function of the instantaneous field of view, *β*, of the scanner and the altitude above ground level (AGL) of the aircraft, *H*, at the time of data collection. The thermal infrared radiant flux recorded by the sensor is a function of the radiant energy emitted from all the terrain materials within the ground resolution element. In this case, that would include energy from healthy pine trees, dead trees, bare soil, and any radiant energy from the intervening atmosphere. The detector also looks at cold and hot calibration sources during each line scan.

flight and drives a 45° scanning mirror facet located at the end of the shaft which has a precise instantaneous field of view (e.g., 2.5 mrad). The mirror scans the terrain at a right angle (perpendicular) to the direction of flight. The mirror normally sweeps out a *total angular field of view* of 90 – 120° during each scan, depending on the sensor system. The mirror also views internal *hot* and *cold calibration sources* (targets) during each scan. The exact temperature of these calibration sources is known.

Photons of thermal infrared radiant flux, Φ, emitted by the terrain, are routed to a mirror that focuses the photons onto

the *detector*. The detector converts the incoming radiant energy into an analog electrical signal. The greater the number of photons impacting the detector, the greater the signal strength. The infrared detectors (Figure 8-11) are usually composed of:

- *In:Sb* (indium antimonide) with a peak sensitivity near 5 μm;

- *Ge:Hg* (mercury-doped germanium) with a peak sensitivity near 10 μm;

Table 8-4.   Aircraft multispectral scanner flight altitudes and pixel size based on an instantaneous field of view of 2.5 mrad.

Flight Altitude AGL, m	Pixel Size, m
1,000	2.5
2,000	5.0
4,000	10.0
6,000	15.0
20,000	50.0
50,000	125.0

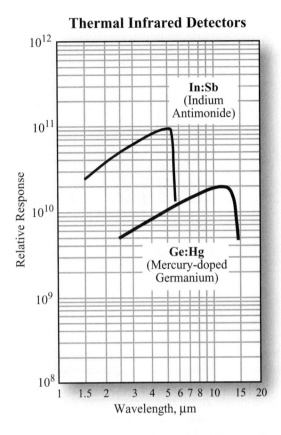

Figure 8-11   The peak spectral sensitivity of an indium antimonide (*In:Sb*) thermal infrared detector is approximately 5.0 μm, while the peak sensitivity of a mercury-doped germanium (*Ge:Hg*) detector is approximately 10 μm (after McDonnell Douglas, 1982).

- *Hg:Cd:Te* (mercury-cadmium-telluride) which is sensitive over the range from 8 – 14 μm.

The detectors are cooled to low temperatures (-196°C; -243 °F; 73 K) using liquid helium or liquid nitrogen. Cooling the detectors ensures that the radiant energy (photons) recorded by the detectors comes from the terrain and not from the ambient temperature of objects within the scanner itself.

The Earth does not emit very much thermal infrared radiation; therefore, the relatively weak signal is usually amplified. The signal is then recorded on magnetic tape or other media for future analog-to-digital (A-to-D) conversion and analysis. If desired, the signal can also be used to modulate a light source that reflects visible light onto a recorder mirror located at the other end of the motorized shaft. Here the process is reversed and visible light radiant flux proportional to the amount of infrared energy received is used to expose photographic film pixel-by-pixel and line-by-line, creating a thermal infrared image of the terrain. To properly expose the photographic film, it must be advanced forward in relation to how fast the shaft is turning. The hard-copy thermal infrared image may be processed as a negative or positive print.

It is important to remember that the infrared radiant flux recorded by the sensor system is an integration of all the radiant flux emitted from the various materials within the IFOV and any radiant flux that the atmosphere might scatter into the IFOV of the sensor. For example, radiant flux emitted from the healthy pine trees, dead trees, bare soil, and the atmosphere would be integrated into a single measurement of the terrain shown in Figure 8-10.

The following factors should be considered when collecting aircraft MSS thermal infrared data:

- There is an inverse relationship between having high spatial resolution and high radiometric resolution when collecting thermal infrared data. The larger the radiometer instantaneous field of view, β, the longer the *dwell time* that an individual detector views the terrain within the IFOV during a single sweep of the mirror. A larger IFOV provides good *radiometric resolution,* which is the ability to discriminate between very small differences in radiant energy exiting the terrain element. In fact, the radiant energy *signal* measured may well be much stronger than any *noise* introduced from the sensor system components. When this takes place, we say that we have a good *signal-to-noise ratio*. Of course, the larger the IFOV, the poorer the ability to resolve fine spatial detail. Selecting a smaller IFOV will increase the spatial resolution, but the sensor

Inverse-Square Law

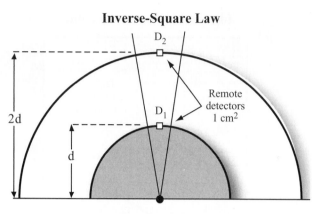

Blackbody Point Source, S

Figure 8-12　The inverse-square law reveals that the intensity of thermal radiation emitted from a blackbody point source, S, varies as the inverse square of the distance, d, between the source and remote detector receivers, D_1 and D_2.

will dwell a shorter time on each terrain element during a sweep of the mirror, resulting in poorer radiometric resolution and perhaps a poorer signal-to-noise ratio.

- Cutting in half the distance of a remote sensing detector from a point source quadruples the infrared energy received by that detector. The *inverse-square law* states that "the intensity of radiation emitted from a point source varies as the inverse square of the distance between source and receiver." Thus, we can obtain a more intense, strong thermal infrared signal if we can get the remote sensor detector as close to the ground as practical. For example, consider a blackbody point source, S, and two remote detectors (D_1 and D_2) of equal sensitive area, say, 1 cm^2. Detector D_1 is a distance d cm from S, and detector D_2 is at a distance $2d$ cm from S (Figure 8-12). From the Stefan-Boltzmann law (Equation 8-1), the total infrared energy radiated by S into a hemisphere is M_b Watts/cm^2 of the radiating surface. Thus, M_b is the total infrared energy radiated into a hemisphere of radius d on which sensor D_1 is located, that is, into a surface area of πd^2 cm^2. But M_b is also the total infrared energy radiated into a hemisphere of radius $2d$ on which sensor D_2 is located, specifically, into a surface area of $4\pi d^2$ cm^2. Therefore,

$$\text{Infrared energy detected by } D_1 = \frac{M_b}{\pi d^2} = M_b'. \quad (8\text{-}15)$$

$$\text{Infrared energy detected by } D_2 = \frac{M_b}{4\pi d^2} = \frac{M_b'}{4}. \quad (8\text{-}16)$$

- Most thermal infrared remote sensing investigations try to maintain good radiometric and spatial resolution by 1) selecting a fairly large IFOV such as 2.5 mrad, and 2) flying at a relatively low altitude to obtain smaller pixel sizes. Unfortunately, at lower altitudes, the high spatial resolution may be outweighed by the fact that more flightlines are required to cover the area compared to more efficient coverage at higher altitudes with larger pixels. The pixel size and the geographic size of the survey are considered, objectives are weighed, and a compromise is reached. Multiple flightlines of aircraft MSS data are difficult to mosaic.

Geometric Correction of Across-Track Thermal Infrared Scanner Data

Thermal infrared scanning systems (actually all scanning systems) introduce numerous types of geometric error that must be understood because they impact 1) the quality of the imagery for visual or digital image processing and analysis, and 2) the creation of planimetric maps from the thermal infrared data. The most important considerations include:

- ground swath width;

- spatial ground resolution cell size;

- one-dimensional relief displacement; and

- tangential scale distortion.

Ground Swath Width (gsw): The ground swath width is the length of the terrain strip remotely sensed by the system during one complete across-track sweep of the scanning mirror. It is a function of the total angular field of view of the sensor system, θ, and the altitude of the sensor system above ground level, H (Figure 8-13). It is computed as

$$gsw = \tan\left(\frac{\theta}{2}\right) \times H \times 2. \quad (8\text{-}17)$$

For example, the ground swath width of an across-track scanning system with a 100° total field of view and an altitude above ground level of 6,000 m would be 14,301 m:

$$gsw = \tan\left(\frac{100}{2}\right) \times 6000 \times 2$$

$$gsw = 1.191753 \times 6000 \times 2$$

$$gsw = 14,301 \text{ m}.$$

β

θ
total field
of view of
one scan line

β
instantaneous
field of view

$H_\phi = H \sec \phi$

scan
angle off-
nadir

H

ϕ

$(H \sec \phi)\beta$

$(H \sec^2 \phi)\beta$

$D = H \times \beta$
at nadir

line of flight

Figure 8-13   The ground resolution cell size along a single across-track scan is a function of a) the distance from the aircraft to the observation, i.e., H is the altitude of the aircraft above ground level (AGL) at nadir and $H \sec \phi$ off-nadir; b) the instantaneous field of view of the sensor, β, measured in milliradians, and c) the scan angle off-nadir, ϕ. Thus, pixels off-nadir have semi-major and semi-minor axes (diameters) that define the resolution cell size. One-dimensional relief displacement and tangential scale distortion also occur in a direction perpendicular to the line of flight and parallel with a line scan.

If the total field of view were 90°, the ground swath width would be 12,000 m:

$$gsw = \tan\left(\frac{90}{2}\right) \times 6000 \times 2$$

$$gsw = 1 \times 6000 \times 2$$

$$gsw = 12,000 \text{ m}.$$

Most scientists utilizing across-track scanner data only use the central 70 percent of the swath width (35 percent on each side of nadir) primarily because ground resolution elements have larger cell sizes the farther they are away from nadir.

Ground Resolution Cell Size (D): The diameter of the circular ground area viewed by the sensor, D, at nadir is a function of the instantaneous field of view, β, of the scanner measured in milliradians (mrad) and the altitude of the scanner above ground level, H, where $D = H \times \beta$. Interestingly, as the scanner's instantaneous field of view moves away from nadir on either side, the circle becomes an ellipsoid. One of the major reasons is that the distance from the aircraft to the resolution cell is increasing, as shown in Figure 8-13. In fact, the distance from the aircraft to the resolution cell on the ground, H_ϕ, is a function of the scan angle off-nadir, ϕ, at the time of data collection and the true altitude of the aircraft, H (Lillesand and Kiefer, 2003):

$$H_\phi = H \cdot \sec\phi. \tag{8-18}$$

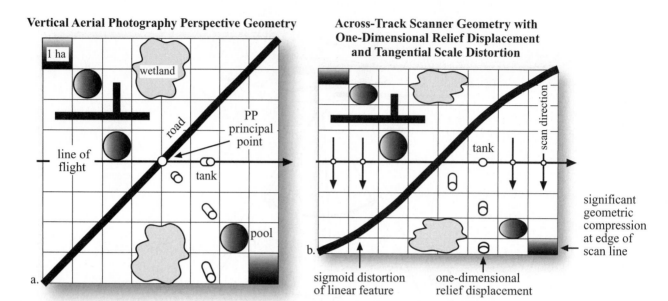

Figure 8-14   a) The hypothetical perspective geometry of a vertical aerial photograph obtained over level terrain. Four 50-ft tanks are distributed throughout the landscape and experience varying degrees of radial relief displacement the farther they are from the principal point (PP). b) Across-track scanning systems introduce one-dimensional relief displacement perpendicular to the line of flight and tangential scale distortion and compression the farther the object is away from nadir. Linear features trending across the terrain are often recorded with s-shaped or sigmoid curvature characteristics due to tangential scale distortion and compression.

Thus, the size of the ground resolution cell increases as the angle increases away from nadir. The nominal (average) diameter of the elliptical resolution cell, D_ϕ, at this angular location from nadir has the dimension:

$$D_\phi = (H \cdot \sec\phi) \cdot \beta \qquad (8\text{-}19)$$

in the direction of the line of flight, and

$$D_\phi = (H \cdot \sec^2\phi) \cdot \beta \qquad (8\text{-}20)$$

in the orthogonal (perpendicular) scanning direction.

Scientists using thermal across-track scanner data usually only concern themselves with the spatial ground resolution of the cell at nadir, D. If it is necessary to perform precise quantitative work on pixels some angle ϕ off-nadir, then it may be important to remember that the radiant flux recorded is an integration of the radiant flux from all the surface materials in a ground resolution cell with a constantly changing diameter. Using only the central 70 percent of the swath width reduces the impact of the larger pixels found at the extreme edges of the swath.

One-Dimensional Relief Displacement: Truly vertical aerial photographs have a single principal point directly beneath the aircraft at nadir at the instant of exposure. This perspective geometry causes all objects that rise above the local terrain elevation to be displaced from their proper planimetric position radially outward from the principal point (discussed in Chapter 6). For example, the four theoretical tanks in Figure 8-14a are each 50 ft high. The greater the distance from the principal point, the greater the radial relief displacement of the top of the tank away from its base.

Thermal infrared images acquired using an across-track scanning system also contain relief displacement. However, instead of being radial from a single principal point, the displacement takes place in a direction that is perpendicular to the line of flight for each and every scan line as shown in Figure 8-14b. In effect, the ground-resolution element at nadir functions like a principal point for each scan line. At nadir, the scanning system looks directly down on the tank, and it appears as a perfect circle in Figure 8-14b. The greater the height of the object above the local terrain and the greater the distance of the top of the object from nadir (i.e., the line of flight), the greater the amount of *one-dimensional relief displacement* present. One-dimensional relief displacement is introduced in both directions away from nadir for each sweep of the across-track mirror.

Vertical Aerial Photograph

science
buildings

dorms

library

parking

soccer field

a.

Predawn Thermal Infrared Image

one-dimensional
relief
displacement

steam
lines

manhole
cover

line of flight

steam
plant

b.

Figure 8-15 a) Vertical panchromatic aerial photography of the University of South Carolina campus obtained on April 26, 1980. Note that the relief displacement of the science buildings is radial away from the principal point. b) Predawn 1 × 1 m thermal infrared imagery. Note the intricate underground steam-line network served by the steam plant. The science buildings exhibit one-dimensional relief displacement caused by the cross-track scanning system.

One-dimensional relief displacement can be beneficial or cause image-interpretation problems. For example, consider the aerial photograph and predawn thermal infrared image of the University of South Carolina campus shown in Figure 8-15. The science buildings exhibit radial relief displacement in the aerial photograph away from the principal point and one-dimensional relief displacement in the thermal infrared image perpendicular to the line of flight. Note how it is easy to view the side (façade) of the science buildings in the thermal infrared image. This is valuable if we want to study the temperature characteristics of the side of these buildings. However, if we wanted to evaluate the thermal characteristics of the road or objects immediately behind the buildings, they are obscured from view.

Aerial photography and predawn thermal infrared imagery of downtown New York City provide an even greater appreciation of one-dimensional relief displacement (Figure 8-16). In this case, the radial relief displacement in the aerial photograph makes it difficult to obtain information about the

Aerial Photograph of New York City   **Thermal Infrared Image of New York City**

a.   b.

Figure 8-16   a) Perspective aerial photograph of downtown New York City. Note the radial relief displacement of the Empire State Building away from the principal point (not shown). b) Predawn thermal infrared image of downtown New York City. Note the one-dimensional relief displacement that is perpendicular to the line of flight. The metal on the top of the Empire State Building shows up as being very cold due to its low emissivity. The building on the left is very cool, while the Empire State Building radiates a significant amount of energy. The metal cars have a low emissivity, as do the circular metal evaporative coolers on top of the nearest building.

street pattern. In fact, the street pattern is almost completely obscured from view. Conversely, because the thermal imagery was obtained along a line of flight that was parallel with the street orientation, the one-dimensional relief displacement creates an excellent view of the temperature characteristics of the buildings and streets, especially those that lie perpendicular to the direction of flight. Notice the significant amount of thermal detail on the side of the Empire State Building that is visible because of the one-dimensional relief displacement. Also note the radiometrically cold metal on the top of the Empire State Building.

While some aspects of one-dimensional relief displacement may be of utility for visual thermal infrared image interpretation, it seriously displaces the tops of objects projecting above the local terrain from their true planimetric position. Maps produced from such imagery contain serious planimetric errors. Thermal infrared imagery must be geometrically rectified in order to produce maps with some semblance of

geometric accuracy. Methods of geometric rectification are summarized in Jensen (2005).

Tangential Scale Distortion: The mirror on a thermal infrared across-track scanning system rotates at a constant speed and typically views from 70° to 120° of terrain during a complete line scan. Of course, the amount depends on the specific sensor system. From Figure 8-13 it is clear that the terrain directly beneath the aircraft (at nadir) is closer to the aircraft than the terrain at the edges during a single sweep of the mirror. Therefore, because the mirror rotates at a constant rate, the sensor scans a shorter geographic distance at nadir than it does at the edge of the image. This relationship tends to *compress* features along an axis that is perpendicular to the line of flight. The greater the distance of the ground resolution cell from nadir, the greater the image scale compression. This is called *tangential scale distortion*. Objects near nadir exhibit their proper shape. Objects near the edge of the flightline become compressed and their shape distorted. For example, consider the tangential geometric distortion and compression of the circular swimming pools and one hectare of land the farther they are from nadir in the hypothetical diagram (Figure 8-14b).

This tangential scale distortion and compression in the far range also causes linear features such as roads, railroads, utility right of ways, etc., to have an *s-shape* or *sigmoid distortion* when recorded on thermal infrared imagery (Figure 8-14b). Interestingly, if the linear feature is parallel with or perpendicular to the line of flight it does not experience sigmoid distortion.

Some sophisticated across-track scanning systems take tangential scale distortion into consideration and sweep the exposing spot of light at the film recorder at a continuously varying speed. Ideally, this is the type of thermal infrared imagery available for analysis. When tangential scale distortion has not been removed, it is best to 1) use the central 70 percent of the field of view for analysis to minimize the effects of distortion, and 2) geometrically rectify the remote sensor data using ground-control point rectification discussed in Jensen (2005).

Even single flightlines of aircraft MSS data are difficult to rectify to standard map projection because of aircraft roll, pitch, and/or yaw during data collection (Jensen et al., 1988). Notches in the edge of a flightline of data are indicative of aircraft roll. Such data require significant human and machine resources to make the data planimetrically accurate. Several firms have placed GPS and inertial navigation systems (INS) onboard the aircraft to obtain the exact location of the plane and its roll, pitch, and yaw at the exact instant of data collection. This information can be used to geometrically correct each line of remote sensor data without the need for the collection of ground control points.

Radiometric Calibration of Thermal Scanner Data

In order to derive surface temperatures from thermal infrared remote sensing data, two important problems must be solved (Coll et al., 2003): (i) compensation for atmospheric absorption and emission in the radiance arriving at the sensor; and (ii) correction for surface emissivity effects, which are coupled with temperature in the radiance emitted by the surface. This radiometric calibration may be performed using 1) internal blackbody source referencing, or 2) external empirical referencing based on *in situ* data collection (Quattrochi and Goel, 1995).

Internal Source Referencing: When an across-track scanning system is used to collect thermal infrared data, the detector first looks at a "cold" reference target, approxi-

mately 120° of terrain, and then a "hot" reference target during each line scan, as demonstrated in Figure 8-10. The true kinetic temperature of these cold and hot targets is constantly monitored by the remote sensing system and is recorded on disk, tape, or some other medium along with the image data for each line scan. If desired, all of the terrain brightness values collected during a scan can then be calibrated (converted) to *apparent* true temperature values based on their relationship to the cold and hot target information stored for each scan line. The radiometric resolution is usually accurate to within $\pm 0.2°C$. This should be the ideal method of radiometric calibration because it involves no field work. Unfortunately, this method does not account for the intervening atmosphere 1) emitting spurious radiant energy into the IFOV of the sensor system, or 2) absorbing energy emitted from the ground before it reaches the detector optics.

External Referencing: To incorporate atmospheric effects, it may be necessary to perform external empirical referencing. This involves taking *in situ* measurements with 1) a thermometer that measures the true kinetic temperature of a material or water body, 2) a handheld *radiometer* that measures the radiant temperature exiting the terrain in a specific instantaneous field of view, or 3) a *radiosonde* (a balloon carrying sensitive meteorological instruments) launched to obtain atmospheric profiles of temperature, barometric pressure, and water vapor.

Thermometers are straightforward *in situ* measurement devices. A radiometer is a handheld remote sensing instrument that you point at the terrain and an apparent radiant temperature reading is returned. Either type of temperature measurement should be obtained at the exact time that the remote sensor data are collected overhead. Ideally, more than 30 *in situ* samples are obtained. Unfortunately, this is not always practical due to the constraints of hiring people, boats, etc., and obtaining a sufficient number of high-quality thermometers and/or radiometers. The exact location of each of the *in situ* thermometer or radiometer measurements are obtained using GPS. The locations of the *in situ* data-collection points are then located in the rectified remote sensor data, and the brightness values at these locations are extracted. The accuracy of the geometric correction of the thermal infrared data becomes very important at this point.

The *n* thermometer or radiometer measurements collected *in situ* are then regressed with the corresponding *n* remote sensing brightness values obtained at the same geographic locations. For example, consider the eight *in situ* water-temperature measurements in Figure 8-17 and the corresponding uncalibrated thermal infrared remote sensing brightness val-

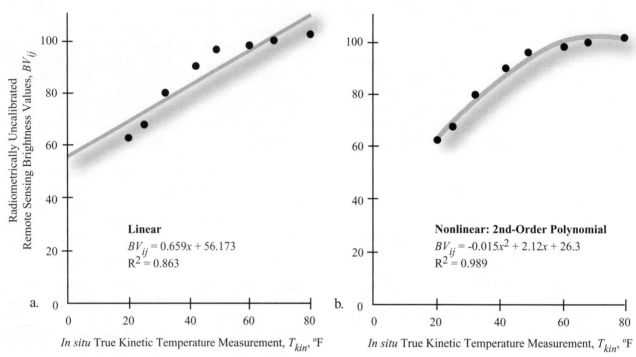

Figure 8-17   a) Linear regression line fit to eight *in situ* temperature measurements and uncalibrated remote sensing brightness values obtained for the same geographic location. b) Nonlinear 2nd-order polynomial equation fit to the same observations.

ues ($BV_{i,j}$) for these same eight locations. Linear and/or nonlinear curves are fit through the observations. The linear equation explained 86 percent of the variance while the 2nd-order polynomial explained 99 percent of the variance. Either of these equations may be used to relate the *in situ* temperature measurements to the remote sensing brightness values, in effect, radiometrically calibrating the remote sensor data. This method does not take into account the intervening atmosphere present at the time of data collection. Note that the linear equation is of the form $BV_{ij} = aT_{kin} + b$. If we want to take into account the emissivity (ε) of the terrain, we utilize the *y*-intercept (b) and slope of the relationship (a) and the equation:

$$BV_{ij} = a \cdot \varepsilon \cdot \mathrm{T}_{kin}^{4} + b. \tag{8-21}$$

Rearranging the equation allows the true kinetic temperature T_{kin} of every pixel, BV_{ij} in the uncalibrated matrix of remote sensor data to be determined:

$$T_{kin} = \left(\frac{BV_{ij} - b}{a \cdot \varepsilon}\right)^{\frac{1}{4}}. \tag{8-22}$$

The radiometrically calibrated remote sensor data may then be used to make temperature maps. This method requires a costly and well-coordinated field data-collection program. Also, all the thermometers, radiometers, and GPS instruments must be carefully calibrated, and the *in situ* readings should be obtained at exactly the same time, if possible.

But what if it is not practical to collect *in situ* thermometer or radiometer measurements in the field over vast or intractable distances at the same time as the remote sensing overflight? How can we then take into account the deleterious effects of atmospheric absorption or emission on our thermal infrared measurements? The answer is the use of 1) radiosondes and atmospheric radiative transfer modeling, or 2) split-window atmospheric correction techniques.

As noted, the intervening atmosphere has a significant impact on the energy emitted by the terrain before it is actually recorded by the thermal infrared remote sensing system. In fact, the maximum atmospheric transmittance of thermal infrared radiation in the spectral region from 8.0 – 14 μm is only about 80 percent. The amount of atmospheric absorption is primarily a function of the atmospheric water vapor content, although there is an ozone absorption band around 9.5 μm. To obtain accurate thermal surface radiance values, radiosonde launches need to be made concurrently with daytime and nighttime overflights. Atmospheric profiles of temperature, barometric pressure, and humidity obtained using

Table 8-5.   Spectral sensitivity and operating temperatures for
selected infrared electro-optical focal plane arrays.

Wavelength (μm)	Detector Material	Operating Temperature (K)
0.3 – 5.5	In:Sb	<90
1.0 – 3.0	PV Hg:Cd:Te	>150
3.0 – 5.0	PV Hg:Cd:Te	≤120
8.0 – 12.0	PV Hg:Cd:Te	50 – 80
12.0 – 25.0	Si:As	10

the radiosonde are transmitted to the ground. These data are then incorporated into an atmospheric transmission model (e.g., LOWTRAN or MODTRAN) to calculate atmospheric transmission characteristics (Quattrochi and Goel, 1995). The output from the LOWTRAN or MODTRAN model is combined with calibrated spectral response curves for the individual bands of the thermal infrared detectors being used and internal blackbody source referencing data recorded during the flight. All these data are then modeled and used to produce a look-up table for converting each pixel's brightness value into true kinetic temperature measurements (e.g., Luvall, 1999; Coll et al., 2003). This is the most rigorous method of calibrating thermal infrared imagery. Unfortunately, very few persons have access to radiosonde technology. Sometimes atmospheric profile meteorological information from a nearby National Weather Service station (or airport) can be used instead of the radiosonde data as input to the atmospheric transmission model (Quattrochi and Goel, 1995).

Scientists attempting to remove atmospheric effects and determine surface temperature (T_s) data have also found that a *split-window* approach works well in certain instances (Czajkowski et al., 1997; Ouaidrari et al., 2002). Basically, the brightness temperatures observed in one thermal channel (e.g., AVHRR band T_4) are corrected for atmospheric effects by the linear difference between the brightness temperatures in other thermal channels (e.g., AVHRR band T_5). An example of a split-window equation is

$$T_s = a + T_4 + b(T_4 - T_5) \qquad (8\text{-}23)$$

where a and b are constants that can be estimated from model simulations (Becker and Li, 1995) or correlation with ground observations. Six split-window equations are summarized in Ouaidrari et al. (2002) and in Czajkowski et al (2004). Coll et al. (2003) describe land-surface temperature

Figure 8-18   A 640 × 480 Hg:Cd:Te (mercury-cadmium-telluride) long-wavelength infrared (LWIR) area array and readout assembly. This detector array is cooled to 50 – 80 K. New arrays have > 1000 × 1000 detectors.

retrieval using split-window methods, TIMS data, and ground measurements.

Thermal Infrared Remote Sensing Based on Area- and Linear-Arrays of Detectors

Until recently, it was difficult to make commercially available CCDs that had spectral sensitivity to mid- and long-wavelength infrared radiation. Therefore, most thermal infrared remote sensor data were collected with a single detector and a scanning mirror as previously discussed. This has changed (FLIR, 2006). It is now possible to make both linear and area-arrays (sometimes referred to as "staring" focal-plane arrays) that are sensitive to mid- and thermal infrared radiation such as those summarized in Table 8-5. For example, a 640 × 480 Hg:Cd:Te long-wavelength infrared area array with its readout assembly attached is shown in Figure 8-18. Detector arrays > 1024 × 1024 elements are now available and offer low noise and state-of-the-art sensitivity for extremely low-background applications. Staring array detectors made of platinum silicide (Pt:Si) are also popular.

Linear and area staring arrays allow improved thermal infrared remote sensing to take place because (FLIR, 2006):

- the solid-state micro-electronic detectors are smaller in size (e.g., 20 × 20 μm) and weight, require less power to operate, have fewer moving parts, and are more reliable;

Figure 8-19   A helicopter with a forward-looking infrared (FLIR) sensor system located under the nose (courtesy FLIR Systems, Inc.).

Figure 8-20   Nighttime thermal infrared image of a break-in (courtesy FLIR Systems, Inc.).

- each detector in the array can view the ground resolution element for a longer time (i.e., it has a longer dwell time), allowing more photons of energy from within the IFOV to be recorded by the individual detector, resulting in improved radiometric resolution (the ability to resolve smaller temperature differences);

- each detector element in the linear or area array is fixed relative to all other elements, therefore, the geometry of the thermal infrared image is much improved relative to that produced by an across-track scanning system;

- some linear and area thermal detectors now use a miniature Sterling closed-cycle cooling system that does not require the compressed gas-cooling apparatus (argon or liquid nitrogen) previously discussed.

Forward-Looking Infrared (FLIR) Systems

During the 1991 Gulf War and the War in Iraq beginning in 2004, the public saw day and nighttime *forward-looking infrared (FLIR)* images of the terrain and various targets. For decades, military organizations throughout the world have funded the development of FLIR-type systems that look obliquely ahead of the aircraft and acquire high-quality thermal infrared imagery, especially at night. In fact, their goal is usually "to own the night." Some FLIR systems collect the infrared energy based on the same principles as an across-track scanner (previously discussed), except that the mirror points *forward* about 45° and projects terrain energy during a single sweep of the mirror onto a linear array of thermal infrared detectors. Some systems use staring focal-plane

array technology. An example of a FLIR system housed under the nose of an aircraft is shown in Figure 8-19.

FLIR and other thermal infrared systems are routinely used by law-enforcement agencies to see nighttime criminal activity and find lost people. For example, a criminal is in the process of breaking into a second-story building in Figure 8-20. The person is approximately 98°F, while the terrain is 70 – 80°F. The moist rooftop is relatively cool (dark), the tree trunk and canopy relatively warm (bright), and the metal components of the ladder are cool (dark). Foresters and firefighters use FLIR to locate the perimeter of forest fires and effectively allocate firefighting resources. Security personnel at Gatwick Airport in London use pole-mounted FLIR surveillance systems to monitor the perimeter fences and make sure no unauthorized people gain access to the airport. Several other FLIR images are presented in Figure 8-21.

The Thermal Airborne Broadband Imager (TABI)

The Itres, Inc. of Canada has developed two of the most innovative airborne thermal infrared remote sensing systems based on linear array pushbroom technology. The *Thermal Airborne Broadband Imager* (TABI) obtains data in a single, broad thermal-infrared channel from 8.0 – 12.0 μm. The linear array contains 320 picture elements that are 50 × 50 micrometers in size. The IFOV is 2.8 mrad with a total field of view of 48°. The data are quantized to 12-bits (0 - 4095). The sensor can function to a maximum altitude of 10,000 ft.

FLIR (Forward-looking Infrared) Imagery

Figure 8-21   Examples of airborne FLIR (forward-looking infrared) imagery. a) Nighttime image of people in a boat being rescued. b) Nighttime image of a person on a metal roof being pursued by another person on the ground. Note the warm transformers on the telephone pole and the low emissivity of the metal telephone lines. c) Helicopter view of policemen making an armed arrest. d) Nighttime image of a Hummer military vehicle with the engine on. e) Storage tank with a relatively cool lower layer of contents. f) Nighttime image of an urban area with water, boat, automobiles, and pedestrians (courtesy FLIR Systems, Inc.).

(3,048 m) ASL. Geometric rectification is performed using onboard GPS and INS data making the collection of ground-control points unnecessary. Characteristics of the Itres TABI-320 are found in Table 8-6 (Itres TABI, 2006).

The Thermal Airborne Spectrographic Imager (TASI)

The *Thermal Airborne Spectrographic Imager* (TASI) is one of the first true hyperspectral thermal-infrared remote sensing systems based on linear array technology. It collects data in 32 thermal channels in the region from 8.0 - 11.5 μm. The pixel size is 30 × 30 μm and the IFOV is 1.25 mRad. The linear array contains 600 pixels. The total field of view is 40°. The data are quantized to 14-bits. Characteristics of the Itres TASI-600 are summarized in Table 8-6 (Itres TASI, 2006).

 Thermal Infrared Environmental Considerations

When interpreting a thermal infrared image, it is useful to understand the diurnal cycle and how it relates to the temperature of objects on the Earth's surface.

Diurnal Temperature Cycle of Typical Materials

The diurnal cycle encompasses 24 hours. Beginning at sunrise, the Earth begins intercepting mainly short-wavelength energy (0.4 – 0.7 μm) from the Sun (Figure 8-22a). From dawn to dusk, the terrain intercepts the incoming short wavelength energy and reflects much of it back into the atmosphere, where we can use optical remote sensors to measure the reflected energy. However, some of the incident short-wavelength energy is absorbed by the terrain and then re-radiated back into the atmosphere as thermal infrared wavelength radiation (3 – 14 μm). The outgoing long-wave radiation peak usually lags two to four hours after the midday peak of incoming shortwave radiation, due to the time taken to heat the soil. The contribution of reflected short-wavelength energy and emitted long-wavelength energy causes an energy surplus to take place during the day, as shown in Figure 8-22a. Both incoming and outgoing short-wave radiation become zero after sunset (except for light from the moon and stars), but outgoing long-wave radiation exiting the terrain continues all night.

The typical diurnal temperature variations for soils and rock, water, vegetation, moist soil, and metal objects are shown in

Table 8-6.   Itres, Inc., Thermal Airborne Broadband Imager (TABI-320) and Thermal Airborne Spectrographic Imager (TASI-600) characteristics.

	TABI-320	**TASI-600**
Spectral sensitivity (nm)	8 - 12 μm	8 - 11.5 μm
Spatial resolution (cm) at Nadir	variable based on the aircraft altitude above ground level (AGL)	
Number of thermal bands	1	32
Sensor	linear array CCD	
IFOV	2.8 mrad	1.25 mrad
Total field-of-view	48°	40°
Swath width	320 pixels	600 pixels
Detector size	50 × 50 μm	30 × 30 μm
Detectors in linear array	320	600
Radiometric resolution	12-bit	14-bit
Geo-referencing	Applanix POS software with GPS and Inertial Navigation System (INS) data. Ground-control points are not required.	

Figure 8-22b. If all of these curves lie exactly on top of one another, then remote sensing in the thermal infrared portion of the spectrum would be of no value because all the phenomena would have the same apparent radiant temperature. There would be no contrast in the imagery between the different phenomena. Fortunately, there are only two times during the day (after sunrise and near sunset) when some materials like soil, rock, and water have exactly the same radiant temperature. During this *crossover* time period it is generally not wise to acquire thermal infrared remotely sensed data.

Fortunately, some materials store heat more efficiently than others, i.e., they have a higher *thermal capacity*. For example, water has a much higher thermal capacity than soil and rock (Table 8-3). Its diurnal temperature range fluctuates very little when compared with the dramatic temperature fluctuation of soil and rock during a 24-hr period.

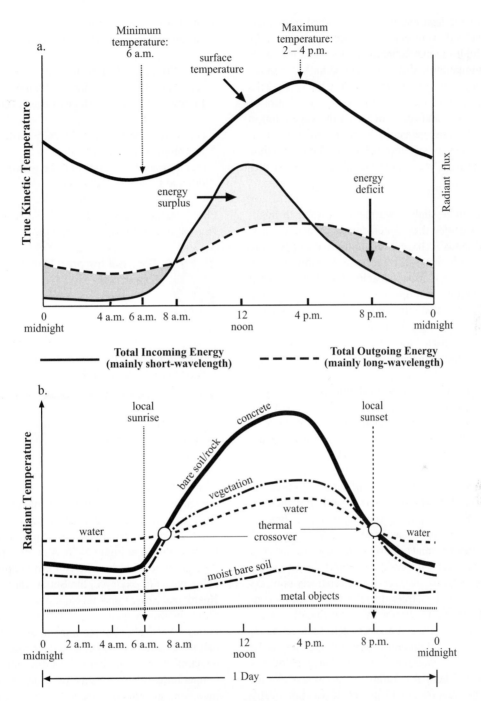

Figure 8-22   a) The diurnal cycle of reflected short-wavelength and emitted long-wavelength energy. Note the peak period of daily outgoing long-wavelength radiation and the general daily maximum temperature. b) The diurnal radiant temperature of bare soil and rock, concrete, vegetation, water, moist bare soil and metal objects.

If we were interested in performing temperature mapping of terrain consisting of just soil, rock, and water, we could predict what the image would look like if we acquired thermal infrared imagery at about 2:00 p.m. and at 4:00 a.m. in the morning. The soil and rock would appear brighter than water in the daytime thermal imagery due to their higher apparent temperature. Rock and soil continue to radiate energy into the atmosphere during the night. During the early evening rock and soil (and concrete) are still warmer than much of the surrounding terrain. By midnight these surfaces have radiated most of the heat energy they absorbed during the day and gradually they recede in brightness. Conversely, water with its high thermal capacity, maintains a relatively stable surface temperature that may well be higher than the soil and rock (and concrete) by about 4:00 a.m. This results in water being brighter on the nighttime thermal infrared imagery than the soil and rock, and even the vegetation in many instances. Because vegetation contains water, it is usually cooler (darker) than soil and rock on 2:00 p.m. daytime imagery and warmer (brighter) than soil and rock on predawn imagery.

Vegetation tends to be slightly warmer than water throughout the day and cooler than water during predawn hours. Very moist soil tends to have a stable diurnal temperature cycle, as shown in Figure 8-22b, because the more water in the soil, the higher its thermal capacity. Metal objects such as cars and aluminum rooftops appear cool (dark) on both day and nighttime thermal infrared imagery due to their low emissivity (remember, good metal reflectors are poor absorbers, and therefore poor emitters). In fact, they are often the darkest objects in the imagery.

Several of the previously mentioned diurnal temperature cycle relationships are demonstrated in Figure 8-23. An ATLAS mission was flown over a large sandbar in the Mississippi River at 5:00 a.m. and 10:30 a.m (Jensen et al., 2004). The data have a spatial resolution of 2.5 × 2.5 m. Daytime thermal bands 10, 11, and 12 reveal a dramatic difference in the temperature properties of sand and gravel on the sandbar. The different materials absorb the incident energy from the Sun differently, resulting in substantial differences in exitance from the sand and gravel surfaces in the three thermal infrared bands. During the day, the water is cooler than much of the surrounding countryside. The vegetation (V) on the bank of the river is also relatively cool. As expected, water is much warmer than the surrounding countryside at night and the vegetation is cooler than the water. There is still a noticeable difference between the large gravel and sand areas in the nighttime imagery, although not to the degree present in the daytime imagery. The goal of this project was to see if thermal infrared imagery was useful for discriminating between sand and gravel. It was practically impossible to discriminate between sand and gravel areas on 1:8,000-scale natural color aerial photography obtained at the same time as the 10:30 a.m. overflight.

Unless scientists are specifically trying to compute thermal inertia which requires both day and nighttime imagery, they often prefer to collect predawn thermal infrared imagery because:

- short-wavelength reflected energy from the Sun can create annoying shadows in daytime thermal infrared imagery;

- by 4:00 a.m., most of the materials in the terrain have relatively stable equilibrium temperatures, as shown in Figure 8-22b, i.e., the slopes are near zero;

- convective wind currents usually settle down by the early morning, resulting in more accurate flightlines (less crabbing of the aircraft into the wind) and no wind smear or wind streaks on the imagery.

Pilots are now able to obtain very accurate flightlines of thermal infrared imagery at night using onboard GPS.

Thermal Infrared Remote Sensing Examples

Water Pollution Monitoring: Identification of Thermal Effluent in the Savannah River

The Savannah River Site in South Carolina at one time extracted Savannah River water and used it to cool nuclear reactor operations. The hot cooling water was then returned to the Savannah River swamp system via Four Mile Creek (Figure 8-24). Sometimes the thermal effluent reached the Savannah River and created a thermal plume as shown in the flightline of predawn thermal infrared imagery in Figure 8-25. A thermal infrared view of the plume at full spatial resolution is shown in Figure 8-26. A number of state and federal laws governed the characteristics of the plume in the Savannah River. For example, the South Carolina Department of Health and Environmental Control (DHEC) required that the plume be less than one-third the width of the river at temperatures >2.8°C above river ambient temperature. The Savannah River after a heavy rain is dangerous, fast moving, and contains substantial debris. It is difficult to obtain accurate temperature measurements by placing people with thermometers in boats that are at the mercy of the current. Furthermore, it is not practical to place *in situ* thermometers in the river because of debris flowing at high rates of speed. Therefore, a remote sensing approach was used to obtain the required spatial temperature information.

Table 8-7 summarizes the user requirements and the sensor system parameters that were used to obtain the necessary temperature data. An across-track multispectral scanning

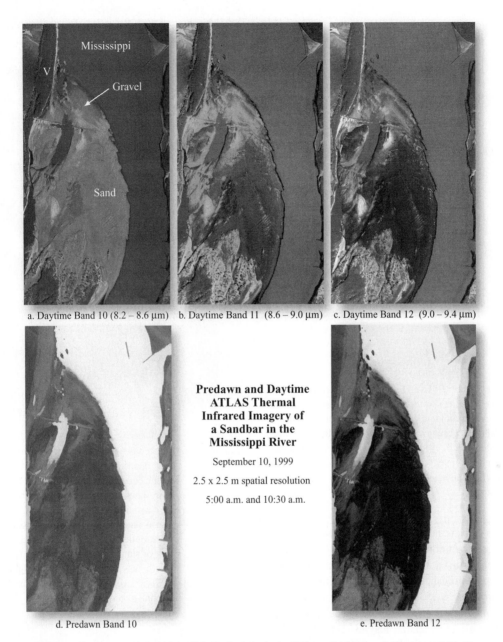

a. Daytime Band 10 (8.2 – 8.6 μm)   b. Daytime Band 11 (8.6 – 9.0 μm)   c. Daytime Band 12 (9.0 – 9.4 μm)

Predawn and Daytime ATLAS Thermal Infrared Imagery of a Sandbar in the Mississippi River

September 10, 1999

2.5 x 2.5 m spatial resolution

5:00 a.m. and 10:30 a.m.

d. Predawn Band 10

e. Predawn Band 12

Figure 8-23   Thermal infrared imagery of a large sandbar in the Mississippi River obtained at 5:00 a.m. and 10:30 a.m. on September 10, 1999. The ATLAS Band 11 detector did not function properly during predawn data collection (Jensen et al., 2004).

system equipped with a thermal infrared detector (Hg:Cd:Te) was used to record emitted radiant energy in the 8 – 14 μm region of the spectrum. This region was selected because the terrain was relatively cool, with a maximum emittance of thermal energy peaking at about 9.7 μm.

The width of the Savannah River was approximately 90 m (300 ft) in the vicinity of the plume. For scientists to obtain

a minimum representative sample of 30 pixels within the river, each pixel had to record the surface temperature for an area $\leq 3 \times 3$ m. To achieve this size of pixel, a sensor with a 2.5 milliradian IFOV was used to produce a nominal spatial resolution at nadir of approximately 3.05×3.05 m when the sensor was flown at 1,220 m (4,000 ft) above ground level.

Savannah River Site

Figure 8-24   The Savannah River Site in South Carolina at one time extracted water from the Savannah River and used it to cool nuclear-reactor operations. The thermal effluent returned to the Savannah River swamp system via Four Mile Creek, where it was to dissipate. However, the thermal effluent on occasion reached the Savannah River, resulting in a thermal plume.

Thermal Infrared Detection of Thermal Water Pollution in the Savannah River

Figure 8-25   A predawn thermal infrared image of thermal effluent entering the Savannah River swamp system from Four Mile Creek on March 31, 1981, at 4:28 a.m. The water migrated through the swamp system and eventually entered the Savannah River, producing a thermal plume. The temperature and spatial distribution of this plume were governed by a number of state and federal statutes. This is a 2× reduction image of a portion of the flightline and consists of 1,065 rows and 632 columns. It provides a regional overview of the spatial distribution of the thermal effluent.

Because an across-track scanning system was used to record the radiant temperature of objects, tangential scale distortion was present. If not removed, the quality of the cartographic products would be diminished. Therefore, the image was geometrically corrected by modeling each pixel as a function of scan angle from nadir. This resulted in each pixel in a scan line having an effective spatial resolution of 2.8 × 2.8 m. The data were then geometrically corrected by selecting ground-control points and rectifying the data using nearest-neighbor resampling, as discussed in Jensen (2005).

For investigators to identify a thermal plume and map its temperature and spatial distribution, the plume, river, and surrounding land all had to be distinguishable on the remotely sensed imagery. Each day the maximum thermal contrast between Savannah River water and other land surfaces is reached just before dawn, as previously discussed when the river is warmer than the land surface. It was assumed that the hot plume would be distinguishable from the ambient Savannah River water and land surface day or

night. An additional factor in favor of a predawn overflight was the fact that at night there are no tree shadows along the bank of the river that can cause spurious temperature measurements of the river or plume. These considerations resulted in the data being collected on March 31, 1981, at 4:28 a.m. under the influence of a high-pressure weather system with an absolute humidity of 27%.

To comply with the ± 0.2°C requirement, the thermal infrared scanner data were calibrated using internal blackbody source referencing. While scanning a single line of data, the detector looked at a cold plate with a known temperature, the terrain, and then a hot plate with a known temperature. By

Thermal Infrared Detection of Thermal Water Pollution in the Savannah River

Figure 8-26 Predawn thermal infrared image (8.0 – 14 μm) of thermal effluent flowing into the Savannah River acquired at 4:28 a.m. on March 31, 1981, using a Daedalus DS-1260 multispectral scanning system.

Table 8-7. User Requirements for monitoring the temperature distribution of a thermal plume in the Savannah River and sensor system parameters.

User Requirement	Remote Sensing System Parameter
The apparent temperature of each pixel must be accurate to within ± 0.2°C of its true temperature. Temperatures in the scene ≥ the ambient river temperature are of primary importance.	Sense in the 8 – 14 μm region using mercury-cadmium-telluride (Hg:Cd:Te) detector with an A-to-D conversion of 8 bits (0 – 255). Radiometrically calibrate to ±0.2°C with pixel values ranging from -3 to 48°C.
At least 30 pixels should fall within the 90-m width of the river. Therefore, a spatial resolution element of ≤ 3 × 3 m is required. Rectify all pixels to be within ±10 percent of their planimetric position.	Use a sensor with an IFOV of 2.5 mrad flown at 1200 m (4,000 ft) AGL. Remove geometric sensor system distortion and resample to 2.8 × 2.8 m picture elements.
Acquire predawn imagery in the spring and analyze results rapidly.	Predawn data acquisition with stable atmospheric conditions. Calibrate and analyze.
Produce isotherm maps and extract quantitative temperature information for selected class intervals.	Apply image processing and cartographic principles to obtain quantitative information from the isothermal map and associated transects.

knowing the relationship between the temperature of these plates, the analyst could calculate the apparent temperature of the terrain for each pixel, from -3 to 48°C. The major drawback of this calibration method was that it did not incorporate the effect of the intervening atmosphere on signal response. However, it appeared that the calibration was fairly successful. When the calibrated data were compared with two *in situ* surface-temperature measurements made in the Four Mile Creek plume during the time of the overflight, both remote sensing measurements were within ± 0.2°C of the *in situ* measurements.

Data Analysis

Environmental scientists and ichthyologists are interested in the temperature and spatial distribution of thermal plumes and how they relate to the ambient river temperature. If a plume exists, it is important to determine where the plume is greater than a specified number of degrees above river ambient temperature. Depending on the time of year, thermal plumes may attract certain species of aquatic organisms and exclude others. Unfortunately, a hot plume extending across

the river at some depth will inhibit certain species from swimming upstream during the spawning season.

Two types of data analysis were performed to extract quantitative temperature information. First, the spatial distribution of selected temperature class intervals > 2.8°C above river ambient temperature were mapped. Next, transects were passed through the isotherm map to document the cross-section surface area of the plume at specific temperature class intervals.

Mapping the spatial distribution of the temperature made it necessary to identify:

- the land (soil and vegetation) to make sure it was not confused with the thermal plume;

- the ambient river temperature;

- the spatial distribution of the plume temperature > 2.8°C above river ambient temperature.

Table 8-8.   Savannah River thermal plume apparent temperature transects and color-coding (density slicing) logic.

		Relationship of Class to Ambient River Temperature, 12°C						
		Class 1 Dark blue ambient	Class 2 Light blue +1°C	Class 3 Green 1.2°–2.8°C	Class 4 Yellow 3.0°–5.0°C	Class 5 Orange 5.2°–10°C	Class 6 Red 10.2°–20°C	Class 7 White >20°C
Transect[a]	Average Width of River[b]	Brightness Value Range for Each Class Interval						
		74–76	77–80	81–89	90–100	101–125	126–176	177–255
A	32 pixels = 89.6 m	15/42[c]	17/47.6	—	—	—	—	—
B	38 pixels = 106.4 m	25/70	1/2.8	1/2.8	2/5.6	1/2.8	5/14	3/8.4
C	34 pixels = 95.2 m	19/53.2	2/5.6	2/5.6	2/5.6	6/16.8	3/8.4	—

[a] Each transect was approximately 285 m in length (66 pixels at 2.8 m/pixel). Transect measurements in the river were made only after the image was rotated, so that the beginning and ending pixels of the transect fell on the same scan line.

[b] Includes one mixed pixel of land and water on each side of the river.

[c] Notation represents pixels and meters; for example, 15 pixels represent 42 m.

At 4:28 a.m. the river is much warmer than the surrounding land and easily distinguishable. Two methods were used to determine ambient river temperature. First, *in situ* temperature measurements of the Savannah River were available which placed the temperature above the plume at 12°C. Second, a sample of 200 pixels just above the Four Mile Creek plume yielded a mean of 12°C and a standard deviation of $\pm 0.2°C$. Thus, the ambient river temperature was determined to be 12°C.

Seven class intervals were then selected. Three of the class intervals were for temperatures < 2.8°C above river ambient temperature and included the land and thermally unaffected river. Four class intervals were used to highlight special ranges of temperature within the thermal plume that were > 2.8°C above river ambient temperature. Each of the seven class intervals were assigned a unique color with yellow, orange, red, and white corresponding to warmer temperatures in the plume (Table 8-8). Analysis of Color Plate 8-2a reveals that the hottest part of the plume was just offshore from the mouth of Four Mile Creek (34.6 °C; 94.28 °F). The temperature of the creek itself was just as hot as the warmest part of the plume; however, the overhanging shrubbery along the banks of the creek dampened the remotely sensed apparent temperature of the creek along its main channel. Also, the northern eddy of Four Mile Creek warmed the water slightly just above the plume as it entered the Savannah River.

A color-coded (density sliced) isotherm map provided valuable information about the spatial distribution of the temperature of the plume as it entered the Savannah River and progressed downstream. However, more quantitative data were obtained by passing three transects (A, B, and C) through the plume, as shown in Color Plate 8-2b. The brightness values encountered along each transect are summarized in Table 8-8. These values were obtained only after the original image was geometrically rotated 16° clockwise so that the end points of each transect fell on the same scan line. This ensured that the number of meters in each temperature class along each transect was accurately measured. If the analyst extracts transects where the end points do not fall on the same scan line (or column), the hypotenuse of stair-stepped pixels must be considered instead of the simple horizontal pixel distance. These relationships are demonstrated in Color Plate 8-2c.

A histogram of transect B is shown in Color Plate 8-2d. The relationship between the original brightness values and the class intervals of the density-sliced map is provided in Table 8-8. By counting the number of pixels along a transect in specific temperature class intervals within the plume and counting the total number of pixels of river, it is possible to determine the proportion of the thermal plume falling within specific temperature class intervals (Jensen et al., 1983 and 1986). For example, in 1981 in South Carolina a thermal plume could not be >2.8°C above river ambient temperature for more than one-third of the width of the river. Transect

Enhancements Applied to Thermal Infrared Imagery of a Thermal Plume in the Savannah River

Figure 8-27   a) Original contrast-stretched image of the Four Mile Creek thermal plume in the Savannah River recorded on March 31, 1981, at 4:28 a.m. b) Low-frequency filtered image. c) High-frequency filtered image. d) Application of a minimum filter. e) Application of a maximum filter. f) An embossed filter gives the impression of shaded relief. g) Roberts edge enhancement. h) Sobel edge enhancement. i) Laplacian edge enhancement.

information extracted from thermal infrared imagery and summarized in Table 8-8 were used to determine if the plume was in compliance.

The fundamental black-and-white thermal infrared image of the plume contained valuable information (Figure 8-27a). However, it was also possible to digitally enhance the original thermal infrared image to visually appreciate the more subtle characteristics of the thermal plume and the surrounding phenomena. For example, Figure 8-27b is a low-frequency filtered image of the original contrast-stretched image It emphasized the slowly varying components within the image. Conversely, the application of a high-frequency filter enhanced the high-frequency detail in the image. Note how several radiometric errors in the scan lines are more pronounced in the high frequency filtered image in Figure 8-27c. A minimum filter enhanced the core of the plume, while a maximum filter enhanced the entire plume (Figure 8-27d,e). An embossed filter created a shaded-relief impression of the plume (Figure 8-27f). Three edge-enhancement algorithms (Roberts, Sobel, and Laplacian) highlighted different parts of the plume and the edges of the plume/land/water boundaries (Figure 8-27g,h,i). The logic and mathematics of these algorithms are summarized in Jensen (2005).

Thermal infrared surveys of the Four Mile Creek plume were routinely collected at least two times per year from 1981 through 1988 to ensure that the plume was in compliance with state and federal water-quality standards. Cooling towers were then built to receive the thermal effluent.

Water Pollution Monitoring: Identification of Leaking Septic Tanks that Contribute to Fecal Coliform Count

---

A high fecal coliform count in fresh and marine waters may indicate the presence of bacteria that can cause typhoid, dysentery, hepatitis A, and cholera. Health departments routinely monitor drinking water, recreational lakes, rivers, ponds, estuaries, and ocean water for fecal coliform contamination. Shellfish are sensitive to fecal coliform contamination because filter-feeding organisms such as oysters, clams, and mussels can accumulate bacteria, protozoa, and viruses. The bacteria concentrated in the shellfish by this method of feeding are not harmful to the shellfish, but can be harmful to people that consume shellfish. Increased fecal coliform levels near shellfish beds lead to shellfish bed closures.

The South Carolina Department of Health and Control (DHEC) has investigated the use of forward-looking infrared (FLIR) remote sensing to identify leaking septic tanks near Charleston, SC. Leaking septic tanks are a common source of fecal coliform in bays, tidal creeks, and estuaries. For example, two leaking septic tank systems are identified in the FLIR imagery shown in Figure 8-28. The imagery was obtained at night in the winter when the warm effluent from the septic tank contrasted with the relatively cool marsh and tidal creek water. Imagery such as this helped officials know where to obtain samples on the ground to bring the septic

Thermal Infrared Detection of Leaking Septic Tanks near Charleston, SC

Figure 8-28   Predawn winter FLIR imagery of two failed septic tanks along a tidal creek near Charleston, SC.

tank into compliance with health regulations (NOAA Coastal Services, 2006)

Thermal Infrared Residential Insulation Surveys

Homes, office buildings, and industries are insulated so that in the winter, expensive heat does not easily exit. It is possible to use remote sensing to monitor the effectiveness of the insulation. However, the results are usually subjective rather than quantitative because many important parameters must be known to quantitatively document the exact insulation characteristics of every home or business in a neighborhood.

Some of the most important parameters to be taken into consideration when performing a residential thermal imagery insulation survey are summarized in Table 8-9. Colcord (1981) suggests the temperature (radiometric) resolution should be $\pm 0.8°C$ for differentiation of an R-10 and R-20 house. This results in the discrimination of approximately four levels of apparent insulation (poor, fair, good, and excellent) based on visual examination. Most houses are

Table 8-9. Emissivity of building materials from 8 – 14 μm (Colcord, 1981; Wolfe, 1985).

Material	Emissivity, ε
asphalt shingle (dry)	0.97
asphalt shingle (wet)	1.00
cedar-shake shingle (dry)	0.95
cedar-shake shingle (wet)	0.99
tar/stone	0.97
aluminum (sheet)	0.09
copper (oxidized)	0.78
iron (sheet-rusted)	0.69
tin (tin-plated sheet iron)	0.07
brick (red-common)	0.93
paint (average of 16 colors)	0.94
sand	0.90
wood (planed oak)	0.90
frost crystals	0.98

Table 8-10. User requirements for conducting a residential thermal imagery insulation survey.

User Requirement	Remote Sensing System Parameter
Temperature (radiometric) resolution should be ± 0.8°C for differentiation of R-10 and R-20 houses (Colcord, 1981).	Sense in the 8–14 μm region using a mercury-cadmium-telluride (Hg:Cd:Te) detector. The true kinetic temperature of several rooftops is required to calibrate to ±0.8°C.
Obtain 15 × 15 cm spatial resolution to get many samples from each roof; however, 1 × 1 m spatial resolution may be sufficient. Rectify data to be within ±1 pixel of their planimetric position.	Use sensor with an IFOV of 2.5 mrad flown at 400 m AGL. This will result in a spatial resolution of 1 × 1 m at nadir. Remove tangential scale distortion if possible and rectify to large-scale planimetric maps using ground-control points and nearest-neighbor resampling.
Acquire predawn imagery in the spring and analyze results rapidly.	Acquire imagery when the sky is clear, wind is < 7 km/hr, there is low humidity and no moisture, snow, or ice on the roof. Predawn imagery has no confusing shadows from surrounding buildings or trees.
Produce isotherm maps and extract qualitative and quantitative temperature information. Reduce the information to temperature class levels that the public can understand (e.g., poor, fair, good, and excellent).	Obtain rooftop emissivity characteristics for the neighborhood. Note differences in rooftop materials, roof pitch and orientation, and building occupancy. Use digital image processing to obtain quantitative information from the thermal images.

approximately 80°F. This approximates a 300 K blackbody with a dominant wavelength of approximately 9.7 μm. Therefore, the scene should be inventoried using an 8 – 14 μm mercury-cadmium-telluride (Hg:Cd:Te) detector. The true rooftop kinetic temperature of several houses is required if the data are to be calibrated to ±0.8°C.

The sensor system must acquire enough pixels on a rooftop to characterize its thermal environment. One pixel per rooftop is not sufficient. Some have recommended a 15 × 15 cm spatial resolution cell, but a 1 × 1 m spatial resolution appears to be sufficient for most homes, excluding trailer parks. A sensor with an IFOV of 2.5 milliradians flown at 400 m above ground level would achieve a nominal ground spatial resolution of 1 × 1 m per pixel at nadir. This will result in problems in hilly terrain as local relief approaches 400 m. Tangential scale distortion should be removed and the data geometrically rectified. The thermal infrared survey should take place under the following conditions, if possible.

- standing water, ice, or snow should not be present on the roof, as this dramatically changes the emissivity characteristics of the roof, as summarized in Table 8-10;

- wind speed should be < 7 km/hr;

- humidity and ground fog should be minimal;

- cloud cover should be minimal, as clouds radiate heat back onto the roof.

It is important to obtain precise rooftop-emissivity characteristics for the residential neighborhood or business districts of interest because different rooftop materials have different emissivities. For example, cedar shake shingles have an emissivity of 0.99 when wet and 0.95 when dry, while an aluminum rooftop has an emissivity of 0.09 (Table 8-10).

Figure 8-29   Predawn thermal infrared image of a residential subdivision in Fort Worth, TX. The image was acquired at an altitude of 250 m above ground level with an across-track scanning system and a 1 mrad field of view at 6:45 a.m. on January 10, 1980. This yielded an effective ground spatial resolution at nadir of 0.25 × 0.25 m per pixel. The sky was clear.

Ideally, all rooftops are flat. Unfortunately, roof pitch (how steep it is) and orientation toward or away from the sensor will impact the accuracy of the apparent temperature values derived. Finally, people tend to lower their house temperature in the winter when they leave town for a few days, from approximately 72°F to approximately 55 – 60°F. Therefore, some rooftops that appear cold in a thermal infrared image (i.e., they appear to have good attic insulation) may in fact be unoccupied with the thermostat set low. Thermal infrared imagery of a residential area in Fort Worth, TX, is shown in Figure 8-29. It was acquired at an altitude of 250 m above-ground-level with an across-track scanning system with a 1.0 milliradian field of view at 6:45 a.m. on January 10, 1980. This yielded an effective ground spatial resolution at nadir of 0.25 × 0.25 m per pixel.

Theoretically, in Figure 8-29, the most poorly insulated home in the subdivision is at (*a*), while the house next door at (*b*) appears to be well insulated. There is a fire in the fireplace at home (*c*). The road network (*d*), concrete sidewalks (*e*), and back-porch concrete pads (*f*) are all much warmer than the surrounding terrain. Some of the backyard terrain is more moist (*g*) than other areas. All metal vehicles (*h*) are dark (cold) because of their low emissivity. Most of the

homes in this subdivision probably received about the same insulation; therefore, it is not surprising that their rooftop apparent temperatures appear similar.

Some utility companies subcontract for thermal infrared data and then attempt to convince homeowners that their insulation might not be as good as their neighbors'. The only problem with this logic is that 1) the neighbor's home might have a different roof material, 2) the orientation of the neighbor's roof toward the sensor might be different, and 3) the neighbor might keep their home much cooler. If these parameters are taken into consideration, then the rooftop temperature information extracted from the thermal infrared data may be of value to the homeowner.

Thermal infrared energy surveys are also performed routinely for public and industrial facilities. Figure 8-30 depicts an aerial photograph and predawn thermal infrared image of the Solomon Blatt fieldhouse on the University of South Carolina campus. The thermal imagery was acquired at 4:30 a.m. on March 10, 1983, at 500 m above-ground-level using a 2.5 mrad IFOV sensor system. This resulted in a spatial resolution of approximately 1.25 × 1.25 m.

a. Vertical Aerial Photograph   b. Predawn Thermal Infrared Image

Figure 8-30 a) Aerial photograph of the Solomon Blatt fieldhouse on the University of South Carolina campus, Columbia, SC. b) Thermal infrared image obtained at 4:30 a.m. on March 10, 1983, with a spatial resolution of 1.25 × 1.25 m. Note 11 cool metal vents, the hot air escaping from nine of the vents, and the cool overhanging eaves. The eaves are exposed on both sides to the cool early morning air.

Thermal Infrared Commercial/Industrial Roof Moisture Surveys

Roof maintenance is a billion-dollar industry. It will always be necessary for a roofing contractor to walk on the roof to determine exactly what repairs need to be made, but it has become increasingly useful to conduct thermal infrared remote sensing studies of rooftops, especially when substantial roof surface and/or subsurface water may be present. This helps the roofing contractor locate all of the roof problems. The process is called roof moisture surveying.

Diurnal expansion and cooling, wind, rain, snow, hail, ice, chemicals, punctures, and time eventually cause roofs to fail. Roof problems manifest themselves in two ways: leakage and entrained water contamination. Leakage is simple, although the leak inside the building rarely reflects the exact location on the roof where there is a hole or tear in the roofing material. Because most types of insulation absorb a certain amount of water, it's difficult to find the entrained water contamination because the roof may not actually show a leak until the sub-roof material and insulation have absorbed all the water that they can hold (Stockton, 2004a).

Three nondestructive methods can be used to locate roof subsurface moisture: 1) nuclear gauges that count neutrons, 2) capacitance meters that measure resistance, and 3) thermal infrared remote sensing that can be used to map roof temperature patterns. A technic an can use a nuclear gauge or capacitance meter to take sample readings on a 1 × 1 m or 5 × 5 m grid on the roof. These point samples are then used to extrapolate where the roof subsurface water is located.

Thermal infrared remote sensing is the preferred method for roof moisture surveying (Stockton, 2004b). During the day, the Sun's incident energy heats the rooftop material. Some of this energy is transferred below the roof surface into any water, wood, or insulation present. During the night, the Sun radiates this stored energy back into the atmosphere. This is known as radiational cooling. Higher mass (wet) areas absorb and dissipate heat at a different rate than the lower mass (dry) areas The wet areas radiate heat for a longer period of time at night because they have greater thermal capacity as previously discussed. Airborne thermal infrared remote sensing can detect these sources of heat and identify the higher mass (wet areas) during this nighttime window of uneven heat dissipation.

For example, consider the commercial building shown in Figure 8-31a. A large-scale vertical serial photograph is

Thermal Infrared Roof Moisture Survey

a. Oblique view of commercial building.

b. Large-scale vertical aerial photograph.

c. Predawn thermal infrared image.

Probable wet area

d. CAD drawing of probable roof moisture.

Figure 8-31   a) Oblique aerial photograph of a commercial building in North Carolina. b) Vertical large-scale aerial photograph. c) Predawn thermal infrared image obtained on May 6, 2000. d) CAD drawing showing polygons of probable roof moisture (courtesy Stockton Infrared Thermographic Services; Stockton Infrared; 2006).

shown in Figure 8-31b. A predawn thermal infrared image is shown in Figure 8-31c and a computer-aided-design (CAD) map of the extent of the subsurface moisture is shown in Figure 8-31d. This information is used to estimate the amount of roof that must be replaced or repaired.

Analysis of the Urban Heat Island Effect

It is well known that an urban heat island exists over most urban areas compared to the relatively cooler nonurban surrounding countryside. Urban heat islands are caused by deforestation and the replacement of the land surface by nonevaporative and nonporous materials such as asphalt and concrete. In addition, air-conditioning systems introduce a significant amount of heat energy into the urban landscape. The result is reduced evapotranspiration and a general increase in urban landscape temperature.

Quattrochi and Ridd (1994) and Lo et al. (1997) evaluated several cities using high spatial resolution thermal infrared

remote sensing systems to document the urban heat island effect. In general, they found that during the daytime hours commercial land cover exhibited the highest temperatures followed by services, transportation, and industrial land uses. The lowest daytime temperatures were found over water bodies, vegetation, and agricultural land use, in that order. Residential housing being composed of a heterogeneous mixture of buildings, grass, and tree cover exhibited an intermediate temperature, as expected.

At night, commercial, services, industrial, and transportation land cover types cooled relatively rapidly. Nevertheless, their temperatures even in the predawn early morning hours were still slightly higher than those for vegetation and agriculture. Water has a high thermal capacity; therefore, it is typically the warmest land cover during the predawn hours as previously discussed. Conversely, agriculture typically exhibits the lowest temperature at night.

Examples of daytime and nighttime thermal infrared images of Atlanta, GA, confirm these observations (Figure 8-32ab).

Atlanta, GA

a. Daytime thermal infrared
image of Atlanta, GA.

b. Nighttime thermal infrared image.

Figure 8-32   a) Daytime ATLAS airborne thermal infrared channel 13 (9.60 – 10.2 μm) image of Atlanta, GA. Prominent urban features are annotated for orientation. b) Nighttime, predawn thermal infrared image of Atlanta, GA (courtesy D. Quattrochi and J. Luvall, *Project ATLANTA*; NASA Marshall Space Flight Center, Global Hydrology Climate Center, Huntsville, AL).

The ATLAS channel 13 imagery (9.60 – 10.2 μm) at 10×10 m spatial resolution was obtained as part of the NASA EOS investigation Project ATLANTA (Quattrochi and Luvall, 1999). Shadows from tall buildings located in the Atlanta CBD are observed on the daytime imagery. The intense thermal exitance from buildings, pavement, and other surfaces typical of the urban landscape, as well as the heterogeneous distribution of these responses, stand in significant contrast to the relative "flatness" of the Atlanta thermal landscape at night. Also, the dampening effect that the urban forest has on upwelling thermal energy response is evident, particularly in the northeast portion of the daytime image where residential tree canopy is extensive. In the nighttime image there is still evidence, even in the very early morning, of the elevated thermal responses from buildings and other surfaces in the Atlanta CBD and from streets and freeways. Thermal energy responses for vegetation across the image are relatively uniform at night, regardless of vegetation type (e.g., grass, trees).

Such thermal information has been used to 1) model the relationship between Atlanta urban growth, land-cover change, and the development of the urban heat island phenomenon through time, 2) model the relationship between Atlanta urban growth and land-cover change and air quality through time, and 3) model the overall effects of urban development on surface-energy budget characteristics across the Atlanta urban landscape. Such data can also be used to recommend tree-planting programs that may be able to substantially decrease the urban heat island effect (Quattrochi and Ridd, 1998).

Weng et al. (2004) analyzed Landsat 7 ETM[+] data to extract land surface temperature information for urban heat island studies of Indianapolis, Indiana. They found the incorporation of the amount of green vegetation, dry soil, and shade within each pixel was useful for estimating the land surface temperature/vegetation abundance relationship for urban heat island studies.

Use of Thermal Infrared Imagery for Forestry Applications

Quantitative information about forest canopy structure, biomass, age, and physiological condition have been extracted from thermal infrared data. Basically, a change in surface temperature can be measured by an airborne thermal infrared sensor (e.g., TIMS or ATLAS) by repeatedly flying over the same area a few times. Usually a separation of about 30 minutes results in a measurable change in surface temperature caused by the change in incoming solar radiation; however, a longer period of time may be required. Average surface net radiation (R_n) is measured *in situ* for the study area and is used to integrate the effects of the nonradiative fluxes. The change in surface temperature from time period t_1 to t_2 (i.e., Δt) is the value that reveals how those nonradiative fluxes are reacting to radiant energy inputs. The ratio of these two parameters is used to compute a surface property defined by Luvall (1997) as a *Thermal Response Number* (TRN):

$$TRN = \frac{\sum_{t_1}^{t_2}[(R_n)(\Delta t)]}{\Delta t}. \tag{8-24}$$

Terrain containing mostly soil and bare rock have the lowest TRN values, while forests have the highest. The TRN is a site-specific property that may be used to discriminate among various types of coniferous forest stands and some of their biophysical characteristics (Luvall, 1997).

Remote Sensing Plant Stress Degree Days

Moran (2004) summarized various *relative* measures of plant ecosystem health that can be monitored using thermal infrared remote sensing. For example, the stress degree day (SDD) was defined by Idso et al. (1977) as:

$$SDD = \sum_{i=1}^{n}(T_c - T_a)_i \tag{8-25}$$

where T_c is plant canopy temperature (a thermal infrared remote sensing measurement) 1 to 1.5 hours after solar noon, T_a is the air temperature 1.5 m above the soil (an *in situ* measurement), and ($T_c - T_a$) is summed over n days beginning at day i. The SDD index is related to plant water consumption and has been used to schedule crop irrigation, monitor grassland stress, and improve yield estimates.

Summary

Thermal infrared remote sensing will become even more important in the future. Temperature and emissivity are powerful biophysical variables critical to many investigations. We now have very sensitive linear- and area-array thermal infrared detectors that can function in broad thermal bands or in hyperspectral configurations. In the future, it will be commonplace to find miniature thermal infrared sensors on unmanned aerial vehicles (UAV) used by the military, scientists and lay persons.

References

Becker, F. and A. L. Li, 1995, "Surface Temperature and Emissivity at Various Scales: Definition, Measurement and Related Problems," *Remote Sensing of Environment*, 12:225–253.

Colcord, J., 1981, "Thermal Imagery Energy Surveys," *Photogrammetric Engineering & Remote Sensing*, 47(2):237–240.

Coll, C., Caselles, V., Valor, E. and E. Rubio, 2003, "Validation of Temperature-Emissivity Separation and Split-Window Methods from TIMS and Ground Measurements," *Remote Sensing of Environment*, 85:232-242.

Czajkowski, K. P., Goward, S. N., Mulhern, T., Goetz, S. J., Walz, A, Shirey, D., Stadler, S., Prince, S. and R. O. Dubayah, 2004, "Estimating Environmental Variables Using Thermal Remote Sensing," in D. A. Quattrochi and J. C. Luvall, (Eds.), *Thermal Remote Sensing in Land Surface Processes*, Boca Raton: CRC Press, 11-32.

Czajkowski, K. P., Goward, S. N., Mulhern, T., Goetz, S. J., Walz, A, Shirey, D., Stadler, S., Prince, S. and R. O. Dubayah, 1997, "Recovery of Environmental Variables from Thermal Remote Sensing," in D. A. Quattrochi and M. F. Goodchild (Eds.), *Scale in Remote Sensing and GIS*, Chelsea: Lewis.

Estes, J. E., 1966, "Some Applications of Aerial Infrared Imagery," *Annals* of the Association of American Geographers, 56(4):673–682.

Fischer, W. A., 1983, "History of Remote Sensing," *Manual of Remote Sensing*, R. Reeves (Ed.), Bethesda: ASP&RS, 27–50.

FLIR Systems, 2006, *Infrared Imaging Systems*, Portland: FLIR Systems, Inc., www.FLIR.com.

Idso, S. B., Jackson, R. D. and R. J. Reginato, 1977, "Remote Sensing of Crop Yields," *Science*, 196: 19–25.

Itres TABI, 2006, *Thermal Airborne Broadband Imager (TABI)*, Alberta, Canada: Itres Research, www.itres.com.

Itres TASI, 2006, *Thermal Airborne Spectrographic Imager (TASI)*, Alberta, Canada: Itres Research, www.itres.com.

Jacob, F., Francois, P., Schmugge, T. Vermote, E., French, A. and K. Ogawa, 2004, "Comparison of Land Surface Emissivity and Radiometric Temperature Derived from MODIS and ASTER," *Remote Sensing of Environment*, 90:137–152.

Janza, F. J., 1975, "Interaction Mechanisms," in *Manual of Remote Sensing*, R. Reeves (Ed.), Bethesda: ASP&RS, 75–179.

Jensen, J. R., 2005, *Introductory Digital Image Processing: A Remote Sensing Perspective*, 3rd Ed., Upper Saddle River: Prentice-Hall, 525 p.

Jensen, J. R., Luders, J. K., Schill, S. R. and G. T. Raber, 2004, "Identifying Riverine Sand and Gravel Deposits Using A High Resolution Day-night Temperature Difference Map and Digitized Aerial Photography," *Geocarto International*, 19(2): 49-56.

Jensen, J. R., Hodgson, M. E., Christensen, E. J., Mackey, H. E., Tinney, L. R. and R. Sharitz, 1986, "Remote Sensing Inland Wetlands: A Multispectral Approach," *Photogrammetric Engineering & Remote Sensing*, 52(1)87–100.

Jensen, J. R., Pace, P. J. and E. J. Christensen, 1983, "Remote Sensing Temperature Mapping: the Thermal Plume Example," *The American Cartographer*, 10:111–127.

Jensen, J. R., Ramsey, E., Mackey, H. E. and M. E. Hodgson, 1988, "Thermal Modeling of Heat Dissipation in the Pen Branch Delta Using Thermal Infrared Imagery," *Geocarto International*, 4:17–28.

Kahle, A. B., Schieldge, J. P., Abrams, M. J., Alley, R. E. and C. J. VeVine, 1981, *Geologic Applications of Thermal Inertia Imaging Using HCMM Data*, #81–55, Pasadena: JPL.

Kahle, A. B., Schieldge, J. P. and R. E. Alley, 1984, "Sensitivity of Thermal Inertia Calculations to Variations in Environmental Factors," *Remote Sensing of Environment*, 16:211–232.

Kirkland, L., Herr, K., Keim, E., Adams, P., Salisbury, J., Hackwell, J. and A. Treiman, 2002, "First Use of An Airborne Thermal Infrared Hyperspectral Scanner for Compositional Mapping," *Remote Sensing of Environment*, 80:447–459.

Lillesand, T. M. and R. W. Kiefer, 2003, *Remote Sensing and Image Interpretation*, N.Y.: John Wiley, 750 p.

Lo, C. P., Quattrochi, D. A. and J. C. Luvall, 1997, "Application of High-resolution Thermal Infrared Remote Sensing and GIS to Assess the Urban Heat Island Effect," *International Journal of Remote Sensing*, 18(2):287–304.

Luvall, 1997, "The Use of Remotely Sensed Surface Temperatures from an Aircraft Based Thermal Infrared Multispectral Scanner (TMS) to Estimate the Spatial Variability of Latent Heat Fluxes from a White Pine (*Pinus strobus* L.) Plantation," in D. A. Quattrochi and M. F. Goodchild (Eds.), *Scale in Remote Sensing and GIS*, Chelsea: Lewis, 169–185.

Luvall, J. C., 1999, personal correspondence, NASA Marshall Space Flight Center.

McDonnell Douglas, 1982, *Reconnaissance Handy Book for the Tactical Reconnaissance Specialist*, St. Louis: McDonnell Douglas, 183 p.

Moran, M. S., 2004, "Thermal Infrared Measurement as an Indicator of Plant Ecosystem Health," in D. A. Quattrochi and J. C. Luvall, (Eds.), *Thermal Remote Sensing in Land Surface Processes*, Boca Raton: CRC Press, 257–282.

NOAA Coastal Services, 2006, *Identifying Sources of Fecal Coliform Bacteria in Charleston, S.C.*, http://www.csc.noaa.gov/crs/rs_apps/issues/septic.htm

Omega, 2005, *Transactions in Measurement Control*, Vol. 1, 2nd Ed., Omega Engineering, LLC, http://www.omega.com, 86 p.

Oregon DEQ, 2003, *Aerial Surveys in the Umpqua River Basin: Thermal Infrared and Color Videography*, Portland: Oregon Department of Environmental Quality, 90 p., www.deq.state.or.us/wq/tmdls/umpquabasin/flir/umpqualir.pdf.

Ouaidrari, H., Goward, S. N., Czajkowski, K. P., Sobrino, J. A. and E. Vermote, 2002, "Land Surface Temperature Estimation from AVHRR Thermal Infrared Measurements: An Assessment for the AVHRR Land Pathfinder II Data Set," *Remote Sensing of Environment*, 81:114–128.

Petitcolin, F and E. Vermote, 2002, "Land Surface Reflectance, Emissivity and Temperature from MODIS Middle and Thermal Infrared," *Remote Sensing of Environment*, 83:112-134.

Peres, L. F. and C. C. DaCamara, 2004, "Land Surface Temperature and Emissivity Estimation Based on the Two-Temperature Methods: Sensitivity Analysis Using Simulated MSG/SEVIRI Data," *Remote Sensing of Environment*, 91:377–389.

Price, J. C., 1985, "On the Analysis of Thermal Infrared Imagery: The Limited Utility of Apparent Thermal Inertia," *Remote Sensing of Environment*, 18:59–73.

Quattrochi, D. A. and N. S. Goel, 1995, "Spatial and Temporal Scanning of Thermal Infrared Remote Sensing Data," *Remote Sensing Reviews*, 12:225–286.

Quattrochi, D. A. and J. C. Luvall, 2004, *Thermal Remote Sensing in Land Surface Processes,* Boca Raton: CRC, 440 p.

Quattrochi, D. A. and J. C. Luvall, 1999, *High Spatial Resolution Airborne Multispectral Thermal Infrared Data to Support Analysis and Modeling Tasks in the EOS IDS Project ATLANTA*, http: //wwwghcc.msfc.nasa.gov/atlanta/.

Quattrochi, D. A. and M. K. Ridd, 1994, "Measurement and Analysis of Thermal Energy Responses from Discrete Urban Surfaces Using Remote Sensing Data," *International Journal of Remote Sensing*, 15(10):1991–2022.

Quattrochi, D. A. and Merrill K. Ridd, 1998, "Analysis of Vegetation within a Semi-arid Urban Environment Using High Spatial Resolution Airborne Thermal Infrared Remote Sensing Data," *Atmospheric Environment*, 32(1):19–33.

Schmugge, T., French, A., Ritchie, J. C., Rango, A. and H. Pelgrum, 2002, "Temperature and Emissivity Separation from Multispectral Thermal Infrared Operations," *Remote Sensing of Environment*, 79:189-198.

Sabins, F. F., Jr., 1997, *Remote Sensing Principles and Interpretation*, NY: Freeman, 494 p.

Salisbury, J. W. and D. M. D'Aria, 1992, "Emissivity of Terrestrial Materials in the 8 – 14 μm Atmospheric Window," *Remote Sensing of Environment*, 42:83–106.

Slater, P. N., 1980, *Remote Sensing: Optics and Optical Systems*, NY: Addison-Wesley, 575 p.

Snyder, W. C., Zhengming, W., Zhan, Y. and Y. Feng, 1997, "Thermal Infrared (3 – 14 μm) Bi-directional Reflectance Measurements of Sands and Soils," *Remote Sensing of Environment*, 60:101–109.

Sobrino, J. A., Jimenez-Munoz, J. C. and L. Paolini, 2004, "Land Surface Temperature Retrieval from Landsat TM 5," *Remote Sensing of Environment*, 90:434–440.

Stockton, G. R., 2004a, "Aerial Infrared Thermography," *Cadalyst*, September (2004):47–53.

Stockton, G. R., 2004b, "Finding Pollution with Aerial Infrared Thermography," *Industrial Physicist*, May (2004):24–27.

Stockton Infrared, 2006, *Stockton Infrared Thermographic Services*, www.stocktoninfrared.com.

Torgersen, C. E., Faux, R. N., McIntosh, B. A., Poage, N. J. and D. J. Norton, 2001, "Airborne Thermal Remote Sensing for Water Temperature Assessment in Rivers and Streams," *Remote Sensing of Environment,* 76:386–398.

Vaughn, R. G., Calvin, W. M. and J. V. Taranik, 2003, "SEBASS Hyperspectral Thermal Infrared Data: Surface Emissivity Measurement and Mineral Mapping," *Remote Sensing of Environment*, 85:48–63.

Wallace, K., 1999, *ATLAS - Airborne Terrestrial Applications Sensor Specifications*, SSC: NASA Stennis Space Ctr., 20 p.

Weng, Q., Dengsheng, L. and J. Schubring, 2004, "Estimation of Land Surface Temperature - Vegetation Abundance Relationship for Urban Heat Island Studies," *Remote Sensing of Environment*, 89:467–483.

Wolfe, W. L., 1985, *Infrared Handbook*, Ann Arbor: Environmental Research Institute of Michigan.

Active and Passive Microwave Remote Sensing

9

assive remote sensing systems record electromagnetic energy that is reflected (e.g., blue, green, red, and near-infrared light) or emitted (e.g., thermal infrared radiation) from the surface of the Earth. There are also *active* remote sensing systems that are not dependent on the Sun's electromagnetic energy or the thermal properties of the Earth. Active remote sensors create their own electromagnetic energy that 1) is transmitted from the sensor toward the terrain (and is largely unaffected by the atmosphere), 2) interacts with the terrain producing a backscatter of energy, and 3) is recorded by the remote sensor's receiver. The most widely used active remote sensing systems include:

- *active microwave* (*RADAR*), which is based on the transmission of long-wavelength microwaves (e.g., 3 – 25 cm) through the atmosphere and then recording the amount of energy backscattered from the terrain;

- *LIDAR* which is based on the transmission of relatively short-wavelength laser light (e.g., 1040 nm) and then recording the amount of light backscattered from the terrain;

- *SONAR* which is based on the transmission of sound waves through a water column and then recording the amount of energy backscattered from the bottom or from objects within the water column.

Of the three, RADAR remote sensing is the most widely used for Earth-resource observations. LIDAR and SONAR are especially useful for topographic and bathymetric mapping, respectively. It is also possible to record passive microwave energy that is naturally emitted from the surface of the Earth using a *passive microwave radiometer*. This chapter provides an overview of active and passive microwave remote sensing. LIDAR remote sensing is discussed in Chapter 10 and SONAR in Chapter 12.

History of Active Microwave (RADAR) Remote Sensing

James Clerk Maxwell (1831– 1879) provided the mathematical descriptions of the magnetic and electric fields associated with electromagnetic radiation. Then, Heinrich R. Hertz (1857 – 1894) increased our knowledge about the creation and propagation of electromagnetic energy in the microwave and radio portions of the spectrum. Hertz also studied the interaction of radio waves with metallic surfaces and initiated some of the early thinking that eventually led to the invention of radios and radars. Building on the fundamental physics principles discovered by Maxwell and Hertz, Guglielmo M.

Marconi (1874 – 1937) constructed an antenna that transmitted and received radio signals. In 1901, he sent radio waves across the Atlantic and in 1909 shared the Nobel Prize in physics for his work.

RADAR as we know it was investigated by A. H. Taylor and L. C. Young in 1922. These scientists positioned a high-frequency radio transmitter on one side of the Anacostia River near Washington, DC and a receiver on the opposite side. Ships passing up or down the river interrupted the long-wavelength radio signal (1 – 10 m) sent between the transmitter and receiver. Such systems provided the first clues that radio signals might be useful for detecting the distance to ships (i.e., the range) at sea. This had implications for ship navigation because the active radio transmission and reception could take place both at night and even in bad weather. The military took an early interest in radar because during a war it is important to know the location of all ships and planes (friendly and enemy). A phrase describing the process was "**ra**dio **d**etection **a**nd **r**anging" or RADAR. Although radar systems now use microwave wavelength energy almost exclusively instead of radiowaves, the acronym was never changed.

By 1935 Young and Taylor (and independently Sir Robert Watson-Watt in Great Britain) combined the antenna transmitter and receiver in the same instrument. Eventually, high-power transmissions in very specific bands of the EM spectrum were possible. These and other electronic advancements laid the groundwork for the development of RADAR during World War II for navigation and target location. By late 1936 experimental radars were working in the United States, Great Britain, Germany, and the Soviet Union. Of these nations, none was more vulnerable to air attack than Britain. London lay within 90 miles of foreign territory and within 275 miles of Germany. Under the direction of Air Chief Marshal Hugh Dowding, the British constructed 21 radars along the south and east coasts of England and the east coast of Scotland. The code name "Chain Home — CH" arose from a plan to build a "home chain" of radars in Britain to be followed by an overseas chain to defend threatened portions of the British empire. The CH radars were the primary British long-range early warning radars well into World War II. The radars allowed incoming planes to be detected out to a maximum range of 50 miles. Without the radar, Britain would not have been able to effectively counter the German Luftwaffe bombers and fighter escorts during the Battle of Britain, which began in July 1940 (Price, 1990). The circularly scanning Doppler radar that we watch every day during television weather updates to identify the geographic location of storms around cities is based on the same circularly scanning radar concept (plan-position indi-

cator radar, PPI) found in the original World War II radars. PPI radars are also used for air-traffic control at airports.

RADAR images obtained from aircraft or spacecraft as we know them today were not available during World War II. The continuous-strip mapping capability of *side-looking airborne radar* (SLAR) was not developed until the 1950s. An important advantage of SLAR is its ability to obtain reconnaissance images over vast regions to the left or right of the aircraft (Sabins, 1997). This is called *long-range standoff* data collection because the pilot can fly along the edge of friendly air space while obtaining detailed RADAR imagery far into unfriendly air space. It also became possible to perform *radargrammetric* measurement—the science of extracting quantitative geometric object information from radar images (Leberl, 1990; Henderson and Lewis, 1998).

The military began using SLARs in the 1950s. By the mid-1960s some systems were declassified. There are two primary types of SLAR: *real aperture radar* (also known as brute-force radar) and *synthetic aperture radar* (*SAR*), where the word "aperture" means *antenna*. Real aperture radars use an antenna of fixed length, e.g., 1 – 2 m. Synthetic aperture radars also use a 1 – 2 m antenna, but they are able to synthesize a much larger antenna (e.g., perhaps 600 m in length), which has improved resolving power. SARs achieve very fine resolution from great distances. For example, an 11-m SAR antenna on an orbital platform can be synthesized electronically to have a synthetic length of 15 km. Later it will be demonstrated how important this capability is to SAR azimuth resolution. Military and civilian commercial RADARS are synthetic aperture radars.

In the 1960s and 1970s, real and synthetic aperture SLARs were used extensively for Earth resource reconnaissance of vast, previously unmapped regions of the Earth. Early studies focused on continents perennially shrouded in cloud cover. The first large-scale SLAR project was the province of Darien, which connects Panama and South America. In 1968, Westinghouse Electric, Inc., in cooperation with Raytheon, Inc., employed a real aperture SLAR that Westinghouse developed for the U.S. Army and successfully made a 20,000 km^2 mosaic of this area. Up to this time, the area had never been seen or mapped in its entirety because of almost perpetual cloud cover.

Goodyear Aerospace, Inc., and the Aero Service Division of Western Geophysical, Inc., adapted a synthetic aperture radar built by Goodyear and installed it in an Aero Service jet for civilian surveys. In 1971 they initiated project RADAM (Radar of the Amazon) to map the Amazon Basin in Venezuela and Brazil. Approximately 4 million km^2 of

Table 9-1.   Characteristics of selected Earth-orbiting Synthetic Aperture Radars (SARs).

Synthetic Aperture Radar (SAR)	Launch	Origin	Wave-length, cm	Depression angle, γ (near to far-range) [Incident angle]	Polari-zation	Azimuth Resolu-tion, m	Range Resolu-tion, m	Swath width, km	Alti-tude, km	Cover-age	Dura-tion
SEASAT	06-26-78	USA	L - (23.5)	73° − 67° [23°]	HH	25	25	100	800	10° − 75° N	105 d
SIR-A	11-12-81	USA	L - (23.5)	43° − 37° [50°]	HH	40	40	50	260	41° N − 36° S	2.5 d
SIR-B	10-05-84	USA	L - (23.5)	75° − 35° [15° − 64°]	HH	17 − 58	25	10 − 60	225 and 350	60° N − 60° S	8 d
SIR-C/X-SAR	04-94 10-94	USA	X - (3.0) C - (5.8) L - (23.5)	variable [15° − 55°]	HH, HV, VV,VH	30	10 − 30	15 − 90	225	57° N − 57° S	10 d
ALMAZ-1	03-31-91	USSR	S - (9.6)	59° − 40° [30° − 60°]	HH	15	15 − 30	20 − 45	300	73° N − 73° S	18 mo
ERS-1,2	1991 1995	ESA	C - (5.6)	67° [23°]	VV	30	26	100	785	polar orbit	--
JERS-1	02-11-92	Japan	L - (23.5)	51° [39°]	HH	18	18	75	568	polar orbit	6.5 yr.
RADARSAT	11-95	Canada	C - (5.6)	70° − 30° [10° − 60°]	HH	8 − 100	8 − 100	50 − 500	798	polar orbit	--
SRTM	02-11-00	USA	X - (9.6) C - (5.3)	--	HH VV,HH	30 30	30 30	X-50 C-225	225	60° N − 56° S	11 d
RADARSAT2	2006	Canada	C - (5.4)	70° − 30° [10° − 60°]	Quad-pol	3 − 100	2.4 − 100	20 − 500	798	polar orbit	--
Envisat ASAR	03-01-02	ESA	C - (5.3)	[15° − 45°]	VV, HH, VV/VH HV/HH VH/VV	30 − 1000	30 − 1000	5 − 150	786	polar orbit 98°	--

land area (about half the size of the United States) were recorded and assembled into radar mosaics. Goodyear and Aero Service eventually surveyed the Amazon rainforest in Brazil, Venezuela, eastern Colombia, Peru, and Bolivia. Subsequent radar investigations mapped Guatemala, Nigeria, Togo, portions of Indonesia, the Philippines, Peru, and other regions (Leberl, 1990).

NASA has launched several successful SARs such as SEASAT and the Space Shuttle Imaging Radar experiments. *SEASAT* (for "sea satellite") was launched in 1978 to obtain L-band (23.5 cm) 25 × 25 m spatial resolution oceano-

graphic information (Table 9-1). This was the first orbital SAR that provided public-domain data (Evans et al., 2005). It also provided valuable land information, but functioned for only 105 days. The *Shuttle Imaging Radar* experiment A (SIR-A) with its L-band (23.5 cm) 40 × 40 m SAR was launched in 1981. SIR-B was launched in 1984 with 17 × 25 m resolution. SIR-C was launched in 1994 with multifrequency and multipolarization capability. SIR-A, SIR-B, and SIR-C missions lasted only 2.5, 8, and 10 days, respectively. The Shuttle Topography Mission (SRTM) was launched on February 11, 2000, and lasted 11 days (NASA SRTM, 2006) (Table 9-1).

Table 9-2.   Advantages of RADAR remote sensing.

Advantages

Primary

- Certain microwave frequencies will penetrate clouds, allowing all-weather remote sensing.

- Synoptic views of large areas for mapping at 1:10,000 to 1:400,000. Satellite coverage of cloud-shrouded countries is possible.

- Coverage can be obtained at user-specified times, even at night.

- Permits imaging at shallow look angles, resulting in different perspectives that cannot always be obtained using aerial photography.

- Senses in wavelengths outside the visible and infrared regions of the electromagnetic spectrum, providing information on surface roughness, dielectric properties, and moisture content.

Secondary

- Certain frequencies of microwave energy penetrate vegetation, sand, and surface layers of snow.

- Based on its own illumination, and the angle of illumination can be controlled.

- Enables resolution to be independent of distance to the object, with the size of a resolution cell being as small as 1×1 m.

- Images can be produced from different types of polarized energy (HH, HV, VV, VH).

- May operate simultaneously in several wavelengths (frequencies) and thus has multi-frequency potential.

- Can measure ocean wave properties, even from orbital altitudes.

- Can produce overlapping images suitable for stereoscopic viewing and radargrammetry.

- Supports interferometric operation using two antennas for 3-D mapping, and analysis of incident-angle signatures of objects.

The former Soviet Union launched the ALMAZ-1 S-band (9.6 cm) radar in 1991. The European Space Agency (ESA) launched the *European Remote Sensing Satellite ERS-1* with its C-band (5.6 cm) imaging radar in 1991 and *ERS-2* in 1995. The ESA launched Envisat on March 1, 2002 with its C-band (5.3 cm) *Advanced Synthetic Aperture Radar* (ASAR) to provide continuity with the *ERS-1,2* radars. Japan launched the L-band (23.5 cm) *Japanese Earth Resources Satellite JERS-1* in 1992. The Canadian government placed the C-band (5.6 cm) *RADARSAT-1* in orbit in 1995. *RADARSAT-2* is scheduled for launch in 2006. The system characteristics of these orbital SAR systems are summarized in Table 9-1 and discussed later in this chapter.

Active and passive microwave remote sensing will continue grow in importance. Such sensors provide the only viable information for the tropical portions of the world where extensive, fragile ecosystems are at risk and under perennial cloud cover. For a detailed history of radar development, see Henderson and Lewis (1998). The primary and secondary

Table 9-3.   RADAR wavelengths and frequencies used in active microwave remote sensing investigations.

RADAR Band Designations (common wavelengths in parentheses)	Wavelength (λ) in cm	Frequency (υ) in GHz
K_a (0.86 cm)	0.75 – 1.18	40.0 – 26.5
K	1.19 – 1.67	26.5 – 18.0
K_u	1.67 – 2.4	18.0 – 12.5
X (3.0 and 3.2 cm)	2.4 – 3.8	12.5 – 8.0
C (7.5, 6.0 cm)	3.9 – 7.5	8.0 – 4.0
S (8.0, 9.6, 12.6 cm)	7.5 – 15.0	4.0 – 2.0
L (23.5, 24.0, 25.0 cm)	15.0 – 30.0	2.0 – 1.0
P (68.0 cm)	30.0 – 100	1.0 – 0.3

advantages of RADAR remote sensing are summarized in Table 9-2.

 Active Microwave System Components

Active microwave imagery is obtained using instruments and principles that are different from those used when acquiring data in the visible, near-, mid-, and thermal infrared portions of the EM spectrum using passive remote sensing techniques. Therefore, it is necessary to understand the active microwave system components and how electromagnetic energy is sent, received, and recorded by the sensor system before an analyst can accurately interpret a radar image. The discussion is based initially on the system components and functions of a *real aperture side-looking airborne radar* (SLAR). The discussion then expands to include *synthetic aperture radars* (SAR) that have improved capabilities.

Sending and Receiving a Pulse of Microwave Electromagnetic Energy – System Components

A typical active microwave SLAR is shown mounted on an aircraft in Figure 9-1a. It consists of a pulse-generating device, a transmitter, a duplexer that carefully coordinates when the active microwave energy is transmitted and received, an antenna, a receiver, a recording device such as a high-density digital tape recorder or hard disk, and typically a CRT monitor so that the technician in the plane can make sure radar imagery is actually being collected (Figure 9-1b).

a. Intermap LearJet 36 Star 3*i*.

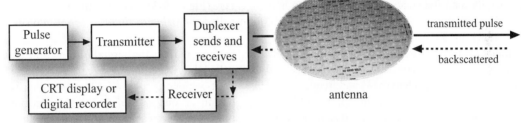

b. Typical active microwave system components.

Figure 9-1   a) A side-looking airborne radar (SLAR) called the Star 3*i* IFSAR (interferometric synthetic aperture radar) is mounted underneath a LearJet 36 aircraft (courtesy Intermap Technologies, Inc.). b) The active microwave system components and flow of energy through a typical SLAR. It consists of a pulse-generating device, transmitter, duplexer, antenna, receiver, and a digital recorder. Some systems have a quick-look CRT display to make sure data is being collected. The 1-m synthetic aperture antenna shown can synthesize an antenna hundreds of meters long.

The radar can also be mounted onboard a satellite. The pulse-generating device sends a pulse of electromagnetic energy at a specific wavelength (frequency) to the transmitter. When in sending mode, the duplexer then sends the polarized pulse of energy through the antenna toward the Earth's surface.

Wavelength, Frequency, and Pulse Length

The pulse of electromagnetic radiation sent out by the transmitter through the antenna is of a specific wavelength and duration (i.e., it has a *pulse length* measured in microseconds, μsec). The wavelengths of energy most commonly used in imaging radars are summarized in Table 9-3. The wavelengths are much longer than visible, near-infrared, mid-infrared, or thermal infrared energy used in other remote sensing systems (Figure 9-2). Therefore, microwave energy is usually measured in centimeters rather than micrometers (Carver, 1988). The unusual names associated with the radar wavelengths (e.g., K, K_a, K_u, X, C, S, L, and P) are an artifact of the secret work on radar remote sensing in World War II when it was customary to use an alphabetic descriptor instead of the actual wavelength or frequency. These descriptors are still used today in much of the radar scientific literature.

The shortest radar wavelengths are designated K-band. K-band wavelengths should theoretically provide the best radar resolution. Unfortunately, K-band wavelength energy is partially absorbed by water vapor and cloud penetration can be limited. This is the reason that most ground-based weather radars used to track cloud cover and precipitation are K-band. X-band is often the shortest wavelength range used for orbital and suborbital imaging radars (Mikhail et al., 2001). Some RADAR systems function using more than one frequency and are referred to as *multiple-frequency radars* (e.g., SIR-C and SRTM) (Table 9-1).

Table 9-3 and Figure 9-2 also provide the radar band designation in units of frequency measured in billions of cycles

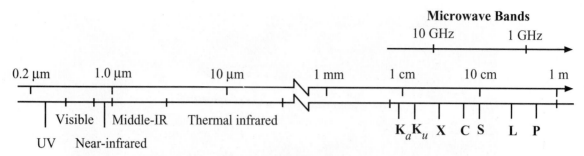

Figure 9-2   The wavelength and frequency of commonly used RADAR bands. RADAR antennas transmit and receive very long wavelength energy measured in centimeters, unlike the relatively short wavelength visible, near-infrared, middle-infrared, and thermal infrared regions measured in micrometers.

per second (Gigahertz or GHz, 10^9 cycles sec^{-1}). Earth-resource image analysts seem to grasp the concept of wavelength more readily than frequency, so the convention is to describe a radar in terms of its wavelength. Conversely, engineers generally prefer to work in units of frequency because as radiation passes through materials of different densities, frequency remains constant while velocity and wavelength change. Since wavelength (λ) and frequency (υ) are inversely related to the speed of light (c), it really does not matter which unit of measurement is used as long as one remembers the following relationships:

$$c = \lambda \upsilon \qquad (9-1)$$

$$\lambda = \frac{3 \times 10^8 \text{m sec}^{-1}}{\upsilon} \qquad (9-2)$$

$$\upsilon = \frac{3 \times 10^8 \text{m sec}^{-1}}{\lambda}. \qquad (9-3)$$

The following simple equation can be used to rapidly convert frequencies into units of radar wavelength:

$$\lambda \text{ in centimeters} = \frac{30}{\upsilon \text{ (in GHz)}}. \qquad (9-4)$$

Azimuth and Range Direction, Depression Angle, Look Angle, Incident Angle, and Polarization

There are additional parameters that must be known about the nature of the pulse of microwave energy that is sent by the radar antenna to the terrain, including aircraft *azimuth direction, radar range* or *look direction, depression angle, look angle, incident angle,* and *polarization.* The following discussion is based on the use of an airborne radar system, although the principles are equally applicable to satellite radar systems.

Azimuth Direction: In a typical SLAR configuration the antenna is mounted beneath and parallel to the aircraft fuselage (Figure 9-3). The aircraft travels in a straight line that is called the *azimuth flight direction.* Pulses of active microwave energy illuminate strips of the terrain at right angles (orthogonal) to the aircraft's direction of travel, which is called the *range* or *look direction.* Figure 9-3 depicts an aircraft equipped with a SLAR system that is illuminating the terrain on one side of the aircraft but not beneath it. The pulses of energy sent out in the range direction only illuminate a certain part of the terrain. The terrain illuminated nearest the aircraft in the line of sight is called the *near-range* (Figure 9-3). The farthest point of terrain illuminated by the pulse of energy is called the *far-range.*

Range Direction: The *range* or *look direction* for any *radar* image is the direction of the radar illumination that is at right angles to the direction the aircraft or spacecraft is traveling. Look direction usually has a significant impact on feature interpretation. The extent to which linear features are enhanced or suppressed on the imagery depends significantly on their orientation relative to a given look direction of radar illumination. Generally, objects that trend (or strike) in a direction that is orthogonal (perpendicular) to the range or look direction are enhanced much more than those objects in the terrain that lie parallel to the look direction. Consequently, linear features that appear dark or are imperceptible in a radar image using one look direction may appear bright in another radar image with a different look direction. A good example of this is demonstrated in Figure 9-4, which shows radar imagery of an area in Nigeria, West Africa, that were obtained using two different look directions. Note how certain terrain features are emphasized and/or de-emphasized in the two images.

Depression Angle: The *depression angle* (γ) is the angle between a horizontal plane extending out from the aircraft

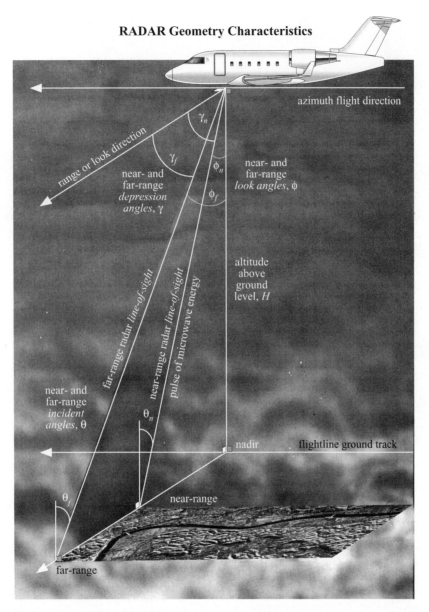

RADAR Geometry Characteristics

azimuth flight direction

γ_n

γ_f

ϕ_n

ϕ_f

range or look direction

near- and far-range *depression angles*, γ

near- and far-range *look angles*, ϕ

far-range radar line-of-sight

near-range radar line-of-sight

pulse of microwave energy

altitude above ground level, H

near- and far-range *incident angles*, θ

θ_n

θ_f

nadir

flightline ground track

near-range

far-range

Figure 9-3   Geometric characteristics of radar imagery acquired by a side-looking airborne radar (SLAR) through cloud cover. All the nomenclature assumes that the terrain is flat.

fuselage and the electromagnetic pulse of energy from the antenna along the radar *line-of-sight* to a specific point on the ground (Figure 9-3). The depression angle within a strip of illuminated terrain varies from the *near-range depression angle* to the *far-range depression angle*. The *average depression angle* of a radar image is computed by selecting a point midway between the near and far-range in the image strip. Summaries of radar systems sometimes only report the average depression angle.

Look Angle: The *look angle* (ϕ) is the angle between the vertical from the antenna to the ground and the radar line of sight (Figure 9-3). The look angle varies from the near- to the far-range and is the complement of the depression angle (Mikhail et al., 2001).

Incident Angle: The *incident angle* (θ) is the angle between the radar pulse of energy and a line perpendicular to the Earth's surface where it makes contact. When the terrain is flat, the incident angle (θ) is the complement ($\theta = 90 - \gamma$) of the depression angle (γ). However, if the terrain is sloped, there is no relationship between depression angle and incident angle. The incident angle best describes the relationship between the radar beam and surface slope. A diagram of the

a. X-band, HH polarization   ↑ look direction

b. X-band, HH polarization   ↓ look direction

Figure 9-4   a) X-band image of the Kaduna State in Nigeria created by flying east-west and having the RADAR
look north. b) This image was obtained by flying
the aircraft east-west with the RADAR looking
south. Generally, it is good practice to always orient
a radar image so that the look direction is toward the
viewer. This causes the shadows to fall toward the
analyst and keeps him or her from experiencing
pseudoscopic illusion (i.e., topographic inversion).

relationship is shown in Figure 9-5. Many mathematical
radar studies assume the terrain surface is flat (horizontal),
therefore, the incident angle is assumed to be the complement of the depression angle.

Polarization: Unpolarized energy vibrates in *all* possible
directions perpendicular to the direction of travel. Radar
antennas send and receive *polarized* energy. This means that
the pulse of energy is filtered so that its electrical wave
vibrations are only in a single plane that is perpendicular to
the direction of travel. The pulse of electromagnetic energy
sent out by the antenna may be *vertically* or *horizontally*

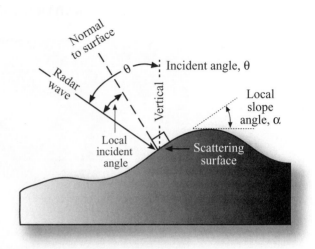

Figure 9-5   The relationship between radar incident angle (θ),
true vertical, and local slope angle (α) for non-level
terrain (after Henderson and Lewis, 1998).

polarized as shown in Figure 9-6. The transmitted pulse of
electromagnetic energy interacts with the terrain and some
of it is backscattered at the speed of light toward the aircraft
or spacecraft, where it once again must pass through a filter.
If the antenna accepts the backscattered energy, it is
recorded. Various types of backscattered polarized energy
may be recorded by the radar. For example, it is possible to

- send vertically polarized energy and receive only
  vertically polarized energy (designated *VV*);

- send horizontal and receive horizontally polarized energy
  (*HH*);

- send horizontal and receive vertically polarized energy
  (*HV*); or

- send vertical and receive horizontally polarized energy
  (*VH*).

HH and *VV* configurations produce *co-* or *like-polarized*
radar imagery. *HV* and *VH* configurations produce *cross-
polarized* imagery. The ability to record different types of
polarized energy from a resolution element in the terrain
results in valuable Earth resource information in certain
instances. For example, Figure 9-7 documents how it is possible to selectively record only vertically or horizontally
polarized visible light (0.4 – 0.7 μm) to our advantage. The
scene consists of water, vegetation, and a floating block of
wood. A vertically polarized filter placed in front of a cam-

Polarization

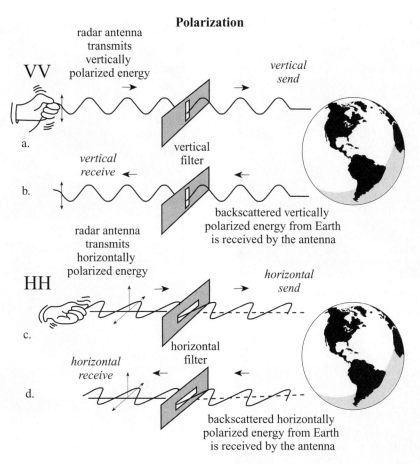

Figure 9-6   a) VV – Polarization. The radar antenna transmits vertically polarized energy toward the terrain. b) Some of the vertically po-larized energy is *not* depolarized by the terrain. A vertical filter in the antenna allows only vertically polarized energy backscat-tered from the terrain to be recorded. c) HH – Polarization. The radar antenna transmits horizontally polarized energy toward the terrain. d) Some of the horizontally polarized energy is *not* depolarized by the terrain. A horizontal filter in the antenna al-lows only horizontally polarized energy backscattered from the terrain to be recorded.

era lens allows only vertically polarized light to pass through to the film. A horizontally polarized filter placed in front of the lens allows only the horizontally polarized energy to pass through to the film. In this example, the horizontally polar-ized light provides more detailed information about the pres-ence and distribution of underwater rocks. Similarly, surface rocks and other materials often have unique polarization characteristics in the microwave region (e.g., K_a band = 0.86 cm). For example, Figure 9-8 demonstrates how a northern Arizona basalt lava flow is much easier to delineate in the HH polarization real aperture K_a-band radar image than in the HV polarization image acquired at the same time using the same look direction. More will be said about how the ter-rain interacts with a pulse of polarized electromagnetic energy in the section on radar environmental considerations.

Slant-Range versus Ground-Range RADAR Image Geometry

Radar imagery has a different geometry than that produced by most conventional remote sensor systems, such as cam-eras, multispectral scanners or area-array detectors. There-fore, one must be careful when attempting to make radargrammetric measurements. First, uncorrected radar imagery is displayed in what is called *slant-range geometry*, i.e., it is based on the actual distance from the radar to each of the respective features in the scene. For example, in Fig-ure 9-9 we see two fields, A and B, that are the same size in the real world. One field is in the near-range close to the air-craft and one is in the far-range. Field A in the near-range is compressed much more than field B in the far-range in a

Use of Vertical and Horizontal Polarizing Filters

a. Vertical filter allows only vertically polarized light to pass.

b. Horizontal filter allows only horizontally polarized light to pass.

Figure 9-7   a) A vertically polarized filter placed in front of a camera lens allows only vertically polarized light to pass through. b) A horizontally polarized filter allows only horizontally polarized light to pass. In this example based on visible light, horizontally polarized light provides much more information about the underwater rocks than the vertically polarized light. Active microwave sensors can be configured to send and receive horizontally and vertically polarized microwave energy.

slant-range display (Ford et al., 1980). It is possible to convert the *slant-range display* (S_{rd}) information into the true *ground-range display* (G_{rd}) on the x-axis so that features in the scene are in their proper planimetric (x,y) position relative to one another in the final radar image. The following equation, based on the Pythagorean theorem applied to a

a. K$_a$-band, HH polarization   ↓ look direction

b. K$_a$-band, HV polarization   north ⟶

Figure 9-8   a) K$_a$-band, HH polarization real aperture radar image of a cinder cone and basalt lava flow in north-central Arizona. b) Simultaneously obtained HV polarization image with the same look direction. The strong response of the lava flow in the direct return image (HH) and the weak return on the cross-polarized image (HV) indicates that the blocky flow is highly polarized. This is due to the direct reflection of blocks that are large relative to the wavelength (courtesy NASA; Carver, 1988).

right triangle, transforms the slant-range distance, S_{rd}, at the very beginning of field A to a corrected ground-range distance, G_{rd}, based on the trigonometric relationship between the altitude of the sensor above-ground datum (H) and the other two sides of the right triangle, S_{rd} and G_{rd}, shown in Figure 9-9:

$$S_{rd}^2 = H^2 + G_{rd}^2 \tag{9-5}$$

$$G_{rd} = \sqrt{S_{rd}^2 - H^2} \; . \tag{9-6}$$

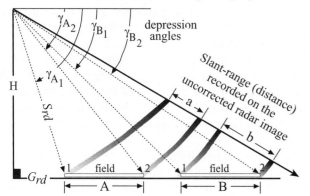

**Slant-Range Display
versus Ground-Range Display**

Ground-range (distance) on corrected radar image

Figure 9-9   Uncorrected radar images have slant-range geometry, where objects in the near-range are compressed
more than objects in the far-range. In this example,
field A and B are the same size and have no local relief. The slant-range image display is a plane connecting the antenna at altitude H, with the farthest
point in the far-range. When the corner of field A is
illuminated with a pulse of energy at a depression
angle of γ_A, it is projected onto the slant-range display as shown. The same thing occurs at the end of
field A and for both sides of field B. This causes
field A to be compressed much more than field B in
the slant-range image display.

It is also possible to transform the slant-range display to a
ground-range display using the relationship between the
height of the antenna above the local ground level, H, and
the depression angle (γ) at the point of interest using the following equation (Ford et al., 1980):

$$G_{rd} = H\sqrt{\frac{1}{\sin^2\gamma} - 1}.$$ (9-7)

If we wanted to measure the true ground-range distance
between two points in a radar image such as between points
1 and 2 in field A, we may use the following relationship
(Henderson and Lewis, 1998):

$$G_{rd_{1-2}} = H(\cot\gamma_2 - \cot\gamma_1)$$ (9-8)

which takes into account the altitude of the sensor, H, and
the depression angle (γ) to points 1 and 2 in the image.

These equations assume that the terrain is flat. It does not
correct for distortion due to radar layover (foreshortening)
caused by topographic relief (to be discussed). Radar image
analysts should always inquire as to whether they are viewing radar imagery that has been converted from slant-range
to ground-range geometry. Most radar systems and data providers now provide the data in ground-range geometry.

Computing Range Resolution

To determine the spatial resolution at any point in a radar
image, it is necessary to compute the resolution in two
dimensions: the *range* and *azimuth resolutions*. Radar is in
effect a ranging device that measures the distance to objects
in the terrain by means of sending out and receiving pulses
of active microwave energy (Figure 9-10). The *range resolution* in the across-track direction is proportional to the
length of the microwave pulse. The shorter the pulse length,
the finer the range resolution. *Pulse length* is a function of
the speed of light (c) multiplied by the duration of the transmission (τ). The length of time that the microwave energy
(e.g., L-band, 23.5 cm) is actually transmitted is measured in
microseconds (10^{-6} sec) and typically ranges from $0.4 - 1.0$
microsecond. This translates into a pulse length ranging
from $8 - 210$ m. The pulse length must travel to the target
and back to the sensor. Therefore, it is necessary to divide by
2 to measure the slant-range resolution. To scale it to
ground-range, it is multiplied by the cosine of the depression
angle (γ). Thus, the equation for computing the *range resolution* becomes

$$R_r = \frac{\tau \cdot c}{2\cos\gamma}.$$ (9-9)

One might ask, Why not select an extremely short pulse
length to obtain an extremely fine range resolution? The reason is that as the pulse length is shortened, so is the total
amount of energy that illuminates the target of interest. Soon
we would have such a weak backscattered return signal that
it would be of no value. Therefore, there is a trade-off
between shortening the pulse length to improve range resolution and having enough energy in the transmitted pulse to
receive a strong signal from the terrain, which is the heart of
microwave remote sensing.

The general rule is that signals reflected from two distinct
objects in the terrain (e.g., two houses) can be resolved if
their respective range distances are separated by at least half
the pulse length. For example, consider Figure 9-10 in which
the terrain is being illuminated with a single pulse of microwave energy that lasts 10^{-7} seconds. This translates to a pulse

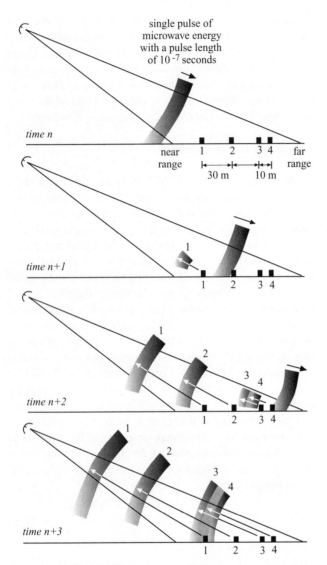

single pulse of
microwave energy
with a pulse length
of 10^{-7} seconds

time n

near   1   2   3 4   far
range   range
      30 m   10 m

time n+1

1
1   2   3 4

time n+2

1
2
3 4
1   2   3 4

time n+3

1
2
3
4
1   2   3 4

Figure 9-10   The distance of an object (its range) from the aircraft antenna is determined by the length of time required for the pulse of microwave energy traveling at the speed of light to reach the object and be reflected back to the antenna. Signals from houses 1 and 2 will arrive sooner than the signals returned from the two houses farther away (3 and 4). In this example, the 10^{-7} second pulse length equates to 30 m. The resolution across-track is equal to half the pulse length, in this case 15 m. Houses 1 and 2 will be resolved as distinct objects on the radar image. Two objects (houses 3 and 4) separated by less than half the radar pulse length will be perceived by the antenna as one broad object.

length of 30 m and therefore has a range resolution of 15 m. The fate of the single pulse of microwave energy is monitored for four brief time periods. At *time n* the pulse has not

impacted any homes yet. At *time n+1* a portion of the pulse has been reflected back toward the antenna while the remaining part of the pulse continues across-track. By *time n+2* homes 2, 3, and 4 have reflected a part of the incident microwave energy back to the antenna. Because houses 1 and 2 were greater than 15 m apart, they will appear as distinct features in the radar imagery. However, houses 3 and 4 were less than 15 m apart; therefore, their two returns will overlap and they will be perceived by the antenna as one broad object. They will probably be difficult to resolve as individual houses in the radar image.

While the pulse length remains constant through the near- and far-range, the range resolution varies linearly from the near- to the far-range. For example, consider Figure 9-11 which depicts towers 1 and 2 that are 30 m apart in the near-range and towers 3 and 4 that are 30 m apart in the far-range. If we use Equation 9-9 to compute the range resolution in the far-range with a depression angle of 40° and a pulse length duration of transmission of 0.1 μsec, the range resolution is:

$$R_r = \frac{(0.1 \times 10^{-6}\,\text{sec}) \cdot (3 \times 10^{8}\,\text{m sec}^{-1})}{2\cos 40°}$$

$$R_r = \frac{\left(0.1 \times \dfrac{1\,\text{sec}}{1,000,000}\right) \times \dfrac{300,000,000\,\text{m}}{\text{sec}}}{2 \times 0.766}$$

$$R_r = \frac{\dfrac{0.1\,\text{sec}}{1,000,000} \times \dfrac{300,000,000\,\text{m}}{\text{sec}}}{2 \times 0.766}$$

$$R_r = \frac{30\,\text{m}}{1.532}$$

$$R_r = 19.58\,\text{m}.$$

Therefore, towers 3 and 4 in the far-range must be separated by more than 19.58 m to resolve the individual towers on the radar image. Because towers 3 and 4 are separated by 30 m, it is possible to identify the individual towers. Conversely, towers 1 and 2, located at a depression angle of 65° in the near-range, would not be resolved because the range resolution in this area would be 35.5 m. The two towers would probably appear as a single bright return.

Computing Azimuth Resolution

Thus far we have only identified the length in meters of an active microwave resolution element at a specific depression angle and pulse length in the range (across-track) direction. To know both the length and width of the resolution element,

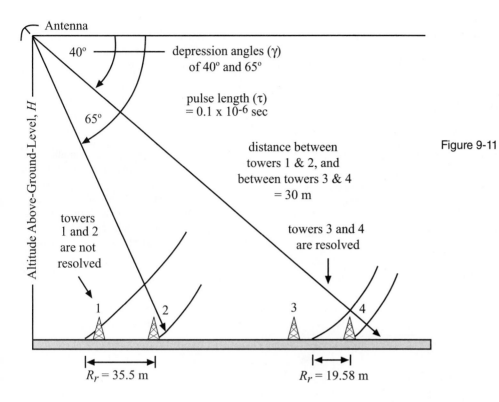

Figure 9-11 Computing the range resolution at two different depression angles (40° and 65°) for a real aperture radar with a pulse length of 0.1×10^{-6} sec. The towers can be resolved in the far-range but not in the near-range (after Sabins, 1997).

we must also compute the width of the resolution element in the direction the craft is flying — the azimuth direction. *Azimuth resolution* (R_a) is determined by computing the width of the terrain strip that is illuminated by the radar beam. Real aperture active microwave radars produce a lobe-shaped beam similar to the one shown in Figure 9-12, which is narrower in the near-range and spreads out in the far-range. Basically, the angular beam width is directly proportional to the wavelength of the transmitted pulse of energy, i.e., the longer the wavelength, the wider the beam width, and the shorter the wavelength, the narrower the beam width. Therefore, in real aperture (brute force) radars a shorter wavelength pulse will result in improved azimuth resolution. Unfortunately, the shorter the wavelength, the poorer the atmospheric and vegetation penetration capability.

Fortunately, the beam width is also inversely proportional to antenna length (L). This means that the longer the radar antenna, the narrower the beam width and the higher the azimuth resolution. The relationship between wavelength (λ) and antenna length (L) is summarized in Equation 9-10, which can be used to compute the *azimuth resolution* (Henderson and Lewis, 1998):

$$R_a = \frac{S \times \lambda}{L} \qquad (9\text{-}10)$$

where S is the *slant-range distance* to the point of interest. The equation can be used to compute the azimuth resolution at any location between the near- and far-range. For example, consider the conditions shown in Figure 9-12 where the near slant-range is 20 km and the far slant-range is 40 km. Tanks 1 and 2 and tanks 3 and 4 are separated by 200 m. If an X-band radar (3 cm) is used with a 500 cm antenna, then Equation 9-10 can be used to compute the near-range *azimuth resolution*:

$$R_a = \frac{20 \text{ km} \times 3 \text{ cm}}{500 \text{ cm}}$$

$$R_a = \frac{20,000 \text{ m} \times 0.03 \text{ m}}{5 \text{ m}}$$

$$R_a = \frac{600 \text{ m}}{5 \text{ m}}$$

$$R_a = 120 \text{ m}.$$

The far-range azimuth resolution at the 40 km distance is:

$$R_a = \frac{40 \text{ km} \times 3 \text{ cm}}{500 \text{ cm}}$$

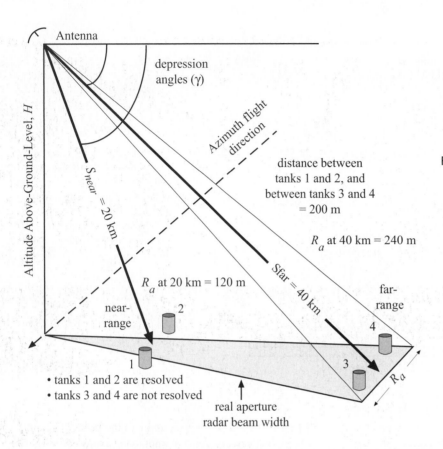

Figure 9-12 Computing the azimuth resolution at two different slant-range distances (20 and 40 km) for a real aperture radar with an X-band wavelength of 3 cm and a 500 cm antenna. The tanks can be resolved in the near-range but not in the far-range.

$$R_a = \frac{40,000 \text{ m} \times 0.03 \text{ m}}{5 \text{ m}}$$

$$R_a = \frac{1200 \text{ m}}{5 \text{ m}}$$

$$R_a = 240 \text{ m}.$$

Tanks 1 and 2 in the near-range would most likely be resolved because the azimuth resolution at this slant-range distance (120 m) is less than the distance between tanks 1 and 2 (200 m). Conversely, tanks 3 and 4 in the far-range would probably not be resolved, because at a slant-range distance of 40 km the azimuth resolution is 240 m, much greater than the 200 m separating the tanks.

There is a trigonometric relationship between the slant-range distance (S) and the depression angle (γ) and the height of the aircraft or spacecraft platform above the local datum, H:

$$S = \frac{H}{\sin\gamma}. \qquad (9\text{-}11)$$

The equation for computing the azimuth resolution becomes:

$$R_a = \left(\frac{H}{\sin\gamma}\right) \cdot \frac{\lambda}{L}. \qquad (9\text{-}12)$$

Ideally, one could build an extremely long antenna and place it on an aircraft to achieve very high azimuth resolution. Unfortunately, there is a practical limit (about 2 m) to the size of antenna that can be mounted on an aircraft (although the size is virtually unlimited in outer space!). Fortunately, there are ways to electronically synthesize a longer antenna, which is the heart of the synthetic aperture radar concept to be discussed shortly.

While we have not yet defined how a synthetic aperture radar (SAR) works, it is instructive to point out at this time a significant difference between the computation of the azimuth resolution for a real aperture radar versus a synthetic aperture radar. The equation for the azimuth or along-track resolution for a point target in a synthetic aperture radar (SAR_a) is

$$SAR_a = \frac{L}{2} \qquad (9\text{-}13)$$

where L is the antenna length. As Henderson and Lewis (1998) point out,

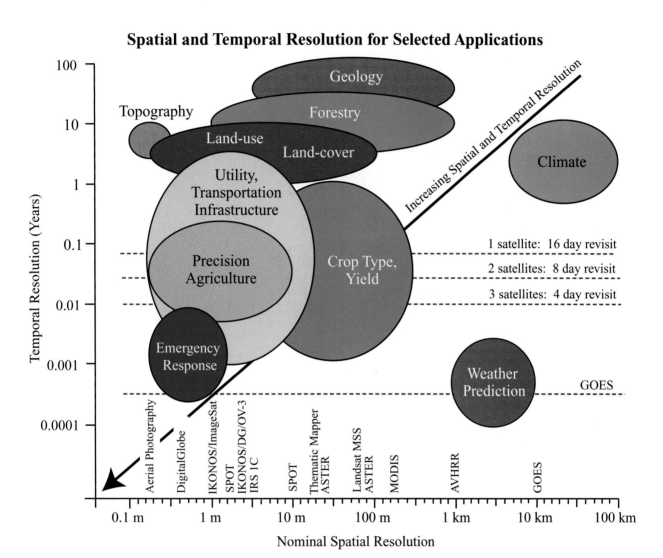

Color Plate 1-1 The nominal spatial and temporal resolution characteristics for selected applications are presented. When conducting a remote sensing project, there is usually a trade-off between spatial and temporal resolution requirements. Generally, as temporal resolution requirements increase, it is usually necessary to lower the spatial resolution requirements so that the amount of remote sensor data collected does not become unmanageable. Fortunately, many applications that require very detailed spatial information (e.g., land-use mapping) normally do not require high temporal resolution data (i.e., data collected once every five to ten years may be sufficient). There are exceptions to these general rules. For example, precision agriculture, crop yield investigations, traffic studies, and emergency response applications sometimes require very high spatial and temporal resolution remote sensor data. The volume of data collected can create serious data management and analysis problems. There are also trade-offs with the other resolutions (e.g., spectral, radiometric, polarization) that may need to be considered.

Solar and Heliospheric Observatory (SOHO) Image
of the Sun Obtained on September 14, 1999

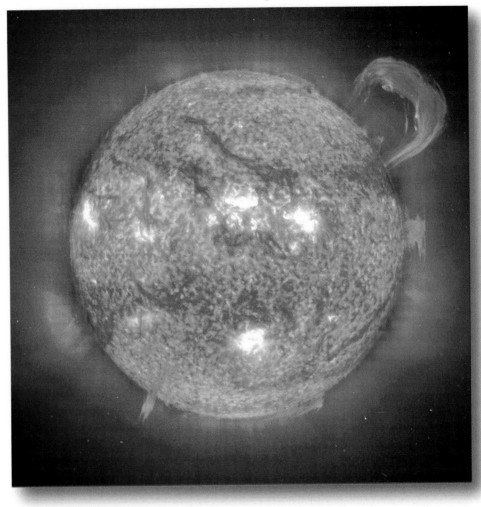

Color Plate 2-1 An image of the Sun obtained by the Solar and Heliospheric Observatory (SOHO) EIT instrument on September 14, 1999 (courtesy Solar Data Analysis Center, NASA Goddard Space Flight Center). The bright features are called active regions. The Sun has a diameter of 1,390,000 km. The temperature of the Sun's core is 15,600,000 K. The temperature of the Sun's surface (the photosphere) is 5800 – 6000 K. Just above the Sun's surface is the chromosphere. The region above the chromosphere is called the corona and extends millions of kilometers into space but is visible only during eclipses. Temperatures in the corona are over 1,000,000 K. A huge erupting prominence is seen escaping the Sun in the upper right. Prominences are relatively "cool" 60,000 K plasma embedded in the much hotter surrounding corona. If an eruption like this is directed toward the Earth it can cause significant geomagnetic activity in the Earth's environment. The Sun is the largest object in the solar system and contains more than 99 percent of its total mass (Jupiter contains most of the remainder). The Sun is 75 percent hydrogen and 25 percent helium by mass. The Sun's energy output is 386 billion billion megawatts per second produced by nuclear fusion reactions. The Greeks called the Sun Helios and the Romans referred to it as Sol. It is the source of much of the electromagnetic energy recorded by remote sensing systems.

Visible Portion of the Electromagnetic Spectrum from 400 to 700 nm

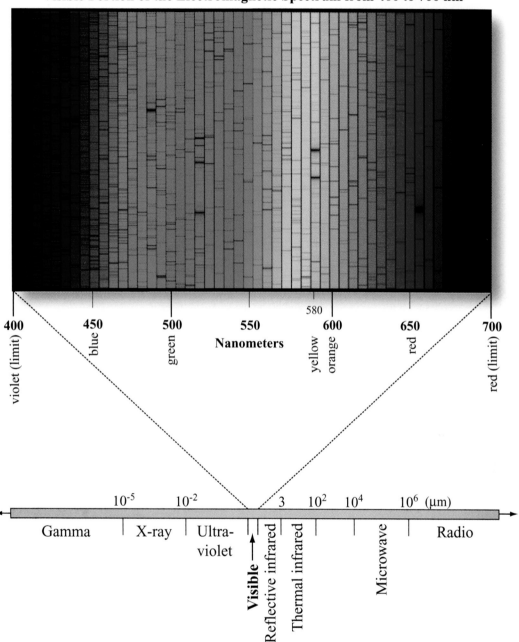

Color Plate 2-2 The portion of the electromagnetic spectrum visible by human beings is generally considered to be from 400 to 700 nm consisting of blue (400 - 500 nm), green (500 - 600 nm), and red (600 - 700 nm) light. The image of the solar spectrum was observed with the Fourier Transform Spectrometer at the McMath-Pierce Solar Facility at Kitt Peak National Observatory, near Tucson, AZ. Each of the 50 slices covers 60 angstroms and wavelength increases from top to bottom along each slice. Note that it contains relatively narrow, dark Fraunhofer lines named for the German physicist Joseph von Fraunhofer (1787-1826) who "mapped" the location of over 570 of them in the solar spectrum. Later scientists discovered that the dark lines are caused by chemical element absorption in the upper layers of the Sun and some absorption by oxygen molecules in the Earth's atmosphere (used with permission of N. A. Sharp, NOAO/NSO/Kitt Peak FTS/AURA/NSF).

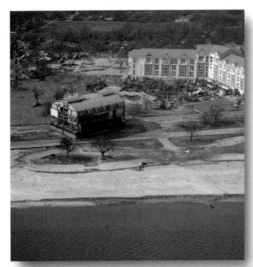

**Hurricane Katrina Damage
at Gulfport, MS**

August 30, 2005

a. Low-oblique aerial photograph of a portion
of the Grand Casino moved by storm surge onto
Highway 90.

b. Vertical aerial photograph of the Grand Casino and surrounding area obtained
at a spatial resolution of 37 x 37 cm (1.2 x 1.2 ft).

Color Plate 4-1 a) Low-oblique aerial photograph of wind and storm surge damage caused by Hurricane Katrina at Gulfport, MS, obtained
on August 30, 2005 (courtesy of USGS Coastal and Marine Geology Program). b) Vertical aerial photograph (courtesy of
NOAA Remote Sensing Division).

Dispersion of White Light Through a Prism

a.

Additive Color Theory

Subtractive Color Theory

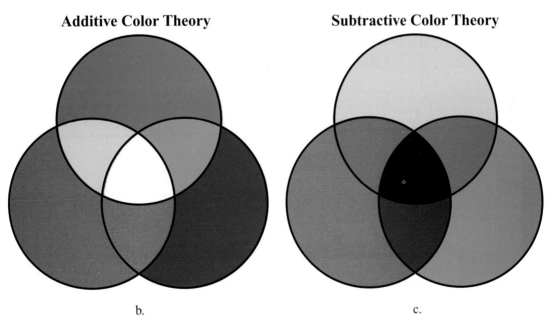

b.

c.

Color Plate 4-2 a) Sir Isaac Newton discovered that white light could be dispersed into its spectral components by passing white light through a prism (© David Parker, Photo Researchers, Inc.). b) *Additive color theory* — equal proportions of blue, green, and red light superimposed on top of one another creates white light, i.e., white light is composed of blue, green, and red light. The complementary colors yellow, magenta, and cyan are created by selectively adding together red and green, blue and red, and blue and green light, respectively. c) *Subtractive color theory* — equal proportions of blue, green, and red pigments yield a black surface. A yellow filter effectively absorbs all blue light, a magenta filter absorbs all green light, and a cyan filter absorbs all red light.

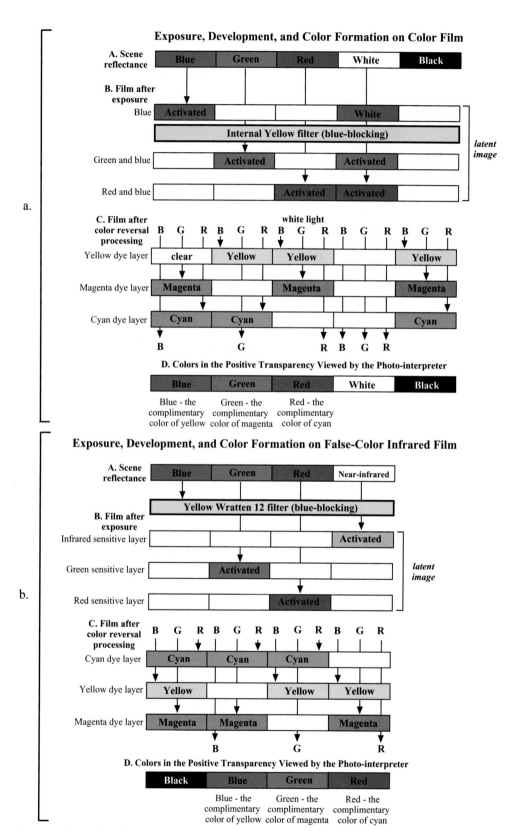

Color Plate 4-3 Process of exposing, developing, and forming color on a) normal color film, and b) false-color infrared film (courtesy Eastman Kodak).

Tivoli North Bay on the Hudson River, NY

a. Normal color aerial photograph.

b. Color-infrared aerial photograph.

Color Plate 4-4 a) A normal (natural) color photograph of Tivoli North Bay along the Hudson River in New York. Color aerial photography records blue, green and red light reflected from the scene. A haze filter (HF3) is usually placed in front of the lens to prevent scattered ultraviolet light from reaching the film plane. Vegetation shows up in green hues because the plants absorb more blue and red incident light than green light. b) A color-infrared aerial photograph records green, red, and near-infrared light reflected from the scene. A yellow (Wratten 12) filter is usually placed in front of the lens to prevent blue light from reaching the film plane. The land-water interface is especially well-delineated because water absorbs most of the incident near-infrared light causing it to appear dark. Vegetation appears in magenta hues because healthy vegetation reflects substantial amounts of near-infrared light while absorbing much of the green and red incident light (after Berglund, 1999).

University of Nebraska, Lincoln, NB

a. Normal color aerial photograph.

b. Color-infrared aerial photograph.

Color Plate 4-5 a) Normal (natural) color terrestrial photograph of a building at the University of Nebraska. b) Color-infrared photograph acquired using a Wratten 12 minus-blue filter (after Rundquist and Sampson, 1988).

Remote Sensing Can Provide Knowledge by Measuring Energy Characteristics in Spectral Regions Beyond Our Human Visual Perception

a. Green reflectance of an agricultural area in Saudi Arabia.

b. Red reflectance.

c. Near-infrared reflectance.

d. Color composite (RGB = near-infrared, red, green).

Color Plate 5-1 Multispectral imagery of center-pivot agriculture in Saudi Arabia. a,b) Vegetation absorbs most of the green and red incident energy causing vegetated fields to appear dark in green and red multispectral images. c) Conversely, vegetation reflects a substantial amount of the incident near-infrared energy causing it to appear bright. In this example, several fields appear dark in the green and red images (possibly due to recent irrigation, stubble from a previous crop, or plowing) suggesting that vegetation is present. Careful examination of the same fields in the near-infrared image reveals that very little vegetation is present. The near-infrared image also provides detailed information about the spatial distribution of the biomass present in each field. d) A color-infrared color composite makes it clear which fields are vegetated (IRS-1C LISS III images courtesy of Indian Space Agency).

A Mangrove Island near Key West, FL

Color Plate 5-2 Calibrated Airborne Multispectral Scanner (CAMS) data of one of the island keys near Key West, FL. The spatial resolution is 2.5 × 2.5 m. The image is a color composite of green, red, and near-infrared bands. The dense stands of healthy red mangrove (*Rhizophora mangale*) are 6 to 12 m tall. The tallest trees are found around the periphery of the island where tidal flushing provides a continuous flow of nutrients. Two areas of bare sand are located in the interior of the island. A coral reef surrounds the island (Davis and Jensen, 1998).

Color Perception Charts

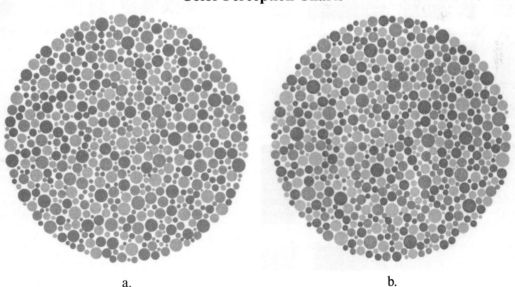

a.   b.

Color Plate 5-3 Some people do not perceive color like the majority of the population. This does not preclude them from becoming excellent image analysts. They only need to understand the nature of their color perception differences and adjust. Two color perception charts are provided. The viewer should see a "3" in (a) and a "6" in (b). If you see a "5", the next time you have an eye examination you might ask the optometrist to give you a color perception test to identify the nature of your color perception (courtesy Smith, J. T., 1968, *Manual of Color Aerial Photography*, American Society for Photogrammetry & Remote Sensing).

Planimetric Information Extraction:
Aero-triangulated Stereopair Displayed Anaglyphically

Color Plate 6-1 Anaglyph stereopair of Columbia, SC, created using soft-copy photogrammetric techniques. Please use anaglyph glasses to view the three-dimensional model. The left aerial photograph is viewed through the red lens while the right photograph is viewed through the blue lens.

Exterior Orientation: Westinghouse Savannah River Site
Stereoscopic Collection of Image Coordinates of Ground Control Points

a.

Point #	Point ID	>	Description	Type	Usage	Active	X Reference	Y Reference	Z Reference
1	5	>	laymax_pip	Full	Control	X	438912.700	3682631.232	91.489
2	6		laymax_pip	Full	Control	X	438907.990	3682623.710	91.483
3	7		laymax_pip	Full	Control	X	438912.870	3682623.110	91.461
4	8		entonite_to	Full	Check	X	438931.951	3682650.304	88.752
5	9		entonite_pip	Full	Control	X	438945.249	3682631.695	91.193
6	10		entonite_pip	Full	Control	X	438945.250	3682623.192	91.056

mage #	Image Name	Active	X File	Y File
1	srs95_fr1684	X	10254.924	6110.706
	srs95_fr1685	X	9202.335	11632.155

Aerial Photograph Draped over the Photogrammetrically-derived Digital Terrain Model

b.

Color Plate 6-2 a) Exterior orientation of a stereopair of the Westinghouse Savannah River Site near Aiken, SC. b) The digital terrain model (DTM) extracted using soft-copy photogrammetric techniques is draped with one of the natural-color aerial photographs from the stereopair and displayed from an oblique vantage point. The four test pads are used to analyze the effectiveness of various materials used to cap hazardous waste sites. Any subsidence due to problems associated with the cap can be documented by comparing DTMs derived from multiple dates of remote sensor data.

Planimetric Information Extraction:
Stereoscopic Analysis of An Anaglyph Stereopair Using the Floating Mark

a. Aero-triangulated anaglyph stereopair analyzed in three-dimensions
to extract planimetric detail. Please view this stereopair using anaglyph glasses.

b. Building footprints, pond perimeter, and street centerline were extracted from the
anaglyph stereopair shown above. Please view this stereopair using a stereoscope.

Color Plate 6-3 a) This anaglyph stereopair was analyzed in three-dimensions using the floating mark to identify three building footprints, the perimeter of a reflecting pond, and a street centerline. b) The features extracted are displayed in the aero-triangulated stereopair and can be viewed with a stereoscope.

USGS *Global Visualization Viewer*

a. Search for images of Charleston, SC, using the criteria: Path 16 and Row 37, Landsat 4-5 TM, May 2005, 100% maximum cloud cover; resampled to 1,000 m pixels.

b. Search for Charleston, SC, using pixels resampled to 240 m.

Color Plate 7-1 Use of the USGS *Global Visualization Viewer* to locate Landsat Thematic Mapper images of Charleston, SC.

Landsat 7 Enhanced Thematic Mapper Plus Imagery of San Diego, CA

Path 40, Row 37

Color-infrared color composite (RGB = bands 4, 3, 2).

Color Plate 7-2 Color composite of Landsat 7 Enhanced Thematic Mapper Plus imagery of San Diego, CA, obtained on April 24, 2000 (courtesy of NASA).

NOAA-16 Advanced Very High Resolution Radiometer (AVHRR) Imagery

Sea-surface temperature (SST) map derived from NOAA-16 AVHRR
band 4 (10.3 - 11.3 μm) imagery obtained on October 16, 2003.

Color Plate 7-3 Sea-surface temperature (SST) map derived from NOAA-16 AVHRR thermal infrared imagery (courtesy of NOAA and the Ocean Remote Sensing Program at Johns Hopkins University; Gasparovic, 2003).

Advanced Very High Resolution Radiometer (AVHRR) Mosaic of the United States

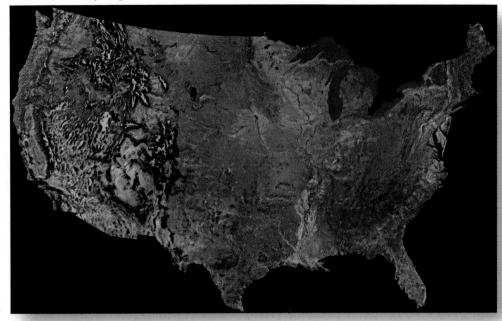

a.

Average August AVHRR Normalized Difference Vegetation Index (NDVI) 1981 - 2000

b.

Color Plate 7-4 a) AVHRR mosaic of the conterminous United States derived from sixteen 1 × 1 km NOAA-8 and NOAA-9 AVHRR images obtained from May 24, 1984, to May 14, 1986, using channels 1 and 2 (image courtesy of NOAA and U.S. Geological Survey. b) Average August Normalized Difference Vegetation Index (NDVI) image of North America derived from AVHRR imagery 1981 to 2000 (image courtesy of NASA Goddard Space Flight Center Scientific Visualization Studio).

Aircraft Multispectral Scanner Data of Four Mile Creek Delta in South Carolina

a. Daedalus band 10 (0.92 – 1.10 μm) on March 31, 1981.

b. Color-infrared composite (RGB = bands 10, 6, 4).

c. Daedalus band 10 (0.92 – 1.10 μm) on April 23, 1992.

b. Color-infrared composite (RGB = bands 10, 6, 4).

Color Plate 7-5 Daedalus DS-1260 aircraft multispectral scanner images of the Four Mile Creek delta on the Westinghouse Savannah River Site, SC. a) Band 10 near-infrared image obtained on March 31, 1981. b) Color-infrared composite. c) Band 10 image obtained on April 23, 1992. d) Color-infrared composite. Note the differences in the wetland vegetation distribution in the sloughs between 1981 and 1992 (images courtesy of Westinghouse Savannah River Company and the Department of Energy).

SPOT 5 Imagery of Los Angeles, CA

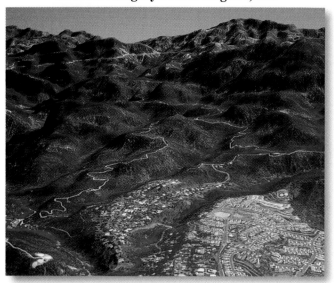

Color Plate 7-6 SPOT 5 imagery of Los Angeles, CA, created by merging 2.5 × 2.5 m panchromatic data with 10 × 10 m multispectral data. The imagery was draped over a 30 × 30 m USGS digital terrain model (© SPOT Image Inc.).

Portion of the First Global Ten-Day Synthesis Image Produced Using the SPOT Vegetation Sensor

Color Plate 7-7 Portion of a Global Ten-Day Synthesis product generated using the SPOT 4 Vegetation sensor. The daily product provides vegetation index information at a spatial resolution of 1 × 1 km over the whole Earth. The Global Ten-Day Synthesis is derived from analysis of Daily Synthesis data (© SPOT Image, Inc.).

May 11-20, 1998

Terra ASTER Optical Imagery of Oahu, Hawaii

a. ASTER 15 x 15 m color composite obtained on June 3, 2000 (RGB = bands 3, 2, 1).

b. Enlargement centered on Pearl Harbor.

Color Plate 7-8 *Terra* ASTER imagery of Oahu, Hawaii (courtesy of NASA/GSFC/MITI/ERSADC/JAROS, U.S./Japan ASTER Science Team, and the California Institute of Technology).

IKONOS Multispectral and Panchromatic Imagery of Columbia, SC

a. Band 2 (0.52 - 0.60 μm) 4 x 4 m.

b. Band 3 (0.63 - 0.69 μm) 4 x 4 m.

c. Band 4 (0.76 - 0.90 μm) 4 x 4 m.

d. Panchromatic band (0.45 - 0.90 μm) 1 x 1 m.

e. Color composite (RGB = bands 4, 3, 2).

f. Brovey merge of bands 4, 3, 2 and panchromatic.

Color Plate 7-9 IKONOS imagery of a business park in Columbia, SC. a–d) Individual 4 × 4 m multispectral bands and the 1 × 1 m panchromatic band are displayed. e) Standard color composite of IKONOS bands 4, 3, and 2. f) Color composite of merged multispectral (4 × 4 m) and panchromatic (1 × 1 m) data (images courtesy of GeoEye, Inc.).

AVIRIS Image of Moffett Field, CA

Image on top of the datacube
is a color composite of
three of the 224 bands

San Francisco Bay

evaporation
ponds

400 nm

1400 nm

1900 nm

2500 nm

400 nm

224
bands at
10 nm
nominal
bandwidth

1400 nm

1900 nm

2500 nm

Color Plate 7-10 An Airborne Visible Infrared Imaging Spectrometer (AVIRIS) dataset of Moffett Field, CA, adjacent to San Francisco Bay. The data were acquired on August 20, 1992, by the NASA ER-2 aircraft (a modified U-2) at an altitude of 20,000 m (65,000 ft.) resulting in 20×20 m nominal spatial resolution. AVIRIS acquires images in 224 bands each 10 nm wide, in the 400 to 2500 nm region. Three of the 224 spectral bands of data were used to produce the color composite on top of the "hyperspectral datacube" to accentuate the structure in the water and in the evaporation ponds on the right. The sides of the cube depict just the edge pixels for all 224 of the AVIRIS spectral channels. The top of the sides are in the visible part of the spectrum (400 nm) and the bottom is in the middle-infrared (2500 nm). The black areas in the datacube represent atmospheric absorption bands, especially at 1400 and 1900 nm. The sides are displayed in a pseudo-color ranging from black and blue (very low response) to red (high response). Of particular interest in the datacube is the small region of high response in the upper right corner of the datacube just below the evaporation ponds. This response is in the red part of the visible spectrum (about 680 nm) and is due to the presence of 1 cm (0.5 in.) long red brine shrimp in the evaporation ponds (courtesy of R. O. Green, NASA Jet Propulsion Lab).

Emerge Digital Sensor System (DSS) Images of New York

a. Color-infrared photograph.

b. Natural-color photograph.

Color Plate 7-11 Emerge Digital Sensor System (DSS) data. a) Color-infrared image (RGB = near-infrared, red, and green) of Dunkirk, NY, at 1 × 1 m obtained on December 12, 1998. The tennis court and track are covered with astroturf. b) Natural color image (RGB = red, green, and blue) of an area adjacent to a N.Y. Power Authority lake. The 1 × 1 ft data were acquired on October 13, 1997. The deciduous hardwood trees were captured senescing in this beautiful image (images courtesy of Don Light).

Thermal Infrared and Color Videography of Segments of the Umpqua River Basin in Oregon

a. Color density sliced thermal infrared image and natural color videography of the confluence of the South Umpqua River (27.5 °C) and the North Umpqua River (25.7 °C) at river mile 109.3.

b. Confluence of Wolf Creek (19.7 °C) and Little River (22.8 °C) at river mile 11.

Color Plate 8-1 Forward-looking infrared (FLIR) imagery of two segments of the Umpqua River Basin displayed along with natural-color videography obtained at the same time. The thermal-infrared data were calibrated to surface temperature and color-coded (density sliced). The remote sensing-derived water surface temperature measurements are generally within ± 0.5 °C of the true kinetic temperature (images courtesy of Oregon Department of Environmental Quality, 2003).

Thermal Infrared Detection of Thermal Water Pollution in the Savannah River

Savannah River

a. Density sliced predawn thermal infrared (8 - 14 µm) data.

A

B

Four Mile Creek Plume

C

b. Rotated 16° and transects extracted.

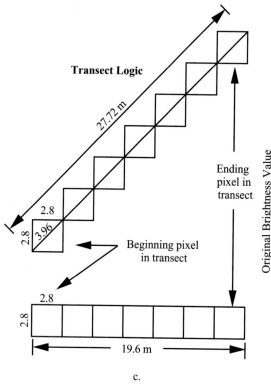

Transect Logic

27.72 m

2.8

2.8

3.96

Ending pixel in transect

Beginning pixel in transect

2.8

2.8

19.6 m

c.

Transect B Histogram

Original Brightness Value

white

200

red

180

160

140

orange

120

yellow

100

green

land

80

60

40

0   10   20   30   40   50   60

Pixel in Transect

>20° C above river ambient temperature

10.2 - 20° C

5.2 - 10° C

3.0 - 5.0° C
1.2 - 2.8° C

swamp

+1° C
12° C
river ambient

d.

Color Plate 8-2 a) Density sliced display of predawn thermal infrared data of the Four Mile Creek thermal plume in the Savannah River on March 28, 1981 (2.8 × 2.8 m spatial resolution). The data are color-coded according to the class intervals summarized in Table 8-8. b). Three transects are passed through the plume after rotating it 16°. c) Why it is important to rotate the image. d) The temperature information in Transect B displayed in a histogram format.

Space Shuttle Photograph and SIR-C/X-SAR Image of the Nile River, Sudan

a. Space Shuttle photograph.

b. SIR-C/X-SAR image.

Color Plate 9-1 a) Space Shuttle photograph of the Nile River, Sudan. b) SIR-C/X-SAR color-composite image of C-band with HV polarization, L-band with HV polarization, and L-band with HH polarization. The data were acquired by the Space Shuttle *Endeavor* in April, 1994 (courtesy NASA Jet Propulsion Lab).

SIR-C/X-SAR Imagery of Rondonia, Brazil

X-band VV   C-band HV   L-band HV

Color Plate 9-2 SIR-C/X-SAR image of a portion of Rondonia, Brazil, obtained on April 10, 1994. It is a color composite of X-band VV polarization, C-band HV polarization, and L-band HV polarization. Heavy precipitation appears as a black cloud in the X-band image, more faintly in the C-band image, and is relatively invisible in the L-band image. The bright pink is rainforest (courtesy NASA Jet Propulsion Lab).

SIR-C L-Band and C-Band Interferograms of Fort Irwin, CA

L-band

C-band

Color Plate 9-3 SIR-C/X-SAR interferograms of Fort Irwin, CA, in the Mojave Desert. The colored bands provide detailed, quantitative elevation information that can be used to construct a digital terrain model (courtesy NASA Jet Propulsion Lab).

SIR-C /X-SAR Imagery of Owens Valley, CA

Color Plate 9-4 Three-dimensional perspective view of Owens Valley, near Bishop, CA, created by combining two spaceborne SIR-C/X-SAR images obtained in October, 1994, using interferometric techniques. The White Mountains are in the center of the image and rise to 3,000 m (10,000 ft). The Owens River and its tributaries are seen (courtesy NASA Jet Propulsion Lab).

Digital Elevation Model of the Wasatch Front in Utah Derived from Shuttle Radar Topography Mission (SRTM) Data Draped with Landsat Thematic Mapper Imagery

Great Salt Lake

Salt Lake City

Wasatch Mountain Range

Provo

Utah Lake

a. SRTM-derived digital elevation model displayed in perspective looking north.

c. Stereopair.

b. Anaglyph stereopair.

Color Plate 9-5 a) SRTM-derived digital elevation model draped with Landsat TM image (RGB = bands 4, 3, 1). The C-band SRTM data were obtained February 12, 2000. b) An anaglyph presentation of the SRTM C-band data (please use anaglyph glasses). c) A stereopair of the Landsat TM data draped over the SRTM-derived DEM (images courtesy NASA Jet Propulsion Lab).

Comparison of GTOPO30 DEM and SRTM-derived DEM
of Mount Kilamanjaro in Tanzania and Kenya

a. GTOPO30 1 x 1 km digital elevation model.

b. SRTM-derived 30 x 30 m digital elevation model.

c. GTOPO30 1 x 1 km digital elevation model
observed from an oblique vantage point looking W-NW.

d. SRTM-derived 30 x 30 m digital elevation model.

Color Plate 9-6 a) GTOPO30 digital elevation model at 1 × 1 m. This is part of the global GTOPO30 DEM database maintained by the USGS. b) SRTM-derived DEM at 30 × 30 m. c) GTOPO30 DEM viewed from the W-NW. d) SRTM-derived DEM viewed from the W-NW. Note the detail associated with the small cinder cones on the flank of the mountain (images courtesy NASA Jet Propulsion Lab).

Advanced Microwave Scanning Radiometer (AMSR-E)

a. Composite of AMSR-E derived sea surface temperature (SST) for June, 2002.

b. Sea surface temperature (SST) derived from AMSR-E data collected August 25-27, 2005.

Color Plate 9-7 a) Composite image of Advanced Microwave Scanning Radiometer (AMSR-E) derived sea surface temperature (SST) in the birthplace of hurricanes, the inter-tropical convergence zone. b) A 3-day average of SST for the Caribbean Sea and the Atlantic Ocean from August 25-27, 2005, derived from the AMSR-E. Warm ocean water fuels hurricanes, and there was plenty of warm water for Hurricane Katrina to build up strength once she crossed over Florida and moved into the Gulf of Mexico. All yellow, orange or red water is ≥ 27.8 C (82 F). A hurricane requires SSTs at about this level or warmer to strengthen. Cloud cover data were extracted from NOAA GOES imagery. Land data were obtained from MODIS imagery (images courtesy of NASA and the National Space Development Agency of Japan).

Sweetgum
(*Liquidambar styraciflua L.*)

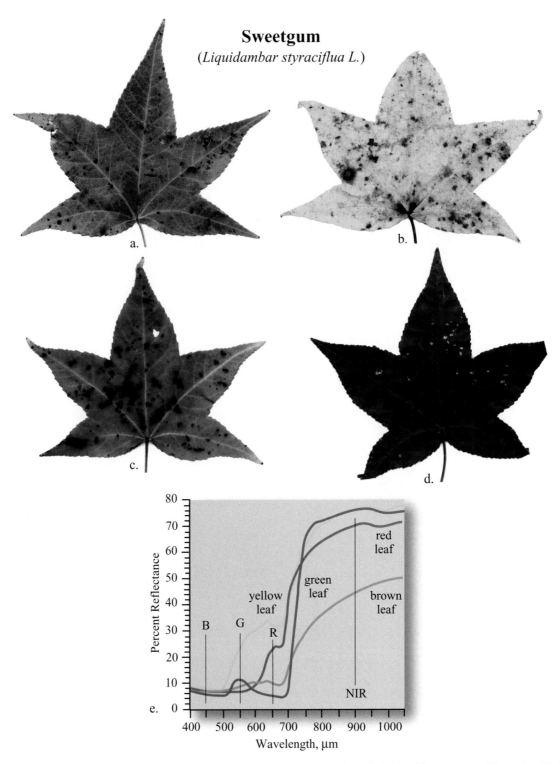

Color Plate 11-1 a) Photosynthesizing green Sweetgum leaf (*Liquidambar styraciflua* L.) obtained from a tree on November 11, 1998. b-c) Senescing yellow and red Sweetgum leaves obtained from the tree. d) Senesced Sweetgum leaf that was on the ground. e) Spectroradiometer percent reflectance measurements over the wavelength interval 400 – 1050 nm.

Ground Reference Information Overlaid on a Single Channel of AVIRIS Imagery San Luis Valley, CO

Vegetation Classification Map September 3, 1993

a.

b.

	alfalfa		barley		canola		pasture
	oat hay		spinach		potato		

	spinach		nothing mapped

Vegetation Senescence/Stress Map

c.

	healthy green vegetation		stressed vegetation
	dry vegetation/ bare ground		

Color Plate 11-2 a) Ground reference information overlaid on a single channel of 20 × 20 m AVIRIS data obtained on September 3, 1993. b) Vegetation species classification map derived from analysis of AVIRIS data. c) Vegetation senescence/stress map (after Clark et al., 1995).

a. CAMS data (RGB = bands 3, 2, 1).

b. CAMS data (RGB = bands 6, 4, 2).

c. Map of biomass.

Biomass in a Portion of Murrells Inlet, SC, Derived from 3 x 3 m Calibrated Airborne Multispectral Scanner (CAMS) Data Obtained on August 2, 1997

Total Biomass (grams/m^2)

	500 - 749
	750 - 999
	1000 - 1499
	1500 - 1999
	2000 - 2499
	2500 - 2999

Color Plate 11-3 a) Natural color composite of a small portion of Murrells Inlet, SC, recorded by the NASA Calibrated Airborne Multispectral Scanner (CAMS) on August 2, 1997. b) Color-infrared color composite. c) Biomass (g/m^2) information extracted from the CAMS data (Jensen et al., 1998).

Phenological Cycle of Crops in California's San Joaquin and Imperial Valleys and Landsat Multispectral Scanner Images of One Field During a Growing Season

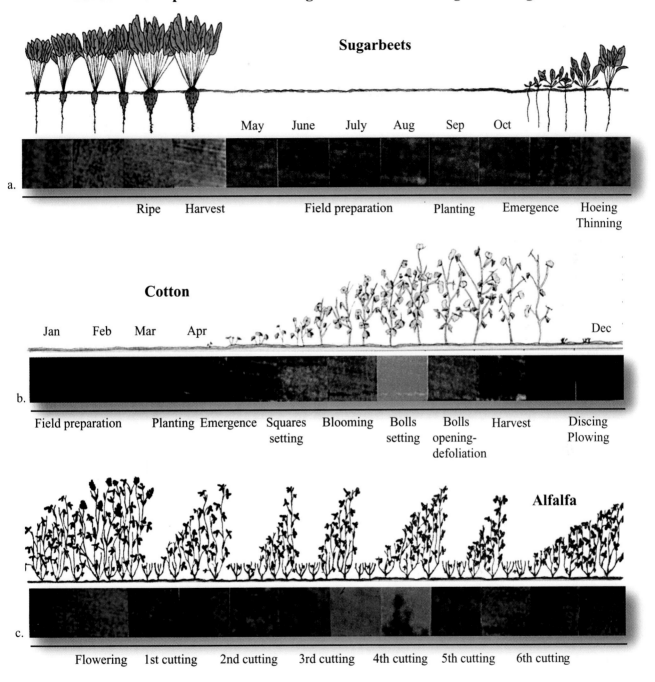

Color Plate 11-4 Phenological cycles of a) sugarbeets, b) cotton, and c) alfalfa grown in the San Joaquin and Imperial Valleys of southern California. Landsat MSS images were obtained over a 12-month period in the San Joaquin Valley. The color composite images (RGB = bands 4, 2, 1) of three fields are extracted and placed below the crop calendar information.

Classification Map of Imperial Valley, CA, on December 10, 1982
Derived from Landsat Thematic Mapper Bands 1-5 and 7

a. TM bands 3, 2, 1 (RGB).

b. TM bands 4, 3, 2 (RGB).

c. TM bands 5, 3, 2 (RGB).

d. TM bands 7, 3, 2 (RGB).

e. Classification map.

Sugarbeets

Alfalfa

Cotton

Fallow

Color Plate 11-5 a) Landsat Thematic Mapper natural color composite image of a portion of the Imperial Valley, CA, obtained on December 10, 1982. b – d) Several other false-color composites. e) Classification map derived by digital image processing.

Agricultural Landcover Along the Mhlathuze River in South Africa

a. Part of Water Management Area #6: Usutu and Mhlathuze River catchments
recorded on SPOT 5 imagery collected on January 9, 2004 (RGB = bands 3, 2, 1).

Legend:
- Sugarcane
- Citrus
- Forest
- Bushveld
- Road
- Water
- Bare soil
- Windbreak
- Banana

b. Classification map derived from 2004 SPOT multispectral and panchromatic data.

Color Plate 11-6 a) SPOT HRVIR color composite image of an area along the Mhlathuze River in the Republic of South Africa. b) Remote sensing-derived classification map of the study area. This type of information is used by the Republic of South Africa Department of Water Affairs & Forestry to inventory the agricultural land cover for input to the Water Authorization and Resource Management System (WARMS) spatial decision support system (courtesy Jensen et al., 2006).

Sea-viewing Wide Field-of-View (SeaWiFS) Imagery

a. Global chlorophyll *a* (g/m³) derived from SeaWiFS imagery obtained from September 3, 1997, through December 31, 1997. The warmer the color, the greater the chlorophyll concentration.

b. True-color SeaWiFS image of the Eastern Seaboard of the U. S. obtained on September 30, 1997.

c. Chlorophyll *a* distribution on September 30, 1997 derived from SeaWiFS data.

Color Plate 12-1 Examples of Sea-viewing Wide Field-of-View (SeaWiFS) remote sensor data [courtesy NASA Goddard Space Flight Center and Orbital Imaging Corporation (ORBIMAGE); used with permission].

MODIS Chlorophyll *a* Product

Color Plate 12-2 Chlorophyll *a* distribution on November 23, 2003, along the southeastern United States derived using the MODIS Chlor_*a*_2 algorithm (courtesy NASA).

Lake Tahoe, California-Nevada

a. Bathymetry derived from LIDAR for depths from 0 to 15 m and from multiple-beam SONAR for depths > 15 m.

b. Backscatter information derived from multiple-beam SONAR.

c. Oblique view of the bathymetric data looking west (arrow in *a* above).

Color Plate 12-3 a) Bathymetry of Lake Tahoe derived using the SHOALS LIDAR system for depths < 15 m and a Kongsberg multiple beam SONAR for depths > 15 m. b) Brighter tones in the acoustic backscatter data indicate harder surfaces such as rock. Darker tones indicate softer surfaces such as sediment. The surrounding land is a shaded-relief image generated from USGS 10-m DEMs. c) The central part of the western margin collapsed sending large blocks across the floor of the lake. Large blocks within the debris tongue are up to 20 m high. This large failure is called a debris avalanche and is very similar in size to collapse features found along ocean margins in tectonically active areas and on volcanoes (courtesy U.S. Geological Survey, 2006).

NOAA AVHRR-derived Sea-Surface Temperature

a. Composite sea-surface temperature (SST) map of the southeastern bight derived from AVHRR data.

b. Worldwide sea-surface temperature (SST) map derived from NOAA-14 AVHRR data.

Color Plate 12-4 a) Sea-surface temperature (SST) map derived from a three-day composite of NOAA AVHRR thermal infrared data centered on March 4, 1999. Each pixel was allocated the highest surface temperature that occurred during the three days (courtesy of NOAA Coastal Services Center). b) Global ocean 50 × 50 km SST (°C) derived from March 9, 1999 through March 13, 1999 (95 hours) NOAA-14 AVHRR data (courtesy of NOAA/NESDIS).

Monthly Sea-Surface Temperature (°C)

a. La Nina in December, 1998.

b. Normal in December, 1990.

c. El Nino in December, 1997.

Color Plate 12-5 Reynolds monthly sea-surface temperature (SST) maps derived from *in situ* buoy data and remotely sensed data (courtesy NOAA/TAO/National Center for Environmental Prediction).

Tropical Rainfall Measurement Mission (TRMM)

a. TRMM Microwave Imager (TMI) data obtained on March 9, 1998.

b. TRMM Precipitation Radar (PR) data
obtained on March 9, 1998.

c. Along-track cross-section of TRMM Precipitation Radar data obtained on March 9, 1998.

Color Plate 12-6 a) Worldwide TRMM Microwave Imager data collected on March 9, 1998. b) Precipitation Radar data of southern Florida. c) Cross-section of Precipitation Radar data (courtesy NASA Goddard TRMM Office and Japan National Space Development Agency).

Selected Databases from
South Carolina's Spatial Data Infrastructure

Digital
Orthorectified
Imagery

Hydrology

Digital Terrain

Transportation

Landcover/
Landuse

Color Plate 13-1 Selected databases in the South Carolina Spatial Data Infrastructure.

Level I Urban/Non-urban Landcover for
Berkeley, Dorchester, and Charleston Counties in South Carolina

a. February 14, 1975.   b. March 26, 1981.   c. February 10, 1985.

d. January 4, 1989.   e. March 17, 1992.   f. February 3, 1994.

Color Plate 13-2 Level I urban/non-urban landcover for Berkeley, Dorchester, and Charleston counties centered on Charleston, SC, derived from Landsat MSS 79 × 79 m data. The landcover information is draped over a near-infrared image (MSS band 4).

Mineral Maps of Cuprite, NV, Derived from AVIRIS Hyperspectral Data

Color Plate 14-1 Mineral maps of Cuprite, NV, derived from low-altitude (3.9 km AGL) and high-altitude (20 km AGL) AVIRIS data obtained on October 11, 1998, and June 18, 1998, respectively (courtesy Swayze et al., 1999).

Shuttle SIR-C/X-SAR Images of Kilauea Volcano in Hawaii

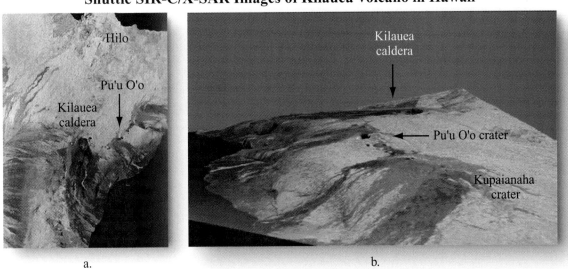

Color Plate 14-2 a) Composite Shuttle SIR-C/X-SAR image of Kilauea volcano, HI, (bands C, X, L) obtained on April 12, 1994. b) The image overlaid on a digital elevation model. The overland flow of lava on this shield volcano is very evident (courtesy NASA Jet Propulsion Lab).

Stereopair of Mount St. Helens, Washington

Color Plate 14-3 U.S. Geological Survey High Altitude Photography (HAP) color-infrared stereopair of Mount St. Helens, WA, on August 6, 1981. The active lava dome in the center of the cone is visible. A sediment-choked radial drainage pattern has developed. North is to the left (courtesy U.S. Geological Survey).

The Grand Canyon

a. Landsat TM color composite (RGB = bands 7, 4, 2).

b. Color composite (RGB = bands 4, 3, 2).

Color Plate 14-4 Landsat Thematic Mapper color composites of a portion of the Grand Canyon (courtesy NASA).

Ugab River, Namibia

Color Plate 14-5 ASTER image of Namibia's Ugab River obtained on December 25, 2000 (RGB = bands 7, 3, 1). The river passes through nearly vertical folded layers of thinly bedded limestone, sandstone, and siltstone (courtesy of NASA/GSFC/METI/ERS-DAC/JAROS and U.S./Japan ASTER Science Team).

Mississippi River

Color Plate 14-6 Landsat Thematic Mapper color composite (RGB = bands 4, 3, 2) of the Mississippi River (courtesy Space Imaging, Inc.).

Morro Bay, CA

a. Landsat TM color composite (RGB = bands 7, 4, 2).

b. Color composite (RGB = bands 4, 3, 2).

Color Plate 14-7 Landsat Thematic Mapper color composites of Morro Bay, CA (courtesy Space Imaging, Inc.).

Sullivan's Island, SC

Isle of Palms, SC

a. NAPP color-infrared orthophotograph.

b. ATLAS multispectral data of the tidal flat behind Isle of Palms, SC.

Color Plate 14-8 a) Portion of a USGS NAPP color-infrared orthophoto quarter quadrangle of Sullivan's Island, SC (courtesy U.S. Geological Survey). b) ATLAS multispectral scanner data (2.5 × 2.5 m) of the tidal flats behind Isle of Palms, SC, (RGB = bands 6, 4, 2) (courtesy NASA Stennis Space Center).

Agassiz Seward
Glacier Glacier

Malaspina Glacier, AL

a. Landsat 7 Enhanced Thematic Mapper Plus imagery of the Malaspina Glacier obtained on August 31, 2000 (RGB = bands 5, 4, 3).

b. High-oblique view of the glacier created by draping a Landsat Thematic Mapper image obtained in February, 2000, (RGB = bands 3, 2, 1), over a digital elevation model extracted from Space Shuttle Radar Topography Mission C-band interferometric SAR data collected in February 2000.

Color Plate 14-9 a) Vertical near-infrared and b) high-oblique natural-color view of the Malaspina Glacier in Alaska. This is Alaska's largest glacier and covers 1,500 sq. mi (3,880 sq. km). It is a compound glacier formed by the merging of valley glaciers, the most prominent of which are the Agassiz Glacier and the Seward Glacier (courtesy NASA and Jet Propulsion Lab).

Remote Sensing Glaciated Landscapes

a. Uinta Mountain Range in Utah.

b. Wasatch Range in the Rocky Mountains of Utah.

c. Lake Tahoe in the Sierra Nevada on the border of California and Nevada.

Color Plate 14-10 Landsat color composites of glaciated landscapes in portions of a) the Uinta Mountains in Utah, b) the Wasatch Range in Utah, and c) the Lake Tahoe region in the Sierra Nevada, CA. Please refer to the black-and-white illustrations in Chapter 14 for annotation (images courtesy U.S. Geological Survey and Space Imaging, Inc.).

Namib Desert, Namibia, Africa

a. Landsat 7 Enhanced Thematic Mapper Plus (ETM+) imagery of the Namib Desert
obtained on August 12, 2000 (RGB = bands 5, 4, 3).

b. Low-oblique view of a portion of the same area in ASTER imagery obtained on October 14, 2002.
The image was draped over a digital elevation model extracted from stereoscopic ASTER data.

Color Plate 14-11 a) Vertical and b) low-oblique views of the Namib Desert in Namib-Naukluft National Park. These are the tallest dunes in the world, some > 300 m (almost 980 ft). The area receives an average of only 63 mm rainfall per year. The orange color develops over time as iron in the sand is oxidized. The older the dune, the brighter the orange color (courtesy USGS and NASA/GSFC/METI/ERSDAC/JAROS and U.S./Japan ASTER Science Team).

This is one of the more incredible equations in the discipline of radar remote sensing. The omission of the slant-range distance (S) from the equation means that the azimuth resolution of a SAR system is independent of range distance or sensor altitude. Theoretically, the spatial resolution in the azimuth direction from a SAR imaging system will be the same from an aircraft platform as it is from a spacecraft. There is no other remote sensing system with this capability!

Equation 9-13, however, is not the only parameter that has an impact in SAR data. The coherent nature of the SAR signal produces speckle in the image. To remove the speckle, the image is usually processed using several *looks*, i.e., an averaging takes place. For example, four looks (N) might be averaged. This dramatically improves the interpretability of the SAR image data. However, the azimuth resolution must be adjusted by the equation:

$$SAR_a = N\left(\frac{L}{2}\right) .\qquad(9\text{-}14)$$

The SIR-C SAR had a 12 m antenna, which would produce a 6 m along-track resolution if Equation 9-13 were used. However, the speckle in the SIR-C SAR data were processed using $N = 4$ looks to improve the interpretability of the data. Thus, the adjusted azimuth resolution is 24 m.

This discussion first summarized how the range and azimuth resolution are computed for specific locations within the real aperture radar beam swath. Note that the resolution element might have different dimensions, i.e., the range resolution could be 10 m and the azimuth resolution 20 m. Therefore, because scale is constantly changing throughout a real aperture radar image, it is important to be careful when making radargrammetric measurements over large regions. When using such data, it is best to compute the range and azimuth resolution for the major region you are interested in and then make measurements. If a new area is selected for study, then new resolution measurements should be computed before area measurements are obtained.

RADAR Relief Displacement, Image Foreshortening, Layover, Shadows, and Speckle

Geometric distortions exist in almost all radar imagery, including foreshortening, layover, and shadowing. When the terrain is flat, it is a straightforward matter to use Equation 9-6 to convert a slant-range radar image into a ground-range radar image that is planimetrically correct in x,y. However, when trees, tall buildings, or mountains are present in the scene, relief displacement in the radar image occurs. In radar relief displacement, the horizontal displacement of an object in the image caused by the object's elevation is in a direction toward the radar antenna (Figure 9-13a). Because the radar image is formed in the range (cross-track) direction, the higher the object, the closer it is to the radar antenna, and therefore the sooner (in time) it is detected on the radar image. This contrasts sharply with relief displacement in optical aerial photography where the relief displacement is radially outward from the principal point (center) of a photograph. The elevation-induced distortions in radar imagery are referred to as *foreshortening* and *layover*.

Foreshortening and Layover

Basically, all terrain that has a slope inclined toward the radar will appear compressed or foreshortened relative to slopes inclined away from the radar. The effect is illustrated in Figure 9-13a. The foreshortening factor, F_f, is approximately

$$F_f = \sin(\theta - \alpha)\qquad(9\text{-}15)$$

where the *incident angle* θ is the angle between the vertical plane at nadir and a line that links the imaging radar antenna to a feature on the ground, and α is the slope angle of the surface. Alpha is positive (α^+) where the slope is inclined toward the radar (foreslope), and negative (α^-) where the slope is inclined away from it (backslope).

For example, consider the relatively low relief mountain in Figure 9-13a. It is perfectly symmetrical with the distance from AB' and B'C being equal when plotted in the ground-range (i.e., on a planimetric map). A single pulse of microwave energy illuminating the terrain will first encounter the base of the mountain (A) and record it in the slant-range image at a. Because the top of the mountain (B) extends above the terrain and is relatively close to the antenna, it is recorded at b. The base of the mountain at C is recorded at c in the slant-range display. The higher the object above the terrain, the more foreshortening that will take place. In this case, the radar image will have a very bright short foreslope and a darker backslope. It would be difficult to make a map of mountainous features using this radar image because the top of the mountain, even though it was not very high, is displaced from its true ground-range planimetric position in the radar image, i.e., ab does not equal bc. Foreshortening is influenced by the following factors:

- *object height*: The greater the height of the object above local datum, the greater the foreshortening.

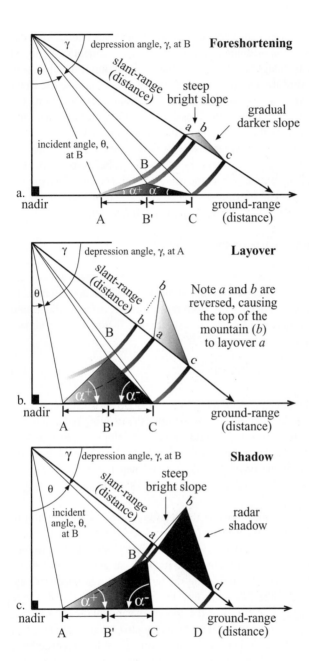

Figure 9-13   a) Radar foreshortening in the slant-range display. Slope *AB* in the ground-range display and *ab* in the slant-range display are supposed to be equal as are *BC* and *bc*. Instead, slope *ab* is shorter and steeper while slope *bc* is more gradual and longer than it should be. b) Image layover of foreslope *AB* in the slant-range display (*ba*) caused by the incident angle *θ* being smaller than the foreslope angle *α⁺*. Slant-range *a* and *b* are reversed, causing layover. c) A substantial shadow is produced at *bd* because the backslope angle (*α⁻*) is greater than the depression angle (*γ*) (after Ford et al., 1980).

- *depression angle* (or *incident angle*): The greater the depression angle (γ) or smaller the incident angle (θ), the greater the foreshortening. A good example is found in Figure 9-14a,b where the ERS-1 sensor with its large 67° depression angle and its 23° incident angle introduces more foreshortening than the JERS-1 radar with its 51° depression angle and 39° incident angle. Also consider the foreshortened cinder cone in Arizona when recorded on a radar image versus a conventional vertical aerial photograph (Figure 9-14c,d).

- *location of objects in the across-track range*: Features in the near-range portion of the swath are generally foreshortened more than identical features in the far-range. Foreshortening causes features to appear to have steeper slopes than they actually have in nature in the near-range of the radar image and to have shallower slopes than they actually have in the far-range of the image.

Image *layover* is an extreme case of image foreshortening. It occurs when the incident angle (θ) is smaller than the foreslope (α⁺), i.e., θ < α⁺. This concept is illustrated in Figure 9-13b. In this case, the mountain has so much relief that the summit (*B*) backscatters energy toward the antenna before the pulse of energy even reaches the base of the mountain (*A*). Remember that in terms of planimetric distance from the nadir point directly beneath the aircraft, the base of the mountain (*A*) is much closer than the summit (*B*), as documented by the ground-range distance display. However, because the mountain summit (*B*) reflects the incident microwave energy sooner than the base of the mountain (*A*), the summit (*b*) in the slant-range radar image actually *lays over* (hence the terminology) the base of the mountain recorded on the radar image at *a*. Once again, the summit of the mountain (*a*) is significantly displaced from its true planimetric position. This distortion cannot be corrected even when the surface topography is known. Great care must be exercised when interpreting radar images of mountainous areas where the thresholds for image layover exist. Bright white ridges in Figure 9-15 represent severe SIR-C L-band (HH) radar layover in the San Gabriel Mountains just east of Los Angeles, CA.

Shadows

Shadows in radar images can enhance the geomorphology and texture of the terrain. Shadows can also obscure the most important features in a radar image, such as the information behind tall buildings or land use in deep valleys. If certain conditions are met, any feature protruding above the local datum can cause the incident pulse of microwave energy to reflect all of its energy on the foreslope of the

a.   **C-band ERS-1**   b.   **L-band JERS-1**   look direction
depression angle 67°   depression angle 51°
incident angle 23°   incident angle 39°   ↓

c.   **X-band**   ↓ look direction   d.   **Aerial photograph**   —N—▷

Figure 9-14   a) The C-band ERS-1 image of the White Mountains exhibits substantial foreshortening due to the satellite SAR's small incident angle (courtesy Alaska SAR Facility; © 1992, European Space Agency). b) The L-band JERS-1 image with a larger incident angle has significantly less image foreshortening (courtesy Alaska SAR Facility; © NASDA). c) X-band aircraft
synthetic aperture radar (SAR) image of a cinder cone in Arizona. The shape of the basically cylindrical cinder cone is distorted
in the slant-range look direction, appearing more elliptical in shape than it does in the aerial photograph. Foreshortening (relief
displacement toward the radar antenna) occurs in the near-range. There is also a dark shadow under the rim of the crater at the
top of the cone. d) Vertical panchromatic aerial photograph of SP Mountain, AZ (courtesy Eric Kasischke, ERIM International,
Inc.).

object and produce a black shadow for the backslope. A
backslope is in radar shadow when its angle (α^-) is steeper
than the depression angle (γ), i.e., $\alpha^- > \gamma$ (Figure 9-13c). If
the backslope equals the depression angle ($\alpha^- = \gamma$), then the
backslope is just barely illuminated by the incident energy.
This is called grazing illumination because the radar pulse
just grazes the backslope. The backslope is fully illuminated
when it is less than the depression angle ($\alpha^- < \gamma$).

Figure 9-13c demonstrates how a large shadow from a relatively steep backslope might be produced. In this case, we
have a backslope of approximately 85° ($\alpha^- = 85°$) and a
depression angle of 45° ($\gamma = 45°$). Because the backslope is
greater than the depression angle, we expect this area to be
in shadow. In fact, this is the case with the terrain surface
BCD in the ground-range being in complete shadow in the
slant-range radar image display (*bd*). In the ground-range
display the distance from the summit at B to the back base of

L-band SIR-C (HH)
July 20, 1995

N

look direction

Figure 9-15   L-band SIR-C (HH) image of the San Gabriel Mountains and a portion of the city of Pasadena. Extensive layover is found in the mountainous terrain, causing the ridge lines to be displaced toward the antenna (courtesy NASA Jet Propulsion Lab).

the mountain at C is relatively short. But in the slant-range radar image, *bd* is very long. This particular radar image would also experience image foreshortening (but not layover because *A* is recorded by the antenna before *B*) and have a very bright return from the foreslope.

Below are some important characteristics of radar shadows.

• Unlike aerial photography, where light may be scattered into the shadow area and then recorded on film, there is no information within the radar shadow area. It is black.

• Two terrain features (e.g., mountains) with identical heights and fore- and backslopes may be recorded with entirely different shadows, depending upon where they are in the across-track. A feature that casts an extensive shadow in the far-range might have its backslope completely illuminated in the near-range.

• Radar shadows occur only in the cross-track dimension. Therefore, the orientation of shadows in a radar image provides information about the look direction and the location of the near- and far-range.

Shadows in radar imagery are valuable when evaluating geomorphic characteristics. Radar shadows often function like low Sun-angle photography, enhancing lineaments and other geologic conditions. Terrain information such as topographic height, slope, etc., can be used to configure radar overflights or, purchase radar data that have the most appropriate depression angles and near- and far-range characteristics to emphasize terrain shadows.

Speckle

Speckle is a grainy salt-and-pepper pattern in radar imagery (Figure 9-16) that is present due to the coherent nature of the radar wave, which causes random constructive and destructive interference, and hence random bright and dark areas in a radar image. The speckle can be reduced by processing separate portions of an aperture and recombining these portions so that interference does not occur (Kasischke et al., 1984). As previously mentioned, this process, called multiple *looks* or noncoherent integration, produces a more pleasing appearance, and in some cases may aid in interpretation of the image but at a cost of degraded resolution. For example, consider the radar imagery in Figure 9-16, which was processed using 1-look, 4-looks, and 16-looks. Most interpreters would prefer working with the 4- or 16-look imagery.

Synthetic Aperture Radar Systems

A major advance in radar remote sensing has been the improvement in azimuth resolution through the development of *synthetic aperture radar* (SAR) systems. Remember, in a real aperture radar system that the size of the antenna (*L*) is inversely proportional to the size of the angular beam width (Equation 9-10). Therefore, great improvement in azimuth resolution could be realized if a longer antenna were used. Engineers have developed procedures to *synthesize* a very long antenna electronically. Like a brute force or real aperture radar, a synthetic aperture radar also uses a relatively small antenna (e.g., 1 m; refer to Figure 9-1) that sends out a relatively broad beam perpendicular to the aircraft. The major difference is that a greater number of additional beams are sent toward the object. Doppler principles are then used to monitor the returns from all these additional microwave pulses to synthesize the azimuth resolution to become one very narrow beam.

The *Doppler principle* states that the frequency (pitch) of a sound changes if the listener and/or source are in motion relative to one another. For example, an approaching train whistle will have an increasingly higher frequency pitch as it

a. 1-look radar image.

b. 4-look radar image.

c. 16-look radar image.

Figure 9-16   Radar speckle reduction using multiple-look techniques (courtesy Eric Kasischke, ERIM International, Inc.).

approaches. This pitch will be highest when it is directly perpendicular to the listener (receiver). This is called the point of zero Doppler. As the train passes by, its pitch will decrease in frequency in proportion to the distance it is from the listener (receiver). This principle is applicable to all harmonic wave motion, including the microwaves used in radar systems.

Figure 9-17 depicts the Doppler frequency shift due to the relative motion of a terrain object at times n, $n+1$, $n+2$, $n+3$, and $n+4$ through the radar beams due to the forward motion of the aircraft. The Doppler frequency diagram reveals that the frequency of the energy pulse returning from the target increases from a minimum at time n to a maximum at point $n+3$, normal (at a right angle) to the aircraft. Then, as the target recedes from $n+3$ to $n+4$, the frequency decreases.

How is a synthetic aperture image actually produced? A long antenna can be synthesized using a short antenna by taking advantage of the aircraft's motion and the Doppler principle. It is assumed that the terrain is stable and *not* moving. It is also assumed that the object of interest remains a fixed distance away from the aircraft's flightline. As the aircraft flies along a straight line, a short antenna sends out a series of microwave pulses at regular intervals. As an object (black dot) enters the antenna's beam (Figure 9-17a), it backscatters a portion of the pulse it receives toward the antenna. At some point in the aircraft's path the object will be an integral number of microwave wavelengths away; between those points it will not be. For example, in (a) we see that the object is first 9 wavelengths away, then 8 (b), then 7 (c), then 6.5 (d), at which point the object is at a right angle to the antenna, i.e., the shortest distance and area of zero Doppler shift. From then on the distance between the aircraft and the object will be increasing, i.e., perhaps 7 wavelengths away at location (e). The antenna receives the series of reflected waves (gray lines in illustrations a – e) and electronically combines them with a train of reference wavelengths (not shown), causing the two series of oscillations to interfere. The interference signal emerges as a voltage that controls the brightness of a spot scanning across the screen of a cathode-ray tube. When the returned pulse coincides with a reference pulse, the interference is constructive; the voltage will be high and the moving spot will be bright. When the phase of the returned wavelength does not coincide with the phase of the reference frequency, the interference is destructive; the voltage will be low and the moving spot will be dim or dark. The moving spot thus tracks out a series of light and dark dashes of unequal length that are recorded on a strip of data film moving at a velocity proportional to the velocity of the aircraft. The series of opaque and transparent dashes on the film are actually a one-dimensional interference pattern; the film on which they are recorded is a *radar hologram*.

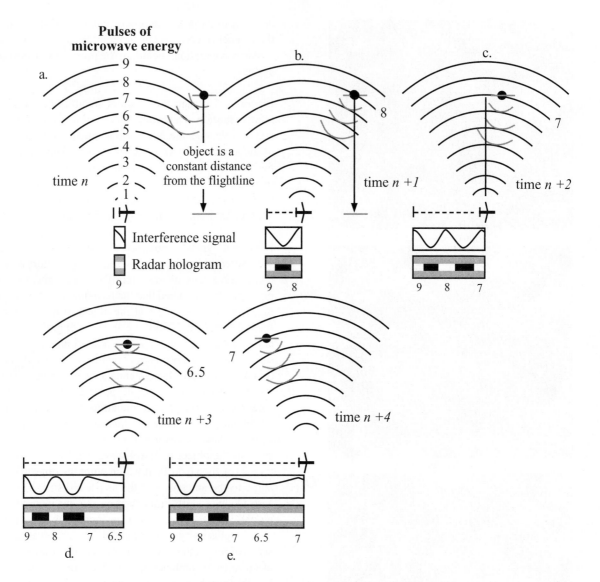

Figure 9-17   A long antenna can be synthesized by a short antenna (e.g., 1 m long) by taking advantage of the aircraft's motion and Doppler principles. As the craft flies along a straight line, a mounted short real antenna sends out a series of pulses at regular intervals. Each pulse consists of a train of coherent microwaves. Although the length of the pulses determines the resolution across the track, it is the wavelength of the microwave radiation that determines the resolution along the track. As an object (black dot) enters the antenna's beam (*a*), it reflects a portion of the pulses it receives back toward the antenna. At some point in the aircraft's path the object is an integral number of microwave wavelengths away; between these points it will not be. For example, in the illustration the object is first 9 wavelengths away at *a*, then 8 at *b*, then 7 at *c*, then 6.5 at *d*, where the object is at right angles to the antenna. From then on, the crest is increasing its distance from the object, e.g., point *e*. The antenna receives the series of reflected waves and electronically combines them with a train of reference wavelengths (not shown), causing the two series of oscillations to interfere. The interference signal emerges as a voltage that controls the brightness of a spot scanning across a cathode-ray tube. At the times that a returned pulse coincides with a reference pulse, the interference is constructive; the voltage will be high and the moving spot will be bright. At the times that the phase of the returned wavelength does not coincide with the phase of the reference frequency the interference is destructive; the voltage will be low and the moving spot will be dim. The moving spot thus traces out a series of light and dark dashes of unequal length that are recorded on a strip of data film moving at a velocity proportional to the velocity of the aircraft (after Kasischke et al., 1984).

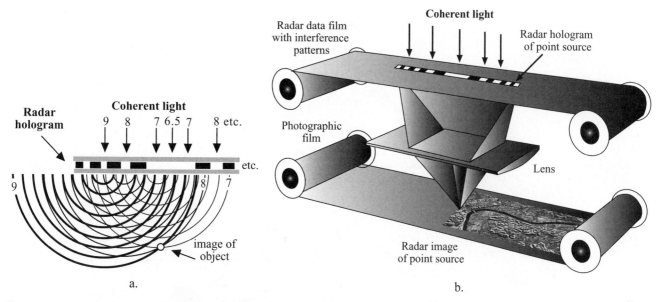

Figure 9-18   a) When the developed hologram is illuminated by laser coherent light, each transparent dash acts as a separate source of coherent light. Below the hologram there is a single point where the resulting light waves all constructively interfere. Here the
9th wavelength of light from the transparent dash created by the 9th microwave will meet the 8th wavelength of light from the
transparent dash created by the 8th microwave, and both will meet with the 7th wavelength of light from the transparent dash
created by the 7th microwave. At that one point light from the entire length of the interference pattern is focused to form a
miniature image of the original object. b) As the data film is advanced through the beam of laser light, the reconstructed image
is recorded on another moving strip of film. Because the data film is holographic only in the along-track coordinate, the images
in the across-track coordinate must be focused with a cylindrical lens.

When the developed hologram is illuminated by a source of
coherent light (Figure 9-18a), each transparent dash functions as a separate source of coherent light. Below the hologram there will be a single point where the resulting light
waves all constructively interfere. In this example, the 9th
wavelength of light (thick curved lines) from the transparent
dash created by the 9th microwave will meet the 8th wavelength of light (regular lines) from the transparent dash created by the 8th microwave, and both will meet the 7th
wavelength of light (thin curves) from the transparent dash
created by the 7th microwave and so on (only the 9th, 8th,
and 7th patterns are shown in this example for clarity). At
that one point, light from the entire length of the interference
pattern is focused to form a miniature image of the original
object. Figure 9-18b demonstrates how the holographic
image is reconstructed and recorded on film. After processing the negative film to become a positive print, the radar
image is ready for analysis.

The record of Doppler frequency enables the target to be
resolved on the image film as though it was observed with an
antenna of length L, as shown in Figure 9-19 (Sabins, 1997).
This synthetically lengthened antenna produces the effect of
a very narrow beam with constant width in the azimuth

direction, shown by the shaded area in Figure 9-19. For both
real and synthetic aperture systems, range resolution is
determined by pulse length and depression angle. Generally,
synthetic aperture images are higher in azimuth resolution
than real aperture radars.

The aforementioned method is often called synthetic aperture radar *optical correlation* because of all the precision
optics utilized. It is also possible to use *SAR digital correlation* techniques to record and process the amplitude and
phase history of the radar returns. Digital correlation does
not produce an intermediate radar film. The digital process is
computation intensive. Major advantages of digital correlation include the ability to 1) perform both radiometric and
geometric corrections rapidly during onboard processing
(good for emergencies such as oil spills, floods, fires, etc.),
2) telemeter the processed radar data directly to the ground
to support real-time decision making, and/or 3) store the digitally processed SAR data on hard disks for subsequent digital processing on the ground. Many of the commercial (e.g.,
Intermap Star 3*i*) and government SARs (e.g., Canadian
RADARSAT, ESA ASAR) use digital SAR correlation techniques.

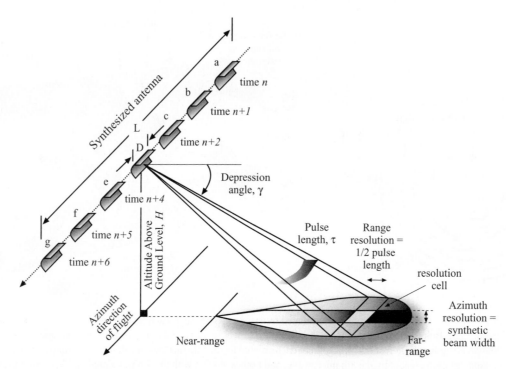

Figure 9-19 A synthetic antenna of length L is produced by optically or digitally processing the phase histories of microwave radar returns sent and received by a real antenna of length D sent at times n. The synthetic aperture radar has a range resolution that is approximately equal to half the pulse length. Note how the azimuth resolution remains constant from the near-range to the far-range (after Sabins, 1997).

The Radar Equation

The following discussion provides additional quantitative information about the radar signal. A radar image is a two-dimensional representation of the power (or voltage) returned to the radar from a specific area on the ground presented as a picture element (pixel). This returned power is usually quantized to a radiometric scale of n bits for presentation on image processing systems (Leberl, 1990). RADARSAT is quantized to 11 bits. To understand how to interpret radar images, we should understand the nature of the power scattered back toward the radar antenna. In the most simple case, this can be stated verbally as suggested by Moore (1983):

$$\begin{pmatrix} \text{Power} \\ \text{received} \end{pmatrix} = \begin{pmatrix} \text{Power per unit} \\ \text{area at target} \end{pmatrix} \times \begin{pmatrix} \text{Effective scattering} \\ \text{area of the target} \end{pmatrix}$$
$$\times \begin{pmatrix} \text{Spreading loss of} \\ \text{reradiated signal} \end{pmatrix} \times \begin{pmatrix} \text{Effective antenna} \\ \text{receiving area} \end{pmatrix}$$

where the power per unit area at the receiver is the energy scattered back from the terrain — *backscatter*. The spreading loss occurs because the signal starts from the backscattering point source on the ground (e.g., a large rock) and spreads out in all directions, so that the power per unit area is less at a greater distance than it would be near the scat-

terer. Therefore, the strength of the backscatter toward the receiver is a product of the power per unit area illuminating the target, times the effective scattering area of the target, and then the retransmission of this wave back toward the receiving antenna. The actual size of the receiving antenna also makes a difference.

The fundamental radar equation is derived by combining these word quantities to create the mathematical expression (Moore, 1983; Kasischke et al., 1984):

$$P_r = \frac{P_t \cdot G_t \cdot \sigma \cdot A_r}{(4\pi)^2 \cdot R^4} \tag{9-16}$$

where P_r is power received, P_t is the power transmitted toward the target, G_t is the gain of the antenna in the direction of the target, R is the range distance from the transmitter to the target, σ is the effective backscatter area of the target (often called the radar cross-section), and A_r is the area of the receiving antenna. This equation can be broken down into seven parts for even more clarity:

$$P_r = P_t \cdot G_t \cdot \frac{1}{4\pi R^2} \cdot \sigma \cdot \frac{1}{4\pi R^2} \cdot A_r \tag{9-17}$$

$$1 = 2 \cdot 3 \cdot 4 \cdot 5 \cdot 6 \cdot 7.$$

The power received [1] by the radar system, P_r, is a function of a pulse of electromagnetic energy, P_t [2], that has been focused down to an angular beam width by the antenna so that the flux becomes higher by a factor [3] of G_t over a spherically expanding wave [4]. The focused energy illuminates an *area* on the ground that has a cross-section of σ [5]. The *radar cross-section* is defined as the equivalent of a perfectly reflecting area that reflects isotropically (spherically). The energy backscattered from the ground once again spherically expands from the source [6]. Finally, the receiving antenna area (A_r) intercepts a portion of the reflected wave and records it [7].

Most radars use the same antennas for transmitting (A_t) and receiving (A_r). Consequently, the gain factors of the antennas may be combined using the relationship between gain and receiving aperture as (Moore, 1983):

$$G = G_t = G_r = \frac{4\pi A_r}{\lambda^2} \qquad (9\text{-}18)$$

where λ is the wavelength (or frequency) of the radar system. Substituting this value in Equation 9-16 or 9-17 results in a modified radar equation (Carver et al., 1985):

$$P_r = \frac{P_t \cdot G^2 \cdot \sigma \cdot \lambda^2}{(4\pi)^3 \cdot R^4}. \qquad (9\text{-}19)$$

Thus, we find that the radar equation can be viewed as a product of system parameters and other environmental terrain parameters that produce the backscatter cross-section, σ. Because the system parameters are well known, their effects are typically removed from the radar images, i.e., the system parameters may be set to unity (1).

It is the effects of terrain on the radar signal that we are most interested in, i.e., the amount of radar cross-section, σ, reflected back to the receiver, per unit area (A) on the ground. This is called the *radar backscatter coefficient* ($\sigma°$) and is computed as

$$\sigma° = \frac{\sigma}{A} \qquad (9\text{-}20)$$

where σ is the radar cross-section. The radar backscatter coefficient determines the percentage of electromagnetic energy reflected back to the radar from within a resolution cell, e.g., 10×10 m. The actual $\sigma°$ for a surface depends on a number of terrain parameters, like geometry, surface roughness, moisture content, and the radar system parame-

ters (wavelength, depression angle, polarization; Leberl, 1990). $\sigma°$ is a dimensionless quantity characteristic of the scattering behavior of all the elements contained in a given ground cell. Because $\sigma°$ can vary over several orders of magnitude, it is expressed as a logarithm with units of decibels (dB) that usually range from -5 to +40 dB.

The total radar cross-section of an area (A) on the ground therefore becomes ($\sigma°A$), and the final form of the radar equation for an area-extensive target becomes (Henderson and Lewis, 1998; Mikhail et al., 2001):

$$P_r = P_t(\sigma°A)\left(\frac{G^2 \cdot \lambda^2}{(4\pi)^3 \cdot R^4}\right). \qquad (9\text{-}21)$$

A digital SAR image is created that consists of a two-dimensional array (matrix) of picture elements (pixels) with the intensity (called the brightness) of each pixel proportional to the power of the microwave pulse reflected back from the corresponding ground cell (Waring et al., 1995). The reflected radar signal is proportional to the backscattering coefficient ($\sigma°$) of a given ground cell.

 RADAR Environmental Considerations

Having defined the radar backscatter coefficient, $\sigma°$, as a quantitative measure of the intensity of energy returned to the radar antenna from a specific area on the surface of the Earth, it is important to identify the environmental parameters within the resolution cell on the ground that are responsible for backscattering the incident energy.

Surface Roughness Characteristics

Surface roughness is the terrain property that strongly influences the strength of the radar backscatter. When interpreting aerial photography, we often use the terminology "rough" (coarse), "intermediate," or "smooth" (fine) to describe the surface texture characteristics (refer to Chapter 5). It is possible to extend this analogy to the interpretation of radar imagery if we keep in mind that the *surface roughness* we are talking about may be envisioned at the *microscale, mesoscale*, and/or *macroscale*.

Microscale surface roughness is usually measured in centimeters (i.e., the height of stones, size of leaves, or length of branches in a tree) and not in hundreds or thousands of meters as with topographic relief or mountains. The amount

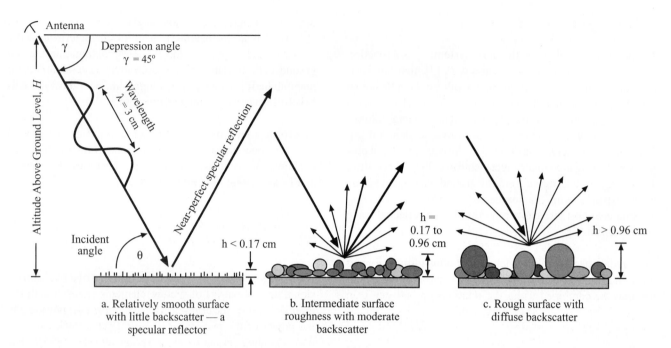

Figure 9-20   Expected surface roughness backscatter from terrain illuminated with 3 cm wavelength microwave energy. a) With a depression angle of 45° and a local relief of < 0.17 cm, the terrain represents a relatively smooth surface and should produce very little backscatter toward the sensor. This specular reflector will appear dark on the radar image. b) Local relief from 0.17 – 0.96 cm represents intermediate surface roughness and should produce a medium gray response on the radar image. c) When the local surface roughness relief is > 0.96 cm, a rough surface exists that will produce very diffuse backscatter. This diffuse reflector will produce a bright return on the radar image due to the large amount of microwave energy reflected back toward the antenna.

of microwave energy backscattered toward the sensor based on microscale components is a function of the relationship between the wavelength of the incident microwave energy (λ), the depression angle (γ), and the local height of objects (h in cm) found within the resolution cell being illuminated. We may use the *modified Rayleigh criteria* to predict what the Earth's surface will look like in a radar image if we know the microscale surface roughness characteristics and the radar system parameters (λ, γ, h) mentioned (Peake and Oliver, 1971). For example, one would expect that an area with *smooth surface roughness* would send back very little backscatter toward the antenna, i.e., it acts like a specular reflecting surface, where most of the energy bounces off the terrain away from the antenna (Figure 9-20a). The small amount of backscattered energy returned to the antenna is recorded and shows up as a dark area on the radar image. The quantitative expression of this *smooth criteria* is:

$$h < \frac{\lambda}{25 \sin \gamma}. \tag{9-22}$$

To illustrate, we will compute what the local height (h) of objects (e.g., grass and rocks in this example) must be in

order to produce a smooth (dark) radar return when an X-band ($\lambda = 3.0$ cm) radar is used with a depression angle (γ) of 45° (Figure 9-20a). Substituting these values into Equation 9-22 yields:

$$h < \frac{3 \, \text{cm}}{25 \sin 45°}$$

$$h < \frac{3 \, \text{cm}}{25 \times 0.70711}$$

$$h < 0.17 \, \text{cm} \, .$$

This means that a resolution cell uniformly filled with dry grass that has a surface height of < 0.17 cm should produce relatively little radar backscatter and will therefore be recorded on the radar image in a dark tone.

A bright return is expected if the following modified Rayleigh *rough criteria* are used:

Table 9-4.   Modified Rayleigh surface roughness criteria for three different radar systems at two different depression angles.

Surface Roughness Category	Aircraft K$_a$-band $\lambda = 0.86$ cm $\gamma = 45°$	Aircraft X-band $\lambda = 3$ cm $\gamma = 45°$	Seasat L-band $\lambda = 23.5$ cm $\gamma = 70°$
Smooth, cm	h < 0.048	h < 0.17	h < 1.00
Intermediate, cm	h = 0.048 to 0.276	h = 0.17 to 0.96	h = 1.00 to 5.68
Rough, cm	h > 0.276	h > 0.96	h > 5.68

$$h > \frac{\lambda}{4.4 \sin \gamma}. \qquad (9\text{-}23)$$

Substituting the same wavelength and depression angle into this equation yields:

$$h > \frac{3\,\text{cm}}{4.4 \sin 45°}$$

$$h > \frac{3\,\text{cm}}{4.4 \times 0.70711}$$

$$h > 0.96 \text{ cm}.$$

Therefore, if the uniformly distributed rocks in the resolution cell had a local relief of > 0.96 cm (Figure 9-20c), then a strong return from the terrain would be expected and would be recorded as a bright tone in the radar image.

If the terrain were composed of rocks with local relief between 0.17 and 0.96 cm, it would be considered to have an *intermediate surface roughness* for this combination of wavelength and depression angle and would produce an intermediate tone of gray on the radar image.

It is important to remember that the radar backscatter is dependent on the wavelength and the depression angle. For example, consider Table 9-4 where the modified Rayleigh criteria are computed for three different radar wavelengths ($\lambda = 0.86$, 3.0, and 23.5 cm) at two different depression angles ($\gamma = 45°$ and 70°). Terrain with a local relief of 0.5 cm would appear very bright on the K$_a$-band imagery and as an intermediate shade of gray on the X-band imagery. In this case the only parameter changed was the wavelength. If we changed both the wavelength and the depression angle to be that of the Seasat SAR (Table 9-4), we see that the same 0.5 cm local relief would produce a smooth (dark) return in this L-band imagery. The significance of this relationship is that the same terrain will appear differently in radar imagery as a function of the sensor's depression angle and wavelength (Henderson and Xia, 1998). Therefore, it is difficult to create "radar image interpretation keys" of selected phenomena. An analyst must constantly keep this in mind when interpreting radar imagery. Later we will see that aircraft or spacecraft look direction also impacts the backscattered energy.

Mikhail et al. (2001) suggest that the following criteria can also be used to predict whether a surface will yield a weak, intermediate, or strong radar return:

- A surface will produce a relatively weak return if its local relief is less than one-eight the incident radar wavelength. In this case much of the incident microwave energy is scattered or specularly reflected away from the antenna resulting in a weak return.

- An intermediate return is produced when the local relief ranges from one-eight to one-half the incident radar wavelength. These diffuse reflectors disperse energy at all angles and a moderate portion is backscattered toward the radar antenna.

- A strong radar return is produced when the local terrain relief is greater than one-half the incident radar wavelength.

The incident radar energy may be scattered according to both *mesoscale* and *macroscale surface roughness* criteria (Henderson and Lewis, 1998). As mentioned, the microscale roughness is a function of the size of leaves, twigs, and perhaps branches within an individual resolution cell. Conversely, mesoscale surface roughness would be a function of the characteristics within numerous resolution cells, perhaps covering the entire forest canopy. Based on this logic, we would expect forest canopies to generally show up with a more coarse texture than grasslands when recorded in the same image at the same depression angle and wavelength. Finally, *macroscale* roughness would be influenced greatly by how the entire forest canopy is situated on the hillside. Macroscale roughness is significantly influenced by the topographic slope and aspect of the terrain with the existence or absence of shadows playing a very important role in creating image surface roughness.

Electrical Characteristics (Complex Dielectric Constant) and the Relationship with Moisture Content

A radar sends out a pulse of microwave energy that interacts with the Earth's terrain. Different types of terrain conduct this electricity better than others. One measure of a material's electrical characteristics is the *complex dielectric constant,* defined as a measure of the ability of a material (vegetation, soil, rock, water, ice) to conduct electrical energy. Dry surface materials such as soil, rock, and even vegetation have dielectric constants from 3 to 8 in the microwave portion of the spectrum. Conversely, water has a dielectric constant of approximately 80. The most significant parameter influencing a material's dielectric constant is its *moisture content.* Therefore, the amount of moisture in a soil, on a rock surface, or within vegetative tissue may have a significant impact on the amount of backscattered radar energy (Mikhail et al., 2001).

Moist soils reflect more radar energy than dry soils, which absorb more of the radar wave, depending on the dielectric constant of the soil material. Radar images may be used to estimate *bare* ground soil moisture content when the terrain is devoid of most other material such as plants and rocks and has a uniform surface roughness. The amount of soil moisture influences how deep the incident electromagnetic energy penetrates into the material. If the soil has a high surface soil moisture content, then the incident energy will only penetrate a few centimeters into the soil column and be scattered more at the surface producing a stronger, brighter return.

The general rule of thumb for how far microwave energy will penetrate into a dry substance is that the penetration should be equal to the wavelength of the radar system. However, active microwave energy may penetrate extremely dry soil several meters. For example, Figure 9-21 depicts four views of a part of the Nile River, near the Fourth Cataract in the Sudan. The top image is a photograph taken by the crew of the Space Shuttle *Columbia* in November 1995. The three radar images were acquired by the Shuttle Imaging Radar C/X-band Synthetic Aperture Radar (SIR-C/X-SAR) onboard Space Shuttle *Endeavor* in April 1994; C-band HV, L-band HV, and L-band HH. Each radar image provides some unique information about the geomorphology of the area. The thick white band in the top right of the radar image is an ancient channel of the Nile that is now buried under layers of sand. This channel cannot be seen in the photograph, and its existence was not known before the radar imagery was processed. The area to the left in both images shows how the

a. Shuttle photograph.

b. SIR-C C-band HV.

c. SIR-C L-band HV.

d. SIR-C L-band HH.

Figure 9-21   Example of radar penetration of dry soil along the Nile River, Sudan. a) Black-and-white version of a color-infrared photograph acquired by Space Shuttle *Columbia* crew in November 1995; b–d) SIR-C/X-SAR images acquired by the Space Shuttle *Endeavor* in April 1994. Subtle, different information is recorded in each of the three radar images. Each reveals an ancient, previously unknown channel of the Nile. Radar brightness values are inverted in these examples (courtesy Jet Propulsion Lab).

Nile is forced to flow through a set of fractures that causes the river to break up into smaller channels, suggesting that the Nile has only recently established this course. The radar images have allowed scientists to develop new theories to explain the origin of the "Great Bend" of the Nile in the Sudan. Color Plate 9-1 is a color composite of the three radar images. It provides more revealing visual information about the regional geomorphology.

Ocean surfaces have a very high dielectric constant with most of the radar energy being reflected at the water's surface. The dielectric constant of snow depends upon how much liquid water is in the snow. Thus, microwave remote sensing can be used to determine snow water content. Similarly, there are differences in the dielectric constants of ice, depending upon age, degree of compaction, and type of ice. Microwave remote sensing has been useful for the extraction of these biophysical variables. Finally, healthy agricultural crops and forest canopy leaves often have relatively large surface areas and high moisture content (high relative turgidity). Therefore, it is not surprising that dense canopies of moist vegetation reflect radar energy well. In effect, a fully turgid vegetated forest canopy acts like a cloud of water droplets hovering above the surface of the Earth. Also, the steeper the depression angle, the greater the sensitivity of the radar to soil and vegetation moisture content.

Vegetation Response to Microwave Energy

Emphasis is being placed on quantitatively monitoring the spatial distribution, biomass, gross and net primary productivity, and condition of global vegetation communities, especially forests (approximately 33 percent of the Earth's land surface), semi-arid ecosystems of grassland/steppe/desert, agricultural land (10 percent), and wetlands. Scientists are interested in how energy and matter move within these vegetated ecosystems. Many of these vegetated areas are shrouded in perennial cloud cover. Synthetic aperture radar imagery may provide some of the following vegetation biophysical parameters (Carver, 1988):

- canopy water content;

- vegetation type;

- biomass by component (foliage, higher-order stems and main stem);

- canopy structure (including green leaf area index), leaf orientation, main stem (trunk) geometry and spatial distribution, stem, branch size, and angle distributions.

Any plant canopy (forest, agriculture, grassland, etc.) may be thought of as a seasonally dynamic three-dimensional water-bearing structure consisting of foliage components (leaves) and woody components (stems, trunk, stalks, and branches). Remote sensing systems such as the Landsat Thematic Mapper, SPOT, or aerial photography sense reflected optical wavelength energy measured in micrometers that is reflected, scattered, transmitted, and/or absorbed by the first few layers of leaves and stems of a healthy vegetation canopy. We typically get little information about the internal characteristics of the canopy, much less information about the surface soil characteristics lying below the canopy. Conversely, active microwave energy can penetrate the canopy to varying depths depending upon the frequency, polarization, and incident angle of the radar system. Microwave energy responds to objects in the plant's structure that are measured in centimeters and even decimeters. It is useful to identify the relationship between the canopy components and how they influence the radar backscattering.

If a radar sends a pulse of vertically or horizontally polarized microwave energy toward a stand of trees, it interacts with the components present and scatters some of the energy back toward the sensor. The amount of energy received is proportional to the nature of the energy sent (its frequency and polarization) and is dependent upon whether or not the canopy components depolarize the signal, how far the signal penetrates into the canopy, and whether it eventually interacts with the ground soil surface.

Penetration Depth and Polarization

Like-polarization backscatter toward the sensor results from single reflections from canopy components such as the leaves, stems, branches, and trunk. These returns are generally very strong and are recorded as bright signals in like-polarized radar imagery (HH or VV). This is often called canopy *surface scattering*. Conversely, if the energy is scattered multiple times within a diffuse volume such as a stand of pine trees (i.e., from a needle, to a stem, to the trunk, to a needle), the energy may become depolarized. This is often called *volume scattering*. A radar can measure the amount of depolarized volume scattering that takes place. For example, it is possible to configure a radar to send a vertically polarized pulse of energy. Some of this energy becomes depolarized in the canopy and exits toward the sensor in the horizontal domain. The depolarized energy may then be recorded by the sensor in VH mode – vertical send and horizontal receive.

Kasischke and Bourgeau-Chavez (1997) provide insight as to how the radar backscattering coefficient, $\sigma°$, from both woody (forested) and non-woody (e.g., brush, scrub, crops) environments is produced when the terrain is impacted by microwave energy.

To understand radar scattering from complex vegetation covers, it is necessary to think in terms of the different canopy layers affecting the radar signature. For wetlands containing shrubs and trees, there are three distinct layers to consider, as shown in Figure 9-22a: 1) a canopy layer that consists of small branches and foliage (leaves), 2) a trunk layer that consists of large branches and trunks or boles, and 3) a surface layer that may or may not be covered by water if wetland is present. For wetlands that do not contain woody plants, a simple two-layer model can be used (Figure 9-22b): 1) a canopy layer consisting of herbaceous vegetation, and 2) a surface layer that may be covered by water or soil. The backscattering coefficient exiting a woody vegetation canopy toward the radar system is σ°_w and can be expressed as (Wang et al., 1995; Dobson et al., 1995; Kasischke and Bourgeau-Chavez; 1997):

$$\sigma^\circ_w = \sigma^\circ_c + \tau^2_c \tau^2_t (\sigma^\circ_m + \sigma^\circ_t \, \sigma^\circ_s + \sigma^\circ_d) \qquad (9\text{-}24)$$

where

- σ°_c is the backscatter coefficient of the canopy layer of smaller woody branches and foliage (i.e., surface scattering);

- τ_c is the transmission coefficient of the vegetation canopy;

- τ_t is the transmission coefficient of the trunk layer;

- σ°_m is the multiple-path scattering between the ground and canopy layer;

- σ°_t is direct scattering from the tree trunks;

- σ°_s is direct surface backscatter from the ground;

- σ°_d is the double-bounce scattering between the trunks and the ground.

By eliminating all terms associated with the trunk layer, it is possible to determine the total radar-scattering coefficient from terrain with non-woody, herbaceous vegetation, σ°_h:

$$\sigma^\circ_h = \sigma^\circ_c + \tau^2_c (\sigma^\circ_s + \sigma^\circ_m). \qquad (9\text{-}25)$$

The terms in Equations 9-24 and 9-25 are dependent on 1) the type of vegetation present (which has an impact on sur-

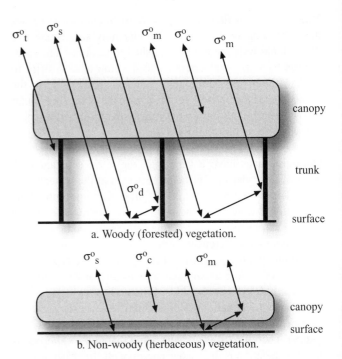

a. Woody (forested) vegetation.

b. Non-woody (herbaceous) vegetation.

Figure 9-22   Major sources of scattering from a) woody and b) herbaceous vegetation canopies where σ°_c is the backscatter coefficient of the crown layer of smaller woody branches and foliage (i.e., surface scattering), σ°_m is the multiple-path scattering between the ground and canopy layer, σ°_t is direct scattering from the tree trunks, σ°_s is direct surface backscatter from the ground, and σ°_d is the double-bounce scattering between the tree trunks and ground (adapted from Kasischke and Bourgeau-Chavez, 1997).

face roughness), 2) the wavelength and polarization of the incident microwave energy, 3) the dielectric constant of the vegetation, and 4) the dielectric constant of the ground surface. The scattering and attenuation in the equations are all directly proportional to the dielectric constant. Live vegetation, with a higher water content (turgidity) has a higher dielectric constant than drier or dead vegetation. The presence of dew or moisture acts to increase the dielectric constant of vegetated surfaces (Kasischke and Bourgeau-Chavez, 1997). Often, the primary quantity governing the attenuation coefficient of a vegetation canopy is the water content per unit volume, not necessarily the actual structure and geometry of the leaves, stems, and trunk of the plants.

The condition of the ground layer is also very important in microwave scattering from vegetation surfaces. There are two properties of this layer that are important, including: 1) the micro- and mesoscale surface roughness (relative to the radar wavelength previously discussed), and 2) the reflection coefficient. In general, a greater surface roughness 1) increases the amount of microwave energy backscattered (increasing $\sigma°_s$), and 2) decreases the amount of energy scattered in the forward direction (decreasing $\sigma°_m$ and $\sigma°_d$). The reflection coefficient is dependent on the dielectric constant (or conductivity) of the ground layer. A dry ground layer has a low dielectric constant and therefore has a low reflection coefficient. As soil moisture increases, so does the dielectric constant and, hence, the reflection coefficient. Given a constant surface roughness, as the soil dielectric constant increases, so does both the amount of backscattered and forward scattered microwave energy (resulting in increases in $\sigma°_m$, $\sigma°_s$, and $\sigma°_d$).

If there is a layer of water over the ground surface of a vegetated landscape such as in wetland environments, two things happen: 1) it eliminates any surface roughness, and 2) it significantly increases the reflection coefficient. In terms of microwave scattering, the elimination of any surface roughness means that all the energy is forward scattered, eliminating the surface backscattering term ($\sigma°_s$) in the equations; and, the increased forward scattering and higher reflection coefficient lead to significant increases in the ground-trunk and ground-canopy interaction terms $\sigma°_d$ and $\sigma°_m$ respectively (Kasischke and Bourgeau-Chavez, 1997).

Penetration Depth and Frequency

The longer the microwave wavelength, the greater the penetration into the plant canopy (ESA ASAR, 2006). For example, Figure 9-23 depicts the response of a hypothetical pine forest to microwave energy. Surface scattering takes place at the top of the canopy as the energy interacts with the leaves (or needles) and stems. Volume scattering by the leaves, stems, branches, and trunk takes place throughout the stand, and surface scattering can occur again at the soil surface. A comparison of the response of X-, C-, and L-band microwave energy incident to the same canopy is presented in Figure 9-24a–c. The shorter wavelength X-band (3 cm) energy is attenuated most by surface scattering at the top of the canopy by foliage and small branches. The C-band (5.8 cm) energy experiences surface scattering at the top of the canopy as well as some volume scattering in the heart of the stand. Little energy reaches the ground. L-band (23.5 cm) microwave energy penetrates farther into the canopy, where volume scattering among the leaves, stems, branches, and trunk cause the beam to become depolarized. Also, numerous pulses may be transmitted to the ground, where surface scattering from the soil-vegetation boundary layer may take place. Longer P-band radar (not shown) would afford the greatest penetration through the vegetation and mainly

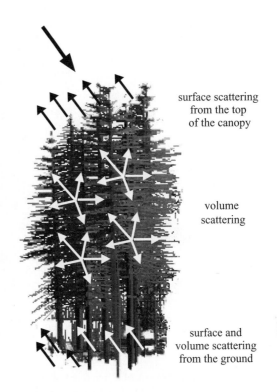

surface scattering
from the top
of the canopy

volume
scattering

surface and
volume scattering
from the ground

Figure 9-23   The types of active microwave surface and volume scattering that might take place in a hypothetical pine forest stand (after Carver, 1988).

reflect off large stems and the soil surface (Waring et al., 1995).

Radar Backscatter and Biomass

Radar backscatter increases approximately linearly with increasing biomass until it saturates at a biomass level that depends on the radar frequency. For example, Dobson et al. (1992) found that the biomass saturation level was about 200 tons/ha of Loblolly pine using P-band and 100 tons/ha for L-band, and that the C-band backscattering coefficient showed much less sensitivity to total aboveground biomass. Wang et al. (1994) evaluated Loblolly pine using ERS-1 SAR backscatter data. They also found that the C-band functioned poorly due to its high sensitivity to soil moisture and the steep local incident angle of the sensor (23°). Generally, backscatter at lower frequencies (P- and L-bands) is dominated by scattering processes involving the major woody biomass components (trunks and branches), while scattering at high frequencies (C- and X-bands) is dominated by scattering processes in the top crown layer of branches and foliage. Radar canopy measurements have also been found to be correlated with leaf-area-index (LAI) measurements (Franklin et al., 1994).

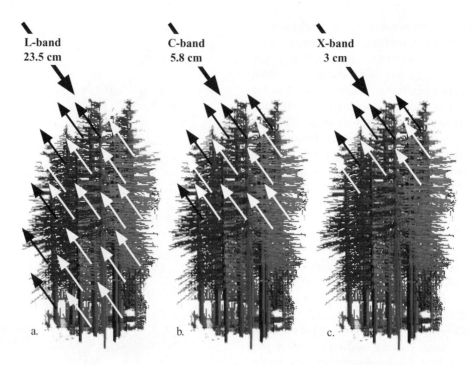

Figure 9-24   Theoretical response of a pine forest stand to X-, C-, and L-band microwave energy. The shorter the wavelength, the greater the contribution from *surface scattering*. The longer the wavelength, the greater the penetration into the material and the greater the *volume scattering*.

Some other general observations about SAR vegetation interpretation include

• Vertically polarized energy is highly attenuated by the vertically oriented canopy components (leaves, stems, branches, and trunk) while horizontally polarized energy is not.

• The brighter the return on like-polarized radar images (HH or VV), the greater the contribution from surface scattering.

• The brighter the return on cross-polarized images (HV or VH), the greater the contribution from volume (internal canopy) scattering.

• When the radar wavelength is approximately the same size as the canopy components (e.g., C- or X-band), substantial surface and volume scattering will take place and little energy may reach the ground. Consequently, shorter wavelength radars (2 – 6 cm) may be preferred when monitoring crop canopies and tree leaves. Longer wavelength radars (9 – 30 cm) exhibit substantial volume scattering as incident energy interacts with larger trunk and branch components. Surface scattering from the underlying soil may also occur, which can cause confusion.

• Cross-polarized images (HV or VH) are less sensitive to slope variations. This suggests vegetation monitoring in mountainous areas may best be performed using cross-polarization techniques. Also, the same row crop planted in different directions can produce like-polarized images that are difficult to interpret. This ambiguity may be reduced when cross-polarized images are available in addition to like-polarized images.

• One of the key variables necessary to model the hydrological cycle is how much water is being stored in vegetation canopies and the rate of evapotranspiration taking place. Generally, the more moisture in the structure, the greater the dielectric constant and the higher the radar backscatter return. Active microwave remote sensing is capable of sensing canopy (or leaf) water content in certain instances.

• Radar imagery can provide some information on landscape-ecology patch size and canopy gaps that are of value when monitoring ecosystem fragmentation and health (Sun and Ranson, 1995).

Shuttle Imaging Radar C- and X-Band SAR (SIR-C/X-SAR) images of the Amazon rain forest obtained on April 10, 1994, are displayed in black and white in Figure 9-25 and as a color composite in Color Plate 9-2. These images may be used to demonstrate several of the concepts mentioned. First, the multifrequency radar data reveals rapidly changing land-use patterns, and it also demonstrates the capability of the different radar frequencies to detect and penetrate heavy rainstorms. The top X-band image has VV polarization, the center C-band image is HV, and the bottom L-band image is HV. A heavy downpour in the lower center of the image appears as a black "cloud" in the X-band image, more faintly in the C-band image, and is invisible in the L-band image. When combined in the color composite, the rain cell appears red and yellow. Although radar can usually penetrate through clouds, short radar wavelength (high frequency), such as X- and C-band, can be changed by unusually heavy rain cells. L-band, at 23.5 cm (9 in.) wavelength, is relatively unaffected by such rain cells. Such information has been used to estimate rainfall rates (NASA JPL, 1996). Of course, there is very little backscatter from the river in all three radar images, causing it to appear black.

The area shown is in the state of Rondonia, in western Brazil. The pink areas in the color composite are pristine tropical rain forests, and the blue and green patches are areas where the forest has been cleared for agriculture. Radar imaging can be used to monitor not only the rain forest modification but also the rates of recovery of abandoned fields. Inspection of the black and white images reveals that as the wavelength of the radar progresses from X- to C- to L-band, it becomes easier to discriminate cleared land from rain forest. In the L-band black-and-white image, the cleared land appears relatively dark while the rain forest appears very bright. Evidently, the cleared land contains sufficient scatterers to cause it to appear bright at X- and C-band frequencies, while at L-band frequencies the incident energy to the cleared fields is reflected away from the radar receiver. The rain forest appears relatively bright in all images because of 1) the tremendous number of leaves, branches, stems, and trunks that can contribute to canopy surface scattering and volume scattering within the canopy (L-band penetrates the greatest distance into the canopy and is brighter), and 2) there is a high moisture content in the rain forest canopy, further increasing the amount of backscatter present in all three bands (NASA JPL, 1996).

Water Response to Microwave Energy

Flooding occurs in many areas, and cloud cover obscures the collection of data using optical and thermal sensors. Fortunately, the smooth surface of standing water reflects almost all the incident microwave radiation away from the sensor (unless there is a strong wind with lots of chop, which can

a. X-band VV

b. C-band HV

c. L-band HV

Figure 9-25 SIR-C/X-SAR images of a portion of Rondonia, Brazil, obtained on April 10, 1994. a) X-band image with VV polarization. b) C-band image is HV. c) L-band image is HV. A heavy rain in the lower center of the image appears as a black "cloud" in the X-band image, more faintly in the C-band image, and is invisible in the L-band image. L-band, at 24 cm (9 in) wavelength, is relatively unaffected by such rain cells. Also, the L-band image differentiates the pristine rain forest (bright) from the darker clear-cut areas because it penetrates farther into the canopy, experiencing greater volume scattering. Water is dark in all three bands. A color composite of the three images is found in Color Plate 9-2 (courtesy NASA Jet Propulsion Lab).

cause confusion), resulting in lower backscatter than a dry surface. When standing water is present under vegetation, a unique corner-reflection backscatter interaction between surface water and tree stems results in an extremely high backscatter and allows inundation to be clearly mapped (Waring et al., 1995). Ford et al. (1986) demonstrated how wetlands along the Savannah River appeared especially bright compared to standing water and other upland land cover such as Loblolly pine, due to this corner-reflector condition between the water and cypress-tupelo tree trunks.

Assessing Soil Moisture

L-band radar penetrates into bare, damp, smooth soil to a maximal depth of approximately 10 cm. Shorter wavelengths penetrate to only 1 – 3 cm. In agricultural fields that have smooth soil surfaces and biomass of less than 1 mg/ha, moisture content of surface layers can be determined fairly accurately (Wang et al., 1994). Once vegetation biomass exceeds a certain limit, the ability of radar to sense surface soil-water conditions decreases rapidly (Waring et al., 1995). Under a dense forest canopy, the amount of moisture held in the leaves is so large that it interferes with any direct assessment of soil-water status (Jackson and Schmugge, 1991).

Thus, the presence of vegetation over the soil surface can add considerable complexity to measuring soil moisture using microwave remote sensing techniques. Because the vegetation's transmissivity decreases with increasing microwave frequency, it is best to use the longest wavelength available for soil moisture mapping (Dobson and Ulaby, 1998). This would be P-band ($\lambda = 68$ cm) for aircraft SARs and L-band ($\lambda = 23$ cm) for satellite SARs. If the vegetation biomass is < 0.5 kg/m^2, the effect of the vegetation backscatter may be ignored for like-polarized L-band data, i.e., $\sigma°_{VV}$ and $\sigma°_{HH}$. Unfortunately, when the biomass is > 0.5 kg/m^2, it is not currently possible to disentangle the separate soil moisture and vegetation backscatter contributions.

Urban Structure Response to Microwave Energy

Urban buildings, cars, fences, bridges, etc., act as corner reflectors that send much of the incident energy back toward the antenna. This generally results in bright signatures for urban phenomena on the radar image. Unfortunately, this plethora of bright returns is often confusing in radar images of urban areas especially when 1) the cardinal effect takes place, and/or 2) the imagery is of relatively low spatial resolution.

SIR-C/X-SAR Image of Greater Los Angeles, CA

Figure 9-26   The cardinal effect is responsible for the bright signature of portions of Santa Monica and San Fernando in this SIR-C/X-SAR image of the Los Angeles basin obtained on October 3, 1994. The look direction is from the top to the bottom of the image (courtesy NASA Jet Propulsion Laboratory).

A SIR-C/X-SAR image of Los Angeles, CA, is found in Figure 9-26. It has approximately 30×30 m resolution. The radar look direction is from the top to the bottom of the image. Only the major terrain features can be resolved, including the freeways, major urban development, the Pacific Ocean, and the mountainous terrain. In general, it is not possible to distinguish between commercial and residential communities. Interestingly, there are some very bright polygonal areas in the image. These are produced by the radar *cardinal effect*. Early in radar remote sensing research it was noted that reflections from urban areas, often laid out according to the cardinal points of the compass, caused significantly larger returns when the linear features were illuminated by the radar energy at an angle orthogonal (perpendicular) to their orientation (Raney, 1998). For example, the residential communities of San Fernando and Santa Monica are for all practical purposes similar in nature to other residential communities in their vicinity (Figure 9-26). However, due to the cardinal effect caused by their unique orientation relative to the SIR-C sensor system at the time of data collection, they appear in the radar image to contain land cover that is dramatically different from the

a. Oblique photograph of the Pentagon.   b. Radar image of the Pentagon.

Figure 9-27   a) Low-oblique aerial photograph of the Pentagon in Washington, DC. b) Synthetic aperture radar image of the Pentagon (courtesy Federation of American Scientists).

other residential communities found throughout the Los Angeles basin.

It has been demonstrated repeatedly that two identical tracts of urban land (e.g., two single-family housing tracts) built at the same time, with the same lot size, and using the same materials will appear dramatically different from one another on radar imagery if one is laid out in a different orientation than the other, e.g., one is laid out with streets trending northeast and one is laid out with streets trending due north. Similarly, the regular spacing of agricultural row crops can produce a similar effect. Also, the same parcel of urban land may appear quite different on radar imagery acquired on two different dates if practically any of the system parameters are changed, especially look direction. This plays havoc when attempting to perform digital change detection using radar imagery.

RADARSAT, JERS-1, ERS-1,2, and Envisat ASAR with their relatively coarse spatial resolutions (Table 9-1) may be of value for obtaining general Level I land cover and land-use information. When attempting to extract Level II and III urban land cover and land-use information, optical remote sensor data is usually superior. However, if high spatial resolution radar is available (e.g., from RADARSAT-2), then detailed urban information may be extracted. For example, a photograph and a high-resolution radar image of the Pentagon are found in Figure 9-27. Henderson and Xia (1998) provide examples of how settlement patterns and socioeconomic information such as population estimates may be extracted from radar data.

 SAR Remote Sensing from Space

The following sections summarize the characteristics of the major satellite SAR sensor systems.

Seasat

Seasat was launched by NASA on June 26, 1978, and functioned for 105 days. It carried an L-band (23.5 cm) active microwave SAR at an altitude of 800 km. The antenna was 10.7×2.16 m in dimension. It collected HH-polarized data at an incident angle of 23°. It had a range resolution of 25 m and an azimuth resolution of 25 m. The swath width was 100 km. The data were processed with "4 looks." Seasat had an orbital repeat cycle of 17 days. The data were processed originally optically and then digitally (Evans et al., 2005).

Shuttle Imaging Radar SIR-A, SIR-B, SIR-C, and Shuttle Radar Topography Mission

Several important scientific radar instruments have been carried aboard NASA's Space Shuttle and operated for a number of days before returning to Earth. SIR-A and SIR-B were launched on November 12, 1981, and October 5, 1984, respectively, and were in orbit for 2.5 and 8 days. Both payloads consisted of an L-band (23.5 cm) SAR.

SIR-A had a 9.4×2.16 m antenna with HH polarization. The incident angle was 50°. The sensor had a range and azimuth resolution of 40 m with 6 looks. The swath width was 50 km. The Shuttle was in orbit at 260 km above the Earth. The data were processed optically.

SIR-B had a 10.7 × 2.16 m antenna with HH polarization. It had an incident angle of 15 – 64°. Its azimuth resolution was 17 – 58 m and its range resolution was 25 m with 4 looks. The swath width was 10 – 60 km. It operated at 225 and 350 km above the Earth. The data were processed both optically and digitally.

SIR-C was a significant breakthrough in radar remote sensing. It was a joint project between the United States (NASA JPL) and a consortium of European groups. SIR-C carried aloft a three frequency SAR: X-band (3 cm), C-band (5.8 cm), and L-band (23.5 cm). The three antennas were placed on a common platform in the Shuttle bay. The L- and C-bands had quad polarization (i.e., HH, HV, VV, and VH) while the X-band had VV polarization. The incident angle was from 15 – 55°. The range resolution was 10 – 30 m, and the azimuth resolution was 30 m with approximately 4 looks. The swath width was 15 – 90 km. It was flown at 225 km above the Earth. The data were processed digitally. *This was the first true multifrequency, multipolarization space-borne SAR.* Data from all three bands are excellent and widely available through JPL and other locations. Several examples of SIR-C data are found in this chapter and in the color plate section.

The Space Shuttle Radar Topography Mission (SRTM) was launched on February 11, 2000, and lasted 11 days. It used C-band and X-band interferometric synthetic aperture radars. SRTM characteristics are discussed in the section on radar interferometry.

RADARSAT and RADARSAT-2

RADARSAT was launched by the Canadian government on November 4, 1995, into a near-polar, Sun-synchronous orbit 798 km above the Earth. It has a dawn-to-dusk orbit, meaning it crosses the equator at dawn and dusk (approximately 6:00 a.m. and p.m. local time) and is rarely in eclipse or darkness. The orbital inclination is 98.6° with a period of 100.7 minutes and 14 orbits per day. It has a single C-band (5.6 cm) active microwave sensor that transmits at 5.3 GHz frequency at a pulse length of 42.0 μs. The antenna size is 15 × 1.5 m. Its polarization is horizontal-send and horizontal-receive (HH).

Unlike many of the other radars, RADARSAT provides a range of spatial resolutions and geographic coverages. There are seven image sizes, termed *beam modes,* summarized in Figure 9-28 and Table 9-5. Each beam mode is defined by the geographic area it covers and the spatial resolution. The beam modes range from *Fine,* which covers a 50 × 50 km area and has a 10 × 10 m spatial resolution, to *ScanSAR Wide,* which covers a 500 × 500 km area at 100 × 100 m spatial resolution.

RADARSAT obtains data using a range of incident angles from less than 20° (steep angle) to almost 60° (shallow angle). Within each beam mode, a number of incident angle positions are available (Figure 9-28 and Table 9-5). These are called *beam positions.* For example, Standard beam mode, which covers a 100 × 100 km area, has seven beam positions. By specifying a beam position, one of seven 100 × 100 km images within a 500 km accessible swath will be collected. Factors influencing the choice of beam mode include the sensitivity of the application to incident angle, type of terrain, stereo requirements, spatial resolution desired, and how often coverage of the area is required.

RADARSAT's orbit has a 24-day cycle, meaning it returns to the same location every 24 days. However, it can be pointed to provide a more frequent revisit cycle. The user also has the option of collecting imagery based on two different look directions. As RADARSAT descends from the North Pole (a descending orbital pass), it views the Earth in a westerly direction. As it ascends from the South Pole (an ascending orbital pass) it views the Earth in an easterly direction. This can be very useful when working in areas with high relief, when we are interested in highlighting features with a particular orientation, and/or when the study requires imagery acquired in the early morning or early evening.

RADARSAT-2 is to be launched in 2006 or 2007 (MDA, 2006a). It is has many specifications that are identical to RADARSAT, however, there are some very significant improvements that make RADARSAT-2 the premier earth observation active microwave remote sensing system. First, it is very important to maintain continuity in a series of dedicated satellite remote sensing systems. This allows mature applications to use the data produced by the new system. Major considerations designed to maintain continuity with RADARSAT-1 include the use of the same altitude (798 km) and orbital path, and all of the C-band imaging modes.

Improved RADARSAT-2 specifications include (MDA, 2006ab):

• < 1 s delay to switch between imaging modes.

Figure 9-28   The beam modes and incident angle options for acquiring RADARSAT data. The size of the footprints provide information
on the spatial resolution. Refer to Table 9-5 for specifications (after MDA, 2006ab).

- Full polarization imaging modes (HH, VV, VH, HV).

- A sensor that can look both left and right of nadir, which is
very important when trying to obtain data in an
emergency.

- *Ultrafine beam mode* (3.0 × 2.4 – 3.4 m; azimuth and
range, respectively), which provides the highest spatial
resolution of any commercial satellite SAR (Table 9-5).
Ground swath coverage is limited to 20 × 20 km to keep
the data rate within the recorder limits. The set of ultra-
fine beams cover any area within the incidence angle
range of 30 to 40°. Images are generated in a selectable,
single polarization. Improved object detection and
recognition enables a variety of new applications
including defense surveillance and urban mapping at
scales of 1:20,000 (Van der Sanden and Thomas, 2004).

- *Multi-look fine beam mode*, which obtains data in the same
swath locations as the Ultra-fine beam mode with a spatial
resolution of (7.9 × 7.4 – 9.1 m; azimuth and range,
respectively). However, these data have improved
radiometric resolution because of multiple looks. The
swath width is 50 × 50 km (Table 9-5). Data are collected
in the incidence angle range from 30 to 50°. Multi-look
fine-beam mode images are generated in a single
polarization.

- Onboard GPS receivers that monitor spacecraft position,
which yields more accurate geometric rectification of the
radar imagery without the need for ground-control points
(GCP).

- First commercial spaceborne SAR to offer quadrature-
polarimetry ("quad-pol") capability where *amplitude* and

Table 9-5.   RADARSAT and RADARSAT-2 beam position characteristics (MDA, 2006ab).

Operational Beam Modes	Beam Position	Incident Angle Positions (Degrees)	Nominal Spatial Resolution (m) [azimuth × range]	Nominal area (km)	Number of Processing Looks
Ultrafine beam and **Multiple-look fine beam** (only on RADARSAT-2)	—	30 – 40 30 – 50	3.0 × 2.4 - 3.4 7.9 × 7.4 - 9.1	20 × 20 50 × 50	1 × 1 >3
Fine (5 positions)	F1 F2 F3 F4 F5	37 – 40 39 – 42 41 – 44 43 – 46 45 – 48	10	50 × 50	1
Standard (7 positions)	S1 S2 S3 S4 S5 S6 S7	20 – 27 24 – 31 30 – 37 34 – 40 36 – 42 41 – 46 45 – 49	30	100 × 100	1 × 4
Wide (3 positions)	W1 W2 W3	20 – 31 31 – 39 39 – 45	30	165 × 165 150 × 150 130 × 130	1 × 4
ScanSAR Narrow (2 positions)	SN1 SN2	20 – 40 31 – 46	50	300 × 300	2 × 2
ScanSAR Wide	SW1	20 – 49	100	500 × 500	2 × 4
Extended High (6 positions)	H1 H2 H3 H4 H5 H6	49 – 52 50 – 53 52 – 55 54 – 57 56 – 58 57 – 59	25	75 × 75	1 × 4
Extended Low	L1	10-23	35	170 × 170	1 × 4

phase information are retained. There are a number of unique polarimetric parameters that can be derived from these data and used to develop an understanding of the features present in the radar imagery (e.g., co-polarized and cross-polarized ratio images; co-polarized correlation coefficient images). Table 9-6 summarizes some of these parameters and the expected response for a specific type of surface (MDA, 2006b).

European Space Agency ERS-1 and ERS-2

The European Space Agency launched the ERS-1 on July 16, 1991. It has a C-band (5.6 cm) SAR with a 10 × 1 m

antenna and VV polarization. It functions with an incident angle of 23°. The range resolution is 26 m and the azimuth resolution is 30 m with 6 looks. It has a swath width of 100 km. It is in orbit approximately 785 km above the Earth. The data are processed digitally. An identical ERS-2 was launched in 1995. ERS-1 and ERS-2 have on occasion been operated in tandem to provide image pairs for SAR interferometry research.

European Space Agency Envisat ASAR

Launched in March 1, 2002, Envisat is the largest Earth Observation spacecraft ever built. It carries 10 sophisticated

Table 9-6.   Several RADARSAT-2 polarimetric parameters and interpretation guidelines (MDA, 2006b).

Parameter	Co- or like-polarized ratio	Cross-polarized ratio	Co-polarized correlation coefficient
Ultra-fine	$\dfrac{\left\|S_{hh}\right\|^2}{\left\|S_{vv}\right\|^2}$	$\dfrac{\left\|S_{hv}\right\|^2}{\left\|S_{vv}\right\|^2}$	$\dfrac{S_{hh}\cdot S_{vv}}{\sqrt{\left(\left\|S_{hh}\right\|^2\right)\cdot\left(\left\|S_{vv}\right\|^2\right)}}$
Use	Measure surface roughness	Discrimination between surface and volume scattering	Detection of depolarization and double bounce
Surface scatter	Low for smooth surfaces tending to 1 as roughness increases	Very low for smooth surfaces, increases with surface roughness	High-amplitude (low depolarization), small phase difference
Volume scatter	approximately 1	Relatively high	Low-amplitude (high depolarization), poorly defined phase difference
Dihedral (double bounce)	> 1	Low	Phase difference close to 180°

Table 9-7.   ENVISAT Advanced Synthetic Aperture Radar (ASAR) specifications (ESA ASAR, 2006b).

ASAR Mode	Characteristics
Image Mode (IM)	30×30 m spatial resolution for the precision product. VV or HH polarization images from any of 7 selectable swaths. Swath width is between 56 km (swath 7) and 100 km (swath 1) across-track.
Alternating Polarization Mode (AP)	30×30 m spatial resolution for the precision product. Two co-registered images per acquisition, from any of 7 selectable swaths. Polarization pairs possible: HH/VV, HH/HV, or VV/VH.
Wide Swath Mode (WS)	150×150 m spatial resolution. Swath width is 400 km. Polarization: VV or HH.
Global Monitoring Mode (GM)	Spatial resolution of 1000×1000 m (azimuth and range, respectively) for nominal product. Up to a full orbit of coverage. HH or VV polarization.
Wave Mode (WV)	Small imagette (dimensions range between 10×5 km to 5×5 km) is acquired at regular intervals of 100 km along-track. Imagette can be positioned anywhere in an Image Mode swath. Polarization: VV or HH. Imagettes are converted to wave spectra for ocean monitoring.

optical and radar instruments that provide continuous observation and monitoring of the Earth's land, atmosphere, oceans and ice caps (Table 9-7). The Advanced Synthetic Aperture Radar (ASAR) instrument on board ENVISAT extends the mission of the Active Microwave Instrument (AMI) Synthetic Aperture Radar (SAR) instruments flown on the ERS-1 and ERS-2 satellites.

ASAR uses an active phased-array antenna, with incidence angles between 15 and 45°. Applications for this sensor include the study of ocean waves, sea-ice extent and motion, and land-surface studies, such as deforestation and desertification. In contrast to the ERS-1 and -2 SARs, which had a fixed swath position (23° mid-swath incidence angle), ASAR Image Mode provides data acquisition in seven dif-

ferent swath positions (i.e., IS1 to IS7), giving incidence angles ranging from 15° to 45° (Table 9-1). One significant advantage of higher incidence angles is that terrain distortion is reduced.

JERS-1

The National Space Development Agency (NASDA) of Japan launched the Japanese Earth Resource Satellite (JERS-1) on February 11, 1992. It was very similar to the original Seasat. It had an L-band (23.5 cm) SAR with an 11.9×2.4 m antenna and HH polarization. The incident angle was 39°. Range and azimuth resolution were both 18 m with 3 looks. The swath width was 75 km. The SAR orbited at approximately 568 km above the Earth. The data were processed digitally. JERS-1 operation was terminated on October 12, 1998.

Almaz-1

This SAR was placed in orbit by the former Soviet Union on March 31, 1991, and functioned for 1.5 years. It consisted of an S-band (9.6 cm) SAR with a 1.5×15 m antenna and HH polarization. The incident angle ranged from $30 - 60°$. Range resolution was $15 - 30$ m, and azimuth resolution was 15 m with greater than 4 looks. The swath width was $20 - 45$ km. It was placed in a 300 km orbit above the Earth. The data were processed digitally.

 ### Radar Interferometry

The previous discussion of synthetic aperture radar systems was primarily concerned with collecting a single image of the terrain. It is possible to acquire multiple SAR images of the terrain from aircraft or spacecraft to extract valuable three-dimensional and velocity information. *Imaging radar interferometry* is the process whereby radar images of the same location on the ground are recorded by antennas at 1) different locations or 2) different times (Madsen and Zebker, 1998; Rabus et al., 2003; Hodgson et al., 2003). Analysis of the resulting two interferograms allow very precise measurements of the range to any specific *x,y,z* point found in each image of the interferometric pair. The precision may be at the sub-wavelength scale.

Interferometric Topographic Mapping

Topographic mapping based on SAR interferometry relies on acquiring data from two different look angles and assumes that the scene imaged did not move between data acquisitions. The two measurements can be from two radars placed on the same platform separated by a few meters. This is called *single-pass interferometry*. The first single-pass interferometric SAR was the Shuttle Radar Topography Mission (SRTM) launched on February 11, 2000. It had a C-band and an X-band antenna in the cargo bay and a C-band and an X-band antenna at the end of a 60 m mast (Figure 9-29a). Interferometry may also be conducted using a single radar that obtains two measurements on different orbital tracks that are closely spaced but a day or so apart. This is the methodology used for the SIR-C and ERS-1,2 interfer-

ometry and is called *multiple-pass* or repeat-pass interferometry.

In many instances, interferometric SAR data can provide topographic information (*x,y,z*) that is just as accurate as digital elevation models derived using traditional optical photogrammetric techniques. However, interferometry can operate through clouds, day or night. This is important for cloud-shrouded tropical or arctic environments or when disasters strike and one cannot wait for an atmospheric window to obtain optical (photographic) data. Digital elevation information derived from SIR-C interferometric data are displayed in Color Plates 9-3 and 9-4. The cartographic community is especially interested in interferometric SAR.

Interferometric SAR obtains digital elevation information in the following manner (JPL, 1999b). First, the two radar images must be precisely registered. Then their geometry is such that the two radars and any object on the ground form a triangle (Figure 9-29b). If we know the distance (range) from each radar to the object on the ground (r_1 and r_2), the distance between the two radars (the baseline *B*) and the angle of that baseline, α (with respect to the horizontal), we can use trigonometry (the cosine rule) to calculate the height, *h*, of one of the radars above the position of the object on the ground. From Figure 9-29b we know that

$$h = r_1 \cos\theta \qquad (9\text{-}26)$$

and by the cosine rule,

$$(r_2)^2 = (r_1)^2 + B^2 - 2r_1 B \cos(90 + \theta - \alpha) \qquad (9\text{-}27)$$

which is the same as

$$(r_2)^2 = (r_1)^2 + B^2 - 2r_1 B \sin(\theta - \alpha). \qquad (9\text{-}28)$$

We solve for θ, then for *h* (using $h = r_1 \cos\theta$). This calculation is repeated for every point on the ground within the image. If we determine the precise height of one of the radars above sea level, we can produce a map of the surface heights. It turns out that we can only precisely measure the relative distance ($r_2 - r_1$) from the phase difference between each pair of radar measurements. This relative distance can be related to the height, *h*, after some additional algebra (SIR-CED home page; JPL, 1999b).

The Space Shuttle Radar Topography Mission (SRTM) used C-band and X-band interferometric synthetic aperture radars (IFSARs) to acquire topographic data over 80 percent of the Earth's landmass between 60° N and 56° S during its 11-day

Interferometry

Shuttle Radar Topography Mission (SRTM)

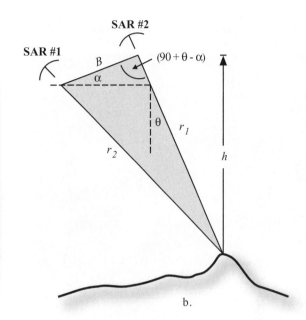

a.

b.

Figure 9-29   a) Location of the C-band and X-band antenna pairs on the space shuttle *Endeavor* used to collect interferometric data for the Shuttle Radar Topography Mission (SRTM). b) The geometric relationship between two SAR systems used for interferometry to extract topographic information (after NASA JPL, 1999b).

mission (NASA SRTM, 2006). NASA JPL processed the C-band interferometric data while the German Aerospace Center processed the X-band interferometric data. The topographic surfaces produced meet Interferometric Terrain Height Data (ITHD-2) specifications. SRTM mapped 99.97% of the targeted land at least once (119,000,000 km²; 46,000,000 mi²). Thus, the first worldwide collection of digital elevation data was acquired using interferometry, not photogrammetry. The interferometer pairs can also be of significant value for many Earth resource applications. Rodriguez et al. (2006) assessed the accuracy of SRTM-derived elevation products.

An example of a SRTM-derived digital elevation model of the Wasatch front in Utah is shown in Figure 9-30a. It is viewed in perspective and is draped with a single band of Landsat TM data. Color Plate 9-5 depicts the SRTM-derived data of the Wasatch front as an anaglyph and as a stereopair.

Prior to the launch of the SRTM, *x,y,z* data were compiled from a variety of *in situ* and remote sensing sources to produce global DEMs with a spatial resolution of 30 arc sec (1 × 1 km) (e.g., GLOBE and GTOPO30). DEMS of 1 arc sec (30 m) or 3 arc sec (90 m) were still lacking for many parts of the world (Rabus et al., 2003). SRTM-derived DEMs are now available for much of the planet at a spatial resolution of 3 arc-sec (90 m). However, a considerable amount of data

have been processed to 30 m. The data can be ordered through the NASA Jet Propulsion Laboratory.

GTOPO30 data are maintained by the USGS. A comparison of a GTOPO30 DEM and a SRTM-derived DEM of Mount Kilimanjaro in Tanzania and Kenya is shown in Color Plate 9-6. Note how much more detail is available in the SRTM-derived dataset for more elevation, slope, and aspect analysis as well as drainage and geomorphic investigations (Jensen et al., 2002).

Private commercial companies also collect interferometric synthetic aperture radar (IFSAR) data. One of the most widely used is the Intermap X-band Star 3*i* system, which generates high-quality 3 × 3 m X-band microwave imagery plus a detailed digital elevation model of the terrain. A good example is shown in Figure 9-31 with the X-band image of Bachelor Mountain, CA, on the left and the digital elevation model on the right. Of course, the radar image and the DEM are registered. Such data are very valuable for land-use and land-cover analysis as well as watershed hydrology studies.

Interferometric Velocity Mapping

If the look angles of multiple data acquisitions are held constant, there is no sensitivity to topography, and the interferometry can be used to extract information about things that

Digital Elevation Model of the Wasatch Front in Utah Derived from Shuttle Radar Topography Mission (SRTM) Data Draped with Landsat Thematic Mapper Imagery

Figure 9-30   Perspective view of an SRTM-derived DEM of the Wasatch Front in Utah draped with Landsat TM band 3 data. The C-band SRTM data were obtained February 12, 2000 (images courtesy NASA Jet Propulsion Lab).

have changed in the scene. Quantitative information about the velocity of objects that moved between the two observations may be made. Interferometry has been successfully applied to measuring movement along fault lines, measuring seismic displacement due to earthquakes, mapping glacier velocity, monitoring ocean currents, and measuring wave spectra. In addition, interferometry can be used to determine if man-made objects in the scene have moved. This is very powerful for change detection purposes.

 Passive Microwave Remote Sensing

There is a great interest in the measurement of passive microwave energy to monitor some of the more important global hydrologic variables such as soil moisture, precipitation, ice water content, and sea-surface temperature. In fact, several sensors onboard NASA's *Aqua* satellite launched on May 4, 2002, include specialized passive microwave radiometers (NASA AMSR-E, 2006).

As discussed in Chapter 2, the Earth approximates a 300 K blackbody with a dominant wavelength of approximately 9.7 μm. While the dominant wavelength may be 9.7 μm, a continuum of energy is emitted from the Earth and the atmosphere. In fact, the Earth passively emits a steady stream of microwave energy. The only difference is that 1) the Earth's materials do not emit a tremendous amount of passive microwave energy, and 2) what energy it does emit is relatively weak in intensity due to its long wavelength. Nevertheless, a suite of radiometers have been developed that can record subtle, passive microwave energy. The instruments measure the *brightness temperature* of the terrain or the atmosphere and its constituents.

Passive Microwave Radiometers

Passive microwave remote sensing devices may be 1) profiling radiometers or 2) scanning radiometers. A profiling radiometer simply stares at the terrain directly beneath the craft at nadir (or off-nadir if desired) and records the radiance from within the instantaneous field of view of the sensor. The output is a profile of the microwave brightness temperature recorded as the aircraft or spacecraft moves forward. A scanning microwave radiometer is much like the scanning thermal infrared radiometer discussed in Chapter 8. It collects data across-track as the craft moves forward. The result is a matrix of brightness temperature values that can be used to construct a passive microwave image.

Passive microwave radiometers generally record energy in the region between 0.15 and 30 cm (between 1 and 200 GHz), well beyond the thermal infrared region (3 – 14 μm). The microwave frequencies most commonly used are centered at 1, 4, 6, 10, 18, 21, 37, 50, 85, 157, and 183 GHz. This means that it is theoretically possible to acquire multispectral passive microwave imagery. The actual bandwidths (range of frequencies) recorded are usually fairly broad so that enough passive microwave energy is available to be recorded by the antenna. Similarly, the spatial resolution of passive microwave radiometers is usually large so that sufficient energy is collected within the IFOV to be recorded by the antenna. Aircraft sensors flying closer to the ground may have spatial resolutions measured in meters while most satellite passive microwave scanning radiometers have a spatial resolution measured in kilometers. The sensor is actually a large antenna sensitive to passive microwave energy.

Special Sensor Microwave/Imager (SSM/I)

One of the first passive microwave sensors was the Special Sensor Microwave/Imager (SSM/I) onboard the Defense

a. Intermap Star 3i X-band image.   b. Intermap digital terrain model derived from IFSAR data.

Figure 9-31   a) Intermap X-band Star 3*i* orthorectified image of Bachelor Mountain, CA. b) Digital elevation model of the same area derived using interferometric synthetic aperture radar (IFSAR) techniques (orthorectified imagery and Global Terrain Digital
Elevation Model provided by Intermap Technologies, Inc.).

Meteorological Satellite Program (DMSP) satellites since
1987. The Department of Defense also releases the data to
the scientific community (Figure 9-32). The SSM/I is a four-
frequency, linearly polarized passive microwave radiometric
system that measures atmospheric, ocean and terrain microwave brightness temperatures at 19.3, 22.3, 36.5, and 85.5
GHz (Table 9-8). The SSM/I rotates continuously about an
axis parallel to the local spacecraft vertical and measures the
upwelling scene brightness temperatures. It is calibrated
using cold sky radiation and a hot reference absorber. The
swath is approximately 1400 km. The data are converted
into sensor counts and transmitted to the National Environmental Satellite, Data, and Information Service (NESDIS).
The SSM/I is an excellent sensor for measuring the brightness temperature of very large regions. For example, a SSM/
I passive microwave image of almost the entire Amazon
Basin is shown in Figure 9-32.

NOAA has developed a SSM/I rainfall algorithm that utilizes the 85.5 GHz channel to detect the scattering of
upwelling radiation by precipitation-size ice particles within
the rain layer. The scattering technique is applicable over the
land and the ocean. Rain rates can be derived indirectly,
based on the relationship between the amount of ice in the
rain layer and the actual rainfall on the surface. In addition,
a scattering-based global land rainfall algorithm has been
developed. Monthly rainfall at 100×100 km and 250×250

km grids have been produced for the period from July, 1987
to the present (Li et al., 1998; NSIDC SSM/I, 2006).

TRMM Microwave Imager (TMI)

The Tropical Rainfall Measuring Mission (TRMM) is sponsored by NASA and the National Space Development
Agency (NASDA) of Japan to study the tropical rainfall and
the associated release of energy that helps to power global
atmospheric circulation (NASA TRMM, 2006). It carries
five instruments and was launched on November 27, 1997.
The TRMM Microwave Imager (TMI) is a passive microwave sensor designed to provide quantitative rainfall information over a 487 mile (780 km) swath. It is based on the
design of the SSM/I. It measures the intensity of radiation at
five frequencies: 10.7 (45 km spatial resolution), 19.4, 21.3,
37, and 85.5 GHz (5 km spatial resolution). Dual polarization at four frequencies provides nine channels. The 10.7
GHz frequency provides a more linear response for the high
rainfall rates common in tropical rainfall.

Calculating the rainfall rates from both the SSM/I and TMI
sensors requires complicated calculations because water
bodies such as oceans and lakes emit only about one-half the
energy specified by Planck's radiation law at microwave frequencies. Therefore, they appear to have only about half
their actual temperature at the surface and appear very

Table 9-8.　Comparative operating characteristics of SSM/I, AMSR-E on *Aqua,* and AMSR on *ADEOS-2* (NSIDC, 2006).

Parameter	SSM/I (DMSP F8, F10, F11, F13)	AMSR-E on *Aqua*	AMSR on *ADEOS-2*
Time period	7-09-87 to present	5-04-02 to present	12-14-02 to present
Frequencies	19.3, 22.3, 36.5, 85.5	6.925, 10.65, 18.7, 23.8, 36.5, 89.0	6.925, 10.65, 18.7, 23.8, 36.5, 50.3, 52.8, 89.0
Footprint sizes (km)	37 × 28 (37 GHz) 15 × 13 (85.5 GHz)	74 × 43 (6.9 GHz) 14 × 8 (36.5 GHz) 6 × 4 (89.0 GHz)	74 × 43 (6.9 GHz) 14 × 8 (36.5 GHz) 6 × 4 (89.0 GHz)

"cold" to a passive microwave radiometer. Fortunately, raindrops appear to have a temperature that equals their real temperature and appear "warm" or bright to a passive microwave radiometer. The more raindrops, the warmer the whole scene appears. Research over the last three decades has made it possible to obtain relatively accurate rainfall rates based on the temperature of the passive microwave scene.

Land is very different from oceans in that it emits about 90 percent of its real temperature at microwave frequencies. This reduces the contrast between the rain droplets and the land. Fortunately, high-frequency microwaves (85.5 GHz) are strongly scattered by ice present in many raining clouds. This reduces the microwave signal of the rain at the satellite and provides a contrast with the warm land background, allowing accurate rainfall rates to be computed over land as well. An example of rainfall measurement using the TRMM Microwave Imager is presented in Chapter 12: Remote Sensing of Water.

Advanced Microwave Scanning Radiometer (AMSR-E)

The Advanced Microwave Scanning Radiometer (AMSR-E) is one of six sensors onboard *Aqua* (NASA AMSR-E, 2006). AMSR-E is a cooperative effort between NASA and the National Space Development Agency of Japan. AMSR-E was modified for *Aqua* based on the design of AMSR on the Japanese *ADEOS-2* satellite.

AMSR-E is flown in a polar, Sun-synchronous orbit. It is a six frequency passive microwave radiometer that measures frequencies at 6.925, 10.65, 18.7, 23.8, 36.5, and 89 (HV polarization). It has a mean spatial resolution of 5.4 km at 89 GHz and 56 km at 6.9 GHz (Table 9-8) with a 1445 km swath width. The Earth-emitted microwave radiation is col-

SSM/I Passive Microwave Image of the Amazon Basin

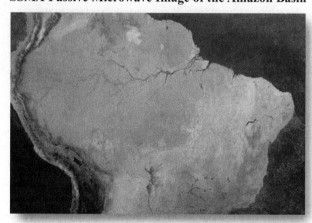

Figure 9-32　SSM/I passive microwave radiometer image of the Amazon Basin obtained at a frequency of 85 GHz with vertical polarization (courtesy DMSP).

lected by a parabolic reflector 1.6 m in diameter that scans the Earth maintaining a constant Earth incidence angle of 55°. Six data collection feedhorns focus the passive microwave radiation onto the six radiometers. Calibration is accomplished by observing cosmic background radiation and an onboard warm target.

AMSR-E measures total water-vapor content, total liquid-water content, precipitation, snow-water equivalent, soil moisture (using the 6.925 and 10.65 GHz frequencies), sea-surface temperature (SST), sea-surface wind speed, and sea-ice extent (NSIDC, 2006). Examples of AMSR-E derived sea-surface temperature (SST) information are provided in Color Plate 9-7. The SST surrounding Hurricane Katrina August 25–27, 2005, is shown in Color Plate 9-7b.

References

Carver, K. R., 1988, SAR *Synthetic Aperture Radar – Earth Observing System*, Vol. IIf, Washington: NASA Instrument Report, 233 p.

Carver, K. R., E. Elachi and F. T. Ulaby, 1985, "Microwave Remote Sensing from Space," *Proceedings of the IEEE*, 7(6):970–996.

Dobson, M. C. and F. T. Ulaby, 1998, "Mapping Soil Moisture Distribution with Imaging Radar," Lewis and Henderson (Eds.), *Principles and Applications of Imaging Radar*, NY: John Wiley, 407–433.

Dobson, M. C., F. T. Ulaby, T. LeToan, A. Beaudoin, E. S. Kasischke and N. Christensen, 1992, "Dependence of Radar Backscatter on Coniferous Forest Biomass," *IEEE Transactions on Geoscience and Remote Sensing*, 30(2):412–415.

Dobson, M. C., F. T. Ulaby, L. E. Pierce, T. L. Sharik, K. M. Bergen, J. Kellndorfer, J. R. Kendra, E. Li and Y. C. Lin, 1995, "Estimation of Forest Biomass Characteristics in Northern Michigan with SIR-C/S-SAR Data," *IEEE Transactions on Geoscience and Remote Sensing*, 33:877–894.

ESA ASAR, 2006, *ASAR User's Guide*, European Space Agency, http://envisat.esa.int/dataproducts/asar/CNTR3-1-2.htm.

Evans, D. L and 12 co-authors, 2005, "Seasat – 25 Year Legacy of Success," *Remote Sensing of Environment*, 94:384–404.

Flood, M. and B. Gutelius, 1997, "Commercial Implications of Topographic Terrain Mapping Using Scanning Airborne Laser Radar," *Photogrammetric Engineering and Remote Sensing*, 63:327.

Ford, J. P., R. G. Blom, M. L. Bryan, M. K. Daily, T. H. Dixon, C. Elachi and E. C. Xenos, 1980, *Seasat Views North America the Caribbean and Western Europe with Imaging Radar*, Publication #80 - 67, Pasadena: NASA Jet Propulsion Lab, 1–6 p.

Ford, J. P., J. B. Cimino and C. Elachi, 1983, *Space Shuttle Columbia View the World with Imaging Radar: the SIR-A Experiment*, Pasadena: NASA Jet Propulsion Lab, 179 p.

Ford, J. P., J. B. Cimino, B. Holt and M. R. Ruzek, 1986, *Shuttle Imaging Radar Views the Earth from Challenger: SIR-B Experiment,* Pasadena: NASA Jet Propulsion Lab, 135 pp.

Franklin, S. E., M. B. Lavigne, B. A. Wilson and E. R. Hunt, 1994, "Empirical Relations Between Balsam Fir Stand Conditions and ERS-1 SAR data in Western Newfoundland," *Canadian Journal of Remote Sensing*, 20:124–130.

Henderson, F. M. and Z Xia, 1998, "Radar Applications in Urban Analysis, Settlement Detection and Population Estimation," in Henderson and Lewis (Eds.) *Principles and Applications of Imaging Radar*, NY: John Wiley, 733–768.

Henderson, F. M. and A. J. Lewis, 1998, "Radar Fundamentals: The Geoscience Perspective," in Henderson and Lewis (Eds.) *Principles and Applications of Imaging Radar*, NY: John Wiley, 131–181.

Hodgson, M. E., Jensen, J. R., Schmidt, L., Schill, S. and B. A. Davis, 2003, "An Evaluation of LIDAR- and IFSAR-Derived Digital Elevation Models in Leaf-on Conditions with USGS Level 1 and Level 2 DEMS," *Remote Sensing of Environment*, 84(2003):295–308.

Jackson, T. J. and T. J. Schmugge, 1991, "Vegetation Effects on the Microwave Emission of Soils," *Remote Sensing of Environment*, 36(3):203–212.

Jensen, J. R., Botchway, K., Brennan-Galvin, E., Johannsen, C., Juma, C., Mabogunje, A., Miller, R., Price, K., Reining, P., Skole, D., Stancioff, A. and D. R. F. Taylor, 2002, *Down to Earth: Geographic Information for Sustainable Development in Africa*, Washington: National Academy Press, 155 p.

JPL, 1999a, *Shuttle Radar Topography Mission*, Pasadena: Jet Propulsion Lab, http://www-radar.jpl.nasa.gov/srtm.

JPL, 1999b, *Module 2: How Radar Imaging Works: Interferometry*, Pasadena: Jet Propulsion Lab, SIR-C Website.

Kasischke, E. S. and L. L. Bourgeau-Chavez, 1997, "Monitoring South Florida Wetlands Using ERS-1 SAR Imagery," *Photogrammetric Engineering & Remote Sensing*, 63(3):281–291.

Kasischke, E. S., G. A. Meadows and P. L. Jackson, 1984, *The Use of Synthetic Aperture Radar to Detect Hazards to Navigation*, Ann Arbor: Environmental Research Institute of Michigan, Publication # 169200-2-F., 194 p.

Leberl, F. W., 1990, *Radargrammetric Image Processing*, Norwood: Artech House, 595 p.

Li, Q., R. Ferraro and N. C. Grody, 1998, "Detailed Analysis of the Error Associated with the Rainfall Retrieved by the

NOAA/NESDIS SSM/I Rainfall Algorithm: Part 1," *Journal Geophysical Research*, 103(11):419–427.

Madsen, S. N. and H. A. Zebker, 1998, "Imaging Radar Interferometry," in Henderson and Lewis (Eds.) *Principles and Applications of Imaging Radar*, NY: John Wiley, 359–380.

McGlone, J. C., 2004, *Manual of Photogrammetry*, 5th Ed., Bethesda: ASP&RS, 1151 p.

Mikhail, E. M., Bethel, J. S. and J. C. McGlone, 2001, *Introduction to Modern Photogrammetry*, NY: John Wiley, 479 p.

MDA, 2006a, *RADARSAT-2: Innovation in Earth Observation*, Canada: MDA Services, http://www. radarsat2.info/.

MDA, 2006b, *Polarimetric Parameters*, Canada: MDA Geospatial Services, http://www.radarsat2.info/.

Moore, R. K., 1983, "Imaging Radar Systems," *Manual of Remote Sensing*, 2nd Ed., R. Colwell, (Ed.), Bethesda: ASP&RS, 429–474.

NASA AMRS-E, 2006, *Advanced Microwave Scanning Radiometer*, Marshal Space Flight Center, http://www.ghcc.msfc.nasa.gov/AMSR/.

NASA JPL, 1996, "Brazil Rainforest," at *Imaging Radar Home Page* (http:/www.jpl.gov).

NASA SRTM, 2006, *Shuttle Radar Topography Mission*, Pasadena: Jet Propulsion Lab, http://www2.jpl.nasa.gov/srtm/index.html.

NASA TRMM, 2006, *Tropical Rainfall Measuring Mission*, Beltsville: Goddard Space Flight Center, http://trmm.gsfc. nasa. gov/.

NSIDC, 2006, *Advanced Microwave Scanning Radiometer (AMSE-E)*, Boulder: National Snow and Ice Data Center, http://nsidc.org/data/docs/daac amsre_instrument.gd.html.

NSIDC SSM/I, 2006, *DMSP SSM/I Pathfinder Daily EASE-Grid Brightness Temperatures*, Boulder: National Snow and Ice Data Center, http://nsidc.org/data/nsidc-0032.html.

Peake, W. H. and T. L. Oliver, 1971, *Response of Terrestrial Surfaces at Microwave Frequencies,* Air Force Avionics Lab Report #TR-70-301, Columbus: Ohio State Univ.

Price, A., 1990, "Eyes of England," *Air & Space,* 5(4): 82–93.

Rabus, B., Eineder, M., Roth, A., and R. Bamler, 2003, "The Shuttle Radar Topography Mission — A New Class of Digital Elevation Models acquired by Spaceborne Radar," *ISPRS Journal of Photogrammetry & Remote Sensing*, 57:241–262.

Raney, K., 1998, "Radar Fundamentals: Technical Perspective," Henderson and Lewis (Eds.), *Principles and Applications of Imaging Radar*, NY: John Wiley, 42–43.

Rodriguez, E., Morris, C. S. and J. E. Belz, 2006, "A Global Assessment of the SRTM Performance," *Photogrrammetric Engineering & Remote Sensing*, 72(3):249-260.

Sabins, F. F., 1997, *Remote Sensing: Principles and Interpretation.* 3nd Ed., NY: Freeman, 494 p.

Sun, G. and K. J. Ranson, 1995, "A Three-dimensional Radar Backscatter Model for Forest Canopies," *IEEE Transactions on Geoscience and Remote Sensing*, 33:372–382.

Van der Sanden, J. J. and S. J. Thomas (Eds.), 2004, *Applications Potential of RADARSAT-2 – Supplement One*, Ottawa: Canada Centre for Remote Sensing, 115 p.

Wang, Y., E. S. Kasischke, J. M. Melack, F. W. Davis and N. L. Christensen, 1994, "The Effects of Changes in Loblolly Pine Biomass and Soil Moisture on ERS-1 SAR Backscatter," *Remote Sensing of Environment*, 49:25–31.

Wang, Y., E. S. Kasischke, F. W. Davis, J. M. Melack and N. L. Christensen, 1995, "The Effects of Changes in Forest Biomass on Radar Backscatter from Tree Canopies," *International Journal of Remote Sensing*, 16:503–513.

Waring, R. H., J. Way, E. R. Hunt, L. Morrissey, K. J. Ranson, J. F. Weishampel, R. Oren and S. E. Franklin, 1995, "Imaging Radar for Ecosystem Studies," *BioScience*, 45(10):715–723.

LIDAR Remote Sensing

10

Elevation information is a critical component of geographic databases used by many agencies (e.g., DOD, 2000; FEMA, 2004; USGS, 2006). It is so important that a National Digital Elevation Program (NDEP) was established to enhance data sharing among federal, state, and local agencies, the commercial sector, and academia to minimize redundant elevation data production (NDEP, 2006). These data are used in numerous practical applications such as flood zone mapping, regional planning, and utility siting.

Most elevation data are stored in a digital elevation model that can be analyzed in conjunction with other spatial data in a GIS. A *digital elevation model* (DEM) is defined as a file or database containing elevation points over a contiguous area (Miller, 2004; Ma, 2005). DEMs may be subdivided into:

1. *digital surface models* (DSM) that contain elevation information about all features in the landscape, such as vegetation, buildings, and other structures; and

2. *digital terrain models* (DTM) that contain elevation information about the bare-Earth surface without the influence of vegetation or man-made structures.

Four major technologies are used to obtain elevation information, including (Bossler et al., 2002):

- *in situ* surveying;

- photogrammetry (Chapter 6);

- Interferometric Synthetic Aperture Radar (IFSAR), (Chapter 9); and

- Light Detection and Ranging (LIDAR).

In situ surveying using conventional surveying (e.g., total station) or GPS instruments can yield accurate x,y,z information. However, field surveys are time-consuming and expensive on a per-point basis. Even with GPS instruments, it is often difficult for surveyors to obtain points in thick undergrowth. Due to these obstacles, the density of x,y,z observations obtained in an area is sometimes low. It then becomes necessary to interpolate between distant points to produce a digital terrain model (DTM) of the area.

Photogrammetric techniques are routinely used to collect x,y,z topographic information. Photogrammetric surveys can map large areas during the leaf-off season and yield a dense collection of points. Chapter 6 summarized how digital elevation models are extracted directly from stereoscopic aerial photography. Photogrammetry can be used to obtain dense elevation information in inhospitable terrain. In addition, an analyst can selectively obtain a greater density of *x,y,z* observations associated with critical features such as ridgelines or steep changes of slope (e.g., along road overpasses) called breaklines.

LIDAR offers an alternative to *in situ* field surveying and photogrammetric mapping techniques for the collection of elevation data (Maune, 2001). LIDAR technology can be used to provide elevation data that is accurate, timely, and increasingly affordable in inhospitable terrain (McGlone, 2004). LIDAR does not, however, allow the analyst to control the placement of individual *x,y,z* measurements on ridgelines or breaklines.

LIDAR Principles

The first optical laser was developed in 1960 by Hughes Aircraft, Inc. Laser instruments were soon used to compute distance by measuring the travel time of light from a laser transmitter to a target and then back to a laser receiver. Early remote sensing LIDAR systems could only collect measurements directly underneath the aircraft, creating a single profile of elevation measurements across the landscape (e.g., Jensen et al., 1987). The synergistic use of kinematic GPS and inertial measurement units (IMUs) on airborne LIDAR scanning systems has allowed the technology to mature rapidly. LIDAR–derived horizontal and vertical accuracies and cost of operation are now similar to that of photogrammetry (McGlone, 2004; Raber et al., 2006).

LIDAR Laser and Scanning System

The LIDAR instrument consists of a system controller and a transmitter and receiver. As the aircraft moves forward along the line-of-flight, a scanning mirror directs pulses of laser light across-track perpendicular to the line-of-flight (Figure 10-1a). Most LIDAR systems used for topographic mapping use eye-safe near-infrared laser light in the region from 1040 to 1060 nm (Boland et al., 2004). Blue-green lasers centered at approximately 532 nm are used for bathymetric mapping due to their water penetration capability (Mikhail et al., 2001). LIDAR data can be collected at night if necessary because it is an active system, not dependent on passive solar illumination.

LIDAR systems can emit pulses at rates >100,000 pulses per sec, often referred to as *pulse repetition frequency*. A pulse of laser light travels at *c*, the speed of light (3×10^8 m s^{-1}). LIDAR technology is based on the accurate measurement of the laser pulse travel time from the transmitter to the target and back to the receiver. The traveling time of a pulse of light, *t*, is (Boland et al., 2004)

$$t = 2\frac{R}{c} \tag{10-1}$$

where *R* is the range (distance) between the LIDAR sensor and the object. The range, *R* can be determined by rearranging the equation (Baltsavias, 1999ab)

$$R = \frac{1}{2}tc . \tag{10-2}$$

The range measurement process results in the collection of elevation data points (commonly referred to as *masspoints*) arranged systematically in time across the flightline, as shown in Figure 10-1b. The example displays masspoints associated with the ground, several powerlines, a pole, and tree canopy.

The *laser footprint* is approximately circular on the ground and varies with the scan angle and the topography encountered (Baltsavias, 1999b). The diameter of the instantaneous laser footprint (Fp_{inst}) on the ground (Figure 10-1a) is computed by

$$Fp_{inst} = \frac{h}{\cos^2(\theta_{inst})}\gamma \tag{10-3}$$

where *h* is the altitude of the aircraft AGL, θ_{inst} is the instantaneous scan angle under investigation, and γ is the divergence of the laser beam (Baltsavias, 1999b; Boland et al., 2004b). For example, if *h* = 750 m AGL, θ_{inst} = 15°, and γ = 1 mrad, then Fp_{inst} = 0.80 m for flat terrain. At a flying height of 1000 m AGL, a typical laser beam divergence of 1 mrad results in a 1 m diameter footprint (Baltsavias, 1999b).

The across-track swath width (*sw*) is given by:

$$sw = 2h\tan\frac{\theta}{2} \tag{10-4}$$

where θ is the scan angle. For example, if *h* = 750 m AGL, and θ = 30°, then *sw* = 402 m.

The *point spacing across-track* ($p_{spacing}$) is dependent upon the pulse repetition frequency (PRF), the altitude of the air-

LIDAR Data Collection

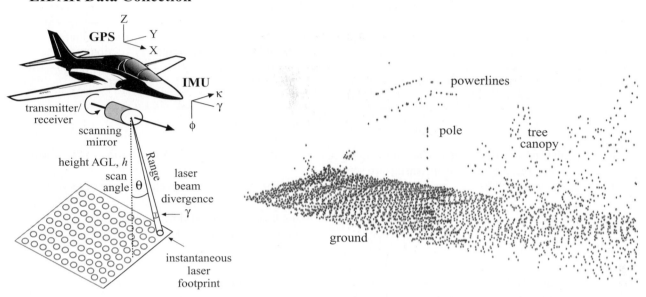

a. LIDAR data collection.

b. LIDAR elevation masspoints for a small area.

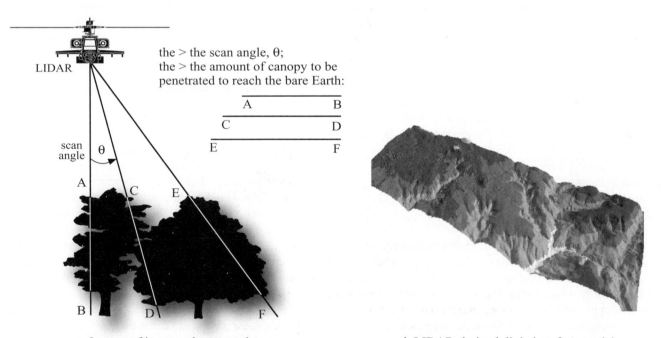

c. Impact of increased scan angle.

d. LIDAR-derived digital surface model.

Figure 10-1 a) LIDAR instruments transmit pulses of laser light toward the ground using a scanning mirror. Some of this energy is scattered back toward the aircraft and recorded by the receiver electronics. Onboard GPS and IMU document the exact location of the aircraft and the roll, pitch, and yaw at the instant the laser pulse is sent and received. b) LIDAR–derived x, y, and z elevation masspoints for an area near Aiken, SC. Some masspoints are located on the Earth's surface while others are located in the tree canopy and even on the power transmission line and pole. c) Assuming that the tree canopy has a uniform height and canopy density, then the greater the scan angle from nadir the greater the amount of canopy that will have to be penetrated to reach the ground to acquire the information necessary to create a bare-Earth digital terrain model. d) LIDAR–derived digital surface model for an area near Aiken, SC (from Cowen et al., 2000).

Table 10-1.   Characteristics of LIDAR data collected on the Savannah River Site on October 10, 2004.

Vendor	LIDAR sensor	Date	Pulses per sec	Wavelength	Scan angle	Nominal post spacing
Sanborn, Inc.	Optech ALTM 2050	10-10-04	50,000 (50 kHz)	1046 nm	± 10	every 30 cm in x and y

craft (h) AGL, the instantaneous angular scanning speed (α_{inst}) in radians per second, and the instantaneous scan angle (θ_{inst}) (Baltsavias, 1999b):

$$p_{spacing} = \frac{h}{\cos^2(\theta_{inst})} \times \frac{\alpha_{inst}}{PRF}. \qquad (10\text{-}5)$$

Actual sampling densities for an *area* also depend on the forward speed of the aircraft. Typical sampling densities on the ground range between 1 point per 20 m^2 to 20 points per m^2 for a flying height of 1000 m (Boland et al., 2004). These higher densities are typically achieved using multiple overlapping flightlines to cover the study area. The observed ground returns are almost always less than the number of emitted pulses.

The maximum off-nadir scan angle can be adjusted to meet the needs of a data-collection mission. The greater the scan angle off-nadir, the more vegetation that will have to be penetrated to receive a pulse from the ground assuming a uniform canopy (Figure 10-1c). LIDAR data may be used to prepare digital terrain or surface models such as the one shown in Figure 10-1d which was used to identify the optimum location for a new railroad line near Aiken, SC (Cowen et al., 2000).

LIDAR remote sensing avoids the problems of aerial triangulation and orthorectification discussed in Chapter 6 because each LIDAR posting is individually georeferenced (Flood and Gutelius, 1997). However, it takes substantial processing to turn the laser range information into georeferenced masspoints. Some of the most important variables used in the processing include:

- the *x,y,z* location of the laser in three-dimensional space at the time of the laser pulse;

- the attitude (roll, pitch, and heading) of the laser at the time of the laser pulse;

- the scan angle of the LIDAR at the time of the laser pulse;

- the effect of atmospheric refraction on the speed of light;

- the laser pulse travel time from the LIDAR instrument to the target (ground) and back.

It is useful to delve more deeply into several of these parameters. For the purposes of discussion, we will use LIDAR data of the Savannah River Site near Aiken, SC, collected on October 10, 2004. The data were acquired using a LIDAR that transmitted 50,000 pulses of near-infrared (1046 nm) laser light per second. The LIDAR data acquisition specifications are summarized in Table 10-1 (Sanborn, 2005).

LIDAR Laser Location

It is important to know the exact location of the LIDAR laser at all times during data collection. This is accomplished using Differential Global Positioning System (DGPS) technology. DGPS is based on the use of two GPS receivers that simultaneously record positional information. A terrestrial GPS base station is located at an accurately surveyed location with well-documented *x*, *y*, and *z*-coordinates. The terrestrial base station records its GPS position for the duration of the LIDAR data collection mission. The second GPS receiver is located on the aircraft and records the LIDAR antenna position at all times during data collection.

After the LIDAR data are collected, the data from both GPS units (one on the ground and one in the aircraft) are post-processed along with the known location of the base station antenna. This process determines the exact location of the aircraft's antenna for the entire flight. The accuracy of the aircraft position is typically < 5-10 cm, and is output in units of latitude, longitude, and ellipsoidal height in a WGS 84 coordinate system.

LIDAR Antenna Attitude (Orientation)

It is necessary to have accurate LIDAR antenna *orientation* information (roll, pitch, and heading) at all times during data collection. This is measured by an inertial measurement unit (IMU). The IMU uses roll, pitch, and yaw gyroscopes and three accelerometers to measure the orientation of the LIDAR antenna at the exact moment every pulse is transmitted and received. After the LIDAR data are collected, data from the IMU is post-processed along with the GPS-derived antenna position data to output a file indicating the trajectory of the aircraft and the laser antenna at all times during the LIDAR mission. The output file documents the position of the aircraft (latitude, longitude, and ellipsoidal height) and sensor orientation (roll, pitch, and heading), indexed by GPS time.

LIDAR Post-Processing of Multiple Returns

Thus far we have collected GPS data, IMU data, and sent and received laser pulses. How are these data turned into digital elevation values at specific x,y and z locations on the surface of the Earth? This is accomplished through LIDAR post-processing which takes place after the aircraft has landed. Post-processing software is used to associate 1) LIDAR antenna x,y,z position, 2) antenna roll, pitch, and yaw orientation, and 3) LIDAR range (distance) information into a set of latitude, longitude, and altitude (x,y,z) coordinates for each LIDAR return. The output is typically in a compact binary format of WGS 84 coordinates, with options for converting the output to ASCII formats, and/or UTM coordinates. Now let us consider the characteristics of the LIDAR data.

LIDAR Returns

As mentioned, a pulse of laser energy exiting the transmitter is directed toward the terrain at a certain angle by the rotating mirror. Depending upon the altitude of the LIDAR instrument AGL and the angle at which the pulse is sent, each pulse illuminates a near-circular area on the ground called the *instantaneous laser footprint*, e.g., 30 cm in diameter (Figure 10-1a). This single pulse can generate one return or multiple returns. Figure 10-2 depicts how multiple returns might be produced from a single pulse. All of the energy within laser pulse *A* interacts with the ground. One would assume that this would generate only a single return. However, if there are any materials whatsoever with local relief within the instantaneous laser footprint (e.g., grass, small

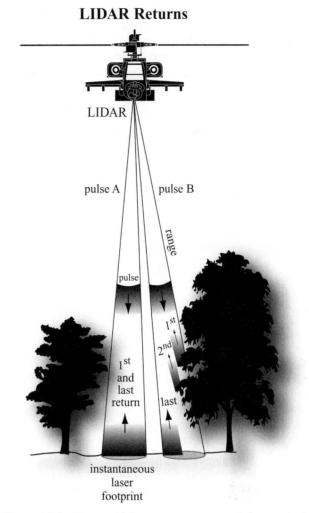

LIDAR Returns

Figure 10-2 How multiple returns are generated from a single pulse of laser light.

rocks, twigs), then there will be multiple returns. The *1st return* will come from these materials that have local relief (even on the order of 3 to 5 cm) and the 2nd and perhaps *last return* will come from the bare-Earth. Although not identical, the range (distance) associated with the first and last returns will be very similar.

Laser pulse *B* encounters two parts of a tree at different elevations and then the bare Earth. In the example, part of pulse *B* encounters a branch at 3 m AGL causing some of the incident laser pulse to be backscattered toward the LIDAR receiver. This is recorded as the *1st return* (Figure 10-2). The remainder of the pulse continues until at 2 m AGL it encounters another branch that scatters energy back toward the LIDAR receiver. This is recorded as the *2nd return*. In this example, approximately one-half of the pulse finally reaches the ground, and some of it is backscattered toward the

LIDAR receiver. This is the *last return*. If we wanted information about the height of the tree and its structural characteristics then we would be interested in the 1st and 2nd returns associated with pulse *B*. If we are only interested in creating a bare-Earth digital terrain model then we would be especially interested in the last return associated with pulses *A* and *B*.

Thus, each laser pulse transmitted from the aircraft can yield multiple returns. This is referred to as *multiple-return LIDAR* data. Post-processing the original data results in several LIDAR files commonly referred to as:

- 1st return;

- possible intermediate returns;

- last return; and

- intensity.

The masspoints associated with each return file (e.g., 1st return) are distributed throughout the landscape at various densities depending upon the scan angle, the number of pulses per second transmitted (e.g., 50,000 pps), aircraft speed, and the materials that the laser pulses encountered. Areas on the ground that do not yield any LIDAR-return data are referred to as *data voids*.

Data collection vendors deliver LIDAR data according to user specifications. For example, the Savannah River Site LIDAR return data were delivered in ASCII format as:

- time of day, *x*-coordinate, *y*-coordinate, *z*-coordinate, and intensity.

The simple ASCII format makes it straightforward to input the LIDAR data into a GIS for examination and analysis. The LIDAR returns were processed by the vendor into separate files of:

- 1st returns;

- last returns; and

- bare-Earth returns.

Bare-Earth returns were derived from the first and last returns using a post-processing procedure (to be discussed).

Extraction of First, Intermediate, and/or Last Return Digital Surface Models (DSMs)

Masspoints associated with the last returns from the LIDAR mission flown over the Savannah River Site on October 10, 2004 are shown in Figure 10-3a. Note the distribution and density of these points. Each masspoint has a unique *x,y* location. Careful examination reveals that the last-return dataset contains points derived from multiple, overlapping flightlines of last-return LIDAR data. In an interactive computer environment, we could click on the individual points in Figure 10-3a and obtain the elevation. This is useful but does not reveal any elevation patterns in the dataset. Therefore, it is common to use several digital image processing techniques to enhance our understanding of the masspoint dataset. For example, the individual masspoints were processed using inverse distance weighting (IDW) interpolation logic to create a raster (grid) of elevation values every 0.25 × 0.25 m (Figure 10-3b) [refer to Jensen (2005) for a discussion of IDW]. The interpolation process creates a *digital surface model* (DSM) which contains the elevation characteristics of all the trees, shrubs, and man-made structures. The brighter the pixel in the DSM, the greater the elevation. For example, the buildings in Figure 10-3b are higher than the surrounding ground therefore the buildings are brighter than the ground. The original masspoints have been overlaid onto the IDW digital surface model for illustrative purposes in Figure 10-3b. Many scientists create DSMs using a triangular irregular network (TIN) data structure (to be shown later).

The LIDAR–derived IDW DSM can be made even easier to interpret by applying a shaded-relief algorithm that highlights the terrain as if it were illuminated by the Sun from a specific direction (e.g., from the northwest). An example is presented in Figure 10-3c. The original masspoints have been overlaid on the shaded-relief display for illustrative purposes. Draping masspoints onto orthophotography is also very useful (Figure 10-3d).

Most LIDAR projects generate so much data that it is necessary to subdivide the dataset into tiles. We have been examining a small portion of one tile. The first- and last-return digital surface models for an entire tile are shown in Figure 10-4ab. This graphic depicts both the first- and last-return data as inverse distance weighted (IDW) digital surface models (DSM). First- and last-return IDW DSMs are then portrayed in shaded-relief format.

The multiple return DSM files are of significant value for many applications such as the extraction of vegetation

Various Displays of Last-Return LIDAR Masspoints

a. Distribution of unedited last-return masspoints. b. IDW Digital surface model overlaid with masspoints.

c. Shaded-relief of IDW model overlaid with masspoints. d. Unedited last-return masspoints overlaid on orthophoto.

Figure 10-3   a) Masspoints can be viewed individually. This procedure is useful for identifying any data voids. One can also see that the data in this last return dataset came from multiple flightlines. Digital surface models (DSM) can be created using inverse distance weighting (IDW) as in (b) or using triangular irregular networks (TIN) data structures (not shown). c) The IDW digital surface model is portrayed in shaded-relief. The terrain is modeled as if the sun were in the northwest casting shadows that fall toward the viewer. Cartographic theory suggests that this is one of the best ways to communicate three-dimensional information. d) Last-return masspoints overlaid on digital orthophotography can also be useful.

height and biomass and building height information (e.g., Lefsky et al., 2005; Ma, 2005). However, what if we desire a bare-Earth digital terrain model (DTM) without the presence of trees and/or man-made structures? How is it created?

Extraction of Bare-Earth Digital Terrain Models (DTMs)

If the purpose of a LIDAR overflight is to collect data to create a digital terrain model, the presence of vegetation (and other surface obstructions) can be a nuisance. In areas covered by dense vegetation, the majority of the LIDAR returns will be from the canopy, with only a few pulses reaching the ground. Hendrix (1999) found that up to 93% of LIDAR pulses never reached the ground in mixed bottomland hardwoods near Aiken, SC. Separating ground returns from vegetation canopy returns can be problematic. Nevertheless, it can be done.

A bare-Earth digital terrain model (DTM) may be created by systematically removing masspoints in the first, intermediate, and/or last return LIDAR data that come from trees, shrubs, and even grass that extend above the bare ground. This procedure is typically performed in two steps: 1) semi-

LIDAR Data of the Savannah River Site Obtained on October 10, 2004

a. Inverse distance weighted first return.   b. Inverse distance weighted last return.   c. IDW bare Earth after editing.

d. Shaded-relief version of (a).   e. Shaded-relief version of (b).   f. Shaded-relief version of (c).

Figure 10-4   a) Inverse distance weighting (IDW) applied to first-return LIDAR data to create a digital surface model (DSM). b) IDW applied to last-return LIDAR data. c) IDW applied to edited masspoints to create a bare-Earth digital terrain model (DTM). d–f) Shaded-relief versions of a–c.

automatic masspoint filtering, and b) manual masspoint editing (Raber et al., 2002; Leonard, 2005).

Semiautomatic Vegetation and/or Man-made Structure Masspoint Filtering

Many landscapes contain dense vegetation. It is imperative to have a technique that removes 90 – 98% of the above-ground vegetation masspoints. This is done using a filtering algorithm that systematically passes through the LIDAR dataset examining each masspoint and the elevation characteristics associated with its *n* nearest-neighbors. The filter then identifies those points that are a) bare ground, b) scrub-shrub, c) trees, and/or d) man-made structures. For example, Raber et al. (2002) developed an adaptive vegetation removal filter that took into account the LIDAR-predicted land-cover type. Sithole and Vosselman (2004) review eight different filters applied to twelve LIDAR datasets. Progress has been made in the development of semiautomatic vegetation and/or man-made object removal algorithms. New algorithms will hopefully be more efficient and accurate.

Manual Vegetation and/or Man-made Structure Masspoint Editing

Unfortunately, semiautomatic filtering rarely identifies all masspoints associated with shrubs, trees, and/or man-made objects (Leonard, 2005). It is usually necessary for a well-trained analyst to visually examine the results of applying the semiautomatic filter and then selectively identify and set aside any above-ground masspoints that were not detected semiautomatically. This approach may involve viewing the point clouds in 3-dimensions or overlaying the masspoints on rectified aerial photography. The manually edited masspoints can then be used to create a bare-Earth *digital terrain model* (DTM). Figures 10-4c and 10-4f depict a bare-Earth DTM in which all of the vegetation and buildings have been removed. Figure 10-5a presents an enlarged view of a small part of the area shown in Figure 10-4c and 10-4f with the filtered bare-Earth masspoints overlaid on the last return IDW DSM. Note the data voids on the building rooftops where the masspoints were deleted by automated and/or manual filtering. The bare-Earth masspoints can then be processed to yield a DTM that contains no buildings as shown in Figure 10-5d. Note also that almost all of the vegetation present in the first- and last-return LIDAR data (Figures 10-4a,b and 10-4d,e) is gone in the bare-Earth DTM shown in Figure 10-4c and 10-4f.

It is important to note that the final bare-Earth DTM may or may not include man-made structures such as buildings. For example, if we are concerned with creating a DTM that will be used to model the flow of water on the ground, then we would not want to use the DTM shown in Figure 10-4f (and Figure 10-5d) because the buildings have been removed. Instead, we would want to remove all above-ground vegetation and then create the DTM. The DTM would contain bare-Earth and building elevation information. A hydrologic model using this DTM would flow water correctly around buildings, not through them.

LIDAR Intensity

Most LIDAR systems provide an intensity file in addition to the multiple return data. For example, the intensity data produced by the Leica ALS40/50 LIDAR system represents the peak voltage of the return signal as recorded by the system controller (Leonard, 2005). The recorded intensity is in most cases not the integration of the returned echo from all the pulse returns, but just its *maximum* (Baltsavias, 1999). There are several factors which influence the intensity value.

First, it is important to remember that laser light is monochromatic and has an extremely small bandwidth (e.g., 2 – 5 nm) usually centered on a near-infrared wavelength (e.g., 1046 nm). Conversely, multispectral remote sensing bandwidths are often 50 – 100 nm wide although hyperspectral sensors might have bandwidths as small as 10 nm (Baltsavias, 1999). Other significant factors include the range to the target, angle of incidence and atmospheric dispersion. In addition, the system controller also records the state of the AGC (automatic gain control). The AGC circuit adjusts the return signal gain in response to changes in target reflectance. The gain and intensity values may vary over a scene and from day to day. This variability in gain control can cause problems when interpreting intensity data (Leonard, 2005).

Theoretically, one would think that because the laser uses near-infrared light that the intensity value at each of the masspoint locations would in effect be similar to the reflectance found when conducting optical remote sensing in the near-infrared portion of the spectrum. Unfortunately, this is not the case. For example, Figure 10-6a presents a DSM produced using IDW applied to last-return LIDAR data. Figure 10-6b is an intensity DSM produced using IDW. The intensity image contains a wealth of detail and in many respects looks like a panchromatic aerial photograph of the study area. However, there are some interesting anomalies when we compare the intensity image to what we would expect to find in a typical black-and-white near-infrared image.

First, consider the large forested area shown at *a* in Figure 10-6b. This is a mature forest that has substantial local relief

Distribution of Masspoints After Digital and Manual Editing

a. Edited masspoints overlaid on IDW last-return elevation.

b. Shaded relief of (a).

c. Edited masspoints overlaid on 1 x 1 m orthophoto.

d. Bare Earth TIN derived from the edited masspoints.

Figure 10-5 a) Edited masspoints overlaid on the original last-return IDW digital surface model. b) Shaded-relief version of (a). c) Edited masspoints overlaid on an orthophoto. d) Bare-Earth digital terrain model in a triangular irregular network (TIN) data structure.

as evidenced by the bright return in the adjacent last return dataset (Figure 10-6a). We would normally expect this area to appear bright in a black-and-white near-infrared image. Conversely, it appears quite dark in the intensity image.

Similarly, consider several individual deciduous trees at location *b* in Figure 10-6b and shown in Figure 10-7. These trees also exhibit less intensity than expected. The natural grass surrounding these trees has a higher intensity range than the trees (Figure 10-7c).

Finally, consider the bright areas at location *c* in the intensity image (Figure 10-6b). The northern half of three of the test-

pad mounds were cleared and sodded with centipede grass. They do not have as much biomass as the trees at *a* or *b*. Nevertheless, they exhibit a very bright return in the intensity image (Figure 10-6b).

In these examples, it appears that when the LIDAR pulses encounter trees, volume scattering within the tree canopy reduces the amount of energy backscattered toward the LIDAR receiver. This causes the trees to appear dark in the intensity imagery. Conversely, LIDAR pulses that hit the natural grass at location *b* and newly sodded centipede grass at location *c* are not absorbed or scattered away. Rather, a

LIDAR Last Return and Intensity Data

a. LIDAR inverse distance weighted (IDW)
last-return elevation.

b. LIDAR intensity.

Figure 10-6   a) Inverse distance weighting applied to LIDAR last-return data. b) IDW applied to LIDAR intensity data. Three areas are highlighted for discussion.

strong return exits these mixed grass and bare-Earth surfaces causing them to appear bright.

Many of the buildings are bright in the intensity image suggesting that most of the incident laser pulse is backscattered resulting in strong voltage readings. The intensity of the return is a function of the rooftop material.

Leonard (2005) states that range and angle of incidence also have an effect on intensity readings over the course of a single flightline. As the LIDAR scanning mirror moves from nadir to the edge of the field of view, more energy is reflected away from the receiver optics. As a result, intensity

readings are higher closer to nadir than at the edge. Range to target is another significant factor in intensity normalization. Shorter ranges have higher intensity readings due to less atmospheric signal lost.

Currently, there are no rigorous guidelines about how to interpret LIDAR intensity images. Nevertheless, it is clear that they contain a wealth of information. For example, consider Color Plate 10-1 which depicts the intensity data draped over the last return elevation data for the deciduous trees shown at location *b* in Figure 10-6b. When these data are color-coded (Color Plate 10-1d) and evaluated obliquely (Color Plate 10-1e), it is clear that the intensity of the returns

LIDAR Last-Return Elevation Compared with LIDAR Intensity

a. Inverse distance weighted last-return elevation.

b. Shaded-relief version of (a).

c. LIDAR intensity.

Figure 10-7   a) Inverse distance weighting applied to last-return data. b) Shaded-relief version of (a). c) LIDAR intensity data.

in the trees is a function of how far the pulses penetrated into the tree canopy. Such information could be related to tree biophysical characteristics (e.g., LAI, percent canopy closure, biomass, basal area).

Contours

Sometimes it is valuable to extract contours (lines of equal elevation) from DSMs or DTMs to highlight subtle differences in the terrain and to identify depressions. For example, consider the testpad mounds located at *c* in Figure 10-6b. These testpads were engineered specifically to test the effectiveness of various clay-cap materials used to protect subsurface hazardous materials. Various types of impermeable polyurethane barriers were placed on the testpads, covered with clay, and planted with centipede grass. The surfaces were then left to the ravages of nature. LIDAR first-return data of the testpads processed using IDW

are presented in Figure 10-8a. A shaded-relief version of the IDW DSM is shown in Figure 10-8b. Note that one-half of the trees were removed in the northern half of three of the testpads and not on the eastern-most control testpad. The cleared areas on the three testpads were sodded with new centipede grass at the time of data collection.

The last-return LIDAR data were filtered using automated and manual procedures to derive the bare-Earth DTM shown in Figure 10-9a. As expected, almost all of the trees present in the first-return data (Figure 10-8a) have been removed. Despite its usefulness, it is not possible to visually extract quantitative information about the nature of the testpads from the bare-Earth DTM shown in Figure 10-9a. It is only when we derive contours or other synthetic 3-D representations from the bare-Earth DTM that we can appreciate the three-dimensional nature of the testpads and

LIDAR Data of the Testpads on the Savannah River Site

a. Inverse distance weighted first-return elevation.

b. Shaded-relief version of (a).

Figure 10-8   a) Inverse distance weighting applied to first-return data. Note the presence of trees on the testpads and in the forest to the east. b) Shaded-relief version of (a).

Contours Extracted from the Bare-Earth Digital Terrain Model of the Testpads

a. Inverse distance weighted bare-Earth elevation.

b. Contours (0.25 m interval) derived from (a) overlaid on a shaded-relief model of (a).

Figure 10-9   a) Inverse distance weighted bare-Earth elevation. Note that the trees have been removed when compared to Figure 10-8. b) Contours with a contour interval of 0.25 m are overlaid on the shaded-relief version of (a) above.

Table 10-2. Vertical accuracy of LIDAR–derived elevation data collected for the Savannah River Site on October 10, 2004, using the *Geospatial Position Accuracy Standards Part 3: National Standard for Spatial Data Accuracy* (NSSDA) (FGDC, 1998).

Vendor	LIDAR sensor	Date	Horizontal datum	Units	Vertical datum	Number of check-points	NSSDA RMSE	NSSDA Vertical accuracy at 95% confidence interval (1.96 × RMSE)
Sanborn, Inc.	Optech ALTM 2050	10-10-04	UTM NAD 83 Zone 17	m	NAVD88	95	6.6 cm	12.9 cm

the subtle depressions, especially on the most eastern testpad. The contour interval was 0.25 m.

Accuracy of LIDAR Measurements

LIDAR is a relatively new technology. As such, there is healthy skepticism about its ability to provide x, y, and z-elevation information as accurately as traditional *in situ* surveying and photogrammetry (e.g., Baltsavias, 1999; McGlone, 2004). Fortunately, there are accuracy assessment standards that can be used to provide an unbiased assessment of the accuracy of LIDAR–derived products.

NSSDA Horizontal and Vertical Accuracy Assessment

To determine the accuracy of a LIDAR–derived digital surface model (DSM) or bare-Earth digital terrain model (DTM), it is customary to identify *in situ* x, y, z checkpoints throughout the study area using a higher accuracy technique, such as total station surveying or differential GPS. Each *in situ* checkpoint is then located in the LIDAR–derived DSM or DTM. The nearest LIDAR–derived information is then compared with the location and elevation information associated with each *in situ* checkpoint. This is performed for a number of checkpoints. The results are then used to compute the horizontal and vertical accuracy of the LIDAR–derived data expressed as the root mean squared error (RMSE).

In 1998, the Federal Geographic Data Committee (FGDC) published the *Geospatial Position Accuracy Standards Part 3: National Standard for Spatial Data Accuracy* (NSSDA) (FGDC, 1998). This standard replaced both the *United States National Map Accuracy Standards* (NMAS) published by the Office of Management and Budget in 1947 and the American Society for Photogrammetry and Remote Sensing (ASPRS) *ASPRS Accuracy Standards for Large-Scale Maps* published in 1990.

The *Geospatial Accuracy Standard* for horizontal and vertical accuracy of spatial products put forth by the FGDC is based on the computation of RMSE where (Miller, 2004)

$$RMSE = \sqrt{\frac{\sum\limits_{i=1}^{n} \Delta^2}{n}} \qquad (10\text{-}6)$$

and Δ is the difference between an *in situ* checkpoint measurement and a remote sensing–derived measurement at the same location. The standard recommends that a minimum of 20 checkpoints be used, i.e., $n > 20$. The checkpoints should be of substantially higher accuracy than the remote sensing-derived product under investigation.

Horizontal accuracy (i.e., circular standard error) assessment at the 95% confidence level is computed using

$$Horizontal = 2.4477 \times 0.5 \times (RMSE_x + RMSE_y). \qquad (10\text{-}7)$$

Vertical accuracy assessment at the 95% confidence level is computed using

$$Vertical = 1.96 \times RMSE_z. \qquad (10\text{-}8)$$

Equations 10-7 and 10-8 assume the error distribution for x, y, and z are normally distributed. FEMA built upon these criteria and provides detailed accuracy assessment procedures for its flood-hazard mapping partners (FEMA, 2004).

The vertical accuracy of the LIDAR–derived elevations on the Savannah River Site was evaluated using 95 checkpoints (Table 10-2). The spatial distribution of the checkpoints on the Mixed Waste Management Area and the testpads is shown in Figure 10-10a,b. The vertical accuracy of the

Checkpoint Locations for LIDAR Accuracy Assessment on the Savannah River Site

a. Checkpoints on the Mixed Waste Management Facility.

a. Checkpoints on the test pads.

c. Relationship Between *In Situ* Ground Survey Elevation and LIDAR-derived Elevation for 95 Checkpoints on the Savannah River Site

$$y = 1.0027x - 0.253$$
$$R^2 = 0.9992$$

In Situ Ground Survey Elevation (meters)

LIDAR-derived Elevation (meters)

Figure 10-10   a) Location of checkpoints on the Mixed Waste Management Facility. This area is clay-capped and covered with Centipede grass. b) Locations of checkpoints on the testpads. c) Statistical relationship between GPS *in situ* ground survey elevation at 95 checkpoints and LIDAR–derived elevation at the same location extracted from a bare-Earth TIN digital terrain model (DTM). The *National Standard for Spatial Data Accuracy* (NSSDA) vertical accuracy assessment statistics are summarized in Table 10-2.

LIDAR data according to the NSSDA specifications was 12.9 cm. As expected, the LIDAR–derived elevation values were highly correlated with the GPS *in situ* elevation measurements (Figure 10-10c).

Miller (2004) stated that LIDAR–derived vertical accuracies are usually in the 5 to 30 cm range. But does this level of accuracy hold up when the terrain is steep? Does it hold up for all land-cover types (e.g., bare soil, grassland, shrub/scrub, forest, desert, water, shallow water, ice, snow, urban infrastructure)? Unlike photogrammetric surveys, can LIDAR be used successfully during leaf-on conditions? It is useful to consider some of these questions.

Influence of Land Cover, Slope, and Leaf-on/off Condition on LIDAR–derived Elevation Accuracy

Many scientists are investigating the effects of land cover, slope, and leaf-on/off condition on LIDAR–derived elevation accuracy. Hodgson et al. (2005) investigated the effects of land cover and surface slope using leaf-off LIDAR–derived elevation data for a watershed in the piedmont of North Carolina. Survey-grade elevation checkpoints (1,225) for seven different land cover classes were obtained. LIDAR elevation error for points associated with the seven land-cover classes ranged from 14.5 cm in low grass to 36.1 cm for dense shrub/scrub. Elevations found in pine and deciduous forest exhibited mean errors of 27.6 cm and 27.3 cm, respectively. On average, the LIDAR–derived elevation under-predicted true elevation regardless of land-cover type. The under-prediction was significant, and ranged up to 23.6 cm under dense pine land cover.

In another study, less-dense pine exhibited elevation errors as low as pavement and low grass (Hodgson and Bresnahan, 2004). The finding that scrub-shrub and deciduous land-cover exhibited the highest error appears to be consistent regardless of leaf-on/leaf-off condition (e.g., Hodgson *et al.*, 2003ab; Hodgson and Bresnahan, 2004). Basically, it appears that the more dense scrub/shrub and deciduous land-cover keeps laser pulses from reaching the ground, causing the interpolated elevation surface to be derived from the few points that do reach the ground. The resulting DSM or DTM may contain error.

Hodgson et al. (2005) found that over the small slope range in their study area (0° to 10°), there was not a statistically significant increase in LIDAR–derived elevation error as the slopes increased. However, they did find that slope errors from LIDAR–derived products increased with increasing surface slope, typically under-predicting true slope by 2°.

LIDAR–derived Vegetation Information

The ability of LIDAR technology to obtain returns from both vegetation and the ground surface has useful applications in measuring vegetation attributes. Weltz et al. (1994) used a profiling LIDAR system to map vegetation height and canopy cover at the Walnut Gulch experimental watershed in Arizona. This study demonstrated the ability of LIDAR to distinguish between different vegetation communities and ground-cover types. Nilsson (1996) examined the usefulness of LIDAR for estimating timber volume as a part of Sweden's National Forest Inventory.

Zimble et al. (2003) were able to extract vertical forest structure information from small footprint LIDAR data. They obtained tree height information by subtracting the bare-Earth DTM from the 1st-return DSM, i.e., tree height DSM = 1st return DSM – bare Earth DTM.

Carbon storage in moderate to high biomass forests is difficult to estimate with conventional optical or radar sensors. Lefsky et al. (2005) compared the relationships between LIDAR-measured canopy structure and coincident field measurements of forest stand structure at five locations in the Pacific Northwest. They obtained Coefficient of Determination values (R^2) for LAI and aboveground biomass of 81% and 92%, respectively.

Morris et al. (2005) used LIDAR data to characterize the elevation habitat of two salt marsh species in North Inlet, SC. Approximately 80% of the *Spartina alterniflora* existed within the narrow elevation range from 0.22 – 0.481 m ASL while 80% of the *Juncus roemerianus* existed within the elevation range of 0.296 – 0.981 m ASL.

LIDAR–derived Urban Information

LIDAR–derived information are often used in urban applications such as:

- identifying the location and elevation of utility transmission lines (Maune, 2001);

- obtaining accurate building elevation (e.g., Ma, 2005);

- measuring the elevation characteristics of transportation infrastructure such as highways, bridges and overpasses (NCRST, 2004);

LIDAR-derived Elevation of New York City, NY, after September 11, 2001

Figure 10-11   Perspective view of New York City, NY, after September 11, 2001. The pile of rubble is associated with the Twin Towers of the World Trade Center (LIDAR data courtesy of EarthData International; Jensen and Hodgson, 2004; 2006).

- identifying optimum routes that require the least cut and fill (e.g., Cowen et al., 2000); and

- urban environment visualization (Queija et al., 2005).

Before and after LIDAR–derived elevation data are especially useful for calculating the amount of material excavated (scoured) or deposited in a region (e.g., Shrestha et al., 2005). For example, LIDAR–derived elevation information were used to compute the volume of rubble associated with the Twin Towers of the World Trade Center in New York City (Figure 10-11) (Jensen and Hodgson, 2006).

The Future of LIDAR Remote Sensing

LIDAR remote sensing technology has revolutionized the collection of digital surface models and digital terrain models. LIDAR sensors obtain hundreds of thousands of masspoints per second that already possess accurate *x*,*y*, and *z*-coordinates without the necessity of aerotriangulation. The accuracy of the onboard GPS and IMUs continue to improve. The number of pulses per second transmitted continues to increase. The ability to measure the pulse range continues to improve. LIDAR sensors are becoming more affordable. All of these factors result in ever improving LIDAR–derived digital elevation information. LIDAR will provide considerable vegetation canopy information in the

future (e.g., LAI, biomass). It will also be used extensively to obtain three-dimensional urban infrastructure information (e.g., building footprints and height). LIDAR instruments will eventually be flown on unmanned aerial vehicles.

 ### References

Baltsavias, E. P., 1999a, "A Comparison between Photogrammetry and Laser Scanning," *ISPRS Journal of Photogrammetry & Remote Sensing*, 54:83–94.

Baltsavias, E. P., 1999b, "Airborne Laser Scanning: Basic Relations and Formulas," *ISPRS Journal of Photogrammetry & Remote Sensing*, 54:199–214.

Boland, J. and 16 co-authors, 2004, "Chapter 8: Cameras and Sensing Systems," in McGlone, J. C., (Ed.), *Manual of Photogrammetry*, 5th Ed., Bethesda: ASP&RS, 629–636.

Bossler, J. D., Jensen, J. R., McMaster, R. B. and C. Rizos, 2002, *Manual of Geospatial Science and Technology*, London: Taylor & Francis, 623 p.

Cowen, D. C., Jensen, J. R., Hendrix, C. J., Hodgson, M.E. and S.R. Schill, 2000, "A GIS-Assisted Rail Construction Econometric Model that Incorporates LIDAR Data," *Photogrammetric Engineering & Remote Sensing*, 66(11):1323–1328.

DOD, 2000, *Performance Specification Digital Elevation Data (DTED)*, MIL-PRF-89020B, Washington: DOD, May 23rd.

FEMA, 2003, "Appendix A: Guidelines for Aerial Mapping and Surveying," in *Map Modernization: Guidelines and Specifications for Flood Hazard Mapping Partners*, Washington: Federal Emergency Management Agency, 59 p., www.fema.gov.

FEMA, 2004, "Appendix N: Data Capture Standards," in *Map Modernization: Guidelines and Specifications for Flood Hazard Mapping Partners*, Washington: FEMA, 115 p., www.fema.gov.

FGDC, 1998, *Geospatial Position Accuracy Standards Part 3: National Standard for Spatial Data Accuracy*, Washington: FGDC, http://www.fgdc.gov/standards/status/sub1_3.html.

Flood, M., and B. Gutelius, 1997, "Commercial Implications of Topographic Terrain Mapping Using Scanning Airborne Laser Radar," *Photogrammetric Engineering & Remote Sensing*, 63:327.

Hendrix, C., 1999, *Parameterization of LIDAR Interaction with Vegetation Canopy in a Forested Environment*, Columbia: University of South Carolina, Masters Thesis, 150 p.

Hodgson, M.E. and P. Bresnahan, 2004, "Accuracy of Airborne Lidar–Derived Elevation: Empirical Assessment and Error Budget," *Photogrammetric Engineering & Remote Sensing*, 70(3):331–339.

Hodgson, M. E., Jensen, J. R., Schmidt, L., Schill, S. and B. A. Davis, 2003a, "An Evaluation of LIDAR- and IFSAR-Derived Digital Elevation Models in Leaf-on Conditions with USGS Level 1 and Level 2 DEMS," *Remote Sensing of Environment*, 84(2003):295–308.

Hodgson, M. E., Jensen, J. R., Tullis, J. A. Riordan, K. D., and C. M. Archer, 2003b, "Synergistic Use of LIDAR and Color Aerial Photography for Mapping Urban Parcel Imperviousness," *Photogrammetric Engineering & Remote Sensing*, 69(9):973–980.

Hodgson, M. E., Jensen, J. R., Raber, G. T., Tullis, J. A., Davis, B. A., G. Thompson and K. Schuckman, 2005, "An Evaluation of LIDAR–derived Elevation and Terrain Slope in Leaf-Off Conditions," *Photogrammetric Engineering & Remote Sensing*, 71(7):817–823.

Jensen, J. R., 2005, *Introductory Digital Image Processing: A Remote Sensing Perspective,* 3rd Ed., Upper Saddle River: Prentice-Hall, 525 p.

Jensen, J. R. and M. E. Hodgson, 2006, "Remote Sensing of Natural and Man-made Hazards and Disasters," in *Manual of Remote Sensing: Settlements*, M. K. Ridd (Ed.), Bethesda: ASP&RS, 401-444.

Jensen, J. R., Hodgson, M. E., Mackey, Jr., H. E. and W. Krabill, 1987, "Correlation Between Aircraft MSS and LIDAR Remotely Sensed Data on a Forested Wetland, *Geocarto International*, 4:39–54.

Jensen, J. R. and M. E. Hodgson, 2004, "Remote Sensing of Selected Biophysical Variables and Urban/Suburban Phenomenon" in *Geography and Technology*, Stanley D. Brunn, Susan L. Cutter and J. W. Harrington, Jr. (Eds.), Dordrecht: Kluwer, 109–154.

Konecny, G., 2003, *Geoinformation: Remote Sensing, Photogrammetry and Geographic Information Systems*, London: Taylor & Francis, 248 p.

Lefsky, M. A., Hudak, A. T., Cohen, W. B. and S. A. Acker, 2005, "Geographic Variability in Lidar Predictions of Forest Stand Structure in the Pacific Northwest," *Remote Sensing of Environment*, 95:532–548.

Leonard, J., 2005, *Technical Approach for LIDAR Acquisition and Processing*, Frederick, MD: EarthData Inc., 20 p., www.earthdata.com.

Ma, R., 2005, "DEM Generation and Building Detection from Lidar Data," *Photogrammetric Engineering & Remote Sensing*, 71(7):847–854.

Maune, D. F., (Ed.), 2001, *Digital Elevation Model Technologies and Applications: The DEM Users Manual*, Bethesda: ASP&RS, 540 p.

McGlone, J. C., 2004, *Manual of Photogrammetry*, 5th Ed., Bethesda: ASP&RS, 1151 p.

Miller, S. B., 2004, "Photogrammetric Products," in McGlone, J., (Ed.), *Manual of Photogrammetry*, 5th Ed., Bethesda: ASP&RS, 983–1013.

Mikhail, E. M., Bethel, J. S. and J. C. McGlone, 2001, *Introduction to Modern Photogrammetry*, NY: John Wiley, 479 p.

Morris, J. T., Porter, D., Neet, M., Noble, P. A., Schmidt, L., Lapine, L. A. and J. R. Jensen, 2005, "Integrating LIDAR Elevation Data, Multispectral Imagery, and Neural Network Modeling for Marsh Characterization," *International Journal of Remote Sensing,* 26(23):5221–5234.

NCRST, 2004, *National Consortium on Remote Sensing in Transportation: Infrastructure*, Santa Barbara: NCGIA Department of Geography, UCSB; Washington: DOT, 2 p., http://www.ncgia.ucsb.edu/ncrst/research/ncgia.html.

NDEP, 2006, *National Digital Elevation Program*, Washington: NDEP, www. ndep. gov.

Nilsson, M., 1996, "Estimation of Tree Heights and Stand Volume using an Airborne LIDAR system," *Remote Sensing of Environment*, 56(1):1–7.

Queija, V. R., Stoker, J. M. and J.J. Kosovich, 2005, "Recent U.S.G.S. Applications of LIDAR," *Photogrammetric Engineering & Remote Sensing*, 71(1):1–5.

Raber, G., Hodgson, M. E. and J. R. Jensen, 2006, "Impact of LIDAR Nominal Posting Density on DEM Accuracy, Hydraulic Modeling, and Flood Zone Delineation," *Photogrammetric Engineering & Remote Sensing*,72: in press.

Raber, G. T., Jensen, J. R., Schill, S. R. and K. Schuckman, 2002, "Creation of Digital Terrain Models Using An Adaptive Lidar Vegetation Point Removal Process," *Photogrammetric Engineering & Remote Sensing*, 68(12):1307–1315.

Sithole, G. and G. Vosselman, 2004, "Experimental Comparison of Filter Algorithms for Bare-Earth Extraction from Airborne Laser Scanning Point Clouds," *ISPRS Journal of Photogrammetry & Remote Sensing*, 59:85–105.

Sanborn, 2005, *LIDAR Technology*, Colorado Springs: Sanborn Map Company., www.sanborn.com.

Shrestha, R. L., Carter, W., Sartori, M., Luzum, B. J. and K. C. Slatton, 2005, "Airborne Laser Swath Mapping: Quantifying Changes in Sandy Beaches over Time Scales of Weeks to Years," *ISPRS Journal of Photogrammetry & Remote Sensing*, 59:222–232.

U.S.G.S., 2006, *Standards for Digital Elevation Models*, Washington: USGS National Mapping Division, http://mapping.usgs.gov/standards/index.html.

Weltz, M. A., J. C. Ritchie, and H. D. Fox, 1994, "Comparison of Laser and Field Measurements of Vegetation Height and Canopy Cover," *Water Resources Research*, 30(5):1311–1319.

Zimble, D. A., Evans, D. L., Carlson, G. C., Parker, R. C., Gradob, S. C. and P. D. Gerard, 2003, "Characterizing Vertical Forest Structure Using Small-footprint Airborne LIDAR," *Remote Sensing of Environment*, 87:171–182.

Remote Sensing of Vegetation 11

Approximately 70 percent of the Earth's land surface is covered with vegetation. Furthermore, vegetation is one of the most important components of ecosystems. Knowledge about variations in vegetation species and community distribution patterns, alterations in vegetation phenological (growth) cycles, and modifications in the plant physiology and morphology provide valuable insight into the climatic, edaphic, geologic, and physiographic characteristics of an area (Weiers et al., 2004; Jackson and Jensen, 2005). Scientists have devoted significant effort to develop sensors and visual and digital image processing algorithms to extract important vegetation biophysical information from remotely sensed data (e.g., Townshend and Justice, 2002). Many of the remote sensing techniques are generic in nature and may be applied to a variety of vegetated landscapes, including:

- agriculture

- forests

- rangeland

- wetland

- manicured urban vegetation.

This chapter introduces the fundamental concepts associated with vegetation biophysical characteristics and how remotely sensed data can be processed to provide unique information about these parameters. It then summarizes some of the vegetation indices developed to extract biophysical vegetation information from digital remote sensor data. Several of the metrics used by landscape ecologists to extract meaningful parameters about vegetation patch shape, size, etc., using remote sensor data are then summarized. Case studies are provided throughout the chapter on remote sensing of agriculture, inland wetland, coastal wetland, and biodiversity.

 Photosynthesis Fundamentals

Oil and coal today provide more than 90 percent of the energy needed to power automobiles, trains, trucks, ships, airplanes, factories, and a myriad of electrically energized appliances, computers, and communication systems. The energy within oil and coal was originally "captured" from the Sun by plants growing millions of years ago that were transformed into fossil fuels by geological forces. Therefore, photosynthesis, at least indirectly, is not only the principal means of enabling a civilized society to function normally but also the sole means of sustaining life — except for a few bacteria that derive their energy from sulfur salts and other inorganic compounds. This unique photosynthetic manufacturing process of green plants furnishes raw material, energy, and oxygen. In photosynthesis, energy from the Sun is harnessed and packed into simple sugar molecules made from water and carbon dioxide (CO_2) with the aid of chlorophyll. Oxygen (O_2) is given off as a by-product of the process.

The naturalist Joseph Priestly discovered in 1772 that when he placed a candle in an inverted jar it would burn out quickly, long before it ran out of wax. He also found that a mouse would die if placed under the same jar. He believed that the air was *injured* by the candle and the mouse but that it could be restored by placing a plant under the jar. Jan Ingen-Housz built on Priestly's experiments and discovered in 1788 that the influence of sunlight on the plant could cause it to rescue the mouse in a few hours. In 1796 the French pastor Jean Senebier discovered that it was the carbon dioxide, CO_2, in the jar that was the injured air and that it was actually taken up by the plants. Finally, Theodore de Saussure demonstrated that the increase in the mass of the plant as it grows was due not only to the uptake of CO_2 but also to the uptake of water, H_2O. In the early twentieth century, scientists found that the oxygen for photosynthesis was derived from the water. In effect, light energy entering the plant splits the water into oxygen and hydrogen. The photosynthetic process is described by the equation

$$6CO_2 + 6H_2O + \text{light energy} \rightarrow C_6H_{12}O_6 + 6O_2 \quad (11\text{-}1)$$

Photosynthesis is an energy-storing process that takes place in leaves and other green parts of plants in the presence of light. The light energy is stored in a simple sugar molecule (glucose) that is produced from carbon dioxide (CO_2) present in the air and water (H_2O) absorbed by the plant primarily through the root system. When the carbon dioxide and the water are combined and form a sugar molecule ($C_6H_{12}O_6$) in a chloroplast, oxygen gas (O_2) is released as a by-product. The oxygen diffuses out into the atmosphere. The photosynthetic process begins when sunlight strikes *chloroplasts*, small bodies in the leaf that contain a green substance called chlorophyll.

Plants have adapted their internal and external structure to perform photosynthesis. This structure and its interaction with electromagnetic energy has a direct impact on how

leaves and canopies appear spectrally when recorded using remote sensing instruments.

 ## Spectral Characteristics of Vegetation

A healthy green leaf intercepts incident radiant flux (Φ_i) directly from the Sun or from diffuse skylight scattered onto the leaf. This incident electromagnetic energy interacts with the pigments, water, and intercellular air spaces within the plant leaf. The amount of radiant flux reflected from the leaf (Φ_ρ), the amount of radiant flux absorbed by the leaf (Φ_α), and the amount of radiant flux transmitted through the leaf (Φ_τ) can be carefully measured as we apply the energy balance equation and attempt to keep track of what happens to all the incident energy. The general equation for the interaction of *spectral* (λ) radiant flux on and within the leaf is

$$\Phi_{i_\lambda} = \Phi_{\rho_\lambda} + \Phi_{\alpha_\lambda} + \Phi_{\tau_\lambda}. \quad (11\text{-}2)$$

Dividing each of the variables by the original incident radiant flux, Φ_{i_λ},

$$\frac{\Phi_{i_\lambda}}{\Phi_{i_\lambda}} = \frac{\Phi_{\rho_\lambda}}{\Phi_{i_\lambda}} + \frac{\Phi_{\alpha_\lambda}}{\Phi_{i_\lambda}} + \frac{\Phi_{\tau_\lambda}}{\Phi_{i_\lambda}} \quad (11\text{-}3)$$

yields

$$i_\lambda = \rho_\lambda + \alpha_\lambda + \tau_\lambda \quad (11\text{-}4)$$

where ρ_λ is spectral hemispherical reflectance of the leaf, α_λ is spectral hemispherical absorptance, and τ_λ is spectral hemispherical transmittance by the leaf. Most remote sensing systems function in the $0.35 - 3.0$ μm region measuring primarily reflected energy. Therefore, it is useful to think of this relationship as

$$\rho_\lambda = i_\lambda - (\alpha_\lambda + \tau_\lambda) \quad (11\text{-}5)$$

where the energy reflected from the plant leaf surface is equal to the incident energy minus the energy absorbed directly by the plant for photosynthetic or other purposes and the amount of energy transmitted directly through the leaf onto other leaves or the ground beneath the canopy.

Dominant Factors Controlling Leaf Reflectance

Pioneering work by Gates et al. (1965), Gausmann et al. (1969), Myers (1970) and others demonstrated the impor-

Spectral Characteristics of Healthy Green Vegetation

Dominant Factors Controlling Leaf Reflectance

Leaf pigments in the palisade mesophyll: chlorophyll *a, b* β-carotene, etc.

Scattering in the spongy mesophyll

Leaf water content

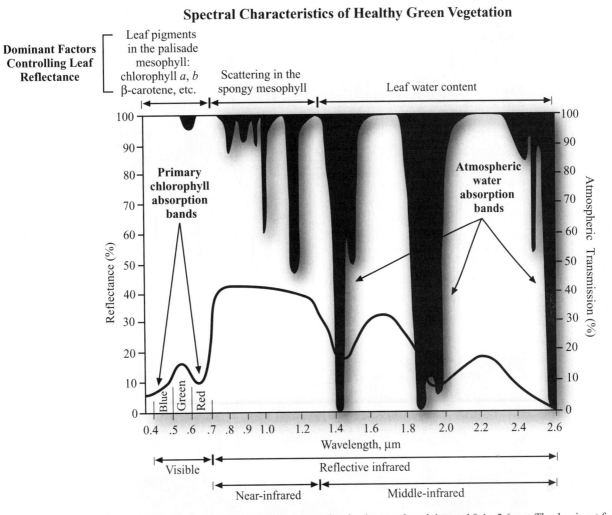

Figure 11-1 Spectral reflectance characteristics of healthy green vegetation for the wavelength interval 0.4 – 2.6 μm. The dominant factors controlling leaf reflectance are the various leaf pigments in the palisade mesophyll (e.g., chlorophyll *a* and *b*, and β-carotene), the scattering of near-infrared energy in the spongy mesophyll, and the amount of water in the plant. The primary chlorophyll absorption bands occur at 0.43 – 0.45 μm and 0.65 – 0.66 μm in the visible region. The primary water absorption bands occur at 0.97, 1.19, 1.45, 1.94, and 2.7 μm.

tance of understanding how leaf pigments, internal scattering, and leaf water content affect the reflectance and transmittance properties of leaves (Peterson and Running, 1989). Dominant factors controlling leaf reflectance in the region from 0.35 – 2.6 μm are summarized in Figure 11-1.

Visible Light Interaction with Pigments in the Palisade Mesophyll Cells

The process of food-making via photosynthesis determines how a leaf and the associated plant canopy actually appear radiometrically on remotely sensed images. A healthy leaf needs three things to make food:

- carbon dioxide (CO_2)

- water (H_2O)

- irradiance (E_λ) measured in W m^{-2}.

The carbon dioxide from the air and the water provided by the root and stem system represent the fundamental raw materials of photosynthesis. Sunlight provides the irradiance (E_λ) that powers photosynthesis.

The leaf is the primary photosynthesizing organ. A cross-section of a typical green leaf is shown in Figure 11-2. The cell structure of leaves is highly variable depending upon

Hypothetical Leaf Cross-section

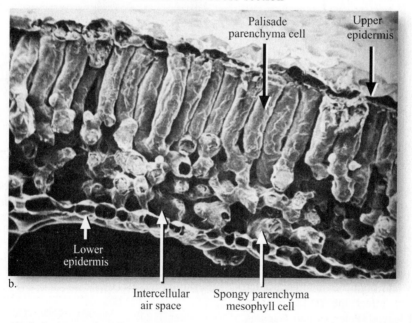

Actual Leaf Cross-section

Figure 11-2   a) Hypothetical cross-section of a typical healthy green leaf, showing both the top and underside of the leaf. The chlorophyll pigments in the palisade parenchyma cells have a significant impact on the absorption and reflectance of visible light (blue, green, and red), while the spongy parenchyma mesophyll cells have a significant impact on the absorption and reflectance of near-infrared incident energy. b) Electron microscope image of a green leaf.

species and environmental conditions during growth. Carbon dioxide enters the leaf from the atmosphere through tiny pores called *stomata* or *stoma,* located primarily on the underside of the leaf on the *lower epidermis.* Each stomata is surrounded by *guard cells* that swell or contract. When they swell, the stomata pore opens and allows carbon dioxide to enter the leaf. A typical sunflower leaf might have 2 million stomata, but they make up only about 1 percent of the leaf's surface area. Usually, there are more stomata on the bottom of a leaf; however, on some leaves the stomata are evenly distributed on both the upper and lower epidermis.

The top layer of leaf *upper epidermis* cells has a *cuticular* surface that diffuses but reflects very little light (Philpott, 1971). It is variable in thickness but is often only 3 – 5 μm thick with cell dimensions of approximately 18 × 15 × 20 μm. It is useful to think of it as a waxy, translucent material similar to the cuticle at the top of your fingernail. Leaves of many plants that grow in bright sunlight have a thick cuticle that can filter out some light and guard against excessive plant water loss. Conversely, some plants such as ferns and some shrubs on the forest floor must survive in shaded conditions. The leaves of many of these plants have a thin cuticle so that the plant can collect as much of the dim sunlight as possible for photosynthesis.

Many leaves in direct sunlight have hairs growing out of the upper (and lower) epidermis, causing them to feel fuzzy. These hairs can be beneficial, as they reduce the intensity of the incident sunlight to the plant. Nevertheless, much of the visible and near-infrared wavelength energy is transmitted through the cuticle and upper epidermis to the palisade parenchyma mesophyll cells and spongy parenchyma mesophyll cells below.

Photosynthesis occurs inside the typical green leaf in two kinds of food-making cells — *palisade parenchyma* and *spongy parenchyma* mesophyll cells. Most leaves have a distinct layer of long palisade parenchyma cells in the upper part of the mesophyll and more irregularly shaped, loosely arranged spongy parenchyma cells in the lower part of the mesophyll. The palisade cells tend to form in the portion of the mesophyll toward the side from which the light enters the leaf. In most horizontal (planophile) leaves the palisade cells will be toward the upper surface, but in leaves that grow nearly vertical (erectophile), the palisade cells may form from both sides. In some leaves the elongated palisade cells are entirely absent and only spongy parenchyma cells will exist within the mesophyll.

The cellular structure of the leaf is large compared to the wavelengths of light that interact with it. Palisade cells are typically 15 × 15 × 60 μm, while spongy mesophyll parenchyma cells are smaller. The palisade parenchyma mesophyll plant cells contain chloroplasts with chlorophyll pigments.

The chloroplasts are generally 5 – 8 μm in diameter and about 1 μm in width. As many as 50 chloroplasts may be present in each parenchyma cell. Within the chloroplasts are long slender *grana* strands (not shown) within which the

chlorophyll is actually located (approximately 0.5 μm in length and 0.05 μm in diameter). The chloroplasts are generally more abundant toward the upper side of the leaf in the palisade cells and hence account for the darker green appearance of the upper leaf surface compared with the bottom lighter surface.

A molecule, when struck by a wave or photon of light, reflects some of the energy or it can absorb the energy and thus enter into a higher energy or excited state (refer to Chapter 2). Each molecule absorbs or reflects its own characteristic wavelengths of light. Molecules in a typical green plant have evolved to absorb wavelengths of light in the visible region of the spectrum (0.35 – 0.70 μm) very well and are called *pigments*. An *absorption spectrum* for a particular pigment describes the wavelengths at which it can absorb light and enter into an excited state. Figure 11-3a presents the absorption spectrum of pure chlorophyll pigments in solution. Chlorophyll *a* and *b* are the most important plant pigments absorbing blue and red light: chlorophyll *a* at wavelengths of 0.43 and 0.66 μm and chlorophyll *b* at wavelengths of 0.45 and 0.65 μm (Farabee, 1997). A relative lack of absorption in the wavelengths between the two chlorophyll absorption bands produces a trough in the absorption efficiency at approximately 0.54 μm in the green portion of the electromagnetic spectrum (Figure 11-3a). Thus, it is the relatively lower absorption of green wavelength light (compared to blue and red light) by the leaf that causes healthy green foliage to appear green to our eyes.

There are other pigments present in the palisade mesophyll cells that are usually masked by the abundance of chlorophyll pigments. For example, there are yellow *carotenes* and pale yellow *xanthophyll* pigments, with strong absorption primarily in the blue wavelength region. The β-carotene absorption spectra is shown in Figure 11-3b with its strong absorption band centered at about 0.45 μm. *Phycoerythrin* pigments may also be present in the leaf which absorb predominantly in the green region centered at about 0.55 μm, allowing blue and red light to be reflected. *Phycocyanin* pigments absorb primarily in the green and red regions centered at about 0.62 μm, allowing much of the blue and some of the green light (i.e., the combination produces cyan) to be reflected (Figure 11-3b). Because chlorophyll *a* and *b* chloroplasts are also present and have a similar absorption band in this blue region, they tend to dominate and mask the effect of the other pigments present. When a plant undergoes senescence in the fall or encounters stress, the chlorophyll pigment may disappear, allowing the carotenes and other pigments to become dominant. For example, in the fall, chlorophyll production ceases, causing the yellow coloration of the carotenes and other specific pigments in the tree foli-

Figure 11-3   a) Absorption spectra of chlorophyll *a* and *b* pigments. Chlorophyll *a* and *b* pigments in a leaf absorb much of the incident blue and red wavelength energy. b) Absorption spectra of β-carotene, which absorbs primarily in the blue. Other pigments that might be found in a leaf include Phycoerythrin which absorbs primarily green light, and Phycocyanin which absorbs primarily green and red light (after Farabee, 1997).

age to become more visible to our eyes. In addition, some trees produce great quantities of *anthocyanin* in the fall, causing the leaves to appear bright red.

The two optimum spectral regions for sensing the chlorophyll absorption characteristics of a leaf are believed to be 0.45 – 0.52 μm and 0.63 – 0.69 μm (Figure 11-3a). The former region is characterized by strong absorption by carotenoids and chlorophylls, whereas the latter is characterized by strong chlorophyll absorption. Remote sensing of chlorophyll absorption within a canopy represents a fundamental biophysical variable useful for many biogeographical investigations. The absorption characteristics of plant canopies may be coupled with other remotely sensed data to identify vegetation stress, yield, and other hybrid variables. Thus, many remote sensing studies are concerned with monitoring what happens to the *photosynthetically active radiation* (PAR) as it interacts with individual leaves and/or the plant canopy. The use of high spectral resolution imaging spectrometers are particularly useful for measuring the absorption and reflectance characteristics of the photosynthetically active radiation.

To demonstrate these principles, consider the spectral reflectance characteristics of four different leaves obtained from a single healthy Sweetgum tree (*Liquidambar styraciflua* L.) on November 11, 1998, in Columbia, SC (Color Plate 11-1).

The green leaf (a), yellow leaf (b), and red leaf (c) were still on the Sweetgum tree at the time of data collection. The dark brown leaf (d) was collected on the ground beneath the tree.

A GER 1500 (Geophysical & Environmental Research, Inc.) handheld spectroradiometer was used to obtain spectral reflectance measurements from each of the leaves. The spectroradiometer obtained spectral reflectance measurements in 512 bands in the ultraviolet, blue, green, red, and near-infrared spectral regions from 350 – 1050 nm. Percent reflectance measurements were obtained in the lab by measuring the amount of energy reflected from the surface of the leaf (the target) divided by the amount of energy reflected from a Spectralon® reflectance reference (percent reflectance = target/reference × 100). The reflectance measurements for each of the leaves from 400 – 1050 nm were plotted in a percent reflectance graph (Color Plate 11-1e).

The green leaf (Color Plate 11-1a) was still photosynthesizing and yielded a typical healthy green reflectance spectra with strong chlorophyll absorption bands in the blue and red regions (approximately 6 percent reflectance at 450 nm and 5 percent at 650 nm, respectively), and a peak in reflectance in the green region of the visible spectrum (11 percent at 550 nm). Approximately 76 percent of the incident near-infrared radiant flux was reflected from the leaf at 900 nm.

The yellow leaf (Color Plate 11-1b) was undergoing senescence. As the influence of the chlorophyll pigments diminished, relatively greater amounts of green (24 percent at 550 nm) and red (32 percent at 650 nm) light were reflected from the leaf, resulting in a yellow appearance. At 750 nm the yellow leaf reflected less near-infrared radiant flux than the healthy green leaf. However, near-infrared reflectance at 900 nm was about 76 percent, very similar to the healthy green leaf.

The red leaf (Color Plate 11-1c) reflected 7 percent of the blue at 450 nm, 6 percent of the green energy at 550 nm, and 23 percent of the incident red energy at 650 nm. Near-infrared reflectance at 900 nm dropped to 70 percent.

The dark brown leaf (Color Plate 11-1d) produced a spectral reflectance curve with low blue (7 percent at 450 nm), green (9 percent at 550 nm), and red reflectance (10 percent at 650 nm). This combination produced a dark brown appearance. Near-infrared reflectance dropped to 44 percent at 900 nm.

It is important to understand the physiology of the plants under investigation and especially their pigmentation characteristics so that we can appreciate how a typical plant will appear when chlorophyll absorption starts to decrease, either due to seasonal senescence or environmental stress. As demonstrated, when a plant is under stress and/or chlorophyll production decreases, the lack of chlorophyll pigmentation typically causes the plant to absorb less in the chlorophyll absorption bands. Such plants will have a much higher reflectance, particularly in the green and red portion of the spectrum, and therefore may appear yellowish or *chlorotic*. In fact, Carter (1993) suggests that increased reflectance in the visible spectrum is the most consistent leaf reflectance response to plant stress. Infrared reflectance responds consistently only when stress has developed sufficiently to cause severe leaf dehydration (to be demonstrated shortly).

Leaf spectral reflectance is most likely to indicate plant stress first in the sensitive 535 – 640 and 685 – 700 nm visible light wavelength ranges. Increased reflectance near 700 nm represents the often reported "blue shift of the red edge," i.e., the shift toward shorter wavelengths of the red-infrared transition curve that occurs in stressed plants when reflectance is plotted versus wavelength (Cibula and Carter, 1992). The shift toward shorter wavelengths in the region from 650 – 700 nm is particularly evident for the yellow and red reflectance curves shown in Color Plate 11-1e. Remote sensing within these spectrally narrow ranges may provide improved capability to detect plant stress not only in individual leaves but for whole plants and perhaps for densely vegetated canopies (Carter, 1993; Carter et al., 1996).

Normal color film is sensitive to blue, green, and red wavelength energy. Color-infrared film is sensitive to green, red, and near-infrared energy after minus-blue (yellow) filtration (refer to Chapter 4). Therefore, even the most simple camera with color or color-infrared film and appropriate band-pass filtration (i.e., only certain wavelengths of light are allowed to pass) can be used to remotely sense differences in spectral reflectance caused by the pigments present in the palisade parenchyma layer of cells in a typical leaf. However, to detect very subtle spectral reflectance differences in the relatively narrow bands suggested by Cibula and Carter (1992) and Carter et al. (1996), it may be necessary to use a high spectral resolution imaging spectroradiometer that has very narrow bandwidths.

Near-Infrared Energy Interaction within the Spongy Mesophyll Cells

In a typical healthy green leaf, the near-infrared reflectance increases dramatically in the region from 700 – 1200 nm. For example, the healthy green leaf in the previous example reflected approximately 76 percent of the incident near-infrared energy at 900 nm. Healthy green leaves absorb radiant energy very efficiently in the blue and red portions of the spectrum where incident light is required for photosynthesis. But immediately to the long wavelength side of the red chlorophyll absorption band, why does the reflectance and transmittance of plant leaves increase so dramatically, causing the absorptance to fall to low values (Figure 11-1)? This condition occurs throughout the near-infrared wavelength range where the direct sunlight incident on plants has the bulk of its energy. If plants absorbed this energy with the same efficiency as they do in the visible region, they could become much too warm and the proteins would be irreversibly denatured. As a result, plants have adapted so they do not use this massive amount of near-infrared energy and simply reflect it or transmit it through to underlying leaves or the ground.

The spongy mesophyll layer in a green leaf controls the amount of near-infrared energy that is reflected. The spongy mesophyll layer typically lies below the palisade mesophyll layer and is composed of many cells and intercellular air spaces as shown in Figure 11-2. It is here that the oxygen and carbon dioxide exchange takes place for photosynthesis and respiration. In the near-infrared region, healthy green vegetation is generally characterized by high reflectance (40 – 60 percent), high transmittance (40 – 60 percent) through the leaf onto underlying leaves, and relatively low absorptance (5 – 10 percent). Notice that a healthy green leaf's reflectance and transmittance spectra throughout the visible and near-infrared portion of the spectrum are almost mirror

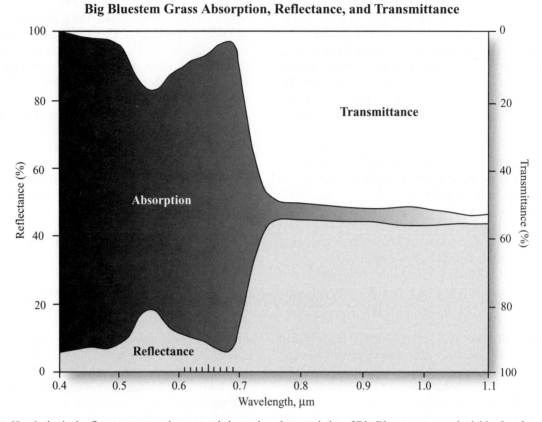

Big Bluestem Grass Absorption, Reflectance, and Transmittance

Figure 11-4   Hemispherical reflectance, transmittance, and absorption characteristics of Big Bluestem grass adaxial leaf surfaces obtained using a laboratory spectroradiometer. The reflectance and transmittance curves are almost mirror images of one another throughout the visible and near-infrared portions of the electromagnetic spectrum. The blue and red chlorophyll in plants absorb much of the incident energy in the visible portion of the spectrum (0.4 – 0.7 μm) (after Walter-Shea and Biehl, 1990). Imaging spectrometers such as AVIRIS are capable of identifying small changes in the absorption and reflection characteristics of plants because the sensors often have channels that are only 10 nm apart, i.e., we could have 10 channels in the region from 0.6 – 0.7 μm (600 to 700 nm).

images of one another, as shown in Figure 11-4 (Walter-Shea and Biehl, 1990).

The high diffuse reflectance of the near-infrared (0.7 – 1.2 μm) energy from plant leaves is due to the internal scattering at the cell wall–air interfaces within the leaf (Gausmann et al., 1969; Peterson and Running, 1989). A water vapor absorption band exists at 0.92 – 0.98 μm; consequently, the optimum spectral region for sensing in the near-infrared region is believed to be 0.74 – 0.90 μm (Tucker, 1978).

The main reasons that healthy plant canopies reflect so much near-infrared energy are:

• the leaf already reflects 40 – 60 percent of the incident near-infrared energy from the spongy mesophyll (Figure 11-4), and

• the remaining 45 – 50 percent of the energy penetrates (i.e., is transmitted) through the leaf and can be reflected once again by leaves below it.

This is called *leaf additive reflectance*. For example, consider the reflectance and transmission characteristics of the hypothetical two-layer plant canopy shown in Figure 11-5. Assume that leaf 1 reflects 50 percent of the incident near-infrared energy back into the atmosphere and that the remaining 50 percent of the near-infrared energy is transmitted through leaf 1 onto leaf 2. The transmitted energy then falls on leaf 2 where 50 percent again is transmitted (25 percent of the original) and 50 percent is reflected. The reflected energy then passes back through leaf 1 which allows half of that energy (or 12.5 percent of the original) to be transmitted and half reflected. The resulting total energy exiting leaf 1 in this two-layer example is 62.5 percent of the incident energy. Therefore, the greater the number of leaf

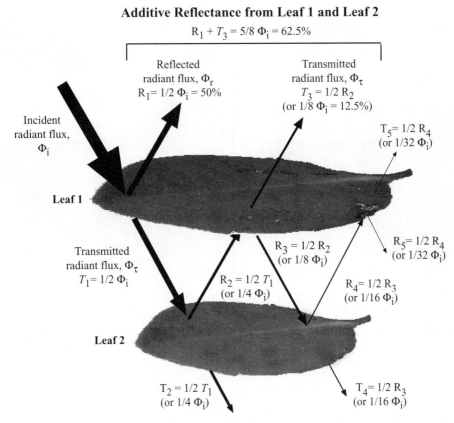

Additive Reflectance from Leaf 1 and Leaf 2

$R_1 + T_3 = 5/8\ \Phi_i = 62.5\%$

Reflected radiant flux, Φ_r
$R_1 = 1/2\ \Phi_i = 50\%$

Transmitted radiant flux, Φ_τ
$T_3 = 1/2\ R_2$
(or $1/8\ \Phi_i = 12.5\%$)

Incident radiant flux, Φ_i

$T_5 = 1/2\ R_4$
(or $1/32\ \Phi_i$)

Leaf 1

$R_5 = 1/2\ R_4$
(or $1/32\ \Phi_i$)

Transmitted radiant flux, Φ_τ
$T_1 = 1/2\ \Phi_i$

$R_3 = 1/2\ R_2$
(or $1/8\ \Phi_i$)

$R_2 = 1/2\ T_1$
(or $1/4\ \Phi_i$)

$R_4 = 1/2\ R_3$
(or $1/16\ \Phi_i$)

Leaf 2

$T_2 = 1/2\ T_1$
(or $1/4\ \Phi_i$)

$T_4 = 1/2\ R_3$
(or $1/16\ \Phi_i$)

Figure 11-5   A hypothetical example of additive reflectance from a canopy with two leaf layers. Fifty percent of the incident radiant flux, Φ_i, to leaf 1 is reflected (R_1), and the other 50 percent is transmitted onto leaf 2 (T_1). Fifty percent of the radiant flux incident to leaf 2 is transmitted through leaf 2 (T_2); the other 50 percent is reflected toward the base of leaf 1 (R_2). Fifty percent of the energy incident at the base of leaf 1 is transmitted through it (T_3) while the remaining 50 percent (R_3) is reflected toward leaf 2 once again. At this point, an additional 12.5 percent (1/8) reflectance has been contributed by leaf 2, bringing the total reflected radiant flux to 62.5 percent. However, to be even more accurate, one would have to also take into account the amount of energy reflected from the base of leaf 1 (R_3) onto leaf 2, the amount reflected from leaf 2 (R_4), and eventually transmitted through leaf 1 once again (T_5). This process would continue.

layers in a healthy, mature canopy, theoretically the greater the infrared reflectance. Conversely, if the canopy is only composed of a single, sparse leaf layer then the near-infrared reflectance will not be as great because the energy that is transmitted through the leaf layer may be absorbed by the ground cover beneath.

Changes in the near-infrared spectral properties of healthy green vegetation may provide information about plant senescence and/or stress. For example, consider the four leaves and their spectral reflectance characteristics shown in Color Plate 11-1. The photosynthesizing green leaf (a) exhibited strong chlorophyll absorption in the blue and red wavelength regions, an understandable increase in green reflectance, and approximately 76 percent reflectance in the near-infrared region. After a certain point, near-infrared reflectance decreased as the leaves senesced (b – d). However, if the

leaves were to dry out significantly during senescence, we would expect to see much higher reflectance in the near-infrared region (to be discussed shortly).

Scientists have known since the 1960s that a *direct* relationship exists between response in the near-infrared region and various biomass measurements. Conversely, it has been shown that an *inverse* relationship exists between the response in the visible region, particularly the red, and plant biomass. The best way to appreciate this is to plot all of the pixels in a typical remote sensing scene in red and near-infrared reflectance space. For example, Figure 11-6a depicts where approximately 10,000 pixels in a typical agricultural scene are located in red and near-infrared multispectral feature space (i.e., within the gray area). Dry bare soil fields and wet bare soil fields in the scene would be located at opposite ends of the soil line. This means that a wet bare

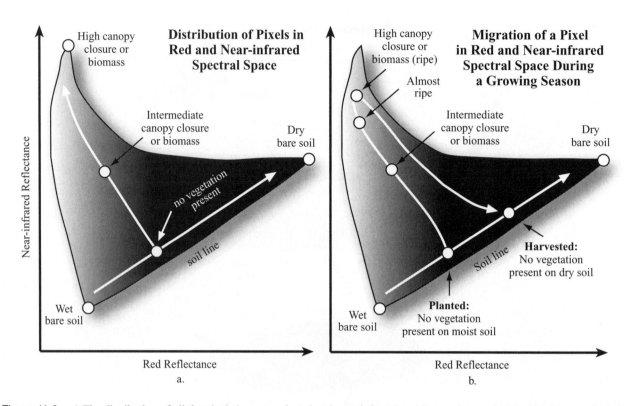

Figure 11-6   a) The distribution of all the pixels in a scene in red and near-infrared multispectral space is found in the gray shaded area. Wet and moist bare soil fields are located along the soil line. The greater the biomass and/or crop canopy closure, the greater the near-infrared reflectance and the lower the red reflectance. This condition moves the pixel's spectral location a perpendicular direction away from the soil line. b) The migration of a single vegetated agricultural pixel in red and near-infrared multispectral space during a growing season is shown. After the crop emerges, it departs from the soil line, eventually reaching complete canopy closure. After harvesting, the pixel will be found on the soil line, but perhaps in a drier soil condition.

soil would have very low red and near-infrared reflectance. Conversely, a dry bare soil area would probably have high red and high near-infrared reflectance. As a vegetation canopy matures, it reflects more near-infrared energy while at the same time absorbing more red radiant flux for photosynthetic purposes. This causes the spectral reflectance of the pixel to move in a perpendicular direction away from the soil line. As biomass increases and as the plant canopy cover increases, the field's location in the red and near-infrared spectral space moves farther away from the soil line.

Figure 11-6b demonstrates how just one agricultural pixel might move about in the red and near-infrared spectral space during a typical growing season. If the field was prepared properly, it would probably be located in the moist bare soil region of the soil line with low red and low near-infrared reflectance at the beginning of the growing season. After the crop emerges, it would depart from the soil line, eventually reaching complete canopy closure. At this point the reflected near-infrared radiant flux would be high and the red reflectance would be low. After harvesting, the pixel would prob-

ably be found once again on the soil line but perhaps in a drier condition.

The relationship between red and near-infrared canopy reflectance has resulted in the development of numerous remote sensing vegetation indices and biomass-estimating techniques that utilize multiple measurements in the visible and near-infrared region (e.g., Richardson and Everitt, 1992; Lyon et al., 1998). The result is a linear combination that may be more highly correlated with biomass than either red or near-infrared measurement alone. Several of these algorithms are summarized in the section on Vegetation Indices in this chapter.

Middle-Infrared Energy Interaction with Water in the Spongy Mesophyll

Plants require water to grow. A leaf obtains water through the plant's roots. The water travels from the roots, up the stem, and enters the leaf through the *petiole*. Veins carry water to the cells in the leaf. If a plant is watered so that it contains as much water as it can possibly hold at a given time, it is said to be fully *turgid*. Much of the water is found in the spongy mesophyll portion of the plant. If we forget to water the plant or rainfall decreases, the plant will contain an amount of water that is less than it can potentially hold. This is called its *relative turgidity*. It would be useful to have a remote sensing instrument that was sensitive to how much water was actually in a plant leaf. Remote sensing in the middle-infrared, thermal infrared (Chapter 8), and passive microwave (Chapter 9) portion of the electromagnetic spectrum can provide such information to a limited extent.

Liquid water in the atmosphere creates five major absorption bands in the near-infrared through middle-infrared portions of the electromagnetic spectrum at 0.97, 1.19, 1.45, 1.94, and 2.7 μm (Figure 11-1). The fundamental vibrational water-absorption band at 2.7 μm is the strongest in this part of the spectrum (there is also one in the thermal infrared region at 6.27 μm). However, there is also a strong relationship between the reflectance in the middle-infrared region from 1.3 – 2.5 μm and the amount of water present in the leaves of a plant canopy. Water in plants absorb incident energy between the absorption bands with increasing strength at longer wavelengths. In these middle-infrared wavelengths, vegetation reflectance peaks occur at about 1.6 and 2.2 μm, between the major atmospheric water absorption bands (Figure 11-1).

Water is a good absorber of middle-infrared energy, so the greater the turgidity of the leaves, the lower the middle-infrared reflectance. Conversely, as the moisture content of leaves decreases, reflectance in the middle-infrared region increases substantially. As the amount of plant water in the intercellular air spaces decreases, this causes the incident middle-infrared energy to be more intensely scattered at the interface of the intercellular walls resulting in greater middle-infrared reflectance from the leaf. For example, consider the spectral reflectance of Magnolia leaf samples at five different moisture conditions displayed over the wavelength interval from 0.4 – 2.5 μm (Figure 11-7). The middle-infrared wavelength intervals from about 1.5 – 1.8 μm and from 2.1 – 2.3 μm appear to be more sensitive to changes in the moisture content of the plants than the visible or near-infrared portions of the spectrum (i.e., the y-axis distance between the spectral reflectance curves is greater as the moisture content decreases). Also note that substantive changes in the visible reflectance curves (0.4 – 0.7 μm) did not begin to appear until the plant moisture in the leaves decreased to about 50 percent. When the relative water content of the plant decreases to 50 percent, almost any portion of the visible, near- and middle-infrared regions might provide some valuable spectral reflectance information.

Leaf reflectance in the middle-infrared region is inversely related to the absorptance of a layer of water approximately 1 mm in depth (Carter, 1991). The degree to which incident solar energy in the middle-infrared region is absorbed by vegetation is a function of the total amount of water present in the leaf and the leaf thickness. If proper choices of sensors and spectral bands are made, it is possible to monitor the relative turgidity in plants.

Most optical remote sensing systems (except radar) are generally constrained to function in the wavelength intervals from 0.3 – 1.3, 1.5 – 1.8, and 2.0 – 2.6 μm due to the strong atmospheric water absorption bands at 1.45, 1.94, and 2.7 μm. Fortunately, as demonstrated in Figure 11-7, there is a strong "carryover" sensitivity to water content in the 1.5 – 1.8 and 2.0 – 2.6 μm regions adjacent to the major water absorption bands. This is the reason that the Landsat Thematic Mapper (4 and 5) and Landsat 7 Enhanced Thematic Mapper Plus (ETM[+]) were made sensitive to two bands in this region: band 5 (1.55 – 1.75 μm) and band 7 (2.08 – 2.35 μm). The 1.55 – 1.75 μm middle-infrared band has consistently demonstrated a sensitivity to canopy moisture content. For example, Pierce et al. (1990) found that this band and vegetation indices produced using it were correlated with canopy water stress in coniferous forests.

Much of the water in a plant is lost via transpiration. *Transpiration* occurs as the Sun warms the water inside the leaf, causing some of the water to change its state to water vapor that escapes through the stomata. The following are several important functions that transpiration performs:

• It cools the inside of the leaf because the escaping water vapor contains heat.

• It keeps water flowing up from the roots, through the stem, to the leaves.

• It ensures a steady supply of dissolved minerals from the soil.

As molecules of water vapor at the top of the leaf in the tree are lost to transpiration, the entire column of water is pulled upward. Plants lose a considerable amount of water through transpiration each day. For example, a single corn plant can lose about 4 quarts (3.8 liters) of water on a very hot day. If the roots of the plant cannot replace this water, the leaves wilt, photosynthesis stops, and the plant dies. Thus, monitoring the moisture content of plant canopies, which is correlated with rates of transpiration, can provide valuable information on the health of a crop or stand of vegetation. Thermal infrared and passive microwave remote sensing

Figure 11-7   Reflectance response of a single magnolia leaf (*Magnolia grandiflora*) to decreased relative water content. As moisture content decreased, reflectance increased throughout the 0.4 – 2.5 μm region. However, the greatest increase occurred in the middle-infrared region from 1.3 – 2.5 μm (after Carter, 1991).

have provided valuable plant canopy evapotranspiration information.

The most practical application of plant moisture information is the regional assessment of crop water conditions for irrigation scheduling, stress assessment, and yield modeling for agriculture, rangeland, and forestry management.

Advanced Vegetation Spectral Reflectance Characteristics: the Bidirectional Reflectance Distribution Function (BRDF)

It would be wonderful if a vegetated canopy such as a mature corn crop or Loblolly pine plantation reflected the same amount of radiant flux toward the sensor irrespective of 1) the solar incidence and azimuth angles, or 2) the sensor viewing geometry. This would mean that the vegetated canopy was a true Lambertian surface, i.e., it reflects incident energy equally well in all directions. Unfortunately, considerable research has demonstrated that this is not the case (e.g., Kimes, 1983). In fact, the spectral radiant flux leaving a vegetation canopy is significantly impacted by a number of factors, many of which are listed in Table 11-1. We briefly discuss the primary vegetation parameters and then delve more deeply into Sun illumination and sensor system geometric characteristics.

The amount and spectral distribution of radiant flux leaving a vegetated canopy is influenced by the type of vegetation

present. For example, grasses hopefully reflect energy differently than a mature stand of trees. Some vegetated canopies have 100 percent *canopy closure* meaning that the understory and soil beneath are not visible through the canopy. Conversely, some canopies have < 100 percent canopy closure, allowing portions of the understory and/or soil to reflect energy into the instantaneous field of view (IFOV) of the sensor, creating a hybrid or mixed pixel. It is possible to disentangle the spectral contribution from the individual land covers, but it requires considerable image processing expertise.

Some vegetation is oriented randomly, while other vegetation is often systematically arranged in rows in a cardinal direction (0 – 360°). Individual tree crowns often have unique shapes (e.g., a conical Ponderosa pine crown or circular Blackjack oak crown) with unique tree crown diameters that may be measured. Tree trunks or plant stems have a certain density (e.g., number of trunks per unit area) with unique *diameter-at-breast-height* (DBH) values.

Leaf-area-index (LAI) is the total one-sided (or one half of the total all-sided) green leaf area per unit ground-surface area. It is an important biological parameter because 1) it defines the area that interacts with solar radiation and provides much of the remote sensing signal, and 2) it is the surface that is responsible for carbon absorption and exchange within the atmosphere (Chen and Black, 1992). Some canopies have substantially higher leaf-area-indices than others.

Figure 11-8   The bidirectional reflectance effect on a field of ryegrass (*Lolium perenne L.*) observed under different viewing angles in the solar principal plane from a FIGOS mounted camera. Solar zenith angle was at 35° as indicated by the dashed arrows. The sensor viewing angles are shown as black lines at nadir (0°) and at ±45° and ±75° off-nadir (after Sandmeier and Itten, 1999).

The *leaf-angle-distribution* (LAD) may change throughout the day as the leaves orient themselves toward or away from the incident radiation. Some leaves lie predominantly in the horizontal plane (e.g., planophile leaves in many broadleaf trees), while others are oriented vertically (e.g., erectophile leaves of coastal cordgrass marsh).

All of these vegetation factors can have a significant impact on the reflection of incident light toward the sensor system. Therefore, it is best to hold as many of them as constant as possible when attempting to extract biophysical information such as primary productivity or biomass using multiple dates of remote sensor data. Even if we held these variables as constant as possible, the zenith and azimuth angles of the incident solar radiation and the azimuth and viewing angle of the sensor system can introduce such dramatic effects that we may not be able to compare our spectral reflectance measurements obtained at one time with those of another. Fortunately, if we know a great deal about the vegetation characteristics of the canopy (just discussed) and obtain some information about the bidirectional reflectance characteristics of the solar angle-of-incidence and the sensor viewing geometry, it is possible to calibrate the remote sensor data to extract useful biophysical information.

Bidirectional Reflectance Distribution Function (BRDF): Early measurements of bidirectional reflectance obtained over various vegetation and soil surfaces demonstrated that most terrestrial surfaces exhibit non-Lambertian (*anisotropic*) reflectance characteristics (Kimes, 1983). In the past, bidirectional reflectance effects of the land surface

did not play a major role in global change and ecological analysis, although they are assumed to be crucial for multi-temporal studies with varying Sun incidence angles. This was mainly due to a lack of bidirectional data available from remote sensors or acquired in field campaigns. The impact of the bidirectional reflectance distribution function is still not well understood despite the fact that we know it exists in much of the remotely sensed data commonly used for Earth observation, especially for sensors with large fields of view such as the NOAA AVHRR 1.1 × 1.1 km data. A great amount of data from sensors with off-nadir viewing capability is becoming available, especially from commercial satellites such as Space Imaging's IKONOS, and NASA's *Terra* MODIS and MISR (Chapter 7). To properly apply these data in land-use change and ecologically relevant studies, the BRDF should be well understood to calibrate the data (Walter-Shea and Biehl, 1990; Abuelgasim et al., 1998).

Most terrain surfaces (soil, vegetation, and even water) reveal a relationship between the amount of reflected radiance, and 1) the geometric characteristics of the Sun's irradiance, and 2) the sensor viewing geometry. Hence the term *bidirectional*. Depending upon the specific irradiance and sensor viewing angles, most surfaces appear brighter or darker. We have all witnessed this effect. If we walk around a patch of grass or a pool of deep water, it may appear brighter or darker depending upon our viewing angle in relation to the constant incident angle of the Sun. For example, Figure 11-8 depicts a portion of a ryegrass lawn (*Lolium perenne L.*) illuminated by sunlight with a constant solar zenith angle of 35°. Individual photographs of the same

Table 11-1.   Major variables affecting the bidirectional reflectance distribution function (BRDF) of a vegetated canopy.

Variables Affecting Bidirectional Reflectance from a Vegetated Canopy

Illumination	• Geometry
	  - Sun angle of incidence (or RADAR)
	  - azimuth
	• Spectral characteristics (λ)

Sensor	• Geometry
	  - angle of view (e.g., 0° nadir)
	  - azimuth look direction (0 – 360°)
	• Spectral sensitivity (λ)
	• IFOV (milliradians)

Vegetation	• Canopy
	  - type (plant or tree nominal class)
	  - closure (%)
	  - orientation
	    - systematic (e.g., rows 0 – 360°)
	    - unsystematic (random)
	• Crown
	  - shape (e.g., circular, conical)
	  - diameter (m)
	• Trunk or Stem
	  - density (units per m^2)
	  - tree diameter-at-breast-height (DBH)
	• Leaf
	  - leaf-area-index (LAI)
	  - leaf-angle-distribution (LAD) (planophile, erectophile)

Understory	same as vegetation
Soil	• Texture • Color • Moisture content

Hot spot

Figure 11-9   A hot spot near the left fiducial mark on a vertical aerial photograph of the Savannah River swamp in South Carolina. The Cypress-Tupelo forested wetland has relatively uniform percent canopy closure in this area. Its tone and texture should appear relatively homogeneous, but they do not. A hot spot is produced when the angle of illumination and sensor viewing angle are nearly identical and in the same plane.

grass canopy are obtained at nadir (0°) and at ±45° and ±75° off-nadir (Sandmeier and Itten, 1999). The terrain is generally brighter when the sensor records back-scattered energy as opposed to forward-scattered energy as the diagram reveals (i.e., the image on the left produced by the +75° back-scattered energy appears to be the brightest). Note, however, that the +75° and -75° images are both much brighter than the image acquired at nadir, 0°.

The bidirectional effect in remote sensing data is most obvious when the angle of illumination and sensor viewing angle are nearly identical and in the same plane. This can produce a *hot spot* (i.e., a shadow is not visible). The solar principal plane is formed when the source, target, and the sensor are in the same plane (in the solar azimuth angle). This is where the

BRDF effects are most pronounced. An example of a hot spot in aerial photography is shown in Figure 11-9.

Fortunately, scientists have developed instruments to measure the bidirectional reflectance distribution function of various surfaces. One of the most useful instruments is the *goniometer*. Here we briefly review the field goniometer system (FIGOS) designed by S. Sandmeier. FIGOS consists of a computer-controlled GER 3700 spectroradiometer sensitive to the wavelength interval from 300 – 2,450 nm in 704 bands with a spectral resolution of 1.5 nm for the interval 300 – 1,050 nm and 8.4 nm for the interval 1,050 – 2,450 nm. It consists of three major parts: a zenith arc rail, an azimuth rail, and a motorized sled where the radiometer is mounted (Figure 11-10a,b). The aluminum goniometer weighs approximately 230 kg and may be transported to field sites as shown. Mounted on the goniometer, the sensor views an area in the very center of the circle approximately 10.5 cm in diameter at nadir and about 41 cm (major axis) in the most extreme view zenith angle position of 75°.

Thus, it is possible in a very short period of time while the Sun is approximately in the same zenith arc (θ_i) and azimuth (φ_i), to vary the position of the radiometer to determine if the amount of radiant flux in very specific wavelength regions

Goniometer

a.

Goniometer

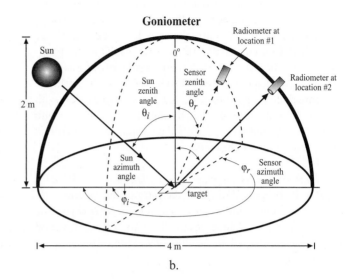

b.

Bidirectional Reflectance Distribution Function (BRDF)

c.

d.

e.

Figure 11-10 a) The field goniometer system (FIGOS) developed by S. Sand-meier. b) In this hypothetical example, the spectroradiometer is located at two locations. Spectral measurements are made at location #1 with a Sun zenith angle of θ_i and Sun azimuth angle of φ_i and a sensor angle of view of θ_r and azimuth angle of φ_r. The GER spectroradiometer records the amount of radiance leaving the target in 704 spectral bands. The radiometer is then moved to location #2 (i.e., perhaps only varying the sensor azimuth angle as shown) where it acquires radiance data again in 704 bands. c) The concept and parameters of the BRDF. A target is bathed in irradiance (dE_i) from a specific zenith and azimuth angle, and the sensor records the radiance (dL_r) exiting the target of interest at a specific azimuth and zenith angle. d) Anisotropy factor for ryegrass (*Lolium perenne L.*) at multiple zenith viewing angles in the solar principal plane for four of the 704 bands. The Sun zenith angle was 35° (after Sandmeier and Itten, 1999). e) Second-generation Sandmeier Field Goniometer (SFG) obtaining smooth cordgrass (*Spartina alterniflora*) readings at North Inlet, SC, on October 21, 1999 (Sandmeier, 1999; Schill et al., 2004).

(e.g., blue, green, red, near-infrared, middle-infrared) leaving the target is influenced by the sensor angle-of-view (θ_r) and azimuth (φ_r). If it is not, then we do not have to worry about BRDF effects. If the reflectance results are not uniform as will be demonstrated, then we may need to concern ourselves with adjusting for BRDF effects if we want to compare remote sensor data obtained on multiple dates (with varying Sun azimuth and zenith angle) or by sensors with multiple look angles (e.g., SPOT data acquired at 0° nadir on day 1 and 20° off-nadir on day 2).

The *bidirectional reflectance distribution function* (BRDF), f_r (sr^{-1}), is formally defined as the ratio of the radiance dL_r (W m^{-2} sr^{-1} nm^{-1}) reflected in one direction (θ_r, φ_r) to the Sun's incident irradiance dE_i (W m^{-2} nm^{-1}) from direction (θ_i, φ_i) (Sandmeier, 1999; Sandmeier and Itten, 1999):

$$f_r(\theta_i, \varphi_i; \theta_r, \varphi_r; \lambda) = \frac{dL_r(\theta_i, \varphi_i; \theta_r, \varphi_r; \lambda)}{dE_i(\theta_i, \varphi_i; \lambda)} . \qquad (11\text{-}6)$$

Note that dL_r and dE_i are directional quantities measured in solid angle steradians (sr^{-1}) as shown in Figure 11-10c.

Normally, a *bidirectional reflectance factor* (*BRF*, also referred to as *R*) is computed, which is the radiance dL_r reflected from a surface in a specific direction divided by the radiance dL_{ref} reflected from a loss-less Lambertian reference panel measured under identical illumination geometry (Sandmeier and Itten, 1999):

$$R(\theta_i, \varphi_i; \theta_r, \varphi_r; \lambda) = \qquad (11\text{-}7)$$

$$\left[\frac{dL_r(\theta_i, \varphi_i; \theta_r, \varphi_r; \lambda)}{dL_{ref}(\theta_i, \varphi_i; \theta_r, \varphi_r; \lambda)} \right] \times R_{ref}(\theta_i, \varphi_i; \theta_r, \varphi_r; \lambda)$$

where R_{ref} is a calibration coefficient determined for the spectral reflectance panel. Bidirectional reflectance factors (BRF) are dimensionless. Both BRF and BRDF take on values from zero to infinity. Values > 1 for BRF and > $1/\pi$ for BRDF are obtained in peak reflectance directions, such as the hot spot, where the reflected flux from a target surface is higher than the flux from a Lambertian surface.

It is also possible to develop an *anisotropy factor*, which is used to analyze the spectral variability in BRDF data. Anisotropy factors (ANIF) allow separation of spectral BRDF effects from the spectral signature of a target. They are calculated by normalizing bidirectional reflectance data R to nadir reflectance, R_o (Sandmeier et al., 1998a; Sandmeier and Itten, 1999):

$$ANIF(\theta_i, \varphi_i; \theta_r, \varphi_r; \lambda) = \frac{R(\theta_i, \varphi_i; \theta_r, \varphi_r; \lambda)}{R_o(\theta_i, \varphi_i; \lambda)} . \qquad (11\text{-}8)$$

So what do these measurements tell us about the BRDF of a typical surface? To answer this question, consider the nadir-normalized BRDF data (i.e., the ANIF data) of perennial ryegrass (*Lolium perenne L.*) shown in Figure 11-10d. During goniometer data collection the Sun zenith angle was 35°. Spectral results from just four of the 704 spectroradiometer bands are presented using just the viewing zenith angle in the solar principal plane. It reveals that BRDF effects were very pronounced in the blue (480 nm) and red (675 nm) chlorophyll absorption bands previously discussed, whereas in the green and particularly in the low absorbing near-infrared range, relatively low BRDF effects were observed. Thus, persons using remote sensor data might consider radiometrically adjusting the brightness values associated with the blue and red bands, but not necessarily the near-infrared bands under investigation. Typical for vegetated surfaces, all four bands exhibit a bowl shape, hot spot, and forward-scattering component.

It is even more interesting to view the ryegrass anisotropy factors for the four wavelengths of interest according to not only the viewing zenith angle of 0° but in a range from $\pm 75°$ (Figure 11-11). Ideally, the entire three-dimensional surface should be relatively flat as with the 750 nm near-infrared example, meaning that measurements in this specific band are relatively free of BRDF effects. Conversely, the 480 nm and 675 nm bands exhibit significant anisotropy factors. In the high-absorbing (i.e., low-reflecting) wavelength ranges, multiple scattering effects are reduced due to the relatively low amount of radiation in the canopy. Thus, the contrast between shadowed and illuminated canopy components are enhanced, which then enhances the BRDF effects. BRDF effects are rather small in the low-absorbing (i.e., high-reflecting) green and near-infrared wavelength ranges where multiple scattering effects are strong and diminish the contrast in the canopy (Sandmeier et al., 1998b).

Future field research using goniometers coupled with off-nadir pointing aircraft and satellite remote sensing systems such as *Terra* MISR (Chapter 7) will eventually document when we should be concerned with BRDF effects. Such information may help us 1) identify and select the bands that are impacted the least by BRDF effects, 2) recognize that there are certain Sun and/or sensor angle-of-view relationships that should be avoided when collecting remotely sensed data, and/or 3) provide us with methods to radiometrically adjust the remote sensor data to minimize BRDF effects.

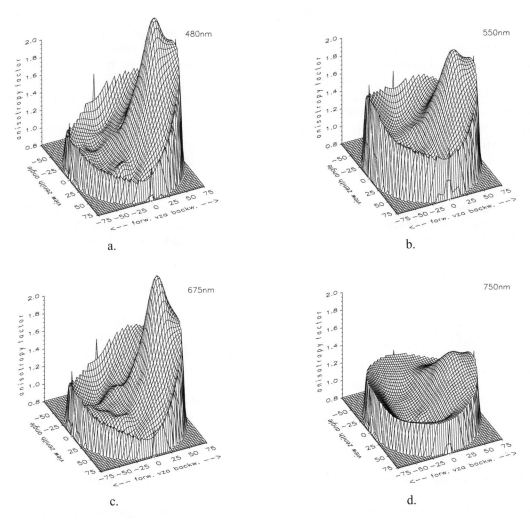

Figure 11-11   Anisotropy factors (nadir-normalized BRDF data) of ryegrass (*Lolium perenne L.*) for four spectral bands acquired with the FIGOS instrument with a Sun zenith angle of 35° (after Sandmeier and Itten, 1999).

One should be careful not to infer that BRDF effects are all bad. Qi et al. (1995) point out that multidirectional remote sensing measurements of the terrain (e.g., at viewing angles off-nadir) can provide complementary information to that provided by the nadir measurement. In fact, a single nadir view remote sensing measurement obtains information about the surface as if the surface had no vertical structures, which is usually not the case in practice, while off-nadir view measurements reveal different aspects of the vertical structures such as vegetation height. *Consequently, to objectively characterize vegetation biophysical parameters, a single viewing geometry (e.g., at nadir) may be insufficient.*

Modeling Canopy Reflectance

Scientists have attempted to predict exactly how much radiant energy in specific wavelengths should be exiting a given leaf or vegetated canopy based on a number of factors, including (Danson, 1998): the leaf-area-index (LAI), soil reflectance properties below the canopy, amount of direct and/or diffuse skylight onto the canopy, the leaf-angle-distribution (ranging from erectophile canopies with vertical leaves at 90° inclination to planophile canopies dominated by horizontal leaves with 0° inclination), and the BRDF influenced by the Sun angle and sensor viewing angle geometry just discussed. For example, the SAIL model (Scattering by Arbitrarily Inclined Leaves) has been widely used in remote sensing research for investigating the spectral and directional reflectance properties of vegetation canopies. It uses radiative transfer equations to model energy interacting with a vegetation canopy in three distinct streams, including a downward flux of direct radiation and a downward and upward flux of diffuse radiation. It assumes that the canopy may be represented by small absorbing and scattering elements (e.g., leaves) with known optical properties, distributed randomly in horizontal layers and with known angular distribution. The model has been used to simulate the effects of off-nadir viewing, to simulate spectral shifts of the red-edge, and to estimate canopy properties directly from remotely sensed data.

Similarly, the Li-Strahler model (1985, 1992) yields estimates of the size and density of trees from remotely sensed images. The signal received by the sensor is modeled as consisting of reflected light from tree crowns, their shadows, and the background within the field of view of the sensor. The reflected signal is modeled as a linear combination of four components and their areal proportions (Woodcock et al., 1997):

$$S = K_g G + K_c C + K_t T + K_z A \qquad (11\text{-}9)$$

where S is the brightness value of a pixel, K_g, K_c, K_t, and K_z stand for the areal proportions of sunlit background, sunlit crown, shadowed crown, and shadowed background, respectively, and G, C, T, and A are the spectral signatures of the respective components. This is called a mixture model, or a scene component model.

The SAIL and Li-Strahler models represent important attempts to explicitly model the energy-matter interactions taking place above, in, and below the vegetative canopy. Scientists strive to invert the models and use the reflectance characteristics recorded by the remote sensor system to predict the characteristics of specific types of structure within the canopy, such as tree height, density, leaf-area-index, etc. Unfortunately, it is often difficult to calibrate such models because so much information must be known about the leaf and canopy characteristics, atmospheric conditions, Sun angle and viewing geometry, and terrain slope and aspect. This is an important area of remote sensing research.

Imaging Spectrometry of Vegetation

As discussed in Chapter 7, an imaging spectrometer collects a continuous reflectance spectrum of the Earth's surface. But instead of having just four (e.g., SPOT) or eight (e.g., Landsat 7 Enhanced Thematic Mapper Plus) spectral channels to characterize the vegetation spectral characteristics, an imaging spectrometer records information in hundreds of spectral channels.

Imaging spectrometry has great potential for monitoring vegetation type and biophysical characteristics (Goetz, 1995). Vegetation reflectance spectra are often quite informative, containing information on the vegetation chlorophyll absorption bands in the visible region, the sustained high reflectance in the near-infrared region, and the effects of plant water absorption in the middle-infrared region.

Geologists have been using hyperspectral imagery for years to discriminate between different rock types based on their spectral reflectance and absorption characteristics. They have prepared an exhaustive spectral library of the most important minerals, soils, and rock types (e.g., Clark et al., 2003). When calibrated airborne hyperspectral imagery are acquired, they often compare the spectra obtained from the airborne data to the spectra stored in the database to determine mineral type and other characteristics. Several examples are provided in Chapter 14 (Remote Sensing of Soils, Minerals and Geomorphology).

Ideally, there would be a vegetation spectral databank available that contains information on: 1) the spectral reflectance and emittance characteristics of every type of vegetation in the region from $0.35 - 14$ μm, 2) how these data appear on different dates throughout the pertinent growing seasons, and 3) what these spectra should look like in the event of stress or insect infestation. Unfortunately, such a database does not yet exist. Instead, it is still usually necessary to collect *in situ* spectroradiometer data of the vegetation of interest at the time of the overflight and then use this data to calibrate the spectral reflectance information derived from the airborne spectroradiometer. When this is performed carefully, it is possible to extract vegetation type and condition information from the hyperspectral data.

For example, consider the AVIRIS data collected over the San Luis Valley of Colorado on September 3, 1993, shown in Color Plate 11-2a (Clark et al., 1995). The 224 spectral channels of 20×20 m AVIRIS data were radiometrically corrected to remove the effects of atmospheric water vapor and geometrically rectified to a standard map projection. Land cover of some of the fields at the time of the overflight are annotated in color on top of the single AVIRIS channel in Color Plate 11-2a. AVIRIS reflectance spectra for some of these fields, including potato, alfalfa, barley, oat hay, canola, and pasture are portrayed in Figure 11-12a,b.

The alfalfa, canola, oat hay, and Nugget potato spectra show the plants to be green and healthy (Figure 11-12a). Barley had lost all of its chlorophyll response at the time of the overflight (Figure 11-12b). Norkotah potatoes were not being irrigated as they were about to be harvested, and con-

AVIRIS Spectral Signatures of Various Crops

Figure 11-12 a) Spectra of healthy green vegetation in the San Luis Valley of Colorado obtained on September 3, 1993, using AVIRIS; 224 channels at 20 × 20 m pixels. b) Most of these other vegetation types exhibited reduced chlorophyll absorption effects. Note that the spectra are offset for clarity (after Clark et al., 1995). Differences in the reflectance and/or absorption characteristics of the vegetation make it possible in certain instances to distinguish one vegetation type from another or to identify if the vegetation is undergoing stress by some agent.

sequently they show weak chlorophyll and cellulose absorptions, with soil (clay) absorption from exposed soil. Field investigation revealed that the potatoes were also being sprayed with a defoliant. Thus, they exhibit decreased chlorophyll absorption along with a shift of the red-edge to shorter wavelengths as we would expect. The Chico and pasture spectra exhibit combinations of chlorophyll and cellulose (dry vegetation) absorption. The spectra from the known fields was used in a special computer program called *Tricorder,* developed at the U.S. Geological Survey, to identify other pixels in the study area with similar spectral response (Clark et al., 1995). The result was a classification map that was approximately 96 percent accurate (Color Plate 11-2b).

Clark et al. (1995) point out that the long-wavelength side of the chlorophyll absorption (approximately 0.68 – 0.73 μm) forms one of the most extreme slopes found in spectra of naturally occurring common materials. The absorption is usually very intense, ranging from a low reflectance of < 5 percent (near 0.68 μm) to a near-infrared reflectance maximum of ≥ 50 percent (at approximately 0.73 μm). When the chlorophyll absorption in the plant decreases, the overall *width* of the absorption band decreases. These factors cause a shift to shorter wavelengths as the chlorophyll absorption decreases. This is known as the "red-edge shift" or the "blue shift of the red edge" previously discussed and can be caused

by natural senescence, water deprivation, or toxic materials. Clark et al. (1995) found that the AVIRIS data could detect red-edge shifts of < 0.1 nm. The red-edge shift information was analyzed using the Tricorder program and used to identify vegetation undergoing any type of stress in the study area (Color Plate 11-2c).

Temporal Characteristics of Vegetation

Timing is very important when attempting to identify different vegetation types or to extract useful vegetation biophysical information (e.g., biomass, chlorophyll characteristics) from remotely sensed data. Selecting the most appropriate date(s) for data collection requires an intimate knowledge of the plants' temporal *phenological* (growth) *cycle*. Plants whose leaves drop seasonally are *deciduous*. Plants whose leaves remain from season to season are *evergreen*.

A noted remote sensing scientist, Dave Simonett, often said "Green is green is green." By this he meant that if two different crops (e.g., corn and cotton) were planted at the same time and had complete canopy closure on the same date as the remotely sensed data were collected, the spectral reflectance characteristics of the two crops would most likely appear to be very similar throughout the visible and near-infrared portion of the spectrum. His comment was based on the use of the relatively broadband sensors available in the 1970s and 1980s. We now have hyperspectral sensors that allow us to sample in relatively narrow portions of the spectrum, hopefully identifying unique absorption features that will allow us to discriminate between one vegetation type and another or to extract biophysical information.

Nevertheless, Simonett's observation is still correct in many instances. Often our only hope of discriminating between two crops using relatively coarse spectral resolution remote sensor data (e.g., Landsat Thematic Mapper, SPOT, IKONOS) is if the crops were:

- planted at slightly different times in the growing season (e.g., 10 days apart), which caused one canopy to be less developed (lower percent canopy closure) than the other;

- one crop received significantly different irrigation than the other, causing it to produce more or less biomass;

- one crop matured more rapidly than the other (e.g., through fertilization or careful weeding);

- the row spacing or field orientation was dramatically different for the two crops; or

- one crop has a different canopy structure.

The difference in crop percent canopy closure, soil moisture, biomass, or the difference in row spacing or orientation might cause one crop to have dramatically different reflectance properties due to the influence of the background soil or understory materials present. Therefore, it is often the proportion of background material present within the instantaneous field of view of the sensor system (e.g., perhaps 1 × 1 ft for high resolution color-infrared aerial photography; 30 × 30 m for Landsat Thematic Mapper imagery) that allows us to discriminate between the two vegetation types. The amount of understory background material present is largely a function of the stage of the plant in its phenological cycle. Therefore, if a scientist is trying to differentiate (classify) between several crops, wetland, or forest types using remote sensor data, it is essential to know the phenological cycle characteristics of all of them. This information is then used to determine the optimum time of year to collect the remotely sensed data to discriminate one land cover or vegetation type from another.

Another important temporal factor cannot be overlooked. Plants require water to grow. Therefore, their most produc-

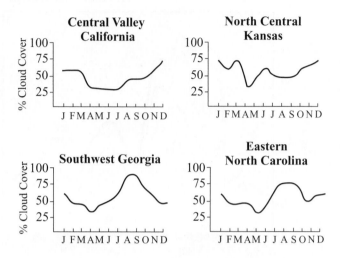

Figure 11-13   Predicted percent cloud cover in four areas in the United States. The greatest amount of cloud cover is often associated with the height of the growing season, except for California. This can significantly complicate the collection of remote sensing data for vegetation analysis.

tive growth period is often associated with the most intense periods of precipitation and associated cloud cover. For example, Figure 11-13 identifies the annual percent cloud cover statistics for several areas in the United States. Except for California, with its low humidity and low cloud cover, most of the areas have their greatest cloud cover at exactly the time when scientists may want to collect the maximum amount of remote sensor data during the growing season. Consequently, scientists must juggle the identification of the optimum date of remote sensor data using the phenological calendar with the likely spectre of considerable cloud cover being present. This is why we need higher temporal resolution sensors.

The development of forests, grassland, and wetland canopies follow relatively predictable cycles each year except when they are under stress from a pathogen or when unseasonably high or low temperatures occur. Similarly, all managed agricultural crops have relatively well-known phenological cycles within a given region, but these may be modified by individual farmers. Therefore, it is useful to review the phenological cycles of both natural vegetation systems and managed agricultural systems to gain insight into how important the cycles are when using remote sensing to extract vegetation biophysical information. Several case studies are provided.

Natural Phenological Cycles

Many plant species found in forests, wetland, rangeland, etc., have relatively unique phenological growth cycles. They leaf-out, grow to maturity, and senesce at approximately the same time each year. However, unseasonably cold or warm temperatures in the spring or fall can shift their annual phenological cycles by as much as thirty days. Therefore, the analyst must be aware of whether the remotely sensed data were collected in a typical or atypical year. If a scientist is attempting to classify vegetation using remote sensor data, it is often wise to collect the data early in the growing season when the vegetation are developing at different rates, which yield different percent canopy closures, hopefully creating differences in their spectral signatures. For example, if one is attempting to discriminate among various inland wetland species (e.g., cattail, willow, buttonbush) in the southeastern United States, it is best to collect imagery from February 15 through May 1. After May 1, most of the wetland vegetation has achieved complete canopy closure, and the spectral signatures of the various canopies become similar, i.e., "Green is green is green."

Conversely, if the goal is to monitor the biomass of the vegetation, it is useful to collect remote sensor data at the height of the growing season when the maximum biomass exists. Monitoring biomass through time can provide important information about the stability of the natural ecosystem and whether significant change is taking place. The following section demonstrates this logic applied to monitoring coastal Smooth Cordgrass wetland (Figure 11-14).

Phenological Cycle of Coastal Smooth Cordgrass in South Carolina

The annual phenological cycle of evergreen Smooth Cordgrass (*Spartina alterniflora*), which grows extensively along the eastern coast of the United States is shown in Figure 11-14. The decomposed *Spartina* organic matter (detritus) is carried out to sea and represents the foundation of the food chain that supports the Eastern Seaboard marine fishery industry. The time of greatest *Spartina* growth and dry weight biomass occurs from July through October. This is the optimum time for remote sensing data collection. It serves little purpose to obtain imagery of the *Spartina* in January through March, when it is in a senesced (dormant) state. Still, one often encounters persons who attempt to use winter imagery collected for some other purpose to monitor the *Spartina* wetland.

Smooth Cordgrass (*Spartina alterniflora*)

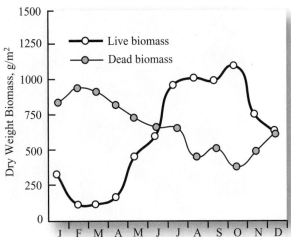

Figure 11-14 Phenological cycle of Smooth Cordgrass (*Spartina alterniflora*) biomass in South Carolina. The greatest amount of live biomass is present from July through October. The greatest amount of dead biomass is present from January through March. The marsh contains approximately equal proportions of live and dead biomass in December and June (after Jensen et al., 1998).

Figure 11-15 displays nine bands of 3 × 3 m NASA Calibrated Airborne Multispectral Scanner (CAMS) data of Murrells Inlet, SC, obtained on August 2, 1997, during the high biomass time of year. *In situ* Smooth Cordgrass total dry biomass (g/m^2) measurements were obtained at 27 locations on August 2 and 3, 1997. The *in situ* data were then correlated with individual band brightness values and vegetation transforms of the original CAMS data (Jensen et al., 1998). One of the most impressive relationships was between the fundamental near-infrared band 6 CAMS data and the *in situ* measurements, which yielded a correlation coefficient of 0.88 and an r^2-value of 0.774 meaning that approximately 77 percent of the variance was accounted for (Figure 11-16).

Color Plate 11-3 is a map of the spatial distribution of total dry biomass for a small portion of Murrells Inlet on August 2, 1997. It was produced using the regression equation:

$$y = 3.4891x - 23.9 \qquad (11-10)$$

where y is total biomass (g/m^2) and x is the CAMS band 6 brightness value. Studies such as this provide important baseline biophysical information that can be compared against future studies to determine if deleterious changes in

Calibrated Airborne Multispectral Scanner (CAMS) Data of Murrells Inlet, SC

Figure 11-15 Nine bands of 3 × 3 m Calibrated Airborne Multispectral Scanner (CAMS) data of Murrells Inlet, SC, obtained on August 2, 1997. The data were obtained at 4,000 ft above ground level.

biomass are occurring in the inlet. Comparative biomass studies would require that most of the system and environmental variables remain constant (e.g., approximately the same sensor system configuration and viewing geometry, altitude above ground level, atmospheric conditions, anni-versary date, time of day, and tidal cycle). The limitations of using a regression approach when predicting biophysical variables using remotely sensed data are summarized in Qi et al. (1995).

$$y = 3.4891x - 23.9$$

$$R^2 = 0.77$$

Figure 11-16 Relationship between Calibrated Airborne Multispectral Scanner (CAMS) band 6 brightness values and *in situ* measurement of *Spartina alterniflora* total dry biomass (g/m^2) at twenty-seven locations in Murrells Inlet, SC, obtained on August 2 and 3, 1997 (Jensen et al., 1998).

Managed Phenological Cycles

Crops associated with mechanized forestry and agriculture are established, grown, and harvested according to relatively predictable phenological cycles. It would be wonderful if all farmers planted the same crop on exactly the same date in a given growing season. This would make the analysis of remotely sensed data straightforward. Fortunately for us, who eat the bread of their labors, they do not do this, but rather plant the crop based on current meteorological conditions, the availability of equipment, and heuristic rules of thumb. Therefore, not all crops of the same species are planted during the same month or harvested during the same month. If two identical corn crops are planted twenty days apart, their spectral signatures will likely be very different throughout their respective phenological cycles. This condition may cause problems when attempting to perform crop identification using remote sensor data.

County agricultural extension agents and land-grant university scientists know the local and regional crop phenological cycles very well. They often represent the most important sources for valuable crop phenology and soils information and any idiosyncrasies associated with local farming practices. It is instructive to provide examples of several agricultural phenological cycles that span the geography of the conterminous United States to demonstrate the importance of vegetation phenological information. This will include case studies concerning hard red winter wheat in the Midwest United States, several major crops in South Carolina, and remote sensing of sugar beets, cotton, and alfalfa in the Imperial Valley, CA. Each example includes detailed phenological information displayed in different formats.

Phenological Cycle of Hard Red Winter Wheat in the Midwest United States

The detailed phenological cycle of hard red winter wheat (*triticale*) grown on the prairie soils in the Midwest United States is shown in Figure 11-17. The crop is established in October and November. It lies dormant under snow cover (if present) until March, when growth resumes. Hard red winter wheat greens up in April and produces heads of grain in May. It matures by mid-June. It is usually ripe and harvested by early July. Remotely sensed data acquired in October and November provide information on the amount of land prepared during the crop establishment period. Imagery acquired during the green-up phase in April and May can be used to extract information on standing-crop biomass and perhaps predict the harvested wheat yield.

Detailed crop calendar information such as this is used by governments, individuals, and private companies to acquire remote sensing data at optimum dates in the phenological cycle in order to model and predict the production of specific food for sale on the world market. Such monitoring may in certain instances also be used to predict future agricultural disasters due to severe drought, hopefully alleviating suffering due to famine.

Phenological Cycle of Agricultural Crops in South Carolina

The phenological cycles of natural and cultivated vegetation varies greatly by region. For example, if a scientist wanted to monitor the condition of the wheat crop in South Carolina, it is *not* possible to use the Midwest wheat phenological information presented in Figure 11-17. The phenological cycles for wheat and South Carolina's other major cash crops are presented in Figures 11-18 and 11-19.

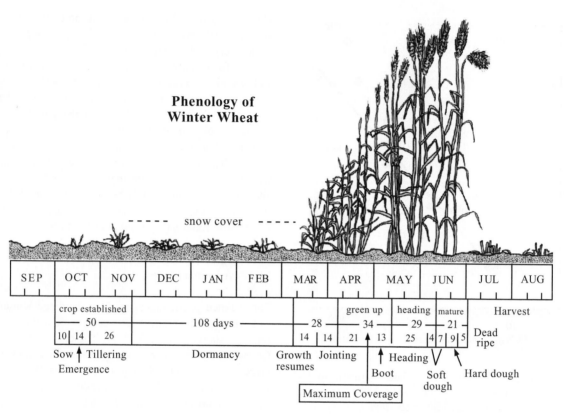

Figure 11-17   The phenological cycle of hard red winter wheat in the Great Plains of the United States. The crop is established in October and November. It lies dormant under snow cover until March, when growth resumes. The plants green up in April, produce heads of grain in May, and mature by mid-June. The wheat is usually dead ripe and harvested by early July. Remotely sensed data acquired in October and November provide information on the amount of land prepared during the crop establishment period. Imagery acquired during the green-up phase in April and May can be used to extract information on standing-crop biomass and perhaps predict the harvested wheat yield.

Wheat heads in late May and early June in the Midwest. Wheat heads in early May in South Carolina, approximately 15 days ahead of the Midwest phenological cycle for wheat, primarily due to an earlier and longer frost-free growing season. Imagery acquired in the Midwest in June might be valuable for inventorying wheat production, whereas imagery obtained in June in South Carolina might well reveal that the wheat crop has already been harvested.

The crop calendars also reveal that it should be straightforward to identify wheat production in South Carolina by acquiring a single image in March or early April when wheat is the only crop in existence with a complete canopy closure. Conversely, it may be difficult to differentiate between tobacco and corn unless we obtain a mid-June image and hope that the tobacco is leafed out more than corn, resulting in a difference in spectral response. Similarly, it may be possible to discriminate between soybeans and cotton if a late July or early August date is selected and soybeans have a greater canopy closure than cotton. If all the phenological

cycles of the major crops were exactly the same for South Carolina, with complete canopy closure at the same time of year, it would be difficult to use remotely sensed data to discriminate between the vegetation types. However, because their phenological cycles are staggered throughout the growing season to some degree, it is possible to discriminate between them if the imagery is collected at optimum times in the growing season (Savitsky, 1986).

Phenological Cycle of Agricultural Crops in the Imperial Valley, CA

The Imperial Valley of California is one of the most productive agricultural ecosystems in the world, producing great quantities of sugarbeets, cotton, and alfalfa. The phenological cycles of several of its major crops are summarized in Color Plate 11-4 (GRSU, 1999). Each calendar depicts a single field monitored throughout a twelve-month period using 79×79 m Landsat Multispectral Scanner (MSS) data (RGB = bands 4,2,1). The brighter red (magenta) the signature, the

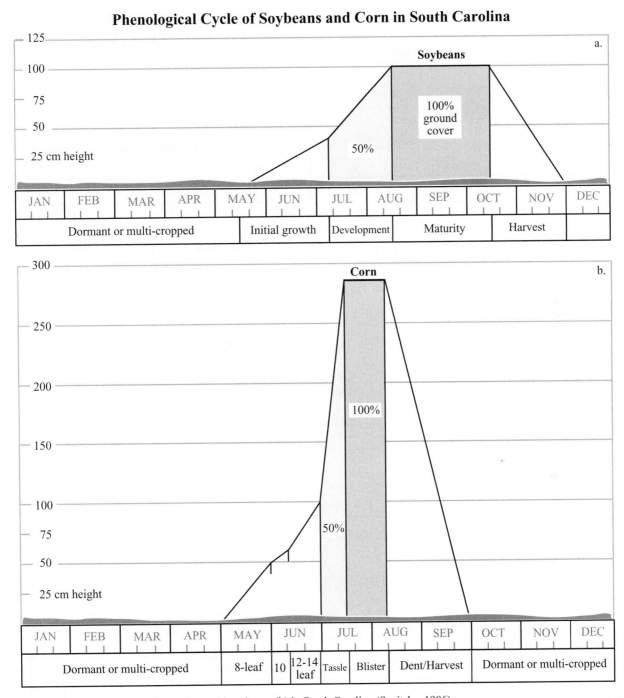

Figure 11-18 Phenological cycles of soybeans (a) and corn (b) in South Carolina (Savitsky, 1986).

greater the amount of biomass and crop canopy closure. Sugarbeets are established in early September, emerge in November, and are harvested in April and early May. By law, cotton must be planted in March and harvested and plowed under in November of each year to control Bole Weevil. In 1982, the plow-down date was January 1. Alfalfa

is planted year-round and may be harvested five or six times per year as depicted in the crop calendar.

It is possible to identify the type of crop in each field if imagery is obtained at times of the year that maximize the spectral contrast between crops. For example, Figure 11-20

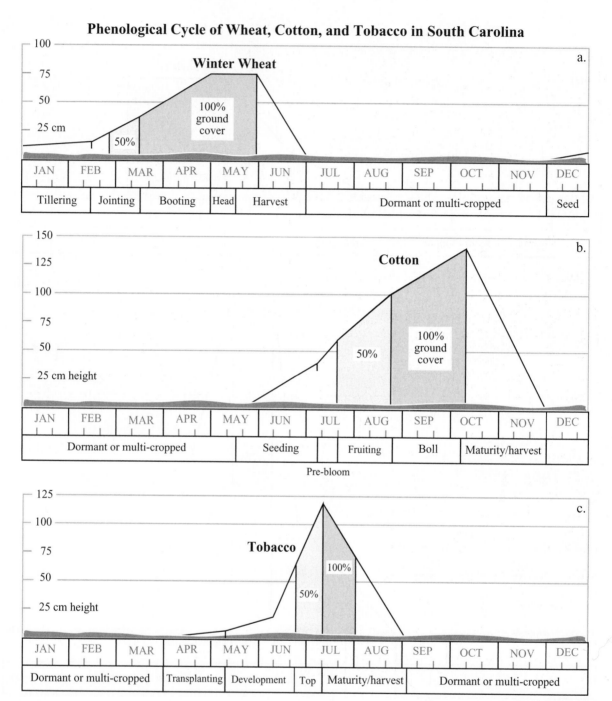

Figure 11-19   Phenological cycles of winter wheat (a), cotton (b), and tobacco (c) in South Carolina. The information was obtained from county agricultural extension agents, Clemson land-grant university extension personnel, and field work (Savitsky, 1986).

depicts seven bands of Landsat Thematic Mapper imagery obtained on December 10, 1982. Six of the bands are at 30 × 30 m, while the thermal infrared band (6) is at 120 × 120 m spatial resolution. A ground reference crop map provided by the Imperial Valley Irrigation Board is also included (Haack and Jampoler, 1995).

Landsat Thematic Mapper Imagery of Imperial Valley, CA, on December 10, 1982

a. Band 1 (blue: 0.45 - 0.52 μm) b. Band 2 (green: 0.52 - 0.60 μm) c. Band 3 (red: 0.63 - 0.69 μm)

d. Band 4 (near-infrared: 0.76 - 0.90 μm) e. Band 5 (mid-infrared: 1.55 - 1.75 μm) f. Band 7 (mid-infrared: 2.08 - 2.35 μm)

feed lot

Legend:

Sugarbeets Cotton

Alfalfa Fallow

g. Band 6 (mid-infrared: 10.4 - 12.5 μm) Ground reference information

Figure 11-20 Individual bands of Landsat Thematic Mapper imagery of a portion of the Imperial Valley, CA, obtained on December 10, 1982. Bands 1 – 5 and 7 have a spatial resolution of 30 × 30 m. Band 6 has a spatial resolution of 120 × 120 m. Sugarbeets are bright in the near-infrared band 4 data and dark in both the middle-infrared bands (5 and 7). Fallow fields are gray in all bands except band 4 (near-infrared), which is dark (images courtesy Space Imaging, Inc.).

The four color composites shown in Color Plate 11-5 demonstrate the importance of the middle-infrared bands when attempting to discriminate among crop types. The fundamental visible (RGB = bands 3,2,1) and near-infrared color composites (RGB = bands 4,3,2) are not nearly as effective for visual determination of the crop types as the visible and middle-infrared composites (RGB = bands 5,3,2 and RGB = bands 7,3,2). Note that all the sugarbeet fields show up as black in the middle-infrared color composite (Color Plate 11-5c; RGB = bands 5,3,2) and in the individual band 5 image (Figure 11-20).

Because of the significant spectral contrast among the crops in this December 10, 1982, Landsat TM image, it is possible to perform a visual classification of the crops based on an examination of the individual bands in Figure 11-20. For

Table 11-2. Tonal characteristics of individual crop types in black-and-white Landsat TM images of Imperial Valley, CA, (Figure 11-20) obtained on December 10, 1982 (D = dark; G = mid-gray; Br = brighter)

Crop Type	TM1	TM2	TM3	TM4	TM5	TM7
Sugarbeets	D	D	D	Br	D	D
Alfalfa	D	D	D	Br	G	G
Cotton	G	G	G	D	Br	Br
Fallow land	G	G	G	D	G	G

example, after careful examination we see that the heuristic rules summarized in Table 11-2 could be used to classify the scene.

It is also possible to program a computer to take into account these same spectral characteristics and perform a classification. Color Plate 11-5e displays a remote sensing–derived digital classification map of the study area. When compared with the ground reference map in Figure 11-20, it is about 85 percent accurate. It was produced using TM bands 1 – 5 and 7 and a supervised maximum likelihood classification algorithm [please refer to Jensen (2005) for information about digital image classification]. Haack and Jampoler (1994, 1995) evaluated this same Landsat TM scene and achieved an overall accuracy of 90 percent using different bands.

The accuracy of a remote sensing derived crop classification map is always dependent upon there being a significant difference in the spectral response between the various crop types. The only way to identify when this maximum contrast among spectral response should take place is to evaluate the phenological crop calendars and select the appropriate dates of imagery for analysis. Then classification maps, such as the one shown in Color Plate 11-5, can be used to predict the amount of acreage in specific crops at a given instant in time. Such information coupled with agricultural-meteorological models can be used to predict crop yield.

 Vegetation Indices

Since the 1960s, scientists have extracted and modeled various vegetation biophysical variables using remotely sensed data. Much of this effort has involved the use of *vegetation indices*—dimensionless, radiometric measures that indicate

relative abundance and activity of green vegetation, including leaf-area-index (LAI), percentage green cover, chlorophyll content, green biomass, and absorbed photosynthetically active radiation (APAR). A vegetation index should (Running et al., 1994; Huete and Justice, 1999):

- maximize sensitivity to plant biophysical parameters, preferably with a linear response in order that sensitivity be available for a wide range of vegetation conditions, and to facilitate validation and calibration of the index;

- normalize or model external effects such as Sun angle, viewing angle, and the atmosphere for consistent spatial and temporal comparisons;

- normalize internal effects such as canopy background variations, including topography (slope and aspect), soil variations, and differences in senesced or woody vegetation (nonphotosynthetic canopy components);

- be coupled to some specific measurable biophysical parameter such as biomass, LAI, or APAR as part of the validation effort and quality control.

There are many vegetation indices. A few of the most widely adopted are summarized in Table 11-3. Many are functionally equivalent (redundant) in information content (Perry and Lautenschlager, 1984), and some provide unique biophysical information (Qi et al., 1995). It is useful to review the historical development of the main indices and provide information about recent advances in index development. Detailed summaries are found in Running et al. (1994) and Lyon et al. (1998). Many indices make use of the inverse relationship between red and near-infrared reflectance associated with healthy green vegetation (Figure 11-21a).

Simple Ratio — SR

Cohen (1991) suggests that the first true vegetation index was the *Simple Ratio* (SR), which is the ratio of red reflected radiant flux (ρ_{red}) to near-infrared radiant flux (ρ_{nir}) as described in Birth and McVey (1968):

$$SR = \frac{\rho_{red}}{\rho_{nir}}. \tag{11-11}$$

The simple ratio provides valuable information about vegetation biomass or LAI (Schlerf et al., 2005). It is especially sensitive to biomass and/or LAI variations in high-biomass vegetation such as forests (Huete et al., 2002b).

Physiological Basis for Developing Vegetation Indices

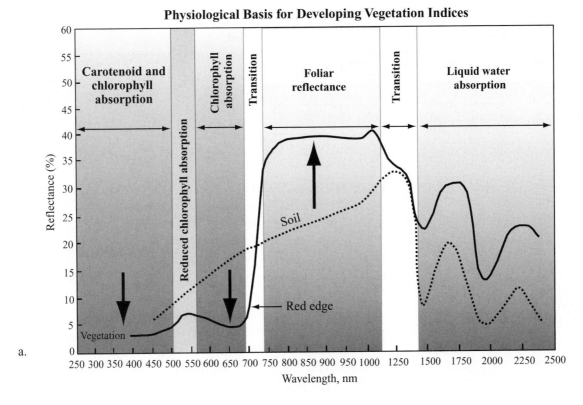

a.

Relationship Between Simple Ratio and NDVI

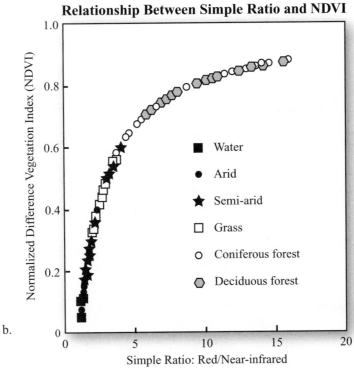

b.

Figure 11-21 a) Physiological basis for developing vegetation indices. Typical spectral reflectance characteristics for healthy green grass and bare dry soil for the wavelength interval from 250 to 1,000 nm. b) The NDVI is a normalized ratio of the near-infrared and red bands. The NDVI is functionally equivalent to and is a nonlinear transform of the simple ratio (adapted from Huete et al., 2002b).

Table 11-3. Selected remote sensing vegetation indices.

Vegetation Index	Equation	References
Simple Ratio (SR)	$$SR = \frac{\rho_{red}}{\rho_{nir}}$$	Birth and McVey, 1968 Colombo et al., 2003 Schlerf et al., 2005
Normalized Difference Vegetation Index (NDVI)	$$NDVI = \frac{\rho_{nir} - \rho_{red}}{\rho_{nir} + \rho_{red}}$$	Rouse et al., 1974 Deering et al., 1975 Huete et al., 2002a Schlerf et al., 2005

Kauth-Thomas Transformation
  Brightness
  Greenness
  Yellow stuff
  None-such
 	
  Brightness
  Greenness
  Wetness
	Landsat Multispectral Scanner (MSS)
	$B = 0.332MSS1 + 0.603MSS2 + 0.675MSS3 + 0.262MSS4$
	$G = -0.283MSS1 - 0.660MSS2 + 0.577MSS3 + 0.388MSS4$
	$Y = -0.899MSS1 + 0.428MSS2 + 0.076MSS3 - 0.041MSS4$
	$N = -0.016MSS1 + 0.131MSS2 - 0.452MSS3 + 0.882MSS4$
	
	Landsat Thematic Mapper (TM)
	$B = 0.2909TM1 + 0.2493TM2 + 0.4806TM3 +$
	$\quad 0.5568TM4 + 0.4438TM5 + 0.1706TM7$
	$G = -0.2728TM1 - 0.2174TM2 - 0.5508TM3 +$
	$\quad 0.7221TM4 + 0.0733TM5 - 0.1648TM7$
	$W = 0.1446TM1 + 0.1761TM2 + 0.3322TM3 +$
	$\quad 0.3396TM4 - 0.6210TM5 - 0.4186TM7$
		Kauth and Thomas, 1976
		Kauth et al., 1979
		
		Crist and Kauth, 1986
		Lunetta et al., 2002
		Price et al., 2002
		Rogan et al., 2002
		Jin and Sader, 2005
		Schlerf et al., 2005

Normalized Difference Moisture or Water Index (NDMI or NDWI)	$$NDMI \text{ or } NDWI = \frac{NIR_{TM4} - MidIR_{TM5}}{NIR_{TM4} + MidIR_{TM5}}$$	Hardisky et al., 1983; Gao, 1996; Jackson et al., 2004 Galvao et al., 2005 Jin and Sader, 2005
Perpendicular Vegetation Index (PVI)	$$PVI = \sqrt{(0.355MSS4 - 0.149MSS2)^2 + (0.355MSS2 - 0.852MSS4)^2}$$	Richardson and Wiegand, 1977
Leaf Relative Water Content Index (LWCI)	$$LWCI = \frac{-\log[1 - (NIR_{TM4} - MidIR_{TM5_{ft}})]}{-\log[1 - NIR_{TM4} - MidIR_{TM5_{ft}}]}$$	Hunt et al., 1987
Soil Adjusted Vegetation Index (SAVI) and Modified SAVI (MSAVI)	$$SAVI = \frac{(1 + L)(\rho_{nir} - \rho_{red})}{\rho_{nir} + \rho_{red} + L}$$	Huete, 1988; Huete and Liu, 1994; Running et al., 1994; Qi et al., 1995
Atmospherically Resistant Vegetation Index (ARVI)	$$ARVI = \left(\frac{\rho^*_{nir} - \rho^*_{rb}}{\rho^*_{nir} + \rho^*_{rb}} \right)$$	Kaufman and Tanre, 1992; Huete and Liu, 1994
Soil and Atmospherically Resistant Vegetation Index (SARVI)	$$SARVI = \frac{\rho^*_{nir} - \rho^*_{rb}}{\rho^*_{nir} + \rho^*_{rb} + L}$$	Huete and Liu, 1994; Running et al., 1994
Enhanced Vegetation Index (EVI)	$$EVI = G \frac{\rho^*_{nir} - \rho^*_{red}}{\rho^*_{nir} + C_1\rho^*_{red} - C_2\rho^*_{blue} + L}(1 + L)$$	Huete et al., 1997 Huete and Justice, 1999 Huete et al., 2002a TBRS, 2003

Table 11-3.   Selected remote sensing vegetation indices.

Vegetation Index	Equation	References
New Vegetation Index (NVI)	$NVI = \dfrac{\rho_{777} - \rho_{747}}{\rho_{673}}$	Gupta et al., 2001
Aerosol Free Vegetation Index (AFRI)	$AFRI_{1.6\mu m} = \dfrac{(\rho_{nir} - 0.66\rho_{1.6\mu m})}{(\rho_{nir} - 0.66\rho_{1.6\mu m})}$ $AFRI_{2.1\mu m} = \dfrac{(\rho_{nir} - 0.5\rho_{2.1\mu m})}{(\rho_{nir} + 0.5\rho_{2.1\mu m})}$	Karnieli et al., 2001
Triangular Vegetation Index (TVI)	$TVI = 0.5(120(\rho_{nir} - \rho_{green})) - 200(\rho_{red} - \rho_{green})$	Broge and Leblanc, 2000
Reduced Simple Ratio (RSR)	$RSR = \dfrac{\rho_{nir}}{\rho_{red}}\left(1 - \dfrac{\rho_{swir} - \rho_{swirmin}}{\rho_{swirmax} + \rho_{swirmin}}\right)$	Chen et al., 2002
Ratio TCARI /OSAVI	$TCARI = 3\left[(\rho_{700} - \rho_{670}) - 0.2(\rho_{700} - \rho_{550})\left(\dfrac{\rho_{700}}{\rho_{670}}\right)\right]$ $OSAVI = \dfrac{(1 + 0.16)(\rho_{800} - \rho_{670})}{(\rho_{800} + \rho_{670} + 0.16)}$ $\dfrac{TCARI}{OSAVI}$	Kim et al., 1994 Rondeaux et al., 1996 Daughtry et al., 2000 Haboudane et al., 2002
Visible Atmospherically Resistant Index (VARI)	$VARI_{green} = \dfrac{\rho_{green} - \rho_{red}}{\rho_{green} + \rho_{red} - \rho_{blue}}$	Gitelson et al., 2002
Normalized Difference Built-up Index (NDBI)	$NDBI = \dfrac{MidIR_{TM5} - NIR_{TM4}}{MidIR_{TM5} + NIR_{TM4}}$ $built\text{-}up_{area} = NDBI - NDVI$	Zha et al., 2003
Red-edge Position	$REP = 700 + 40\left[\dfrac{\rho_{(red\ edge)} - \rho_{(700nm)}}{\rho_{(740nm)} - \rho_{(700nm)}}\right]$ where $\rho^*_{rb} = \rho^*_{red} - \gamma(\rho^*_{blue} - \rho^*_{red})$	Clevers, 1994 Dawson and Curran, 1998 Baranoski, 2005

Normalized Difference Vegetation Index — NDVI

Rouse et al. (1974) developed the generic *Normalized Difference Vegetation Index* (NDVI):

$$NDVI = \frac{\rho_{nir} - \rho_{red}}{\rho_{nir} + \rho_{red}}. \tag{11-12}$$

The NDVI is functionally equivalent to the simple ratio; that is, there is no scatter in an SR vs. NDVI plot, and each SR value has a fixed NDVI value. When we plot the mean NDVI and SR values for various biomes, we find that the NDVI approximates a nonlinear transform of the simple ratio (Figure 11-21b) (Huete et al., 2002b). The NDVI is an important vegetation index because:

• Seasonal and inter-annual changes in vegetation growth and activity can be monitored.

- The ratioing reduces many forms of *multiplicative* noise (Sun illumination differences, cloud shadows, some atmospheric attenuation, some topographic variations) present in multiple bands of multiple-date imagery.

Following are the disadvantages of the NDVI (Huete et al., 2002a; Wang et al., 2005):

- The ratio-based index is nonlinear and can be influenced by *additive* noise effects such as atmospheric path radiance (Chapter 6).

- NDVI is usually highly correlated with vegetation leaf area index (LAI). However, the relationship may not be as strong during periods of maximum LAI, apparently due to the saturation of NDVI when LAI is very high (Wang et al., 2005). For example, Figure 11-21b reveals that the NDVI dynamic range is stretched in favor of low-biomass conditions and compressed in high-biomass, forested regions. The opposite is true for the Simple Ratio, in which most of the dynamic range encompasses the high-biomass forests with little variation reserved for the lower-biomass regions (grassland, semi-arid, and arid biomes).

- It is very sensitive to canopy background variations (e.g., soil visible through the canopy). NDVI values are particularly high with darker-canopy backgrounds.

The NDVI index was widely adopted and applied to the original Landsat MSS digital remote sensor data. Deering et al. (1975) added 0.5 to the NDVI to avoid negative values and took the square root of the result to stabilize the variance. This index is referred to as the *Transformed Vegetation Index* (TVI). These three indices (SR, NDVI, and TVI) respond to changes in the amount of green biomass.

Many scientists continue to use the NDVI (e.g., Miura et al., 2001; Trishchenko et al., 2002; Galvao et al., 2005). In fact, two of the standard MODIS land products are sixteen-day composite NDVI datasets of the world at a spatial resolution of 500 m and 1 km (Huete et al., 2002a). An NDVI image derived from Charleston Landsat Thematic Mapper dataset is shown in Figure 11-22.

Kauth-Thomas Tasseled Cap Transformation

Kauth and Thomas (1976) produced an orthogonal transformation of the original Landsat MSS data space to a new four-dimensional feature space. It was called the *tasseled cap* or *Kauth-Thomas transformation*. It created four new axes: the soil brightness index (*B*), greenness vegetation

NDVI Image of Charleston, SC, Derived from Landsat Thematic Mapper Data

Figure 11-22 Normalized difference vegetation index (NDVI) image of Charleston, SC derived using Landsat Thematic Mapper bands 3 and 4.

index (*G*), yellow stuff index (*Y*), and none-such (*N*). The names attached to the new axes indicate the characteristics the indices were intended to measure. The coefficients are (Kauth et al., 1979):

$$B = 0.332\,MSS1 + 0.603\,MSS2 + 0.675\,MSS3 + 0.262\,MSS4 \quad (11\text{-}13)$$

$$G = -0.283\,MSS1 - 0.660\,MSS2 + 0.577\,MSS3 + 0.388\,MSS4 \quad (11\text{-}14)$$

$$Y = -0.899\,MSS1 + 0.428\,MSS2 + 0.076\,MSS3 - 0.041\,MSS4 \quad (11\text{-}15)$$

$$N = -0.016\,MSS1 + 0.131\,MSS2 - 0.452\,MSS3 + 0.882\,MSS4\,. \quad (11\text{-}16)$$

Crist and Kauth (1986) derived the visible, near-infrared, and middle-infrared coefficients for transforming Landsat Thematic Mapper imagery into brightness, greenness, and wetness variables:

$$B = 0.2909\,TM1 + 0.2493\,TM2 + 0.4806\,TM3 + \quad (11\text{-}17)$$
$$0.5568\,TM4 + 0.4438\,TM5 + 0.1706\,TM7$$

$$G = -0.2728\,TM1 - 0.2174\,TM2 - 0.5508\,TM3 + \quad (11\text{-}18)$$
$$0.7221\,TM4 + 0.0733\,TM5 - 0.1648\,TM7$$

$$W = 0.1446\,TM1 + 0.1761\,TM2 + 0.3322\,TM3 + \quad (11\text{-}19)$$
$$0.3396\,TM4 - 0.6210\,TM5 - 0.4186\,TM7\,.$$

It is called the tasseled cap transformation because of its cap shape (Figures 11-23a–c). Crist and Cicone (1984) identified

Table 11-4. Tasseled cap coefficients for use with Landsat 7 Enhanced Thematic Mapper Plus data (Huang et al., 2002).

Index	TM1	TM2	TM3	TM4	TM5	TM7
Brightness	.3561	.3972	.3904	.6966	.2286	.1596
Greenness	-.334	-.354	-.456	.6966	-.024	-.263
Wetness	.2626	.2141	.0926	.0656	-.763	-.539
Fourth	.0805	-.050	.1950	-.133	.5752	-.777
Fifth	-.725	-.020	.6683	.0631	-.149	-.027
Sixth	.400	-.817	.3832	.0602	-.109	.0985

a third component that is related to soil features, including moisture status (Figure 11-23d). Thus, an important source of soil information is available through the inclusion of the middle-infrared bands of the Thematic Mapper (Price et al., 2002).

The Charleston Landsat TM scene is transformed into brightness, greenness, and wetness (moisture) content images in Figure 11-24 based on the use of the TM tasseled cap coefficients (Equations 11-17 through 11-19). Urbanized areas are particularly evident in the brightness image. The greater the biomass, the brighter the pixel value in the greenness image. The wetness image provides subtle information concerning the moisture status of the wetland environment. As expected, the greater the moisture content, the brighter the response. As expected, water bodies are bright. Crist and Kauth (1986) identified the fourth tasseled cap parameter as being haze.

The tasseled cap transformation is a global vegetation index. Theoretically, it can be used anywhere in the world to disaggregate the amount of soil brightness, vegetation, and moisture content in individual pixels in a Landsat MSS or Thematic Mapper image. Practically, however, it is better to compute the coefficients based on local conditions. Jackson (1983) provides a computer program for this purpose.

The Kauth-Thomas tasseled cap transformation continues to be widely used (e.g., Lunetta et al., 2002; Price et al., 2002; Rogan et al., 2002). Huang et al. (2002) developed tasseled cap coefficients for use with Landsat 7 Enhanced Thematic Mapper Plus at-satellite reflectance data. These coefficients are most appropriate for regional applications where atmospheric correction is not feasible. The coefficients are summarized in Table 11-4.

Normalized Difference Moisture or Water Index — NDMI or MDWI

Information about vegetation water content has widespread utility in agriculture, forestry, and hydrology (Galvao et al., 2005). Hardisky et al. (1983) and Gao (1996) found that the *Normalized Difference Moisture* or *Water Index* (NDMI or NDWI) based on Landsat TM near- and middle-infrared bands,

$$NDMI \text{ or } NDWI = \frac{NIR_{TM4} - MidIR_{TM5}}{NIR_{TM4} + MidIR_{TM5}}, \quad (11\text{-}20)$$

was highly correlated with canopy water content and more closely tracked changes in plant biomass and water stress than did the NDVI. Jackson et al. (2004) found that the NDVI saturated while the NDWI continued to document changes in corn and soybean vegetation water content. Jin and Sader (2005) compared the NDMI and the tasseled cap wetness transformation for detecting forest disturbance.

Perpendicular Vegetation Index — PVI

Richardson and Wiegand (1977) used the perpendicular distance to the "soil line" as an indicator of plant development. The "soil line," which is a two-dimensional analog of the Kauth-Thomas soil brightness index, was estimated by linear regression. The *Perpendicular Vegetation Index* (PVI) based on MSS band 4 data was

$$PVI = \sqrt{(0.355 MSS4 - 0.149 MSS2)^2 + (0.355 MSS2 - 0.852 MSS4)^2} \quad (11\text{-}21)$$

Leaf Water Content Index — LWCI

Hunt et al. (1987) developed the *Leaf Water Content Index* (LWCI) to assess water stress in leaves:

$$LWCI = \frac{-\log[1 - (NIR_{TM4} - MidIR_{TM5})]}{-\log[1 - (NIR_{TM4_{ft}} - MidIR_{TM5_{ft}})]}, \quad (11\text{-}22)$$

where *ft* represents reflectance in the specified bands when leaves are at their maximum relative water content (fully turgid; RWC) defined as:

$$RWC = \frac{\text{field weight} - \text{oven dry weight}}{\text{turgid weight} - \text{oven dry weight}} \times 100 \quad (11\text{-}23)$$

Soil Adjusted Vegetation Index — SAVI

The utility of the normalized difference vegetation index and related indices for satellite and airborne assessment of the

Characteristics of the Kauth-Thomas Tasseled Cap Transformation

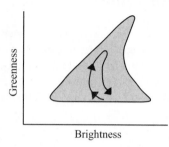

a. Crop development in the tasseled cap brightness-greenness transformation.

b. Location of land cover when plotted in the brightness-greenness spectral space.

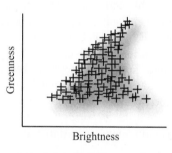

c. Actual plot of brightness and greenness values for an agricultural area. The shape of the distribution looks like a cap.

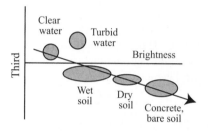

d. Approximate direction of moisture variation in the plane of soils. The arrow points in the direction of less moisture.

Figure 11-23 Characteristics of the Kauth-Thomas tasseled cap transformation (adapted from Crist, 1985; Crist and Kauth, 1986). The tasseled cap transformation is applied correctly if the tasseled cap's base is parallel with the brightness axis.

Earth's vegetation cover has been demonstrated for almost three decades. The time series analysis of seasonal NDVI data have provided a method of estimating net primary production over varying biome types, of monitoring phenological patterns of the Earth's vegetated surface, and of assessing the length of the growing season and dry-down periods (Huete and Liu, 1994; Ramsey et al., 1995). For example, global vegetation analysis was initially based on linearly regressing NDVI values (derived from AVHRR, Landsat MSS, Landsat TM, and SPOT HRV data) with *in situ* measurements of LAI, APAR, percent cover, and/or biomass. This empirical approach revolutionized global science landcover biophysical analysis in just one decade (Running et al., 1994). Unfortunately, studies have found that the empirically derived NDVI products can be unstable, varying with soil color and moisture conditions, bidirectional reflectance distribution function (BRDF) effects (refer to Chapter 1), atmospheric conditions, and the presence of dead material in the canopy itself (Qi et al., 1995). For example, Goward et al. (1991) found errors of ±50% in NDVI images used for global vegetation studies derived from the NOAA Global Vegetation Index product. What is needed are globally accurate NDVI-related products that do not need to be calibrated

by *in situ* measurement within each geographic area yet will remain constant under changing atmospheric and soil background conditions (Huete and Justice, 1999).

Therefore, emphasis has been given to the development of improved vegetation indices that take advantage of calibrated sensor systems such as the Moderate Resolution Imaging Spectrometer (MODIS) (Running et al., 1994). Although the NDVI has been shown to be useful in estimating vegetation properties, many important external and internal influences restrict its global utility. The improved indices typically incorporate a soil background and/or atmospheric adjustment factor.

The *Soil Adjusted Vegetation Index* (SAVI) is

$$\text{SAVI} = \frac{(1 + L)(\rho_{nir} - \rho_{red})}{\rho_{nir} + \rho_{red} + L} \tag{11-24}$$

where L is a canopy background adjustment factor that accounts for differential red and near-infrared extinction through the canopy (Huete, 1988; Huete et al., 1992; Karnieli, et al., 2001). An L value of 0.5 in reflectance space was

a. Brightness.

b. Greenness.

c. Wetness.

**Kauth-Thomas (Tasseled Cap)
Brightness, Greenness, and Wetness
Transformation of Landsat Thematic
Data for Charleston, SC**

Figure 11-24   Brightness, greenness, and wetness images derived by applying Kauth-Thomas tasseled cap transformation coefficients to the Charleston, SC, Thematic Mapper data (6 bands).

found to minimize soil brightness variations and eliminate the need for additional calibration for different soils (Huete and Liu, 1994). The utility of SAVI for minimizing the soil "noise" inherent in the NDVI has been corroborated in many studies. Qi et al. (1995) developed a modified SAVI, called MSAVI, that uses an iterative, continuous L function to optimize soil adjustment and increase the dynamic range of SAVI.

Atmospherically Resistant Vegetation Index — ARVI

SAVI was made less sensitive to atmospheric effects by normalizing the radiance in the blue, red, and near-infrared bands. This became the *Atmospherically Resistant Vegetation Index* (ARVI):

$$ARVI = \left(\frac{\rho^*_{nir} - \rho^*_{rb}}{\rho^*_{nir} + \rho^*_{rb}} \right) \tag{11-25}$$

where

$$\rho^*_{rb} = \rho^*_{red} - \gamma(\rho^*_{blue} - \rho^*_{red}) . \tag{11-26}$$

The technique requires prior correction for molecular scattering and ozone absorption of the blue, red, and near-infrared remote sensor data, hence the term p^*. ARVI uses the difference in the radiance between the blue channel and the red channel to correct the radiance in the red channel and thus reduce atmospheric effects. Unless the aerosol model is known a priori, gamma (γ) is normally equal to 1.0 to minimize atmospheric effects. Kaufman and Tanre (1992) provide guidelines where different gammas might be used over continental, maritime, desert (e.g., the Sahel), or heavily vegetated areas.

Soil and Atmospherically Resistant Vegetation Index — SARVI

Huete and Liu (1994) integrated the L function from SAVI and the blue-band normalization in ARVI to derive a *Soil and Atmospherically Resistant Vegetation Index* (SARVI) that corrects for both soil and atmospheric noise, as would a modified SARVI (MSARVI):

$$\text{SARVI} = \frac{\rho^*_{nir} - \rho^*_{rb}}{\rho^*_{nir} + \rho^*_{rb} + L} \quad (11\text{-}27)$$

and

$$\text{MSARVI} = \frac{2\rho^*_{nir} + 1 - \sqrt{\left[(2\rho^*_{nir} + 1)^2 - \gamma(\rho^*_{nir} - \rho^*_{rb})\right]}}{2}. \quad (11\text{-}28)$$

Huete and Liu (1994) performed a sensitivity analysis on the original NDVI and improved vegetation indices (SAVI, ARVI, SARVI, MSARVI) and drew these conclusions:

- If there were a total atmospheric correction, then there would mainly be "soil noise," and the SAVI and MSARVI would be the best equations to use, and the NDVI and ARVI would be the worst.

- If there were a partial atmospheric correction to remove the Rayleigh and ozone components, then the best vegetation indices would be the SARVI and MSARVI, with the NDVI and ARVI being the worst.

- If there were no atmospheric correction at all (i.e., no Rayleigh, ozone, or aerosol correction), the SARVI would become slightly worse but still would have the least overall noise. The NDVI and ARVI would have the most noise and error.

Aerosol Free Vegetation Index — AFRI

Karnieli et al. (2001) found that under clear sky conditions the spectral bands centered on 1.6 and 2.1 μm are highly correlated with visible spectral bands centered on blue (0.469 μm), green (0.555 μm), and red (0.645 μm). Empirical linear relationships such as $\rho_{0.469\mu m} = 0.25\rho_{2.1\mu m}$, $\rho_{0.555\mu m} = 0.33\rho_{2.1\mu m}$ and $\rho_{0.645\mu m} = 0.66\rho_{1.6\mu m}$ were found to be statistically significant. Therefore, based on these and other relationships, two *Aerosol Free Vegetation Indices* (AFRI) were developed:

MODIS Enhanced Vegetation Index versus Normalized Difference Vegetation Index over Dense Vegetation

Figure 11-25 A comparison of the sensitivity of the MODIS-derived Normalized Difference Vegetation Index (NDVI) values and MODIS Enhanced Vegetation Index (EVI) values for dense vegetation (Didan, 2002).

$$\text{AFRI}_{1.6\mu m} = \frac{(\rho_{nir} - 0.66\rho_{1.6\mu m})}{(\rho_{nir} + 0.66\rho_{1.6\mu m})} \text{ and} \quad (11\text{-}29)$$

$$\text{AFRI}_{2.1\mu m} = \frac{(\rho_{nir} - 0.5\rho_{2.1\mu m})}{(\rho_{nir} + 0.5\rho_{2.1\mu m})}. \quad (11\text{-}30)$$

Under clear sky conditions, the AFRIs (especially $\text{AFRI}_{2.1\mu m}$) result in values very similar to NDVI. However, if the atmospheric column contains smoke or sulfates then the AFRIs are superior to NDVIs. This is because the electromagnetic energy centered at 1.6 and 2.1 μm is able to penetrate the atmospheric column better than red wavelength energy used in the NDVI. Therefore, the AFRIs have a major application in assessing vegetation in the presence of smoke, anthropogenic pollution, and volcanic plumes. Limited success is expected in the case of dust due to presence of larger particles that are similar in size to the wavelength, and therefore not transparent at 2.1 μm (Kaufman et al., 2000). The AFRIs can be implemented using any sensor that incorporates bands centered on 1.6 and/or 2.1 μm such as the Landsat TM, and ETM$^+$, MODIS, ASTER, and the Japanese Earth Resource Satellite-Optical System (JERS-OPS), SPOT 4-Vegetation, and IRS-1C/D.

MODIS Enhanced Vegetation Index Map of the World

Figure 11-26 MODIS Enhanced Vegetation Index (EVI) map of the world obtained over a 16-day period beginning on day 193 of 2003. The brighter the area, the greater the amount of biomass (courtesy Terrestrial Biophysics and Remote Sensing MODIS Team, University of Arizona and NASA).

Enhanced Vegetation Index — EVI

The MODIS Land Discipline Group developed the *Enhanced Vegetation Index* (EVI) for use with MODIS data:

$$EVI = G \frac{\rho^*_{nir} - \rho^*_{red}}{\rho^*_{nir} + C_1 \rho^*_{red} - C_2 \rho^*_{blue} + L}(1 + L). \quad (11\text{-}31)$$

The EVI is a modified NDVI with a soil adjustment factor, L, and two coefficients, C_1 and C_2, which describe the use of the blue band in correction of the red band for atmospheric aerosol scattering. The coefficients, C_1, C_2, and L, are empirically determined as 6.0, 7.5, and 1.0, respectively. G is a gain factor set to 2.5. This algorithm has improved sensitivity to high-biomass regions and has improved vegetation monitoring through a decoupling of the canopy background signal and a reduction in atmospheric influences (Huete et al., 1997; Huete and Justice, 1999; Huete et al., 2002). A comparison of MODIS-derived NDVI values and MODIS-derived EVI values for dense vegetation is shown in Figure 11-25 (Didan, 2002). The EVI has improved sensitivity to high biomass. Two MODIS vegetation indices, the MODIS NDVI and the EVI, are produced at 1×1 km and 500×500 m resolutions (and occasionally at 250×250 m) using sixteen-day compositing periods. A black-and-white MODIS EVI map of the Earth obtained starting on day 193 of 2003 is shown in Figure 11-26.

Triangular Vegetation Index — TVI

Broge and Leblanc (2000) developed a *Triangular Vegetation Index* (TVI), which describes the radiative energy absorbed by pigments as a function of the relative difference between red and near-infrared reflectance in conjunction with the magnitude of reflectance in the green region, where the light absorption by chlorophyll *a* and *b* is relatively insignificant. The index is calculated as the area of the triangle defined by the green peak, the chlorophyll absorption minimum, and the near-infrared shoulder in spectral space. It is based on the fact that both chlorophyll absorption causing a decrease of red reflectance and leaf tissue abundance causing increased near-infrared reflectance will increase the total area of the triangle. The TVI index encompasses the area spanned by the triangle ABC with the coordinates given in spectral space:

$$TVI = \frac{1}{2}(120(\rho_{nir} - \rho_{green})) - 200(\rho_{red} - \rho_{green}) \quad (11\text{-}32)$$

where ρ_{green}, ρ_{red}, and ρ_{nir} are the reflectances centered at 0.55 μm, 0.67 μm, and 0.75 μm, respectively.

Reduced Simple Ratio — RSR

Chen et al. (2002) modified the simple ratio algorithm to include information from the short-wavelength infrared (SWIR) band found in the SPOT *Vegetation* sensor. They used the sensor to map the leaf-area-index of Canada. The *Reduced Simple Ratio* (RSR) is:

$$RSR = \frac{\rho_{nir}}{\rho_{red}}\left(1 - \frac{\rho_{swir} - \rho_{swirmin}}{\rho_{swirmax} + \rho_{swirmin}}\right) \quad (11\text{-}33)$$

where $\rho_{swirmin}$ and $\rho_{swirmax}$ are the minimum and maximum SWIR reflectance found in each image and defined as the 1 percent minimum and maximum cutoff points in the histograms of SWIR reflectance in a scene. The major advantages of RSR over SR are a) the difference between land-cover types is much reduced so that the accuracy for leaf area index retrieval for mixed cover types can be improved or a single LAI algorithm can be developed without resorting to coregistered land-cover maps as the first approximation, and b) the background (understory, litter, and soil) influence is suppressed using RSR because the SWIR band is most sensitive to the amount of vegetation containing liquid water in the background (Chen et al., 2002).

Chlorophyll Absorption in Reflectance Index — CARI; Modified Chlorophyll Absorption in Reflectance Index — MTCARI; Optimized Soil-Adjusted Vegetation Index — OSAVI; Ratio TCARI/OSAVI

Many scientists are interested in the amount of chlorophyll in vegetation (e.g., Daughtry et al., 2000). Kim et al. (1994) developed the *Chlorophyll Absorption in Reflectance Index* (CARI). It was modified to become the *Transformed Absorption in Reflectance Index* (TCARI):

$$TCARI = 3\left[(\rho_{700} - \rho_{670}) - 0.2(\rho_{700} - \rho_{550})\left(\frac{\rho_{700}}{\rho_{670}}\right)\right] \quad (11\text{-}34)$$

It uses bands corresponding to the minimum absorption of the photosynthetic pigments, centered at 550 and 700 nm, in conjunction with the chlorophyll *a* maximum absorption band, around 670 nm. The choice of 700 nm is due to its location at the boundary between the region where vegetation reflectance is dominated by pigment absorption and the beginning of the red edge portion where vegetation structural characteristics (i.e., the spongy mesophyll) have more influence on reflectance (Kim et al., 1994).

Unfortunately, TCARI is still sensitive to the underlying soil reflectance properties, particularly for vegetation with a low leaf area index. Therefore, Daughtry et al. (2000) proposed that TCARI be combined with a soil line vegetation index like the *Optimized Soil-Adjusted Vegetation Index* (OSAVI) (Rondeaux et al., 1996):

$$OSAVI = \frac{(1 + 0.16)(\rho_{800} - \rho_{670})}{(\rho_{800} + \rho_{670} + 0.16)}. \quad (11\text{-}35)$$

The ratio became

$$\frac{TCARI}{OSAVI} \quad (11\text{-}36)$$

and is highly correlated with vegetation chlorophyll content (Haboudane et al., 2002).

Visible Atmospherically Resistant Index — VARI

Many resource managers would like vegetation fraction information (e.g., 60%) (Rundquist, 2002). Building upon the Atmospherically Resistant Vegetation Index, scientists developed the *Visible Atmospherically Resistant Index* (VARI) computed as (Gitelson et al., 2002):

$$VARI_{green} = \frac{\rho_{green} - \rho_{red}}{\rho_{green} + \rho_{red} - \rho_{blue}}. \quad (11\text{-}37)$$

The index was minimally sensitive to atmospheric effects, allowing estimation of vegetation fraction with an error of <10% in a wide range of atmospheric optical thickness.

Normalized Difference Built-up Index — NDBI

Many professionals working on urban/suburban problems are interested in monitoring the spatial distribution and growth of urban built-up areas. These data can be used for watershed runoff prediction and other planning applications. Zha et al. (2003) calculated a *Normalized Difference Built-up Index* (NDBI):

$$NDBI = B_u - NDVI \quad (11\text{-}38)$$

where

$$B_u = \frac{NIR_{TM4} - MidIR_{TM5}}{NIR_{TM4} + MidIR_{TM5}}. \quad (11\text{-}39)$$

This resulted in an output image that contained only built-up and barren pixels having positive values while all other land cover had a value of 0 or −254. The technique was reported to be 92 percent accurate.

New Vegetation Index — NVI

The near-infrared bands found on Landsat TM, NOAA AVHRR, and the Indian Remote Sensing Linear Imaging Self Scanning (LISS) sensor are in the region 770 to 860,

760 to 900, and 725 to 1100 nm, respectively. Unfortunately, these near-infrared regions include water vapor absorption bands. Therefore, to improve the biomass monitoring capability of the NDVI, Gupta et al. (2001) chose to exclude water vapor absorption bands in the creation of a *New Vegetation Index* (NVI):

$$NVI = \frac{\rho_{777} - \rho_{747}}{\rho_{673}} \qquad (11\text{-}40)$$

where ρ_{777}, ρ_{747}, and ρ_{673} are the reflectances centered at 777, 747, and 673 nm, respectively, in 3- to 10-nm bandwidth hyperspectral data. Other narrow-band vegetation indices are found in Jensen (2005).

Red-edge Position Determination — REP

The abrupt change in the 680–800 nm region of reflectance spectra of leaves caused by the combined effects of strong chlorophyll absorption and leaf internal scattering is called the red edge. The *red edge position* (REP) is the point of maximum slope on a vegetation reflectance spectrum between the red and near-IR wavelengths. The red edge was first described by Collins (1978), and is perhaps the most studied feature on the vegetation spectral curve according to Schlerf and Atzberger (2001). The REP is useful because it is strongly correlated with foliar chlorophyll content and can be a sensitive indicator of vegetation stress. Determining the red edge position using remote sensing data usually requires the collection of hyperspectral data.

A linear method proposed by Clevers (1994) can be implemented that makes use of four narrow bands and is computed as follows:

$$REP = 700 + 40\left[\frac{\rho_{(red\ edge)} - \rho_{(700nm)}}{\rho_{(740nm)} - \rho_{(700nm)}}\right], \qquad (11\text{-}41)$$

where

$$\rho_{(red\ edge)} = \frac{\rho_{(670nm)} + \rho_{(780nm)}}{2}. \qquad (11\text{-}42)$$

Dawson and Curran (1998) and Baranoski and Rokne (2005) summarize additional methods used to determine the red edge position.

Scientists throughout the world are studying the role of terrestrial vegetation in large-scale global processes. This is necessary in order to understand how the Earth functions as a system. Vegetation indices are being used to accurately inventory the global distribution of vegetation types as well

as their biophysical (e.g., LAI, biomass, APAR) and structural (e.g., percent canopy closure) properties. Monitoring these characteristics through space and time will provide valuable information for understanding the Earth as a system (Townshend and Justice, 2002).

Landscape Ecology Metrics

Vegetation indices are useful for monitoring the condition and health of vegetated pixels. However, such per-pixel analysis does not provide any information about the nature of surrounding pixels, including their site and association characteristics. Landscape ecology principles have been developed that increasingly incorporate remote sensor data to assess the health and diversity of vegetation and other variables within entire ecosystems. This has resulted in the development of numerous *landscape ecology metrics* or *indicators* that are of significant value when analyzing rangeland, grassland, forests, and wetland (Frohn, 1998; McGarigal et al., 2006). Numerous government agencies, such as the Environmental Protection Agency, base much of their environmental modeling and landscape characterization on these metrics and indicators (EPA Landscape Ecology, 2006). Therefore, it is useful to provide a brief review of their origin and summarize several of the more important landscape ecology metric variables that can be extracted from remotely sensed data.

The term *landscape ecology* was first introduced by the German geographer Carl Troll (1939) who made widespread use of the then new technique of aerial photography. Troll intended for the term landscape ecology to distinguish his approach for using such imagery to interpret the interaction of water, land surfaces, soil, vegetation, and land use from that of conventional photographic interpretation and cartography. Landscape ecology has been intensively practiced in Europe for many decades and became generally recognized in the United States in about 1980. Since then, landscape ecology has rapidly evolved as a discipline, spurred by the synergistic interactions between remote sensing and GIS techniques and advances in ecological theory.

Landscape ecology is the study of the structure, function, and changes in heterogeneous land areas composed of interacting organisms (Bourgeron and Jensen, 1993). It is the study of the interaction between landscape patterns and ecological processes, especially the influence of landscape pattern on the flows of water, energy, nutrients, and biota. What distinguishes landscape ecology from the many separate disciplines that it embraces (e.g., geography, biology, ecology, hydrology) is that it provides a hierarchical framework for interpreting ecological structure, function, change, and resiliency at multiple scales of inquiry.

Traditional measures to protect the environment, such as preventing water pollution or protecting biodiversity, often focused on specific effluent discharges or fine-scale habitat requirements. This method has been described as the fine-filter approach. In contrast, the coarse-filter approach to resource conservation states that "by managing aggregates (e.g., communities, ecosystems, landscapes), the components of these aggregates will be managed as well" (Bourgeron and Jensen, 1993). In other words, the most cost-effective strategy to maintain the resiliency and productivity of ecological systems is to conserve (or restore) the diversity of species, ecosystem processes, and landscape patterns that create the systems (Jensen and Everett, 1993). Applying this coarse-filter management method requires that landscape patterns be evaluated at multiple spatial and temporal scales rather than simply at the traditional scales of stream reach or forest stand.

Hierarchy theory allows landscape ecologists to integrate multiple scales of information to determine whether landscape patterns are sufficient to allow ecological processes to operate at the necessary scales. The objective is to investigate changes in the distribution, dominance, and connectivity of ecosystem components and the effect of these changes on ecological and biological resources. For example, ecosystem fragmentation has been implicated in the decline of biological diversity and ecosystem sustainability at a number of spatial scales (Forman, 1995; Jones et al., 1998). Determining status and trends in the pattern of landscapes is critical to understanding the overall condition of ecological resources. Landscape patterns thus provide a set of indicators (e.g., pattern shape, dominance, connectivity, configuration) that can be used to assess ecological status and trends at a variety of scales.

A hierarchical framework also permits two important types of comparisons: 1) to compare conditions within and across landscapes, and 2) to compare conditions across different types of ecological risks. Such ecological risks include the risk of erosion, loss of soil productivity, loss of hydrologic function, and loss of biodiversity.

Scalable units are needed to address landscape ecology issues at multiple scales within a hierarchical framework. Examples of scalable units include patches, patterns, and landscapes. A *patch unit* is a set of contiguous measurement units (e.g., pixels) that have the same numerical value. A

pattern unit is a collection of measurement units and/or patch units that have the property of being the minimum unit descriptor of a larger spatial area. The scales of assessment questions and indicators suggest two types of landscape units: watersheds and landscape pattern types (LPT) (Wickham and Norton, 1994). Watersheds and LPTs capture or bound four important flow processes operating within and among landscapes: flows of energy, water, nutrients, and biota. Scales of watersheds and LPTs range from approximately 10^3 to 10^6 units in extent, and from 1 to 100 ha.

Landscape Indicators and Patch Metrics

Jones et al. (1998) suggest that landscape integrity can be monitored by carefully watching the status of the following indicators:

- land-cover composition and pattern

- riparian extent and distribution

- ground water

- greenness pattern

- degree of biophysical constraints

- erosion potential.

Monitoring these landscape indicators requires precise, repeatable measurements of terrain *patches* such as individual forest stands, rangeland, wetland, and/or agricultural fields (Table 11-5). It is also important to identify patches of pure urban structure such as residential and commercial land use. These measurements of terrain patches are routinely referred to as *landscape pattern* and *structure metrics* (McGarigal et al., 2006).

Numerous landscape structure metrics have been developed (e.g., Ritters et al., 1995; Schuft et al., 1999; Weiers et al., 2005). Many of these landscape ecology metrics are available in *FRAGSTATS - A Spatial Pattern Analysis Program for Categorical Maps* (McGarigal and Marks, 1994; McCarigal et al., 2006). Many scientists use this program to extract patch, class, and landscape indices (e.g., Millington et al., 2003; Jackson and Jensen, 2005).

O'Neill et al. (1997) suggest that the health of an ecosystem could be monitored if the following three landscape ecology metrics were monitored through time: dominance, contagion, and fractal dimension.

Table 11-5.   Societal values, example indicators, and candidate landscape ecology metrics.

Societal Value	Indicator	Candidate Metrics
Biodiversity	wildlife habitat suitability	patch statistics (number, total area, average size, largest size, distance between, ratio perimeter to area, shape, fractal dimension, square pixel model, etc.), fragmentation, contagion, zone fragmentation index, patch per-unit-area index, dominance, adjacency of land-cover types, Shannon diversity, biophysical attribute patterns
	stream biological condition	diversity, square pixel model, dominance, fragmentation, zone fragmentation index, patch per-unit-area index, adjacency of land-cover types, slope, elevation, diffusion rates, percolation threshold, erosion index, texture, biophysical attribute patterns, geochemical attributes
	forest plant species richness	diversity, dominance, fragmentation, zone fragmentation index, patch per-unit-area index, slope, erosion index, texture, patch statistics, square pixel model, biophysical attribute patterns
	landscape sustainability	patch statistics, contagion, zone fragmentation index, patch per-unit-area index, fragmentation, texture, dominance, fractal dimension, square pixel model, biophysical attribute patterns
Watershed integrity	water quality	patch statistics, erosion index, hydrologic modification, adjacency of land-cover types, dominance, contagion, zone fragmentation index, patch per unit area index, fractal dimension, square pixel model, elevation, slope, biophysical attribute patterns, geochemical attributes
	vulnerability to flooding	patch statistics, adjacency of land-cover types, erosion index, dominance, contagion, zone fragmentation index, patch per-unit-area index, fractal dimension, square pixel model, hydrologic modification, elevation, slope, texture, biophysical attribute patterns
Landscape resilience	landscape sustainability	patch statistics, dominance, contagion, zone fragmentation index, patch per-unit-area index, fragmentation, fractal dimension, square pixel model, biophysical attribute patterns

Dominance, D, is the information theoretic index that identifies the extent to which the landscape is dominated by a single land-cover type. The metric, $0 < D < 1$, is computed as

$$D = 1 - \left[\sum_k \frac{(-P_k \cdot \ln P_k)}{\ln(n)} \right] \qquad (11\text{-}43)$$

where $0 < P_k < 1$ is the proportion of land-cover type k, and n is the total number of land-cover types present in the landscape.

Contagion, C, expresses the probability that land cover is more "clumped" than the random expectation. The index, $0 < C < 1$, is

$$C = 1 - \left[\sum_i \sum_j \frac{(-P_{ij} \cdot \ln P_{ij})}{2 \ln(n)} \right] \qquad (11\text{-}44)$$

where P_{ij} is the probability that a pixel of cover type i is adjacent to type j.

The *fractal dimension, F,* of patches indicates the extent of human reshaping of landscape structure (O'Neill et al., 1997). Humans create simple landscape patterns; nature creates complex patterns. The fractal dimension index is calculated by regressing the log of the patch perimeter against the log of the patch area for each patch on the landscape. The index equals twice the slope of the regression line. Patches of four or fewer pixels are excluded because resolution problems distort their true shape.

O'Neill et al. (1997) suggest that this set of three indices may capture fundamental aspects of landscape pattern that influence ecological processes. Significant changes in these indices for an ecosystem might indicate that perhaps deleterious processes are at work in the environment. For example,

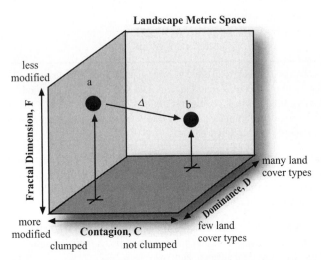

Figure 11-27 A hypothetical depiction of a three-dimensional landscape metric feature space. Point *a* represents where a stable, unmodified ecosystem might reside in dominance, contagion, and fractal dimension feature space. When the landscape is modified by man or natural forces, it may move to point *b* with many land-cover types being introduced (dominance change), be less clumped (contagion), and exhibit a more modified fractal dimension. This could be good or bad. Remote sensing of vegetation is critical to the extraction of landscape patch metrics.

consider a small ecosystem that exhibits a less modified fractal dimension, is highly clumped, and has relatively few land-cover types within it. It might appear in a three-dimensional landscape metric space at location *a* in Figure 11-27. If this small ecosystem were subdivided with several new roads and fragmented, its location might move in three-space to *b* with many land-cover types being introduced (dominance change), be less clumped (contagion), with a more modified fractal dimension. This could be good or bad. In fact, the relationships between the metric values and how they actually relate to ecological principles are still being determined. *Remote sensing of vegetation within these patches is very important and constitutes one of the major factors responsible for whether or not the metrics are robust and useful for ecological modeling.*

Ritters et al. (1995) reviewed 55 patch metrics and concluded that the following metrics accounted for most of the variance, including: the number of attribute cover types in a region (*n*); contagion (previously discussed); average perimeter-area ratio; perimeter-area scaling; average large-patch density-area scaling; standardized patch shape; and patch perimeter-area scaling. The algorithms for all 55 metrics are provided in the paper.

Applying the principles of landscape ecology requires an understanding of the natural variability of landscape patterns and processes across both space and time. Estimates of this variability are essential to determining whether the current condition of landscape is sustainable, given its historic patterns and processes (Jensen and Everett, 1993). Moreover, the estimates are extremely useful in both broad-level assessment of risk to resources, as well as to finer-scale assessments.

The goal is usually to

• estimate, on a regional basis and with known confidence, the current status, trends, and changes in selected *indicators* of the important landscapes;

• estimate with known confidence the geographic coverage and extent of the landscapes' patterns and types;

• seek associations between selected indicators of natural and anthropogenic stressors and indicators of landscape condition; and

• provide statistical summaries and periodic assessments of the condition of the landscapes.

Monitoring the vegetation in landscape patches using remotely sensed data and deriving accurate metrics from these data is becoming a very important component of many ecosystem monitoring programs. For example, Arroyo-Moya et al. (2005) used remote sensing-derived landscape ecology metrics to analyze forest fragmentation and regrowth characteristics in Costa Rica from 1960 – 2000. Jackson and Jensen (2005) used remote sensing-derived landscape ecology metrics to perform reservoir shoreline environmental monitoring in South Carolina.

It is also important to monitor the health of ecosystems at the regional and national levels to identify significant problem areas that may be experiencing a loss of biodiversity. One of the most useful methods of maintaining biodiversity that is based on the use of remote sensing derived vegetation data is Gap Analysis.

 Biodiversity and Gap Analysis

The rapid loss of biological diversity (biodiversity) remains one of mankind's greatest threats. The conventional approach to maintaining biodiversity generally has been to proceed species by endangered species and threat by threat. Within the United States, the primary means of stemming this loss is the Endangered Species Act (ESA). Some have criticized the ESA for several reasons, including a backlog of unaddressed petitions, failure to develop and implement recovery plans in a timely fashion, and lack of adequate funding to meet objectives. A primary cause of these problems is that the Act focuses on individual species. Effort expended on this species-by-species approach is inefficient, expensive, and biased toward *charismatic megafauna* having broad public appeal (i.e., a cute animal is more likely to be protected than a not-so-cute animal). Last-minute protection efforts also contribute to economic conflict because they fail to provide a reasonable planning framework for economic interests (e.g., between biologists and logging companies). While necessary, these efforts need to be complemented with more proactive methods that attempt to maintain species and ecosystems while they are still common (Edwards et al., 1995; Millington et al., 2003).

One approach for assessing the current status of biodiversity at all levels, not just endangered species, is called *Gap Analysis*. As an evaluation process, it provides a systematic approach for determining the protection afforded biological diversity in given areas. Its fundamentals are presented here because it is heavily based on the use of remote sensing and geographic information systems (GIS) to identify gaps in biodiversity protection that may be filled by the establishment of new preserves or changes in land-use practices (Scott et al., 1993; Edwards et al., 1995).

A typical Gap Analysis includes the creation and analysis of four primary GIS layers:

- the distribution of actual vegetation-cover types delineated from satellite and aircraft remotely sensed data and ancillary information;

- land ownership;

- land-management status;

- distribution of terrestrial vertebrates as predicted from the distribution of vegetation and *in situ* observations.

Gap Analysis provides an overview of the distribution and conservation of several components of biodiversity. It uses the distribution of actual vegetation types and terrestrial vertebrate and, when available, invertebrate taxa as indicators (surrogates) for biodiversity. Digital map polygon overlays in a GIS are used to identify individual species, species-rich areas, and cover types that are unrepresented or under-repre-

sented in existing management areas. It functions as a preliminary step to the more detailed studies needed to establish actual boundaries for potential biodiversity management areas. Gap Analysis, by focusing on higher levels of biological organizations, is likely to be less expensive and more likely to succeed than conservation programs focused on single species or populations (Scott et al., 1993).

One of the first Gap Analyses focused on endangered forest birds in Hawaii (Scott et al., 1987). When compared with a map of the existing reserves, < 10 percent of the geographic ranges of endangered forest birds were protected, i.e., there were significant gaps in the existing biodiversity protection scheme. Several areas with highly endangered bird species richness have since been protected by the Nature Conservancy and state and federal agencies. Successful Gap Analysis has taken place in many locations, including Hawaii, Utah, Puerto Rico, and California. Gap Analysis is under way in many other states and countries (Savitsky, 1998).

 Remote Sensing of Vegetation Change

Analyzing an individual date of remote sensor data to extract meaningful vegetation biophysical information is often of value. However, to appreciate the dynamics of the ecosystem, it is necessary to monitor the vegetation through time and determine what changes in succession are taking place. Relatively high temporal resolution satellite data is often of value when conducting such successional studies. The following case study documents the usefulness of SPOT remote sensor data and the methods used to monitor wetland change in a freshwater reservoir.

Remote Sensing Inland Wetland Successional Changes

Inland wetlands assimilate pollutants, control floods, and serve as breeding, nursery, and feeding grounds for fish and wildlife. Unfortunately, the conterminous United States lost 53 percent of its wetlands to agricultural, residential, and/or commercial land use from the 1780s to 1980s. Accurate and timely information on wetland distribution is essential for effective protection and management. This case study documents the use of multiple-season and multiple-year SPOT satellite multispectral data for monitoring inland wetland successional changes in Par Pond on the Savannah River Site, SC, caused by the draw-down of the reservoir's water level from June 1991 through October 1994 (Jensen et al.,

1997). The study brings to bear many of the considerations discussed in this chapter.

Par Pond Reservoir Study Area

Par Pond is a 1012-ha reservoir originally created in 1958 to receive reactor cooling water, but now receives flow-through water from the Savannah River. Par Pond has a stable, predictable cattail and waterlily macrophyte community when water levels fluctuate < 1 m per year.

Approximately 30 macrophyte plant species are found in Par Pond, but only a few dominate:

- Cattails (*Typha* spp.) are persistent emergent macrophytes that exist year-round in Par Pond. They begin to "green up" in early to mid-April and often form a dense, green canopy by late May, as documented in the phenological calendar shown in Figure 11-28. Cattails senesce in late September or early October, remaining brown throughout the winter months. Cattails are normally found in shallow water < 1 m in depth adjacent to the reservoir shoreline.

- Waterlilies (*Nymphaea odorata*) are dominant floating macrophytes that do not persist through the winter (Figure 11-28). They appear at the outer edge of the cattails by late April or early May and reach full growth six to eight weeks later. Waterlily beds expand into the open-water areas of the reservoir through September and typically extend to depths of 2 – 5 m. The beds persist until mid-October to mid-November and then disappear.

Before June, 1991, the water level of Par Pond was relatively stable at 200 ft above mean sea level (msl), fluctuating < 0.15 m in most years. Extensive beds of cattails and waterlilies from 20 – 150 m in width extended out into the lake. From June 23, 1991, through September 15, 1991, the water level of Par Pond was lowered to 181 ft above msl for reservoir repair. The 19 ft (approximately 6 m) drop in water level turned a submerged zone along the shoreline about 50 – 250 m in width into upland. This exposed the emergent and non-emergent aquatic macrophyte beds of the Par Pond shoreline to drying conditions for about four years, until the water level was raised to 200 ft once again in 1995.

Remote Sensor Data

Seven SPOT multispectral (XS) images acquired in the spring and fall of 1992, 1993, and 1994 were analyzed (Table 11-6). Near-anniversary dates were selected whenever possible to minimize changes in reflectance caused by

Phenological Cycle of Cattails and Waterlilies in Par Pond, SC

Figure 11-28 Annual phenological cycle of inland wetland cattails and waterlilies in Par Pond on the Savannah River Site in South Carolina. Cattails are persistent emergent marsh vegetation that green up earlier than waterlilies in late March and early April. They begin senescing in late October and lie dormant over the winter. Waterlilies disappear completely by November each year and reappear in late April. Both cattail and waterlily have a full, dense canopy throughout the summer months.

Table 11-6. SPOT HRV images used to document the vegetation succession of Par Pond.

Date	SPOT Data Type	Reference Color-Infrared Aerial Photography
March 14, 1992	HRV-XS	January 31, 1992
May 5, 1992	HRV-XS	May 2, 1992
October 13, 1992	HRV-XS	August 10, 1992
May 10, 1993	HRV-XS	April 23, 1993
October 28, 1993	HRV-XS	January 22, 1994
April 2, 1994	HRV-XS	April 18, 1994
October 25, 1994	HRV-XS	

Sun angle and seasonal soil moisture (Jensen et al., 1995). Color-infrared aerial photography was also available for this experiment. SPOT image-to-map rectification was based on the use of approximately 50 ground-control points per date and nearest-neighbor resampling to a Universal Transverse Mercator projection. Total geometric root-mean-square-

Figure 11-29 a) Perspective view of the percentage of land cover in Par Pond from March 14, 1992, through October 25, 1994, derived from analysis of seven dates of SPOT HRV XS data. Extensive areas of dead vegetation were present in early 1992. Spikerush dominated in late 1992 and early 1993. Old field natural succession dominated in late 1993 and throughout 1994. Pine and hardwood seedlings were colonizing extensively by October 25, 1994. b) The percent land cover of spikerush, old field, and pine/hardwood vegetation on draw-down areas of Par Pond from June 23, 1991, through October 25, 1994, are presented as raw data and modeled with the use of fifth-order polynomial equations (Jensen et al., 1997).

error (RMSE) was $\pm$ 0.5 pixel ($\pm$ 20 m) for each date of imagery. All terrain > 200 ft above msl was masked out of the study area.

Image Classification and Change Detection

Fifty spectrally unique clusters were extracted from each date of imagery using an Interactive Self Organizing Data Analysis Technique — ISODATA (refer to Jensen, 2005). The clusters were labeled to produce seven land-cover maps. The overall accuracy for two of the classification maps (October 28, 1993, and April 2, 1994) was 75.6 percent and 77.9 percent, respectively. This level of accuracy is about what is expected from simple three-band 20 × 20 m SPOT multispectral data (Jensen et al., 1995).

The goal of this study was to document the ecological change in vegetation cover in the draw-down areas using multiple dates of remote sensor data. This required the use of a change-detection algorithm that provided "from-to" map information, i.e., a 20 × 20 m pixel of Dead Vegetation in 1992 changed into a pixel of Spikerush by 1993. Twenty-one change detection maps were produced and may be seen in Jensen at al. (1997).

Results

The percentage of land-cover in each category on each date from March 14, 1992, through October 25, 1994, for the Par Pond study area are graphed in Figure 11-29a. This information documents the four year vegetation succession trends on the Par Pond exposed shoreline. On March 14, 1992, Dead Vegetation occupied approximately 35.48 percent of the study area. The cattail and waterlily aquatic macrophytes did not survive the draw-down. Also present were 7.46 percent Bare Soil and 8.74 percent early invasion by Spikerush. As time passed, the Dead Vegetation areas were colonized by other species. By October 25, 1994, no Dead Vegetation was present.

Two major types of successional development took place on the Dead Vegetation and Bare Soil draw-down areas. First, there was a significant increase in Spikerush in 1992. By May 5, 1992, Spikerush occupied 20.28 percent of the study area. As the soil dried out in subsequent years, Spikerush coverage declined to about 10.75 percent by October 25, 1994.

Spikerush and Bare Soil were succeeded primarily by Old Field species as expected, and at a very rapid rate. The most dramatic increase was from May 10, 1993 to October 28, 1993, which saw Old Field land cover increase from almost nothing to 25.62 percent of the study area. Pine/Hardwood seedling coverage also increased steadily from almost no cover in 1992 to 10.25 percent by October 25, 1994. If the draw-down areas were unmolested for approximately 75 years, oak/hickory climax vegetation would be present in all the draw-down areas.

Using the land-cover statistics derived from multiple-date remote sensor data, it was possible to develop predictive successional models for each of the land-cover vegetation types in the draw-down area. For example, Figure 11-29b depicts the actual and predicted percent of the scene occupied by Spikerush at different dates since the initiation of draw-down on June 23, 1991. The Spikerush succession was modeled using a fifth-order polynomial where

$$\%Spikerush = -0.0129x^5 + 0.4018x^4 - 4.0989x^3 + 15.383x^2 - 14.425x,$$

with $R^2 = 0.992$. Using this equation it was possible to predict the percent of Spikerush that would be present at user-specified dates after a serious reservoir draw-down. Similarly, Old Field and Pine/Hardwood land-cover succession were modeled using the equations

$$\%Old\ Field = -0.0416x^5 + 0.6421x^4 - 3.0604x^3 + 5.1495x^2 - 2.3275x,\ and$$

$$\%Pine/Hardwood = 0.00762x^5 - 0.1543x^4 + 1.0971x^3 - 2.9539x^2 + 2.4909x.$$

The polynomial model for Old Field ($R^2 = 0.89$) did not explain the succession as well as the Pine/Hardwood predictive model ($R^2 = 0.94$). The first derivative of these models can be used to predict the rate-of-change (gradient) in percent of land cover on a specific day since draw-down.

The analysis of the land-cover and change-detection maps derived from satellite remote sensor data quantitatively documented 1) the spatial distribution of the successional changes in land cover, and 2) the rate of successional change in Par Pond caused by the draw-down.

Numerous state and government agencies recognize the importance of utilizing remote sensor data for monitoring their sensitive vegetation resources. For example, the first comprehensive inventory of the vegetation at the plant community level of the Everglades in Florida was completed using a combination of satellite imagery, aerial photography, GPS, and GIS technologies (Doren et al., 1999).

Remote Sensing South African Agriculture

Many countries use remote sensing to obtain critical biophysical information to help protect its citizens from famine during sustained drought. For example, the Republic of South Africa Department of Water Affairs & Forestry is mandated by the *South African National Water Act* (1998) to obtain information on the spatial distribution of all agriculture that uses water each year. All agricultural water users must become "registered water users" in order to legally utilize the nation's scarce water resources. The DWAF is responsible for conducting the water use accounting using the Water Authorization and Resource Management System (WARMS) spatial decision support system. Remote sensing is one of the primary methods used to obtain the agricultural water use information. A NASA-sponsored study assisted the DWAF to develop remote sensing-assisted techniques to obtain the necessary agricultural water use information (Jensen et al., 2006).

Study Area

The study was conducted in Water Management Area #6: Usutu and Mhlathuze River catchments (Figure 11-30a). The area contains natural vegetation consisting of forest and scrub-shrub (bushveld), non-irrigated dryland farming, and irrigated agriculture consisting primarily of sugarcane and citrus. SPOT remote sensor data and *in situ* ground reference information were used to train and classify the imagery into agricultural land-cover classes.

Remote Sensor Data

Many of the agricultural fields in the *Usutu* and *Mhlathuze* River Water Management Area are < 100 ha. Therefore, it was necessary to use remote sensing imagery with a spatial resolution of $< 30 \times 30$ m to identify the crop type within such small fields. Sensors like MODIS (NASA), MERIS (European Space Agency) and AVHRR (NOAA) were of no value due to their coarse spatial resolutions (e.g., 250 m, 300 m, and 1100 m, respectively). After analyzing crop calendar information, DWAF obtained the following remote sensor data:

- SPOT 5 High Resolution Visible Infrared (HRVIR) imagery obtained on January 9, 2004 [multispectral bands 1-4 at 10×10 m; panchromatic band 5 at 5×5 m].

A SPOT image of the study area is shown in Figure 11-30a. A SPOT color composite is shown in Color Plate 11-6a.

Table 11-7.   Object-oriented image segmentation classification of SPOT 2004 multispectral (10 × 10 m) and panchromatic (5 × 5 m) data of an area in the Republic of South Africa.

Class	Sugar cane	Citrus	Forest	Bushveld	Road	Water	Fallow	Wind-break	Banana	Row Total	User's Accu-racy
Sugarcane	1261	J61	0	4	2	0	1	2	4	1335	94.46
Citrus	27	744	5	0	1	2	4	10	0	793	93.82
Forest	13	23	87	13	0	2	1	1	0	140	62.14
Bushveld	28	25	8	82	1	0	5	0	0	149	55.03
Road	6	1	1	1	45	0	2	1	0	57	78.95
Water	0	0	0	0	0	44	2	0	0	46	95.66
Fallow	7	0	2	2	0	1	39	0	0	49	79.59
Windbreak	2	4	0	0	1	1	0	36	0	44	81.82
Banana	4	4	1	1	0	0	0	0	60	69	86.97
Column Total	1348	862	101	103	50	50	54	50	64	2682	
Producer's Accuracy	93.55	86.31	86.14	79.61	90	88	72.22	72	93.75		
Overall Accuracy	89.4%										
Kappa Coefficient	83.7%										

In Situ Ground Reference Data

DWAF obtained *in situ* ground reference information on the type of agriculture present in 2682 fields in the study area. The centroids of many of the fields are overlaid on 1:10,000 large-scale orthophotography in Figure 11-30b.

Data Analysis

The SPOT data were subjected to an object-oriented image segmentation classification that takes into account both the spectral and spatial information inherent in the SPOT data (please refer to Jensen, 2005, for information about object-oriented image segmentation).

Results

The object-oriented segmentation classification using both SPOT 2004 multispectral radiance (10 × 10 m) and panchromatic (5 × 5 m) data yielded the classification map shown in

Color Plate 11-6b. The overall accuracy was 89.4 percent with a Kappa coefficient of agreement of 83.7 (Table 11-7). The user's accuracy for sugarcane and citrus were 94.46 percent and 93.82 percent, respectively while the producer's accuracies were 93.55 percent and 86.31 percent, respectively. This type of agricultural land cover information is used by the Republic of South Africa Department of Water Affairs & Forestry as input to the Water Authorization and Resource Management System (WARMS) spatial decision support system (courtesy Jensen et al., 2006).

 ### References

Abuelgasim, A. A., Gopal, S. and A. H. Strahler, 1998, "Forward and Reverse Modeling of Canopy Directional Reflectance Using a Neural Network," *Remote Sensing of Environment*, 59:453–471.

SPOT Data of the Mhlathuze River in South Africa

a. Part of Water Management Area #6: Usutu and Mhlathuze River catchments
recorded on SPOT 5 red band imagery collected on January 9, 2004.

b. Ground reference points overlaid on orthophotography.

Figure 11-30 a) SPOT HRVIR image of an area along the Mhlathuze River in the Republic of South Africa. b) A portion of the 2,682 fields
used to assess the agricultural land cover classification accuracy are overlaid on 1:10,000-scale orthophotography (Jensen et
al., 2006).

Arroyo-Moya, J. P., Sanchez-Azofeifa, G. A., Rivard, B., Calvo, J. C. and D. H. Janzen, 2005, "Dynamics in Landscape Structure and Composition for Chorotega Region, Costa Rica from 1960 to 2000," *Agriculture, Ecosystems and Environment*, 106:27–39.

Baranoski, G. V. and J. G. Rokne, 2005, "A Practical Approach for Estimating the Red Edge Position of Plant Leaf Reflectance," *Intl. Journal of Remote Sensing*, 26(3):503–521.

Birth, G. S. and G. McVey, 1968, "Measuring the Color of Growing Turf with a Reflectance Spectrophotometer," *Agronomy Journal*, 60:640–643.

Bourgeron, P. S. and M. E. Jensen, 1993, "An Overview of Ecological Principles for Ecosystem Management," in Jensen, M. and P. Bourgeron (Eds.), *Eastside Forest Ecosystem Health Assessment,* Vol. II, *Ecosystem Management: Principles and Applications*, Missoula: USDA Forest Service, 49–60.

Broge, N. H. and E. Leblanc, 2000, "Comparing Prediction Power and Stability of Broadband and Hyperspectral Vegetation Indices for Estimation of Green Leaf Area Index and Canopy Chlorophyll Density," *Remote Sensing of Environment*, 76:156–172.

Carter, G. A., 1991, "Primary and Secondary Effects of the Water Content on the Spectral Reflectance of Leaves," *American Journal of Botany,* 78(7):916–924.

Carter, G. A., 1993, "Responses of Leaf Spectral Reflectance to Plant Stress," *American Journal of Botany*, 80(3):2310–243.

Carter, G. A., Cibula, W. G. and R. L. Miller, 1996, "Narrowband Reflectance Imagery Compared with Thermal Imagery for Early Detection of Plant Stress," *Journal of Plant Physiology*, 148:515–522.

Chen, T. M. and T. A. Black, 1992, "Defining Leaf Area Index for Non-flat Leaves," *Plant, Cell and Environment*, 15:421–429.

Chen, J. M., Pavlic, G., Brown, L., Cihlar, J., Leblanc, S. G., White, H. P., Hall, R. J., Peddle, D. R., King, D. J., Trofymow, J. A., Swift, E., Van der Sanden, J., and P. K. Pellikka, 2002, "Derivation and Validation of Canada-wide Coarse-resolution Leaf Area Index Maps Using High-resolution Satellite Imagery and Ground Measurements," *Remote Sensing of Environment*, 80:165–184.

Cibula, W. G. and G. A. Carter, 1992, "Identification of a Far-Red Reflectance Response to *Ectomycorrhizae* in Slash Pine," *International Journal of Remote Sensing*, 13(5):925–932.

Clark, R. N., King, T. V., Ager, C. and G. A. Swayze, 1995, "Initial Vegetation Species and Senescence/Stress Indicator Mapping in the San Luis Valley, CO, Using Imaging Spectrometer Data," *Proceedings*, Summitville Forum; H. Posey, J. Pendelton, and D. Van Zyl, (Eds.), Colorado Geological Survey Publication #38, 64–69; http://speclab.cr.usgs.gov.

Clark, R. N., Swayze, G. A., Wise, R., Livo, K. E., Hoefen, T. M., Kokaly, R. F. and S. J. Sutley, 2003, *USGS Digital Spectral Library splib05a*, USGS Open File Report 03–395.

Clevers, J. G., 1994, "Imaging Spectrometry in Agriculture: Plant Vitality and Yield Indicators," in Hill, J. and J. Megier (Eds.), *Imaging Spectrometry: A Tool for Environmental Observations*, Dordrecht, Netherlands: Kluwer Academic, 193–219.

Cohen, W. B., 1991, "Response of Vegetation Indices to Changes in Three Measures of Leaf Water Stress," *Photogrammetric Engineering & Remote Sensing*, 57(2):195–202.

Collins, W., 1978, "Remote Sensing of Crop Type and Maturity," *Photogrammetric Engineering & Remote Sensing*, 43–55.

Colombo, R., Bellingeri, D., Fasolini, D. and C. M. Marino, 2003, "Retrieval of Leaf Area Index in Different Vegetation Types Using High Resolution Satellite Data," *Remote Sensing of Environment*, 86:120–131.

Crist, E. P., 1985, "A Thematic Mapper Tasseled Cap Equivalent for Reflectance Factor Data," *Remote Sensing of Environment,* 17:301–306.

Crist, E. P. and R. C. Cicone, 1984, "Application of the Tasseled Cap Concept to Simulated Thematic Mapper Data," *Photogrammetric Engineering & Remote Sensing*, 50:343–352.

Crist, E. P. and R. J. Kauth, 1986, "The Tasseled Cap De-mystified," *Photogrammetric Engineering & Remote Sensing*, 52(1):81–86.

Danson, F. M., 1998, "Teaching the Physical Principles of Vegetation Canopy Reflectance Using the SAIL Model," *Photogrammetric Engineering & Remote Sensing*, 64(8):8010–812.

Daughtry, C. S. T., Walthall, C. L., Kim, M. S., Brown de Colstoun, E. and J. E. McMurtrey III, 2000, "Estimating Corn Leaf Chlorophyll Concentration from Leaf and Canopy Reflectance," *Remote Sensing of Environment*, 74:229–239.

Dawson, T. P. and P. J. Curran, 1998, "A New Technique for Interpolating the Reflectance Red Edge Position," *International Journal of Remote Sensing*, 19(11): 2133–2139.

Deering, D. W., Rouse, J. W., Haas, R. H. and J. A. Schell, 1975, "Measuring Forage Production of Grazing Units from Landsat MSS Data," *Proceedings*, 10th Intl. Symposium on Remote Sensing of Environment, Ann Arbor: ERIM, 2:1169–1178.

Didan, K., 2002, *MODIS Vegetation Index Production Algorithms*, MODIS Vegetation Workshop, Missoula, MO, July 15–18; Terrestrial Biophysics and Remote Sensing MODIS Team, Tucson: Univ. of Arizona, www.ntsg.umt.edu/MODISCon/index.html.

Doren, R. F., Rutchey, R. and R. Welch, 1999, "The Everglades: A Perspective on the Requirements and Applications for Vegetation Map and Database Products," *Photogrammetric Engineering and Remote Sensing*, 65(2):155–161.

Du, Y., Teillet, P. M. and J. Cihlar, 2002, "Radiometric Normalization of Multitemporal High-resolution Satellite Images with Quality Control for Land Cover Change Detection," *Remote Sensing of Environment*, 82:123–134.

Edwards, T. C., Homer, C. G., Bassett, S. C., Falconer, A., Ramsey, R. D. and D. W. Wight, 1995, *Utah Gap Analysis: An Environmental Information System*, Logan: Utah Cooperative Fish & Wildlife Research Unit, 68 pp.

EPA Landscape Ecology, 2006, *Landscape Ecology*, Washington: Environmental Protection Agency, http://www.epa.gov/nerlesd1/land-sci/default.htm.

Farabee, M. J., 1997, *Photosynthesis*, http://gened.emc.maricopa.edu/bio/bio181/BIOBK/BioBookPS.html

Forman, R. T., 1995, "Some General Principles of Landscape and Regional Ecology," *Landscape Ecology*, 10(3):133–142.

Frohn, R. C., 1998, *Remote Sensing for Landscape Ecology*, Boca Raton: Lewis, 99 p.

Gates, D. M., J. J. Keegan, J. C. Schleter and V. R. Weidner, 1965, "Spectral Properties of Plants," *Applied Optics*, 4(1):11–20.

Galvao, L. S., Formaggio, A. R. and D. A. Tisot, 2005, "Discrimination of Surface Varieties in Southeastern Brazil with EO-1 Hyperion Data," *Remote Sensing of Environment*, 94:523–534.

Gao, B. C., 1996, "NDWI—A Normalized Difference Water Index for Remote Sensing of Vegetation Liquid Water from Space, *Remote Sensing of Environment*, 58:257–266.

Gausmann, H. W., Allen, W. A. and R. Cardenas, 1969, "Reflectance of Cotton Leaves and their Structure," *Remote Sensing of Environment*, 1:110–22.

Gitelson, A. A., Kaufman, Y. J., Stark, R. and D. Rundquist, 2002, "Novel Algorithms for Remote Estimation of Vegetation Fraction," *Remote Sensing of Environment*, 80:76–87.

Goetz, A. F. H., 1995, "Imaging Spectrometry for Remote Sensing: Vision to Reality in 15 Years," *Proceedings*, SPIE International Society for Optical Engineers, 2480:12 pp.

Goward, S. N., Markham, B., Dye, D. G., Dulaney, W. and J. Yang, 1991, "Normalized Difference Vegetation Index Measurements from the Advanced Very High Resolution Radiometer," *Remote Sensing of Environment*, 35:257–277.

GRSU, 1999, *Phenological Calendars of Selected Crops*, Santa Barbara: Geography Remote Sensing Unit.

Gupta, R. K., Vijayan, D. and T. S. Prasad, 2001, "New Hyperspectral Vegetation Characterization Parameters," *Advances in Space Research*, 28(1):201–206.

Haack, B. and S. Jampoler, 1994, "Agricultural Classification Comparisons Using Landsat Thematic Mapper Data," *ITC Journal*, 1994(2):113–118.

Haack, B. and S. Jampoler, 1995, "Colour Composite Comparisons for Agricultural Assessments," *Intl. Journal of Remote Sensing*, 16(9):15810–1598.

Hardisky, M. A., Klemas, V. and M. Smart, 1983, "The Influence of Soil Salinity, Growth Form, and Leaf Moisture on the Spectral Radiance of Spartina Alterniflora Canopies," *Photogrammetric Engineering & Remote Sensing*, 49:77–83.

Haboudane, D., Miller, J. R., Tremblay, N., Zarco-Tejada, P. J. and L. Dextraze, 2002, "Integrated Narrow-band Vegetation Indices for Prediction of Crop Chlorophyll Content for Application to Precision Agriculture," *Remote Sensing of Environment*, 81:416–426.

Huang, C., Wylie, B., Yang, L., Homer, C. and G. Zylstra, 2002, "Derivation of a Tasseled Cap Transformation based on Landsat 7 at-satellite Reflectance," *Intl. Journal of Remote Sensing*, 23(8):1741-1748.

Huete, A. R., 1988, "A Soil-adjusted Vegetation Index (SAVI)," *Remote Sensing of Environment*, 25:295–309.

Huete, A. R. and C. Justice, 1999, *MODIS Vegetation Index (MOD 13) Algorithm Theoretical Basis Document*, Greenbelt: NASA Goddard Space Flight Center, http://modarch.gsfc.nasa.gov/MODIS/LAND/#vegetation-indices, 129 p.

Huete, A. R., Didan, K., Miura, T., Rodriguez, E. P., Gao, X. and G. Ferreira, 2002a, "Overview of the Radiometric and Biophysical Performance of the MODIS Vegetation Indices," *Remote Sensing of Environment*, 83:195–213.

Huete, A. R., Didan, K, and Y. Yin, 2002b, *MODIS Vegetation Workshop*, Missoula, Montana, July 15–18; Terrestrial Biophysics and Remote Sensing (TBRS) MODIS Team, University of Arizona, http://utam.geophys.utah.edu/ebooks/gg527/modis/ndvi.html.

Huete, A. R., Hua, G., Qi, J., Chehbouni A. and W. J. Van Leeuwem, 1992, "Normalization of Multidirectional Red and Near-infrared Reflectances with the SAVI," *Remote Sensing of Environment*, 40:1–20.

Huete, A. R. and H. Q. Liu, 1994, "An Error and Sensitivity Analysis of the Atmospheric and Soil-Correcting Variants of the NDVI for the MODIS-EOS," *IEEE Transactions on Geoscience and Remote Sensing*, 32(4):897–905.

Huete, A. R., Liu, H. Q., Batchily, K. and W. J. van Leeuwen, 1997, "A Comparison of Vegetation Indices Over a Global Set of TM Images for EOS-MODIS," *Remote Sensing of Environment*, 59:440–451.

Hunt, E. R., Rock, B. N. and P. S. Nobel, 1987, "Measurement of Leaf Relative Water Content by Infrared Reflectance," *Remote Sensing of Environment*, 22:4210–435.

Jackson, R. D., 1983, "Spectral Indices in *n*-Space," *Remote Sensing of Environment*, 13:409–421.

Jackson, M. W. and J. R. Jensen, 2005, "An Evaluation of Remote Sensing-derived Landscape Ecology Metrics for Reservoir Shoreline Environmental Monitoring," *Photogrammetric Engineering & Remote Sensing*, 71(12):1387–1397.

Jackson, T. J., Chen, D., Cosh, M., Li, F., Anderson, M., Walthall, C., Doriaswamy, P. and E. R. Hunt, 2004, "Vegetation Water Content Mapping using Landsat Data derived Normalized Difference Water Index for Corn and Soybeans," *Remote Sensing of Environment*, 92:475–482.

Jensen, J. R., 2005, *Introductory Digital Image Processing: A Remote Sensing Perspective*, Upper Saddle River: Prentice-Hall, 325 p.

Jensen, J. R., Coombs, C., Porter, D., Jones, B., Schill, S. and D. White, 1998, "Extraction of Smooth Cordgrass (*Spartina alterniflora*) Biomass and Leaf Area Index Parameters from High Resolution Imagery," *Geocarto International*, 13(4):25–46.

Jensen, J. R., Huang, X. and H. E. Mackey, 1997, "Remote Sensing of Successional Changes in Wetland Vegetation as Monitored During a Four-Year Drawdown of a Former Cooling Lake," *Applied Geographic Studies*, 1(1):31–44.

Jensen, J. R., Garcia-Quijano, M., Hadley, B., Im, J., Wang, Z., Nel, A. L., Teixeira, E. and B. A. Davis, 2006, "Remote Sensing Agricultural Crop Type for Sustainable Development in South Africa," *Geocarto International- A Multidisciplinary Journal of Remote Sensing and GIS*, in press.

Jensen, J. R., Rutchey, K., Koch, M. S. and S. Narumalani, 1995, "Inland Wetland Change Detection in the Everglades Water Conservation Area 2A Using a Time Series of Normalized Remotely Sensed Data," *Photogrammetric Engineering & Remote Sensing*, 61(2): 199–209.

Jensen, M. E. and P. S. Bourgeron (Eds.), 1993, *Eastside Forest Ecosystem Health Assessment,* Vol. II, *Ecosystem Management: Principles and Applications*, Missoula, MT: USDA Forest Service, PNW-GTR-318.

Jensen, M. E. and R. Everett, 1993, "An Overview of Ecosystem Management Principles," in Jensen, M. E., and P. S. Bourgeron (Eds.), *Eastside Forest Ecosystem Health Assessment,* Vol. II, *Ecosystem Management: Principles and Applications*, Missoula: USDA Forest Service, 7–16.

Jin, S. and S. A. Sader, 2005, "Comparison of Time Series Tasseled Cap Wetness and the Normalized Difference Moisture Index in Detecting Forest Disturbances," *Remote Sensing of Environment*, 94:364–372.

Jones, K. B., Ritters, K. H., Wickham, J. D., Tankersley, R. D., O'Neill, R. V., Chaloud, D. J., Smith, E. R. and A. C. Neale, 1998, *Ecological Assessment of the United States: Mid-Atlantic Region,* Washington: EPA, 103 p.

Karnieli, A., Kaufman, Y. J., Remer, L. and A. Wald, 2001, "AFRI: Aerosol Free Vegetation Index," *Remote Sensing of Environment*, 77:10–21.

Kaufman, Y. J. and D. Tanre, 1992, "Atmospherically Resistant Vegetation Index (ARVI) for EOS-MODIS," *IEEE Transactions on Geoscience and Remote Sensing*, 30(2):261–270.

CHAPTER 11

Kaufman, Y. J., Karnieli, A. and D. Tanre, 2000, "Detection of Dust Over Deserts Using Satellite Data in the Solar Wavelengths," *IEEE Transactions Geoscience & Remote Sensing*, 38:525–531.

Kauth, R. J. and G. S. Thomas, 1976, "The Tasseled Cap — A Graphic Description of the Spectral-Temporal Development of Agricultural Crops as Seen by Landsat," *Proceedings*, Machine Processing of Remotely Sensed Data, West Lafayette: Laboratory for the Applications of Remote Sensing, 41–51.

Kauth, R. J., Lambeck, P. F., Richardson, W., Thomas, G. S. and A. P. Pentland, 1979, "Feature Extraction Applied to Agricultural Crops as Seen by Landsat," *Proceedings*, LACIE Symposium, Houston: NASA, 705–721.

Kim, M., Daughtry, C. S., Chappelle, E. W., McMurtrey III, J. E. and C. L. Walthall, 1994, "The Use of High Spectral Resolution Bands for Estimating Absorbed Photosynthetically Active Radiation (APAR)," *Proceedings*, 6th Symposium on Physical Measurements and Signatures in Remote Sensing, January 17–21, Val D'Isere, France, 299–306.

Kimes, D. S., 1983, "Dynamics of Directional Reflectance Factor Distributions for Vegetation Canopies," *Applied Optics*, 22(9):1364–1372.

Li, X. and A. H. Strahler, 1985, "Geometric-optical Modeling of a Conifer Forest Canopy," *IEEE Transactions Geoscience Remote Sensing*, 23(5):705–721.

Li, X. and A. H. Strahler, 1992, "Geometric-optical bidirectional Reflectance modeling of the Discrete-Crown Vegetation Canopy: Effect of Crown Shape and Mutual Shadowing," *IEEE Transactions Geoscience Remote Sensing*, 30:276–292.

Lunetta, R. S., Ediriwickrema, J., Johnson, D. M., Lyon, J. G. and A. McKerrow, 2002, "Impacts of Vegetation Dynamics on the Identification of Land-cover Change in a Biologically Complex Community in North Carolina, USA," *Remote Sensing of Environment*, 82:258–270.

Lyon, J. G., Yuan, D., Lunetta, R. S. and C. D. Elvidge, 1998, "A Change Detection Experiment Using Vegetation Indices," *Photogrammetric Engineering & Remote Sensing*, 64(2):143–150.

Millington, A. C., Velez-Liendo, X. M. and A. V. Bradley, 2003, "Scale Dependence in Multitemporal Mapping of Forest Fragmentation in Bolivia: Implications for Explaining Temporal Trends in Landscape Ecology and Applications to Biodiversity Conservation," *ISPRS Journal of Photogrammetry & Remote Sensing*, 57: 289–299.

Miura, T., Huete, A. F., Yoshioka, H. and B. N. Holben, 2001, "An Error and Sensitivity Analysis of Atmospheric Resistant Vegetation Indices Derived from Dark Target-based Atmospheric Correction," *Remote Sensing of Environment*, 78:284–298.

McGarigal, K. and B. J. Marks, 1994, *FRAGSTATS: Spatial Pattern Analysis Program for Quantifying Landscape Structure*, Version 2, Washington: USDA Forest Service, 141 p.

McGarigal, K., Cushman, S. A., Neel, M. C. and E. Ene, 2006, *FRAGSTATS: Spatial Pattern Analysis Program for Categorical Maps*, Version 3, Amherst: Univ. of Massachusetts, http://www.umass.edu/landeco/research/fragstats/faq/fragstats_faq.html.

Myers, V. I., 1970, "Soil, Water and Plant Relations," *Remote Sensing with Special Reference to Agriculture and Forestry*, Washington: National Academy of Sciences, 253–297.

O'Neill, R. V., Hunsaker, C. T., Jones, K. B., Ritters, K. H., Wickham, J. D, Schwarz, P., Goodman, I. A., Jackson, B. and W. S. Baillargeon, 1997, "Monitoring Environmental Quality at the Landscape Scale," *BioScience*, 47(8):513–519.

Perry, C. R. and L. F. Lautenschlager, 1984, "Functional Equivalence of Spectral Vegetation Indices," *Remote Sensing of Environment*, 14:169–182.

Peterson, D. L. and S. W. Running, 1989, "Applications in Forest Science and Management," in *Theory and Applications of Optical Remote Sensing*, NY: John Wiley, 4210–4273.

Pierce, L. L., Running, S. W. and G. A. Riggs, 1990, "Remote Detection of Canopy Water Stress in Coniferous Forests Using NS001 Thematic Mapper Simulator and the Thermal Infrared Multispectral Scanner," *Photogrammetric Engineering & Remote Sensing*, 56(5):5710–586.

Price, K. P., Guo, X. and J. M. Stiles, 2002, "Optimal Landsat TM Band Combinations and Vegetation Indices for Discrimination of Six Grassland Types in Eastern Kansas," *Intl. Journal of Remote Sensing*, 23:5031–5042.

Qi, J., Cabot, F., Moran, M. S. and G. Dedieu, 1995, "Biophysical Parameter Estimations Using Multidirectional Spectral Measurements," *Remote Sensing of Environment*, 54:71–83.

Ramsey, R. D., A. Falconer and J. R. Jensen, 1995, "The Relationship between NOAA-AVHRR NDVI and Ecoregions in Utah," *Remote Sensing of Environment*, 53:188–198.

Richardson, A. J. and C. L. Wiegand, 1977, "Distinguishing Vegetation from Soil Background Information," *Remote Sensing of Environment*, 8:307–312.

Richardson, A. J. and J. H. Everitt, 1992, "Using Spectral Vegetation Indices to Estimate Rangeland Productivity," *Geocarto Intl.*, 1:63–77.

Ritters, K. H., R. V. O'Neill, C. T. Hunsaker, J. D. Wickham, D. H. Yankee, S. P. Timmins, K. B. Jones and B. L. Jackson, 1995, "A Factor Analysis of Landscape Pattern and Structure Metrics," *Landscape Ecology*, 10(1):23–39.

Rondeaux, G., Steven, M. and F. Baret, 1996, "Optimization of Soil-adjusted Vegetation Indices," *Remote Sensing of Environment*, 55:95–107.

Rogan, J., Franklin, J. and D. A. Roberts, 2002, "A Comparison of Methods for Monitoring Multitemporal Vegetation Change Using Thematic Mapper Imagery," *Remote Sensing of Environment*, 80:143–156.

Rouse, J. W., Haas, R. H., Schell, J. A. and D. W. Deering, 1974, "Monitoring Vegetation Systems in the Great Plains with ERTS," *Proceedings*, Third Earth Resources Technology Satellite-1 Symposium, Greenbelt: NASA SP-351, 3010–3017.

Rundquist, B. C, 2002, "The Influence of Canopy Green Vegetation Fraction on Spectral Measurements over Native Tallgrass Prairie," *Remote Sensing of Environment*, 81:129–135.

Running, S. W., Justice, C. O., Solomonson, V., Hall, D., Barker, J., Kaufmann, Y. J., Strahler, A. H., Huete, A. R., Muller, J. P., Vanderbilt, V., Wan, Z. M., Teillet, P. and D. Carneggie, 1994, "Terrestrial Remote Sensing Science and Algorithms Planned for EOS/MODIS," *Intl. Journal of Remote Sensing*, 15(17):3587–3620.

Sandmeier, S. R., 1999, *Guidelines and Recommendations for the Use of the Sandmeier Field Goniometer for the NASA Stennis Space Center*, Washington: NASA, 47 p.

Sandmeier, S. R. and K. I. Itten, 1999, "Field Goniometer System (FIGOS) for Acquisition of Hyperspectral BRDF Data," *IEEE Transactions on Geoscience & Remote Sensing*, 37(2):978–986.

Sandmeier, S., Muller, C., Hosgood, B. and G. Andreoli, 1998a, "Physical Mechanisms in Hyperspectral BRDF Data of Grass and Watercress," *Remote Sensing of Environment*, 66:222–233.

Sandmeier, S., Muller, C., Hosgood, B. and G. Andreoli, 1998b, "Sensitivity Analysis and Quality Assessment of Laboratory BRDF Data," *Remote Sensing of Environment*, 64(2):176–191.

Savitsky, B. G., 1986, *Agricultural Remote Sensing in South Carolina: A Study of Crop Identification Capabilities Utilizing Landsat Multispectral Scanner Data*, unpublished masters thesis, Columbia: Univ. of South Carolina Geography Dept., 78 pp.

Savitsky, B. G., 1998, "Overview of Gap Analysis" *GIS Methodologies for Developing Conservation Strategies*, B. G. Savitsky and T. E. Lacher, Jr., (Eds.), NY: Columbia Univ. Press, 151–158.

Schill, S., Jensen, J. R., Raber and D. E. Porter, 2004, "Temporal Modeling of Bidirectional Reflection Distribution Function (BRDF) in Coastal Vegetation," *GIScience & Remote Sensing*, 41(2):116–135.

Schlerf, M. and Atzberger, C., 2001, "Imaging Spectrometry and Vegetation Science," in *Imaging Spectrometry: Basic Principles and Prospective Applications*, F. van der Meer and S. de Jong, Eds., Dordrecht: Kluwer, 111–155.

Schlerf, M., Atzberger, C. and J. Hill, 2005, "Remote Sensing of Forest Biophysical Variables Using HyMap Imaging Spectrometer Data," *Remote Sensing of Environment*, 95:177–194.

Schuft, M. J., Moser, T. J., Wiginton, P. J., Stevens, D. L., McAllister, L. S., Chapman, S. S. and T. L. Ernst, 1999, "Development of Landscape Metrics for Characterizing Riparian-Stream Networks," *Photogrammetric Engineering & Remote Sensing*, 65(10):1157–1167.

Scott, J. M., Jacobi, J. J. and J. E. Estes, 1987, "Species Richness: A Geographic Approach to Protecting Future Biological Diversity," *BioScience*, 37:782–788.

Scott, J. M., Davis, F., Csuti, B., Noss, R., Butterfield, B., Groves, C, Anderson, H., Caicco, S., D'Erchia, F., Edwards, R., Ulliman, J. and R. G. Wright, 1993, "Gap Analysis: A Geographic Approach to Protection of Biological Diversity," *Wildlife Monographs*, 123:1–41.

TBRS, 2003, *Enhanced Vegetation Index*, Terrestrial Biophysics and Remote Sensing Lab, Univ. of Arizona, http://tbrs.arizona.edu/project/MODIS/evi.php.

Townshend, J. R. G. and C. O. Justice, 2002, "Towards Operational Monitoring of Terrestrial Systems by Moderate-resolution Remote Sensing," *Remote Sensing of Environment*, 83:351–359.

Troll, C., 1939, Luftbildplan and okologische Bodenforschung, *A. Ges. Erdkunde*, Berlin: 241–298.

Tucker, C, 1978. "A Comparison of Satellite Sensor Bands for Vegetation Monitoring, "Photogrammetric *Engineering & Remote Sensing*, 44:13610–1380.

Walter-Shea, E. A. and L. L. Biehl, 1990, "Measuring Vegetation Spectral Properties," *Remote Sensing Reviews*, 5(1):179–205.

Wang, Q., Adiku, S., Tenhunen, J. and A. Granier, 2005, "On the Relationship of NDVI with Leaf Area Index in a Deciduous Forest Site," *Remote Sensing of Environment*, 94:244–255.

Weiers, S., Bock, M., Wissen, M. and G. Rossner, 2004, "Mapping and Indicator Approaches for the Assessment of Habitats at Different Scales using Remote Sensing and GIS Methods," *Landscape and Urban Planning*, 67:43–65.

Wickham, J. D. and D. J. Norton, 1994, "Mapping and Analyzing Landscape Patterns," *Landscape Ecology*, 9(1):7–23.

Woodcock, C. E., Collins, J. B., Jakabhazy, V. D., Li, X., Macomber, S. A. and Y. Wu, 1997, "Inversion of the Li-Strahler Canopy Reflectance Model for Mapping Forest Structure," *IEEE Transactions Geoscience and Remote Sensing*, GE-35(2):405–414.

Zha, Y., Gao, J. and S. Ni, 2003, "Use of Normalized Difference Built-up Index in Automatically Mapping Urban Areas from TM Imagery," *Intl. Journal of Remote Sensing*, 24(3):583–594.

Remote Sensing of Water

W ater covers approximately 74 percent of the Earth's surface. Nowhere else in the known universe is such an abundance of liquid water found. Almost 97 percent of the Earth's volume of water is in the great saline oceans. Only about 0.02 percent of the Earth's water is found in freshwater streams, rivers, lakes, and reservoirs. The remaining water is contained in underground aquifers (0.6 percent), the Earth's atmosphere in the form of water vapor (0.001 percent), and the icecap (approximately 2.2 percent).

Water exists in various states on Earth, including freshwater, saltwater, water vapor, rain, snow, and ice. Meteorologists, oceanographers, hydrologists, some geographers, and others devote their lives to measuring, monitoring, and predicting the spatial distribution, volume, and movement of water as it progresses through the hydrologic cycle.

It is possible to obtain *in situ* measurements of various hydrologic (water) parameters such as precipitation, water depth, temperature, salinity, velocity, volume, etc., at very specific locations. For example, the U.S. Geological Survey maintains a dense network of *in situ* river-flow gauges on major streams and rivers that provides continuous records of river stage (height) and velocity. Major cities and airports collect *in situ* precipitation (rain and snow) information. Departments of health and environmental control are often mandated to collect water-quality samples from rivers, lakes, reservoirs, and estuaries. These point measurements are very important. If enough of the point observations are collected throughout a region, it is possible to interpolate between the point observations and infer regional geographic patterns. Unfortunately, there are usually not enough point observations to create a statistically significant distribution map (Kroger and Law, 2005). In fact, it is often difficult to obtain regional spatial information using *in situ* point observations for a number of the most important hydrologic variables, including:

- water-surface area (streams, rivers, ponds, lakes, reservoirs, and seas),

- water constituents (organic and inorganic),

- water depth (bathymetry),

- water-surface temperature,

- snow-surface area,

- snow-water equivalent,

- ice-surface area,

- ice-water equivalent,

- cloud cover,

- precipitation, and

- water vapor.

Therefore, a significant amount of research has taken place to develop remote sensing methods that can obtain quantitative, spatial measurements of these important hydrologic variables (e.g., Gregg and Casey, 2004; Karaska et al., 2004). This chapter introduces the fundamental principles associated with remote sensing surface water and its constituents: clouds, water vapor, precipitation, and snow. It concludes with a nonpoint source pollution water-quality study.

Remote Sensing Surface Water Biophysical Characteristics

This section reviews how remote sensing can be used to inventory and monitor the spatial extent, organic/inorganic constituents, depth, and temperature of water in rivers, lakes, reservoirs, seas, and oceans. It is important to first obtain an appreciation for the energy-matter interactions that may impact our ability to perform an accurate aquatic remote sensing investigation.

Water Surface, Subsurface Volumetric, and Bottom Radiance

The total radiance, (L_t) recorded by the sensor onboard the aircraft or satellite is a function of the electromagnetic energy from the four sources identified in Figure 12-1 (Bukata et al., 1995; Legleiter and Roberts, 2005):

$$L_t = L_p + L_s + L_v + L_b \qquad (12\text{-}1)$$

where

- L_p is the portion of the radiance recorded by a remote sensing instrument resulting from the downwelling solar (E_{Sun}) and sky (E_{sky}) radiation that never actually reaches the water surface. This is atmospheric noise and may be considered to be the unwanted *path radiance* discussed in Chapter 2.

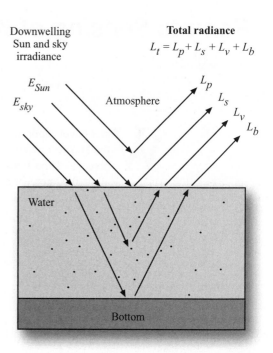

Figure 12-1   Water bodies receive irradiance from the Sun (E_{Sun}) and atmosphere (E_{sky}). The total radiance upwelling (exiting) from a water body toward the remote sensor is a function of the radiance from atmospheric scattering (L_p), water-surface radiance (L_s), subsurface volumetric radiance (L_v), and radiance from the bottom (L_b) of the water body (after Bukata et al., 1995).

- L_s is the radiance from the downwelling solar and sky radiation that reaches the air-water interface (sometimes called the *free-surface layer* or *boundary layer*), but only penetrates it a millimeter or so and is then essentially *reflected* from the water surface. This reflected energy contains valuable spectral information about the near-surface characteristics of the water body (Figure 12-2a). Unfortunately, if the solar zenith angle and sensor viewing angle are almost identical, then we may get a purely specular reflection from the surface of the water body, which provides very little useful spectral information (Figure 12-2b). Such *sunglint* is to be avoided whenever possible. This is why we rarely collect remotely sensed data at nadir within one or two hours of local noon.

- L_v is the radiance from the downwelling solar and sky radiation that actually penetrates the air-water interface, interacts with the water and organic/inorganic constituents and then exits the water column without encountering the bottom (called *subsurface volumetric radiance*). This radiance provides valuable information about the internal bulk characteristics of the water column.

Sunglint

a. Black-and-white infrared photograph   b. Black-and-white infrared photograph with sunglint.
of water bodies in Florida.

Figure 12-2   a) A black-and-white infrared photograph of an area in Florida containing natural vegetation, a golf course, and fairly deep
nonturbid water bodies. b) In this example, specularly reflected sunglint is present that seriously impedes our ability to extract
useful information. The only data of value are the wind-streak patterns on the water surface that provide some information
on wind direction at the time of data collection.

- L_b is that portion of the recorded radiance resulting from the downwelling solar and sky radiation that penetrates the air-water interface, reaches the bottom of the water body, is propagated back through the water column, and then exits the water column. If we want to obtain information about the bottom, such as when performing bathymetric (depth) mapping or coral reef mapping, then this radiance from the bottom may be of significant value (Mumby et al., 1997; Pasqualini et al., 1997). However, radiance from the bottom makes it very difficult to properly characterize the water column above it. Basically, it is difficult to disentangle or disaggregate L_v and L_b (Jensen et al., 1989; Islam et al., 2004).

The goal of most aquatic remote sensing is to extract the radiance of interest from all the other radiance components being recorded by the sensor system (Doxaran et al., 2002). For example, the scientist interested in identifying the organic and inorganic constituents in the water column (e.g., suspended sediment or chlorophyll a) is most concerned with isolating the subsurface volumetric radiance (L_v) computed as:

$$L_v = L_t - (L_p + L_s + L_b).$$ (12-2)

This usually involves careful radiometric correction of the remote sensor data to remove atmospheric attenuation (L_p),

surface sun-glint and other surface reflection (L_s), and bottom reflectance (L_b) (e.g., Mustard et al., 2001).

Scientists interested in monitoring water depth (bathymetry) or bottom characteristics such as reef structure are most interested in precise measurements of bottom radiance (L_b), so they must attempt to remove atmospheric, surface, and subsurface volume radiance.

Spectral Response of Water as a Function of Wavelength

When conducting a remote sensing investigation on water bodies, it is first useful to understand how *pure* water selectively absorbs and/or scatters the incident, downwelling sunlight in the water column. Later we will consider how the incident light is affected when the water column is not pure, but contains organic and inorganic materials.

Pure water is free from organic and inorganic matter. Bukata et al. (1995) summarized the absorption coefficient $a(\lambda)$, the scattering coefficient $b(\lambda)$, and the total attenuation coefficient $c(\lambda)$ of pure water molecules at wavelengths from 250 – 800 nm from a number of studies (Table 12-1). Several important relationships are observed when the absorption and scattering data are graphed, as shown in Figure 12-3.

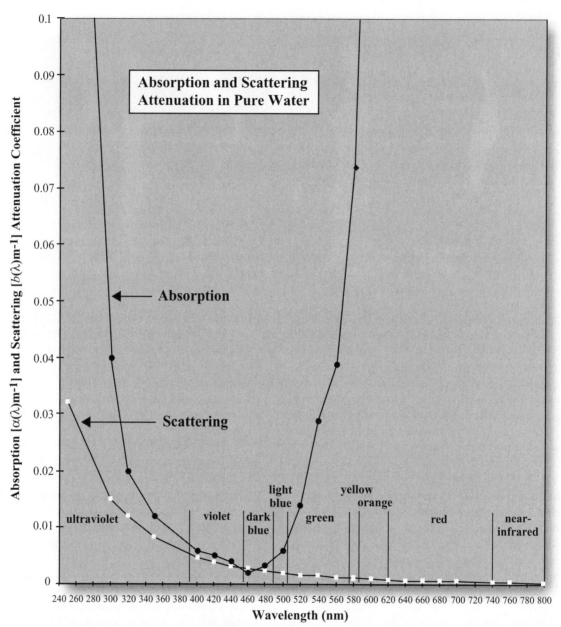

Figure 12-3   Absorption and scattering of light in pure water. Molecular water absorption dominates in the ultraviolet (< 400 nm) and in the yellow through the near-infrared portion of the spectrum (> 580 nm). Almost all of the incident near- and middle-infrared (740 – 2500 nm) radiant flux entering a pure water body is absorbed with negligible scattering taking place. This is why water is so dark on black-and-white infrared or color-infrared film. Scattering in the water column is especially important in the violet, dark blue, and light blue portions of the spectrum (400 – 500 nm). This is the reason water appears blue to our eyes. These data were derived from a variety of sources by Bukata et al. (1995). The graph truncates the absorption attenuation information in the ultraviolet and in the yellow through near-infrared regions because the attenuation is so great. Refer to Table 12-1 for absorption attenuation information in these regions.

The most noticeable characteristic is that the *least* amount of absorption *and* scattering of incident light in the water column (therefore the best transmission) takes place in the blue wavelength region from approximately 400 – 500 nm, with the minimum located at approximately 460 – 480 nm. These wavelengths of violet to light blue light penetrate further than any other type of light into the water column (Clark et al., 1997).

Palancar Reef on Cozumel Island, Quintana Roo, Mexico

a. SPOT Band 1 (0.50 - 0.59 μm).　　b. SPOT Band 2 (0.61 - 0.68 μm).　　c. SPOT Band 3 (0.791 - 0.89 μm).

Figure 12-4　Individual bands of SPOT XS data of Cozumel, Mexico. Incident green and red radiant flux (bands 1 and 2) penetrate through the water column and are reflected by the sand, hard-bottom, and Palancar Reef. Conversely, much of the green and red energy is absorbed by terrestrial vegetation, causing it to appear dark. It is difficult to identify the land-water interface (boundary) using these visible bands. Water absorbs almost all of the incident near-infrared radiant flux, while the terrestrial vegetation reflects most of the incident near-infrared energy. The result is that water appears almost black on the near-infrared (band 3) data, while the upland vegetation appears bright. This makes it easy to identify the land-water boundary (© SPOT Image, Inc.).

Incident green and yellow light from 520 – 580 nm is absorbed very well by the water column with relatively little scattering taking place. Similarly, scattering of orange and red wavelength energy (580 – 740 nm) by water molecules becomes insignificant when compared to absorption by water molecules. Almost all of the incident near- and middle-infrared (740 – 3000 nm) radiant flux entering a deep, pure water body is absorbed with negligible scattering taking place (Figure 12-3).

Thus, molecular scattering of violet and blue light (< 520 nm) in a water column and significant absorption of green, yellow, orange, and red wavelength light (520 – 700 nm) in the same water column, cause pure water to appear blue to our eyes. The blue color is especially evident in pure mid-ocean water and deep nonturbid inland water bodies.

Monitoring the Surface Extent of Water Bodies

The best wavelength region for discriminating land from pure water is the near-infrared and middle-infrared regions at wavelengths between 740 – 2500 nm. In the near- and middle-infrared regions, water bodies appear very dark, even black, because they absorb almost all of the incident radiant flux, especially when the water is deep and pure with no suspended sediment or organic matter in it (e.g., Figure 12-2a). Conversely, land surfaces are typically composed of vegetation and bare soil that reflect significant amounts of near- and middle-infrared energy, as summarized in previous chapters. This causes the land surfaces to appear relatively bright in near- and middle-infrared bands of imagery.

For example, consider the green (band 1), red (band 2), and near-infrared (band 3) SPOT HRV XS images of a portion of the Palancar Reef just offshore the island of Cozumel, Quintana Roo, Mexico (Figure 12-4). Underwater visibility in this portion of the Caribbean Sea is usually 80 – 120 ft. Incident green and red wavelength energy from the Sun penetrate through the water column and are reflected by bottom sand, hard-bottom, and coral reef. Note how the details of the reef are more clearly seen in the green (band 1) and red (band 2) images than in the near-infrared band 3 image (Figure 12-4).

Table 12-1. Optical properties of pure water (derived from various sources by Bukata et al., 1995).

Wavelength (nm)	Absorption $\alpha(\lambda)$ (m^{-1})	Scattering $b(\lambda)$ (m^{-1})	Total Attenuation $c(\lambda)$ (m^{-1})
250 – ultraviolet	0.190	0.032	0.2200
300 – ultraviolet	0.040	0.015	0.0550
320 – ultraviolet	0.020	0.012	0.0320
350 – ultraviolet	0.012	0.0082	0.0202
400 – violet	0.006	0.0048	0.0108
420 – violet	0.005	0.0040	0.0090
440 – violet	0.004	0.0032	0.0072
460 – dark blue	0.002	0.0027	0.0047
480 – dark blue	0.003	0.0022	0.0052
500 – light blue	0.006	0.0019	0.0079
520 – green	0.014	0.0016	0.0156
540 – green	0.029	0.0014	0.0304
560 – green	0.039	0.0012	0.0402
580 – yellow	0.074	0.0011	0.0751
600 – orange	0.20	0.00093	0.2009
620 – orange	0.24	0.0082	0.2408
640 – red	0.27	0.00072	0.2707
660 – red	0.310	0.00064	0.3106
680 – red	0.38	0.00056	0.3806
700 – red	0.60	0.0005	0.6005
740 – near-infrared	2.25	0.0004	2.2504
760 – near-infrared	2.56	0.00035	2.5604
800 – near-infrared	2.02	0.00029	2.0203

Mississippi River Delta

Figure 12-5   Photograph of the Mississippi River delta just below New Orleans, LA, obtained by Space Shuttle astronauts during Mission STS 51. The suspended sediment in the water is reflecting much of the incident radiant flux back into the atmosphere. Conversely, the more pure water, farther offshore, is absorbing most of the incident radiant flux, causing it to appear dark. The boundary (front) between the sediment-laden water and the nonturbid water is quite evident (courtesy K. Lulla, NASA Johnson Space Center).

red radiant flux (as shown in Figure 12-4c), causing the SPOT HRV band 3 sensor to record ocean water as black, while the upland vegetation reflects much of the near-infrared radiant flux, causing it to appear bright. Therefore, it is a straightforward task to delineate the land-water interface using the SPOT near-infrared band 3 image.

Care must be exercised, however, when there are organic and inorganic constituents in the water column (especially those near the surface) because these materials will cause near-infrared surface reflection and subsurface volumetric scattering to take place, dramatically increasing the amount of near-infrared radiant flux leaving the surface of the water body. For example, consider the Space Shuttle photograph of the Mississippi River delta in Figure 12-5. The suspended sediment in the water column causes significant scattering and reflection of radiant flux back toward the sensor system, causing the water to appear almost as bright as the few land features in this bird's-foot delta.

While the subsurface information available in the green and red SPOT images (bands 1 and 2) is useful for many reasons, it is not of much value when trying to identify the land-water interface, i.e., the actual edge of the island of Cozumel. Fortunately, the water absorbs almost all the incident near-infra-

Spectral Response of Water as a Function of Organic and Inorganic Constituents — Monitoring Suspended Minerals (Turbidity), Chlorophyll, and Dissolved Organic Matter

Thus far we have mainly considered the spectral response of pure water. Most natural water bodies, however, contain a variety of organic (e.g., phytoplankton chlorophyll *a*) and inorganic (e.g., suspended minerals) constituents. When natural waters contain a mixture of these materials, one of the most difficult remote sensing problems is to disentangle (extract) quantitative information about these specific constituents from the remotely sensed data (Doxaran et al., 2002; Karaska et al., 2004).

When conducting water-quality studies or trying to predict water productivity using remotely sensed data, we are usually most interested in the *subsurface volumetric radiance*, L_v (Figure 12-1), which is the radiance from the downwelling solar and sky radiation that actually penetrates the air-water interface, interacts with the water and organic/inorganic constituents, and then exits the water column toward the sensor without encountering the bottom. The subsurface volumetric radiance exiting the water column toward the sensor (L_v) is a function of the concentration of pure water (w), inorganic suspended minerals (SM), organic chlorophyll *a* (*Chl*), dissolved organic material (DOM), and the *total* amount of absorption and scattering attenuation that takes place in the water column due to each of these constituents, $c(\lambda)$, i.e.,

$$L_v = f\left[w_{c(\lambda)}, SM_{c(\lambda)}, Chl_{c(\lambda)}, DOM_{c(\lambda)}\right]. \qquad (12\text{-}3)$$

It is instructive to look at the effect that each of these constituents has on the spectral reflectance characteristics of a water column.

Suspended Minerals

Minerals such as silicon, aluminum, and iron oxides are found in suspension in most natural water bodies. The particles range from fine clay particles ($3 - 4$ μm in diameter), to silt ($5 - 40$ μm), to fine-grain sand ($41 - 130$ μm), and coarse grain sand ($131 - 250$ μm). The sediment comes from a variety of sources including upland agricultural cropland erosion, weathering of mountainous terrain, shoreline erosion caused by natural waves or boat traffic, and volcanic eruption (ash). Most of the suspended mineral sediment is concentrated in the inland and nearshore water bodies (Bukata et al., 1995). Clear, deep ocean (referred to as Case I water) far from shore rarely contains suspended minerals greater

than 1 μm in diameter. Thus, suspended mineral concentration is usually of no significance to deep ocean remote sensing studies. This is important since the contributions from suspended minerals can often be ruled out when conducting a deep ocean remote sensing investigation. Conversely, inland and nearshore water bodies may carry a significant load of suspended sediment that can dramatically impact the spectral reflectance characteristics of the water bodies (Miller and McKee, 2004).

Monitoring the type, amount, and spatial distribution of suspended minerals in inland and nearshore water bodies is very important. For example, soil erosion in a watershed contributes sediment loads to surface waters, which results in faster filling of major rivers, reservoirs, farm ponds, flood-control impoundments, and estuaries. This can shorten the useful life of reservoirs, ponds, and flood-control devices and require dredging of rivers and estuaries. For example, the reduction in storage capacity in reservoirs in the United States caused by the infusion of suspended sediment results in a loss of >$100 million annually. Sediment also affects water quality and its suitability for drinking, recreation, and industrial purposes. It serves as a carrier and storage agent of pesticides, absorbed phosphorus, nitrogen, and organic compounds and can be an indicator of pollution. Suspended sediments can impede the transmission of solar radiation and reduce photosynthesis in submerged aquatic vegetation and near-bottom phytoplankton. The aquatic vegetation and phytoplankton play a vital role in the food chain of the aquatic ecosystem.

Fortunately, remote sensing can be used to monitor the suspended mineral concentrations in water bodies. This usually requires obtaining *in situ* measurements of suspended mineral concentrations and relating it to the remote sensor data to derive a quantitative relationship. It is good practice to collect both the remote sensor data and the *in situ* suspended sediment measurements on days that have little wind. Wind-roughened surface water creates specular reflections, which can be deleterious to remote sensing of suspended sediment concentrations (Han and Rundquist, 1998).

The spectral reflectance of suspended sediment in surface waters is a function of both the quantity and characteristics (particle size, absorption) of the material in the water. Suspended sediment concentration is measured *in situ* using either a secchi disk or a nephelometric turbidity unit (NTU) detector. The secchi disk (Figure 12-6) is a circular plate that is lowered into the water until it cannot be seen. This secchi depth is inversely correlated with the amount of suspended material in the water. The greater the amount of suspended

Seechi Disk

Figure 12-6   A secchi disk is used to measure suspended sediment in water bodies by lowering it into the water column and determining the depth at which it disappears. The accuracy of the measurement is a function of the visual acuity of the observer, which can vary dramatically.

Spectroradiometer

Figure 12-7   A spectroradiometer may be used to measure *in situ* spectral reflectance characteristics of pure water and water with various suspended sediment and chlorophyll *a* concentrations (after Lodhi et al., 1997).

material, the less the secchi depth, i.e., it will disappear from view relatively quickly. This method relies on human visual perception, which is not constant among scientists. The more rigorous nephelometer detector passes light through a sample of water obtained at various depths to determine its transmission characteristics that are related to suspended material in the water.

Research has documented the general impact of increased suspended mineral concentration in water bodies in the visible and near-infrared portion of the spectrum. For example, consider the experiment conducted by Lodhi et al. (1997), which determined the spectral reflectance characteristics of two Nebraska soil types (clayey and silty) at various suspended sediment concentrations in water.

Figure 12-7 depicts the experimental design where a spectroradiometer was used to collect spectral reflectance data from a height of 165 cm directly above a controlled water surface. The spectral measurements were made in 252 discrete spectral bands between 368 and 1,114 nm. The spectroradiometer was calibrated with a gray card (18 percent reflectance) that was cross-referenced to a BaSO$_4$ calibration panel. Reflectance, ρ_T, was calculated as a simple ratio using the equation (Markham and Barker, 1986):

$$\rho_T = \frac{L_T}{L_r} \times k \qquad (12\text{-}4)$$

where L_T is the radiance measured from the water surface, L_r is the radiance from the BaSO$_4$ panel, and k is a calibration factor for the BaSO$_4$ panel (refer to Chapter 15).

Figures 12-8a,b depict the spectral reflectance of clear water and water with varying suspended sediment concentrations of clayey and silty soil. Spectral reflectance of the clear water drops continuously after about 580 nm due to absorption in the water column, as previously described (refer to Figure 12-3). As the suspended sediment concentration is increased, reflectance increases at all wavelengths for both clayey and silty soils. The clayey soil (Figure 12-8a) had more organic matter and was darker in color, which resulted in approximately 10 percent lower volume reflectance at all wavelengths than the light-colored silty soil (Figure 12-8b). Reflectance increased in the 580 – 690 nm region and in the near-infrared region as more minerals were suspended in the water bodies. *Thus, the peak reflectance shifts toward longer wavelengths in the visible region as more suspended sediments are added.* More green, red, and near-infrared radiant flux is reflected from the water body and recorded by the remote sensing system. A water body with suspended sedi-

Figure 12-8   a) *In situ* spectral reflectance measurements of clear water and clear water with various levels of *clayey* soil suspended sediment concentrations from 0 – 1000 mg/l. b) *In situ* spectral reflectance measurements of clear water and clear water with various levels of *silty* soil suspended sediment concentrations. Note that the silty soil had approximately 10 percent more volume reflectance at all wavelengths when compared with the clayey soil. In both cases, the peak reflectance shifted toward longer wavelengths in the visible region as more suspended sediments were added (after Lodhi et al., 1997).

ment in it will generally appear brighter in imagery than a nearby waterbody without any suspended sediment.

Correlation coefficients (*r*) were computed to describe the relationship between suspended sediment concentration and the reflectance at each of the 252 spectral sampling points. For the clayey soil, the values ranged from 0.28 – 0.97, and

for the silty soil, the range was 0.78 – 0.98. For both soils the values of *r* > 0.90 occurred in the near-infrared region between 714 and 880 nm, as expected (Han and Rundquist, 1994).

These results suggest that:

Diatom   **Algae Cell**

Figure 12-9   a) A micrograph of blue reflected light from a photosynthesizing diatom. b) A micrograph of blue reflected light from a green algae cell (*Micrasterias sp.*). The chloroplast material in the center shows up in dark tones because the chlorophyll *a* pigments absorb much of the incident blue light during photosynthesis.

- the visible wavelength range of 580 – 690 nm may provide information on the *type* of suspended sediments (soil) in surface waters; and

- the near-infrared wavelength range of 714 – 880 nm may be a useful wavelength range for determining the *amount* of suspended minerals in surface waters where suspended minerals are the predominant constituent.

These relationships may not be universally applicable. Scientists attempting to remotely sense suspended sediment concentrations should investigate the spectral absorption and scattering characteristics of the water and soils in their geographic area of interest.

For example, Doxaran et al. (2002) used SPOT satellite data to measure suspended sediment in the Gironde estuary in southwest France. The best correlations between SPOT data and suspended sediment were obtained using a) the atmospherically-corrected NIR band XS3 (790 – 890 nm) when suspended sediment concentration was high, and b) the ratios [XS3 / XS1] or [XS3 / XS2] when the suspended sediment concentration was low. The suspended sediment concentrations within the surface waters in the estuary were estimated up to 2000 mg/l.

Similarly, Binding et al. (2005) found that the successful use of NIR wavelengths for measuring suspended sediment was dependent on the fact that in highly turbid waters there is sufficient scattering to overcome the strong absorption by water at these wavelengths. In less turbid waters, such as the Irish Sea, the wavelengths suitable for mineral suspended sediment algorithm development may be restricted to only

the visible region (e.g., 665 nm). They suggested that any study attempting to monitor suspended sediment using remote sensing should be based on a thorough knowledge of the particle grain size and composition being examined.

Miller and McKee (2004) found a robust linear relationship ($R^2 = 0.89$) between MODIS band 1 (620 – 670 nm) 250 × 250 m data and *in situ* total suspended sediment measurements in the Gulf of Mexico.

Chlorophyll

Plankton is the generic term used to describe all the living organisms (plant and animal) present in a water body that cannot resist the current (unlike fish). Plankton may be subdivided further into algal plant organisms (*phytoplankton*), animal organisms (*zooplankton*), bacteria (*bacterioplankton*), and lower plant forms such as algal fungi. *Phytoplankton* are single-celled plants smaller than the size of a pinhead. Micrographs of the blue radiant energy reflected from a diatom and a single cell of green algae are shown in Figure 12-9a,b. In both instances, the chlorophyll *a* pigments in the plants absorb most of the incident blue light, causing the photosynthetic portion of each to appear dark.

Phytoplankton, like plants on land, are composed of substances that contain carbon. Phytoplankton sink to the ocean or water-body floor when they die. Zooplankton migrate to the surface at night to feed on live phytoplankton and then sink to greater depths during the day. When zooplankton die, they also sink to the bottom, carrying their carbon with them. The carbon in the dead phytoplankton and zooplankton is soon covered by other sediments. Also, phytoplankton use carbon dioxide and produce oxygen during the photosynthetic process. In this way, the water bodies and oceans act as a *carbon sink*, a place that disposes of global carbon, which otherwise can accumulate in the atmosphere as carbon dioxide. Other global sinks include land vegetation and soil. However, the carbon in these sinks frequently is returned to the atmosphere as carbon dioxide by burning or decomposition. No one knows exactly how much carbon the inland water bodies and ocean accumulate. As such, the characteristics of phytoplankton and zooplankton are very important to our knowledge of the global carbon cycle as it represents a significant carbon sink for increased levels of carbon dioxide in the atmosphere (Bukata et al., 1995).

All phytoplankton in water bodies contain the photosynthetically active pigment chlorophyll *a*. However, chlorophyll *c*, *d*, and even *e* may also be present and at various depths. There are two other phytoplankton photosynthesizing agents: *carotenoids* and *phycobilins*. Bukata et al. (1995) suggest, however, that chlorophyll *a* may be considered a reasonable surrogate for the organic component of optically complex natural waters. Fortunately, because different types of phytoplankton have different concentrations of chlorophyll, they appear as different colors to sensitive remote sensors. Thus, recording the color of an area of the ocean or other water body allows us to estimate the amount and general type of phytoplankton in that area and tells us about the health and chemistry of the water body. Comparing images taken at different times tells us about changes that occur over time and the processes at work.

Chlorophyll *a* introduced to pure water changes its spectral reflectance characteristics, i.e., its color. For example, Figure 12-10a depicts the spectral reflectance characteristics of clear water and the same water laden with algae consisting primarily of chlorophyll *a* (Han, 1997). Clear water reflected approximately 2 percent between 400 and 500 nm and dropped gradually to less than 1 percent at wavelengths beyond 710 nm, as expected. Conversely, four pronounced scattering/absorption features of chlorophyll are evident in the algae-laden water:

- strong chlorophyll *a* absorption of blue light between 400 and 500 nm;

- strong chlorophyll *a* absorption of red light at approximately 675 nm (Gitelson, 1992);

- reflectance maximum around 550 nm (green peak) caused by *relatively* lower absorption of green light by algae (Gitelson, 1992);

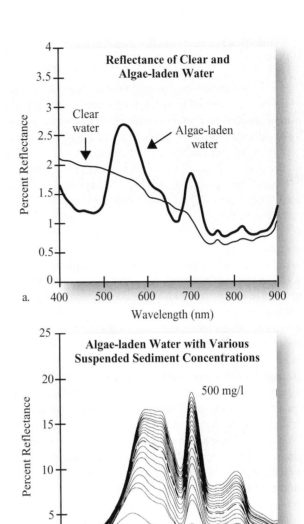

Figure 12-10 a) Percent reflectance of clear and algae-laden water based on *in situ* spectroradiometer measurement. b) Percent reflectance of algae-laden water with various concentrations of suspended sediment ranging from 0 – 500 mg/l (after Han, 1997).

- prominent reflectance peak around 690 – 700 nm caused by an interaction of algal-cell scattering and a minimum combined effect of pigment and water absorption. The height of this peak above the baseline (absorption trough) can be used to accurately measure chlorophyll amount (Rundquist et al., 1995).

Basically, as chlorophyll concentration increases in the water column, there is a significant decrease in the relative amount of energy reflected in the blue and red wavelengths, but an increase in green wavelength reflectance.

When both suspended mineral sediment and chlorophyll are present in the water body at the same time, a dramatically different spectral response is produced. For example, Figure 12-10b demonstrates what happens to the spectral response of water as red loam sediment concentrations from 0 – 500 mg/l are added to water that contains algae (chlorophyll *a*). For algae-laden water, the peak reflectance in the visible region shifted from 547 nm (green light) at 0 mg/l to 596 nm (orange) at 500 mg/l.

Rundquist et al. (1996) and Han and Rundquist (1997) obtained accurate estimates of algal chlorophyll pigment amount in surface water using a simple near-infrared (705 nm) / red (670 nm) ratio when the concentration of chlorophyll was relatively low. Computing the first derivative of reflectance around 690 nm produced the best results when algal chlorophyll concentration was relatively high.

Chlorophyll in Ocean Water: It is assumed that a remote estimate of near-surface chlorophyll concentration constitutes a remote estimate of near-surface biomass (or primary productivity) for deep ocean (Case I) water where there is little danger of suspended mineral sediment contamination. Numerous studies have documented a relationship between selected spectral bands and ocean chlorophyll concentration (*Chl*) using the equation (e.g., Gordon et al., 1983):

$$Chl = x\left[\frac{L(\lambda_1)}{L(\lambda_2)}\right]^y \qquad (12\text{-}5)$$

where $L(\lambda_1)$ and $L(\lambda_2)$ are the upwelling radiances at selected wavelengths recorded by the remote sensing system and x and y are empirically derived constants. Bukata et al. (1995) point out that this approach does not take into account the absorption and scattering characteristics of the inorganic and organic constituents of the study area. The only reason it is successful is that in the mid-ocean, chlorophyll pigments are the principal colorant of Case I waters.

Satellite remote sensing systems used to measure ocean color are summarized in Table 12-2. The operational algorithms used to extract phytoplankton abundance information from several of these satellite remote sensing systems are summarized in Table 12-3. Hopefully, the operational algorithms are spatially and temporally invariant, meaning that they will work most anywhere, anytime.

The Sea-viewing Wide Field of View Sensor (SeaWiFS) launched in 1997 was designed by NASA and industry

(ORBIMAGE, Inc.) to build on what was learned using the Coastal Zone Color Scanner. SeaWiFS has additional bands at 412 nm (to identify yellow substances through their blue wavelength absorption), at 490 nm (to increase sensitivity to chlorophyll concentration), and in the 765 and 865 nm near-infrared (to assist in removing atmospheric attenuation).

Because of the role of phytoplankton in the global carbon cycle, SeaWiFS data are used to assess the ocean's role in the global carbon cycle and to examine oceanic factors that affect global climate change (Pinkerton et al., 2005). Sea-WiFS data are used to identify the magnitude and variability of the annual cycle of primary production by marine phytoplankton and to determine the distribution and timing of spring blooms. The observations help scientists understand the dynamics of ocean and coastal currents, the physics of mixing, and the relationships between ocean physics and large-scale patterns of productivity.

SeaWiFS research incorporates atmospheric correction and improved algorithms applied to ocean imagery. The most important SeaWiFS operational algorithms involve the use of bands 443, 490, or 510 nm divided by the band at 555 nm (Table 2-3) (Falkowski et al., 1998; Pinkerton et al., 2005). For example, consider several examples of the oceanic applications of SeaWiFS data in Color Plate 12-1. The first image is a global map of chlorophyll *a* derived from a composite of images obtained from September 3, 1997, through December 31, 1997 (Color Plate 12-1a). The next image is a true-color composite of a portion of the eastern United States at a spatial resolution of 1.13×1.13 km (Color Plate 12-1b). The true-color image is a color composite of bands 670, 555, and 412 nm (RGB) and highlights vegetation associated with land morphology. White areas are clouds and dense aerosols. Color Plate 12-1c presents the chlorophyll *a* concentrations on September 30, 1997. Red colors reveal high concentrations of chlorophyll, yellows and greens indicate intermediate concentrations of chlorophyll, and blues and purples document low chlorophyll concentrations .

Tang et al. (2003) found good agreement with SEAWiFS data and *in situ* chlorophyll *a* measurements in the Gulf of Tonkin, South China Sea. Gregg and Casey (2004) compared the SeaWiFS chlorophyll data set to global and regional archives of NASA and NOAA *in situ* chlorophyll data. The global comparison yielded a root mean square (RMS) log error of 31%, with a coefficient of determination (R^2) of 0.76, using 4,168 data points. RMS log error for open ocean (defined as bottom depth >200 m) was 27.7% with R^2 = 0.72, compared to 33% RMS log error and R^2 = 0.60 near the coasts.

Table 12-2.   Satellite remote sensing systems used to measure ocean color (courtesy of International Ocean Colour Coordinating Group, http://www.ioccg.org/sensors).

Sensor	Agency	Satellite	Operating Dates	Spatial Resolution (m)	Number of bands	Spectral coverage (nm)
CZCS	NASA (USA)	Nimbus-7 (USA)	10/24/78-06/22/86	825	6	433-12500
OCTS	NASD (Japan)	ADEOS (Japan)	08/17/96-07/01/97	700	12	402-12500
Polder-1	CNES (France)	ADEOS (Japan)	08/17/96-07/01/97	6000	9	433-910
MOS	DLR (Germany)	IRS P3 (India)	Launch 03/21/96	500	18	408-1600
SeaWiFS	NASA (USA)	OrbView-2	Launch 08/01/97	1100	8	402-885
OCI	NEC (Japan)	ROCSAT-1 (Taiwan)	Launch 01/99	825	6	433-12500
OCM	ISRO (India)	IRS-P4 (India)	Launch 05/26/99	350	8	402-885
MODIS-*Terra*	NASA (USA)	*Terra* (USA)	Launch 12/18/99	250/500/1000	36	405-14385
OSMI	KARI (Korea)	KOMPSAT (Korea)	Launch 12/20/99	850	6	400-900
MERIS	ESA (Europe)	ENVISAT-1 (Europe)	Launch 03/01/02	300/1200	15	412-1050
MODIS-*Aqua*	NASA (USA)	*Aqua* (USA)	Launch 05/04/02	250/500/1000	36	405-14385
CMODIS	CNSA (China)	Shen Zhou-3 (China)	03/25/02-09/15/02	400	34	403-12500
COCTS	CNSA (China)	HaiYang-1 (China)	Launch 05/15/02	1100	10	402-12500
CZI	CNSA (China)	HaiYang-1 (China)	05/15/02-10/25/03	250	4	420-890
GLI	NASDA (Japan)	ADEOS-II (Japan)	12/14/02-10/25/03	250/1000	36	375-12500
Polder-2	CNES (France)	ADEOS-II (Japan)	12/14/02-10/25/03	6000	9	443-910

The primary instrument for assessing ocean productivity on the EOS *Terra* spacecraft is MODIS. MODIS bands 8 – 16, ranging from 405 – 877 nm at 1000 × 1000 m spatial resolution, are particularly well suited to the collection of information on ocean color, phytoplankton concentration, and biogeochemistry (MODIS, 2006). *Terra* MODIS product number MOD 21, parameter numbers 26 and 27 (Chlor_a_2 and a_3, respectively) can be used to extract Chlorophyll *a* concentration information from Case 1 and Case 2 water using the algorithms found in Table 12-3 (O'Reilly et al., 2000; Carder et al., 2002; Darecki and Stramski, 2004). A map of the chlorophyll *a* distribution along the southern United States on November 23, 2003, derived using the MODIS Chlor_a_2 algorithm is shown in Color Plate 12-2.

Unfortunately, even these algorithms sometimes have difficulty extracting chlorophyll *a* information from optically-complex water such as the Baltic Sea (Darecki and Stramski,

2004). Maritorena and Siegel (2005) presented a way to merge satellite ocean color data streams from MODIS and SeaWiFS to improve global daily coverage information.

Chlorophyll in Coastal and Inland Water: For inland and near-coastal water bodies, it is often difficult to disentangle the information about the phytoplankton pigments in the remote sensor data from the effects of suspended inorganic materials or dissolved organic matter (DOM). This normally requires the use of sophisticated atmospheric correction techniques applied to the remote sensor data (e.g., Ramsey et al., 1992) and a complex multiple-component extraction methodology (e.g., Bukata et al., 1995). For example, Karaska et al. (2004) documented the use of Airborne Visible/Infrared Imaging Spectrometer (AVIRIS) hyperspectral data to map chlorophyll *a* in coastal and estuarine Case 2 water associated with the Neuse River in North Carolina.

Table 12-3. Selected operational band-ratio algorithms for remote sensing phytoplankton abundance.

Sensor and Algorithm	Algorithm	References
OCTSc SeaDAS 4.3 **Chl_octsc**	$C_a = 10^{(-0.55006 + 3.497R)}$ where $$R = \log_{10}\left[\frac{L_{wn}(520) + L_{wn}(565)}{L_{wn}(490)}\right]$$	O'Reilly et al., 1998 Pinkerton et al., 2005
SeaWiFS OC2v4 SeaDAS 4.3 **chl_oc2**	$C_a = a_4 + 10^{(a_0 + a_1 R + a_2 R^2 + a_3 R^3)}$ where $$R = \log_{10}\left[\frac{R_{rs}(490)}{R_{rs}(555)}\right]$$ where $a = [0.319, -2.336, 0.879, -0.135, -0.071]$	O'Reilly et al., 2000
SeaWiFS OC4v4 SeaDAS4.3 **chl_oc4** **chlor_a**	$C_a = 10^{(a_0 + a_1 R + a_2 R^2 + a_3 R^3 + a_4 R^4)}$ where $$R = \log_{10}\left[\frac{max[R_{rs}(443), R_{rs}(490), R_{rs}(510)]}{R_{rs}(555)}\right]$$ $a = [0.366, -3.067, 1.930, 0.649, -1.532]$	O'Reilly et al., 2000
MODIS DAAC-V4 **Chlor_a_2** (based on SeaWiFS OC4 algorithm with modified bands)	$C_a = 10^{(a_0 + a_1 R + a_2 R^2 + a_3 R^3 + a_4 R^4)}$ where $R = \log_{10}[max(r_{25}, r_{35})]$ and $r_{25} = \dfrac{L(443)}{L(551)}$ and $r_{35} = \dfrac{L(448)}{L(551)}$ where $a = [0.283, -2.753, 1.457, 0.659, -1.403]$	O'Reilly et al., 2000; Morel and Antoine, 2000; Darecki and Stramski, 2004; Pinkerton et al., 2005
MODIS DAAC-V4 **chlor_MODIS**	$C_a = 10^{0.07}\left[\dfrac{L_{wn}(443)}{L_{wn}(550)}\right]^{-1.40}$	Clark, 1997; Pinkerton et al., 2005
MODIS DAAC-V4 **Chlor_a_3** (for Case 1 and Case 2 waters)	$C_a = 10^{(a_0 + a_1 R + a_2 R^2)}$ where $R = \log_{10}\dfrac{L(448)}{L(551)}$. $a = [0.289, -3.2, 1.2]$	Darecki and Stramski, 2004; Carder et al., 2002
MERIS-C **ESA Processor Chl1**	$C_a = 10^{(a_0 + a_1 R + a_2 R^2 + a_3 R^3 + a_4 R^4 + a_5 R^5)}$ where $$R = \log_{10}\left[\frac{R_{rs}(\lambda_0)}{R_{rs}(560)}\right]$$ if $C_a < 1$, then $\lambda_0 = 443$ and $a = [0.2115, -1.844, 1.282, -2.747, 2.865, -1.204]$. if $1 \leq C_a \leq 10$, then $\lambda_0 = 490$ and $a = [0.3866, -2.626, 2.572, -7.156, 10.15, -6.895]$. if $C_a > 10$, then $\lambda_0 = 510$ and $a = [0.390, -3.864, 3.168, -20.33, 63.21, -113.9]$.	Morel and Antoine, 2000
GLI OC4-GLI **NASDA Processor Chla**	$C_a = a_4 + 10^{(a_0 + a_1 R + a_2 R^2 + a_3 R^3)}$ where $$R = \log_{10}\left[\frac{max[L_{wn}(443), L_{wn}(460), L_{wn}(520)]}{L_{wn}(545)}\right]$$ $a = [0.512, -3.65, 4.457, -2.175, -0.167]$	Mitchell, 2001

Dissolved Organic Material

Sunlight penetrates into the water column a certain *photic depth* (the vertical distance from the water surface to the 1 percent subsurface irradiance level). Phytoplankton within the photic depth of the water column consume nutrients and convert them into organic matter via photosynthesis. This is called *primary production*. Zooplankton eat the phytoplankton and create organic matter. Bacterioplankton decompose this organic matter. All this conversion introduces *dissolved organic matter* (DOM) into oceanic, nearshore, and inland water bodies. In certain instances, there may be sufficient dissolved organic matter in the water to reduce the penetration of light in the water column (Bukata et al., 1995).

The decomposition of phytoplankton cells yields carbon dioxide, inorganic nitrogen, sulfur, and phosphorus compounds. The more productive the phytoplankton, the greater the release of dissolved organic matter. In addition, *humic substances* may be produced. These often have a yellow appearance and represent an important colorant agent in the water column, which may need to be taken into consideration. These dissolved humic substances are called *yellow substance* or *Gelbstoffe* and can 1) impact the absorption and scattering of light in the water column, and 2) change the color of the water.

There are sources of dissolved organic matter other than phytoplankton. For example, the brownish-yellow color of the water in many rivers in the northern United States is due to the high concentrations of tannin from the eastern hemlock (*Tsuga canadensis*) and various other species of trees and plants grown in bogs in these areas (Hoffer, 1978). These tannins can create problems when remote sensing inland water bodies.

Water Penetration and Bathymetry

Bathymetric mapping can be performed using passive optical (e.g., aerial photography) or active remote sensing systems. Most geographically extensive and accurate bathymetric charting is performed using active SONAR (which reflects sound waves off the bottom) and LIDAR (which reflects laser light off the bottom).

Bathymetric Mapping Using Passive Optical Remote Sensing

The optimum optical wavelengths to obtain bathymetric (depth) information are from approximately 0.44 to 0.54 μm based on Figure 12-3. In fact, Kodak, Inc., scientists developed a special water penetration film in the 1970s that was sensitive to the wavelength interval 0.44 – 0.54 μm after filtration. This information was also used when specifying the blue-sensitive band (0.45 – 0.52 μm) on the Landsat Thematic Mapper sensor system, which is often called the water penetration band. If the water column is exceptionally clear, it is possible to see subsurface features to a depth of 10 – 30 m by sensing in these wavelengths.

Bathymetric mapping in the 0.44 – 0.54 μm portion of the spectrum requires that the water be almost free from organic and inorganic constituents such as chlorophyll and suspended sediments that would cause scattering and/or absorption to take place and obscure the bottom topography. However, as demonstrated in the Cozumel, Mexico, example (Figure 12-4), even the green and red wavelength region are of value for water penetration when the water is clear.

But what happens when water is not clear? Perhaps the blue-green portion of the spectrum may not be the most ideal region to use if the water contains suspended sediment. Legleiter et al. (2004) documented the effects of channel water depth on remote sensing reflectance (Figure 12-11). They found that at wavelengths <560 nm, remote sensing reflectance from the bottom substrate increased as depth increased from 5 to 80 cm. Conversely, as depth increased, the remote sensing measured reflectance in the red and particularly the near-infrared spectral regions decreased due to the strong absorption by pure water as previously discussed. A crossover point of equal reflectance for all depths separated scattering and absorption-dominated regimes (Legleiter et al., 2004).

Based on these results, Legleiter et al. (2004) found that the natural log of the simple ratio of two bands of hyperspectral data centered on 560 nm and 690 nm correlated well with stream water depth (*wd*):

$$wd = \ln\left[\frac{L(560)}{L(690)}\right]. \qquad (12\text{-}6)$$

across a broad range of stream conditions ($R^2 = 0.79$). This technique outperformed the linear transform method used in historical remote sensing bathymetric mapping studies in shallow coastal waters. Legleiter and Roberts (2005) subsequently studied the effects of channel morphology and sensor spatial resolution on depth estimates derived using the same ratio. Lafon et al. (2002) found that accurate bathymetric measurements in CASE II tidal inlet water could be obtained using SPOT multispectral data if the data were calibrated using *in situ* depth measurements.

Figure 12-11 The effects of water depth on remote sensing reflectance from a stream channel. This particular stream channel bottom substrate was covered with Periphyton (a complex matrix of algae and heterotrophic microbes attached to submerged material in almost all aquatic ecosystems) which has a strong absorption feature at 675 nm. The stream water contained approximately 2 mg/l of suspended sediment. Note that as the depth of the water increased from 5 to 80 cm, the red and near-infrared reflectance from the water column decreased. Conversely, as the depth of the water increased, the greater the reflectance from wavelengths < 560 nm. A reflectance cross-over occurred at approximately 560 nm (Legleiter et al., 2004; reprinted with permission from Elsevier).

When conducting bathymetric surveys using photogrammetric techniques, it important to take into account the fact that the light from the Sun is bent from its true course in *both* the atmosphere and in the water column, causing bathymetric (*z*) information in the imagery to *not* be in its proper planimetric (*x,y*) position. Therefore, the index-of-refraction previously discussed in Chapter 2 must be taken into consideration whenever remote sensing methods are used to perform bathymetric mapping.

Bathymetric Mapping Using Active SONAR

The acronym SONAR (SOund, NAvigation and Ranging) is an American term first used in World War II. The British call SONAR, ASDICS (Anti-Submarine Detection Investigation Committee). SONAR technology may be used to collect measurements of the seafloor. When active SONAR is used to measure the distance to the bottom, it is known as *echo sounding*. Echo sounders are used to measure water depth by sending acoustic pulses via a transducer. The acoustic pulses are reflected by the sea floor and the reflected echoes are received by the transducer. The elapsed time between the outgoing pulse and the return echo can be used to accurately determine depth. There are three main types of SONAR systems (NOAA CSC, 2006):

- single beam SONAR;

- multiple beam SONAR;

- side-scan SONAR.

Single-beam SONARs use a transducer that emits a single sound pulse into the water column at specific intervals with a narrow acoustic footprint (Figure 12-12). The sound wave reflects off the seafloor (or fish!) and the return is captured by the transducer (NOAA CSC, 2006). Single-beam SONAR data are collected along transect lines and typically cannot provide continuous coverage of the seafloor. The output resolution of the data are determined by the acoustic footprint size, sampling interval, sampling speed, and distance between transects. Single-beam sensors may collect bathymetry data or be calibrated to identify seafloor habitats, fish, or subsurface sediments. Many of the systems are designed to be easily transported and used in small boats that can access shallow areas. Recreational fisherman use relatively inexpensive single beam SONAR systems.

Multiple-beam SONARs provide users with two kinds of data: bathymetric (depth) and "acoustic backscatter" (Gardner, 2002). They can be used in extremely deep water (up to 11,000 m). These sensors emit multiple sound pulses (beams) that cover large overlapping swaths of the seafloor, enabling continuous coverage of the seafloor. The backscatter signals can be used for habitat or feature mapping.

Multiple beam acoustic sensors are typically mounted on the ship's hull. The sonar sends a fan of sound energy toward the seafloor, then records the reflected sound through a set of narrow receivers aimed at different angles (Figure 12-12). It can detect features as small as one meter across and locate them to within one meter of their true geographic location (NOAA Office of Coast Survey, 2006). For example, the Kongsberg EM 3002 multiple beam echo sounder has a nominal frequency of 300 kHz with an angular coverage below the ship of $\geq 200°$ and up to 508 acoustic soundings per ping as narrow as 1 degree (Kongsberg, 2005). Achievable swath width on a flat bottom is normally up to 10 times the water depth. The spacing between soundings as well as the acoustic footprints can be set nearly constant over the

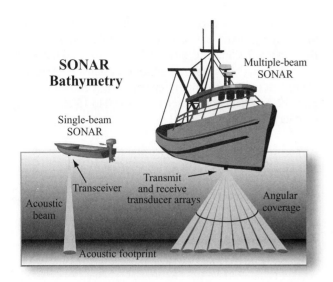

SONAR Bathymetry

Single-beam SONAR

Multiple-beam SONAR

Transmit and receive transducer arrays

Acoustic beam

Transceiver

Angular coverage

Acoustic footprint

Figure 12-12 Single- and multiple-beam SONARs. A Single-beam SONAR transceiver sends out a single pulse of acoustic (sound) energy usually directly beneath the boat. The acoustic energy (referred to as a *ping*) is reflected off the bottom (or any intervening object such as a school of fish) and is recorded by the transceiver. It is then possible to determine the depth to targets based on how long it took the pulse to return to the surface and a knowledge of the speed of sound in water. Multiple-beam SONARs transmit many beams per acoustic ping (e.g, >500) as narrow as 1 degree and can keep track of exactly when they encounter the bottom or other features in the water.

swath in order to provide a uniform and high detection and mapping performance. An acoustic image of the seabed floor can also be produced from the data (Kongsberg, 2005).

Color Plate 12-3a depicts the shaded relief bathymetry of Lake Tahoe, California-Nevada. It was created by combining bathymetric data obtained by a Kongsberg multiple-beam SONAR system (for depths > 15 m) and data obtained from the SHOALS LIDAR system (for depths from 0 to 15 m) (Gardner et al., 2000; U.S.G.S., 2006). The combined bathymetry data is color-coded for depth and merged with Landsat-7 imagery (RGB = bands 3, 2, 1) of the surrounding land. An acoustic backscatter map of Lake Tahoe is shown in Color Plate 12-3b. The brighter tones indicate harder surfaces such as rock, while the darker tones indicate softer surfaces such as sediment. The surrounding land is a shaded relief image generated from USGS 10-m DEMs. An oblique view of the bathymetry is shown in Color Plate 12-3c (USGS, 2006). The lake is about 34 km (21 miles) long, north to south (Gardner et al., 2000).

Side-scan SONARs are usually towed behind a boat (referred to as tow-fish). They are used primarily for searching for and detecting objects on the seafloor, not for bathymetric mapping. The tow-fish SONAR points multiple beams at angles and covers large overlapping swaths, enabling continuous coverage of the seafloor. The transmitted energy is formed into the shape of a fan that sweeps the seafloor from directly under the towfish to either side. Like other SONARS, a side scan transmits sound energy and analyzes the return signal (echo) reflected off the seafloor or other objects. The strength of the return echo is continuously recorded creating a 'picture' of the ocean bottom where objects that protrude from the bottom create a dark image (strong return) and shadows from these objects are light areas (little or no return). While the shape of the seafloor and objects on it can be well depicted, most side-scan systems do not provide depth information. However, they can be deployed in waters up to 11,000 meters deep and provide data with a resolution of a few centimeters.

Bathymetric Mapping Using Active LIDAR

Bathymetric information can be accurately extracted using LIDAR (Light Detection And Ranging) remote sensing (refer to Chapter 10 for information on terrestrial LIDAR applications). Guenter et al. (2000) provided a detailed history of airborne LIDAR bathymetric mapping. The Scanning Hydrographic Operational Airborne Lidar Survey (SHOALS) system is the most advanced airborne LIDAR bathymetric mapping system (West and Wiggins, 2000; Optech, 2006; USACE, 2006). SHOALS typically operates on a Twin Otter or a helicopter. SHOALS simultaneously fires a near-infrared (1064 nm) pulse and a green (532 nm) pulse of laser light toward the water surface at the same instant in time (Figure 12-13) (Optech, 2006). Much of the near-infrared pulse is absorbed by the water as expected, but a certain amount is reflected back toward the LIDAR receiver (Figure 12-13). A portion of the green laser pulse is also absorbed by the water column, but hopefully a sufficient amount reaches the bottom and is reflected back toward the LIDAR receiver. Green light is selected for bottom detection because that is the wavelength within the 0.44 to 0.54 μm region that penetrates typical coastal waters with the least attenuation. As the light travels through the water column and reflects off the seafloor, scattering, absorption, and refraction all combine to limit the strength of the bottom return, and therefore the system's maximum extinction depth. Its maximum water penetration depth is approximately 60 m or 2 to 3 times the Secchi depth. The water depth is calculated from the time difference between the

Simultaneous
transmission of →
near-infrared (1064 nm)
and green (532 nm)
laser pulses toward
the water body

**Airborne
LIDAR
Bathymetry**

← Near-infrared
surface return

Water

← Green
bottom return

Bottom

Figure 12-13 Airborne LIDAR bathymetry sends two laser pulses (near-infrared and green) toward the water at exactly the same instant. The near-infrared pulse is reflected from the water surface toward the LIDAR receiver. The green pulse penetrates into the water column. If the water is clear enough and the depth is not too great, a portion of the green pulse is reflected by the bottom substrate back toward the receiver. The difference in time between the reception of the near-infrared return and the green return is used to compute the water depth. The intensity of the backscattered LIDAR pulse can also be recorded (adapted from SHOALS, 2005).

near-infrared surface return and the green bottom return and the speed of light in air and water (Guenther et al., 2000).

Terrestrial LIDAR systems send out >100,000 pulses per second typically with a 10 nanosecond pulse length. LIDAR bathymetry systems operate at a much slower rate, e.g., 1,000 pulses (i.e., soundings) per second with a 250 nanosecond pulse length. The longer pulse length is necessary because a short laser pulse would be scattered in the water column (Optech, 2006).

The SHOALS system can survey up to 32 km² in one hour, at an altitude of 200 to 400 m above sea level (ASL) with speeds up to 115 knots. The SHOALS system, depending on altitude and speed, is capable of collecting depth soundings on a 4 m grid. Using a kinematic global positioning system, SHOALS references each depth measurement to a horizon-

tal position accurate to 3 m and a vertical position accurate to 15 cm. SHOALS is managed by the Joint Airborne Lidar Bathymetry Technical Center of Expertise based out of the US Army Engineer District in Mobile, Alabama. An example of SHOALS bathymetric information is shown in Color Plate 12-3a where it was used to collect 0 to 15 m bathymetric data of Lake Tahoe.

Airborne LIDAR bathymetry overcomes the limitations of acoustic (SONAR) sensors in some key areas:

• It performs especially well in shallow water where boats might run aground.

• It allows surveys of hazardous coastal areas (e.g., with high surf) to be conducted in comparative safety.

• It is mobile, allowing rapid response in emergency situations.

• Large areas can be inventoried quickly providing economic advantages.

LIDAR and SONAR bathymetric mapping compliment one another. LIDAR is especially useful for high resolution bathymetric mapping in shallow water environments whereas single and multiple beam SONAR are useful for bathymetric mapping in intermediate and deep water.

Water Surface Temperature

We know from Chapter 8 (Thermal Infrared Remote Sensing) that we may obtain the temperature of inland water bodies and sea-surface temperature (SST) during daylight hours and at night using thermal infrared remote sensing techniques. Unlike land surfaces, however, water bodies transfer energy primarily through convection. Therefore, any heat energy introduced into the system may be transferred hundreds of meters into the water body. This mixing is responsible for the relatively uniform surface temperature of a water body both day and night. Thus, water bodies generally have high thermal inertia, meaning that there may only be a few degrees difference between the daytime and nighttime water-surface temperature. This results in water bodies appearing relatively cooler than the land during daylight hours and relatively warmer than surrounding land during predawn hours, as demonstrated in the thermal infrared daytime and nighttime images of Atlanta, GA, in Chapter 8.

Water has an emissivity (ε) very close to 1. Therefore, it is possible to obtain relatively accurate water-surface tempera-

Sea-surface Temperature Maps Derived from NOAA AVHRR Data

a.   b.

Figure 12-14 Sea-surface temperature (SST) maps derived from a three-day composite of NOAA AVHRR thermal infrared data centered on March 4, 1999. Each pixel was allocated the highest surface temperature that occurred during the three days. a) This temperature map has been adjusted to highlight the cool water adjacent to the Georgia and South Carolina coastline. b) This map has been adjusted to highlight the circulation patterns associated with the warm offshore Gulf Stream. A color composite of these images that provides additional information is found in Color Plate 12-4a (courtesy NOAA).

ture measurements because the remote sensor radiant temperature measurement (T_{rad}) is approximately equal to the true kinetic temperature (T_{kin}), assuming the effects of the intervening atmosphere are accounted for. It is important to remember, however, that it is only a surface measurement. If there is a significant change in temperature just a few meters below the surface (i.e., a thermocline), it may not be detected by the thermal infrared detectors.

Two of the most important thermal infrared remote sensing instruments used to measure sea-surface temperature are found on the NOAA Advanced Very High Resolution Radiometer (AVHRR) and the NOAA Geostationary Operational Environmental Satellite (GOES). Their spatial, spectral, and temporal characteristics are summarized in Chapter 7. Data from these sensors is in the public domain, available at a reasonable price, and are used heavily to obtain sea-surface temperature (SST) measurements over large oceanic surfaces usually at a spatial resolution of $> 1 \times 1$ km. For example, SST measurements off the southeastern coast of the United States obtained on March 4, 1999, by the NOAA AVHRR are shown in Figure 12-14 and in Color Plate 12-4a.

The SST images depict the cool water adjacent to the Georgia and South Carolina shoreline caused by the influx of cool water from the major rivers. Also note the location of the warm Gulf Stream winding its way northward from the tip of Florida along the Eastern Seaboard. Tang et al. (2003b) examined 342 NOAA AVHRR thermal infrared images to monitor the temperature characteristics of the thermal plume associated with China's first nuclear power station (Daya Bay) and the largest foreign investment project in China. GOES-8 sea surface temperature measurements have been documented to be accurate to within 1.0° for Brazilian coastal water (Franca and Carvalho, 2004).

El Nino and La Nina

A global sea-surface temperature (SST) map is shown in Color Plate 12-4b. It was created using day/night NOAA-14 thermal data collected from March 9–13, 1999. SST maps such as this have been indispensable in mapping the distribution and movement of the El Nino Southern Oscillation (ENSO). In non-El Nino conditions, the trade winds blow toward the west across the tropical Pacific. These winds pile up warm surface water in the western Pacific, so that the sea surface is about 0.5 m higher near Indonesia than near Ecuador, South America. The sea-surface temperature is also about 8°C higher in the west. Cooler temperatures exist off South America due to an upwelling of cold water from deeper levels. This cold water is nutrient-rich, supporting high levels of primary productivity, diverse marine ecosystems, and major fisheries. Rainfall is found in rising air over the warmest water, and the eastern Pacific is relatively dry.

Every two to seven years the trade winds relax in the central and western Pacific Ocean. When this occurs, the ocean currents and winds off the western coast of South America shift, bringing warm water eastward, displacing the nutrient-rich cold water that normally wells up from deep in the ocean. This invasion of warm water disrupts the marine food chain and the economies of coastal communities. Fishermen named the phenomenon El Nino (the Christ Child) because it occurred during the Christmas season. Thus, El Nino is characterized by unusually warm ocean temperatures, especially in the eastern equatorial Pacific Ocean. La Nina is characterized by unusually cool ocean temperatures in the equatorial Pacific Ocean. Tree ring analysis has documented that these southern oscillations are not new, but have been occurring systematically for more than a hundred years (NOAA, 1999).

El Nino can be seen in measurements of the sea-surface temperature. For example, Color Plate 12-5 depicts three different sea-surface temperature conditions for selected Decembers: a La Nina in 1988, a normal month in 1990, and an El Nino in 1997. These are Reynolds Monthly SST (°C) maps derived by analyzing satellite thermal infrared data in conjunction with daily temperature measurements telemetered from an array of NOAA-sponsored buoys strategically placed in the equatorial Pacific Ocean (Reynolds and Smith, 1994). This is a good example of the utility of using both *in situ* and remote sensing to improve the accuracy of a product. Unfortunately, despite our ability to accurately map the sea-surface temperature of the El Nino Southern Oscillation, a substantial amount of life and property is still lost when they occur, mainly due to poor preparation by local authorities (Suplee, 1999). ASTER and MODIS thermal infrared data are also useful for monitoring water surface temperature.

When high spatial resolution sea-surface temperature data is needed, it may be possible to use Landsat Thematic Mapper (TM) data. For example, Thomas et al. (2002) used 23 Landsat TM band 6 thermal infrared images obtained from 1986 – 1996 to monitor the sea surface temperature along the coast of Maine. When it is necessary to map the temperature of very small reservoirs, lakes, ponds, etc., then thermal infrared sensors flown on suborbital aircraft may be required. An example of using an aircraft-mounted thermal infrared multispectral scanner to map the temperature of water in the Savannah River is found in Chapter 8 (Thermal Infrared Remote Sensing).

 Precipitation

Precipitation may be measured operationally on a local or regional basis using ground-based (terrestrial) active microwave remote sensing. For example, there are more than 150 Next Generation Radar (NEXRAD) WSR-88D systems operating in the United States (NOAA ROC, 2005). The NEXRAD antenna is enclosed in a radome (Figure 12-15a) and emits a burst of microwave energy ($\lambda = 11.1$ cm). If the energy strikes an object (raindrop, snowflake, hailstone, bug, bird, etc.), the energy is scattered. A small fraction of the scattered energy may be directed back toward the radar antenna. Computers analyze the strength of the returned pulse (measured in decibels, dBZ), the time it took to travel to the object and back, and the doppler phase shift of the pulse. This process of emitting a pulsed microwave signal, listening for any returned signal, then emitting the next signal, takes place very fast, up to approximately 1300 times each second. The maximum range of the radar is approximately 250 miles.

NEXRAD functions in two modes: *clear air* mode and *precipitation* mode. In Clear Air mode it produces an updated image every 10 or 11 minutes. In Precipitation mode it produces an updated image every 6 minutes. In both modes, what is important to look at is the level of reflectivity measured in dBZ. The stronger the return, the higher the dBZ. In Clear Air mode, the scale goes as low as -28, which is why it is possible to occasionally detect dust or pollen in the air. In Precipitation mode, the scale goes as high as 75, which indicates severe weather. The reflectivity information is used to make precipitation maps. For example, a map of precipitation in South Carolina on July 30, 2005, at 9:00 p.m. produced from NEXRAD WSR-88D Precipitation mode data is shown in Figure 12-15b.

National Weather Service Forecast Offices use the NEXRAD WSR-88D systems to provide real-time information on the location of dangerous weather and the amount of precipitation. In the United States, The Weather Channel uses NEXRAD information to provide real-time local and regional weather forecasts. Similar systems function in Europe (Bagella et al., 2005).

NEXRAD Terrestrial Remote Sensing System

a. Next generation Radar (NEXRAD)
weather radar antenna.

b. NEXRAD precipitation map of South Carolina
obtained at 9:00 p.m. on July 30, 2005.

Figure 12-15 a) A 28 ft. diameter Next generation Radar (NEXRAD) weather radar antenna is housed inside a protective 30 ft. diameter spherical radome situated on a 100 ft. tower. In precipitation mode, the volume scan of the radar is comprised of 11 full 360° sweeps at nine unique elevation angles (between 0.5° and 19.5° above the local horizon). NEXRAD data is updated every 6 minutes when the radar is in precipitation mode. b) Reflectivity precipitation map of South Carolina obtained from the NEXRAD in Columbia, SC, obtained on July 30, 2005, with the NEXRAD WSR-88D in precipitation mode. The reflectivity data were processed to highlight precipitation associated with a severe thunderstorm passing from west to east. Intense thunderstorm precipitation is highlighted in white.

Unfortunately, the majority of the world's land surfaces do not enjoy such coverage by weather radars or even rain gauge networks, and will not for the foreseeable future. Meteorological satellites are the only systems capable of acquiring rainfall data over vast distances. Fortunately, it is now possible to monitor selected continents on a continuous basis and to do so for a small fraction of the cost of a surface-based observing network of equivalent spatial density (Smith et al., 2004). Unfortunately, instead of directly measuring rainfall using a rain gauge, remote sensing methods must rely on using *indirect* remotely sensed data on cloud reflectance, cloud-top temperature, and/or the presence of frozen precipitation aloft to estimate the surface rain rate.

Visible – Infrared Techniques

The earliest precipitation estimation methods were based on the assumption that the brightness of reflected sunlight in the visible and near-infrared portion of the spectrum from clouds might give an indication of their thickness and thus of their likelihood of bearing rain. Unfortunately, not all bright clouds produce precipitation. Similarly, the temperature of the top of clouds was used to estimate precipitation. Gener-

ally, the colder the cloud top, the greater the likelihood of precipitation. Unfortunately, not all cold cloud tops produce precipitation. Nevertheless, some very useful precipitation estimation techniques were developed based on visible and infrared data, including 1) twice-daily estimates of outgoing longwave radiation from polar-orbiting infrared sensors onboard AVHRR to estimate monthly rainfall, and 2) a GOES Precipitation Index (GPI) based on an analysis of thermal infrared data, where pixels were classified as "raining" at the rate of 3 mm hr^{-1} if the cloud-top temperature was < 235 K. At one time, the GPI was the standard climatological rainfall product that provided a climate-scale precipitation record throughout the tropics and subtropics. Algorithms have also been developed that utilize multiple images obtained through time to refine the rate of rainfall estimates.

Active and Passive Microwave Techniques

A major milestone in satellite precipitation-rate estimation was the development of microwave remote sensing instruments that respond in a more physically *direct* way to the presence of precipitation-size water and/or ice particles

Tropical Rainfall Measurement Mission (TRMM) Imagery on March 9, 1998

Figure 12-16 Orbital data-collection coverage by the Tropical Rainfall Measurement Mission (TRMM) Microwave Imager (TMI) on March 9, 1998. Notice that the 35° angle of inclination obtains good coverage of the tropical and subtropical portions of the Earth. An intense frontal system is over southern Florida. Also please refer to Color Plate 12-6a (courtesy NASA Goddard TRMM Office and Japan National Space Development Agency).

within the clouds, although remaining insensitive to non-precipitating clouds. The first Special Sensor Microwave Imager (SSM/1) was launched in 1987 and included a high-frequency channel at 85.5 GHz. Basically, precipitation-size ice particles and large raindrops significantly reduce the emissivity of the cloud and thus depress its brightness temperature to below a nominal background level. This allows a rainfall estimate to be computed. This sensor could distinguish rainfall over land with reasonable consistency at 15 × 15 km spatial resolution (Atlas et al., 2005). The SSM/I has provided worldwide precipitation estimates since 1987.

The single most important event in rainfall estimation (especially for tropical and subtropical regions) was the launch of the Tropical Rainfall Measuring Mission (TRMM) on November 27, 1997, by NASA and the National Space Development Agency (NASDA) of Japan (TRMM, 2006). The satellite is in a 350 km orbit at an inclination of 35°. TRMM carries five instruments onboard: the Precipitation Radar (PR), the TRMM Microwave Imager (TMI), Visible Infrared Scanner (VIRS), Lightning Imaging Sensor (LIS), and the Clouds and Earth's Radiant Energy System (CERES).

The TRMM Microwave Imager (TMI) is a passive microwave sensor designed to provide quantitative rainfall information on the integrated column precipitation content over a

487 mile (780 km) swath. It is best suited for rainfall estimates over oceans where data are needed most for climate model verification. It measures the intensity of radiation at five frequencies: 10.7 (45 km spatial resolution), 19.4, 21.3, 37, and 85.5 GHz (5 km spatial resolution). It has dual polarization at four of the frequencies. The 10.7 GHz frequency provides a more linear response for high rainfall rates common in tropical rainfall. A composite image of all the TRMM Microwave Imager orbital paths on March 9, 1998, are presented in Figure 12-16 and Color Plate 12-6a. Note the intense line of thunderstorms over southern Florida.

The TRMM Precipitation Radar (PR), the first in space, measures the three-dimensional rainfall distribution over both land and oceans. It defines the layer depth of the precipitation and provides information about the rainfall actually reaching the surface, which is used to determine the latent heat of the atmosphere (TRMM, 2006). Passive microwave sensors have difficulty measuring rain over land areas (discussed in Chapter 9), whereas the scanning radar operating at 13.8 GHz (HH) provides accurate precipitation data over land at a spatial resolution of 4.3 km at nadir and a swath width of 220 km. TRMM Precipitation Radar data over Florida obtained on March 9, 1998 is shown as an oblique image in Color Plate 12-6b and as an along-track cross-section in Color Plate 12-6c. The cross-section provides a significant amount of information about cloud height and precipitation.

Gabella et al. (2005) found that the TRMM Precipitation Radar estimates were good agreement with ground-based radar precipitation measurements. Atlas et al. (2005) found that precipitation estimates from TRMM Precipitation Radar and the Special Sensor Microwave/Imager (SSM/I) provided excellent rainfall information for use in hurricane analysis and forecasting.

The TRMM Visible Infrared Scanner (VIRS) provides high resolution information on cloud coverage, type, and cloud-top temperatures. It is a five channel cross-track scanning radiometer operating at 0.63, 1.6, 3.75, 10.80, and 12 μm. It obtains 2.1 km data at nadir with a swath width of approximately 720 km.

The TRMM Lightning Imaging Sensor (LIS) inventories the global incidence of lightning using an optical remote sensing system operating at 0.777 μm with a spatial resolution of 5 × 5 km at nadir and a 590 km swath width. The Clouds and Earth's Radiant Energy System (CERES) is a visible and infrared sensor designed especially to measure emitted and reflected radiative energy from the surface of the Earth, and from the atmosphere and its constituents (e.g., clouds, aerosols). It is a broadband scanning radiometer with a total channel spectral range of 0.3 – 50 mm.

TRMM's unique combination of sensor wavelengths, coverages, and resolving capabilities with its low-altitude, non-Sun-synchronous orbit provide monthly precipitation amounts to a 500 × 500 km grid (TRMM, 2006). Unfortunately, as can be seen in Figure 12-16, since the TRMM mission was meant to cover tropical latitudes, its 35 degree latitude limit embraces little of Europe and numerous other important land areas in the Northern Hemisphere (Gabella et al., 2005).

International Global Precipitation Measurement (GPM) Program and Mission

Improved microwave-derived rainfall data will likely become available in the future. In particular, the proposed International Global Precipitation Measurement (GPM) mission consisting of a constellation of eight or more satellites may provide approximately three-hour sampling of rainfall for any given Earth coordinate approximately 90 percent of the time (Smith et al., 2004; Atlas et al., 2005). One of these satellites will be similar to the TRMM satellite, referred to as the GPM Core satellite. The GPM Core will have for its main payload two rain-measuring instruments:

- a dual-frequency (Ka/Ku-band) precipitation radar referred to as the DPR, and

- an advanced, large aperture, multichannel passive microwave rain radiometer referred as the GPM Microwave Imager (GMI).

This effort is primarily a function of the Earth Observation Summit (EOS) held in July, 2003, in Washington, D.C., where more than thirty nations met to address global environmental and economic challenges. As a result of the EOS Summit, an ad hoc Group on Earth Observations was established to prepare a ten-year implementation plan for a coordinated, sustained Earth Observation System.

The Global Precipitation Measurement mission will hopefully be launched in 2009 – 2010. The GMI's spatial resolution will exceed that of any current or planned microwave radiometer, including the new Conical-scanning Microwave Imager-Sounder (CMIS) radiometer being developed for the NOAA Polar Orbiter Environmental Satellite System (NPOESS) (Smith et al, 2004).

 ### Aerosols and Clouds

NASA Earth Science Enterprise scientists state that more information is needed on the following variables if we are to be able to model and understand the forces modifying the Earth's global climate system (Kahn, 1999):

- the amount and type of atmospheric particles (aerosols), including those formed by nature and by human activities;

- the amount, type, and height of clouds; and

- the distribution of land-surface cover, including vegetation canopy structure.

Aerosols

Aerosol particles may be solid or liquid and range in size from 0.01 μm to several tens of micrometers. Cigarette smoke particles are in the middle of this size range; typical cloud drops are 10 or more micrometers in diameter. Aerosols tend to cool the surface below them because most aerosols reflect Sunlight back to space reducing the amount of solar radiation that can be absorbed by the surface below (Kahn, 1999). The magnitude of this effect depends on the size and composition of the aerosols, and on the reflecting properties of the underlying surface. Some believe that the cooling effect due to increased amounts of aerosols may actually reduce or delay some of the expected warming due to increases in the amount of atmospheric carbon dioxide. Perhaps more smog is good! Others disagree.

Although remote sensing aerosol retrieval science has been taking place for more than twenty years, until recently there was no method of obtaining information on aerosol amount and distribution (called optical depth) and particle properties (Bojinski et al., 2004). The EOS *Terra* Multi-angle Imaging SpectroRadiometer (MISR) collects such information using four spectral bands (blue, green, red, and near-infrared) and nine look angles (refer to Chapter 7 for a detailed description). Four of the linear-array sensors look forward at an oblique angle (26.1°, 45.6°, 60°, and 70.5°, respectively), one looks directly down (nadir), and four look aft at the same oblique angles. This provides multiple looks at exactly the same part of the atmosphere (and the surface of the Earth), allowing quantitative information to be derived about aerosol location and content (Diner et al., 2005). Aerosol characteristics of the entire Earth may be obtained every nine days. Tang et al. (2005) reported on the extraction of aerosol information using Terra and Aqua MODIS data. Bojinski et al. (2004) provides a history of aerosol mapping and a new aerosol retrieval algorithm using hyperspectral data.

Clouds

Clouds play a major role in controlling the Earth's climate. More than any other component of the climate system, clouds affect the flow of energy within the Earth's atmosphere. A cloud may warm or cool the Earth, depending upon its thickness and height above the surface. Low, thick clouds reflect incoming solar radiation back to space, which causes cooling. High clouds trap outgoing infrared radiation and produce greenhouse warming. Because cloud type, height, moisture content, and location are so variable, their effect on global climate is very difficult to measure. Some believe that the largest uncertainty in climate prediction models is how to characterize and parameterize the radiative and physical properties of clouds.

Clouds in Visible Imagery

The first meteorological satellites only measured visible energy reflected from clouds (0.4 – 0.7 μm). GOES sensors provide data in both the visible and thermal infrared portion of the spectrum, as summarized in Chapter 7. In the daylight hours, visible imagery provides detailed views of the cloud patterns that closely match our visual senses, i.e., clouds usually appear bright while land and water appear darker on the images. Consider typical visible images of both the western and eastern United States obtained by GOES-East and GOES-West on April 17, 1998 (Figure 12-17a,b). Note the large low-pressure frontal system approaching Washington state from the Bay of Alaska. There is also a significant low-pressure system located along the Eastern Seaboard and throughout the southeast. GOES was first launched on October 16, 1975. Since that time many new GOES satellites have been parked at 35,790 km in geostationary orbit to obtain visible and infrared imagery. They are used to monitor frontal systems, intense thunderstorm activity associated with tornadoes, and hurricanes.

Europe has a system similar to GOES, called METEOSAT. METEOSAT 4 is a geostationary meteorological satellite located at 0° of latitude and longitude. It continuously observes the whole Earth disk. The images gathered by METEOSAT in three channels—visible (0.4 – 1.14 μm), thermal infrared (10.5 – 12.5 μm) and water vapor (5.1 – 7.1 μm)—are transmitted every 30 min. The METEOSAT images available in high resolution mode have 5000 × 5000 pixels for the visible channel and 2500 × 2500 pixels for the infrared and water vapor channels (Ameur et al., 2004).

Most remote sensing systems operating in the visible portion of the spectrum can only collect data during the daytime. However, a light-sensitive instrument onboard the Defense Meteorological Satellite Program (DMSP) can obtain visible images at night. This is done by recording the features illuminated at night by moonlight.

In general, clouds do not reflect solar radiation equally well in all directions. Therefore, a single measurement of reflectivity from a single direction (e.g., at nadir) makes it difficult to determine the total amount of light reflected by the cloud (its *albedo*) relative to the incident energy. One solution is to obtain multiple images of a cloud from different vantage points. This introduces stereoscopic parallax. The EOS *Terra* MISR collects stereoscopic cloud information by viewing each cloud from nine angles previously discussed. The stereoscopic data can be analyzed to yield three-dimensional quantitative information about cloud height, structure, thickness, shape, and roughness of cloud tops. Accurate albedo information can also be computed.

Clouds in Thermal Infrared Imagery

The most common thermal infrared band used in meteorological investigations is 10 – 12.5 μm. The atmosphere is relatively transparent to this wavelength energy upwelling from the Earth's surface and clouds. Also, thermal infrared images can be obtained at night, so we can have a continuous 24-hr record of cloud patterns that includes important meteorological events taking place at night.

GOES-West and GOES-East Images

a. GOES-West visible.

b. GOES-East visible.

c. GOES-West thermal infrared.

d. GOES-East thermal infrared.

e. GOES-West water vapor.

f. GOES-East water vapor.

Figure 12-17 Examples of GOES-East and GOES-West images of the United States and portions of Central America obtained on April 17, 1998 (courtesy NOAA).

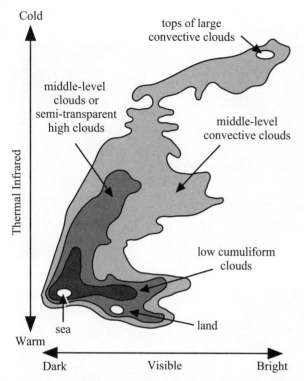

Figure 12-18 It is possible to determine cloud type based on multispectral measurements in the visible and thermal infrared portion of the electromagnetic spectrum (after Desbois et al., 1982).

In satellite meteorology, infrared images are normally inverted, i.e. the greater the radiance of a pixel, the darker the pixel. This way, clouds, which are usually colder than the surface of the Earth, appear white, and the warmer ground or ocean surfaces appear darker than clouds, as in visible images. Thermal infrared GOES-west and GOES-east images of the United States are shown in Figure 12-17c,d. It is easier to identify the edge of frontal systems using the thermal infrared imagery. Also, the land-water boundaries are more distinct.

It has been known for some time that it is possible to extract information on the type of clouds and their height using multispectral remote sensing. One of the earliest relationships is demonstrated in Figure 12-18 where visible and infrared data can be used to differentiate between the sea, land, cumuliform clouds, semitransparent high clouds, and convective clouds (like thunderstorms). Tall convective cumulonimbus clouds are cold and bright. The sea and land surface are warm and dark (Desbois et al., 1982). The analyst extracts the pixel value in the visible and thermal infra-

red bands, locates it in the diagram, and identifies the nature of the cloud under investigation.

Numerous scientists have attempted to detect the presence of clouds using a thermal infrared cloud temperature threshold value. The goal is to identify low-temperature pixels that correspond to medium- or high-altitude clouds. The main problem is usually the specification of the exact temperature threshold value. For example, Franca and Cracknell (1995) developed five different cloud-detection algorithms based on visible and infrared spectral information using NOAA AVHRR daytime data for tropical areas. Feidas et al. (2000) developed a cloud classification scheme where thick opaque convective clouds correspond to classes with the lower brightness temperatures (T_b) in both the thermal infrared (below 223 K) and water vapor band (below 223 K). Feidas and Cartalis (2005) built upon this work and developed an automatic algorithm for tracking convective cloud cells. The *Normalized Difference Snow Index* (NDSI) discussed in the Snow section of this chapter can also be used to identify clouds using sensors that contain a visible and mid-infrared band (e.g., Landsat TM bands 2 and 5) (Hall et al., 1995; Choi and Bindschadler, 2004).

Cloud-height information extracted from thermal infrared data can be used to generate pseudo three-dimensional oblique images of major storm events. For example, the AVHRR visible and infrared images in Figure 12-19a,b capture Hurricane Andrew moving toward Louisiana in 1992. In these examples, the visible (band 1; 0.58 – 0.68 μm) and near-infrared images (band 2; 0.725 – 1.10 μm) are draped over the cloud-height information derived from thermal infrared data (band 5; 11.5 – 12.5 μm).

The TRMM Visible Infrared Scanner (VIRS) launched in 1997 provides information on cloud type and cloud-top temperatures using a five channel cross-track scanning radiometer. The *Terra* MISR sensor collects information in only the visible and near-infrared portions of the spectrum, while the *Terra* Clouds and Earth's Radiant Energy System (CERES) sensor collects data from just one look angle, but across the entire solar spectrum. CERES measures both solar-reflected and Earth-emitted radiation from the top of the atmosphere to the surface. It also determines cloud properties including amount, height, thickness, and particle size. Thus, the VIRS, MISR, and CERES instruments complement one another in the collection of cloud information. MODIS obtains cloud-top information from bands 33 – 36 in the thermal infrared region from 13.185 – 14.385 μm.

AVHRR Imagery of Hurricane Andrew on August 25, 1992

a. Visible band 1 (0.58 - 0.68 μm).   b. Near-infrared band 2 (0.725 - 1.10 μm).

Figure 12-19 a) AVHRR visible band 1 (0.58 – 0.68 μm) imagery of Hurricane Andrew bearing down on Louisiana obtained on August 25, 1992, at 20:20 UT. The apparent height of the perspective rendering is inversely proportional to the cloud temperatures observed in band 5 (11.5 – 12.5 μm). b) AVHRR band 2 (0.725 – 1.10 μm) near-infrared data (courtesy of F. Hasler, K. Palaniappan, M. Manyin and H. Pierce, NASA Goddard Space Flight Center).

 Water Vapor

Water vapor is essential for precipitation. Not only is water vapor the most important greenhouse gas in the atmosphere (Chylek et al., 20004), but this variable also plays a key role for the understanding and prediction of most important weather processes from convection, to the formation of clouds and precipitation, to the formation of severe storms (Wulfmeyer et al., 2005). It is possible to map water vapor by sensing in water-vapor absorption bands. Several wavelengths can be used, but one of the most common is centered around 6.7 μm. METEOSAT 1, launched in 1978 by the European Space Agency, was the first geostationary satellite to obtain images of mid- to upper-troposphere water vapor in the 6.7 μm region in addition to visible and 10 – 12 μm infrared images. METEOSAT 4 collects water vapor information in the spectral region from 5.1 – 7.1 μm (Ameur et al., 2004).

GOES sensors routinely provide water-vapor images obtained in the 6.7 μm region. At this wavelength, most of the radiation sensed by the satellite comes from the atmospheric layer between 300 and 600 km, i.e., from the middle layers of the troposphere. Figures 12-17e,f depict water-vapor images of the western and eastern United States. Note the tremendous amount of water vapor in the approaching storm in the western United States and the departing storm in the eastern United States. Also note the lack of water vapor on this particular date around Hawaii and in the lower por-

tion of the Gulf of Mexico near Cozumel, Mexico. The weather was very comfortable on this day in these areas. The relative humidity is likely to be higher in bright areas than in dark areas on a water-vapor image. Bright and dark areas may also indicate rising and sinking motions, respectively.

It is also possible to extract water vapor information in the near-infrared portion of the spectrum which has several features that make it attractive for remote water vapor retrieval. It is a region where the solar radiation is still quite strong; and there are spectral bands within the near-infrared where interference from other atmospheric gases is weak. Within the wavelength span from about 890 to 990 nm there are three major water vapor absorption bands: a strong absorption band centered near 942 nm and weaker absorption bands centered around 906 and 977 nm (Chylek et al., 2004). MODIS has several bands that are sensitive to atmospheric water vapor in these regions, including band 17 (890 – 920 nm), 18 (931 – 941 nm), and 19 (915 – 965 nm). In the future, the European Space Agency may launch the Water Vapour Lidar Experiment in Space (WALES) using a differential absorption Lidar (Wulfmeyer et al., 2005).

 Snow

Snow runoff can be both beneficial and disastrous. Spring runoff is the lifeblood for most agricultural activities in semiarid climates and provides much of the potable drinking water (in addition to groundwater). Unfortunately, a tremendous buildup of snow or an untimely thaw can cause severe flooding and consequently loss of life and property. Therefore, meteorologists, hydrologists, and glaciologists work to identify 1) the geographic extent of the snowpack on a seasonal basis, and 2) its water equivalent.

Snow in the Visible Spectrum

When clouds are not present in a visible and/or near-infrared image, it is a relatively straightforward task to identify the spatial distribution of snow because it is generally much brighter than the vegetation, soil, or water nearby that is not covered with snow. Also, one of the major diagnostic clues that can be used to distinguish between clouds and snow is to obtain numerous images of the terrain in a relatively short time period using AVHRR or GOES data. Clouds move–snow does not.

Snow in the Middle-Infrared and Microwave Regions

However, if one can only obtain a single image of the terrain and it contains both clouds and snow, then it is possible to utilize the middle-infrared portion of the spectrum to differentiate between snow and cloud cover. The reason is made clear in Figure 12-20. Throughout the visible and near-infrared portions of the spectrum, clouds and snow reflect approximately equal amounts of radiant flux. In the middle-infrared portion of the spectrum (especially 1.5 – 2.5 μm) clouds continue to reflect a substantial amount of energy, while snow reflectance drops to near zero (Hall et al., 1995; Choi and Bindschadler, 2004). This relationship was first documented by Skylab sensors operating in the region 1.55 – 1.75 μm. In this band, clouds have a very high reflectance and appear white in the imagery, while the snow has a very low reflectance and appears black. Thus, the Landsat Thematic Mapper band 5 (1.55 – 1.75 μm) and the SPOT 4 sensor with its middle-infrared band can be used to discriminate between clouds and snow cover. The decrease in reflectance of snow from 80 – 90 percent at wavelengths < 1 μm to < 10 percent at approximately 1.5 μm is extraordinary when compared with the reflectance characteristics of other Earth surface materials. Based on these relationships, Hall et al. (1995) developed the *Landsat TM Normalized Difference Snow Index* (NDSI) based on band 2 and band 5 data:

$$\text{Landsat TM NDSI} = \frac{(\text{TM } 2 - \text{TM } 5)}{(\text{TM } 2 + \text{TM } 5)}. \quad (12\text{-}7)$$

Figure 12-20 Reflectance of clouds and snow in the wavelength interval 0.4 to 2.5 μm (after Avery and Berlin, 1992).

A threshold applied to the NDSI image is used to separate cloud pixels from non-cloud (snow) pixels. The automatic cloud cover assessment algorithm (ACCA) developed by Irish (2000) used 0.7 as the threshold value (Choi and Bindschadler, 2004). However, the optimum threshold can vary from one NDSI-derived image to another.

MODIS data have been used to produce global snow-cover map products since September 13, 2000 at 500×500 m resolution (Dozier and Painter, 2004). The automated MODIS snow-mapping algorithm uses at-satellite reflectances in MODIS bands 4 (0.545 – 0.565 μm) and 6 (1.628 – 1.652 μm) to calculate the *MODIS Normalized Difference Snow Index* (Hall et al., 2002; Salomonson and Appel, 2004):

$$\text{MODIS NDSI} = \frac{(\text{MODIS } 4 - \text{MODIS } 6)}{(\text{MODIS } 4 + \text{MODIS } 6)}. \quad (12\text{-}8)$$

A pixel in a non-densely forested region will be mapped as snow if the MODIS NDSI is ≥ 0.4 and reflectance in MODIS band 2 is $\geq 11\%$. However, if the MODIS band 4 reflectance is < 10%, then the pixel will not be mapped as snow even if the other criteria are met (Hall et al., 2002). This prevents pixels containing very dark targets such as black spruce forests from being mapped as snow. Salomonson and Appel (2004) reported that the relationship between fractional snow cover and MODIS NDSI is reasonably robust when applied locally and over large areas like North America. Zhou et al. (2005) found MODIS snow-cover products to be of value for modeling streamflow.

The microwave region of the spectrum offers promise for retrieving important snow information (Foster et al., 2005). Depending upon the wavelength, estimates of the depth, snow water equivalent, and the amount/presence of liquid water in the snow pack are possible. In addition, clouds are transparent to many of the microwave frequencies so that mapping of the snow area and properties is possible even in regions where clouds are common and at night (Foster et al., 2005). Aircraft and Space Shuttle Synthetic Aperture Radar (SAR) measurements have shown that SAR can be used to discriminate between snow and glaciers and between wet and dry snow (Shi et al., 1994). Derksen et al. (2005) documented how data from the passive microwave Special Sensor Microwave/Imager (SSM/I) were used to successfully extract snow water equivalent information over Canada. In general, optical remote sensing used to map snow areal extent (SAE) has much higher spatial resolution when compared with microwave remote sensing which can be used to map snow water equivalent (SWE) (Zhou et al., 2005).

Water Quality Modeling

State and county agencies are required to reduce point source discharges to surface waters. Point sources are relatively easy to control and regulate because of their known location. A phenomenon more common and more difficult to detect is *nonpoint source pollution* (NPS). NPS is defined as pollution originating from urban runoff, construction, hydrologic modification, silviculture, mining, agriculture, irrigation return flows, solid-waste disposal, atmospheric deposition, stream bank erosion, and sewage disposal. Impacts of NPS include decreased recreational water use; reduction of water storage capacity in streams, lakes and estuaries; clogging of drainage ditches and irrigation canals; excessive enrichment and sedimentation of surface waters that contribute to the loss of fish, shellfish and wildlife habitat; and a reduction in the aesthetic qualities of the aquatic environment (Corbitt, 1990).

Traditional *in situ* measurement techniques have had a limited effect on identifying and modeling nonpoint source pollution. Scientists have found that an integration of *in situ* and remote sensing data collection with GIS modeling techniques can provide some very useful water-quality information. In fact, remote sensing data collection combined with GIS modeling offers a means of identifying and ranking NPS potential in surface waters with output that is relatively easy for the general public to understand. Schultz (1988) provides examples of how remotely sensed biophysical vari-

ables such as slope, soils and land-cover characteristics can be used as input to hydrological models.

An Integrated Remote Sensing and GIS Water-Quality Model

This brief case study evaluates water-quality degradation and impacts from urban runoff within the Withers Swash watershed in Myrtle Beach, SC. Withers Swash is an urbanized watershed (34% impervious) that covers approximately 4.2 mi^2 and includes industrial, residential, commercial, agricultural, recreational, and undeveloped land uses. The watershed is divided into two subbasins (Figure 12-21a), which drain into a tidal pool designed to capture sediment before emptying into the Atlantic Ocean. A study was conducted using remotely sensed data to derive land-cover information which was then integrated with other spatial information such as hydrology, topography, and soil data, into a GIS database. The data were then analyzed using the *Agricultural NonPoint Source* (AGNPS) pollution model to generate water-quality predictions (USDA, 1998; Schill and Jensen, 1998).

The AGNPS water-quality model uses eight directional flow attributes determined by topography and channel characteristics to route stormwater throughout the watershed to a single outlet cell (Cronshey and Theurer, 1998). Approximately one third of the model input parameters are derived from topography (Figure 12-21b). These parameters include terrain slope, the Universal Soil Loss Equation (USLE) slope length, and slope shape. Soil attributes including erodability (USLE K factor) and texture, were verified through field sampling and sieve tests (Figure 12-21c). Land-cover information was derived from 1:10,000-scale color-infrared aerial photography and from NASA's Airborne Terrestrial Applications Sensor (ATLAS) 3 × 3 m multispectral data (Figure 12-21d). From these land-cover attributes, the SCS curve number (assigned by soil hydrologic group), Manning roughness coefficient, USLE C (cropping) and P (practice) factors, soil condition constant, and chemical oxygen demand (COD) were derived based on lookup tables in Young et al. (1994).

Before executing the model, all spatial data were assigned to a 30 × 30 m cell (0.2 × 0.2 acre) for each of the two subbasins. A single storm event was simulated using the following parameters: storm energy intensity (*r* factor) of 95.39, storm duration of 24 hours, total storm rainfall equal to 2.68 in., and rainfall nitrogen concentrations of 0.80 ppm. An example of two model output parameters include concentrated peak flow in ft^3 per second (cfs) (Figure 12-21e) and

Nonpoint Source Pollution Modeling

a. Hydrology

0 - 10
10 - 50
50 - 100
100 - 200

e. Peak flow (cfs)

b. Topography

0.0 - 0.05
0.05 - 0.15
0.15 - 0.25
0.25 - 0.50

f. Cell erosion
(tons per acre)

c. Soils

d. Land cover derived from remote sensor data

Figure 12-21 Nonpoint source pollution modeling using the Agricultural NonPoint Source (AGNPS) pollution water-quality model applied to two sub-basins. a) Hydrology of the Withers Swash watershed in Myrtle Beach, SC. Modeling was based on 30×30 m cells and a 2.68 in. rain in 24 hours. b) Topography was derived photogrammetrically and used to extract slope information. c) Soils data were sampled in the field. d) Land cover characteristics were derived from analysis of aerial photography and ATLAS multispectral scanner data. e) Modeled concentrated peak flow in ft^3 per second. f) Modeled total cell erosion in tons per acre (Schill and Jensen, 1998).

total cell erosion in tons per acre (Figure 12-21f). Model results have been compared with *in situ* measurements and used to locate areas of increased sediment and nutrient loss and to indicate the spatial extent of nutrient and sediment transport, deposition and accumulation. This information has been used to spatially allocate vegetated buffer strips and other best management practices (BMPs). King et al (2005) report on the use of remote sensing land-use inputs from Landsat TM data and surface roughness inputs from RADARSAT to the STREAM runoff and erosion model.

References

Ameur, Z., Ameur, S., Adane, A., Sauvageot, H. and K. Bara, 2004, "Cloud Classification using the Textural Features of Meteosat Images," *Intl. Journal of Remote Sensing*, 25(21):4491–4503.

Atlas, R., Hou, A. Y. and O. Reale, 2005, "Application of SeaWinds Scatterometer and TMI-SSM/I Rain Rates to Hurricane Analysis and Forecasting," *ISPRS Journal of Photogrammetry & Remote Sensing*, 59:233–243.

Avery, T. E. and G. L. Berlin, 1992, *Fundamentals of Remote Sensing and Airphoto Interpretation*, NY: Macmillan.

Binding, C. E., Bowers, D. G. and E. G. Mitchelson-Jacob, 2005, "Estimating Suspended Sediment Concentrations from Ocean Colour Measurements in Moderately Turbid Waters: The Impact of Variable Particle scattering Properties," *Remote Sensing of Environment*, 94:373–383.

Bojinski, S., Schlapfer, D., Schaepman, M., Keller, J. and K. Itten, 2004, "Aerosol Mapping over Land with Imaging Spectroscopy using Spectral Autocorrelation," *Intl. Journal of Remote Sensing*, 25(22):5025–5047.

Bukata, R. P., J. H. Jerome, K. Y. Kondratyev and D. V. Pozdnyakov, 1995, *Optical Properties and Remote Sensing of Inland and Coastal Waters*, NY: CRC, 362 p.

Carder, K. L., Chen, F. R., Lee, Z., Hawes, S. K. and J. P. Cannizzaro, 2002, *MODIS Ocean Science Team Algorithm Theoretical Basis Document, Case 2 Chlorophyll a*, MODIS Algorithm Theoretical Basis Doc. 19, Ver. 7, Greenbelt: NASA GSFC, http://modis.gsfc.nasa.gov/data/atbd/atbd_mod19.pdf.

Carder, J. V., Mayer, L. A. and J. E. H. Clarke, 2000, "Morphology and Processes in Lake Tahoe (California-Nevada)," *Geological Society of America Bulletin*, 112(5):736–746.

Choi, H. and R. Bindschadler, 2004, "Cloud Detection in Landsat Imagery of Ice Sheets using Shadow Matching Technique and Automatic Normalized Difference Snow Index Threshold Value," *Remote Sensing of Environment*, 91:237–242

Chylek, P., Clodius, W. B., Bender, S. C., Atkins, W. H. and L. K. Balick, 2004, Sensitivity of Near Infrared Total Water Vapour Estimate to Calibration Errors," *Intl. Journal of Remote Sensing*, 25(21):4457–4470.

Darecki, M. and D. Stramski, 2004, "An Evaluation of MODIS and SeaWiFS Bio-optical Algorithms in the Baltic Sea," *Remote Sensing of Environment*, 89:326–350.

Desbois, M., G. Seze and G. Szejwach, 1982, "Automatic Classification of Clouds on METEOSAT Satellite Imagery: Applications to High-level Clouds," *Journal of Applied Meteorology*, 21:401–412.

Diner, D. J., Martonchik, J. V., Kahn, R. A., Pinty, B., Gobron, N., Nelson, D. L. and B. N. Holben, 2005, "Using Angular and Spectral Shape Similarity Constraints to Improve MISR Aerosol and Surface Retrievals Over Land," *Remote Sensing of Environment*, 94:155–171.

Derksen, C., Walker, A. and G. Boodison, 2005, "Evaluation of Passive Microwave Snow Water Equivalent Retrievals Across the Boreal Forest/Tundra Transition of Western Canada," *Remote Sensing of Environment*, 96:315–327.

Dozier, J. and T. H. Painter, 2004, "Multispectral and Hyperspectral Remote Sensing of Alpine Snow Properties," *Annual Reviews Earth Planetary Science*, 32:465–494.

Doxaran, D., Froidefond, J. M., Lavender, S. and P. Castaing, 2002, "Spectral Signature of Highly Turbid Waters: Application with SPOT Data to Quantify Suspended Particulate Matter Concentrations," *Remote Sensing of Environment*, 81:149–161.

Falkowski, P. G., M. J. Behrenfeld, W. E. Esaias, W. Balch, J. W. Campbell, R. L. Iverson, D. A. Kiefer, A. Morel and J. A. Yoder, 1998, *Satellite Primary Productivity Data and Algorithm Development: A Science Plan for Mission to Planet Earth*, Greenbelt: NASA Goddard Space Flight Center, 42:36 p.

Feidas, H., Cartalis, C. and A. P. Cracknell, 2000, "Use of METEOSAT Imagery to Define Clouds Linked with Floods in Greece," *Intl. Journal of Remote Sensing*, 21:1047–1072.

Feidas, H. and C. Cartalis, 2005, "Application of an Automated Cloud-tracking Algorithm on Satellite Imagery for Tracking and Monitoring Small Mesoscale Convective Cloud Systems," *Intl. Journal of Remote Sensing*, 26(8):1677–1698.

Foster, J. L., Sun, C., Walker, J. P., Kelly, R., Chang, A., Dong, J. and H. Powell, 2005, "Quantifying the Uncertainty in Passive Microwave Snow Water Equivalent Observations," *Remote Sensing of Environment*, 94:187–203.

Franca, G. B. and A. P. Cracknell, 1995, "A Simple Cloud Masking Approach Using NOAA AVHRR Daytime Data for Tropical Areas," *Intl. Journal of Remote Sensing*, 16(9):1697-1705.

Franca, G. B. and W. S. Carvalho, 2004, "Sea Surface Temperature GOES-8 Estimation Approach for the Brazilian Coast," *Intl. Journal of Remote Sensing*, 25(17):3439–3450.

Gabella, M., Michaelides, S. and G. Perona, 2005, "Preliminary Comparison of TRMM and ground-based Precipitation Radars for A European Test Site," *Intl. Journal of Remote Sensing*, 26(5):997–1006.

Gardner, J., 2002, "Using Multibeam Sonar to Map MPSs: Tool of the Future for Planning and Management," *Marine Protected Areas News*, 4(2):1–2.

Gitelson, A. A., 1992, "The Peak Near 700 nm on Radiance Spectra of Algae and Water: Relationships of its Magnitude and Position with Chlorophyll Concentration, *Intl. Journal of Remote Sensing*, 13:3367–3373.

Gordon, H. R., D. K. Clark, J. W. Brown, O. B. Brown, R. H. Evans and W. W. Broenkow, 1983, "Phytoplankton Pigment Concentrations in the Middle Atlantic Bight: Comparison of Ship Determinations and CZCS Estimates," *Applied Optics*, 22:20–36.

Gregg, W. W. and N. W. Casey, 2004, "Global and Regional Evaluation of the SeaWiFS Chlorophyll Data Set," *Remote Sensing of Environment*, 93:463–479.

Guenther, G. C., Cunningham, A. G., LaRocque, P. E. and D. J. Reid, 2000, "Meeting the Accuracy Challenge in Airborne LIDAR Bathymetry," *EARSEL eProceedings of the LIDAR Workshop,* Dresden; June 16-17, 1:1–27.

Hall, D. K., Riggs, G. A. and V. V. Salomonson, 1995, "Development of Methods for Mapping Global Snow Cover using Moderate Resolution Imaging Spectrometer Data," *Remote Sensing of Environment*, 54:127–140.

Hall, D. K., Riggs, G. A., Salomonson, DiGirolamo, N. E. and K. J. Bayr, 2002, "MODIS Snow-cover Products," *Remote Sensing of Environment*, 83:181–194.

Han, L., 1997, "Spectral Reflectance with Varying Suspended Sediment Concentrations in Clear and Algae-Laden Waters," *Photogrammetric Engineering & Remote Sensing*, 63(6):701–705.

Han, L. and D. C. Rundquist, 1994, "The Response of Both Surface Reflectance and the Underwater Light Field to Various Levels of Suspended Sediments: Preliminary Results," *Photogrammetric Engineering & Remote Sensing*, 60:1463–1471.

Han, L. and D. C. Rundquist, 1997, "Comparison of NIR/RED Ratio and First Derivative of Reflectance in Estimating Algal-Chlorophyll Concentration: A Case Study in a Turbid Reservoir," *Remote Sensing of Environment*, 62:253–261.

Han, L. and D. C. Rundquist, 1998, "The Impact of a Wind-roughened Water Surface on Remote Measurements of Turbidity," *Intl. Journal of Remote Sensing*, 19(1):195–201.

Hoffer, R., 1978, "Spectral Characteristics of Water and Snow," *Remote Sensing: The Quantitative Approach*, P. Swain and S. Davis (Eds.), NY: McGraw Hill.

Irish, R., 2000, "Landsat-7 Automatic Cloud Cover Assessment Algorithms for Multispectral, Hyperspectral, and Ultraspectral Imagery," *SPIE*, 4049:348–355.

Islam, A., Gao, J., Ahmad, W., Neil, D. and P. Bell, 2004, "A Composite DOP Approach to Excluding Bottom Reflectance in Mapping Water Parameters of Shallow Coastal Zones from TM Imagery," *Remote Sensing of Environment*, 92:40–51.

Jensen, J. R., Kjerfve, B. K., Ramsey III, E. W., Magill, K. E., Medeiros,C. and J. E. Sneed, J. E, 1989, "Remote Sensing and Numerical Modelling of Suspended Sediment in Laguna de Terminos, Campeche, Mexico," *Remote Sensing of Environment*, 28:33–44.

Kahn, R., 1999, *General Introduction to MISR*, Pasadena: NASA JPL, http://www-misr.jpl.nasa.gov/miintro.html.

Karaska, M. A., Huguenin, R. L., Beacham, J. L., Wang, M., Jensen, J. R. and R. S. Laufmann, 2004, "AVIRIS Measurements of Chlorophyll, Suspended Minerals, Dissolved Organic Carbon, and Turbidity in the Neuse River, NC," *Photogrammetric Engineering & Remote Sensing*, 70(1):125–133.

King, C., Lecomte, V., Bissonnais, Y. L., Gaghdadi, N., Souchere, V. and O. Cerdan, 2005, "Remote-sensing Data as an Alternative Input for the STREAM Runoff Model," *Catena*, 62:125–135.

Kongsberg, 2005, *EM120 Multibeam Echo Sounder*, Horten, Norway: Kongsberg, Inc., http://www.knogsberg.com.

Kroger, S. and R. J. Law, 2005, "Sensing the Sea," *Trends in Biotechnology*, 23(5):250-256.

Lafon, V., Froidefond, J. M., Lahet F. and P. Castaing, 2002, "SPOT Shallow Water Bathymetry of A Moderately Turbid Tidal Inlet Based on Field Measurements," *Remote Sensing of Environment*, 81:136–148.

Legleiter, C. J. and D. A. Roberts, 2005, "Effects of Channel Morphology and Sensor Spatial Resolution on Image-derived Depth Estimates," *Remote Sensing of Environment*, 95:231–247.

Legleiter, C. J. Roberts, D. A., Marcus, W. A and M. A. Fonstad, 2004, "Passive Optical Remote Sensing of River Channel Morphology and in-stream Habitat: Physical Basis and Feasibility," *Remote Sensing of Environment*, 93:493–510.

Lodhi, M. A., D. C. Rundquist, L. Han and M. S. Juzila, 1997, "The Potential for Remote Sensing of Loess Soils Suspended in Surface Waters," *Journal of the American Water Resources Association*, 33(1):111–127.

Maritorena, S. and D. A. Siegel, 2005, "Consistent Merging of Satellite Ocean Color Data Sets Using A Bio-optical Model," *Remote Sensing of Environment*, 94:429–440.

Markham, B. L. and J. C. Barker, 1986, "Landsat MSS and TM Post-Calibration Dynamic Ranges, Exoatmospheric Reflectance and at Satellite Temperature," *Landsat Technical Notes*, 1:3–8.

Miller, R. L. and B. A. McKee, 2004, "Using MODIS *Terra* 250 m Imagery to Map Concentrations of Total Suspended Matter in Coastal Waters," *Remote Sensing of Environment*, 93:259–266.

Mitchell, B. G., 2001, *Global Imager (GLI) Standard Algorithms (Chl-a, CDROM and K490): Calibration and Validation*, http://sharaku.eorc.nasda.go.jp/GLI/meet/2001/O_05.pdf.

MODIS, 2006, *MODIS Home Page*, Greenbelt: NASA Goddard Space Flight Center, http://modis.gsfc.nasa.gov/.

Morel, A. and D. Antoine, 2000, *Pigment Index Retrieval in Case 1 Waters*, MERIS Algorithm Theoretical Basis Document ATBD2, 9, p. 25, http://envisat.esa.int/instruments/meris/pdf/atbd_2_09.pdf.

Mumby, P. J., E. P. Green, A. J. Edwards and C. D. Clark, 1997, "Coral Reef Habitat-mapping: How Much Detail can Remote Sensing Provide?" *Marine Biology*, 130:193–202.

Mustard, J. F., Staid, M. I. and W. J. Fripp, 2001, "A semianalytical Approach to the Calibration of AVIRIS Data to Reflectance Over Water Application in a Temperate Estuary," *Remote Sensing of Environment*, 75:335–349.

NOAA, 1999, *What is an El Nino?*, Washington: NOAA.

NOAA CSC, 2006, *Single Beam Sonar*, Charleston: NOAA Coastal Services Center, www.csc.noaa.gov/crs/rs_appps/sensors/single_beam.htm.

NOAA Office of Coast Survey, 2006, *Side Scan and Multibeam SONAR*, Washington: NOAA Office of Coast Survey, http://chartmaker.ncd.noaa.gov/HSD/wrecks.html.

NOAA ROC, 2005, NEXRAD WSR-88D, Washington: NOAA Radar Operations, http://www.roc.noaa.gov.

Optech, 2006, *SHOALS–1000*, Toronto: Optech, Inc., Http://www.optech.on.ca.

O'Reilly, J. E., Maritorena, S., Mitchell, G., Siegel, D.A., Carder, K. L., Garver, S. A., et al., 1998, "Ocean Colour Algorithms for SeaWiFS," *Journal of Geophysical Research*, 103(C11):24937–24953.

O'Reilly, J. E. and 21 co-authors, 2000, "Ocean Color Chlorophyll Algorithms for SeaWiFS, OC2, and OC4: Version 4," in S. B. Hooker and E. R. Firestone (Eds.), *SeaWiFS Postlaunch Calibration and Validation Analyses, Part 3*, NASA Technical Memorandum, 2000-206892, Greenbelt: GSFC, 11:9–27.

Pasqualini, V., C. Pergent-Martine, C. Fernandez and G. Pergent, 1997, "The Use of Airborne Remote Sensing for Benthic Cartography: Advantages and Reliability," *Intl. Journal of Remote Sensing*, 18(5):1167–1177.

Pinkerton, M. H., Richardson, K. M., Boyd, P. W., Gall, M. P., Zeldis, J. L., Oliver, M. D. and R. J. Murphy, 2005, "Intercomparison of Ocean Colour Band-ratio Algorithms for Chlorophyll Concentration in the Subtropical Front East of New Zealand," *Remote Sensing of Environment*, 97:382–402.

Ramsey, E. W., J. R. Jensen, H. E. Mackey and J. Gladden, 1992, "Remote Sensing of Water Quality in Active to Inactive Cooling Water Reservoirs," *Intl. Journal of Remote Sensing*, 13(18):3465–3488.

Reynolds, R. W. and T. M. Smith, 1994, "Improved Global sea-surface Temperature Analyses Using Optimum Interpolation," *Journal of Climate*, 7:929–948.

Rundquist, D. C., J. F. Schalles and J. S. Peake, 1995, "The Response of Volume Reflectance to Manipulated Algal Concentrations Above Bright and Dark Bottoms at Various Depths in an Experimental Pool," *Geocarto Intl.*, 10:5–14.

Rundquist, D. C., L. Han, J. F. Schalles and J. S. Peake, 1996, "Remote Measurement of Algal Chlorophyll in Surface Waters: The Case for the First Derivative of Reflectance Near 690 nm," *Photogrammetric Engineering & Remote Sensing*, 62(2):195–200.

Schill, S. R. and J. Jensen, 1998, "Modeling to Predict Stormwater Runoff in Withers Swash Myrtle Beach, SC," *Proceedings*, Urban and Regional Information Systems, Charlotte: 83–93.

Schultz, G. A., 1988, "Remote Sensing in Hydrology," *Journal of Hydrology*, 100:239–265.

Shi, J., J. Dozier and H. Rott, 1994, "Snow Mapping in Alpine Regions with Synthetic Aperture Radar," *IEEE Transactions on Geoscience and Remote Sensing*, 32(1).

Smith, E. A. and 18 others, 2004, "International Global Measurement (GPM) Program and Mission: An Overview," in V. Levizzani and F. J. Turk, Eds., *Measuring Precipitation from Space: EURAINSAT and the Future*, London: Kluwer, 43 p., http://gpm.gsfc.nasa.gov/gpm_overview.pdf

Suplee, C., 1999, "El Nino and La Nina: Nature's Vicious Cycle," *National Geographic*, 195(3):73–95.

Tang, D., Kawamura, H., Lee, M. and T. V. Dien, 2003, "Seasonal and Spatial Distribution of Chlorophyll *a* Concentrations and Water Conditions in the Gulf of Tonkin, South China Sea," *Remote Sensing of Environment*, 85:475–483.

Tang, D., Kester, D. R., Wang, Z., Lian, J. and H. Kawamura, 2003b, "AVHRR Satellite Remote Sensing and Shipboard Measurements of the Thermal Plume from the Daya Bay, Nuclear Power Station, China," *Remote Sensing of Environment*, 84:506–515.

Tang, J., Xue, Y., Yu, T. and Y. Guan, 2005, "Aerosol Optical Thickness Determination by Exploiting the Synergy of TERRA and AQUA MODIS," *Remote Sensing of Environment*, 94:327–334.

Thomas, A., Byrne, D. and R. Weatherbee, 2002, "Coastal Sea Surface Temperature Variability from Landsat Infrared Data," *Remote Sensing of Environment*, 81:262–272.

TRMM, 2006, *Tropical Rainfall Measurement Mission (TRMM) Home Page*, Greenbelt: GSFC, http://trmm.gsfc.nasa.gov/.

USACE, 2006, *SHOALS*, Mobile: U. S. Army Corps of Engineers, http://shoals.sam.usace.army.mil/default.htm.

USDA, 1998, *AGNPS 98 Agricultural Non-Point Source Pollution Model (Version 5.0)*, Morris: USDA Agricultural Research Service), hwww.ftc.nrcs.usda.gov/huwq/agnps.htm.

USGS, 2006, *Pacific Sea-Floor Mapping Images*, Washington: U.S. Geological Survey Western Region Coastal & Marine Geology, http://walrus.wr.usgs.gov/pacmaps/.

West, G. R. and C. E. Wiggins, 2000, "Airborne Lidar Bathymetry in the Management of Florida's Coastal Zone," *Integrated Costal Zone*, Autumn:69–72.

Wulfmeyer, V., Bauer, H., Girolamo, P. D. and C. Serio, 2005, "Comparison of Active and Passive Water Vapor Remote Sensing from Space: An Analysis Based on the Simulated Performance of IASI and Space Borne Differential Absorption LIDAR," *Remote Sensing of Environment*, 95:211–230.

Young, R. A., C. A. Onstad, D. D. Bosch, and W. P. Anderson, 1994, *Agricultural Non-Point Source Pollution Model (AGNPS)*, St. Joseph: U.S. Dept. of Agriculture.

Zhou, X., Xie, H. and J. M. H. Hendrickx, 2005, "Statistical Evaluation of Remotely Sensed Snow-cover Products with Constraints from Streamflow and SNOTEL Measurements," *Remote Sensing of Environment*, 94:214–231.

Remote Sensing the Urban Landscape 13

Urban landscapes are composed of a diverse assemblage of materials (concrete, asphalt, metal, plastic, shingles, glass, water, grass, shrubbery, trees, and soil) arranged by humans in complex ways to build housing, transportation systems, utilities, commercial and industrial facilities, and recreational landscapes (e.g., Small, 2004). The goal of this construction is hopefully to improve the quality of life. In many instances, *urbanization* is taking place at a dramatic rate (e.g., Auch et al., 2004), often without planned development (referred to as *urban sprawl*) (Clapham, 2003; Wilson et al., 2003a). A significant number of businessmen and women, public organizations, and scientists constantly require up-to-date information about the city and suburban infrastructure (Carlson, 2003; Herold et al., 2003, 2004). For example, detailed urban information is required by:

- city, county, and regional councils of governments that a) legislate zoning regulations to hopefully improve the quality of life in urbanized areas, and b) would like to model (predict) the location of future urban growth;

- city and state departments of commerce, which are mandated to stimulate development, often to increase the tax base;

- tax-assessor offices, which must maintain legal geographic descriptions of every parcel of land, assess its value, and levy a tax-millage rate;

- county and state departments of transportation that maintain existing facilities, hopefully build new facilities without damaging the environment, and prepare for future transportation demand;

- public and private utility companies (e.g., water, sewer, gas, electricity, telephone, cable) to predict where new demand will occur and plan for the most efficient and cost-effective method of delivering the services;

- public service commissions, which are mandated to make the utility services available economically to the consumer;

- departments of parks, recreation, and tourism to improve public recreation facilities;

- departments of emergency management/preparedness and homeland security, which are responsible for removing the threat and/or mitigating destruction and allocating resources in the event of a disaster or terrorism;

- real estate companies that are paid to find the ideal location for industrial, commercial, and residential development;

- developers who build residential, commercial, and industrial facilities; and

- scientists who model urban hydrologic conditions, climate change, and net primary productivity (e.g., Imhoff, 2005).

These are educated, professional users of urban information (Cullingworth, 1997; Jensen and Hodgson, 2004). The urban/suburban land they manage or develop is of significant monetary value. Therefore, it is not surprising that city, county, and state agencies as well as private companies invest hundreds of millions of dollars each year obtaining aerial photography and other forms of remotely sensed data to extract the required urban information. Ideally, most of this spatial urban information is placed in a national spatial data infrastructure where it can be accessed by a variety of public and private users.

National Spatial Data Infrastructure

In the United States (and many other nations), an interagency Federal Geographic Data Committee (FGDC) promotes the coordinated development, use, sharing, and dissemination of the nation's digital geographic information resources (FGDC, 2006). This nationwide data publishing effort is known as the *National Spatial Data Infrastructure* (NSDI).

The NSDI consists of three *framework foundation spatial databases* (Figure 13-1a):
- geodetic control
- digital terrain (elevation and bathymetry)
- digital orthorectified imagery

four *framework thematic databases*:
- cadastral
- boundaries or political units
- hydrology
- transportation

and several *other thematic databases,* such as urban/suburban landuse and landcover.

The FGDC develops geospatial data standards for implementing the NSDI, in cooperation with state, local, and tribal

governments, the private sector and academic community, and the international community (Jensen et al., 2002). The *Developing Spatial Data Infrastructure: The SDI Cookbook* (GSDI, 2006) introduces regional, national and multinational organizations to the concepts and standards for implementing a globally compatible Spatial Data Infrastructure. Selected databases from the South Carolina Spatial Data Infrastructure are shown in Figure 13-1b and Color Plate 13-1.

Urban Resolution Considerations

Many of the detailed urban/suburban attributes that businesses and agencies require are summarized in Table 13-1 and Figure 13-2. This chapter reviews how remotely sensed data may be of value for the collection of these attributes. To remotely sense urban phenomena, it is first necessary to appreciate the urban attribute's temporal, spectral, and spatial resolution characteristics.

Urban/Suburban Temporal Resolution Considerations

Three types of temporal resolution should be considered when monitoring urban environments using remote sensor data. First, urban/suburban phenomena often progress through an identifiable *developmental cycle* much like vegetation progresses through a *phenological cycle* (Chapter 11). For example, Jensen and Toll (1983) documented a 10-stage single-family residential housing development cycle at work in suburban Denver, CO, that progressed from stage-1 (undeveloped rangeland) to stage-10 (fully landscaped residential housing), often within just six months (Figure 13-3a). The stages of residential development were identified based on the presence or absence of five major factors, including:

1. parcel clearing (partial or complete)

2. land subdivision (i.e., parcels surveyed and graded)

3. roads (dirt or paved)

4. buildings (presence or absence)

5. landscaping (partial or complete).

Single-family residential housing developments in San Diego, CA, undergo a similar cycle (Figures 13-3bc). It is

National Spatial Data Infrastructure

Additional themes
Cultural and Demographic
Geology — Other thematic Data
Soils
Vegetation
Landcover/Landuse

Cadastral
Hydrology — Thematic Framework Data
Boundaries (governmental units)
Transportation

Digital Orthorectified Imagery — Framework Foundation Data
Digital Terrain/Bathymetry
Geodetic Control

a.

Selected Databases from South Carolina's Spatial Data Infrastructure

Digital Orthorectified Imagery
Hydrology
Digital Terrain/Bathymetry
Transportation
Landcover/Landuse

b.

Figure 13-1   a) National Spatial Data Infrastructure framework foundation, thematic and other layers (after FGDC, 2006).
b) Selected databases from South Carolina's spatial data infrastructure (please see Color Plate 13-1).

imperative that the image analyst understand the temporal development cycle of the urban phenomena being analyzed. If it is not understood, embarrassing and costly image interpretation mistakes can be made.

The second type of temporal resolution is how often it is possible for the remote sensor system to collect data of the urban landscape, e.g., every 8 days, 16 days, or on demand. Current information is critical for most urban applications. Generally, sensors that can be pointed off-nadir (e.g., Geo-Eye IKONOS; DigitalGlobe QuickBird) have higher temporal resolution than sensors that only sense the terrain at nadir (e.g., Landsat Thematic Mapper). Urban applications are usually not as time-sensitive as those dealing with highly dynamic phenomena such as vegetation where a life cycle might take place during a single season. For these reasons, most urban applications (except traffic count transportation studies and emergencies) only require that imagery be collected every year or so. Orbital characteristics of the satellite platform and the latitude of the study area also impact the revisit schedule. Remote sensor data may be collected on demand from suborbital aircraft (airplanes, helicopters) if weather conditions permit.

Finally, temporal resolution may refer to how often land managers require a certain type of information. For example, local planning agencies may need precise population estimates every 5 – 7 years in addition to the estimates provided by the decennial census. The temporal resolution requirements for many urban applications are summarized in Table 13-1 and shown graphically in Figure 13-2.

Urban/Suburban Spectral Resolution Considerations

Most image analysts would agree that when extracting urban/suburban information from remotely sensed data, it is often more important to have high spatial resolution (e.g., ≤ 5 × 5 m) than high spectral resolution (i.e., a large number of bands). For example, local population estimates based on building unit counts usually require a minimum spatial resolution of 0.25 – 5 m (0.82 ft – 16.4 ft) to detect, distinguish between, and/or identify the type of individual buildings. Practically any visible band (e.g., green or red) or near-infrared spectral band at this spatial resolution will do. Of course, there must be sufficient spectral contrast between the object of interest (e.g., a building) and its background (e.g., the sur-

Figure 13-2 The clear polygons represent the spatial and temporal requirements for selected urban attributes listed in Table 13-1. Gray boxes depict the spatial and temporal characteristics of the major remote sensing systems that may be used to extract the required urban information (adapted from Jensen and Cowen, 1999; Jensen and Hodgson, 2004; Jensen, 2005).

Table 13-1. Relationship between urban/suburban attributes and the minimum remote sensing resolutions required to provide such data (adapted from Jensen and Cowen, 1999; Jensen and Hodgson, 2004; Jensen, 2005).

Attributes	Minimum Resolution Requirements		
	Temporal	Spatial	Spectral
Landuse/Landcover			
L1 – USGS Level I	5 – 10 years	20 – 100 m	V - NIR - MIR - Radar
L2 – USGS Level II	5 – 10 years	5 – 20 m	V - NIR - MIR - Radar
L3 – USGS Level III	3 – 5 years	1 – 5 m	Pan - V - NIR - MIR
L4 – USGS Level IV	1 – 3 years	0.25 – 1 m	Panchromatic (Pan)
Building and Property-Line Infrastructure			
B1 – building perimeter, area, volume, height	1 – 5 years	0.25 – 0.5 m	Stereo Pan - V, LIDAR
B2 – cadastral information (property lines)	1 – 6 months	0.25 – 0.5 m	Stereo Pan - V
Socioeconomic Characteristics			
S1 – local population estimation	5 – 7 years	0.25 – 5 m	Pan -V - NIR, LIDAR
S2 – regional/national population estimation	5 – 15 years	5 – 20 m	Pan -V - NIR
S3 – quality of life indicators	5 – 10 years	0.25 – 30 m	Pan -V - NIR
Energy Demand and Conservation			
E1 – energy demand and production potential	1 – 5 years	0.25 – 1 m	Pan -V - NIR, LIDAR
E2 – building-insulation surveys	1 – 5 years	1 – 5 m	TIR
Transportation Infrastructure			
T1 – general road centerline	1 – 5 years	1 – 30 m	Pan -V - NIR
T2 – precise road width	1 – 2 years	0.25 – 0.5 m	Pan - Visible
T3 – traffic count studies (cars, airplanes, etc.)	5 – 10 min	0.25 – 0.5 m	Pan - Visible
T4 – parking studies	10 – 60 min	0.25 – 0.5 m	Pan - Visible
Utility Infrastructure			
U1 – general utility line mapping and routing	1 – 5 years	1 – 30 m	Pan -V - NIR
U2 – precise utility line width, right-of-way	1 – 2 years	0.25 – 0.6 m	Pan - V, LIDAR
U3 – location of poles, manholes, substations	1 – 2 years	0.25 – 0.6 m	Panchromatic, LIDAR
Digital Elevation Model (DEM)			
D1 – digital surface model and/or digital terrain model	5 – 10 years	0.25 – 0.5 m	Stereo Pan -Visible, LIDAR
D2 – slope map (derived from DTM)	5 – 10 years	0.25 – 0.5 m	Stereo Pan -Visible, LIDAR
Meteorological Data			
M1 – weather prediction	3 – 25 min	1 – 8 km	V - NIR - TIR
M2 – current temperature	3 – 25 min	1 – 8 km	TIR
M3 – clear air and precipitation mode	5 – 10 min	1 – 4 km	WSR-88D Radar
M4 – severe weather mode	5 – 10 min	1 – 4 km	WSR-88D Radar
M5 – monitoring urban heat island effect	12 – 24 hr	5 – 30 m	TIR
Urban Hydrology			
H1 – impervious surface mapping	1 – 5 years	0.25 – 10 m	V - NIR - MIR, LIDAR
H2 – stream centerline and flood plain delineation	1 – 5 years	1 – 30 m	V - NIR - MIR, LIDAR
Critical Environmental Area Assessment			
C1 – stable sensitive environments	1 – 2 years	1 – 10 m	V - NIR - MIR
C2 – dynamic sensitive environments	1 – 6 months	0.25 – 2 m	V - NIR - MIR - TIR
Disaster Emergency Response			
DE1 – pre-emergency imagery	1 – 5 years	1 – 5 m	Pan -V - NIR
DE2 – post-emergency imagery	12 hr – 2 days	0.25 – 2 m	Pan - NIR - Radar
DE3 – damaged housing stock	1 – 2 days	0.25 – 1 m	Pan - NIR
DE4 – damaged transportation	1 – 2 days	0.25 – 1 m	Pan - NIR
DE5 – damaged utilities, services	1 – 2 days	0.25 – 1 m	Pan - NIR

a. Single-family residential development near Denver, CO. A parcel of land may progress from rangeland (stage 1) to fully developed, landscaped, single-family residential housing (stage 10). The stages of residential development were identified based on the presence or absence of five factors, including parcel clearing, subdivision, roads, buildings, and degree of landscaping (Jensen and Toll, 1983).

b. Cleared, subdivided, terraced lots with dirt roads on steep slopes being developed for single-family residential housing in San Diego, CA.

c. Single-family residential housing in various stages of construction. Rows one and two in the top left have foundation footings. Rows three and four have subflooring and are being framed.

Figure 13-3   Examples of stages in the development cycle of single-family residential housing in Denver, CO, and San Diego, CA. Remote sensor data may capture urban land use in one of many stages of development.

rounding landscape) in order to detect, distinguish between, and identify the object from its background.

There are relatively coarse regions of the electromagnetic spectrum that are very useful for extracting certain types of

urban/suburban information (Table 13-1). For example, USGS Level III landcover is best acquired using the visible (0.4 – 0.7 μm; V), near-infrared (0.7 – 1.1 μm; NIR), middle infrared (1.5 – 2.5 μm; MIR), and/or panchromatic (0.5 – 0.7 μm) portions of the spectrum. Building perimeter, area, and

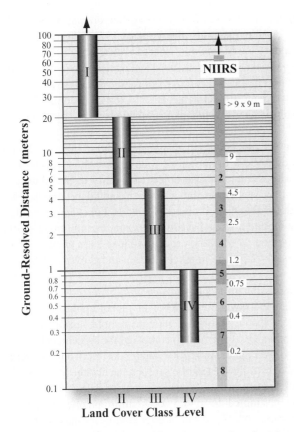

Figure 13-4   Relationship between USGS Land-use/Land-cover Classification System and the spatial resolution of the remote sensing system (often referred to as ground-resolved distance in meters). The National Image Interpretability Rating Scale (NIIRS) is provided for comparison. A NIIRS zero rating suggests that the interpretability of the image is precluded by obscuration, degradation, or poor resolution.

height information are best acquired using black-and-white panchromatic (0.5 – 0.7 µm) or color imagery (0.4 – 0.7 µm). The thermal infrared portion of the spectrum (3 – 12 µm; TIR) may be used to obtain urban surface temperature measurements (Dousset and Gourmelon, 2003). Active microwave sensors may obtain imagery of cloud-shrouded urban areas (e.g., Canadian RADARSAT C-band, and European Space Agency ERS-1,2 C-band). Near-infrared laser light (e.g., 1024 nm) can be used to collect LIDAR digital surface model first, intermediate and last returns plus intensity data (Hodgson et al., 2005).

Many urban applications can also be conducted using hyperspectral remote sensor data. Herold et al. (2004) collected field urban spectral measurements from 350 to 2400 nm and found potential for high spectral-resolution remote sensing for detailed mapping of urban *materials* and their condition.

Unfortunately, they also encountered considerable spectral confusion between some urban land cover types (e.g., specific roof and road types). Small (2002; 2003; 2004) summarized how different materials found within the IFOV of a sensor system (referred to as *endmembers* such as concrete, asphalt, vegetation, etc.) contribute to urban spectral reflectance and how they can be disentangled using spectral mixture analysis [refer to Jensen (2005) for a discussion of spectral mixture analysis]. Sometimes the only way to extract the urban information required is to utilize multiple types of remote sensor data in the same investigation referred to as *multi-sensor data fusion* (Gamba et al., 2003).

Urban/Suburban Spatial Resolution Considerations

Trained image analysts do rely on black-and-white tone or color in aerial photography or other types of imagery to extract useful urban information. However, in many instances the *geometric* spatial elements of image interpretation such as object shape, size, texture, orientation, pattern, and shadow silhouette are often just as useful if not more useful (all elements are discussed in Chapter 5). Generally, the higher the spatial resolution of the remote sensor data, the more detailed information that can be extracted in the urban environment. But how do we know what spatial resolution imagery to use for a specific urban application? Fortunately, there are several methods available for assessing the interpretability of imagery for urban applications based largely on the spatial resolution of the remote sensor data.

NIIRS Criteria

One solution might be to use the military and/or civilian versions of the National Image Interpretation Rating Scales (NIIRS) developed by the Image Resolution Assessment and Reporting Standards Committee (IRARS). NIIRS is the metric used by the intelligence community to characterize the usefulness of imagery for intelligence purposes (Leachtenauer et al., 1998; Logicon, 1997; Pike, 2006; FAS, 2006). Many urban image-interpretation tasks are similar to those required for intelligence applications. The NIIRS criteria consist of 10 rating levels (0 – 9) for a given type of imagery arrived at through evaluation by trained image analysts. The IRARS committee makes it clear that spatial resolution (ground-resolved distance) is only one of the measures of the interpretability of an image, although it is a very important one (Figure 13-4). Other factors such as film quality, atmospheric haze, contrast, angle of obliquity, and noise can reduce the ability of a well-trained analyst to detect, distinguish between, and identify military and civilian objects in an image. While it would be useful to use the NIIRS criteria, it is not optimum for this discussion because 1) the civil NIIRS criteria were only recently made available (Leachtenauer et al., 1998; FAS, 2006), 2) there has not been sufficient time for the civilian community to familiarize itself with the concept, and consequently 3) the civilian community has rarely reported its collective experiences in urban/suburban information extraction using the NIIRS system.

Area Weighted Average Resolution

A figure of merit for measuring the resolvability of a film camera system often used by photogrammetrists is called the *area weighted average resolution* (AWAR), measured in line-pairs-per-millimeter (lp/mm) (Light, 1993). A line pair is the width of one black bar and one white space as contained on resolution targets recorded in an aerial photograph. Together, they form a pair and serve as a measure of image quality for the aerial film camera industry. The five essential elements that make up the system AWAR are the lens, film, image blur (smear) on the film due to aircraft forward velocity, angular motion, and the resolution of the duplicating film. Also, scene contrast of the Earth and atmosphere play a role in system resolution.

Scientists have studied the general relationship between aerial photography scale and AWAR. For example, Light (1993, 1996) documented that if we assume that the Earth is a low-contrast scene, the 1:40,000-scale National Aerial Photography Program (NAPP) photography exhibits approximately 39 lp/mm and yields approximately 25 μm for the size of 1 lp in the image. At 1:40,000 scale, 25 μm equate to a ground resolution of 1×1 m for low-contrast scenes. Therefore, a minimum of 1 m ground resolution can be expected throughout the photographic mission. In fact, the USGS digital orthophoto quarter-quad files produced from 1:40,000-scale NAPP photography are provided at 1×1 m (3.28 × 3.28 ft) spatial resolution by scanning the photography with a pixel size of 11 μm. Light (1998) suggests that there is a general linear relationship for larger scale aerial photography obtained using metric cameras, i.e., 1:20,000-scale photography equates to approximately 0.5 × 0.5 m (1.64 × 1.64 ft); 1:10,000-scale photography to 0.25 × 0.25 m (0.82 × 0.82 ft); and 1:5,000-scale photography to 0.125 × 0.125 m (0.41 × 0.41 ft). Unfortunately, most persons using remote sensor data for practical urban applications do not report their findings in terms of AWAR criteria.

Nominal Spatial Resolution (Ground-Resolved Distance)

The civilian user community usually reports the utility of a given type of imagery for extracting urban information based on the comparatively easy-to-understand concept of *nominal spatial resolution* (also called *ground-resolved distance*). For example, the Landsat 7 Enhanced Thematic Mapper Plus has six multispectral bands at 30 × 30 m nominal spatial resolution, a 15 × 15 m panchromatic band, and a 60 × 60 m thermal infrared band. The SPOT 3 HRV sensor has three multispectral bands with 20 × 20 m nominal spatial resolution and a single panchromatic band at 10 × 10 m.

Another general, nominal spatial resolution rule is that there needs to be a minimum of four spatial observations (e.g., pixels) within an urban object to identify it. Stated another way, the sensor spatial resolution should be one-half the width of the smallest object of interest. For example, to identify mobile homes that are 5 m wide, the minimum spatial resolution of imagery needed without haze or other problems is ≤ 2.5 × 2.5 m pixels (Jensen and Cowen, 1999).

It is possible to identify *urban/builtup* land cover versus all other land cover for regional or global applications using coarse spatial resolution satellite remote sensor data. For example, Giri et al. (2005) compared the agreement between the Global Land Cover 2000 dataset derived from the French SPOT 1 × 1 km Vegetation sensor data, and the MODIS Global Land Cover dataset derived from MODIS 250 and 500 m data resampled to 1 × 1 km. Unfortunately, the percent agreement for the *urban/builtup* land-cover class at the per pixel level was poor in 15 of the 16 biomes examined.

The temporal, spectral, and spatial resolution requirements for selected urban attributes summarized in Table 13-1 and Figure 13-2 were synthesized from practical experience reported in journal articles, symposia, chapters in books, and government and society manuals [e.g., Avery and Berlin, 1992; Light (1993, 1996); Philipson, 1997; Haack et al., 1997; Jensen and Cowen, 1999; Donnay et al., 2001; Jensen et al., 2002, Jensen and Hodgson, 2004; Jensen, 2005; Pike, 2006 and others in the individual sections]. Hopefully, there will always be a remote sensing system available that can obtain images of the terrain that satisfy the urban attributes' resolution requirements (Table 13-1).

Remote Sensing Land Use and Land Cover

The term *land use* refers to how the land is being used by human beings. *Land cover* refers to the biophysical materials found on the land. For example, a state park may be used for recreation but have a coniferous forest cover. Urban land-use or land-cover information is required for a great variety of applications including residential-industrial-commercial site

Table 13-2. Four levels of the *USGS Land-Use/Land-Cover Classification System for Use with Remotely Sensed Data* and representative types of remotely sensed data typically used to provide the information.

Classification Level	Remote Sensor Data Characteristics
I	NOAA AVHRR (1.1 × 1.1 km), MODIS (250 × 250 m; 500 × 500 m), Landsat MSS (79 × 79 m), Thematic Mapper (30 × 30 m), Indian LISS (72.5 × 72.5 m; 36.25 × 36.25 m; 23.5 × 23.5 m), SPOT Vegetation (1 × 1 km), RADARSAT (100 × 100 m), aerial photography 1:120,000 to 1:240,000 scale.
II	Landsat 7 Enhanced Thematic Mapper Plus pan (15 × 15 m), SPOT HRV pan (10 × 10 m), SPOT HRV XS (20 × 20 m), Indian IRS pan (5.8 × 5.8 m), GeoEye IKONOS (1 × 1 m pan; 4 × 4 m multispectral), RADARSAT (11 × 9 m), aerial photography 1:60,000 to 1:120,000 scale.
III	Indian IRS pan (5.8 × 5.8 m), GeoEye IKONOS (1 × 1 m pan), DigitalGlobe QuickBird pan (0.61 × 0.61 m), OrbView 3 (1 × 1 m pan), aerial analog or digital photography 1:20,000 to 1:60,000 (1 – 3 m).
IV	DigitalGlobe QuickBird pan (0.61 × 0.61 m), aerial analog or digital photography 1:6,000 to 1:20,000 (0.25 – 1 m).

selection, population estimation, tax assessment, development of zoning regulations, etc. (Cullingworth, 1997; Wilson et al., 2003b). This means that urban information collected for one application might be useful in another.

Land-Use/Land-Cover Classification Schemes

The best way to ensure that urban information derived from remote sensor data is useful in many applications is to organize it according to a standardized *land-use* or *land-cover classification scheme* (Chen and Stow, 2003).

American Planning Association "Land-Based Classification Standard"

The most comprehensive hierarchical classification system for urban/suburban landuse is the *Land-Based Classification Standard* (LBCS) developed by the American Planning Association (2006). This standard updates the 1965 *Standard Land Use Coding Manual*, which is cross-referenced

with the *1987 Standard Industrial Classification* (SIC) *Manual* and the updated *North American Industrial Classification Standard* (NAICS, 2006). The LBCS requires extensive input from *in situ* data collection, aerial photography, and satellite remote sensor data to obtain information at the parcel level on the following five characteristics: activity, function, ownership, site, structure (Figure 13-5) (American Planning Association, 2006). The system provides a unique code and description for almost every commercial and industrial land-use activity. The LBCS does not provide information on land-cover or vegetation characteristics in the urban environment, as it relies on the Federal Geographic Data Committee standards on this topic. Users are encouraged to keep abreast of the LBCS and to utilize it for intensive urban studies that require detailed commercial/industrial classification codes.

U.S. Geological Survey "Land-Use/Land-Cover Classification System for Use with Remote Sensor Data"

The U.S. Geological Survey *Land-Use/Land-Cover Classification System* (Anderson et al., 1976; USGS, 1992) was specifically designed to be resource-oriented (land cover) in contrast with various people or activity (land-use) oriented systems, such as the *Standard Land-Use Coding Manual*. The USGS rationale was that "although there was an obvious need for an urban-oriented land-use classification system, there was also a need for a resource-oriented classification system whose primary emphasis would be the remaining 95 percent of the United States land area." The USGS classification system was designed to be driven primarily by the interpretation of remote sensor data obtained at various scales and resolutions (Figure 13-4; Table 13-2) and not data collected *in situ*. The USGS system addresses this need with eight of the nine level I categories treating land area that is not in urban or *built-up* categories (Tables 13-3 to 13-6). The classification system was initially developed to include land-use data that was visually photointerpreted, although it has been widely used for digital multispectral remote sensing classification studies.

While the USGS *Land-use/Land-cover Classification System* was not originally designed exclusively to incorporate detailed urban attribute information, it nevertheless has been used extensively for more than two decades for urban land-use studies. This is usually performed by embellishing the classification system with detailed Level III, IV, and V urban class definitions such as those shown in Tables 13-4 to 13-6. The utility of the modified system is that it may be adapted to include as many levels as desired yet is upwardly compatible with all the USGS Level I and II land-use and land-cover data compiled by neighboring cities, counties, states,

American Planning Association Land-Based Classification System

Table 1: Activity

Parcel ID	Activity	Description
10-5-100	2100	shopping
10-5-100	2200	restaurant
10-5-100	6600	social, religious
10-5-100	2100	furniture
10-5-100	5210	vehicular parking
etc.	etc.	etc.

Table 2: Function

Function	Description
2110	retail sales and services
2510	full-service restaurant
6620	religious institutions
2121	furniture
5200	parking facilities
etc.	etc.

Table 3: Structure

Parcel ID	Structure	Description
10-5-100	2500	mall, shopping center

Figure 13-5   The American Planning Association developed the *Land-Based Classification System* (LBCS) that contains detailed definitions of urban/suburban landuse. The system incorporates information derived *in situ* and using remote sensing techniques. This is an oblique aerial photograph of a mall in Ontario, CA. Hypothetical activity and structure codes associated with this large parcel are identified. Site development and ownership information attribute tables are not shown (concept courtesy American Planning Association).

or nations. For example, Table 13-7 documents how the *Land-Use/Land-Cover Classification System for Use with Remote Sensor Data* has been modified to include 22 classes for the *National Land-Cover Characterization* based on continuous production of a U.S. National Land-Cover Dataset (NLCD) derived from Landsat Thematic Mapper or other data. Many of the Level II classes are best derived using aerial photography (USGS, 2006).

The same classes are also used by the NOAA Coastal Services Center in the creation of the Coastal Change Analysis Program (C-CAP) products. NOAA C-CAP products include land cover for the most current date available, a five-year retrospective land-cover product, and a product that documents changes between the two dates (NOAA, 2004).

Urban Land-Use/ Land-Cover Classification (Levels I to IV) Using Remote Sensor Data

The general relationship between USGS land-cover classification system levels (I – IV) and the nominal spatial resolution of the sensor system (ground-resolved distance in meters) was presented in Figure 13-4. Generally, USGS Level I classes may be inventoried effectively using sensors with a nominal spatial resolution of 20 – 100 m such as the Landsat Multispectral Scanner (MSS) with 79 × 79 m nominal spatial resolution, the Thematic Mapper (TM) at 30 × 30 m, SPOT HRV XS at 20 × 20 m, and Indian LISS 1-3 (72.5 × 72.5 m; 36.25 × 36.25 m; 23.5 × 23.5 m, respectively). Of course, any sensor system with higher spatial resolution can

Table 13-3. U.S. Geological Survey *Land-Use/Land-Cover Classification System for Use with Remote Sensor Data* (Anderson et al., 1976).

Classification Level

1 **Urban or Built-up Land**
  11 Residential
  12 Commercial and Services
  13 Industrial
  14 Transportation, Communications, and Utilities
  15 Industrial and Commercial Complexes
  16 Mixed Urban or Built-up
  17 Urban or Built-up Land

2 **Agricultural Land**
  21 Cropland and Pasture
  22 Orchards, Groves, Vineyards, Ornamental Horticulture
  23 Confined Feeding Operations
  24 Other Agricultural Land

3 **Rangeland**
  31 Herbaceous Rangeland
  32 Shrub–Brushland Rangeland
  33 Mixed Rangeland

4 **Forest Land**
  41 Deciduous Forest Land
  42 Evergreen Forest Land
  43 Mixed Forest Land

5 **Water**
  51 Streams and Canals
  52 Lakes
  53 Reservoirs
  54 Bays and Estuaries

6 **Wetland**
  61 Forested Wetland
  62 Nonforested Wetland

7 **Barren Land**
  71 Dry Salt Flats
  72 Beaches
  73 Sandy Areas Other Than Beaches
  74 Bare Exposed Rock
  75 Strip Mines, Quarries, and Gravel Pits
  76 Transitional Areas
  77 Mixed Barren Land

8 **Tundra**
  81 Shrub and Brush Tundra
  82 Herbaceous Tundra
  83 Bare Ground Tundra
  84 Wet Tundra
  85 Mixed Tundra

9 **Perennial Snow or Ice**
  91 Perennial Snowfields
  92 Glaciers

Table 13-4. Urban Residential, Commercial and Services – Level I and II are from the *USGS Land-Use/Land-Cover Classification System for Use with Remote Sensor Data*. Levels III, IV, and V are logical extensions.

Classification Level

1 Urban or Built-up Land

11 Residential
  111 Single-Family Residential
    1111 House, houseboat, hut, tent
    1112 Mobile home
  112 Multiple-Family Residential
    1121 Duplex
    1122 Triplex
    1123 Apartment complex or Condominium ($\geq$ 4 attached units)
    1124 Mobile home (trailer) park

12 Commercial and Services
  121 Commercial
    1211 Automotive (cars, trucks, motorcycles)
      12111 Dealership
      12112 Gas station
      12113 Junkyard
      12114 Service or repair
    1212 Boat
      12121 Dealership
      12122 Marina
      12123 Service or repair
    1213 Department Store
      12131 Isolated department store (e.g., WalMart, K-Mart)
      12132 Mall complex
    1214 Financial and Construction
      12141 Banking, Financial Services
      12142 Insurance
      12143 Real estate
      12144 Construction
    1215 Food and Drug
      12151 Drugstore
      12152 Market
      12153 Restaurant
    1216 Funeral
      12161 Mortuary
      12162 Cemetery-mausoleum
    1217 Housing (temporary)
      12171 Hotel
      12172 Motel
      12173 Campground
    1218 House & Garden
    1219 Recreation (commercial and public)
      12191 Amusement park (e.g., Go-carts, miniature golf)
      12192 Baseball
      12193 Football/soccer/track
      12194 Golf
      12195 Swimming
      12196 Stadiums (multipurpose)
      12197 Tennis
      12198 Theater (includes multiplex and drive-in)
    1220 Utility
      12201 Gas – electric
      12202 Water – sewer
      12203 Waste disposal [solid (e.g., landfill) or liquid]
    1221a Warehousing/shipping
  122 Services (Public and Private)
    1221 Public Buildings and Facilities
      12211 Administration
      12212 Education
        122121 Elementary
        122122 Middle
        122123 High School
        122124 University or college
      12213 Fire, Police, Rescue
      12214 Health and fitness centers
      12215 Library
      12216 Post Office
      12217 Parks
    1222 Medical
      12221 Hospital
      12222 Offices
    1223 Religion (churches)

Table 13-5.   Urban Industrial Land-use Classification – Level I and II are from the *USGS Land-Use/Land-Cover Classification System for Use with Remote Sensor Data.* Levels III, IV, and V are logical extensions.

Classification Level

13 Industrial
  131 Extraction
    1311 Dredging
      13111 Sand and gravel
      13112 Placer mining for gold, platinum, and diamonds
    1312 Open pit mine
      13121 Copper
      13122 Gravel pit – rock quarry
      13123 Iron ore
      13124 Salt
    1313 Shaft mine
      13131 Bauxite
      13132 Coal
      13133 Lead
      13134 Limestone
    1314 Strip mining
      13141 Coal
      13142 Oil shale
    1315 Well
      13151 Petroleum
      13152 Natural gas
      13153 Water
    1316 Cutting field
      13161 Forest clear-cutting or high-grading
    1317 Dredging or mining spoil piles
  132 Processing
    1321 Mechanical
      13211 Agriculture grain mill
      13212 Aluminum
      13213 Hydroelectric power plant
      13214 Ore concentration
      13215 Sewage treatment
      13216 Sugarbeet refining
      13217 Textile
      13218 Timber sawmill
      13219 Water treatment and purification
    1322 Chemical
      13221 Chlorine and caustic soda
      13222 Fertilizer
      13223 Hazardous-waste disposal
      13224 Petroleum refinery
      13225 Pharmaceutical
      13226 Magnesia – magnesium
      13227 Textile (synthetic)
      13228 Wood treatment
    1323 Heat
      13231 Calcium carbide
      13232 Cement
      13233 Clay products
      13234 Copper smelting
      13235 Glass
      13236 Iron and steel
      13237 Lead
      13238 Lime
      13239 Nuclear facilities
      13240 Thermal electric power plant
      13241 Zinc
  133 Fabrication
    1331 Heavy
      13311 Heavy machinery (bulldozers, etc.)
      13312 Railroad machinery manufacture and repair
      13313 Shipbuilding
      13314 Structural steel fabrication
    1332 Light
      13321 Aircraft assembly
      13322 Automotive assembly (e.g., cars, trucks, motorcycles)
      13323 Boat building and repair
      13324 Canneries
      13325 Electronics
      13326 Housing (prefabricated)
      13327 Meat packing
      13328 Plastic products

Table 13-6.   Urban Transportation, Communications, Utilities, and Industrial/Commercial Complexes – Level I and II classes are from the *USGS Land-Use/Land-Cover Classification System for Use with Remote Sensor Data.* Levels III, IV, and V are logical extensions.

Classification Level

14 Transportation, Communications, and Utilities
  141 Transportation
    1411 Roads and Highways
      14111 Dirt
      14112 Paved
      14113 Limited access (freeway, toll)
      14114 Interchange
      14115 Parking
      14116 Bridge
    1412 Railroad
      14121 Track
      14122 Marshalling (classification) yard
      14123 Terminal
      14124 Bridge
    1413 Airport
      14131 Runway, tarmac
      14132 Hangar
      14133 Terminal
    1414 Boats/Ships
  142 Communications
    1421 Sign, billboard
    1422 Radio/Television/Cable
  143 Utility Facilities
    1431 Electricity
    1432 Natural gas
    1433 Petroleum
    1434 Water

15 Industrial and Commercial Complexes
  151 Industrial complex (park)
  152 Commercial complex (mall)

16 Mixed Urban or Built-up Land

17 Other Urban or Built-up Land

be used to acquire Level I data. Color Plate 13-2 presents Level I urban vs. non-urban land-cover information for three counties centered on Charleston, SC, extracted from Landsat MSS data acquired on six dates from February 14, 1975, to February 3, 1994. For global projects, Level I landcover may be extracted from more coarse spatial resolution imagery (e.g., AVHRR at 1.1×1.1 km; MODIS 250×250 m or 500×500 m) (e.g., Giri et al., 2005).

Typical urban spectral reflectance signatures are summarized in Figure 13-6. Urban land cover typically appears "steel–gray" on color-infrared color-composite images because the urban terrain, consisting primarily of concrete and asphalt roads, parking areas, shingles, and bare soil, typically reflects high proportions of the incident green, red, and near-infrared radiant flux. The steel-gray urban signature contrasts sharply with vegetated surfaces that appear bright red or magenta on color-infrared images because the vegetation reflects substantial amounts of near-infrared

Table 13-7. U.S. Geological Survey *Land-Use/Land-Cover Classification System for Use with Remote Sensor Data* modified for the National Land-cover Dataset and the NOAA Coastal Change Analysis Program (NOAA, 2006).

Classification Level

1 Water
   11 Open Water
   12 Perennial Ice/Snow

2 Developed
   21 Low-Intensity Residential
   22 High-Intensity Residential
   23 Commercial/Industrial/Transportation

3 Barren
   31 Bare Rock/Sand/Clay
   32 Quarries/Strip Mines/Gravel Pits
   33 Transitional

4 Forested Upland
   41 Deciduous Forest
   42 Evergreen Forest
   43 Mixed Forest

5 Shrubland
   51 Shrubland

6 Non-Natural Woody
   61 Orchards/Vineyards, Other

7 Herbaceous Upland Natural/Seminatural Vegetation
   71 Grasslands/Herbaceous

8 Herbaceous Planted/Cultivated
   81 Pasture/Hay
   82 Row Crops
   83 Small Grains
   84 Fallow
   85 Urban/Recreation
   86 Grasses

9 Wetland
   91 Woody Wetlands
   92 Emergent Herbaceous Wetlands

energy while absorbing much of the incident green and red wavelength energy, as discussed in Chapter 4. Water absorbs most of the incident radiant flux, causing it to appear dark on color-infrared imagery, easily distinguishable from the urban landscape. Color composites of downtown Charleston, SC, produced using many of the major sensor systems, are provided in Chapter 7.

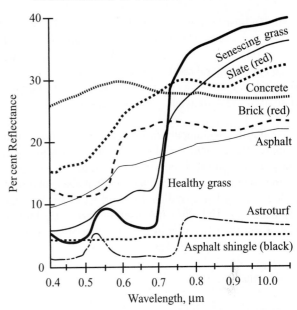

Laboratory Spectroradiometer Reflectance Characteristics of Common Urban Materials

Figure 13-6 Percent reflectance curves for common materials found in urban environments. The reflectance spectra were obtained in a controlled laboratory environment using a GER 1500 spectroradiometer.

Sensors with a minimum spatial resolution of 5 – 20 m are generally required to obtain Level II information. SPOT 5 sensor systems provide 10×10 m multispectral and 5×5 m panchromatic data. RADARSAT provides 11×9 m spatial resolution data for Level I and II land-cover inventories even in cloud-shrouded tropical landscapes. The Landsat 7 ETM[+] has a 15×15 m panchromatic band.

More detailed Level III classes may be inventoried using a sensor with a spatial resolution of approximately $1 – 5$ m such as: IRS-1CD pan (5.8×5.8 m data resampled to 5×5 m); large-scale aerial photography; IKONOS (1×1 m pan; 4×4 m multispectral); QuickBird (0.61×0.61 m pan; 2.44×2.44 m multispectral); and OrbView 3 (1×1 m pan; 4×4 m multispectral). The synergistic use of high spatial resolution panchromatic data (e.g., 1×1 m), merged with lower spatial resolution multispectral data (e.g., 4×4 m), provides an image-interpretation environment that is superior to using panchromatic data alone (Gamba et al., 2003; Jensen, 2005).

Level IV classes and cadastral (property-line) information is best monitored using high spatial resolution panchromatic sensors, including QuickBird 0.61×0.61 m panchromatic data, stereoscopic analog aerial photography ($\leq 0.25 – 1$ m) and stereoscopic digital frame camera imagery [e.g., Leica GeoSystems Airborne Digital Sensor ADS40 ($\leq 5 \times 5$ cm)].

Urban land-use and land-cover classes in Levels I through IV have temporal attribute requirements ranging from $1 - 10$ years (Table 13-1 and Figure 13-2). All of the sensors mentioned have temporal resolutions of < 55 days, so the temporal resolution of the land-use/land-cover attributes is satisfied by the current and proposed sensor systems.

As previously mentioned, urban landscapes are notoriously complex (Ridd, 1995; Small, 2001; 2004). While simple binary urban/non-urban information is valuable for assessing growth trends (e.g., Color Plate 13-2), it does not provide the level of land-use/land-cover detail necessary to make wise decisions. Therefore, it is useful to provide additional detail about the extraction of the following urban land-use information from remote sensor data:

- residential housing

- commercial and services

- industrial

- transportation infrastructure

- communications and utilities.

 Residential Land Use

The *home* in almost all cultures is the single most important possession a person acquires in a lifetime. It shelters the family and usually represents the household's greatest economic investment. Inventorying the location, type, condition, and number of residences is one of the most important tasks performed using remote sensor data. Discriminating between Level II residential, commercial, industrial, and transportation land-uses requires a sensor system with a spatial resolution from $5 - 20$ m. Identifying Level III residential land-use classes requires a spatial resolution from $1 - 5$ m (Table 13-1; Figure 13-4).

Prior to investigating how residential land use appears on remotely sensed data, it is important to introduce the concept of *form* and *function*. Basically, the function of a building often dictates its form. In fact, we often say "form follows function" or "form is dictated by function." For example, the need to house a single family often results in the creation of

a detached dwelling composed of several attached rooms (this is not true in the central business district). Conversely, a company producing finished steel typically requires a very large, linear building (form) that can house the various steel processing activities (function). An analyst must be careful, however, because a building's use may change over time. Thus, a knowledge of the sequence-of-occupance of buildings in a region becomes important.

Single-Family versus Multifamily Residential

A *single-family residence* is detached from any other housing structure. In developed countries it usually has a single driveway, a sidewalk leading to the front door, a front yard, a backyard (often fenced), and a garage or carport. It is usually ≤ 3 stories in height. Several single-family homes in Richmond, CA, are in the process of being framed in Figure 13-7a. Completely landscaped single-family residences in Boca Raton, FL, and San Diego, CA, are shown in Figure 13-7b,c. Mobile homes (trailers) in developed countries are usually much smaller than other single-family homes (approximately 5 m single-wide; 10 m double-wide), rarely have a garage (but may have a carport), and may or may not have a paved driveway or sidewalk (Figure 13-7d).

In developing countries, single-family residences may consist of a building, hut, tent, or lean-to depending upon the culture. For example, traditional housing associated with a pastoral village in Kenya is shown in Figure 13-8a. It may be permanent or temporary (seasonal). The structures may contain individual or extended families. Sometimes it is difficult to identify such structures in developing countries because they are made of the same materials found in the surrounding countryside, resulting in low object-to-background contrast (Jensen et al., 2002).

In developed countries, *multiple-family residences* usually have more than one sidewalk, may have large collective aboveground or belowground parking garages or parking lots, may be ≥ 2 stories in height, and share front and back yards (Figure 13-9a). There may be a community pool and/or tennis court. *Duplexes* (two attached housing units) and *triplexes* (three attached housing units) usually have two or three walkways and driveways, respectively. Sometimes a common parking area is provided. It is often difficult to determine how many families occupy single residences in developing countries. When there are more than three attached housing units, it is usually called an apartment or condominium (Figure 13-9b,c). Such complexes often occupy an extensive area of land, rise many stories above the ground, and in large cities may be high-rise buildings.

a. Several single-family residences being framed in Richmond, CA (level 1111) (courtesy Cris C. Benton).

b. A single-family residence in Boca Raton, FL, with one sidewalk, one driveway, and a fenced backyard (level 1111).

c. Closely spaced single-family residences in San Diego, CA, with one sidewalk, one driveway, and occasionally a small swimming pool (level 1111).

d. Single-family trailers (mobile homes) in a trailer park near Myrtle Beach, SC (level 1112).

Figure 13-7   Examples of single-family residential housing (level 111).

Nairobi, Kenya

a. Traditional Kenyan village with individual dwellings surrounded by a communal livestock corral.

b. Multiple-family informal (squatter) housing on the outskirts of Nairobi, Kenya.

Figure 13-8   Examples of single- and multiple-family residential housing in a developing country.

In developing countries, rural people flock to the city to hopefully obtain an improved quality of life only to find few jobs and no available housing (Da Costa and Cintra, 1999). This often results in the creation of *informal* (squatter) *developments* where multiple families live in buildings constructed with any materials that can be found (Jensen et al., 2002). An informal settlement on the outskirts of Nairobi, Kenya, is shown in Figure 13-8b.

It is usually possible to determine if a building is a single or multiple-family residence using imagery with a spatial resolution from 1 – 5 m. Sometimes it is necessary to merge multiple types of imagery such as Landsat Thematic Mapper and radar imagery to identify villages in tropical parts of the world (Haack and Slonecker, 1994).

On numerous occasions visual, heads-up on-screen image interpretation of urban structure outperforms computer-assisted digital image processing (Jensen, 2005). Visual image interpreters are able to take into account the *site, situation*, and *association* of structures. These elements of image interpretation have been very difficult to incorporate into analytical methods of digital image classification. Only now are analytical neural network-based image interpretation systems beginning to include the fundamental elements of visual image interpretation into the image-analysis process (Jensen et al., 2001; Qiu and Jensen, 2004). For example, Figure 13-10 depicts the user interface associated with a neural network-based digital image processing system that

assists an image interpreter to arrive at the correct land-cover or land-use classification for a particular parcel of land. In the past, the interpreter had to rely on complex branching dichotomous keys to assist in the interpretation process. Now, the image analyst can draw a polygon around a feature such as the one shown in Figure 13-10a and then enter in the characteristics observed in the polygon of interest using a simple point-and-click menu such as the one shown in Figure 13-10b. The characteristics might include the number of sidewalks, the presence of a driveway and/or garage, parking lots, multiple stories, and general size of the phenomena. The neural network then processes this information and reports back to the image analyst what it believes is the most appropriate land-use or land-cover code. The neural network is trained using real-world empirical examples, and in effect it *learns*. The neural network classification system can also be used to photointerpret other land uses, including commercial, services, and industrial (Jensen et al., 2001). Neural network or expert system-assisted interpretation systems will most likely be the means in the future by which most detailed urban information is eventually extracted from digital remote sensor data (Jensen and Hodgson, 2004).

Building and Cadastral (Property-Line) Infrastructure

In addition to fundamental nominal scale land-use and land-cover information (i.e., identifying whether an object is a

a. Condominium residential housing in Colorado (level 1123). The number of driveways, sidewalks, and roof structures indicates how many housing units are present. The top-left condo is a fourplex. The top right is a sevenplex. The bottom two condos are fiveplex.

b. Two-story apartment complex in Columbia, SC (level 1123), with numerous sidewalks and parking around the perimeter.

c. Twelve-story student-housing apartment complex at the University of South Carolina (level 1123).

Figure 13-9 Examples of multiple-family residential housing (level 112).

**Neural Network
Image Interpretation System**

a. User-interface.

b. Data collection interface.

Figure 13-10 a) The user-interface of a Neural Network Image Interpretation System that assists an image analyst in the classification of difficult urban landuse. In this example, the analyst used visual on-screen digitizing to outline an unknown object of interest. b) The interpreter then used a point-and-click dialog box to identify primitive attributes
within the object of interest, such as the existence of parking lot(s), sidewalks, etc. The empirical information is
passed to the neural network, which then compares this pattern with previous patterns (actually weights) and attempts to identify the unknown object as a multiple family residential housing complex (Jensen et al., 2001).

single-family residence or a commercial building), transportation planners, utility companies, tax assessors, and others
require more detailed information on building footprint
perimeter, area, height, driveways, patios, fences, pools,
storage buildings, and the distribution of landscaping every
one to five years (Table 13-1).

It is possible to identify the presence of individual buildings
using monoscopic images and several elements of image
interpretation such as size, shape and context (Tullis and
Jensen, 2003; Jensen et al., 2005). However, detailed building footprints and property lines are best obtained using stereoscopic (overlapping) panchromatic aerial photography or
other remote sensor data with a spatial resolution of ≤ 0.25 –
0.5 m (e.g., Ehlers et al., 2003). For example, panchromatic
stereoscopic aerial photography with 0.25 × 0.25 m (1 ft)
spatial resolution was used to extract the exact dimensions
of individual houses and outbuildings, trees, pools, driveways, fences, and contours for the single-family residential
area in Figure 13-11a,b. In many instances, the fence lines
are the cadastral property lines. If the fence lines are not visible or are not truly on the property line, the property lines

can be identified by a surveyor in the field and the information overlaid onto the orthophotograph or planimetric map
database to represent the legal cadastral (property) map.
Many municipalities in the U.S. use high spatial resolution
imagery as the source for some of the cadastral information
and as an image backdrop upon which surveyed cadastral
and tax information are portrayed (Cowen and Craig, 2004).

Detailed building height and volume data can also be
extracted from high spatial resolution (0.25 – 0.5 m) stereoscopic imagery (Chapter 6). For example, consider the wire-
frame three-dimensional outline of the United States Capitol
shown in Figure 13-11c. The perimeter, area, volume, and
square footage of this building are now available. Such
detailed building information may allow the exact type of
roof to be identified (e.g., ridge, mansard, hipped, lean-to,
flat, etc.), which may be used to infer the date of building
construction (Lo and Yeung, 2002). Building information
can also be obtained from high spatial resolution satellite
imagery. For example, Fraser et al. (2002) found that buildings could be extracted from stereo IKONOS 1 × 1 m imag-

a. Vertical panchromatic 1:2,400-scale aerial photography.

b. Planimetric (x,y) cadastral information extracted from the stereoscopic aerial photography, including 2-ft contours, building footprints, fences, retaining walls, and large trees (Jensen and Cowen, 1999).

c. Wire-frame information of the U.S. Capitol extracted from large-scale vertical aerial photography, including precise building footprint, height, area, perimeter, and volume. The two photographs may be viewed stereoscopically (courtesy LH Systems, LLC, San Diego, CA, and Marconi Integrated Systems, Inc.; © SOCET Set).

Figure 13-11 Extraction of cadastral information from stereoscopic vertical aerial photography using a) traditional, and b,c) soft-copy photogrammetric techniques (please refer to Chapter 6). Such information is vital to the accurate inventory of residential, commercial, and industrial landuse. It can be acquired using traditional surveying or photogrammetric techniques.

ery with an *x,y* accuracy of 0.3 – 0.6 m and height accuracy of 0.5 – 0.9 m.

Socioeconomic Characteristics Derived from Single- and Multifamily Residential-Housing Information

Selected socioeconomic attributes may be extracted directly from remote sensor data or via surrogate information derived from the imagery. Two of the most important are *population estimation* and *quality of life* indicators, derived primarily from single- and multiple-family residential-housing stock information.

Population Estimation

Knowing how many people live within a specific geographic area or administrative unit (e.g., city, county, state, country) is very powerful information (Jensen et al., 2002). In fact, some have suggested that the global effects of increased population density and ecosystem land-cover conversion may be much more significant than those arising from climate change (Skole, 1994). Population estimation can be performed at the local, regional, and national level based on:

• counts of individual dwelling units (requires 0.25 – 5 m nominal spatial resolution (Jensen et al., 2005);

- measurement of urbanized land areas (often referred to as settlement size) (e.g., Tatem et al., 2004);

- estimates derived from land-use/land-cover classification (Lo, 1995; Sutton et al., 1997; Sutton, 2003).

Remote sensing techniques may provide population estimates that approach the accuracy of traditional census methods if sufficiently accurate *in situ* data are available to calibrate the remote sensing model. Unfortunately, the ground-based population estimations are often woefully inaccurate. In many instances, the remote sensing methods may be superior to the ground-based methods.

The most accurate remote sensing method of estimating the population of a local area is to *count individual dwelling units* based on the following assumptions (Lindgren, 1985; Lo, 1995; Holz, 1988; Haack et al., 1997):

- The imagery must have sufficient spatial resolution to allow identification of individual structures even through sparse tree cover and to determine whether the structures are residential, commercial, or industrial.

- Some estimate of the average number of persons per dwelling unit must be available.

- Some estimate of the number of homeless, seasonal, and migratory workers is required.

- It is assumed all dwelling units are occupied, and only *n* families live in each unit (calibrated using *in situ* information).

This is usually performed every five to seven years and requires high spatial resolution remotely sensed data (0.25 – 5 m). For example, individual dwelling units in Irmo, SC, were extracted from 2.5 × 2.5 m aircraft multispectral data. Correlation of the remote sensing–derived dwelling unit data with Bureau of the Census dwelling unit data for the 32 census block area yielded an $R^2 = 0.81$, which accounted for 81 percent of the variance. These findings suggest that the new high spatial resolution panchromatic sensors may provide a good source of information for monitoring the housing stock of a community on a routine basis. This will enable local governments to anticipate and plan for schools and other services with data that has a much more frequent temporal resolution than the decennial census. This data will also be of value for real estate, marketing, and other business applications. Unfortunately, the dwelling-unit approach is not suitable for a regional/national census of population because it

is too time-consuming and costly (Sutton et al., 1997). Some have suggested that this method requires so much *in situ* data to calibrate the remote sensor data that it can become operationally impractical. Therefore, other methods have been developed.

There is a relationship between the urbanized built-up area (settlement size) extracted from a remotely sensed image and settlement population (Olorunfemi, 1984):

$$r = a \times P^b \qquad (13\text{-}1)$$

where *r* is the radius of the populated area circle, *a* is an empirically derived constant of proportionality, *P* is the population, and *b* is an empirically derived exponent. Estimates of these parameters are fairly consistent at regional scales, but the estimate of the *a* parameter varies between regions. Sutton et al. (1997) used Defense Meteorological Satellite Program Operational Linescan System (DMSP-OLS) visible near-infrared nighttime 1 × 1 km imagery to inventory urban extent for the entire United States (Figure 13-12). When the data were aggregated to the state or county level, spatial analysis of the clusters of the saturated pixels predicted population with an $r^2 = 0.81$ once again. Sutton (2003) used nighttime DMSP-OLS imagery to measure per capita land use consumption as an aggregate index for the spatially contiguous urban areas of the conterminous United States with population of $\geq 50,000$. Tatum et al. (2004) delineated settlements in Kenya using Landsat TM and JERS-1 synthetic aperture radar (SAR) data with the goal of producing medium-scale population maps for improved public health planning.

Another widely adopted population estimation technique is based on the use of the Level I – III land-use information. This approach assumes that land use in an urban area is closely correlated with population density. Researchers establish an empirical value for the population density for each land use by field survey or census data (e.g., multiple-family residential housing may contain five persons per pixel when using 30 × 30 m Thematic Mapper data; rural forested areas might have only 0.20 persons per pixel). Then, by measuring the total area for each land-use category, they estimate the total population for that category. Summing the estimated totals for each land-use category provides the total population projection (Lo, 1995). The urban built-up area and land-use data method can be based on more coarse spatial resolution multispectral remote sensor data (5 – 20 m) every 5 – 15 years. Henderson and Xia (1998) discuss how radar imagery can be of value when conducting population estimates using these same techniques.

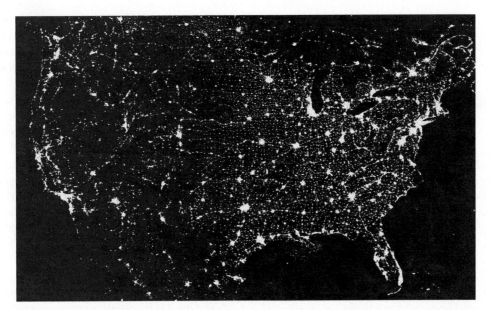

Figure 13-12 Defense Meteorological Satellite Program Operational Linescan System (DMSP-OLS) visible near-infrared nighttime 1×1 km imagery of the conterminous United States. When the data were aggregated to the state or county level, spatial analysis of the clusters of the saturated pixels predicted population with an $R^2 = 0.81$ (Sutton et al., 1997; courtesy American Society for Photogrammetry & Remote Sensing).

Quality-of-Life Indicators

Adequate income, decent housing, good education, health services, and a good physical environment (e.g., no hazards or refuse) are important indicators of social well-being and *quality of life* (Lo and Faber, 1998; Lo and Yeung 2002). Evaluating the quality of life of a population on a continuing basis is important because it helps planners and government agencies involved with the delivery of human services to be aware of problem areas.

Most quality-of-life studies have been based on the analysis of socioeconomic information derived from *in situ* enumeration. Remotely sensed images have rarely been used in social studies. Only recently have factor analysis studies documented how quality of life indicators (such as house value, median family income, average number of rooms, average rent, and education) can be estimated by extracting the urban attributes found in Table 13-8 from relatively high spatial resolution (0.25 – 30 m) imagery (Lindgren, 1985; Holz, 1988; Avery and Berlin, 1992; Haack et al., 1997; Lo and Faber, 1998; Da Costa and Cintra, 1999; Lo and Yeung, 2002; Jensen et al., 2005). Note that the attributes are arranged by *site* (building and lot) and *situation*. The site may be situated in positive and negative surroundings.

These remote sensing–derived attributes must be correlated with *in situ* census observations to compute the quality-of-

life indicators. Lo and Faber (1998) suggest that in many cities green vegetation is very precious and costly, and that only the wealthy can afford to include grass and trees in their living environment. They found that greenness information derived from TM data was highly correlated with quality of life. Quality-of-life indicators are usually collected every 5 – 10 years.

Environmental Influences Associated with Zoning

Zoning is a tool urban planners use to control physical characteristics of developing landscapes by imposing restriction on variables such as maximum building height and density, extent of impervious surface and open space, and land use types and activities. These variables, in turn, influence environmental processes of atmosphere/surface energy exchange, surface and subsurface hydrologic systems, and micro- to mesoscale weather and climate regimes. Wilson et al. (2003b) found that there were significant differences in radiant surface temperature and the Normalized Difference Vegetation Index (NDVI) derived from Landsat 7 ETM[+] imagery for different types of zoning in Indianapolis, IN. This suggests that the integration of a city's GIS zoning data with remote sensing measurements of surface temperature and NDVI can help planners gain insight into the environmental consequences of zoning decisions.

Table 13-8.　Urban/Suburban attributes that may be extracted from remote sensor data and used to assess housing quality and/or quality of life.

Factor	Attributes
Site	**Building** • single or multiple-family • size (m²) • height (m) • age (derived by convergence of evidence) • garage (attached, detached) **Lot** • size (m²) • front yard (m²) • backyard (m²) • street frontage (m) • driveway (paved, unpaved) • fenced • pool (inground, aboveground) • patio, deck • outbuildings (sheds) • density of buildings per lot • percent landscaped • health of vegetation (e.g., NDVI greenness) • fronts paved or unpaved road • abandoned autos • refuse
Situation	**Adjacency to Community Amenities** • schools • shopping • churches • hospitals • fire station • open space, parks, golf courses • marsh **Adjacency to Nuisances or Hazards** • heavy street traffic • railroad or switchyard • airports and/or flightpath • freeway • located on a floodplain • sewage-treatment plant • industrial area • power plant or substation • overhead utility lines • steep terrain

Energy Demand and Conservation

Local urban/suburban energy demand (sometimes referred to as heat-load density) may be estimated using remotely sensed data. First, the square footage (or m²) of individual buildings is determined from high spatial resolution imagery. Local ground-reference information about energy consumption is then obtained for a representative sample of homes in the area. Regression relationships are then derived to predict the anticipated energy consumption for the region. It is also possible to predict how much solar photovoltaic energy potential a geographic region has by modeling the individual rooftop square footage and orientation with known photovoltaic generation constraints. Both applications require high spatial resolution imagery (0.25 – 1 m) (Curran and Hobson, 1987; DOE, 2005). Regional and national energy consumption may also be predicted using DMSP imagery (e.g., Elvidge et al., 1997; Sutton et al., 1997).

Numerous studies have documented how high spatial resolution (1 – 5 m) predawn thermal infrared imagery (8 – 14 μm) can be used to inventory the relative quality of housing insulation if 1) the rooftop material is known (e.g., asphalt versus wood shingles), 2) no moisture is present on the roof, and 3) the orientation and slope of the roof are known (Eliasson, 1992). If energy conservation or the generation of solar photovoltaic power were important in countries, these variables would probably be collected every one to five years.

Active and passive use of solar energy can be used to reduce electrical energy consumption in buildings and thereby reduce the consumption of fossil fuels. Hammer et al. (2003) demonstrated how the amount of surface solar irradiance can be determined from geostationary satellite remote sensor data and then used in a predictive model (HELIOSAT) for solar energy predictions. The goal is to provide near real-time intelligent building temperature-control techniques that take into account systems that use solar energy (e.g., solar panels, skylights).

 Commercial and Services Land Use

People engage in commerce (business) to provide food, clothing, and shelter for themselves and their extended families. Many work in the service industries providing education, medical, sanitation, fire and police protection, etc., to improve the quality of life for the entire community. Interestingly, many of these activities are visible in the landscape. Commercial and service activities often have a unique *cultural signature* defined as the assemblage of specific materials and structures that characterize certain activities in a particular culture. Each culture has commerce and services that are unique to that culture. For example, there are many types of commerce found in Asiatic urban environments that simply are not found in Western urban landscapes. Therefore, an interpreter must understand well the cultural milieu of the country of interest before he or she attempts to inventory and map the commerce and service activities using remote sensor data. In fact, if an interpreter has not actually seen the phenomena on the ground and does not understand completely the cultural significance and purpose of the commerce or activity, then it is unlikely that a correct interpretation will be made (Jensen et al., 2002).

The author extended the *USGS Land-Use/Land-Cover Classification System* in Tables 13-4 to 13-6 to include Level IV and sometimes Level V classes. This extension is primarily for U. S. and Western European landscapes but may be generally applicable to other developed nations. It represents a common-sense hierarchical system that can be aggregated back to Levels I and II. It is not complete in content; rather, it reflects an attempt to categorize the land use that accounts for approximately 95 percent of the urban land use encountered in most developed cities. It uses common terminology whenever possible. A detailed definition of each class is not provided here. However, it is instructive to provide numerous examples of the various types of land use and comment on unique characteristics that allow one type of commerce or service to be distinguished from others.

The Central Business District

The *central business district* (CBD) of many cities is the most distinctive urban landscape with its high-rise buildings and the highest density of land use. Many CBDs have changed considerably in recent years, particularly in the United States. Urban geographers use the term *deindustrialization* to describe the loss of manufacturing industries in cities today, and the rise of *tertiary* and *quaternary* commercial activities, notably financial services, insurance, and real estate. There are also many other professional services that characterize the modern cities and the phenomenon of *suburbanization*. Downtown Seattle, WA, is an excellent example of a central business district (Figure 13-13). Numerous high-rise office buildings house the financial services, insurance, and real estate firms. However, some of the towers also contain single- and multiple-family residences.

The CBDs of some cities in the United States have been 1) transformed into poverty areas, 2) gentrified, meaning that only older people live in them and that very few young families are present, and/or 3) reconstructed with new buildings

Figure 13-13 A portion of the *central business district* (CBD) of downtown Seattle, WA. The region has been deforested. It is primarily composed of impervious high-rise buildings that contain financial services, insurance, and real estate companies. Numerous multiple-story apartment complexes are also present. A rectangular, systematically surveyed transportation system serves the area. A new office building is under construction in the top-left portion of the image. This is a 1×1 m ADAR 5500 digital frame camera image (courtesy Positive Systems, Inc.).

to attract professional middle-class people and their small families (usually ≤ 2 children) to return and stay. These and other conditions make it difficult to identify the exact land use of a particular parcel of land in the central business district. *In situ* observation is usually required in addition to the analysis of remotely sensed data.

Commercial Land Use

Automotive and Boat

The developed world and especially the United States is enamored with the automobile and other forms of personal transportation, including trucks, sport utility vehicles, recreation vehicles, and boats. Much of our disposable income is spent acquiring and maintaining these vehicles. Therefore, the urban landscape is littered with dealerships to sell them, service stations to fill them up with petroleum distillate

a. Automobile dealership in Columbia, SC (level 12111). Cars for sale are arranged systematically for buyer inspection and ease of exit and return.

b. Typical automotive service station with multiple islands for dispensing gasoline or diesel fuel and a main building consisting of service bays or mini-market (level 12112). The circular tops of the underground petroleum storage tanks are visible in this example.

Figure 13-14 Examples of commercial automobile establishments (level 1211).

products, and businesses that repair them. Finally, their remains can often be seen in automobile junkyards.

New and used car, truck, and boat dealerships are relatively easy to identify due to the myriad of vehicles or boats parked systematically in large parking lots, the presence of large showrooms, and attached service centers (Figure 13-14a). Isolated service and repair businesses are more difficult to identify but usually have smaller parking lots with many cars present in the lot in various states of repair.

Gas service stations often have a separate island structure with its own canopy where the fuel is dispensed, a detached building that is usually a mini-market, and occasionally another building where automobiles are washed. Gas stations are often located on corner lots (Figure 13-14b). Numerous cars are often located around the periphery of the service stations.

Junkyards have a large number of vehicles arranged in a haphazard fashion, many missing hoods and trunks. Also, grass may be growing between the cars, making it straightforward to distinguish it from a dealership.

Department Stores

Department stores (e.g., WalMart, K-Mart, Sears-Roebuck) contain many types of products located within a single building (e.g., sporting goods, garden, clothing, market, toys). The building usually is one of the largest (m^2) in the region (Figure 13-15 location *a*). A very large parking lot is adjacent to it. Isolated department stores often have storage locations in the parking lot where carts can be returned.

Conversely, a mall is a collection of multiple department stores (Figure 13-15 locations *bcd*). It is usually the largest building complex in the region (m^2). The department stores within the mall are usually connected by a central walkway that is often enclosed. Skylights are often located above the central walkway. The mall is often completely encircled with single or multiple levels of parking, with perhaps some underground parking. Department stores and malls are often located at the junction of major highways. They are usually straightforward to identify because of their extreme size and situation.

Finance and Construction

There are often a tremendous number of banks, brokerage firms, insurance agencies, construction firms, and real estate offices in the central business district of a major city (e.g., refer to the image of Seattle, WA, in Figure 13-13). Parking may be visible aboveground, in multiple-story aboveground parking structures, or underground. For example, consider the Affinity Office Building located in downtown Columbia, SC (Figure 13-16a). It is the tallest building in the city and

Figure 13-15 Example of department store and mall commercial land use in some developed countries. An isolated Toys-R-Us, Inc., department store (a) is independent of the mall (level 12131). The mall commercial complex (level 12132) is anchored by three
major chains: Sears-Roebuck, Inc. (b), Belk, Inc. (c), and Dillards, Inc. (d). The central walkway of the mall is enclosed with
occasional skylights. The mall is completely surrounded by parking, some of which is asphalt (dark) and some concrete
(light). The roof of Whites department store (e) is asphalt, barely distinguishable from the parking lot. Note the numerous air-
conditioning units on each building.

houses banking, brokerage firms, insurance agencies, and an
airline ticket office. Parking for this building is primarily
aboveground. It is impossible to identify the specific activities taking place within this office building. Therefore, it is
often necessary to simply label the building or parcel as
being *commercial* land use. Field work or census enumeration can provide some of the information.

Large multiple-story buildings like this are commonplace in
major cities of the world. On large-scale aerial photography,
relief displacement of the top of the building is a function of
1) how high the building is above the local datum, 2) the altitude of the aircraft at the instant of exposure, and 3) how far
the building is located away from the principal point of the
photograph. Also, the buildings tend to cast long shadows.
The relief displacement and the shadows can both hinder
and help in building identification. In this example, relief
displacement aids in the identification, as the sides (façade)
of buildings are actually visible providing a good impression
of building vertical relief. Shadows help to set the buildings
off from the remainder of the scene. When using such aerial
photography, it is necessary to use photogrammetric techniques to compensate for the relief displacement if one

desires to map the exact perimeter of the building in its
proper planimetric (x,y) position. One cannot simply draw a
polygon around the top of the building and assume that it is
in its proper planimetric position. One might think that relief
displacement would be absent from high spatial resolution
imagery obtained from sensors placed in orbit. Unfortunately, this is not the case, as demonstrated in the GeoEye
IKONOS 1×1 m image of downtown Seattle, WA, where
building displacement is still apparent (Figure 13-13).

Isolated, detached banking facilities typically have a drive-
through portico structure where financial transactions take
place (Figure 13-16b). One parking area is usually associated with walk-in banking, while the other portion of the
parking area services the drive-through traffic. One must be
careful not to confuse the drive-through bank building with
a gas station that also has a drive-through portico.

Construction companies often have heavy equipment (back-
hoes, dump trucks, etc.) located on the property, as well as
stacks or piles of raw materials (e.g., lumber, cement, palettes of bricks).

a. Commercial financial services, including banking (level 12141), insurance (level 12142), and real estate transactions (level 12143), take place in the Affinity Office Building in downtown Columbia, SC. Building relief displacement in this original 1:6,000-scale vertical aerial photography provides detailed information about the façade (side) of the building and aids in building image interpretation.

b. An isolated small banking establishment (level 12141), complete with the main building and attached drive-through teller system. In this example, one parking area is associated with walk-in banking while the other portion of the parking area services the drive-through customers.

Figure 13-16 Examples of commercial financial services (level 1214).

Food and Drug

The urban landscape is replete with many food and drug business establishments. Chainstore markets and drugstores occupy large buildings with extensive parking facilities. They may be isolated, but are often found attached. For example, consider the attached drugstore (1) and supermarket (2) shown in Figure 13-17a. The food and drug establishments share a large common parking lot. The exact land-use code of such buildings is often difficult to determine without *in situ* inspection because the large extended building and parking facilities may appear similar to other commercial enterprises.

Fast-food restaurants generally occupy relatively small, single-story buildings. They are located on both major and minor roads, have modest but adequate parking facilities, and often exhibit a single automobile drive-through window lane that encircles the building (Figure 13-17bc). This is important because drive-through banks typically have multiple drive-through lanes. There are usually numerous entrances and exits to the lot. The lot is usually modestly landscaped. Sometimes there are elaborate playgrounds with gymnasium equipment present in front of the restaurants.

More expensive, upscale restaurants are larger in size (m²), have well-manicured lawn and tree landscaping, occupy a much larger lot, and have larger parking areas than fast-food restaurants. Both fast-food and upscale restaurants have a large number of vents on their rooftops. Sometimes the smoke discharged from the vents discolors the rooftops.

Funeral and Cemetery

Having traveled the equivalent of 100 times around the Earth in our automobiles during a lifetime (the Earth is only 25,000 miles in circumference) and eaten at far too many fast-food restaurants, American and Western Europeans finally reside in a cemetery or mausoleum. Interestingly, cemeteries often confound image interpreters, yet they have unique characteristics (Figure 13-18). In developed nations they generally occupy an expensive and extensive tract of land, often adjacent to churches. The landscaping is usually meticulous. The road network is intricate, with many narrow roads. Often, the roads follow the contour of the land. But most important, there are hundreds of systematically spaced small white dots on the landscape. There is often a shed on the property with heavy equipment (backhoe) and vaults located in the shed or on the ground out of sight from the

a. Example of a drugstore (level 12151) on the left (1) attached to a supermarket (level 12152) on the right (2). The small one-story building in the parking lot is a one-hour Fotomat (3).

b. A fast-food restaurant (level 12153). Note the ample parking and the drive-through window lane.

c. Another fast-food restaurant (level 12153). Again note the drive-through window lane loaded with cars.

Figure 13-17 Examples of food and drug commercial businesses (level 1215).

public. A mortuary or funeral home may be located on the property. It is often a single large building with substantial parking.

Housing (Temporary)

People throughout the world love to go on overnight and extended vacations. This has resulted in a plethora of small motels that cater to the motorist and larger hotels that are used for both business and pleasure. One of the major differ-

ences between a motel and a hotel is size. Motels tend to be smaller and have fewer stories (Figure 13-19a). Hotels are often imposing structures rising ≥ 4 stories (Figure 13-19b). Both motels and hotels usually have large parking areas and swimming pool(s). Motel parking is often directly in front of the rental unit, whereas hotels often have large underground parking areas that are difficult to detect. Hotels often have tennis courts, whereas motels usually do not. Large hotels tend to be located near major thoroughfares, airports, and limited-access highways while motels may be located just

Figure 13-18 A portion of the Elmwood Cemetery in Columbia, SC (level 12162), originally recorded on 1:10,000-scale aerial photography. The small white dots are individual headstones. The larger white structures with shadows are mausoleums. The building at the right is a large mausoleum.

a. The Heart of Carolina Motel (level 12172). Many of the cars are parked directly in front of the patron's rooms.

b. Temporary housing at the Columbia Plaza Hotel (level 12171). Note the high-rise hotel complex, the large parking area, and attached convention hall with its own elevated ramp for loading and unloading convention exhibits. The pool is inside the building.

Figure 13-19 Commercial motel and hotel temporary housing (level 1217).

about anywhere. Hotels usually have more luxuriant land-scaping.

There are also specialized campgrounds that cater specifically to the motorist. These have unique attributes, including individual driveways where the car is parked and concrete pads where a tent or recreational vehicle may be placed. There is often a centrally located swimming pool and other recreational amenities. The campground is often quite large and located near a major highway. Campgrounds are usually well-manicured.

House and Garden

This type of business (e.g., Lowes, Home Depot) usually has a single large building, attached greenhouse(s), and rows of shrubbery outside or adjacent to the building. Also visible are palettes of fertilizer and other materials, yard machinery such as lawn tractors and trailers sitting outside the building, and a large parking lot. It is usually situated in the city on a major highway or in suburbia near residential housing.

Recreation

The post-WWII baby boomers are maturing rapidly and so are their children and grandchildren. Many of the adults are out of shape, have substantial disposable income, and will obtain more and more leisure time, especially when they retire. Therefore, it is not surprising that a great deal of time, money, and landscape is devoted to public and commercial recreation facilities, especially golf. Also, numerous amusement parks are now routinely available that contain a variety of recreational facilities. These are relatively easy to identify on remotely sensed data due to their unique shapes and curvilinear patterns. For example, consider the roller-coaster, go-cart track and water-slide facilities located in Myrtle Beach, SC, in Figure 13-20.

It is a relatively straightforward task to identify baseball diamonds, football or soccer fields, running tracks, and tennis courts using remotely sensed data, because they are laid out according to strict specifications. Examples of these activities and the dimensions of the diamonds, fields, and courts are summarized in Figures 13-21 and 13-22.

Perhaps the greatest consumer of prime land in and around cities are commercial and public golf courses. They come in an endless variety of fairway shapes and sizes and may include 9, 18, and even 36 holes. They are especially easy to identify on remotely sensed data because they always include a clubhouse with ample parking, long curvilinear fairways, numerous bright sand traps, and well-manicured greens (Figure 13-23).

Swimming pools are easy to identify if one remembers that there is usually a darker deep end to the pool and that many pools have parallel swimming lane lines on the bottom of the pool. They come in a tremendous variety of shapes and sizes but almost always have a significant amount of deck surrounding them. Two examples are shown in Figure 13-24.

Many communities use tax dollars to support public stadiums, coliseums, or concert halls. A good example is the University of South Carolina Coliseum (Figure 13-25). These facilities are used for an amazing variety of recreational and/or ceremonial purposes, including indoor basketball, circuses, tractor-pulls, rock concerts, graduations, etc. They are usually extremely large multiple-story buildings that are well-landscaped and have extensive parking.

It is practically impossible to identify a movie theater located within a mall. However, many theater complexes are now built unattached with numerous large screening rooms and ample parking. Some communities still have a drive-in movie theater that at times has been one of the most difficult features to identify in the urban landscape. The drive-in movie theater site is usually fan-shaped, with a large vertical structure at the apex of the fan (the screen). It typically casts a long shadow, depending on the time of day. A fan-shaped road network is found within the site, and there are poles placed systematically about every 5 m on the ground that hold the detachable speaker. The concession stand is almost always located in the center of the site. There is usually no landscaping, the terrain being composed of asphalt or gravel.

Warehousing/Shipping

Warehouses receive goods and then reorganize the materials according to a bill of laden for shipment via trucks, railroad, or ship to an interim or final destination. They are usually very large buildings with modest landscaping. In the case of truck transport, there is usually a large parking lot with an elevated (5 ft high) docking system on one or multiple sides of the building. Forklifts are often present in the yard. Occasionally, some nonperishable commodities or container cargo boxes may be seen outside in the parking area. The most diagnostic features are the tractor trailers that are situated with their back doors against the building or arranged systematically throughout the site out of harm's way.

Shipping and warehousing along harbor waterfronts usually involves extremely large warehouses situated on the docks or immediately inland. For example, consider the aerial pho-

a. Roller-coaster recreational facility at the Myrtle Beach, SC, Pavilion amusement park (level 12191).

b. A typical go-cart track. The overpass and most of the corners are concrete, while the remainder of the track is asphalt.

c. A water-slide ride that empties into a pool.

Figure 13-20 Commercial recreation amusement-park facilities (level 12191).

tograph of a portion of the Seattle, WA, harbor in Figure 13-26. Here, piles of minerals are being loaded onto the ship from the conical storage structures using a sophisticated overhead conveyor-belt system. Extensive large warehouses are also located nearby. The minerals were brought to the shipping facility most likely by the adjacent railroad.

Other Commercial

Finally, there is the matter of what to do about all the other myriad of commercial landuses that are found in almost every urban landscape (hobby shops, Fotomats, clothing stores, bowling alleys, etc.). It is impossible to determine what type of business is being conducted from their remote sensing attributes. Only *in situ* investigation will allow them to escape from land-use class *17, Other Urban or Built-up Land* category in Table 13-3.

Services (Public and Private)

In developed countries, mankind has organized a sophisticated infrastructure of public services to improve the quality of life. Some of the buildings and grounds associated with these services have relatively diagnostic characteristics that allow them to be identified in remote sensor data.

Public Buildings and Facilities (Administration, Fire, Police, Rescue, Postal, Libraries, Prisons)

Preeminent public facilities such as state capitols (Figure 13-27a) and major administrative buildings (Figure 13-27b) are relatively easy to identify in remote sensor data. They usually have striking architecture because they are supposed to last and inspire us for many generations. They typically have

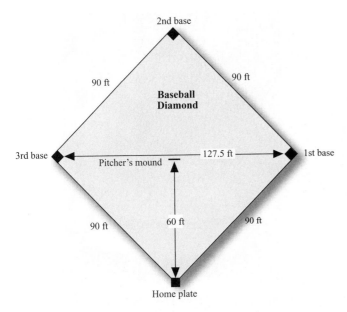

a. Dimensions of a baseball diamond.

b. Capitol City Bombers professional minor-league baseball stadium, Columbia, SC (originally 1:6,000 scale).

c. High school baseball diamond.

d. Atlanta Fulton County stadium, home of the professional Atlanta Braves baseball team (originally 2×2 m).

Figure 13-21 Examples of high school, minor-league, and professional baseball diamonds (level 12192). The baseball-diamond dimensions remain the same while the facilities for the team and spectators change dramatically.

well-landscaped grounds and ample parking, which may be located in multiple-story parking garages. National monuments such as the Statue of Liberty in New York and L'Arc de Triomphe in Paris, France, often have very diagnostic architecture (Figure 13-27cd, respectively).

Conversely, the only way to identify fire-and-rescue services is if the engines, rescue vehicles, and/or firefighting or rescue equipment are temporarily visible outside the building. Also, some fire stations have a two or three-story tower nearby where they practice firefighting. It is possible to

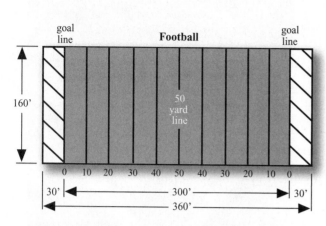

a. Dimensions of a football field.

b. University of South Carolina football stadium (level 12193).

c. Dimensions of a soccer field.

d. High school soccer field and track (level 12193). The goal posts reveal that the field is also used for football.

e. Dimensions of a tennis court.

f. Multiple tennis courts at a tennis club (level 12197).

Figure 13-22 Examples of a university football stadium, high school soccer field and track, and community tennis club.

Figure 13-23 Low-oblique aerial photography of a golf course near Atlanta, GA, with multiple long, curvilinear fairways, bright sandtraps, and manicured greens (level 12194).

Figure 13-24 Large-scale vertical aerial photography of swimming pools (level 12195) and associated decking. The pool on the left has swimming lanes. The pool on the right has debris in it. Both pools have shallow and deep ends.

identify a police station if the cars parked outside have numbers on their roofs.

Regional post offices are large buildings with extensive parking and a significant number of postal trucks and smaller postal vehicles present. Local post offices are difficult to identify because they are small, well-landscaped buildings, with ample parking nearby. In fact, they may be indistinguishable from local libraries.

Prisons are usually easy to identify because they have large buildings, ample parking, guard towers, multiple rows of very high fencing, and restricted access. Exercise fields are usually present within the fenced perimeter.

Figure 13-25 Multipurpose coliseum (stadium) associated with the University of South Carolina (level 12196). A recreational or cultural event is taking place on this date.

Figure 13-26 Shipping and warehousing of grain and other commodities in the Seattle, WA, harbor. Note the conical storage structures, the elaborate conveyor-belt system, the large rectangular warehouses, and railroad spurs that transport raw materials to and from the facility.

a. State Capitol of South Carolina in Columbia (level 12211).

b. Public office building with multiple-story parking garage.

c. Ellis Island and the Statue of Liberty National Park in New York, NY (12217).

d. Low-oblique aerial photograph of L'Arc de Triomphe in Paris (12217).

Figure 13-27 Public administration buildings, national parks, and monuments.

a.   b.

Figure 13-28 Examples of university public education facilities (level 122124). a.) Vertical aerial photograph of the manicured "horseshoe" of the University of South Carolina, founded in 1801. At one time most of the buildings surrounding the horseshoe were for administrators, classes, and students and their attendants. Now most of the buildings facing the horseshoe are administrative. This type of architectural design is common on other southern campuses such as the University of Georgia. b.) Low-oblique photograph of the University of California at Santa Barbara in Goleta, CA, with its varied architectural forms, well-manicured landscape, and intricate network of walkways between buildings.

Education

Universities are relatively easy to identify in high spatial resolution remote sensor data. Except in the downtown areas of the largest cities (e.g., New York, Chicago, Atlanta), universities usually exhibit a geographically extensive collection of buildings, open spaces, recreational facilities (swimming pools, baseball, football, soccer, track, tennis), gymnasiums, coliseums, and stadiums. Many of the education buildings are multiple-story. Also, large dormitories are often present. The problem is usually deciding where the formal university begins and ends. Two different examples of universities are provided in Figure 13-28. The University of South Carolina was founded in 1801. All of the original administration buildings, student dormitories, and teaching facilities were systematically oriented to look out upon a landscaped park called the "horseshoe." Conversely, the oblique aerial photograph of a portion of the University of California at Santa

Barbara reveals the unsystematic placement of facilities interspersed with a complex network of sidewalks.

Most high schools and middle schools have large buildings, ample parking, and outdoor fields for baseball, football, soccer, track, and/or tennis. High schools usually have a football stadium and a pool, whereas middle schools do not. In cold climates, the pool may be indoors. High schools and middle schools are often fenced.

Elementary schools have relatively smaller buildings, ample parking, and a few outdoor fields. They rarely have a pool or a football stadium. Elementary schools may be fenced.

Medical

It is difficult to identify medical business offices or health and fitness clubs. They appear similar to many other nondescript commercial activities that can take place in a building.

However, it is relatively easy to identify hospitals. They are usually very large multistory buildings with many additions. Ample parking is necessary. Walkways, tramways, etc., often connect one building to another. Sometimes it is possible to detect the emergency-room automotive entrance. Some hospitals have a helicopter landing pad on the ground or on top of a building.

 Industrial Land Use

This section introduces the concept of image interpretation for the analysis of aerial photography and other types of high spatial resolution remote sensor data for extracting industrial land-use information. The heuristic rules discussed are designed to enable the interpreter who is not a specialist in industrial image interpretation to identify those industries primarily from their images alone. If the image analyst cannot specifically identify an industry, however, he or she should still be able to place it in a category that limits the kind of industry it may be. The term industry includes those businesses engaged in the *extraction* of raw materials, the *processing* of materials, and the *fabrication* of intermediate and finished products.

First, however, it should come as no surprise that military intelligence gathering organizations throughout the world are very interested in monitoring the industrial land-use of foreign countries. They have developed excellent industrial target image analysis manuals that may also be used for civilian applications. Several of the more important manuals focus on 1) general industrial target analysis logic (DoD, 1976), 2) the coke, iron and steel industries (DoD, 1978a), 3) power and water facilities (DoD, 1977), and 4) the petroleum industry (DoD, 1978b). Copies of these manuals may be viewed at the Federation of American Scientists home page (Appendix A).

Industrial Land-Use Classification Logic

Industries often have unique assemblages of raw materials, equipment, final products and waste, as well as buildings that characterize the industry. Sometimes these *industry-related components* are visible on the photography or imagery; at other times the equipment or material may be sheltered in buildings or otherwise obscured from view. In this case, images of other components, not significant in themselves but closely associated with the hidden essential components, may still be recorded on the imagery. Moreover,

from their distinctive forms, patterns, and relationships, one frequently can infer the kind of material or equipment obscured from view. For instance, unusually shaped buildings may be specially designed to house specific kinds of equipment. Similarly, chimneys and stacks by their number and arrangement may indicate the specific kind of furnace or oven from which smoke or noxious gases are being vented.

Identification of the various industry-related components in an image may allow the industry to be placed into one of three major industrial categories (e.g., NPIC, 1961; Avery and Berlin, 1992):

- extraction

- processing

- fabrication.

When attempting to categorize an industry using remote sensing data, it is recommended that the analyst 1) first decide whether it is an extraction, processing, or fabrication industry; 2) if it is a processing industry, decide whether it is chemical, heat, or mechanical processing, in that order; and 3) if it is a fabrication industry, decide whether it is light or heavy fabrication. There are a tremendous variety of industries. It is only possible to provide a few representative examples of each major industry category.

Extraction Industries

Extractive industries extract the natural resources of the Earth with the minimum handling required to accumulate raw materials in a form suitable for transportation or processing. The extraction usually takes place from a shaft mine, open-pit mine, well, or cutting field, e.g., forestry clear-cutting or high-grading (Table 13-4). Extractive industries typically have several of the following diagnostic recognition features:

- the presence of open-pit excavations, ponds, mine openings or derricks;

- normal and oversized handling equipment including bulldozers, dump trucks, cranes, power shovels, dredges, tree-cutters, and mine cars;

- an elaborate transportation system to move the raw material about the site, including conveyors, pipelines, railroads, drag lines, and/or road network;

Kennecott Ridgeway Mining Company, Ridgeway, SC

Open-pit mine

Tailings pile

Open-pit mine

Tailings pile

Catchment basin

Figure 13-29 Panchromatic stereopair of the Kennecott Ridgeway Mining Company located near Ridgeway, SC, on March 16, 1992 (level 1312) (courtesy Kennecott Ridgeway Mining Company).

- extracted bulk materials stored in piles, ponds, hoppers, or tanks; and

- piles or ponds of waste that have been separated from the valuable material of interest.

All the extractive industry features are rarely seen at any one location. Often the surrounding countryside shows evidence of reclamation, reforestation, etc., in an attempt to repair the effects of the extractive industry.

A 1:19,200-scale black-and-white panchromatic stereopair of the Kennecott Ridgeway Mining Company located near Ridgeway, SC, obtained on March 16, 1992, is shown in Figure 13-29. Two very large open-pit mines are visible in the photographs. They are surrounded primarily by loblolly pine plantations. The raw ore-bearing material is extracted from the earth using hugh shovels within the open-pit mines. It is transported to the central facility by dump trucks using a complex network of roads. The gold is separated from the ore-bearing rock using a chemical process at the mine central processing facility. Two elevated piles of material that have already been processed (often referred to as tailings piles) are visible. Note that a substantial amount of the processed rock material also has been used to construct a catchment basin. The mine was decommissioned in 1999. The process of restoration is underway.

The largest *open-pit mine* in the world is the Bingham Canyon Copper Mine near Salt Lake City, UT, operated by Ken-

a. Low-oblique aerial photograph of the Bingham Canyon Copper Mine near Salt Lake City, UT, the largest open-pit mine in the world.

b. Terrestrial photograph of the mine reveals ore trains, huge shovels, and a dense network of power-line poles.

Figure 13-30 The Bingham Canyon Copper Mine near Salt Lake City, UT, is an extractive industry (level 13121).

necott Copper Corporation. It is 4 km in length and has a depth of 800 m. It is one of the few features that man has created that can be seen from space (another is the Great Wall of China). More than 3 billion tons of copper have been extracted, more than five times the material excavated for the Panama Canal. Figure 13-30a is a low-oblique aerial photograph of the mine. Figure 13-30b is a terrestrial photograph taken within the mine. The terrestrial photograph provides information on just how large the mine is when we discover that the dark linear features located along the farthest wall are actually four long trains that are used to remove material from the mine. The railroad tracks spiral down to the lowest level.

An open-pit gravel mine in South Carolina is shown in Figure 13-31a. Material is dynamited from the walls of the excavation and carried by truck to a collection point at the base of the mine. A conveyor belt then carries the unsorted material up 100 m to the surface, where it is sorted into various grades of gravel and sand. The sorted material is then transported by trucks to final destinations. Precipitation cannot escape the pit, resulting in an internal drainage pond at the base of the mine.

Figure 13-31b depicts a small portion of a petroleum extraction well field situated in the mountainous terrain behind Ventura, CA. Each clearing contains one or more oil wells that are connected via an intricate system of dirt roads. The oil is actually pumped by pipeline from the wells to collecting points not present in the photograph.

Processing Industries

Processing industries subject the accumulated raw materials to mechanical, chemical, or heat treatment to render them suitable for further processing, or to produce materials from which goods and equipment can be made.

The processing industries are characterized by the presence of facilities for the storage and handling of large quantities of bulk materials. These materials may be stored in the open in piles, ponds or reservoirs, or in storage containers such as silos, bins, hoppers, bunkers, and open or closed tanks. The materials are handled by pipelines, conveyors, and fixed cranes, as well as railroad cars and other mobile equipment. The conveyor-belt system used to sort the gravel in the open-pit mine previously discussed is a good example. In addition, other outdoor equipment are commonly present, including blast furnaces, kilns, chemical-processing towers, large chimneys or many stacks, indicative of the kind of processes being carried out.

The processes involved usually require large quantities of power, which is indicated by the presence of coal piles, fuel tanks, boiler houses, or transformer yards if electric power is employed or produced. The buildings that house the processing equipment frequently are large or at least complex in outline and roof structure. Since the processing of raw materials usually involves refinement, piles and ponds of waste

a. An open-pit gravel mine in South Carolina (level 13122). The gravel is transported from the base of the mine via a single conveyor belt to several sorters at the top of the mine. Trucks transport the sorted material to final destinations.

b. Petroleum extraction from oil wells in the mountains behind Ventura, CA (level 13151). Each clearing contains one or more oil wells that are connected via an intricate system of dirt roads.

Figure 13-31 Examples of open-pit mining and petroleum extraction industries.

a. Low-oblique photograph of mechanical processing at an agricultural mill where wheat is processed into baking flour for human consumption and animal feed. The grain is stored in the silos, and the processing takes place in the buildings. Flour and feed are transported to market by truck (level 13211).

b. Low-oblique photo of mechanical processing taking place at a sewage-treatment plant in Newport Beach, CA (level 13215). Solids are mechanically separated from liquids, hence the term mechanical processing. Note the settling and aeration ponds. Purified effluent is pumped 500 m to a submerged outfall in the Pacific Ocean (level 13215).

Figure 13-32 Examples of mechanical-processing industries (level 1321).

are common, and care is required to distinguish between waste versus useful raw materials stored in a similar manner.

The processing industries can be subdivided on the basis of image components into three subcategories:

- mechanical

- chemical

- heat processing.

These subcategories have functional significance.

Mechanical-Processing Industries

Mechanical-processing industries size, sort, separate, or otherwise change the physical form or appearance of the raw materials. The image components that characterize the mechanical-processing industries are the bulk materials stored in piles, ponds, or reservoirs, or outdoor equipment such as silos, bins, bunkers or open tanks, as well as an abundance of handling equipment such as conveyors, launders, cranes, rail cars, and other mobile equipment. The processing involved may require large quantities of power, as indicated by the presence of boiler houses with their fuel supply, or transformer yards when electric power is employed. Many of the buildings may be large or at least complex in outline and roof structure. Piles or ponds of waste are quite common. The industries in the mechanical-processing category differ from the other processing industries in that they have few pipelines, closed or tall tanks, or stacks other than on boiler houses. Furthermore, there is an absence of the kilns associated with the heat-processing industries.

Mechanical processing takes place at the Adluh Flour Meal and Feed Company (level 13211) in Columbia, SC (Figure 13-32a). Grain arrives at the mill by way of a railroad spur behind the buildings or by truck. Three large concrete silos store the unprocessed wheat. Wheat milling takes place in the large, complex building to the right. Tractor trailers backed up to the building transport the finished flour to local and regional market.

Figure 13-32b depicts a low-oblique photograph of mechanical processing taking place at a sewage-treatment plant (level 13215) in Newport Beach, CA. The sewage is routed from the nearby communities to the plant via underground pipelines. Mechanical processing at the plant separates the solids from the liquid. Various settling ponds, enclosed tanks, aerators, and anerobic bacteria facilities purify the sewage so that it can eventually be pumped 500 m to a sub-

merged outfall in the Pacific Ocean. Water purification installations are also mechanical-processing industries.

A hydroelectric power plant utilizes the "hydrostatic head" of water stored in the lake to spin turbines that create electricity. For example, consider the hydroelectric generating facilities at the Lake Murray Dam near Columbia, SC (Figure 13-33). Four intake towers transport water down 200 ft to the turbine house where the water pressure spins the turbines generating electricity. The water then enters the Saluda River. The electricity is transported to the regional power grid via a large substation complex. There is also a thermal-electric power plant at this location which will be discussed in the heat-processing industry section

Chemical-Processing Industries

Chemical-processing industries utilize chemicals or chemical processes to separate or rearrange the chemical constituents of the raw materials. Petroleum refineries and fertilizer plants are good examples. Pressure, heat and catalysts, or other chemicals may be employed. Closed vessels for holding or handling fluids, gases, or suspensions of solids in liquids, and the use of fluid flow in processing the raw materials are typical of these industries. Consequently, the image components that characterize the chemical-processing industries are an abundance of closed tanks, pipelines, and such large processing equipment as towers for cracking or distillation. As with all processing industries, those in this category have facilities for the storage and handling of bulk materials and require large quantities of power. The buildings frequently are complex, and extensive waste piles or ponds are common.

The Al Qaim superphosphate fertilizer plant in Iraq is shown in Figure 13-34. Note the existence of numerous enclosed storage tanks, a large conical pile of raw materials, a circular pond, and large processing buildings (several with recent bomb damage). Phosphates may be used to produce explosives such as those used in the Oklahoma federal building bombing. That is why they are targeted during war.

Industry generates a significant amount of chemical hazardous wastes. Some companies use remote sensing data to document the initial conditions under which the materials are stored and then monitor the surface expression after they are stored. For example, Figure 13-35a is a low-oblique aerial photograph of the Low-Level Waste Management site at the Department of Energy's (DOE) Nevada Test Site. This location provides efficient disposal of radioactive chemical wastes generated in the nuclear weapons program. Waste is trucked to the site in Department of Transportation approved

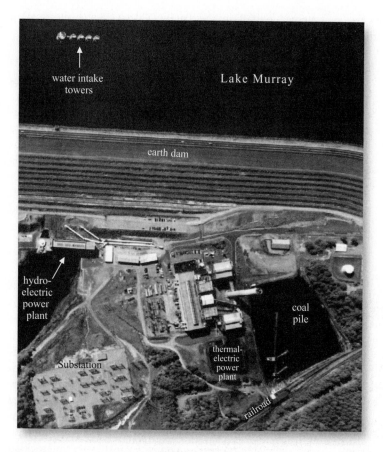

Figure 13-33 Aerial photograph of the Lake Murray Dam, SC, hydroelectric power plant (level 13213) and the Saluda thermal-electric power plant. Four intake towers provide the high-pressure water that drives the turbines that produce the electricity. The thermal-electric power plant burns coal that produces steam to drive the separate turbines that produce electricity. The coal is delivered by railroad and transported from the railroad cars to the pile via conveyor belts. The electricity is distributed from the site onto the regional power grid via the substation.

containers (in this example drums) and buried in the large trench. As it fills, the pit is covered with soil.

Figure 13-35b is a low-oblique photograph of radioactive waste being stored in a tank farm at the Hanford, WA, DOE facility. The tank farms store by-product materials left over from plutonium extraction operations prior to permanent disposal. This by-product material has no useful purpose and is stored in approximately 177 underground storage tanks with a cumulative total of 55 million gallons capacity, with some individual tanks ranging up to 1,000,000 gallons. The trailers and cars in the lower right of the photograph provide some idea of the size of the tanks.

Petroleum chemical-processing industries convert raw crude oil into gasoline, diesel fuel, synthetic textiles, and plastic. The plants usually have an extensive, often dense network of above or underground pipelines such as those found in a petroleum refinery in Texas (Figure 13-36a). Storage tanks are often completely surrounded by Earth dike revetments that can contain the chemical if a spill occurs. For example, Figure 13-36b depicts a tank farm at the petroleum refinery at Al Basrah, Iraq, in 1991. Several of the tanks are on fire

and leaking. The Earth dikes are containing much of the spill. Figure 13-36c depicts the Atlantic Richfield petroleum storage and transport facilities in Seattle, WA.

Heat-Processing Industries

The heat-processing industries utilize primarily heat to refine, separate, or re-form the raw materials, or to derive energy from them. Iron and steel production and thermal-electric power plants that burn oil or coal are good examples. The image components that distinguish this category are the large quantities of coal or other fuel, large chimneys, large numbers of stacks, flues and kilns of various kinds. Although pipelines and tanks frequently are employed in these industries, they are usually not abundant, and their presence is outweighed by the image components evidencing the use of heat. As with the other processing industries, these have extensive facilities for handling and storing bulk materials and require large quantities of fuel or power. Large outdoor equipment, such as blast furnaces and kilns, often are employed. The buildings frequently are complex, and piles of waste are common.

Figure 13-34 Imagery of the Al Qaim superphosphate fertilizer plant in Iraq, obtained during the Gulf War in 1991 (U.S. Navy TARPS image released under the Freedom of Information Act to William M. Arkin).

a. Radioactive hazardous waste being stored in drums at the Department of Energy's Nevada Test Site Low-Level Waste Management facility. The waste is eventually covered with sand (courtesy Department of Energy, ID WB212; John Pike, FAS).

b. Hazardous-waste storage tanks at the DOE Hanford, WA facility. At one time approximately 2,300 personnel supported the total tank farm activities at Hanford (courtesy Department of Energy, ID 303-058-001; John Pike, Federation of American Scientists).

Figure 13-35 Examples of chemical hazardous waste–site monitoring using large-scale low-oblique aerial photography (level 13223).

Iron and steel heat-processing industries are usually straightforward to identify on remote sensor data. For example, Figure 13-37 is a historical aerial photograph of a small portion of the Bethlehem Steel Company at Sparrows Point, MD, in 1952. The photograph records numerous long buildings that house the steel rolling mills and blast furnaces, many support buildings, and considerable smoke from large stacks.

Nuclear facilities are also relatively easy to identify. The Fast Flux Test Facility at the Hanford site in eastern Washington is a 400-megawatt thermal reactor cooled by liquid

a. Large-scale vertical aerial photograph of a dense network of pipes at a petroleum refinery in Texas.

b. Petroleum refinery at Al Basrah, Iraq. Note the large tanks and Earth dikes that separate them. This photograph was acquired in 1991 during the Gulf War, and some tanks are on fire and leaking (U.S. Navy TARPS image released under the Freedom of Information Act to William M. Arkin).

c. Chemical processing, storage, and transportation of petroleum products at the Atlantic Richfield facilities in Seattle, WA. Note the petroleum tank farms, tanker ships, and railroad tank cars.

Figure 13-36 Examples of chemical-processing industries (level 1322).

Figure 13-37 Bethlehem Steel Company at Sparrows Point, MD, on April 23, 1952.

sodium (Figure 13-38a). It was built in 1978 to test plant equipment and fuel for the U.S. Government's liquid metal reactor development program. This program demonstrated the technology of commercial breeder reactors. The containment dome is easily identified in the image. Aboveground testing of nuclear materials has been banned for decades. However, underground testing continues. Figure 13-38b depicts the results of one test at the Nevada Test Site where a 100 kiloton explosive was buried under 635 feet of desert alluvium and detonated on July 6, 1962, displacing 12 million tons of earth. Underground tests conducted by nations throughout the world often leave craters such as these as evidence of what has taken place.

Below are some additional fundamental image-recognition features that may be used to distinguish among the three types of processing industries:

Mechanical-Processing Industries:
- few pipelines or closed tanks
- little fuel
- conveyor-belt systems often present
- few stacks
- no kilns.

Chemical-Processing Industries:
- many closed or tall tanks, including gasholders
- many pipelines
- much large outdoor processing equipment.

Heat-Processing Industries:
- a few pipelines or tanks
- large active chimneys and/or stacks
- large quantities of fuel
- kilns.

Fabrication Industries

Fabrication industries assemble the mechanical and chemical subcomponents into finished products such as automobiles, trucks, boats and ships, trailers, heavy equipment (e.g., bulldozers), plastic products, and electronic devices. The fabrication industries may be subdivided into heavy and light fabrication. *Heavy-fabrication* industries often have tall heavy steel frame one-story buildings, storage yards with heavy lifting equipment and cranes, and occasionally rail lines entering buildings. *Light-fabrication* industries often make use of light steel and wood-frame buildings (perhaps multistory) and typically lack heavy-lifting equipment.

Heavy Fabrication

Below are some of the fundamental image-recognition features that may be used to distinguish among the major heavy-fabrication industries. Heavy fabrication is almost universally associated with large buildings.

a.

b.

Figure 13-38 a) The Department of Energy Fast Flux Test Facility at Hanford, WA, has a 400-megawatt thermal reactor. The dome is the containment structure. b) The Sedan Crater was formed when a 100-kiloton explosive buried 635 ft beneath the desert was detonated on July 6, 1962, at the Nevada Test Site, displacing 12 million tons of earth. The massive crater is 320 ft deep and 1,280 ft in diameter. Two craters formed from other detonations are seen in the distance (courtesy Dept. of Energy).

Heavy Machinery Fabrication:

- storage yard for heavy metal shapes
- piles of structural steel and other materials
- finished heavy machinery.

Railroad Machinery Manufacture and Repair:

- railroad locomotives and/or cars under construction
- railroad track network.

Shipbuilding:

- storage yards for steel plate
- extensive dry-dock and ramp facilities
- ships under construction
- large cranes.

Structural Steel Fabrication:

- storage yards of structural steel
- structural steel fabrication yard.

A structural steel fabrication industry is shown in Figure 13-39a. The actual fabrication of the iron-and-steel girders used in building and bridge construction takes place inside the large elongated buildings. The iron and steel stock materials are sorted by size and type in the central open-air storage yard. Two overhead cranes (not visible) move the iron-and-steel raw materials back and forth (left to right) into the buildings for fabrication. Raw and finished materials are transported to and from the business either by truck or via the railroad spur.

A vertical perspective of how such structural steel is used in the building of a power plant is presented in Figure 13-39b. Aerial images such as this can document the progress on specific dates in the construction process.

Light Fabrication

Below are some of the fundamental image-recognition features that can be used to distinguish among the major light-fabrication industries.

Aircraft Assembly:

- aircraft or seaplane ramps or taxiways adjacent to or connected to the fabrication building
- hangars
- finished aircraft stored in the open.

Automobile Assembly (cars, trucks, motorcycles):

- single- or multiple-story buildings
- finished parked vehicles arranged in an orderly system
- test track(s)
- tractor-trailer shipping facilities with cars on trailers
- rail shipping container cars.

Boat Building and Repair:

a. Heavy fabrication at the Kline Iron and Steel Co. in Columbia, SC. Iron and steel stock is sorted by size and type in the two open-air storage yards. Fabrication takes place inside the buildings. Raw materials and finished products are transported to and from the facility via trucks or the railroad spur (level 13314).

b. A detailed vertical aerial photograph of a power plant under construction. The project contains significant quantities of prefabricated structural iron and steel (courtesy Eastman Kodak Company).

Figure 13-39 Examples of iron and steel fabrication and the use of such materials in the building of a power plant.

- boat components stored in the open
- fiberglass forms
- building where fiberglass application takes place
- boat trailers
- finished boats in open storage.

Meat Packing:
- holding pens
- a few tanks
- railroad siding
- tractor-trailer parking.

Textiles:
- tractor-trailer shipping and receiving
- bales of cotton or other raw material
- pipelines and tanks when creating synthetic fabrics.

An automotive-component manufacturing company in Columbia, SC, is shown in Figure 13-40a. Raw materials that are not susceptible to damage by the Sun or precipitation are sorted and stored by type in the open air. Fabrication takes place inside the building. The finished products are shipped from the shipping bays located at the back of the building.

A fish-canning plant located in San Pedro Harbor, CA, is found in Figure 13-40b. The ships moor at the dock next to the plant. The manufacturing process (canning) takes place within the large building. Numerous tanks hold chemicals used during the canning process. The plant is discharging effluent into the harbor using an underground pipe.

 Transportation Infrastructure

Transportation planners often use remote sensor data to update transportation network maps, evaluate road and railroad conditions, study urban traffic patterns at choke points such as tunnels, bridges, shopping malls, and airports, and conduct parking studies (Haack et al., 1997). The most comprehensive work to date on remote sensing applied to transportation problems was conducted by the National

a. Light fabrication company. Some of the raw materials are sorted by type in the open air. Fabrication takes place in the building.

b. A fish-canning plant (level 13324) in San Pedro Harbor, CA, and its associated pollution.

Figure 13-40 Examples of light fabrication (manufacturing).

Consortium on Remote Sensing in Transportation sponsored by NASA and the Department of Transportation (NCRST, 2006). Thirumalai (2003) reviewed NCRST research performed by consortium members on: a) environmental impact for multimodal corridor impact planning, b) hazard and disaster preparedness and lifelines security, c) infrastructure asset management and security, and d) multimodal transportation flow management.

Roads and Highways

Roads may be unpaved or paved. Paved highways may be accessible to motorists at almost any location along their route or they may be limited-access (freeways, toll). The general updating of a road or highway network centerline map is a fundamental task. This is often done every one to five years and, in areas with minimum tree density, can be accomplished using imagery with a spatial resolution of 1 – 30 m (Lacy, 1992). If more precise road dimensions are required such as the exact width of the road and sidewalks, then a spatial resolution of 0.25 – 0.5 m is required (Jensen et al., 1994). Currently, only large-scale aerial photography can provide such planimetric information. In addition, consider the aerial photograph of the intersection shown in Figure 13-41a. The stoplights suspended across the street, individual cars and trucks, and even the right-of-way white road markings are visible. Future unmanned aerial vehicles (UAVs) carrying lightweight digital cameras might well provide similar high-resolution information at a reasonable cost (NCRST, 2006).

Highway interchanges are used to route traffic onto and off of limited-access highways. They come in an endless variety of shapes and sizes and are relatively easy to distinguish on remote sensor data. For example, consider the classic limited-access interchange shown in Figure 13-41b. It is gently sloping and has very long entrance and exit ramps that allow traffic to merge carefully with other traffic.

Next to meteorological investigations, traffic count studies of automobiles, airplanes, boats, pedestrians, and people in groups require the highest temporal resolution data, often ranging from 5 to 10 minutes. It is difficult to resolve the type of car or boat using even 1×1 m data. This task requires high spatial resolution imagery from 0.25 – 0.5 m. Such information can only be acquired using aerial photography or video sensors that are 1) located on the top edges of buildings looking obliquely at the terrain, or 2) placed in aircraft or helicopters and flown repetitively over the study areas. Figure 13-41c captures the parking characteristics of a parking lot at a single instant in time. When such information is collected at an optimum time of day, future parking and traffic movement decisions can be made. Parking studies require the same high spatial resolution (0.25 – 0.5 m) but slightly lower temporal resolution (10 – 60 minutes).

Road, railroad, and bridge conditions (cracks, potholes, etc.) are routinely monitored both *in situ* and occasionally using high spatial resolution remote sensor data. For example, Figure 13-41d presents a panchromatic image of a highway and railroad bridge. Careful inspection by a trained analyst pro-

a. High spatial resolution image of an intersection acquired using kite aerial photography.

b. A limited-access highway interchange in Texas.

c. Large-scale aerial photography of a parking lot.

d. Aerial photography of highway and railroad bridges.

Figure 13-41 Examples of roads, limited-access highways, interchanges, parking, and bridges recorded using various scales of aerial photography (level 1411).

vides significant information about the condition of the road and bridge. Road and bridge condition can be documented using high spatial resolution aerial photography (< 0.25 × 0.25 m).

Railroads

Railroad tracks, locomotives, cars, and terminal facilities are relatively easy to distinguish on high spatial resolution imagery because

Zurich Train Station, Switzerland

Figure 13-42 The train station in Zurich, Switzerland, recoorded on August 16, 2002 (courtesy DigitalGlobe, Inc.).

- the tracks rarely move from year to year;

- the tracks have no right angles, only gradual curvilinear turns;

- railroad bridges become narrow as opposed to highway bridges, which generally stay the same width as the road;

- railroad crossing signposts and their attendant shadows are located at almost every major intersection with a road;

- railroad tracks rarely cross other railroad tracks; and

- locomotives and railroad cars are large, long, and linear, making them fairly easy to identify.

Much of the actual work of a railroad takes place in the railroad marshalling (classification) yard. It is here that railroad cars going the same direction down the line or railroad cars going to the identical destination are placed in a specific train. Figure 13-42 depicts the Zurich Train Station terminal and railroad marshalling yard.

Airport

Civilian and military airports usually have:

- long runway(s)

- flight-traffic control tower(s)

- terminal(s) [civilian]

- hangars

- an extensive tarmac road network connecting the hangars and terminals with the runways

- large parking areas

- visible aircraft.

Rural airports often have a single runway, a windsock, visible planes, and a hangar or two.

Figure 13-43a depicts a biplane recorded at approximately 6 × 6 in. spatial resolution. The biplane photograph was acquired using kite aerial photography. A portion of the El Toro Marine Air Station, CA (1 × 1 ft), recorded using a conventional camera is found in Figure 13-43b. An IKONOS (1 × 1 m) image of the Ronald Reagan National Airport is presented in Figure 13-43c. The civilian airport has an extensive terminal system to move people to and from the plane.

a. A vintage biplane recorded at approximately 6 × 6 in. spatial resolution using kite aerial photography (courtesy Cris C. Benton).

b. Jet aircraft at the El Toro Marine Air Station, CA, recorded at approximately 1 × 1 ft spatial resolution.

c. Panchromatic 1 × 1 m image of Ronald Reagan National Airport in Washington, DC, recorded by the IKONOS satellite in 1999 (courtesy Space Imaging/GeoEye, Inc.).

Figure 13-43 Examples of aircraft, tarmac, hangars, and military and civilian terminals (level 1413).

a. Sailboats at anchor in a harbor near San Francisco, CA (courtesy Cris C. Benton).

b. High spatial resolution vertical aerial photograph of twelve barges on the Mississippi River being pushed by two tugboats.

c. Four battleships, two troop transports, and an aircraft carrier in storage at National City, San Diego, CA.

Figure 13-44 Examples of boats and ships recorded on remotely sensed imagery (level 1414).

Conversely, some military airports do not have terminals, as the crews are ferried to and from the aircraft by vehicle.

Boats and Ships

Small boats and large ships are relatively easy to identify using high spatial resolution remote sensor data for the following reasons:

• They are often seen as isolated man-made reflective objects on a relatively uniform water surface, which can increase their object-to-background contrast.

• Boats and ships have relatively unique shapes (i.e., the length is almost always at least two times greater than the width, with a relatively pointed bow) because they need to

Military Applications

a. Submarine base at Andreyeva Guba, Russia, obtained on February 9, 2004.

b. Aircraft graveyard at Davis-Monthan Airforce Base near Tucson, AZ, on August 11, 2002.

Figure 13-45 a. Two submarines are present at the Andreyeva Guba, Russia, submarine base (image courtesy of ImageSat International, Inc.). b. The Aerospace Maintenance and Regeneration Center (AMRC) at Davis-Monthan Air Force Base in Tucson, AZ, stores more than 5,000 aircraft and provides spare parts for older aircraft in service throughout the world. The center annually takes in about 400 aircraft for storage and exports about the same number for return to active service. Note the B-52 bombers in the lower left portion of the image that have been cut into pieces using a large blade dropped by a crane. The pieces are left out in the open for an expended period of time for satellite national technical means (i.e., spy satellite) verification as part of the strategic arms limitation treaty (SALT II Treaty and Protocol) agreements between the United States and the Union of Soviet Socialist Republics (image courtesy of DigitalGlobe, Inc.).

hydrodynamically slice through the water with a minimum amount of friction.

Boats and ships range in size from the small family rowboats, power boats, and sailboats from 3 – 10 m in length (Figure 13-44a) to extremely large cruise ships, passenger liners, oil tankers, battleships, submarines, and aircraft carriers. Figure 13-44b depicts twelve commercial barges on the Mississippi River being pushed by two tugboats. Numerous large military ships are in storage in National City, San Diego, CA, in Figure 13-44c.

Military organizations throughout the world use high spatial and spectral resolution remote sensing technology to monitor the location and disposition of military equipment. For example, Figure 13-45a depicts two submarines at the Andreyeva Guba, Russia submarine base. Aircraft in storage at the Aerospace Maintenance and Regeneration Center at Davis-Monthan Air Force Base in Tucson, AZ, are shown in Figure 13-45b.

 ## Communications and Utilities

Urban/suburban environments are enormous consumers of electrical power, natural gas, telephone service, and potable water (Haack et al., 1997). In addition, they create great quantities of refuse, wastewater, and sewage. The removal of storm water from urban impervious surfaces is also a serious problem (Schultz, 1988; Jensen et al., 2005a). Automated mapping/facilities management (AM/FM) and geographic information systems (GIS) have been developed to manage extensive right-of-way corridors for various utilities, especially pipelines (Jadkowski et al., 1994). The most fundamental task is to update maps to show a general centerline of the utility of interest such as a power line right-of-way. This is relatively straightforward if the utility is not buried and 1 – 30 m spatial resolution remote sensor data is available. It is also often necessary to identify prototype utility (e.g., pipeline) routes (Feldman et al., 1995). Such studies require more geographically extensive imagery, such as Landsat Thematic Mapper data. Therefore, the majority of

a. Fenced water tower with adjacent maintenance building.

b. WIS TV-10 television station is in the building on the right. It has a tall antenna on the roof and a large circular satellite dish in the parking lot. A government office building is on the left.

Figure 13-46   Examples of utility transmission facilities (level 143).

the actual and proposed right-of-way may be observed well on imagery with 1 – 30 m spatial resolution obtained once every one to five years.

When it is necessary to inventory the exact location of the utility footpads, transmission towers, utility poles, manhole covers, the true centerline of the utility, the width of the utility right-of-way, and the dimensions of buildings, pumphouses, and substations, then it is necessary to have a spatial resolution of from 0.25 – 0.6 m (Jadkowski et al, 1994). For example, Figure 13-46 depicts a water-storage tower and the satellite transmission facilities associated with a television station.

 Urban Digital Elevation Model Creation

A digital model of the earth's elevation is referred to generically as a digital elevation model (DEM). If the DEM contains building and vegetation elevation information, then it is called a *digital surface model* (DSM). If all the building and vegetation information have been removed, then it is called a bald earth *digital terrain model* (DTM). Most GIS used for socioeconomic or environmental planning in the urban environment include a DSM or DTM (Jensen et al., 2005). Two sensors that can provide such information are

- analog or digital stereoscopic large-scale metric aerial photography with a spatial resolution of $\leq 0.25 - 0.5$ m;

- LIDAR data with masspoint post spacing of < 1 m and vertical accuracy of approximately 15 cm.

Digital soft-copy photogrammetry based on high spatial resolution remote sensor data has revolutionized the creation and availability of digital surface and digital terrain models (Jensen, 1995; Linder, 2003) (Chapter 6). The extraction of high resolution bald-earth DTMs using LIDAR is becoming less expensive and may be the technology of choice in the future (Haala and Brenner, 1999; Hodgson et al., 2005) (Chapter 10).

Terrain elevation in urban environments does not change very rapidly. Therefore, a DTM of an urbanized area need only be acquired once every 1 to 5 years unless there is significant development and the analyst desires to compare two different date DSMs or DTMs to determine change in terrain elevation, identify unpermitted additions onto buildings, or identify changes in building heights. Figure 13-47 depicts 1) a large-scale aerial photograph of downtown Columbia, SC, 2) a digital surface model of the same area extracted from the stereoscopic photography depicting the height of every building, 3) the orthophotograph draped over the DSM, creating a virtual reality representation of a major street, and 4) use of the DSM for modeling the optimum location for locat-

a. Panchromatic vertical aerial photograph.

b. Digital Surface Model (DSM).

c. Orthophoto draped over DSM.

d. Cellular phone transceiver location model.

Figure 13-47 a) Vertical aerial photography of downtown Columbia, SC. b) Digital surface model derived using soft-copy photogrammetry and stereoscopic imagery. c) Digital orthophoto draped over the DSM. d) The application of a GIS intervisibility model to identify dead zones that would be produced if a cellular phone transceiver were located on this particular building.

ing a cellular phone transceiver (Jensen, 1995; Cowen and Jensen, 1998). Architects, planners, engineers, and real estate personnel are beginning to use such information for a variety of purposes.

A digital surface model of a portion of the Westinghouse Savannah River Company near Aiken, SC, is shown in Figure 13-48. This three-dimensional scene was derived from LIDAR last return data obtained at a posting density of approximately 10 cm. Note that this is not a bald-earth digi-

tal terrain model as the building and vegetation elevation information are still present in the scene.

Venderhoe (2005) compared and contrasted elevation information derived using photogrammetric versus lidargrammetric methods for transportation engineering design purposes as part of the U.S. Department of Transportation's National Consortium on Remote Sensing in Transportation research program (NCRST, 2006). He found that elevation information derived from lidargrammetry was just as effec-

Figure 13-48 Analytically hill-shaded digital surface model (DSM) of a portion of the Westinghouse Savannah River Company near Aiken, SC, derived from last return LIDAR data obtained in November, 2004, at a posting density of approximately 10 cm. The elevation of buildings, trees, transportation features and the terrain can be extracted from the DSM.

tive as that derived photogrammetrically for transportation applications.

Meteorological Data

Daily weather in urban environments affects people, schools, businesses, telecommunication, and transportation systems. Great expense has gone into the development of near real-time monitoring of frontal systems, temperature, precipitation, and especially severe storm-warning systems. These important meteorological parameters are monitored almost exclusively by sophisticated airborne and ground-based remote sensing systems. For example, two Geostationary Operational Environmental Satellites (GOES) are positioned at 35,790 km above the equator in geo-synchronous orbits. GOES West obtains information about the western United States and is parked at 135° west longitude. GOES East obtains information about the Caribbean and eastern United States and is parked at 75° west longitude. Every day millions of people watch the progress of frontal

systems that sometimes generate deadly tornadoes and hurricanes. Full hemispheric disk images may be obtained every 25 minutes. Intense thunderstorms in relatively smaller regions may be imaged every 3.1 minutes. The spatial resolution is 1×1 km for the visible band and $4 - 8$ km for the thermal infrared bands. European nations use METEOSAT with visible near-infrared bands obtained at 2.5×2.5 km and thermal infrared data collected at 5×5 km every 25 minutes. Early hurricane monitoring and modeling based on these data have saved thousands of lives in recent history. For example, in 1989 Hurricane Hugo caused approximately $1 billion in damage to residential, commercial, and industrial facilities, but no lives were lost because of remote sensing assisted early warning and evacuation.

The public also relies on ground-based National Weather Service Weather Surveillance Radar (WSR-88D) for precipitation mapping and timely severe storm warning (Chapter 12). The maximum range of the NEXRAD radar is approximately 250 nautical miles. The NEXRAD network provides significant improvements in severe weather and flash flood warnings, air-traffic safety, flow control for air traffic, resource protection at military bases, and management of water, agriculture, forest, and snow removal (NOAA ROC, 2006). The Doppler radar "composite reflectivity" product is projected onto a Cartesian geographical map with a 1×1 km resolution out to 230 km or at a 4×4 km resolution out to 460 km. The data are obtained every 5 minutes in severe weather mode, every 6 minutes in precipitation mode, and every 10 minutes in clear air mode.

High spatial resolution $(5 - 30$ m) day- and nighttime thermal infrared data may be used to obtain detailed quantitative spatial information on the urban heat island effect (Lo et al., 1997). Landsat 7 Enhanced Thematic Mapper Plus, with its 60×60 m spatial resolution and ASTER with its 90×90 m spatial resolution are particularly useful. The spatial information can then be used to develop "greening" campaigns to ameliorate the urban heat island effect.

Urban Hydrology

Civil and hydrologic engineers and urban planners constantly require up to date information about urban hydrology. Two useful measurements that can be remotely sensed include impervious surface area and floodplain delineation.

Extraction of Impervious Surface Information

a. USGS NAPP digital orthophoto 1 x 1 m (red band).

b. Extraction of impervious surface materials.

Figure 13-49 Impervious surfaces were extracted from U.S.G.S. 1 × 1 m National Aerial Photography Program (NAPP) color-infrared digital orthophoto quarter quad (DOQQ) imagery of an area in North Carolina [courtesy Tom Tribble and Frank Obusek; North Carolina Center for Geographic Information and Analysis; Jensen and Hodgson (2004)].

Impervious Surface Mapping

Impervious surfaces such as asphalt, concrete, and building roof materials keep precipitation from percolating into the ground. The greater the amount of impervious surface material in a watershed, the greater the runoff and the higher the peak flow of tributaries that collect the increased runoff. Significant work has been conducted to develop methods to extract impervious surface information from remote sensor data. Impervious surfaces such as parking lots, highways, buildings, etc. can be readily identified on large scale remote sensor data using the fundamental elements of image interpretion. In addition, the spectral characteristics of selected impervious surface materials can be collected and used to train digital image processing programs to automatically identify impervious surface cover and quantify its extent (Ridd, 1995). For example, Figure 13-49 demonstrates how U.S. Geological Survey National Aerial Photography Program (NAPP) 1 × 1 m orthophotography was used to extract impervious surfaces associated with a large mall in North Carolina (Jensen and Hodgson, 2004).

Impervious surfaces can be inventoried most accurately using multispectral remote sensor data that has a spatial resolution of 0.25 – 10 m (Ridd, 1995; Ji and Jensen, 1999;

Jensen et al., 2005ab). Urbanization is taking place at a rapid pace in many countries (Jensen et al., 2002). It is necessary to collect impervious surface information every one to five years in such environments (Table 13-1).

Floodplain Delineation

The geographic extent of floodplains can be identified using multispectral remote sensor data in conjunction with digital terrain model (DTM) information derived from terrestrial surveying, soft-copy photogrammetry, LIDAR or IFSAR. Vegetation cover and soil associations are often in transition at the floodplain boundary. Therefore, it is possible to utilize multispectral data to identify changes in vegetation type or soil association and use this information in conjunction with elevation and slope data to identify the boundary of the floodplain.

Multispectral or hyperspectral remote sensor data with a spatial resolution of 1 – 30 m is usually sufficient for floodplain delineation when used in conjunction with elevation data. In dynamic areas, floodplain delineation should be updated every one to five years (Table 13-1).

Critical Environmental Area Assessment

Urban/suburban environments often include very sensitive areas such as wetlands, endangered-species habitat, parks, land surrounding treatment plants, and the land in urbanized watersheds that provides the runoff for potable drinking water. Relatively stable sensitive environments only need to be monitored every one to two years using a multispectral remote sensor collecting 1 – 10 m data. For extremely critical areas that could change rapidly, multispectral remote sensors (including a thermal infrared band) should obtain ≤ 0.25 – 2 m spatial resolution data every one to six months (Ehlers et al., 2003).

Disaster Emergency Response

The Federal Emergency Management Agency (part of the U.S. Department of Homeland Security) is utilizing remote sensing data as it conducts the Multi-Hazard Flood Map Modernization program. This includes mapping and analyzing data for all types of hazards. The program requires geodetic control, aerial imagery, elevation, surface water extent and other thematic data which are used to produce digital flood maps and other hazard-related products. All data are served via the *Geospatial One-Stop* portal and *The National Map* (Lowe, 2003; FEMA, 2006).

Flooding (e.g., Mississippi River in 1993; Rhine and Danube in 1993), hurricanes (e.g., Hugo in 1989; Andrew in 1991; three in Florida in 2004; Katrina in 2005), tornadoes (every year), fires, tanker spills, earthquakes (e.g., Saugus, CA, in 1971; Northridge, CA, in 1994), and the 2004 Indian Ocean tsunamis demonstrated that a rectified, predisaster remote sensing image database is indispensable (Jensen and Hodgson, 2006). The predisaster data only needs to be updated every one to five years. However, it should be high spatial resolution (1 – 5 m) multispectral data if possible (Jensen and Cowen, 1999).

When disaster strikes, high resolution (≤ 0.25 – 2 m) panchromatic and/or near-infrared data should be acquired within 12 hours to two days (Schweitzer and McLeod, 1997). For example, Figure 13-50a records the spill associated with an overturned tanker truck immediately after an accident. Aerial photography acquired immediately after the lethal Saugus, CA, February 1971 earthquake is shown in Figure 13-50b. Tornado damage near Mechanicsville, NY, is

captured in digital frame imagery in Figure 13-50c. If the terrain is shrouded in clouds, imaging radar often provides the most useful information. Post-disaster images are registered to the predisaster images, and manual and digital change detection takes place (Jensen, 2005). If precise, quantitative information about damaged housing stock, disrupted transportation arteries, the flow of spilled materials, and damage to aboveground utilities are required, it is advisable to acquire post-disaster 0.25 – 1 m panchromatic and near-infrared data within one to two days. Such information were indispensable in assessing damages and allocating scarce cleanup resources.

Mayors and governors often use pre- and post-disaster remote sensor data to obtain a rapid assessment of the devastation. For example, before and after tsunami images of BandAceh and Gleebruk, Indonesia, reveal destroyed homes, washed-out roads and bridges, and deforestation (Figure 13-51). When this information is correlated with the parcel property value stored in a GIS for the same square km, a quantitative dollar damage assessment can be made, which is indispensable when requesting disaster assistance.

Observations

Table 13-1 and Figure 13-2 reveal that there are a number of remote sensing systems that currently provide some of the desired urban/socioeconomic information when the required spatial resolution is poorer than 4 × 4 m and the temporal resolution is between 1 and 55 days. However, very high spatial resolution data (< 1 × 1 m) is required to satisfy many of the socioeconomic data requirements. In fact, as shown in Figure 13-2, the only sensor that currently provides such data on demand is analog or digital aerial photography (0.25 – 0.5 m). GeoEye's IKONOS, with its 1 × 1 m panchromatic data; GeoEye's OrbView 3, with its 1 × 1 m panchromatic data; and DigitalGlobe's QuickBird, with its 0.61 × 0.61 m panchromatic data, may still not satisfy all of the data requirements. None of the sensors can provide the 5- to 60-minute temporal resolution necessary for traffic and parking studies. The GOES satellite constellation (East and West) and the European METEOSAT provide sufficient national and regional weather information at reasonable temporal (3 – 25 minutes) and spatial resolutions (1 – 8 km and 2.5 – 5 km, respectively). Ground-based National Weather Service Weather Surveillance Radar provides sufficient spatial resolution (1 × 1°) and temporal resolution (5 – 10 min) for precipitation and intense storm tracking in urban environments.

a. Overturned tractor-trailer tanker in Alaska (Jensen and Cowen, 1999).

b. Aerial photography of collapsed spans of a freeway near Saugus, CA, immediately after an earthquake in 1971. This span was supported by pillars that could not withstand the quake.

c. Tornado damage near Mechanicsville, NY. This is 1 × 1 ft spatial resolution imagery obtained on June 1, 1998, using a digital frame camera. Fire and medical rescue workers are on the scene (courtesy Litton Emerge, Inc.).

Figure 13-50 Examples of high spatial resolution imagery acquired immediately after disasters using traditional metric aerial photography (a,b) and a digital frame camera (c).

References

American Planning Association, 2006, *Land-Based Classification Standards*, Chicago: American Planning Association Research Department, http://www.planning.org/lbcs.

Anderson, J. R., Hardy, E., Roach, J. and R. Witmer, 1976, *A Land-Use and Land-Cover Classification System for Use with Remote Sensor Data*, Washington: USGS Paper #964, 28 p.

Auch, R., Taylor, J. and W. Acevedo, 2004, *Urban Growth in American Cities: Glimpses of U.S. Urbanization*, Washington: USGS Circular #1252, 52 p.

Avery, T. E. and G. L. Berlin, 1992, *Fundamentals of Remote Sensing & Airphoto Interpretation*, NY: Macmillan, 377–404.

Batistella, M., Robeson, S. and E. F. Moran, 2003, "Settlement Design, Forest Fragmentation, and Landscape Change in Rondonia, Amazonia," *Photogrammetric Engineering & Remote Sensing*, 69(7):805–812.

Tsunami Impacts at Banda Aceh and Gleebruk, Indonesia

a. QuickBird 60 x 60 cm image of Banda Aceh, Indonesia, obtained on June 23, 2004.

b. QuickBird 60 x 60 cm image obtained on Dec. 28, 2004, revealing massive piles of debris.

c. QuickBird 60 x 60 cm image of Gleebruk, Indonesia, obtained on April 12, 2004.

d. QuickBird 60 x 60 cm image obtained on January 2, 2005 revealing massive erosion and vegetation denudation.

Figure 13-51 Before- and after-images of BandAceh and Gleebruk, Indonesia, revealing destroyed homes, washed-out roads and bridges, and deforestation (images courtesy of DigitalGlobe, Inc.).

Carlson, T., 2003, "Applications of Remote Sensing to Urban Problems," *Remote Sensing of Environment*, 86:273–274.

Chen, D. and D. Stow, 2003, "Strategies for Integrating Information from Multiple Spatial Resolutions into Land-use/Land-cover Classification Routines," *Photogrammetric Engineering & Remote Sensing*, 69(11):1279–1287.

Clapham, W. B., 2003, "Continuum-based Classification of Remotely Sensed Imagery to Describe Urban Sprawl on A Watershed Scale," *Remote Sensing of Environment*, 86:322–340.

Cowen, D. J. and W. Craig, 2004, "A Retrospective Look at the Need for a Multipurpose Cadastre," *Surveying and Land Information Science*, 63(4):205–214.

Cowen, D. C. and J. R. Jensen, 1998, "Extraction and Modeling of Urban Attributes Using Remote Sensing Technology," *People and Pixels: Linking Remote Sensing and Social Science*, Washington: National Academy Press, 164–188.

Cullingworth, B., 1997, *Planning in the USA: Policies, Issues and Processes*, London: Routledge, 280 pp.

Curran, P. J., and T. A. Hobson, 1987, "Landsat MSS Imagery to Estimate Residential Heat-Load Density," *Environment and Planning*, 19:1597–1610.

Da Costa, S. M. F. and J. P. Cintra, 1999, "Environmental Analysis of Metropolitan Areas in Brazil," *ISPRS Journal of Photogrammetry & Remote Sensing*, 54:41–49.

DoD, 1976, *Industrial Target Analysis Supplemental Reading*, Washington: U.S. Army Intelligence Center, SupR 62810–11.

DoD, 1977, *Power/Water Facility Clue Sheet*, Fort Belvoir: Defense Mapping School, DMS No. 527.

DoD, 1978a, *Coke, Iron and Steel Industries*, Fort Belvoir: Defense Mapping School, DMS No. 555.

DoD, 1978b, *Petroleum Industries*, Fort Belvoir: Defense Mapping School, DMS No. 553.

DOE, 2005, *Solar Photovoltaics*, Washington: Dept. of Energy, http://www.eia.doe.gov/cneaf/solar.renewables/page/solar-photv/solarpv.html.

Donnay, J. P., Barnsley, M. J. and P. A. Longley, 2001, *Remote Sensing and Urban Analysis*, NY: Taylor & Francis, 268 p.

Dousset, B. and F. Gourmelon, 2003, "Satellite Multi-sensor Data Analysis of Urban Surface Temperatures and Landcover," *ISPRS Journal of Photogrammetry & Remote Sensing*, 58:43–54.

Eliasson, I., 1992, "Infrared Thermography and Urban Temperature Patterns," *Intl. Journal of Remote Sensing*, 13(5):869–879.

Elvidge, C. D., Baugh, K. E., Kihn, E. A., Kroeh, H. W. and E. R. Davis, 1997, "Mapping City Lights with Nighttime Data from the DMSP Operational Linescan System," *Photogrammetric Engineering & Remote Sensing*, 63(6):727–734.

Ehlers, M., Janowsky, R. and M. Gohler, 2003, "Ultra High Resolution Remote Sensing for Environmental Monitoring," *Earth Observation Magazine*, 12(9):27–32.

FAS, 2006, *National Image Interpretation Rating Scales*, Washington: Federation of American Scientists, Intelligence Resource Program, www.fas.org/irp/imint/niirs.

Feldman, S. C., Pelletier, R. E., Walser, E., Smoot, J. R. and D. Ahl, 1995, "A Prototype for Pipeline Routing Using Remotely Sensed Data and Geographic Information System Analysis," *Remote Sensing of Environment*, 53:123–131.

FEMA, 2005, *National Flood Map Web Service*, Washington: FEMA, http://hazards.fema.gov/resources/flood_map_svc.htm.

FGDC, 2006, *Federal Geographic Data Committee*, Washington: FGDC, http://www.fgdc.gov/.

Fraser, C. S., Baltsavias, E. and A. Gruen, 2002, "Processing of IKONOS Imagery for Submeter 3D Positioning and Building Extraction, *Remote Sensing of Environment*, 56:177–194.

Friedl, M. A. and 13 co-authors, 2002, "Global Land Cover Mapping from MODIS: Algorithms and Early Results," *Remote Sensing of Environment*, 83:287–302.

Gamba, P., Hellwich, O. and P. Lombardo, 2003, "Algorithms and Techniques for Multi-source Data Fusion in Urban Areas," *ISPRS Journal of Photogrammetry & Remote Sensing*, 58:1–3.

Giri, C., Zhu, Z. and B. Reed, 2005, "A Comparative Analysis of the Global Land Cover 2000 and MODIS Land Cover Data Sets," *Remote Sensing of Environment*, 94:123-132.

GSDI, 2006, *Global Spatial Data Infrastructure (GSDI)*, Washington: Federal Geographic Data Committee, www.gsdi.org/.

Haack, B. K. and E. T. Slonecker, 1994, "Merged Spaceborne Radar and Thematic Mapper Digital Data for Locating Villages in Sudan," *Photogrammetric Engineering & Remote Sensing*, 60(10):1253–1257.

Haack, B. K., Guptill, S., Holz, R., Jampoler, S., Jensen, J. R. and R. Welch, 1997, "Chapter 15: Urban Analysis and Planning," *Manual of Photographic Interpretation*, Bethesda: ASP&RS.

Haala, N. and C. Brenner, 1999, Extraction of Buildings and Trees in Urban Environments, *ISPRS Journal of Photogrammetry & Remote Sensing*, 54:130–137.

Hammer, A., Heinemann, D., Hoyer, C., Kuhlemann, R., Lorenz, E., Muller, R. and H. G. Beyer, 2003, "Solar Energy

Assessment using Remote Sensing Technologies," *Remote Sensing of Environment*, 86:423–432.

Henderson, F. M. and Z. Xia, 1998, "Radar Applications in Urban Analysis, Settlement Detection and Population Estimation," *Principles and Applications of Imaging Radar*, 3rd Ed., *Manual of Remote Sensing*, NY: John Wiley, 733–768.

Herold, M., Goldstein, N. C. and K. C. Clarke., 2003, "The Spatiotemporal form of Urban Growth: Measurement, Analysis and Modeling," *Remote Sensing of Environment*, 86:286–302.

Herold, M., Roberts, D. A., Gardner, M. E. and P. E. Dennison, 2004, "Spectrometry for Urban Area Remote Sensing - Development and Analysis of a Spectral Library from 350 to 2400 nm," *Remote Sensing of Environment*, 91:304–319.

Herold, M., Scepan, J. and K. C. Clarke, 2002, "The Use of Remote Sensing and Landscape Metrics to Describe Structures and Changes in Urban Land Uses," *Environment and Planning A*, 34:1443–1458.

Hodgson, M. E., Jensen, J. R., Raber, G., Tullis, J., Davis, B., Thompson, G. and K. Schuckman, 2005, "An Evaluation of LIDAR derived Elevation and Terrain Slope in Leaf-off Conditions," *Photogrammetric Engineering & Remote Sensing*, 71(7):817–823.

Holz, R. K., 1988, "Population Estimation of Colonias in the Lower Rio Grande Valley Using Remote Sensing Techniques," *Annual Meeting of the Association of American Geographers*, Phoenix, AZ.

Imhoff, M. L., Bounoua, L. DeFries, R., Lawrence, W. T., Stutzer, D., Tucker, C. J. and T. Ricketts, 2005, "The Consequences of Urban Land Transformations on Net Primary Productivity in the United States," *Remote Sensing of Environment*, 89:434–443.

Jadkowski, M. A., P. Convery, R. J. Birk and S. Kuo, 1994, "Aerial Image Databases for Pipeline Rights-of-Way Management," *Photogrammetric Engineering & Remote Sensing*, 60(3):347–353.

Jensen, J. R., 1995, "Issues Involving the Creation of Digital Elevation Models and Terrain Corrected Orthoimagery Using Soft-Copy Photogrammetry," *Geocarto Intl.*, 10(1): 1–17.

Jensen, J. R., 2005, *Introductory Digital Image Processing: A Remote Sensing Perspective*, 3rd Ed., Upper Saddle River, Prentice-Hall, 525 p.

Jensen, J. R. and D. C. Cowen, 1999, "Remote Sensing of Urban/Suburban Infrastructure and Socio-Economic Attributes," *Photogrammetric Engineering & Remote Sensing*, 65:611–622.

Jensen, J. R. and M. E. Hodgson, 2004, "Chapter 6: Remote Sensing of Selected Biophysical Variables and Urban/Suburban Phenomena," in *Geography and Technology*, S. Brunn, S. Cutter and J. Harrington, Jr. (Eds.), Boston: Kluwer, 109–154.

Jensen, J. R. and M. E. Hodgson, 2006, "Remote Sensing of Natural and Man-made Hazards and Disasters," in *Manual of Remote Sensing: Settlements*, M. K. Ridd (Ed.), Bethesda: ASP&RS, 401-429.

Jensen, J. R., Qiu, F. and K. Patterson, 2001, "A Neural Network Image Interpretation System to Extract Rural and Urban Land Use and Land Cover Information from Remote Sensor Data," *Geocarto Intl.*, 16(1):19–28.

Jensen, J. R. and D. L. Toll, 1983, "Detecting Residential Land-Use Development at the Urban Fringe," *Photogrammetric Engineering & Remote Sensing*, 48:629–643.

Jensen, J. R., Botchway, K., Brennan-Galvin, E., Johannsen, C., Juma, C., Mabogunje, A., Miller, R., Price, K., Reining, P., Skole, D., Stancioff, A. and D. R. F. Taylor, 2002, *Down to Earth: Geographic Information for Sustainable Development in Africa*, Washington: National Academy Press, 155 p.

Jensen, J. R., Cowen, D., Halls, J., Narumalani, S., Schmidt, N., Davis, B. A. and B. Burgess, 1994, "Improved Urban Infrastructure Mapping and Forecasting for BellSouth Using Remote Sensing and GIS Technology," *Photogrammetric Engineering & Remote Sensing*, 60(3):339–346.

Jensen, J. R., Halls, J. and J. Michel, 1998, "A Systems Approach to Environmental Sensitivity Index (ESI) Mapping for Oil Spill Contingency Planning and Response," *Photogrammetric Engineering & Remote Sensing*, 64(10):1003–1014.

Jensen, J. R., Huang, X., Graves, D. and R. Hanning, 1996, "Cellular Phone Transceiver Site Selection," *Raster Imagery in Geographic Information Systems*, S. Morain and S. Baros, (Eds.), Santa Fe: OnWard Press, 117–125.

Jensen, J. R., Hodgson, M. E., Tullis, J. A. and G. T. Raber, 2005a, "Chapter 2: Remote Sensing of Impervious Surfaces and Building Infrastructure", in *Geo-Spatial Technologies in Urban Environments*, Berlin: Springer, 5–21.

Jensen, J. R., Saalfeld, A., Broone, F., Cowen, D., Price, K., Ramsey, D., Lapine, L., and E. Lynn Usery, 2005b, "Chapter 2: Spatial Data Acquisition and Integration", in *A Research Agenda for Geographic Information Science*, Boca Raton: CRC, 17–60.

Ji, M. and J. R. Jensen, 1999, "Effectiveness of Subpixel Analysis in Detecting and Quantifying Urban Imperviousness from Landsat Thematic Mapper Imagery," *Geocarto Intl.*, 14(4):39–49.

Lacy, R., 1992, "South Carolina Finds Economical Way to Update Digital Road Data," *GIS World,* 5(10):58–60.

Leachtenauer, J. C., Daniel, K. and T. Vogl, 1998, "Digitizing Satellite Imagery: Quality and Cost Considerations," *Photogrammetric Engineering & Remote Sensing,* 64:29–34.

Light, D. L., 1993, "The National Aerial Photography Program as a Geographic Information System Resource," *Photogrammetric Engineering & Remote Sensing*, 59(1):61–65.

Light, D. L., 1996, "Film Cameras or Digital Sensors? The Challenge Ahead for Aerial Imaging," *Photogrammetric Engineering & Remote Sensing*, 62(3):285–291.

Light, D. L., 1998, personal communication, Emerge, Inc.

Lindgren, D. T., 1985, *Land-use Planning and Remote Sensing.* Boston, Martinus Nijhhoff.

Lo, C. P., 1986, "The Human Population," *Applied Remote Sensing*, NY: Longman, 40–70.

Lo, C. P., 1995, "Automated Population and Dwelling Unit Estimation from High-Resolution Satellite Images: A Geographic Information System Approach," *Intl. Journal of Remote Sensing*, 16(1):17–34.

Lo, C. P. and A. K. Yeung, 2002, *Concepts and Techniques of Geographic Information Systems*, Upper Saddle River: Prentice-Hall, 492 p.

Lo, C. P. and B. J. Faber, 1998, "Integration of Landsat Thematic Mapper and Census Data for Quality of Life Assessment," *Remote Sensing of Environment,* 62(2):143–157.

Lo, C. P., Quattrochi, D. A. and J. C. Luvall, 1997, "Application of High-Resolution Thermal Infrared Remote Sensing and GIS to Assess the Urban Heat Island Effect," *Intl. Journal of Remote Sensing,* 18(2):287–304.

Logicon, 1997, *Multispectral Imagery Reference Guide*, Fairfax: Logicon Geodynamics, 100 p.

Lowe, A. S., 2003, "The Federal Emergency Management Agency's Multi-Hazard Flood Map Modernization and The National Map," *Photogrammetric Engineering & Remote Sensing,* 69(10): 1133–1135.

NAICS, 2006, *North American Industry Classification System,* Washington: Bureau of the Census, http://www.census.gov/epcd/www/naics.html.

NPIC, 1961, *Photographic Interpretation Keys: Major Industries,* Washington: NPIC, 90 p.

NCRST, 2006, *National Consortium on Remote Sensing for Transportation*, Washington: Department of Transportation, www.ncgia.ucsb.edu/ncrst/.

NOAA, 2006, *Coastal National Land Cover Dataset Classification Scheme*, Charleston: NOAA Coastal Services Center, www.csc.noaa.gov/crs/lca/tech_cls.html.

NOAA ROC, 2006, *Radar Operations Center WSR 88D*, Washington: NOAA, http://www.roc.noaa.gov/.

Olorunfemi, J. F., 1984, "Land-use and Population: A Linking Model," *Photogrammetric Engineering & Remote Sensing*, 50:221–227.

Philipson, W., 1997, *Manual of Photographic Interpretation.* Bethesda: ASP&RS, 830 p.

Pike, J., 2006, *Space Policy Project of the Federation of American Scientists,* Washington: Federation of American Scientists, http://www.fas.org/irp/imint/niirs.htm. Source of the spatial resolutions associated with the National Image Interpretability Rating System (NIIRS).

Qui, F. and J. R. Jensen, 2004, "Opening the Neural Network Black Box and Breaking the Knowledge Acquisition Bottleneck of Fuzzy Systems for Remote Sensing Image Classification," *Intl. Journal of Remote Sensing*, 25(9):1749-1768.

Ridd, M. K., 1995, "Exploring a V-I-S (Vegetation-Impervious Surface-Soil) Model for Urban Ecosystems Analysis through Remote Sensing: Comparative Anatomy for Cities," *Intl. Journal of Remote Sensing*, 16(12):2165–2186.

Schultz, G. A., 1988, "Remote Sensing in Hydrology," *Journal of Hydrology*, 100(1988):239–265.

Schweitzer, B. and B. McLeod, 1997, "Marketing Technology that Is Changing at the Speed of Light," *Earth Observation Magazine,* (6):7, 22–24.

Skole, D. L., 1994, "Data on Global Land-Cover Change: Acquisition, Assessment, and Analysis," *Changes in Land-use and Land-cover: A Global Perspective,* W. B. Meyer and B. L. Turner (Eds.), Cambridge: Cambridge Univ. Press, 437–472.

Slonecker, E. T., Shaw, D. M. and T. M. Lillesand, 1998, "Emerging Legal and Ethical Issues in Advanced Remote Sensing Technology," *Photogrammetric Engineering & Remote Sensing,* 64(6):589–595.

Small, C., 2001, "Multiresolution Analysis of Urban Reflectance," *Proceedings, IEEE/ISPRS Workshop on Remote Sensing and Data Fusion over Urban Areas,* Rome, Nov., 8, 1–6.

Small, C., 2002, "Multitemporal Analysis of Urban Reflectance," *Remote Sensing of Environment,* 81:427-442.

Small, C., 2003, "High Spatial Resolution Spectral Mixture Analysis of Urban Reflectance," *Remote Sensing of Environment,* 88:170–186.

Small, C., 2004, "The Landsat ETM+ Spectral Mixing Space," *Remote Sensing of Environment,* 93:1–17.

Sutton, P., Roberts, D., Elvidge, C. and H. Meij, 1997, "A Comparison of Nighttime Satellite Imagery and Population Density for the Continental United States," *Photogrammetric Engineering & Remote Sensing,* 63(11):1303–1313.

Sutton, P. C., 2003, A Scale-adjusted Measure of "Urban sprawl" using Nighttime Satellite Imagery," *Remote Sensing of Environment,* 86:353–369.

Thirumalai, T. K., 2003, *Remote Sensing and Geospatial Information Technologies Applicable to Multimodal Transportation,* Washington: Dept. of Transportation, www.ncgia.ucsb.edu/ncrst/research/cip/ncgia.html, 6 p.

Tullis, J. A. and J. R. Jensen, 2003, "Expert System House Detection in High Spatial Resolution Imagery Using Size, Shape, and Context," *Geocarto Intl.,* 18(1):5–15.

USGS, 2006, *USGS National Land Cover Data,* Sioux Falls: EROS Data Center, http://landcover.usgs.gov/classes.asp.

Venderhoe, A. P., 2005, *Technical brief: Evaluating Lidar for Engineering and Design,* Washington: U.S. Department of Transportation Research and Special Programs Administration, www.ncgia.ucsb.edu/ncrst/research.html, 2 p.

Wilson, E. H., Hurd, J. D., Civco, D. L., Prisloe, M. P. and C. Arnold, 2003, "Development of A Geospatial Model to Quantify Describe and Map Urban Growth," *Remote Sensing of Environment,* 86:275–285.

Wilson, J. S., Clay, M., Martin, E., Stuckey, D. and K. Vedder-Risch, 2003, "Evaluating Environmental Influences of Zoning in Urban Ecosystems with Remote Sensing," *Remote Sensing of Environment,* 86:303–321.

Remote Sensing of Soils, Minerals, and Geomorphology

14

O nly 26 percent of the Earth's surface is exposed land. The remaining 74 percent is covered by water (including inland seas, lakes, reservoirs, and rivers). Very few people actually live on boats or on structures located in water. Almost all of humanity lives on the terrestrial, solid Earth comprised of bedrock and the weathered bedrock we call *soil*. Humankind is able to obtain a relatively abundant harvest in certain parts of the world from this soil. They are also able to extract important minerals from the bedrock and derivative materials that we use in industrial/commercial processes, hopefully to improve the quality of life on Earth. It is important to have accurate information about the location, quality, and abundance of soils, minerals, and rocks in order to conserve these often irreplaceable natural resources.

Bedrock is continually weathered and eroded by the combined effects of water, wind, and/or ice. The sandstone monolithic dome Ayers Rock in Australia is a good example of tilted sedimentary bedrock being weathered and eroded (Figure 14-1). The once buried sandstone slab was exposed to surface erosion approximately 70 million years ago. The eroded materials have been moved to other locations via mass transport. These unconsolidated sedimentary materials are called surficial deposits. Remote sensing can play a role in the identification, inventory, and mapping of soils that are on the surface of the Earth, especially when *surficial* soils are not covered with dense vegetation. This chapter reviews the fundamental issues associated with remote sensing the spectral characteristics of soils. The impact of soil grain size, organic matter, and water content on soil spectral reflectance are identified. Remote sensing may also assist in the modeling of soil erosion, providing biophysical information for the Universal Soil Loss Equation and other hydrologic models (e.g., King et al., 2005).

Remote sensing can provide information about the chemical composition of rocks and minerals that are on the Earth's surface and not completely covered by dense vegetation. Emphasis is placed on understanding unique *absorption bands* associated with specific types of rocks and minerals, as recorded using imaging spectroscopy (Dalton et al., 2004; Hook et al., 2005). In certain instances, remote sensing can be used to identify geobotanical relationships and to identify soil geochemistry or rock type (e.g., Rowan et al., 2000).

The chapter concludes with an overview of how general geologic information may be extracted from remotely sensed data, including information on lithology, structure, drainage patterns, and geomorphology (landforms) (Walsh et al., 1998; Bocco et al., 2005). Remote sensing data are generally of limited value for detecting deep subsurface geologic features unless they have a surface surrogate expression that can be extrapolated to depth.

Ayers Rock (Uluru), Australia

QuickBird image obtained February 4, 2002.

Figure 14-1   Ayers Rock (Uluru in aborigine) in Australia is a monolithic slab of Arkose sandstone that rises more than 348 m (1,100 ft) above the desert. The once horizontal sedimentary slab was tilted (folded) so that it protrudes through the surface at an angle of approximately 85°. It continues below the ground for 5 to 6 km. Some layers of Arkose sandstone are less resistant than others and erode more rapidly. This produces the parallel ribs or ridges. Ayers Rock has an area of 3.33 sq. km with a circumference of 9.4 km and lies 1,395 km (538 mi) south of Darwin (courtesy DigitalGlobe, Inc.).

Soil Characteristics and Taxonomy

Soil is unconsolidated material at the surface of the Earth that serves as a natural medium for growing plants. Plant roots reside within this material and extract water and nutrients. Soil is the weathered material between the atmosphere at the Earth's surface and the bedrock below the surface to a maximum depth of approximately 200 cm (USDA, 1998). A mature, fertile soil is the product of centuries of physical and chemical weathering of rock, combined with the addition of decaying plants and other organic matter (Loynachan et al., 1999). Soil is essential to the Earth's life-support system on the land. Agronomists refer to this as the *solum*.

Soil is comprised of solid particles (minerals and organic matter) of varying size and composition that make up about 50 percent of the soil's volume. Soils also contain liquid and gases. A soil is characterized by one or both of the following: horizons, or layers, that are distinguishable from the initial material as a result of additions, losses, transfers, and transformations of energy and matter or the ability to support rooted plants in a natural environment (USDA, 2003). The weathered, unconsolidated organic and inorganic mineral material that lies on top of the bedrock shown in Figure 14-2 varies greatly in composition and thickness throughout the Earth. In the heartland of continents such as North America, it may be 25 – 300 m deep. On steep mountain slopes or in deserts it may be almost completely absent. Permafrost soils may exist in arctic climates.

Soil Horizons

Biological, chemical, and physical processes create vertical zonation within the upper 200 cm or so of soils in which there is comparatively free movement of gravity water and groundwater capillary moisture. This results in the creation of relatively horizontal layers, or *soil horizons*. There are several standard horizons in a typical soil profile situated above the bedrock, including (Figure 14-2) *O, A, E, B, C, R,* and *W*, that may be distinguishable from one another based on their color (hue, value, chroma), texture, and chemical properties (USDA, 1998; 2003).

The *epipedon* (Greek *epi*, over, upon and *pedon*, soil) is a horizon that forms at or near the surface and in which most of the rock structure has been destroyed (USDA, 2003). The humus-rich topsoil, or *O horizon,* in the epipedon contains more than 20 percent partially decayed organic matter. Thus, it is a complex mixture of inorganic soil particles and decaying organic matter. *O* horizon soils typically have a dark brown or even black surface layer ranging in thickness from a few centimeters to several meters in areas where dense plant cover exists. This horizon is created by the interaction of water, other chemicals, heat, organic material, and air among the soil particles. Plant root systems extract much of their water and nutrients from within this "zone of life" (Marsh and Dozier, 1981).

The *A horizon* is a *zone of eluviation* or *leaching* formed at the surface or below an *O* horizon, where water moving up and down in the soil column leaches out minerals in solution (ions) and clay colloids from within the soil and relocates them to other horizons below. *A* horizons exhibit obliteration of all or much of the original rock structure. In a humid (wet)

Soil Profile

Humus-rich
Fine-textured humus
Light-colored mineral particles
Colloid accumulation; hardpan likely
Transition to subsoil
Subsoil; usually a deposit but may be minerals weathered from bedrock
Bedrock

Zone of Eluviation
Zone of Illuviation

Figure 14-2   A standard soil profile as defined by the U.S. Department of Agriculture. The major soil horizons are *O*, *A*, *E*, *B*, *C*, situated on top of bedrock, *R*. The *O* horizon contains partially decayed organic matter. Subhorizons are transitional to and between the horizons (U.S. Department of Agriculture, 1998).

climate, clay minerals, iron oxides, and dissolved calcite are usually moved downward. This leaching may cause the *A* horizon to be pale and sandy, but the uppermost part is often darkened by humus (decomposed plant material) that collects at the top of the soil. The *A* horizon soils may be influenced by cultivation or animal pasturing in agricultural environments.

Sometimes an *E horizon* exists between the zone of eluviation (*A* horizon) and the zone of illuviation (*B* horizon). *E* horizons have mineral layers in which the main feature is the loss of silicate clay, iron, aluminum, or some combination of these materials, leaving a concentration of sand and silt particles. All or much of the original rock structure is usually obliterated in an *E* horizon.

The downward movement of ions and colloids in the soil often terminates at a certain location in the soil column where the upward pressure of capillary groundwater offsets that of the percolating gravity water from surface precipita-

tion or irrigation. Here the materials may be deposited in the *B horizon*, the *zone of illuviation*. This layer is commonly rich in clay and is colored red or yellow by iron oxides (Loynachan et al., 1999). Over time, the zone of illuviation can collect such a mass of colloids and minerals that the interparticle air spaces become clogged, cementing the particles together. This can lead to the development of an impervious *hardpan*.

The *C horizon* is simply weathered parent material, lying below the *B* horizon. Most are mineral layers. The parent material is commonly subjected to physical and chemical weathering from frost action, roots, plant acids, and other agents. In some cases, the *C* horizon is transitional between unweathered bedrock (*R*) below and the developing soil above. Sometimes there is a water layer within or beneath the soil. The water may be permanently frozen (*Wf*) or not permanently frozen (*W*).

Soil Grain Size and Texture

The average diameter of grains of soil in a soil horizon is one of the major variables used to identify the taxonomy of a soil. There are three universally recognized soil grain size classes: *sand*, *silt*, and *clay*. Figure 14-3 identifies three different scales used to classify soil particles based on their diameters. The U.S. Department of Agriculture scale is (Loynachan et al., 1999):

- Sand: a) soil particle between 0.05 and 2.0 mm in diameter; b) soil composed of a large fraction of sand-size particles.

- Silt: a) soil particle between 0.002 and 0.05 mm in diameter; b) soil composed of a large fraction of silt-size particles.

- Clay: a) soil particle < 0.002 mm in equivalent diameter; b) soil that has properties dominated by clay-size particles.

Sand, silt, and clay-size particles play different roles in the soil formation process. Soil particles with sand-size diameters enhance soil drainage because water can percolate freely in the large air spaces between the large soil particles. Conversely, silt and clay-size soil particles enhance the movement and retention of soil capillary water. The very small clay-size soil particles carry electrical charges that attract and hold minute particles of dissolved minerals (ions) such as potassium and calcium. Because the ions are attached to the clay particles, they are not readily washed away. Thus, some clay in a soil horizon helps to maintain soil fertility by

a. Soil Science Society of America and U.S. Department of Agriculture Soil Particle Size Scale

Clay	Silt	Sand					Gravel
		v. fine	fine	medium	coarse	v. coarse	

0.002   0.05 0.1 0.25 0.5   1   2 mm   76.2

Clay

Silt

Particle size relative
to a grain of sand
0.15 mm in diameter

0.15 mm

Sand

b. MIT and British Standards Institute

Clay	Silt			Sand			Gravel	Stones
	fine	medium	coarse	fine	medium	coarse		

0.002   0.006   0.02   0.06   0.2   0.6   2 mm

c. International Society of Soil Science

Clay	Silt	Sand		Gravel
		fine	coarse	

0.002   0.02   0.2   2 mm

Figure 14-3   Three soil particle size scales: a) Soil Science Society of America, b) Massachusetts Institute of Technology and British Standards Institute, and c) International Society of Soil Science scale (after Marsh and Dozier, 1981; Loynachan et al., 1999).

retaining nutrient-rich potassium, calcium, and other dissolved minerals (Marsh and Dozier, 1981).

Soil horizons contain various proportions of sand, silt, and clay-size particles. *Soil texture* is the relative proportion of sand, silt, and clay in a soil. A typical soil's *texture* is the percentage in *weight* of particles in various size classes. The USDA soil-texture triangle shown in Figure 14-4 identifies the percentages of sand, silt, and clay, comprising standard soil types. For example, a loam soil found in the lower center of the diagram consists of 40 percent sand, 40 percent silt, and 20 percent clay. If a soil has a greater concentration of sand – say, 60 percent – and 10 percent clay and 30 percent silt, it is called a sandy loam soil, etc.

Soil Taxonomy

We no longer identify a "soil type." Rather, soil scientists determine the *soil taxonomy* (Petersen, 1999). *Keys to Soil Taxonomy* have been used by the USDA Natural Resources Conservation Service (USDA, 2003) since 1975 to qualitatively and quantitatively differentiate between soil taxa. The highest category of the U.S. Soil Taxonomy is Soil Order. Each order reflects the dominant soil-forming processes and the degree of soil formation. The 12 dominant U.S. Soil Orders are: Alfisols (high-nutrient soils), Andisols (volcanic soils), Aridisols (desert soils), Entisols (new soils), Gelisols

(tundra soils), Histosols (organic soils), Inceptisols (young soils), Mollisols (prairie soils), Oxisols (tropical soils), Spodosols (forest soils), Ultisols (low-nutrient soils), and Vertisols (swelling-clay soils). Additional dichotomous keys allow the scientist to classify the soil into Suborders, Great Groups, Subgroups, Family Level, and Soil Series.

The Soil Taxonomy is a relatively complex system based on the use of elimination dichotomous keys that allow the soil scientist to evaluate various characteristics in a soil profile, including soil color (hue, value, chroma), soil-texture class, moisture content, bulk density, porosity, and chemistry. For example, the first entry into the Key to Soil Orders is, Does the soil have 1) permafrost within 100 cm of the soil surface, *or* 2) gelic materials within 100 cm of the soil surface and permafrost within 200 cm of the soil surface? If it does, it is a Gelisol. If it does not, the analyst then progresses through the dichotomous key until he or she finds that the soil meets all the conditions identified.

 Remote Sensing of Soil Properties

From the previous discussion it is clear that most of the information used by soil scientists to map a soil series in a

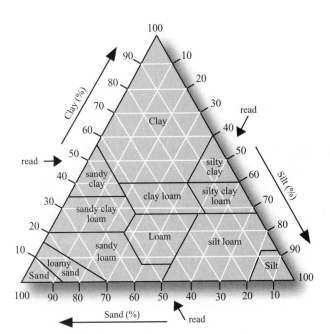

Figure 14-4   Percentages of clay (< 0.002 mm), silt (0.002 – 0.05 mm), and sand (0.05 – 2.0 mm) in the basic soil-textural classes (USDA, 1998).

region is obtained by direct observation in the field. It is essential that subsurface soil profiles be examined and careful biological, chemical, and physical measurements be obtained within each soil horizon. Thus, it is not realistic to expect remote sensing to be a universal panacea that can be used to perform soil taxa mapping without *in situ* data collection. Rather, we should understand that under certain conditions it is possible for remotely sensed data to be of value to the soil scientist as he or she collects all the pertinent material necessary to perform accurate soil classification. In fact, many soil scientists find that remotely sensed images of the terrain under investigation are essential to the soil mapping process (Petersen, 1999). Fortunately, some soil property characteristics may be measured remotely under ideal conditions.

For example, optical remote sensing instruments such as aerial photography, multispectral scanners, and hyperspectral remote sensors can record the spectral reflectance characteristics of the surface properties of soils if they are not totally obscured by dense multiple-story shrubs or trees with their coalescing canopies. Theoretically, the total upwelling radiance from an exposed soil recorded by the sensor onboard the aircraft or satellite, L_t, is a function of the electromagnetic energy from the sources identified in Figure 14-5 and summarized as

$$L_t = L_p + L_s + L_v \qquad (14-1)$$

where,

- L_p: This is the portion of the recorded radiance resulting from the downwelling solar (E_{sun}) and sky (E_{sky}) radiation that never actually reaches the soil surface. This is unwanted atmospheric path radiance noise and should ideally be removed from the imagery prior to trying to extract information about surficial soils or minerals.

- L_s: Some of the downwelling solar and sky radiation reaches the air-soil interface (*boundary layer*) and penetrates it approximately 1/2 wavelength (λ) deep into the soil. If the major wavelength of light being investigated is green light, the depth of penetration into the soil column would be approximately 0.275 µm (i.e., 1/2 of 0.55 µm). The amount of radiant flux exiting the soil column based on the reflection and scattering taking place at this depth is L_s. The characteristics of the soil organic matter (decomposed vegetation or animal material) and inorganic (mineral) constituents and the amount of soil moisture present have a significant impact on the amount of energy absorbed, scattered, and/or reflected from this surficial portion of the soil/rock matrix. Figure 14-5a depicts a soil with well-developed O and A horizons. It is likely that most of the energy reflected from this soil will be representative of the constituents of the O horizon. As the amount of humus (organic matter) diminishes, the surface reflectance will be more representative of the characteristics of the A horizon (Figure 14-5b). If both the O and A horizons are almost nonexistent, as shown in Figure 14-5c, the surface reflectance may be a function of the weathered subsoil (regolith) or even the bedrock if it is completely exposed, as on a steep mountain slope.

- L_v: Some of the incident downwelling solar and sky radiation may be able to penetrate perhaps a few millimeters or even a centimeter or two into the soil column. This may be referred to as *volume scattering, L_v.* Unlike water (refer to Chapter 12), however, there may be very little volumetric visible, near- and middle-infrared radiant flux exiting the soil profile from any appreciable depth. In fact, almost all the specular boundary layer reflectance (L_s) and volumetric scattering (L_v) may take place in the first few millimeters of the soil profile. The amount of volumetric radiant flux scattered or reflected back into the atmosphere is a function of the wavelength of incident energy (i.e., longer wavelength active microwave energy may penetrate farther into the soil, as has been demonstrated in the Sahara), the type and amount of organic/inorganic constituents, the shape and density of the minerals, the degree of mineral compaction, and the amount of soil moisture present. Figures 14-5b,c suggest

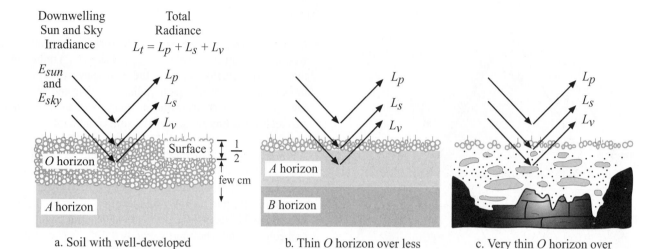

Figure 14-5   Soils and rocks receive irradiance from the Sun (E_{sun}) and atmosphere (E_{sky}). The total radiance upwelling from a soil/rock matrix toward the remote sensor is a function of radiance from unwanted atmospheric scattering (L_p), a substantial amount of radiance (L_s) reflected or scattered from the upper surface of the soil/rock matrix (approximately 1/2 wavelength deep), a small amount of subsurface volumetric radiance (L_v), and a very small amount of radiance from subsequent soil/rock substrate.

that as the O horizon becomes less well-developed, perhaps more incident energy interacts with the soil particles in the A horizon (Figure 14-5b) or even the subsoil and bedrock (Figure 14-5c).

The goal of most soil and mineral remote sensing is to extract the radiance of interest from all the other radiance components being recorded by the sensor system. For example, the scientist interested in identifying the organic and inorganic (mineral) constituents in the very top layers of the soil profile is most concerned with measuring the integrated spectral response of the surface and subsurface radiance, i.e., L_s and L_v:

$$L_s + L_v = L_t - L_p. \qquad (14\text{-}2)$$

This involves careful radiometric correction of the remote sensor data to remove atmospheric attenuation (L_p). Ideally we could disentangle the individual contribution of L_s and L_v to the reflected radiant flux. Unfortunately, this is difficult, and usually we must be content analyzing an integration (summation) of these two radiance constituents. Nevertheless, it is possible to make some general observations about how surficial soils appear in remote sensor data based on their spectral reflectance properties.

The spectral reflectance characteristics of soils are a function of several important characteristics, including:

- soil texture (percentage of sand, silt, and clay)
- soil moisture content (e.g., dry, moist, saturated)
- organic matter content
- iron-oxide content
- soil salinity
- surface roughness.

Generally, a dry soil that contains almost no organic matter exhibits a relatively simple, less complex spectral reflectance curve than those associated with terrestrial vegetation (Chapter 11) or algae-laden water bodies (Chapter 12). For example, Figure 14-6 depicts lab spectroradiometer reflectance of dry silt and dry sand soils. This demonstrates one of the most consistent characteristics of dry soils: *increasing reflectance with increasing wavelength, especially in the visible, near- and middle-infrared portions of the spectrum.* However, as a soil gains moisture, and/or additional organic content or iron oxide, its spectral response may depart from the simple curve. Therefore, it is useful to review how these parameters influence the spectral response of surficial soils. A goal of remote sensing is to disentangle the spectral response recorded from a surficial soil and be able to identify the proportions and/or influence of the characteristics within the instantaneous field of view of the sensor system.

Soil Texture and Moisture Content

There is a relationship between the size of the soil particles found in a mass of soil (e.g., m^3) and the amount of moisture that the soil can store. Figure 14-7a depicts several theoreti-

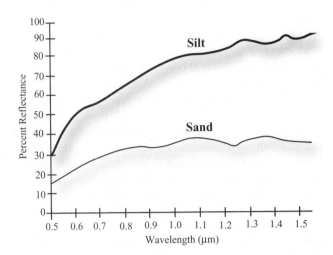

Figure 14-6 *In situ* spectroradiometer reflectance curves for dry silt and sand soils. Reflectance generally increases with increasing wavelength throughout the visible, near- and middle-infrared portions of the spectrum.

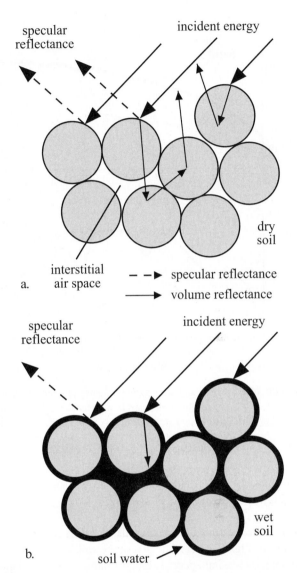

Figure 14-7 a) Incident radiant energy may be reflected from the surface of the dry soil, or it may penetrate into the soil particles, where it may be absorbed or scattered. The total reflectance leaving the soil is a function of the specular reflectance and the internal volume reflectance. b) As soil moisture increases, each particle may be encapsulated with a thin membrane of capillary water. The interstitial spaces may also fill with water. The greater the amount of water in the soil, the greater the absorption of incident electromagnetic energy and the lower the soil reflectance.

cal grains of sand. Incident radiant flux may be reflected from the surface of the sand grains yielding specular reflectance, or the incident energy may penetrate some distance into the sand particle. The energy may then be absorbed by the particle and turned into heat energy or exit the particle and be scattered or absorbed by other particles. The void between the soil particles is called the *interstitial air space*. The total reflectance per wavelength (e.g., blue, green, red, near-infrared light) for a given dry soil with no organic content and no iron oxides is a function of the average of both the *soil specular reflected energy* from the soil particle surfaces and the *soil volume reflectance* taking place due to internal scattering.

The finer clay soils have particles that are packed very closely to one another. The interstitial air spaces between the soil particles are very small. Conversely, sand particles are very large and contain relatively large interstitial air spaces. When precipitation occurs or groundwater rises into the soil horizon, the individual particles may become surrounded by a thin membrane of capillary water. Water may also occupy the interstitial air spaces (Figure 14-7b). The densely packed clayey soil, with each of its small particles holding a membrane of water around it, can hold a tremendous amount of water. Conversely, sandy soils with their significantly larger soil particles and large air spaces 1) drain much more rapidly than clayey soils and 2) are dried out much more rapidly by evaporation than the clayey soils. So what does this have to do with the spectral reflectance characteristics of soils? Basically, the amount of moisture held in the surficial soil

layer is a function of the soil texture. The finer the soil texture, the greater the soil's ability to maintain a high moisture content in the presence of precipitation. The greater the soil

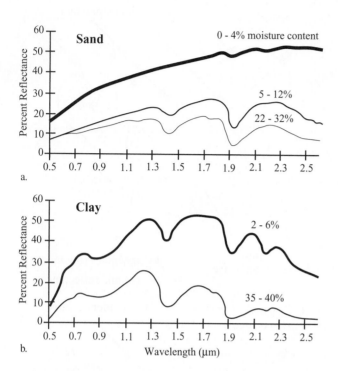

Figure 14-8   Higher moisture content in sandy (a) and clayey soil (b) results in decreased reflectance throughout the visible and near-infrared region, especially in the water-absorption bands at 1.4, 1.9, and 2.7 μm (Hoffer, 1978).

moisture, the more incident radiant energy absorbed and the less reflected energy.

This is demonstrated for sandy soils with little surface vegetation in Figure 14-8a. The dry sandy soil has a relatively simple, increasing spectral response throughout the region from 0.5 – 2.6 μm. However, as soil moisture increases, the water in the upper few centimeters of the soil begins to selectively absorb significant amounts of incident energy in the water-absorption bands at 1.4, 1.9, and 2.7 μm. The soil moisture also absorbs more incident radiant energy in the spectral regions adjacent to these absorption bands. The result is a much more complex spectral response, with characteristic dips in the reflectance curve at the 1.4, 1.9, and 2.7 μm atmospheric water-absorption bands. Also note that the amount of reflected green, red, near- and middle-infrared radiant energy is dramatically reduced as the moisture content increases.

This same relationship holds for clayey soils as demonstrated in Figure 14-8b, where increased moisture in the upper few centimeters of the soil dampens the entire spectral response throughout the wavelength interval from 0.5 – 2.6

μm and deepens the absorption around the water-absorption bands. Notice, however, that the water-absorption bands appear to be much more active. This is because clayey soils with their fine soil texture manage to hold more moisture in the upper portion of the soil horizon, which allows the moisture to absorb some of the incident radiant flux, creating significant dips in and around the water-absorption bands.

Experienced image analysts know that remote sensor data of exposed soil surfaces obtained after a major precipitation event such as a thunderstorm or prolonged frontal activity will be noticeably *darker* than if the imagery was acquired after many days without precipitation. This is because the water in the surficial soil absorbs much of the incident radiant energy, especially in the visible and near-infrared portions of the spectrum, resulting in less radiance exiting toward the sensor system.

If high spectral and radiometric resolution sensors are available, it may be possible to differentiate between soils with different soil textures. This is because almost all soils that have a moderate to large proportion of clay-size particles exhibit strong *hydroxyl absorption bands* at approximately 1.4 and 2.2 μm. The spectral response of the sandy soil in Figure 14-8a was not influenced by the hydroxyl absorption band at 2.2 μm whereas the spectral response of the clayey soil in Figure 14-8b exhibited significant absorption at both 1.4 and 2.2 μm. Of course, it is only possible to differentiate between the clayey and silt/sandy soil texture characteristics if the soils are almost dry and contain very little organic matter that could mask the relationship, and the sensor is sensitive to very specific wavelength intervals (e.g., centered at 1.4 and 2.2 μm). Chabrillat et al. (2002) mapped expansive clay soils (smectite, illite, and kaolinite) along the Colorado front range using hyperspectral data and the information found in the absorption bands in the 1900 - 2400 nm spectral region. Okin and Painter (2004) used hyperspectral data to map the spatial distribution of sand grain size (from 1.0 to 0.2 mm) in windblown abandoned agricultural fields in a desert environment. They found that the larger the grain size, the lower the short-wave infrared reflectance, especially in the regions centered on 1.7 and 2.2 μm.

It is also possible to differentiate between some soil textures using active microwave (RADAR) backscatter characteristics. For example, the 1981 NASA Shuttle Imaging Radar mission (SIR-A) mission revealed buried river channels beneath the sand of the Sahara (McCauley et al., 1982) (Figure 9-21). River channels appeared as dark, dendritic patterns in L-band (24 cm) images, and sand-covered bedrock returned a bright radar signal. The radar signals were able to penetrate up to 1–2 m of very dry sand to return a signal

Figure 14-9   The greater the amount of organic content in a soil, the greater the absorption of incident energy and the lower the spectral reflectance.

from buried bedrock and calcium carbonate-cemented nodules. Williams and Greeley (2004) found that relatively short wavelength C-band (5.8 cm) signals were best for identifying the extend of dry sand while longer wavelength L-band signals penetrated thin moist sand best.

Soil Organic Matter and Biological Soil Crusts

Plants and animals decompose and become organic humus in the upper portions of the soil horizon. The amount of organic matter in the soil has a significant impact on the spectral reflectance characteristics of exposed soils. Generally, the greater the amount of organic content in the upper portions of the soil, the greater the absorption of incident energy and the lower the spectral reflectance. Figure 14-9 summarizes the curvilinear relationship for soils with 0 to 100 percent organic matter.

Sometimes soils have crusts of biological matter (Wassenaar et al., 2005). *Biological soil crusts* are communities of mosses, lichens, liverworts, algae, fungi, cyanobacteria and bacteria. They can survive desiccation and extreme temperatures (up to 70 °C), high pH, and high salinity. Biological soil crusts have been found in deserts throughout the world and play an important role in desert ecosystems in soil formation, stability and fertility and prevent soil erosion (Belnap, 2003). The spectral reflectance characteristics of three types of biological soil crust are shown in Figure 14-10.

Based on these relationships, Chen et al. (2005) developed a biological soil crust index (BSCI) using Landsat Enhanced Thematic Mapper Plus (ETM[+]) imagery:

$$BSCI = \frac{1 - L \times |R_{red} - R_{green}|}{\left[\dfrac{(R_{green} + R_{red} + R_{nir})}{3}\right]} \qquad (14\text{-}3)$$

where R_{green} and R_{red} are reflectance in ETM+ bands 2 and 3 and L is an adjustment factor to amplify the absolute difference between R_{green} and R_{red}. If the numerator gets a negative value, L is restricted to a range from 2 to 4. The denominator is the mean reflectance of green, red and the near-infrared reflectance in ETM+ bands 2, 3, and 4.

Iron Oxide

In the southeastern U.S., Southeast Asia, and several other parts of the world, iron oxides are present in the soil. The existence of iron oxides generally causes an increase in reflectance in the red portion of the spectrum (600 – 700 nm), and hence its reddish color (Figure 14-11). There is also a noticeable decrease in the blue and green reflectance in the iron-oxide soil. The iron-oxide soil also exhibits an absorption band in the 850 – 900 nm region when compared with a sandy loam soil with no iron oxide.

Soil Salinity

Soil salinity is a major environmental hazard. The global extent of primary salt-affected soils is about 955 M ha, while secondary salinization affects some 77 M ha, with 58 percent of these in irrigated areas. Nearly 20 percent of all irrigated land is salt-affected, and this proportion is increasing despite land reclamation efforts. Salts tend to concentrate on the soil surface in arid and irrigated areas.

Metternicht and Zinck (2003) provide an overview of the considerations that must be made when trying to extract salinity information from remote sensing systems. Figure 14-12a depicts the spectral characteristics of saline versus non-saline crusts for silty loam soils in Bolivia. Note that some spectral confusion occurs between salty crusts and silt loam bright crusts in the blue and green portions of the spectrum (450 – 550 nm). Reflectance generally increases with increasing surface soil salt concentrations. Salt-affected soils show relatively higher spectral response in the visible and near-infrared regions of the spectrum than nonsaline soils do, and strongly saline-sodic soils exhibit higher spectral responses than moderately saline-sodic soils (Rao et al.,

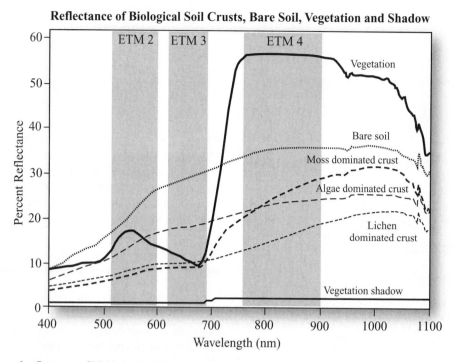

Figure 14-10 Spectral reflectance of biological soil crusts compared with bare soil, vegetation, and vegetation shadow reflectance (Chen et al., 2000; reprinted with permission from Elsevier).

1995). This graph makes it clear why saline deposits typically show up as bright white in color-infrared aerial photography or multispectral false-color composites (due to relatively high proportions of green, red, and near-infrared reflectance), while vegetation shows up as bright red (due to chlorophyll absorption in the green and red portions of the spectrum and high reflectance in the near-infrared). Indicators used to distinguish saline/alkaline soils are spotty growth of crops, presence of dead trees, a blue-green tinge, and moisture stress condition. Spectroradiometer measurements of halite with various salinity concentrations are shown in Figure 14-12b (Farifteh et al., 2006).

Surface Roughness

As we learned in Chapter 9 (Active Microwave Remote Sensing), the smaller the local surface roughness relative to the size of the incident radiation, the greater the specular spectral reflectance from the terrain. We may utilize this information to make some general statements about the amount of spectral reflectance that *should* be exiting certain soil textures, assuming they contain no moisture, organic content, or iron oxides. Dry, fine-texture clayey soil should

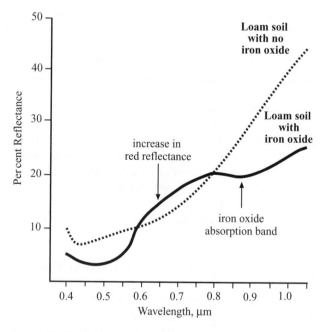

Figure 14-11 Iron oxide in a sandy loam soil causes an increase in reflectance in the red portion of the spectrum (0.6 – 0.7 μm) and a decrease in reflectance in the near-infrared region (0.85 – 0.90 μm).

Figure 14-12 a) Spectral reflectance of saline and non-saline soil crusts (Metternicht and Zinck, 2003; reprinted with permission from Elsevier) versus the spectral response of vegetation. b) Spectroradiometer reflectance measurements of halite (NaCL) at varying salt concentrations (Farifteh et al., 2006; reprinted with permission from Elsevier).

produce a higher spectral response throughout the visible and near-infrared portions of the spectrum due to the near-specular reflection that should take place from its surface versus silt or sand surfaces. Unfortunately, this is often not the case because as moisture, organic content, or iron oxide is added to the clayey soil, it begins dramatically to absorb incident radiant flux, and quite possibly to appear like silt or perhaps even sand on remote sensor data. This can cause interpretation problems.

Conversely, dry sand with its well-drained large grains should diffusely scatter the incident wavelengths of visible and near-infrared energy more than clayey and silt soils. Therefore, it is common for more coarse-grained sand areas to be among the brightest terrain in the landscape while clayey soils are among the darkest. King et al. (2005) and Zribi et al. (2005) describe how surface roughness can be measured using active microwave remote sensing systems.

Remote Sensing of Rocks and Minerals

Rocks are assemblages of *minerals* that have interlocking grains or are bound together by various types of cement (usually silica or calcium carbonate). When there is minimal vegetation and soil present and the rock material is visible directly by the remote sensing system, it may be possible to differentiate between several rock types and obtain information about their characteristics. Most rock surfaces consist of several types of minerals. Clark (1999) suggests that it is possible to model the reflectance from an exposed rock consisting of several minerals or a single mineral based on Hapke's (1993) equation:

$$r_\lambda = \left[\frac{w'}{4\pi} \times \frac{\mu}{\mu + \mu_o}\right] \times [(1 + B_g)P_g + H_\mu H_{\mu_o}^{-1}] \qquad (14\text{-}4)$$

where r_λ is the reflectance at wavelength λ, μ_o is the cosine of the angle of incident light onto the rock or mineral of interest, μ is the cosine of the angle of emitted light, g is the phase angle, w' is the average single scattering albedo from the rock or mineral of interest, B_g is a backscatter function, P_g is the average single particle phase function, and H is a function for isotropic scatterers.

Armed with this advanced reflectance theory and the known optical constants of the minerals involved, it is possible to compute the theoretical reflectance spectra for 1) pure minerals that have a single grain size, 2) a pure mineral with a variety of grain sizes, and 3) mineral mixtures with varying grain sizes (Clark, 1999). This is important because it means that we may be able to predict what the reflectance curves of specific types of minerals or rocks should look like at various wavelengths. This can be important when we attempt to interpret the information content of imaging spectroscopy remote sensor data of mineralized terrain.

Imaging Spectroscopy of Rocks and Minerals

Imaging spectrometry instruments are useful for obtaining quantitative information about rock type and mineral composition. Chapter 7 reviewed the fundamental characteristics of imaging spectroscopy sensor systems such as the Airborne Visible Infrared Imaging Spectrometer (AVIRIS). This section reviews the fundamental characteristics that impact our ability to determine rock type and mineral composition remotely using imaging spectrometry.

Imaging Spectrometers

Imaging spectrometers may be used in the lab, field, or in an airborne remote sensing mission. Reflectance and emittance spectroscopy of natural surfaces are sensitive to specific chemical bonds in materials. In the past, one of the problems associated with spectroscopy was that it was too sensitive to small changes in the chemistry and/or the structure of the material. This resulted in the creation of very complex spectral reflectance curves that were often unintelligible. Fortunately, significant strides have been made in 1) the quality of the imaging spectrometer sensors, and 2) our ability to understand and disentangle the information content of the spectroradiometer information. Thus, what was once a drawback (i.e., very complex spectral reflectance curves) is now an advantage as it allows scientists to extract more valuable information about the chemistry of the natural environment.

Energy-Matter Interactions — Reflection and Absorption Processes

As previously discussed, photons of light incident on a mineral or rock are 1) reflected from grain surfaces onto other grain surfaces, 2) passed through the grain onto other grains, and/or 3) absorbed within a grain. The photons that are reflected from the grain surfaces or refracted through a particle are said to be *scattered*. The scattered photons of light may encounter another grain or be scattered away from the surface, perhaps toward a remote sensing system where the amount and properties can be detected and measured. We are primarily concerned here with photons of light that originated at the Sun and then interacted with the mineral of interest. It is important to remember, however, that photons may also be *emitted* from a mineral because (as discussed in Chapter 8) all objects above absolute zero emit radiation.

If every mineral absorbed and scattered the incident photons of light in an identical manner, then there would be no basis for mineralogical remote sensing. The amount of energy leaving each type of mineral at each specific wavelength would be identical. Fortunately, certain types of minerals absorb and/or scatter the incident energy differently.

There are a number of processes that determine how a mineral will absorb or scatter the incident energy. Also, the processes absorb and scatter light differently depending on the wavelength (λ) of light being investigated. The variety of absorption processes and their wavelength dependence allow us to derive information about the *chemistry* of a mineral from its reflected or emitted energy. The ideal sensor is the imaging spectrometer because it can record much of the absorption information, much like using a handheld spectro-

Figure 14-13 Comparison of a laboratory spectra of alunite (an aluminum sulfate), a simulated Landsat Thematic Mapper spectra (resampled from the lab spectrum), and spectra obtained using an airborne 63-channel Geophysical and Environmental Research Imaging Spectrometer (GERIS) at Cuprite, NV. Symbols and channel numbers on the TM and GERIS spectra identify band centers. Most characteristic absorption band information is lost with the TM spectrum (with the exception of low reflectance in TM band 7 at 2.2 μm) while much of the spectral information is preserved in the GERIS spectrum. The spectra are offset vertically for clarity (after Kruse et al., 1990).

Figure 14-14 a) The index of refraction and extinction coefficient of quartz for the wavelength interval 6 – 16 μm. b) The spectral reflectance characteristics of powdered quartz obtained using a spectroradiometer (after Clark, 1999).

radiometer. For example, consider the spectral reflectance curves for alunite shown in Figure 14-13. The most detailed spectral reflectance information is obtained using a handheld spectroradiometer. Airborne spectral reflectance data obtained using the 63-channel Geophysical and Environmental Research Imaging Spectrometer (GERIS) retains much of the spectral information. Unfortunately, much of the spectral absorption information is lost when the spectral data are obtained from six bands of simulated Landsat Thematic Mapper data (Kruse et al., 1990).

All materials have a complex *index of refraction*. For example, the vacuum of outer space, the atmosphere, quartz, and water all have different indexes of refraction. If we illuminate a plane surface with photons of light from directly overhead, the light, *R*, will be reflected from the surface according to the Fresnel equation:

$$R = \frac{(n-1)^2 + K^2}{(n+1)^2 + K^2} \qquad (14\text{-}5)$$

where n is the index of refraction, and K is the extinction coefficient. Both the index of refraction and the extinction

coefficient of quartz are shown in Figure 14-14a (Clark, 1999). From this illustration it is clear that the optical constants of n and K for quartz vary strongly with wavelength. Note that the index of refraction (n) reaches a minimum just before 8.5 μm and 12.6 μm. The relative reflectance of powdered quartz measured by a spectrometer for the wavelength interval from 6 – 16 μm is shown in Figure 14-14b (Clark, 1999). The reflectance spectra of quartz throughout the visible and near-infrared region is effectively zero and is therefore not shown. However, from 8 – 9.5 μm and at 12.6 μm there is a dramatic increase in reflectance. If quartz is to be detected at all using imaging spectrometry, it may be necessary to sense in the region from 8 – 10 μm as shown. But what causes the reflectance spectra of quartz to appear as it does? Why are certain parts of the spectrum absorbed more completely than others? The answer lies at the heart of using imaging spectrometry for mineral analysis. It is because of the specific types of absorption that take place within the minerals.

Absorption Processes

As demonstrated in the previous illustration, a typical spectral reflectance curve obtained by an imaging spectrometer exhibits various maxima and minima. The minima are caused by strong absorption bands. For example, laboratory and AVIRIS remote sensing derived spectra for three minerals, kaolinite clay, aluminum sulfate (alunite), and buddingtonite (an ammonium feldspar) are shown in Figure 14-15 (Van der Meer, 1994). Scientists have documented that specific minerals exhibit relatively unique absorption spectra. For example, key absorption features associated with kaolinite are typically found at 2.17, 2.21, 2.32, and 2.38 μm. If a spectra exhibits minima at these locations, it may well be kaolinite. It is important to point out here that only a hyperspectral sensor with a spectral bandwidth resolution of approximately 10 nm could capture such information. Spectroradiometers with 20 nm bandwidths might miss the important minima or maxima entirely. This diagram also suggests that the differences in maxima, minima, and the slope between maxima and minima might allow these three minerals to be differentiated one from another using hyperspectral remote sensor data. The absorption bands in these minerals are caused by electronic and vibrational processes.

Electronic Processes: The most common electronic process revealed in the spectral reflectance curves of minerals is due to unfilled electron shells of transition elements such as Ni, Cr, Co, Fe, etc. (Clark, 1999). This is called a crystal field effect. Absorption bands can also be caused by *charge transfer absorptions*, *conduction bands*, and *color centers*. Please refer to Clark (1999) or the *Manual of Remote Sensing – Geosciences* (Rencz, 1999) for detailed information about these electronic absorption processes.

Vibrational Processes: The bonds in a molecule or crystal lattice are like springs with weights attached that cause the system to vibrate. The frequency of vibration depends on the bonds in the molecule and their masses. For a molecule with N atoms, there are 3N–6 normal modes of *vibrations* called *fundamentals*. Each vibration can also occur roughly at multiples of the original fundamental frequency. The additional vibrations are called *overtones* or *combinations*. Hunt (1977) devised a spectral signature diagram that summarizes the impact of both *electronic* and *vibrational* absorption processes for selected minerals (Clark, 1999). Water, hydroxyl, carbonates, phosphates, borates, arsenates, and vanadates have diagnostic vibrational absorption bands.

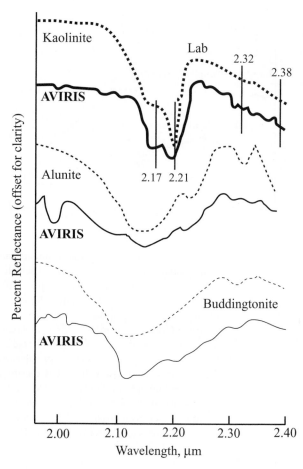

Figure 14-15 Spectra of three minerals derived from the Airborne Visible Infrared Imaging Spectrometer (AVIRIS) using a lab spectroradiometer (after Van der Meer, 1994). The vertical lines identify key absorption features useful for identifying kaolinite. The discrepancy between the lab and AVIRIS curves for each mineral is due to 1) the lab spectra are produced from pure samples whereas the AVIRIS data are extracted from relatively large 20 × 20 m areas that contain a mixture of materials within the IFOV of the sensor system, 2) the AVIRIS data are recorded through an atmosphere, and 3) the lab samples are dry whereas the real-world terrain may exhibit slight differences in soil moisture. The spectra are offset vertically for clarity.

Spectral Reflectance Libraries

The U.S. Geological Survey has compiled a Digital Spectral Library (USGS Speclab, 2006). The California Institute of Technology Jet Propulsion Laboratory compiled a spectral library that includes the Salisbury et al. (1991) library, the JPL ASTER Spectral Library and the aforementioned USGS Digital Spectral Library (JPL, 2006). These are accurate spectral libraries. One should only use spectral libraries that have been created using rigorous imaging spectroscopy calibration standards.

Creating Mineral Maps Using Hyperspectral Data

If we obtain high spectral resolution remote sensing spectra for an unknown surficial rock material, remove the atmospheric effects and convert the brightness values to percent reflectance (or exitance), then it may be possible to search a spectral library and identify the type of mineral that has an identical or very similar spectra. In this manner, imaging spectroscopy can be used to derive significant mineralogical information with little or no *in situ* field work. In certain instances, mineralogical maps can be made.

For example, consider the two mineral maps shown in Color Plate 14-1 derived from low- and high-altitude AVIRIS data obtained over Cuprite, NV (Swayze et al., 1999). USGS scientists first removed atmospheric effects in the data and then used a computer program (Tetracorder) that compared the calibrated spectra obtained for each pixel in the AVIRIS data with the spectra contained in a mineral spectral reflectance library. Twenty-four mineral categories were found in the 2 – 2.5 μm region using Tetracorder. Note the detail present in the 2.3 × 7 m low-altitude AVIRIS data when compared with the results from the coarse 18 × 18 m high-altitude AVIRIS data (Swayze et al., 1999).

Rowan et al. (2000) conducted a geobotanical study using AVIRIS data in North and South Carolina to identify hydrothermally altered rocks based on the spectral reflectance characteristics of forest vegetation growing on them. Sparse monocultures of chestnut oak resulted in lower apparent canopy density at the altered sites, which was expressed in the AVIRIS data by lower O-H absorption intensity. Galvao et al. (2005) used well-defined 2200 nm hydroxyl absorption features to identify hydrothermally altered minerals (kaolinite and muscovite) using ASTER data. They caution, however, that substantial vegetation cover lowers the probability of correct identification in tropical environments. Ellis and Scott (2004) used HyMap hyperspectral data and absorption band features to differentiate between biotite and topaz granite mineral units in the United Kingdom. Absorption bands associated with remote sensing salt-affected soils are summarized in Farifteh et al. (2006).

ASTER Reflectance and Absorption Characteristics of Value for Rock and Mineral Discrimination

Many of the spectral diagnostic characteristics of rocks and minerals are found in the thermal infrared portion of the spectrum. The most important satellite sensor that provides such information is ASTER — the Advanced Spaceborne Thermal Emission and Reflection Radiometer launched onboard EOS *Terra*. Six ASTER bands (4 – 9) cover the short-wavelength infrared (SWIR) range. Band 6 is centered on a clay-absorption feature often associated with hydrothermal alteration and mineral potential. Band 8 is centered on a carbonate-absorption feature, allowing global discrimination of limestones and dolomites from other rocks. Five ASTER bands (10 – 14) cover the thermal infrared (TIR) range. Bands 10, 11, and 12 are designed to detect sulfates and silica spectral features. Evaluating reflectance in SWIR band 6 with the thermal infrared band 10 allows discrimination between common minerals such as alunite (a sulfate mineral important to precious metal deposits) and anhydrite (an evaporative sulfate common in arid regions). Band 14 is centered on a carbonate-absorption spectral feature complementing the SWIR band 8 (NASA ASTER, 2006).

 Geology

The Earth is not rock solid. It is constantly changing, moving, and being rearranged. *Geology* is the science of rocks. It reveals the immense history of the Earth and explains its geological formations. From a geological viewpoint, humanity is a relatively recent arrival, and it is one species among many millions that share an Earthly heritage.

Paleontology is the study of fossils, the remains of ancient organisms that have been turned to stone. William Smith's (1769 – 1839) work marked the beginning of geology. He studied English coal mines and observed that the same layers of sedimentary rocks were revealed in cuttings over large geographic areas. He identified and correlated the strata by their fossil content. He discovered that the order in which rock units were deposited did not vary across their geographic extent. He concluded that rock units at the bottom were older than those above.

Scottish geologist James Hutton (1726 – 1797) built upon Smith's findings and concluded that the history of rocks occurs in *cycles* (Trefil and Hazen, 1995). Rocks are decomposed into sediment by *weathering*. The sediment is often moved by forces of *erosion* and *mass transport* to accumulate in new locations. The sediment is often consolidated into a new type of rock that could be buried under more rock, until, heated to its melting point, it may flow back under pressure to the surface as extrusive lava or to some interior location as intrusive magma. The lava or magma cools into rock. Weathering begins immediately on extruded lava, whereas it may take millennia to expose the intrusive magma at the surface, where it may be weathered. In either case, the cycle begins again. Hutton raised the possibility that the Earth was much older than previously thought, and was continually changing and recycling itself (Selby, 1989).

Charles Lyell (1797 – 1875), the founder of modern geology, introduced *uniformitarianism* — the concept that past events occurred at the same rate as they do today. For example, sedimentary rock was formed by the same processes of sediment deposition and cementation, and at the same rate as those that can be observed today. He was the first to suggest that the Earth was millions of years old.

Finally, Alfred Wegener (1880 – 1930) proposed the theory of *continental drift*. Then, during the cold war of the 1950s, U.S. and Russian scientists set up sensitive vibration detectors to monitor each other's atomic tests. Such instruments also allowed them to identify the epicenter of thousands of earthquakes throughout the world. These epicenters usually occurred along distinctive lines. It became clear to geologists that these lines were edges of enormous plates that covered the Earth's surface and that earthquakes were the result of friction between the plates. Continents on the plates were slowly moving on the surface of the Earth. The concept of *plate tectonics* is related but not identical to continental drift. Today, most geological phenomena can be explained by these two concepts.

Most of our geologic information comes from detailed *in situ* investigation by geologists and paleontologists. They excavate, bore holes in the Earth to extract geologic core samples, and interpret fossil remains. In addition, some of these scientists use airborne and satellite remote sensing technology to supplement their *in situ* investigations. Remotely sensed images are routinely interpreted to identify lithology, structure, drainage-pattern characteristics, and landforms. In addition, remote sensing is often used in mineral and petroleum exploration.

Lithology

Geologists use remote sensing in conjunction with *in situ* observation to identify the *lithology* of a rock type, i.e., its origin. The different rock types are formed by one of three processes:

- *igneous* rocks are formed from molten material;

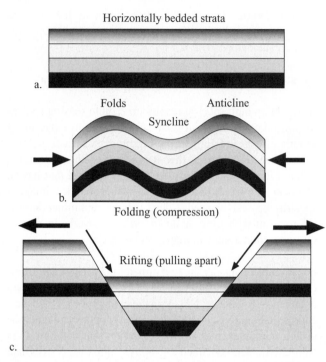

Figure 14-16 a) Horizontally bedded strata. b) Horizontally bedded strata that has been compressed, creating folds in the rock, including anticlines and synclines. c) An example of faulting where a block has been downthrown relative to two stable blocks of rock. This is referred to as rifting.

- *sedimentary* rocks are formed from the particles of pre-existing rocks and plant and animal remains; or

- *metamorphic* rocks are formed by applying heat and pressure to previously existing rock.

Rocks are weathered from the parent material and transported by erosion and mass transport. The unconsolidated sedimentary materials are called *surficial deposits*. The surficial deposits may be transported and eventually deposited by water, in which case we call them *alluvial deposits*. Examples include alluvial fans, sandbars, spits, and river terraces. If the unconsolidated material is transported by ice, we have glacial till, including eskers and drumlins. When the unconsolidated material is transported by the wind, we encounter eolian landscapes, including sand dunes and glacial loess deposits. Finally, the mass wasting of rocks by gravity can produce talus or scree deposits. Under certain favorable circumstances, all these materials may be visible on remote sensor data.

Structure

The major mountain ranges of the world are of volcanic or folded origin. The mountains in the ocean (islands) are usually volcanic. The vast majority of mountains found on the continents were created by folding. Basically, most of the *orogenesis* (mountain building) takes place at the margins of the continents where continental plates push against one another and where rock is compressed and forced to move upwards, perhaps thousands of meters.

Rock type determines how much stress (or compression) it can withstand. When a rock (Figure 14-16a) is subjected to compression, it may experience 1) *elastic* deformation, in which case it may return to its original shape and size after the stress is removed, 2) *plastic* deformation of rock called *folding,* which is irreversible (i.e., the compressional stress is beyond the elastic limit) (Figure 14-16b), or 3) *fracturing,* where the plastic limit is exceeded and the rock breaks into pieces (the pieces can be extremely large!) (Figure 14-16c). Basalt has an average compressive rupture or fracture strength (kg/cm^2) of approximately 2750, quartzite (2020), granite (1480), slate (1480), marble (1020), limestone (960), and sandstone (740). Thus, if more than 2750 kg/cm^2 of compression is applied to a piece of basalt, it may fracture. Sandstone fractures under far less compressional stress.

Folding

Folding takes place when horizontally bedded materials are compressed. The compression results in wavelike undulations imposed on the strata. There are four basic types of folds. A *monocline* is a single fold on horizontally bedded material. It is like a rounded ramp (Figure 14-17b). Monoclines are usually asymmetrical. Archlike upfolds are called *anticlines* (Figure 14-17c). Anticlines typically have a convex upward fold with the oldest rocks in the core or center. The beds of sedimentary strata dip in opposite directions away from the central axis of the anticline. The downward counterpart of an anticline is a *syncline* (a troughlike downfold) (Figure 14-17d). Synclines have a concave downward fold that typically has the youngest rocks in the center. A syncline is a fold in which the sedimentary beds dip inward from both sides toward the axis, forming a synclinal valley.

Anticlines and synclines can be symmetrical or asymmetrical. They can also plunge below the surface, creating a plunging anticline or syncline. Sometimes the compressive forces are so great that the sedimentary beds are *overturned* on top of one another (Figure 14-17e). Anticlines and syn-

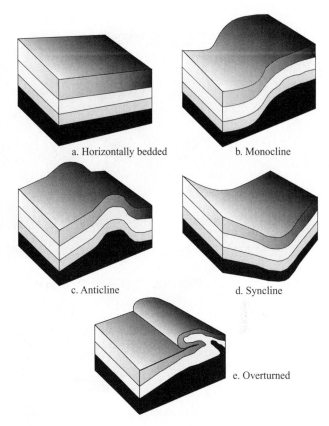

a. Horizontally bedded   b. Monocline

c. Anticline   d. Syncline

e. Overturned

Figure 14-17 a) Horizontally bedded terrain. b) A gently sloping monocline. c) A gently sloping anticline. d) A syncline. e) An overturned fold.

clines are continuously weathered, giving rise to a variety of landforms such as hogbacks, etc. Examples of folded strata in remotely sensed images are found in the section on "Landforms Developed on Folded Strata."

Faults

Rock fractures may be divided into two categories: joints and faults. A *joint* is a crack in rock along which there is no appreciable displacement (i.e., sideways or vertical movement). A *fault* is a crack along which displacement has occurred. Faulting involves the movement of massive blocks of rock and usually generates earthquakes. It is only the fractures along which rock bodies actually move relative to one another that we call *faults*. This displacement can be local in nature or occur for hundreds of kilometers. In certain instances, the direction and amount of movement (displacement) along these *lineaments* can be measured from passive (e.g., Won-In and Charusiri, 2003; Fernandes da Silva, 2005) and active remotely sensed data (Hooper et al., 2003).

Faults

a. Normal dip-slip fault

b. Strike-slip fault

c. Oblique-slip fault

Figure 14-18 a) A normal dip-slip fault, where the hanging wall is displaced from the footwall. Note that the angle of the fault plane is called the dip and that the fault line trends across the terrain in a certain azimuth direction called the strike. The displacement causes a fault scarp. b) A strike-slip fault. c) An oblique-slip fault.

There are three major types of displacement in faults, including: dip-slip, strike-slip, and oblique-slip, as shown in Figure 14-18. Displacement occurring up or down the fault-plane walls produces a *dip-slip* fault (Figure 14-18a). If the displacement occurs parallel with the fault line, a *strike-slip* fault is created (Figure 14-18b). Displacement up and down and along the fault line creates a hybrid *oblique-slip* fault (Figure 14-18c). The diagram also highlights several param-

eters that may be identified using a combination of field work and remote sensor data. The *fault scarp* and the *fault line* may be identifiable. They may trend in a certain direction on the terrain, called a *strike direction*. The angle that the surface is tilted from a horizontal plane is the *dip angle* (θ). It is measured downward from a horizontal plane and ranges between 0° and 90°. The *dip direction* (0° to 359°) is measured orthogonal to the strike direction. There are several

Table 14-1. Drainage morphometric parameters (adapted from Reddy et al., 2004).

Morphometric parameter	Algorithm	Description
Drainage density (D_d)	$$D_d = \frac{\sum_{i=1}^{n} L_i}{A}$$	Total length (L) of n stream channels in a drainage basin divided by the surface area (A) of the basin.
Bifurcation ratio (R_b)	$$R_b = \frac{N_u}{(N_u + 1)}$$	Ratio between the number of streams of any given order to the number of streams in the next higher order.
Basin relief (B_h)	$$B_h = h_{max} - h_{min}$$	Maximum vertical distance between the lowest and highest points of a sub-basin.
Ruggedness number (R_n)	$$R_n = B_h \times D_d$$	Product of the basin relief and its drainage density.
Stream frequency (F_u)	$$F_u = \frac{N}{A}$$	Ratio between the total number of streams and the area of the basin.
Texture ratio (T)	$$T = N\left(\frac{1}{P}\right)$$	Ratio between the total number of first-order streams and perimeter of the basin.
Form factor (R_f)	$$R_f = \frac{A}{(L_b)^2}$$	Ratio between the basin area and the square of the basin length.
Elongation ratio (R_e)	$$R_e = \left(\frac{2}{L_b}\right) \times (A/(\sqrt{A/\pi}))$$	Ratio between the diameter of the circle having the same area (as that of the basin) and the maximum length of the basin.

types of dip-slip faults. These are discussed in the section on "Fault-Controlled Landforms."

Drainage Morphometry and Patterns

River and stream basin surface drainage parameters may be measured using remote sensor data. Some of the most heavily used *morphometric parameters* are summarized in Table 14-1 (Reddy et a., 2004).We will focus here primarily on *drainage density,* which is defined as the total length of *n* stream channels in a drainage basin divided by the surface area of the basin, and how it relates to texture visible in remote sensor data. We will then look at the characteristic of various drainage patterns.

Drainage Density and Texture

The permeability of a soil or rock (i.e., how easily water passes through the material) has a great deal to do with drainage density. Permeability is inversely proportional to runoff. Where permeability is low and runoff is high, many gullies typically form. Conversely, when permeability is high and runoff is low, much of the water infiltrates into the ground and a larger surface area is required to provide sufficient runoff for the creation and maintenance of a channel (tributary). Weak, relatively impermeable clays and shales produce the highest drainage density. Regions with extremely high drainage density are called badlands, e.g., the Badlands National Monument in South Dakota.

The drainage patterns visible in remotely sensed data discussed below are often adjectivally described as having fine, medium, or coarse textures. *Texture* is formally defined as the ratio between the total number of first order streams and the perimeter of the basin (Table 14-1). A *fine-textured* drainage pattern typically exhibits relatively short distances between the first-order tributaries, i.e., it has a high density of closely spaced channels per unit area (or perimeter). This texture indicates high runoff from easily eroded formations such as those composed of shale and clay-rich soils of low permeability. A *medium-textured* drainage pattern typically has moderate tributary spacing and density. It is produced by moderate runoff from relatively permeable soils or bedrock. Thin beds of sandstone can produce such a texture. A *coarse-textured* drainage pattern has a low drainage density with widely spaced channels. It suggests the presence of hard, resistant rock types (e.g., granite, gneiss) and/or very permeable soils. Such materials typically absorb a great deal of water, and there is little runoff. Also, the bedrock may be fractured.

Drainage Pattern

The drainage pattern developed through time on a landscape provides clues about the bedrock lithology (e.g., igneous, sedimentary, metamorphic), topography (slope, aspect), the texture of the soil and/or bedrock materials, the permeability of the soil (how well water percolates through it), and the type of landform present (e.g., alluvial, eolian, glacial). While *in situ* observations are essential, physical scientists often use the synoptic bird's-eye view provided by remote sensing to appreciate a regional drainage pattern. Therefore, it is important to be able to recognize the major drainage patterns present in remote sensor data.

Sometimes there may be a variety of drainage patterns within a watershed or region, in which case it is called a *mixed* pattern. Whenever possible, it is better to stratify the mixed pattern into multiple areas with unique drainage patterns. Also, it is important to remember that the drainage pattern may be composed of wet and dry channels, especially in arid environments.

Dendritic: The most common type of drainage pattern is dendritic (Figure 14-19a). It is characterized by a treelike or fernlike pattern with branches that intersect primarily at acute angles (i.e., < 90°). There are few abrupt bends in the stream channels. Dendritic drainage patterns are developed by random headward erosion of insequent streams on rocks of uniform resistance with little or no structural control caused by folding or faulting. It is often found on landforms composed of relatively homogeneous, horizontally bedded sedimentary rock (e.g., shale), glacial till (e.g., loess), volcanic tuff, on sandy coastal plains, on tidal marshes (Figure 14-19a), or on glaciated outwash plains.

Pinnate: A pinnate drainage pattern is a variation of the dendritic pattern where the streams or gullies have a featherlike branching pattern that typically intersects at acute angles. This pattern indicates that the landform has a high silt content usually consisting of loess, silty alluvium, or very soft erodible sedimentary materials.

Trellis: The trellis drainage pattern is a modified dendritic pattern that resembles a vine on a garden trellis. It often exhibits straight, parallel primary tributaries and shorter secondary tributaries that join the larger branches at right angles. This drainage pattern often indicates that the bedrock structure is tilted, folded, or faulted. It is often found on interbedded sedimentary rocks. Trellis drainage is especially common in areas of folded sedimentary beds of differing resistance such as the Ridge and Valley topography of the Appalachian Mountains from New York to Alabama, or associated with the hogbacks of the Uinta Mountains of Utah (Figure 14-19b). The primary parallel channels follow the less resistant beds. The shorter tributaries flow down the sides of the more resistant upturned beds.

Rectangular: A rectangular drainage pattern also exhibits a treelike pattern; however, the main channels have more abrupt bends. Streams often join at approximately right angles. It develops in areas where joints or faults have developed in the bedrock. Rectangular drainage is present where the bedrock is fractured, jointed, or foliated and is likely to occur on metamorphic slate, schist, or gneiss. It can also form in hard, resistant sandstone in arid climates (Figure 14-19c). The stronger and more rectangular the pattern, generally the thinner the soil covering the bedrock (Rasher and Weaver, 1990). Although both trellis and rectangular drainage patterns have right-angle tributary junctions, individual streams in rectangular patterns generally have right-angle bends in their channels.

Parallel: A parallel drainage pattern consists of tributaries that flow nearly parallel to one another. All the tributaries join the main channel at approximately the same angle. Parallel drainage suggests that the area has a gentle, uniform slope. The tributaries follow beds in the less resistant rock. Parallel and radial drainage on a volcano in Maui, HI, are shown in Figure 14-20a.

Radial (Centrifugal) and Centripetal: A radial drainage pattern forms when water flows downward or outward from a hill or dome. The radial drainage pattern of channels produced can be likened to a wheel consisting of a circular network of parallel channels (i.e., the spokes) flowing away from a central high point or dome (i.e., the hub). In this case, however, the hub of the wheel is elevated above the spokes. Radial drainage is often found on the slopes of volcanic cones such as Diamond Head Crater on Oahu, HI, or on steep mountainous terrain (Figure 14-20b). It is also referred to as *centrifugal* radial drainage.

A radial drainage pattern can also develop on circular areas that drain *into* a common, enclosed central basin or depres-

Figure 14-19   a) Top: Hypothetical dendritic drainage pattern. Bottom: Vertical aerial photograph of a dendritic drainage pattern found on Tivoli North Bay of the Hudson River National Estuarine Research Reserve in New York. b) Trellis drainage pattern found on the hogback ridges on the edge of the Uinta Mountains in Utah (Landsat MSS data). c) Structurally controlled rectangular drainage pattern of the Colorado River in the Grand Canyon (Thematic Mapper band 4 data; courtesy Space Imaging, Inc.).

sion. This is referred to as *centripetal* radial drainage. In this case, the hub of the wheel is lower than the spokes. A cinder cone volcano may have centrifugal radial drainage on the exterior sides of the volcano and centripetal radial drainage on the interior walls of the crater (Figure 14-20b).

Annular: An annular drainage pattern is similar to the radial pattern except that ringlike tributaries intercept radial streams at right angles (Rasher and Weaver, 1990). These develop when stream courses adjust to flow around resistant domes. Granite or sedimentary domes often develop this pattern, as do some meteorite craters (Figure 14-20c).

Dichotomic: A dichotomic drainage pattern may be found on alluvial fans or on alluvial deltas at the mouth of streams or rivers (Figure 14-21a). The water and suspended sediment usually enter the alluvial fan or deltaic area through a single channel. The flow is then redistributed throughout the fan or

delta via a number of *distributaries*. On an alluvial fan, the coarse materials (cobbles, gravel, etc.) are deposited at the apex of the fan while the smaller and lighter minerals are deposited toward the terminus of the fan. In certain instances, the dichotomic drainage pattern may completely disappear near the terminus of the fan, i.e., the material is so porous that all of the water percolates underground. Similar dichotomic distributaries may form on river deltas (e.g., the Mississippi and Nile). The dichotomic drainage patterns can only form on deltas when the suspended sediment is very fine-grained and can be transported great distances.

Braided: A braided stream pattern may develop on floodplains where stream velocity is not sufficient to move the suspended sediment load downstream. The result is the deposition of suspended material in intertwined channels that appear much like braided hair. For example, consider the braided stream pattern found at the mouth of the Betsiboka

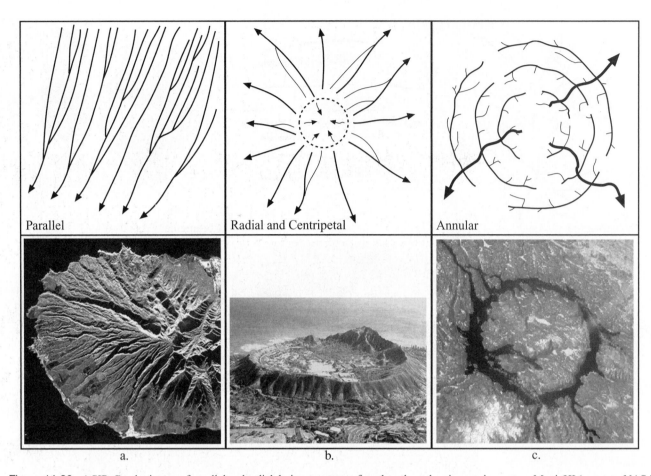

Figure 14-20   a) SIR-C radar image of parallel and radial drainage patterns found on the volcanic cone in western Maui, HI (courtesy NASA Jet Propulsion Laboratory). b) Low-oblique aerial photograph of Diamond Head Crater on Oahu, HI, revealing both radial and centripetal interior drainage patterns. c) Space Shuttle photograph of the annular drainage pattern on the Manicouagan Crater on the Canadian Shield in Quebec Province, Canada (courtesy NASA Johnson Space Center; STS009-48-3139).

River in Madagascar (Figure 14-21b). The sands and gravels that are systematically sorted and deposited in a braided stream drainage network may eventually have significant economic importance.

Deranged: A deranged drainage pattern exhibits streams that wander in disorder in swamps and among water bodies (not shown). It is mainly found on very young landscapes that have almost level topography and a high water table. It can occur in young glacial till plains, moraines, low coastal plains, and floodplains. Usually swamps, marshes, bogs, and lakes or ponds are present.

Anastomotic: An anastomotic drainage pattern may be found on mature floodplains where there is ample homogeneous sediment and reduced stream flow or velocity. Basically, the hydrologic system does not have sufficient energy to remove the sediment from the area. The major stream

channel adjusts to the decrease in energy by depositing its sediment load and increasing the length of the stream channel. Typical diagnostic features include meandering streams and remnant meander scars and/or oxbow lakes such as those found on the Mississippi River (Figure 14-21c).

Sinkhole (doline): A sinkhole drainage pattern consists of isolated lakes or ponds that do not appear to be connected by any systematic surface drainage. If short stream segments are present, they often end abruptly or disappear into karst topography sinkholes or depressions. This drainage pattern develops on sedimentary limestone ($CaCO_3$) where the sinkholes and depressions have formed by chemical dissolution (Trefil and Hazen, 1995). Sometimes the subterranean drainage network that is not visible in the imagery emerges to form springs and lakes. An example of sinkhole topography in Florida is shown in the section on karst topography.

Figure 14-21 a) ASTER image of an dichotomic drainage pattern on an alluvial fan between the Kunlun and Altun mountain ranges on the southern border of the Taklimakan Desert in China's XinJiang Province. The images was acquired on May 2, 2002 (courtesy of NASA/GSFC/METI/ERSDAC/JAROS and U.S./Japan ASTER Science Team). b) Space Shuttle photograph of the braided stream pattern on Betsiboka River mouth in Madagascar (courtesy Kamlesh Lulla; NASA Johnson Space Center). c) SIR-C/ X-SAR L-band radar image of Mississippi River anastomotic drainage pattern (courtesy NASA Jet Propulsion Laboratory).

Geomorphology

A *landform* is a three-dimensional feature on the Earth's surface formed by natural processes. Typical landforms include volcanoes, plateaus, folded mountain ranges, stream channels, etc. *Geomorphology* is the science that studies the nature and history of landforms and the processes of weathering, erosion, and deposition that created them (Bocco et al., 2005). At one time it was known as the science of *physiography*. The study of geomorphology involves an appreciation for five major *processes* at work that constantly erode or deposit materials on the Earth's crust, including running water (fluvial), glacial ice (glacial), wind (eolian), groundwater, and wind-driven water waves. Geologists, physical geographers, and other scientists routinely use the synoptic

view associated with remotely sensed data to identify and interpret geomorphic features on the Earth's surface. In fact, identifying, understanding, and appreciating the nature of landforms present on remotely sensed imagery is one of the great benefits of remote sensing science. One should take time to appreciate the tremendous beauty and variety of landforms on the Earth and how ecosystems associated with the various landforms interact with one another.

A word of caution is in order. In the late nineteenth and early twentieth century, principles of landform evolution were put forth such that any current landform could be interpreted as being in a stage of evolution (e.g., that stream valleys progress from youth, maturity, to old age). Landforms do evolve, but the complexity of climatic change and its influences, the effects of tectonism, and man's impact as a geomorphological agent may be so overriding that few geomorphologists now believe that there is a simple and direct sequence of landforms that can be recognized as developing according to a particular *in*variable pattern (Walsh et al., 1998; James, 2006).

Fortunately, many landforms do exhibit observable, repeatable attributes such as size, shape, height, three-dimensional topography, composition, slope, aspect, etc. These attributes may be used to classify landforms into the following classes:

- igneous landforms

- landforms developed on horizontal strata

- landforms developed on folded strata

- fault-controlled landforms

- fluvial landforms

- karst landforms

- shoreline landforms

- glacial landforms

- eolian landforms.

The Soil Conservation Service produced an excellent manual that summarized the major landforms and how they appear on aerial photography (Rasher and Weaver, 1990). Several of their examples are used in this chapter. Short and Blair (1999) provided a thorough review of how to perform geomorphological analysis from space especially for obtaining information about large, regional landforms. They provide examples from around the world.

One might ask, What is the ideal type of remote sensor data and/or image presentation format to study, appreciate, and classify the Earth's landforms? First, no single sensor is ideal for the study of all landforms. Aerial photography may be ideal for studying landforms in the relatively cloud-free arid western United States, while radar data may be ideal for studying landforms in cloud-shrouded environments (e.g., Lu et al., 2004). Many analysts can extract landform information by viewing a single aerial photograph or image. However, many scientists prefer to study landforms while viewing the terrain in a three-dimensional presentation using 1) stereoscopic photography or imagery (e.g., Baldi et al., 2005), 2) draping a monoscopic single image over a digital elevation model and then manipulating the observer's viewing position to enhance topography, and/or 3) obtaining oblique imagery that generally enhances the third dimen-

sion. Also, imagery obtained at a relatively low Sun angle often enhances the analyst's ability to identify landforms. Finally, one has to be very careful when interpreting images in the Northern Hemisphere because shadows in images fall toward the north, away from the viewer. Therefore, experienced image analysts often rotate the Northern Hemisphere images so that south is toward the top to prevent pseudoscopic illusion from taking place (i.e., confusing mountain ridges with valleys). Some of the illustrations in this chapter are oriented in this manner to facilitate interpretation.

It is beyond the scope of this chapter to provide examples of all the possible geomorphic landscapes present on the surface of the Earth. Therefore, only representative examples of some of the most important landforms are presented. Please refer to books on geomorphic analysis from aerial platforms such as Shelton (1966), Short and Blair (1999), Rasher and Weaver (1990), Strain and Engle (1993), and Way and Everett (1997).

Igneous Landforms

The Earth's upper mantle is partly molten, with many minerals in it that were forced deep by tectonic forces and eventually melted. Igneous rocks are produced from this molten, fluid magma. If the magma is extruded onto the Earth's surface, *extrusive* igneous (volcanic) rock is created. If the magma is extruded into some subsurface portion of the Earth's crust, *intrusive* igneous rock is formed. If the molten lava flows or is exploded out of a central vent and produces a mound or cone-shaped feature, we have a *volcano*.

Cinder cone volcanoes are built entirely of pyroclastic materials that are ejected into the air and fall to Earth nearby. Their accumulation eventually builds the cone. For example, consider one of the Menan Butte volcanic craters (Figure 14-22). It was created when a dike of magma intruded into a shallow aquifer. The water in the aquifer turned to steam and explosively fragmented the basaltic magma into volcanic glasslike particles called tuff. The volcano was built up layer by layer, by aerial deposits of tuff and larger pyroclastic (bomb) material. The two Menan Butte volcanoes also extruded lava onto the surface covering a large portion of eastern Idaho. Diamond Head Crater on Oahu, HI, is also a tuff cone volcano.

Lava dome (shield) volcanoes are created when the lava pours out onto the landscape. Generally this results in more gently sloping volcanoes depending upon the viscosity of the lava (i.e., the more highly fluid the lava, the more gentle the slope; the more viscous the lava, the steeper the slope

Menan Butte Cinder Cone Volcano in Idaho

Figure 14-22 Panchromatic stereopair of the southern Menan Butte cinder cone volcano in Idaho obtained on June 24, 1960. It is one of the world's largest tuff cone volcanoes, with a volume of 0.07 mi^3 (0.30 cubic km^3). The tuff is made of volcanic glass that has been altered by the addition of water. The cone is late Pleistocene in age. The Menan Butte volcano also extruded lava on the surface to the west (bottom of the photographs) for many kilometers. There is sunglint on Henry's Fork in the left photograph as it flows into the Snake River. Note the radial and parallel drainage on the flanks of the crater and the centripetal drainage in the interior of the crater.

produced). Common extrusive rocks are basalt, andesite, dacite, and rhyolite. Basalt is the most widely distributed volcanic rock. It is also the most common rock on the Earth's surface. It flows out from volcanic vents over other rock in great lobes or thin sheets (Lu et al., 2004). It can travel hundreds of miles. It is fine-grained, black, rich in silica and the ferromagnesian minerals. Basalt normally appears dark on panchromatic film because it contains feldspar, hornblende, and micas. Dacite and rhyolite appear in light tones on panchromatic images.

The Shuttle SIR-C/X-SAR radar image in Color Plate 14-2a reveal several components of the Kilauea shield volcano located on Hawaii. The massive volcano continues to be created by the extrusion of two types of lava: *a'a* is extremely rough and spiny and will quickly tear shoes when walking on it, and *pa'hoehoe* is a ropey lava that looks like burned pancake batter. The volcano is enlarged by the extrusion of lava onto the surface. Numerous individual flows are easily seen in the three-dimensional radar-perspective view in Color Plate 14-2b. Note especially the dark materials from the Kupaianaha crater. Kilauea has been erupting almost constantly since 1983.

Isla Isabela of the Galapagos Islands

Figure 14-23 A three-dimensional perspective view of Isla Isabela of the Galapagos Islands. This is a Space Shuttle SIR-C/X-SAR image draped over a digital elevation model. Rough *a'a* lava flows produce a bright radar-return on the side of the dome volcano. Ash deposits and relatively smooth *pa'hoehoe* lava appear as dark areas on the side of the volcano (courtesy NASA JPL).

Isla Isabela is one of the Galapagos Islands located off the west coast of Ecuador, South America. It is also an active lava dome (shield) volcano. Figure 14-23 depicts a Space

Mount St. Helens

a. High oblique photograph.   b. Stereopair.

Figure 14-24 a) High-oblique photograph of Mount St. Helens erupting on May 18, 1980 (courtesy of U.S. Geological Survey). b) USGS High Altitude Photography (HAP) stereopair of Mount St. Helens in Washington on August 6, 1981 (U.S. Geological Survey photos 109-84, 85). The active lava dome in the center of the cone is visible. Steam is rising from within the crater. A sediment choked radial drainage pattern has developed. North is to the left. Please refer to Color Plate 14-3.

Shuttle SIR-C/X-SAR image of Isla Isabela draped over a digital elevation model. The bright areas are very rough lava flows.

Lava flows on the Westdahl Volcano, Alaska, were mapped using radar and Landsat Thematic Mapper imagery (Lu et al., 2004). Baldi et al. (2005) used aerial photography and photogrammetry to map geomorphic changes on the Stromboli volcano in Italy. Thermal infrared AVHRR data were used to map hot spots on Mt. Etna and Stromboli volcanoes (Pergola et al., 2004).

Composite cone or strato volcanoes are created from both pyroclastic materials and extruded lava. The world's most

impressive volcanoes are composite cones. For example, Mount St. Helens in Washington (8,360 ft; 2,548 m) is a composite cone volcano (Figure 14-24 and Color Plate 14-3). It erupted on May 18, 1980, at 8:32 a.m. Pacific time. A series of moderate-to-severe earthquakes preceded the eruption, sending the north side of the mountain cascading downward toward Spirit Lake. This avalanche, the largest ever observed in the Western Hemisphere, weakened the magma chambers within the volcano, causing a northward lateral and vertical explosion that destroyed over 270 mi^2 (7,000 km^2) of forest in five seconds and sent a billowing cloud of ash and smoke 70,000 ft (21,000 m) into the atmosphere. A pumice plain was produced, consisting of volcanic mud, ash, and debris that buried the original Toutle River Valley to a

Mount Taranaki, New Zealand

Figure 14-25 ASTER image of Mount Taranaki on New Zealand's North Island obtained on May 27, 2001. This composite cone volcano is made of andesite lava and a mixture of solid pyroclastic rocks (tephra), and ash. The unique shape of the Egmont National Park results from its protection in 1881, which specified that a forest reserve would extend in a 9.6 km radius from the summit of Mount Taranaki. A series of montane habitats occur in procession from the pasture farmlands up the flanks of the volcano—from rain forest, to shrubs, to alpine, and finally snow cover (courtesy of NASA/GSFC/METI/ERSDAC/JAROS and U.S./Japan ASTER Science Team).

depth of 1,000 ft (300 m). The stereoscopic photography obtained on August 6, 1981, after the eruption revealed another lava dome developing in the center of the crater.

Mount Taranaki in Egmont National Park on New Zealand's North Island is a composite cone volcano made of andesite lava (Figure 14-25). The rocks are a mixture of solid pyroclastic rocks (tephra), ashes, and lava flows, which lack cohesion and are easily carved by erosion. Mount Taranaki stands at 2518 m. The volcano began forming 70,000 years ago, and last erupted in 1755.

Intrusive igneous rock is formed when the molten magma cools and crystallizes within the Earth's crust. The material lying above this plutonic rock may eventually be eroded. Large igneous intrusive rock bodies, or *batholiths,* often form the foundation for entire mountain systems such as the Sierra Nevada in the western United States or the Andes in South America. Smaller dome-shaped intrusive rock bodies are called *lacoliths* or *stocks.*

The most common intrusive (plutonic) rocks include granite, diorite, diabase, and gabbro. Granite is a light-colored igne-ous rock, rich in quartz and feldspar but poor in the ferromagnesian (iron and magnesium) minerals. Granite usually appears in light tones on panchromatic imagery, whereas diabase and gabbro are dark. Granite cools for an exceptionally long time underground; therefore, it has large crystals that are visible to the naked eye. When granite is exposed at the surface, exfoliation may take place where concentric shells of rock material break off due to chemical or thermal weathering and from the release of internal stress when overburden is removed. For example, Figure 14-26a depicts Stone Mountain in Georgia, which is a relatively small intrusive stock. It is approximately 360 m high and 11 km in circumference. An image of the top of the dome reveals linear striations that radiate away from the apex (Figure 14-26b). Another excellent example is the Brandberg Massif in Namibia (Figure 14-27). This massive granite intrusion is exposed at the surface and covers approximately 650 sq. km (250 sq. mi) and rises 2573 m (1.6 mi) above the desert.

Sometimes the less resistant material surrounding the intrusive volcanic neck (or plug) of a volcano erodes, exposing the more resistant volcanic neck materials. For example, Devil's Tower in Wyoming is an intrusive volcanic neck

Stone Mountain, GA

a.   b.

Figure 14-26 a) Panchromatic aerial photograph of Stone Mountain, GA. It is a granite exfoliation dome created by intrusive magma solidifying at great depth. It is believed that more than 10,000 ft. of less resistant material was eroded to expose the intrusive rock.
b) Large-scale photograph of the top of Stone Mountain. Note the gondola facility, linear striations in the rock, and the parallel and annular drainage pattern present.

(Figure 14-28). Occasionally the magma is extruded away from the main body via cracks and fissures in the Earth's crust to form dikes and sills. Later the material above these dikes and sills may be eroded, exposing wall-like resistant dikes.

If the granite at the surface is largely unfractured, a dendritic drainage pattern will result on the domelike hills. Conversely, if the granite has been extensively fractured, this can result in a rectangular or trellis drainage pattern.

Landforms Developed on Horizontal Strata

Many landforms have developed on flat, horizontally layered strata. A layer is considered to be horizontal when the dip (inclination from a horizontal plane) of the strata has not significantly affected the development of the topography. The type of rock, its erodability, amount of precipitation available to the region, and local stream gradient control landform development.

Horizontally bedded sedimentary rocks are formed by cementing the loose particles produced by weathering. The most common cementing agents are quartz, calcite, and iron oxide. Most sediments settle in water in relatively horizontal layers called *strata*. Sediments can collect in slow-moving streams, in swamps, lakes, and shallow seas. Sediments also collect in oceans, where they are deposited on continental shelves or flood down submarine canyons and are deposited.

Brandberg Massif, Namibia, Africa

Figure 14-27 The Brandberg Massif in Namibia is a dome-shaped granite intrusion covering 650 sq. km (250 sq. mi). It rises 2573 m (1.6 mi) above the desert. This image was recorded by the Landsat 7 ETM+ on September 10, 2002 (courtesy of NASA).

The flows first deposit coarse sand, then finer clays and silts, depending upon grain size. The most common rock outcrop formations on the continents are composed of mudstone,

Devil's Tower, WY

a.

b.

Figure 14-28 a) Panchromatic stereopair of the Devil's Tower, WY, volcanic neck obtained on September 15, 1953, with south at the top. Vertical joints divide the rock mass into polygonal columns that extend from just above the base to the top of the Tower. Devil's Tower is 865 ft. tall, with an elevation of 5,117 ft. The Tower is more resistant to erosion than the surrounding sedimentary rock. It was formed by the intrusion of magma into sedimentary rock. Weathered rock (talus) lies at the base. b) Geologic cross-section: Tp = volcanic neck consisting of crystalline intrusive black igneous rock – phonolite porphyry with crystals of white feldspar; Tas = red siltstone and sandstone with interbedded red shale; Jgs = gypsum with interbedded red mudstone; Jsb = gray-green shale with sandstone; Jsh = yellow sandstone; Jsr = gray-green shale with sandstone and limestone.

sandstone, shale, and limestone. Greater water force is required to move larger pebble-size particles that are deposited as *conglomerate*.

Mudstone has no visible grains, being composed of extremely fine sediments deposited on quiet floodplains, in lakes, or deep oceans. Thick deposits of mudstone are present in most deltas, where rivers enter still water. Layering occurs in thick mud deposits because the clay particles, being flat, align themselves horizontally. They are usually gray to shades of red in color. The Mississippi delta contains mudstone.

Sandstone has visible grains like coarse sugar, up to almost 1/12 in. (2 mm) in diameter. Most sandstone is composed of rounded particles of quartz, but it can contain feldspar and fragments of unweathered rock. To move particles of this size, the water or wind must have force. In arid environments, sandstone is a weather-resistant, hard rock. Outcrops

often appear in light tones on black-and-white photography unless covered by desert varnish, in which case they may have dark tones. Sandstone in arid and semiarid environments often produces *plateaus*, *mesas*, and *buttes*. Cliffs may occur where sandstone caps (overlays) less resistant sedimentary rock such as shale. Sandstone is not as plastic (pliable) as shale and other sedimentary rocks. Therefore, it is broken easily when stressed. The fractures become visible and may produce angular, dendritic, trellis, or rectangular drainage patterns.

Limestone is formed from calcium carbonate. Lime precipitates onto shallow sea floors in warm tropical waters where almost no other sediments cloud the water. Limestone rocks formed in the broad centers of flooded ancient continents and along the edge of continental shelves where reefs were built. As a result, many limestones are composed of the skeletons of reef-forming animals. Limestone (and related dolomites) are susceptible to solution, meaning that they can be dissolved. In humid environments, where there is plenty of water, we may find karst topography, named after the Karst region of the former Yugoslavia, created by limestone dissolution. The undulating to hummocky terrain often contains no visible surface drainage pattern but instead may exhibit oval sinkholes called dolines. Limestone in arid environments behaves quite differently. Because there is little water for dissolution, limestone may be very resistant to weathering. In fact, like sandstone, it may form caprocks with vertical faces when underlain by less resistant sedimentary rock such as shale.

Shale sedimentary rock is produced by the deposition and compaction of silt and clay particles. It is relatively impervious to water moving through it; therefore, more water is available for surface erosion. In arid environments the soft impervious shale is easily eroded. It generally does not produce as steep slopes as sandstone or limestone. When no caprock is present, it can yield a rugged badlands topography consisting of a fine-textured pinnate dendritic drainage pattern with sharp ridgelines and steeper slopes. Generally, shale is more pliable than sandstone and, therefore, is not as affected by geologic stress. This results in a greater number of fine- to medium-textured dendritic drainage patterns. Shale generally appears in light tones on panchromatic imagery.

The Grand Canyon in Arizona is a good example of various landforms developed on horizontally bedded sedimentary rock. A diagram of the bedding structure is found in Figure 14-29a. This region contains a nearly continuous sedimentary record of the Earth's history, with approximately 5,000 ft. (1,500 m) of bedded horizontal layers, all stacked in sequence with the oldest at the bottom. One of the top layers, the Kaibab limestone, formed from the remains of corals, sponges, and other marine animals and is one of the youngest in the sequence — about 240 million years old. At the bottom of the canyon lies the Vishnu Schist which is about 1.7 billion years old, some of the planet's oldest rock. The in-between layers of rock reveal a turbulent history of mountain upheaval, lava deposits, erosion by wind and water, and past environmental changes as deserts were replaced by rivers, then lakes and shallow inland seas. Finally, the land was uplifted, forming the Colorado Plateau. Five to six million years ago, the young Colorado River began to cut into the layers of rock, and the Grand Canyon was created.

A panchromatic stereopair of the area near the South Rim of the Grand Canyon and Bright Angel shale butte is found in Figure 14-29b. The South Rim is composed of sandstone caprock of the Coconino Plateau and is resistant to weathering, forming a flat dissected plateau (mesa). Both sandstone and limestone form steep cliffs in arid environments. The less resistant shale is easily eroded, creating more gently sloping eroded surfaces (Figure 14-29b). Also present in the stereopair is the entrenched Colorado River with its deep, V-shaped inner gorge and rapids.

Another example of horizontally bedded strata is found a few hundred miles downstream. Figure 14-30a depicts the Colorado as it dissects the Shivwits Plateau as recorded by the Landsat Thematic Mapper. This band 4 near-infrared image provides significant detail but not nearly the information content of the color composites found in Color Plate 14-4ab, where the more resistant sandstone caprock and the more gently sloping shale sedimentary rocks are evident from their vibrant colors. Image analysts often find it very useful to have information about the elevation and slope of an area when interpreting landforms. Figure 14-30b is a representation of USGS digital elevation data. Note the higher elevation of the Shivwits Plateau and the lower elevation of the Colorado River. An analytical shaded relief image of the digital elevation data is shown in Figure 14-30c. It provides additional information of value to the image analyst about the three-dimensional nature of the terrain. Finally, the slope map (Figure 14-30d) makes it clear that the terrain with the greatest slope occurs adjacent to the Colorado and its major tributaries. Plateaus and mesas are flat, having little slope; therefore, they appear dark in the slope image.

Another excellent example of a plateau or mesa is found in the Phang Hoei Range of north central Thailand about 40 km northeast of the city of Lom Sak (Figure 14-31). The resistant plateau rises majestically above the surrounding countryside in this Shuttle Imaging Radar C-band HV polarized image.

Sometimes we encounter horizontally bedded metamorphic rocks that were created by exposure to heat and pressure. This *metamorphism* causes sandstone to become quartzite, which is very resistant to weathering and can form sharp-crested ridges in all climates. Similarly, metamorphosed shale becomes very resistant slate. Orthogneiss is metamorphosed granite and has a similar appearance. A variety of paragneiss rocks may be developed from metamorphosed sandstone, shale and other sediments. Schists are medium-grained crystalline rocks derived from sedimentary rocks. They are easily weathered in humid climates but resistant in arid environments.

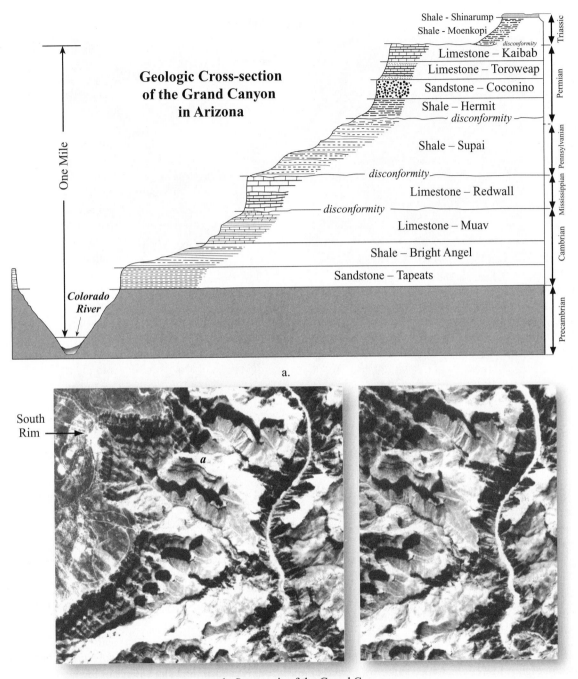

a.

b. Stereopair of the Grand Canyon.

Figure 14-29 a) Geologic cross-section of the Grand Canyon in Arizona. b) Panchromatic stereopair of the Grand Canyon. Viewing and camping facilities are located on the sandstone caprock of the Coconino Plateau on the South Rim. The images were rotated so that south is to the left to avoid pseudoscopic illusion. The resistant sandstone and limestone form steep cliffs. The less resistant shale is easily eroded, causing more gradual slopes to form. The Bright Angel butte is located at "*a*".

Landforms Developed on Folded Strata

Diastrophism (mountain building) can eventually tilt or fold the horizontal sedimentary or metamorphic rock layers. The folded terrain may produce monocline domes such as Ayers

Grand Canyon in Arizona

a. Landsat Thematic Mapper band 4.

b. Digital elevation model.

c. Shaded relief map of DEM.

d. Slope map derived from DEM.

Figure 14-30 a) Another section of the Grand Canyon on the Colorado River in Arizona, as recorded by the Landsat Thematic Mapper (band 4; 30 × 30 m). North is to the left to improve interpretability. Color Plate 14-4 provides additional information. b) Digital elevation model derived from 30 × 30 m USGS data (displayed as 3 arc second, 93 × 93 m cells). Note the higher elevation Shivwits Plateau. c) Shaded relief map. d) Slope map where the brighter the pixel, the steeper the slope (after Mojave Desert Ecosystem Program, 1998; Landsat imagery courtesy of Space Imaging, Inc.).

Rock in Figure 14-1, anticlines, and synclines, as previously illustrated in Figure 14-17. The folded strata are eroded differentially, depending upon the type of sedimentary rock present. For interbedded sandstone and shale in arid environments, the more resistant sandstone forms *anticlinal ridges,* whereas the shales are eroded to *synclinal valleys*. This can form nearly parallel systems of resistant ridges separated by eroded valleys. The resistance of carbonate rocks such as limestone and dolomite depends on the climatic conditions. In arid regions with little water, limestone is resistant to weathering and tends to create ridges. Conversely, in humid

regions limestone may be dissolved and therefore is more easily eroded, forming valleys or lowlands.

Depending upon the steepness of the dipping terrain, the ridges may be asymmetrical or symmetrical. A *hogback* is a sharp-crested ridge formed by differential erosion of a resistant bed of steeply dipping rock. For example, a folded landscape near Maverick Spring, WY, is shown in Figure 14-32a. It is a dissected asymmetrical dome with a series of resistant hogback ridges surrounding the structure. Figure 14-32b depicts an Appalachian Mountain range synclinal valley bounded by the Brush Mountain ridgeline to the west and the

SIR C-band RADAR Image of the Phang Hoei Range in Thailand

Figure 14-31 Shuttle Imaging Radar C-band image of the Phang Hoei Range of north central Thailand on October 3, 1994. The resistant plateau is part of the Phu Kradung National Park (courtesy NASA Jet Propulsion Lab).

Canoe Mountain ridgeline to the east near Tyrone, PA. In this environment, the sandstone is more resistant to erosion than the soluble limestone. Figure 14-33 depicts the San Rafael swell in southern Utah. This is a flat-topped upwarped monocline that slopes to the east/southeast. The more resistant sandstone ridges form hogbacks on the southern and eastern flanks of the monocline.

Sometimes the folding is so severe that the layers are nearly vertical. For example, Figure 14-34 depicts the Ugab River in Namibia in Africa which flows through nearly vertical folded layers of thinly bedded limestone, sandstone, and siltstone (Color Plate 14-5).

The angle of dip of a resistant bed such as a hogback may be calculated using photogrammetric measurement techniques and the formula:

$$\tan\theta = \frac{h}{d} \qquad (14\text{-}6)$$

where θ is the dip angle, h is the vertical distance, and d is the horizontal distance, as shown in Figure 14-35.

Fault-Controlled Landforms

There are three major types of displacement in faults, including: dip-slip, strike-slip, and oblique-slip, as previously shown in Figure 14-18. The dip-slip faults may be further subdivided into normal, reverse, thrust, graben, and horst, as shown in Figure 14-36a-e). In a *normal fault*, the hanging wall is moved (displaced) downward from the footwall, exposing a *fault scarp* on the footwall (Figure 14-36a). The

Wasatch fault along the Wasatch Mountains in Utah is a good example (Figure 14-37). The Wasatch Mountain Range of the Rocky Mountains is the upthrown block, and the downthrown block is the valley floor. Fault scarps at the base of the Wasatch Range generally face west.

A *reverse* fault is created when the displacement takes place in the opposite direction (Figure 14-36b). If the reverse faulting causes the slab of strata to move horizontally on top of the landscape as shown in Figure 14-36c, we have a *thrust* or *over-thrust* fault. A fault can also involve two fault planes. A *graben* is produced when a block of material gets displaced downward between two normal faults (Figure 14-36d). These often produce rift valleys, such as the great East African rift system. A *horst* is created when a block of material is thrust upward between parallel fault planes (Figure 14-36e).

When the displaced rock material occurs parallel with the fault line, a *strike-slip* fault is created (Figure 14-18b). For example, the San Andreas Fault in California is a classic strike-slip fault. It passes through southern California east of Los Angeles, through the southern portion of the great Central Valley, runs through San Francisco, and then out to sea. A portion of the fault line in southern California is shown in Figure 14-38. The west side of the San Andreas Fault is moving northwest, while the east side of the fault is moving southeast; hence the oft quoted observation that Los Angeles and parts of San Francisco, because they are on the western side of the fault, are moving out to sea albeit ever so slowly. Figure 14-38 also identifies the Garlock normal fault. The intersection of the Garlock and San Andreas faults is a very interesting place to live with minor tremors occurring almost every day. Notice that the San Andreas Fault has created a

Maverick Spring, WY

a.

Synclinal Valley in the Appalachian Mountains near Tyrone, PA

b.

Figure 14-32 a) USGS High Altitude Photography (HAP) of an eroded folded landscape near Maverick Spring, WY. It is a dissected asymmetric dome, which is an erosional remnant of a plunging anticline. Note the fine-textured topography, the strike of the ridges and valleys, and the radial and trellis drainage controlled by the anticlinal structure (north is to the right). Also note the prominent hogback ridges. b) HAP of a synclinal valley in the Appalachian Mountains near Tyrone, PA, obtained on May 4, 1981. The ridge lines of Brush Mountain on the west and Canoe Mountain on the east are composed of more resistant sandstone, while the less resistant, soluble limestone sedimentary rock has been eroded. The interbedding of the sandstone and limestone results in hogbacks at the periphery of the syncline with a trellis drainage pattern (courtesy U.S. Geological Survey).

small rift valley that is relatively flat when compared with the terrain on either side of the fault line. This is most evident when viewing the shaded relief version of the digital elevation model (Figure 14-38d). Natural lakes often occur in these rift valleys where the drainage pattern has been disrupted.

Fluvial Landforms

Fluvial landforms are created by the weathering, erosion, transportation, and deposition of materials by flowing water.

Fluvial processes create erosional and depositional landforms in virtually every region of the globe. Consequently, there are a tremendous variety of fluvial landforms.

It is important when interpreting fluvial landforms in an image to understand the characteristics of the rock types within the drainage basin, the drainage pattern, its density, and the gradient (slope) of the watershed. Streams or rivers flowing over terrain with steep gradients have greater velocity (energy) and hence greater bedload and suspended sediment carrying capacity. They can even entrain pebbles, cobbles, and boulders in the flow. Conversely, when the

San Rafael Swell in Southern Utah

Figure 14-33 Landsat Thematic Mapper band 4 image of the San Rafael Swell in southern Utah. It is a large flat-topped monoclinal upwarp bounded by hogback ridges on the southern and eastern flanks. The more resistant hogback ridges are composed of sandstone, while the less resistant shale beds have been eroded (courtesy Doug Ramsey, Utah State Univ. and Space Imaging, Inc.).

Ugab River, Namibia

Figure 14-34 ASTER image of Namibia's Ugab River obtained on December 25, 2000 (band 3). This river only flows aboveground for a few days each year. The river passes through nearly vertical folded layers of thinly bedded limestone, sandstone, and siltstone. See Color Plate 14-5 for a color version of this image (courtesy of NASA/GSFC/METI/ERSDAC/JAROS and U.S./Japan ASTER Science Team).

Figure 14-35 The angle of dip, θ, of a monoclinal ridge such as a hogback may be calculated from stereoscopic imagery using Equation 14-6.

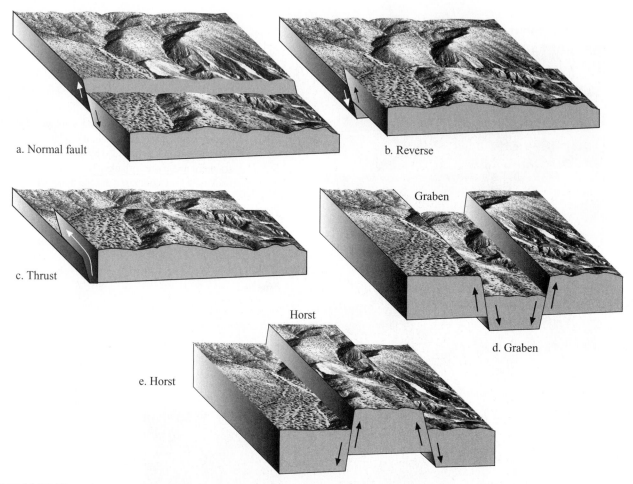

Figure 14-36 The major types of dip-slip faults: a) Normal fault. b) Reverse fault. c) Thrust fault. d) Graben associated with a downthrown block between two normal faults. e) Horst associated with an upthrown fault block.

stream gradient decreases, streams or rivers may no longer be able to hold the sediment in suspension and then deposit the material in the channel bottom or as bars, spits, deltas, or alluvial fans. In effect, the velocity of the stream determines what kind of materials (sand, silt, clay, gravel, cobbles) can be moved along by the stream. Even in deserts, where storms are infrequent, sheet-wash can induce massive erosion because of its intense velocity and transport the eroded materials great distances. It is important to also point out that the greater the velocity, the greater the amount of in-stream or river abrasion, corrosion, and quarrying that will take place. Thus, streams and rivers are active agents of geomorphic change that produce both erosional and depositional landforms. Some of the more important landforms include stream-cut valleys, floodplains, terraces, deltas, alluvial fans, pediments, and playas.

Stream Valleys, Floodplains, and Relic Terraces

Uplifted terrain may be eroded by sheet erosion, creating gullies. Over time and given continued precipitation, these gullies may turn into small tributary channels or streams that have greater erosional power, especially if the terrain has substantial gradient (slope). If the source of water continues unabated, the stream or river may incise and erode both the base and walls of the rock material creating deep, steep-sided V-shaped valleys with minimal floodplains and fairly straight water courses.

If the gradient is sufficient, the stream or river may continue to deepen and widen the river channel. However, at some point the river may erode the river bottom to the point that the river gradient is substantially reduced. If this occurs, the river loses velocity and begins to deposit some of its suspended sediment load. This can be the beginning of a river

Figure 14-37 Landsat Thematic Mapper band 4 image of the Wasatch fault along the Wasatch Mountains to the east of Springville and Spanish Fork, UT. This is a classic normal fault, with the downthrown block being the valley floor and the upthrown block being the Wasatch Mountain Range. There are numerous fault escarpments all along the fault line (courtesy Space Imaging, Inc.).

valley with floodplains, wide meanders, and occasional oxbow lakes. There are a number of factors that can cause the river or stream to have a reduced gradient including land subsidence.

Sometimes a stream or river loses a substantial amount of its source water or its gradient is diminished greatly. When this occurs, even more sediment may be deposited. This can result in extremely broad and shallow valleys that have meanders, meander scars, cut-off meanders, point bars, numerous oxbow lakes, and natural levees. It is important to point out, however, that this process does not always happen. Unusual climatic events and man-induced impacts can inter-

vene and dramatically alter what was once thought of as the youth-maturity-old-age stream-valley erosion cycle (Selby, 1989).

Way and Everett (1997) suggest that there are three types of river floodplains: meander, covered, and composite. *Meander floodplains* are caused by a low stream or river gradient and the deposition of some of the suspended sediment load. Thus, it is a low-energy hydrologic system. Moreover, most of the sediment load in a meander floodplain stays within the confines of the riverbanks, even during flood stage. The greatest erosion in the floodplain takes place at the outer edges of the meanders. The greatest deposition takes place

San Andreas and Garlock Faults in California

a. Landsat Thematic Mapper Band 4.

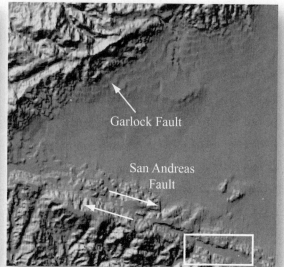

b. Shaded-relief display of a digital elevation model.

c. Enlargement.

d. Enlargement.

Figure 14-38 a) Intersection of the San Andreas and Garlock Faults in southern California recorded on a mosaic of Landsat Thematic Mapper band 4 images. The San Andreas Fault is a northwest-southeast trending strike-slip fault. b) Shaded-relief display of a USGS digital elevation model (3 arc second, 93.3 m cell). c) Enlargement of the Landsat data highlighting a portion of the San Andreas rift zone in the lower right portion of (a). d) The relatively narrow rift zone as portrayed by the shaded-relief model (after the Mojave Desert Ecosystem Program, 1998; courtesy Space Imaging, Inc.).

along inside edges of the meanders. This process can lead to the creation of sand and gravel point bars in the river. *Covered floodplains* are created when additional energy is provided during flooding and the floodwaters routinely overflow the riverbanks. This may create natural levees consisting of coarse sediments (usually sand) which are formed adjacent to the original stream channel. Over time, the natural levees are built up by continued accretion. Any water flowing over the natural levees has less velocity and can usually only entrain fine particle silts and clays that are then transported to distant slackwater locations beyond the natural river levees. Hopefully, these sediments improve the fertility of the soil. A *composite floodplain* contains features common to both meander and covered floodplains. Composite floodplains are the most common type of floodplain.

A small section of the Mississippi River shown in Figure 14-39 provides examples of several composite floodplain features. Both natural and man-made levees are present. The Landsat TM near- and middle-infrared bands are ideal for identifying the oxbow lakes and the more geologically recent flooded meander scars, which are barely discernible in the green and red band images. The sand and point bars are visible on all images. The middle-infrared band 5 provides more detailed soil moisture information than any of the other bands. As expected, vegetated areas are dark in the green and red bands and much brighter in both the near- and middle-infrared bands. A color-composite of TM bands 4, 3, and 2 (RGB) is found in Color Plate 14-6.

Floodplains sometimes have relic terraces associated with them. *A terrace* is a gently inclined, elevated benchlike rem-

Floodplain Landforms on the Mississippi River

a. Thematic Mapper Band 2 (green)

b. Thematic Mapper Band 3 (red)

c. Thematic Mapper Band 4 (near-infrared)

d. Thematic Mapper Band 5 (mid-infrared)

Figure 14-39 Floodplain landforms on the Mississippi River recorded by the Landsat Thematic Mapper on January 13, 1983 (see Color Plate 14-6) (images courtesy of NASA Observatorium and Space Imaging, Inc.).

nant floodplain that lies at a higher elevation than the present day floodplain. Terraces may parallel the existing floodplain on both sides, although this is not always the case. A terrace typically has an escarpment on the side facing the present-day floodplain. Terraces may be produced by glacial activity or by dramatic changes in climate where more water and energy is made available to cut through the existing flood-plain. Several layers of terraces present in a region indicate that dramatic erosional-depositional changes have occurred through time.

Deltas

When a stream or river with a substantial sediment load enters a standing body of water such as a lake, reservoir, sea, or ocean, its velocity is slowed dramatically. This causes the stream or river to deposit its base and suspended sediment load, with the more coarse materials deposited at the apex and the silt and clay particles moving farther out into the water body. The continued deposition of these materials over time may create a delta. These tend to have triangular shapes in planimetric view somewhat similar to the Greek letter Delta (Δ), but many other shapes also occur. There are five common delta forms.

Elongated or *digitate* deltas are created when the system is dominated by the continuous input of a large volume of water and sediment that is *not* impacted severely by waves. This can produce what is called a *bird's-foot* delta, such as the Mississippi River delta as it enters the Gulf of Mexico near New Orleans, LA (Figure 14-40a). Continued input of sediment causes the deltaic stream channels to migrate back and forth across the delta. Sometimes, however, the sediment deposited on the banks of the input channel create natural levees similar to that on the covered floodplain previously discussed. Intense flooding may cause the input stream to break free from its confinement within the natural levee and migrate to a nearby location. This process can lead to the creation of an entirely new lobe of sediment deposition in the delta. There may be several main input distributary channels present in a bird's-foot delta.

On a *lobate* delta the river builds into the sea, but relatively intense wave action redistributes much of the sediment along coastal barriers, causing it to have a convex edge or fan shape facing the water body. It is a wave-dominated system. The Niger River emptying into the Gulf of Guinea in Africa has created a classic lobate delta (Figure 14-40b). The Nile River emptying into the Mediterranean Sea is also a lobate delta (Figure 14-40c).

A *crenulate* delta forms where tidal currents help create numerous sandy islands separated by tidal channels along the delta front. It is a tide-dominated system. The Irrawaddy River delta in Burmah is a good example (Figure 14-40d). Note the extensive mangroves visible as bright areas on the Irrawaddy delta.

If the waves or along-shore currents are extremely strong, they may move the sediment away from the mouth of the river, and only cuspate sand ridges will be formed that parallel the beach. This creates a *cuspate* delta.

Streams entering coastal estuaries also experience a reduced gradient, causing them to deposit much of their suspended sediment load. An *estuarine* delta will form if these deposits are not removed by currents in the estuary or by diurnal tidal flushing.

Alluvial Fans and Bajadas

Streams emerging from a mountain valley into a lowland area often encounter reduced slopes and are forced to deposit material in order to maintain an adequate gradient. An *alluvial fan* is a low, conical-shaped deposit with the apex of the cone at the mouth of the valley from which the fan-building stream issues. The more coarse materials such as gravel, cobbles, and even boulders are deposited near the apex while the sand, silt, and clay particles are transported to more distant parts of the fan. These materials are collectively called alluvium, hence the term alluvial fan.

The conical mound of material is deposited as distributary streams swing back and forth across the fan. This can create dichotomic or braided drainage patterns. Multiple alluvial fans that coalesce at the base of a mountain range create a *bajada*.

A Landsat Thematic Mapper band 4 image of an alluvial fan adjacent to the Salton Sea in California is shown in Figure 14-41a. Note the input stream channel in the upper right and the conical-shaped alluvial fan that has resulted due to millions of years of suspended sediment deposition from very sporadic rainfall events. It is clear that the bright-tone materials deposited on this fan had their origin in a different location when compared with the darker materials found at the base of the nearby mountain range. Also note the dichotomic drainage pattern. Another large alluvial fan in the White Mountains is shown in Figure 14-41b.

An ASTER image of the Taklimakan alluvial fan in the Taklimakan Desert in China's XinJiang Province is shown in Figure 14-42. This fan is located in the landscape between the Kunlun and Altun mountain ranges.

Pediments

A *pediment* is a gently inclined erosional surface carved in bedrock, thinly veneered with gravel, and developed at the base of mountains. Pediments are most prevalent in very arid environments. A pediment may look like an alluvial fan on an image, but it is dramatically different. A pediment is an erosional surface, while an alluvial fan is a depositional surface. In fact, many of the conical features in the Great Basin of North America, in southern Africa, and in Australia are actually pediments and not alluvial fans. Examples of pediments at the base of the White Mountains in California are shown in Figure 14-41b. The headward erosion of the bedrock strews sorted material downslope. The mountain may eventually drown in its own eroded debris.

Playas

Playas are shallow lake basins formed in arid desert regions that are intermittently filled with water that evaporates rela-

Mississippi River Delta, United States

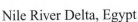

a.

Niger River Delta, Africa

b.

Nile River Delta, Egypt

c.

Irrawaddy River Delta, Burmah

d.

Figure 14-40 a) Space Shuttle photography of the Mississippi River *elongated* bird's-foot delta, consisting of several outlets and their associated natural levees (courtesy K. Lulla, NASA Johnson Space Center). b) Space Shuttle photograph of the Niger River *lobate* delta emptying into the Gulf of Guinea in Africa (NASA Johnson Space Center; STS61C-42-0072). c) MODIS image of the Nile River lobate delta in Egypt (courtesy NASA Goddard Space Flight Center). d) Irrawaddy River *crenulate* delta in Burmah, recorded on a Thematic Mapper band 4 near-infrared image (courtesy Space Imaging, Inc.).

Little San Bernadino Mountains, CA

a.

White Mountains near Death Valley, CA

b.

Figure 14-41 a) Alluvial fan issuing from between the Little San Bernadino Mountains and Chocolate Mountain northeast of the Salton Sea, CA. The California Aqueduct is also visible in this Landsat Thematic Mapper band 4 image. Its bowed shape gives some indication of how the fan has increased in volume over time. b) A large alluvial fan, several pediments, and a playa associated with an area in the White Mountain Range northwest of Death Valley, CA. There are several locations in these images where alluvial fans are coalescing, creating a bajada (courtesy Mojave Desert Ecosystem Program and Space Imaging, Inc.).

Taklimakan Illuvial Fan in XinJiang Province, China

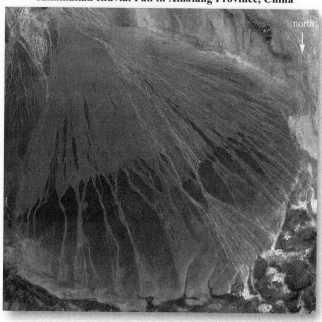

Figure 14-42 The Taklimakan alluvial fan extends across the desolate landscape between the Kunlun and Altun mountain ranges that form the southern border of the Taklimakan Desert in China's XinJiang Province. Note the intricate dichotomic drainage pattern. This ASTER image was obtained on May 2, 2002. North is at the bottom of the illustration to aid visual interpretation (courtesy of NASA/GSFC/METI/ERSDAC/JAROS and U.S./Japan ASTER Science Team).

tively quickly, leaving a residue of fine-textured surface materials. Many playas are saline and exhibit bright tones in imagery. Most playas are barren with little vegetation. A playa is present at the base of the alluvial fan in the White Mountains of California (Figure 14-41b).

Karst Landforms

Landforms created in limestone are generally referred to as *karst topography*. To be a true limestone at least half of the rock consists of carbonate minerals of which calcite ($CaCO_3$) is the most common (Selby, 1989). Dolomite ($CaMg)CO_3$, another carbonate rock, is also susceptible to dissolution but is not as readily affected by slightly acidic precipitation as limestone. This lowers solubility and increases resistance to weathering and erosion. Therefore, dolomite rock does not produce the same karst topography as limestone rock in humid climates.

Surface water derived from precipitation is actually somewhat acidic. When acidic surface water percolates downward through limestone, the chemical process of carbonation takes place along joints and bedding planes, gradually enlarging the openings by dissolution, until subsurface caverns, caves, and subterranean channels are created. Collapse of the roof of a subsurface cavern or downward dissolution of limestone from the surface produces depressions at the surface commonly referred to as *dolines* (the English term is *sinkhole*). *Dolines* are formally defined as closed hollows or depressions that are cone or bowl-shaped, with rocky or vegetated sides, circular or elliptical in shape, 2 – 100 m in depth, and 10 – 1000 m in diameter. The major karst topography is found in the former Yugoslavia along the Adriatic Coast, southern France, Greece, Cuba, Jamaica, the Yucatan Peninsula in Mexico, Tennessee, Kentucky, southern Indiana, northern Florida, and Puerto Rico.

Limestone with > 40 percent calcium carbonate is very susceptible to weathering in humid climates and may result in the development of an extensive network of doline features. For example, consider the black-and-white infrared photograph of karst topography found near Orlando, FL, shown in Figure 14-43 These dolines (sinkholes) were produced by dissolution occurring at favorable sites such as the intersection of major joints in the limestone. The limestone is dissolved and the solution and residue move downward, constantly widening the fissures until a surface depression is created. This permits the collection of more of the surface water runoff and therefore the progressive enlargement of the doline. Sometimes, the dolines collapse, leaving large surface depressions. In this example, there are few surface drainage channels, only subterranean ones.

In areas of high local relief and abundant tropical precipitation such as in Puerto Rico, the weathering is greatly accelerated, causing the sinkholes to coalesce, creating haystack

Karst Topography

Figure 14-43 Calibrated Airborne Multispectral Scanner (CAMS) near-infrared 3 × 3 m data of karst doline (sinkhole) topography north of Orlando, FL. The depressions are created by dissolution of limestone rock. There is no surface drainage pattern in this area because all drainage is subterranean. The darkest areas are deep water. The bright area is sand. Many of these dolines are heavily vegetated.

karst topography. Conversely, dolomitic limestone such as that found in parts of Kentucky is less susceptible to erosion. Consequently, it develops a hill and valley topography along with a surface drainage pattern.

Limestone is especially difficult to identify in extremely arid deserts or cold Arctic environments because the small amount of precipitation does not produce the weathering features discussed above. Instead, limestone is likely the resistant caprock on escarpments or plateaus.

Shoreline Landforms

More than two-thirds of the world's population lives within 100 miles of the coast. Scientists are interested in understanding coastal processes associated with the dynamic coastal plain, beaches, spits, bars, terraces, estuaries, etc. Three major agents are responsible for the creation of specific types of landforms in the coastal zone: 1) energy from the Sun, which produces waves and ocean currents, 2) the gravitational pull of the moon and the Sun combine to create tides, and 3) man is a very active geomorphic agent.

Kuwait City, Kuwait

Figure 14-44   SPOT panchromatic 10 × 10 m image of Kuwait City. Several man-made groins (jetties) influence the deposition or erosion of sediment carried by the longshore current, which generally flows from southeast to northwest. A large spit or hook is present at the northernmost part of the city. This is a good example of how mankind is an active agent of geomorphic change in the coastal zone (© SPOT Image, Inc.).

Radiant energy from the Sun differentially heats the Earth's surface, creating variations in air pressure, which produce weather disturbances. Wind-generated surface waves are the main source of energy along a coast and are responsible for many of the erosional landforms along the coast. Conversely, waves also generate coastal currents that are responsible for the longshore movement of sediment along beaches (Selby, 1989). A wave of height (h) is produced by a sustained wind blowing over a waterbody for a certain *fetch* distance (F) where $h = 0.36\sqrt{F}$. Wave height may also be estimated if only the sustained wind velocity (V) is known using the relationship $h = 0.0024 \cdot V^2$. Sustained prevailing winds blowing over great expanses of water may create ocean currents that are also influenced by variations in water temperature and salinity. These oceanic currents such as the Gulf Stream can impact shoreline landform development because they influence the temperature and amount of suspended sediment in the water, which in turn can impact the development of coral reef and mangrove ecosystems.

The gravitational pull of both the moon and the Sun on the Earth creates tides defined as the periodic rise and fall of sea level that occurs in most seas twice daily. Tides are impor-

tant because they increase the vertical height over which wave action can affect the coast. Currents produced by tidal ebb and flow can erode as well as transport sediment. The tides also flush much of the nutrient-laden detritis from within estuaries and lagoons into the open ocean. The detritis is very important in the aquatic food chain.

Finally, mankind is an active geomorphic agent constantly engineering and reengineering the coastal zone. He routinely diverts major streams, rivers, and longshore currents from their normal courses, significantly impacting their traditional suspended sediment loads. He dredges rivers and streams, changing their gradient; fills wetlands; builds jetties, revetments, and seawalls that redirect wave energy; and builds man-made structures on top of fragile coastal landforms such as dunes. Figure 14-44 depicts groins (jetties) revetments, seawalls, and marinas present on the Kuwait City shoreline that impact the deposition and erosion of suspended sediment carried by the longshore current.

The *coast* is formally defined as the geographical upland region adjacent to a sea. The *shoreline* (or *coastline*) separates the coast from the sea, and is the geographical area

Farewell Spit, South Island, New Zealand

Figure 14-45 Farewell Spit at the northwestern point of South Island, New Zealand, stretches for 30 km east from Cape Farewell. This AS-TER image was obtained on February 13, 2001. The north side of the spit is composed of sand dunes, while the south side facing Golden Bay is wet soil. Submerged tidal mudflats are visible south of the spit (courtesy of NASA/GSFC/METI/ERS-DAC/JAROS and U.S./Japan ASTER Science Team).

affected by wave action. The *shore* may be further subdivided into (Selby, 1989):

- the *offshore* area below the low-tide wave-breaking line;

- the *nearshore* between the low-tide breaker line and the high-tide shoreline;

- the *foreshore* between high- and low-tide shorelines;

- the *backshore* between the highest high-tide line with storm effects, to the base of the permanent land, which could be a cliff or sand dune.

Depositional and erosional coastal landforms are discussed below.

Beaches, Spits, Hooks, Bars, Barrier Islands, Lagoons, and Estuaries

A *beach* is composed primarily of inorganic sediment deposited by longshore currents in the shore zone. Continuous wave action causes beaches to be composed of well-sorted sand-sized particles, although this can vary. For example, some beaches adjacent to major rivers are composed of gravel materials or even cobbles transported along shore from the bedload of the river. The source of beach sed-

iment may be from upland weathered materials adjacent to the beach such as a cliffwall, from the offshore zone where wave action moves sediment shoreward, from fragments of decomposing shell and/or coral, and from suspended sediment or bedload material from rivers and streams.

The continued buildup of sediment on gently sloping shorelines may cause the beach to enlarge in size, move inland, move up or down the shoreline, or move seaward. A major storm can cause any of these effects in a very short time. In any case, *beach ridges* may be formed. If the beach ridge ends in open water, it is called a *spit*. The Farewell Spit in New Zealand is shown in Figure 14-45. The beach ridge or spit in front of Morro Bay, CA (Figure 14-46; Color Plate 14-7), creates an excellent harbor if it is dredged routinely. Spits tend to develop in the predominant direction of the longshore sediment transport under wave action. If the spit curves back toward the land, it is called a *hook*. The hook along the shore of Kuwait City is a good example (Figure 14-44). A spit connecting two headlands is called a *bar*. Behind the sand spit or sandbar a shallow *lagoon* or *bay* may form. Some bars are so extensive that they are called *barrier islands*. On the eastern coast of the United States they extend for as much as 100 km and may be several kilometers wide. For example, Sullivans Island is a coastal barrier island near Charleston, SC (Figure 14-47 and Color Plate 14-8a). Barrier islands take the brunt of hurricane storm surf and storm

Morro Bay, CA

a. Part of Morro Bay,
CA, digital orthophoto
quarter quadrangle
(DOQQ) May 28, 1994.

b. Landsat Thematic Mapper Band 4.   c. Landsat Thematic Mapper Band 7.

Figure 14-46 a) Panchromatic USGS DOQQ of Morro Bay, CA. The bay or lagoon is separated from the ocean by an extensive beach ridge
(courtesy U.S. Geological Survey). b) The beach bar also has numerous dunes, some of which are vegetated. Wave activity in
the shore zone is visible in Landsat TM band 4. c) Band 7 data can be used to clearly identify the beach ridge shoreline. Wave-
cut terraces are also present in this area (courtesy NASA Observatorium and Space Imaging, Inc.). Two color composites of
this area are shown in Color Plate 14-7.

Sullivan's Island, SC

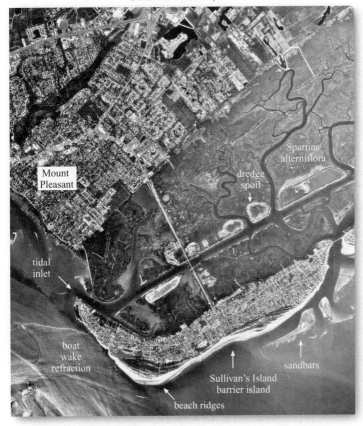

Figure 14-47 Panchromatic version of the USGS digital orthophoto quarter quadrangle of Sullivan's Island, SC. Numerous coastal landforms are
present, including a barrier island created by
accretion of multiple beach ridges, sandbars,
tidal inlets, and tidal marsh (*Spartina alterniflora*). A color-infrared composite of the area
is found in Color Plate 14-8a.

surge. Sullivan's Island experienced a 15 ft storm surge during Hurricane Hugo in 1989. A *tombolo* is formed when a
bar or spit joins an offshore island to the mainland. An *estuary* is a bay at the mouth of a river where the tide influences
the river current.

Tidal Flats, Tidal Marshes, and Mud Flats

A curious thing about the nearshore zone is that the continuous wave action causes clay and silt-size particles to floccu-
late (meaning that the particles stick together) into loose,
relatively large aggregates in seawater. Often this material is
moved from the shore zone into lagoons or estuaries by
wave or tidal action. When the wave or tidal action ceases or
becomes slack, the flocculated material is deposited on the
lagoon or estuary bottom. Thus, one might think of the shore
zone as a place that scrubs the water column, depositing the
more coarse sand and gravel-size material on the beach and
preparing the clay and silt-size particles for deposition else-
where. This flocculation and movement of materials
explains why there is such a difference between beach and
estuarine environments.

Tidal flats are coastal areas that are protected from direct
wave action by spits, hooks, bars, and beach barrier ridges or
islands. Tidal flats have very little slope and may be completely exposed at low tide and are totally submerged at high
tide. Tidal flats may contain a significant amount of vegetation such as smooth cordgrass (*Spartina alterniflora*), mangrove forest (*Rhizophora mangale*), or other species of
vegetation in which case they are called *tidal marshes*
(Davis and Jensen, 1998). If no vegetation is present, the
area is simply called a *mud flat*. Tidal marshes and mud flats
usually have a well-developed dendritic drainage pattern
because the clay-silt mud is homogeneous such as that
shown south of Farewell Spit in New Zealand (Figure 14-
45). The tidal marsh behind the barrier island Isle of Palms,
SC, is shown in Figure 14-48. It consists of tidal channels
containing saline and brackish water, unvegetated tidal mud
flats, and tidal marsh vegetation (smooth cordgrass). Color
Plate 14-8b provides more dramatic information about the
characteristics of the tidal flats and tidal marsh behind Isle of
Palms, SC.

Marsh behind Isle of Palms, SC

a. NASA ATLAS Band 10 (thermal infrared).

b. NASA ATLAS Band 6 (near-infrared).

Figure 14-48 a) NASA ATLAS band 10 thermal infrared 3 × 3 m image of the marsh behind Isle of Palms, SC, obtained in August, 1998. b) Enlarged ATLAS Band 6 near-infrared image of a portion of the area. The images were obtained at mid-tide. Some of the tidal mudflats (which have no vegetation) are exposed, while others are inundated by the incoming tide flowing through the tidal channels. The more dense patches of smooth cordgrass (*Spartina alterniflora*) show up as bright areas along the natural levees in the near-infrared image. The dendritic drainage pattern associated with homogeneous soil materials is evident. It is much easier to discriminate between the tidal flats, tidal water, and smooth cordgrass when viewing the color-infrared color composite of the same area in Color Plate 14-8b.

Marine Terraces

When a coastline is uplifted through tectonic activity, the original wave-cut platform is raised to a higher altitude and wave action begins to create a new wave-cut platform. The result is the presence of a relic marine terrace and associated steep bluffs or cliffs. During the Pleistocene many tectonic uplifts occurred. The dramatic marine terraces along the southern California coast near Palos Verdes were produced in this manner.

Coral Reefs

Coral reefs are created by living coral polyps and calcareous algae (Figure 14-49). Coral reefs develop best in seawater between 77° and 86°F with normal levels of salinity. Thus, coral is generally restricted to the tropical regions of the

Great Barrier Reef, Queensland, Australia

lagoon

Coral Sea of the Pacific Ocean

reef →

↗ Cape York shoreline

a.

Palancar Reef, Cozumel, Mexico

Cozumel Upland

upland

Palancar Reef →

wetland

outer reef wall →

Caribbean

b.

French Mururoa Atoll in the South

Pacific Ocean

lagoon

reef

c.

Figure 14-49 a) The Great Barrier Reef of Australia extends for approximately 1,200 miles (2,000 km) along the Queensland coast. The reefs captured on this Shuttle photograph (STS 046-077-031) are located along the eastern coast of Cape York Peninsula (courtesy Kamlesh Lulla, NASA Johnson Space Center). b) SPOT HRV XS band 1 (green) image of a portion of the Palancar fringing reef on the Island of Cozumel, Mexico (© SPOT Image, Inc.). c) Landsat TM band 2 (green) image of the Mururoa coral atoll in the Pacific Ocean (courtesy Space Imaging, Inc.).

world between 30° N and 25° S latitude. Coral requires water that is <100 m deep and generally free of sediment so that photosynthesis can take place. Coral reefs often grow best on the windward side of islands where wave energy provides food and oxygen for the coral polyps. However, significant reef development can also take place on the leeward side of islands if a strong ocean current is present. For example, the Palancar reef system on the leeward side of the Island of Cozumel, Mexico, is bathed by a strong, clear upwelling ocean current. Fresh water and significant amounts of sediment can prevent coral reef growth. Coral cannot grow out of seawater; therefore, mean low-tide determines the upper limit of coral growth. Sea levels may rise due to global warming. Consequently, we may expect a commensurate growth in coral reef building during the coming century if pollution does not destroy the coral first.

The coral polyps literally create a built-up limestone surface anchored to hardbottom materials. There are three primary types of reefs. A *barrier reef* lies offshore and is separated from the land by a lagoon. The Great Barrier Reef is an excellent example (Figure 14-49a). It extends for more than 1,200 miles along the north coast of Australia. If a reef is attached to the land and extends out into the sea, it is called a *fringing reef*. A portion of the Palancar fringing reef on Cozumel is shown in Figure 14-49b. There may be gaps in fringing reefs due to the input of fresh water or suspended sediment. Finally, we may encounter an *atoll,* which is a ringlike reef enclosing a lagoon with a central island (which may or may not be present either due to subsidence or sea-level rise). The Mururoa atoll in the Pacific Ocean does not have a central island (Figure 14-49c).

Coastal Erosional Landforms

Coastal erosional landforms are created when the combined transporting capacity of wave and longshore current energy exceeds the supply of sediment. Most coastal erosion takes place during storm events due to the destructive force of wave energy. Depending upon how a coastline is oriented, wave refraction may cause either a spreading out or a convergence of wave energy. Headlands or points experience a convergence of wave energy and are therefore more susceptible to erosion. Conversely, adjacent bays generally experience a divergence of wave energy and less erosion.

The erosive power of water may produce *beach cliffs* with varying gradients. Rock fragments in the surf and the tremendous air pressure produced when surf breaks may combine to erode beach cliffs, creating *caves, arches,* and even *blowholes*.

Glacial Landforms

A *glacier* is defined as a body of ice, firn (compacted granular snow), and snow, originating on land and showing evidence of past or present flow. Glacial landforms are created by the moving ice, nivation (a combination of freezing and thawing processes), and glacial meltwater. There are two general categories of glaciation: *continental* and *alpine*. During the Pleistocene Epoch (sometimes referred to as the Great Ice Age), layer upon layer of snow accumulated from year to year, creating vast glaciers that covered much of the Northern Hemisphere. There were approximately four major advances and retreats of the continental glaciers, with the most recent continental glacier retreating (disappearing) approximately 10,000 years ago from the northern United States. Only portions of Antarctica and Greenland currently

Antarctic Ice Sheet near Mount Vinson

Figure 14-50 A small portion of the Antarctic ice sheet near Mount Vinson recorded by ASTER on November 15, 2003. Note the compression ridges formed on the edge of the glacial ice as it passes between two mountain ridges. Antarctica contains 90% of the world's ice. The Antarctic ice sheet has an amazing average thickness of 2,160 m (7,090 ft) (courtesy of NASA/GSFC/METI/ERSDAC/JAROS and U.S./Japan ASTER Science Team).

have continental glaciation. For example, Figure 14-50 is an ASTER image of a small portion of the Antarctic ice sheet near Mount Vinson. Note the compression ridges along the edges of the ice as it flows through the valley between two mountain ridges.

Alpine glaciation is still present in many of the great mountain ranges of the world, such as the Alps, Himalayas, Rocky Mountains, Sierra Nevada, Andes, etc. Both alpine and continental glaciation produce unique landforms based on erosion and deposition.

We focus our attention first on the major landforms that can be identified through analysis of remotely sensed data. It should be stressed that the optimum method of studying glacial landforms is often to evaluate them in three-dimensions, i.e., stereoscopically or in pseudo three-dimensions. For example, Color Plate 14-9 contains both vertical and three-dimensional views of a small part of the Malaspina Glacier in Alaska. The three-dimensional high oblique view was created by draping a Landsat Thematic Mapper image on top of a Space Shuttle RADAR Topography Mission (SRTM) derived digital elevation model. The three-dimensional view provides unique information about the relative height, depth, and slope of the glacial features.

Erosional Glacial Landforms: Cirques, Tarns, Aretes, Horns, U-Shaped Valleys, Hanging Valleys

A *cirque* is a semicircular steep-walled amphitheater-shaped valley in the upper reaches of an alpine glacier. An alpine glacier erodes headward, downward, and along the sides of the original V-shaped stream valley using the immense pressure of the overlying ice, nivation, and glacial quarrying of rock. Interestingly, after the alpine glacier retreats, there may be relatively little material in the bottom of the cirque. In fact, the base of the cirque may have been eroded more severely in certain areas creating a depression. After glacial retreat, the depression may fill with runoff, creating a glacial lake or *tarn*. Numerous cirques in the Uinta Mountains of Utah are shown in Figure 14-51ab. A color image of a Landsat MSS image of the area draped over a digital elevation model is found in Color Plate 14-10a.

An *arete* is formed when cirques on opposite sides of a divide erode headward creating a narrow, serrated mountain ridge. If the two cirques continue to erode headward, they may eventually produce a *col*, which is a sag in the serrated ridge. Numerous aretes and cols are visible in the oblique image of the Uinta Mountains shown in Figure 14-51b. A *horn* is formed when more than two cirques erode headward, leaving only a spire-shaped pinnacle. The most famous is the Matterhorn in Switzerland (Figure 14-52). Horns often appear triangular-shaped when viewed from above in vertical remote sensor data.

During glaciation in mountainous terrain, the glaciers do not normally produce new valleys. Rather, the accumulation and movement of glacial ice in existing valleys erodes headward, erodes the sides of the valley, and scours and deepens the valley floor. This process creates a *U-shaped glacial trough* or valley that is very impressive once the glacier retreats. Wallsburg Canyon in the Wasatch Mountains of Utah is a good example of a glaciated U-shaped valley (Figure 14-53; Color Plate 14-10b). Some glaciers scour extremely deep valleys that become lakes after the glacier retreats. For example, Lake Tahoe shown in Figure 14-54 and Color Plate 14-10c was created in this manner. Lake Tahoe is 23 miles long (37 km) and 12 miles wide (19 km). It lies 6,228 feet (1,898 m) above sea level and is 1,640 feet (500 m) deep. It is one of the deepest lakes in the continental United States.

If the major glacial U-shaped valley erodes downward more extensively than the tributary glacier flows, it may create *hanging valleys*. Hanging valley streams often enter the main valley as waterfalls.

Depositional Glacial Landforms: Moraines, Morainal Lakes, Till Plains, Eskers, Kettles, and Drumlins

Impressive glacial landforms are also created by the deposition of glacially eroded material. Rock material that accumulates on the side or edge of a glacier is called a *lateral moraine*. If two glaciers meet, a *medial* (or middle) *moraine* may be created at their confluence. This often appears as a dark ribbon on aerial photography or other imagery. For example, Figure 14-55a depicts the Barnard Glacier in Alaska with several tributary glaciers. Numerous lateral moraines are present. Several medial moraines are created by the confluence of the lateral moraines.

If a glacier reaches equilibrium between accumulation and ablation (melting), it produces a *terminal moraine* at the end of the glacier (Figure 14-55b). If the equilibrium lasts for a considerable time, the glacier continues to move rock and debris downslope, causing the terminal moraine to increase in volume. If ablation exceeds accumulation, the glacier retreats, leaving the terminal moraine as a remnant landform. If the glacier reaches equilibrium once again, a *recessional moraine* may be produced. It is possible for a number of recessional moraines to be present. Sometimes water is impounded behind the terminal moraine, forming a *morainal lake*.

Sometimes continental or even alpine glaciation deposits a mixture of clay, sand, gravel, and even large boulders over the entire countryside. After retreat, the material is referred to collectively as *glacial till* and the landform as a *glacial till plain*. There are typically no major landforms with any significant elevation in the till plain, only gently rolling terrain. Much of the midwestern United States, the bread basket of the world, is covered with glacial till deposits of varying thickness. Such till often has a mottled texture, with the lighter areas being composed of well-drained sandy soil and the darker area being composed of poorly drained clay-rich soil.

Several glacial landforms are produced directly within, upon, and even under the glacial ice. For example, a subglacial stream running within stagnant (non-moving) glacial ice may deposit well-sorted sand and gravel within the stream tunnels. After glacial retreat, the deposited material appears as a sinuous, linear ridge or *esker* on the terrain, often reaching heights from 5 – 30 m high and 50 – 60 m wide. Eskers may run for many kilometers. For example, Figure 14-56a is a low-oblique aerial photograph of a sinuous, gravel esker

**Uinta
Mountains
in Utah**

a. Landsat Thematic Mapper band 3 (red) image.

b. Landsat TM band 3 image draped over a USGS 30 x 30 m digital elevation model.

Figure 14-51 a) Landsat Thematic Mapper band 3 (red) image of a portion of the Uinta Mountain Range of the Rocky Mountains in Utah.
b) Landsat TM band 3 image draped over a 30 × 30 m USGS digital elevation model (vertical exaggeration is 4x) and viewed
obliquely. A color version is found in Color Plate 14-10a (courtesy Space Imaging, Inc.).

Matterhorn in Switzerland

Figure 14-52 High-oblique aerial photograph of the Matterhorn in Switzerland formed by the headward erosion of more than two cirques. The cirque on this side of the Matterhorn still contains a glacier and to the left is a hanging glacier clinging to the side of the peak. Glacial crevasses are readily apparent (courtesy American Geographical Society Collection archived at the National Snow and Ice Data Center, University of Colorado, Boulder, CO).

Wasatch Range of the Rocky Mountains in Utah

Figure 14-53 Landsat Thematic Mapper band 4 (near-infrared) image of a portion of the Wasatch Mountain Range of the Rocky Mountains in Utah draped over a 30 × 30 m USGS digital elevation model (vertical exaggeration is 5x) highlighting the Mount Timpanogos glacial cirques and aretes and U-shaped Wallsburg Canyon. Several cols are present on the serrated Mount Timpanogos ridgeline. A color version is found in Color Plate 14-10b (courtesy Space Imaging, Inc.).

trending across the terrain in Northern Manitoba, Canada. This esker also deposited sandy alluvium into the glacial lake as the glacier retreated.

Sometimes a large block of glacial ice that contains an assortment of glacial debris becomes detached from the main glacier. The block of ice and materials may be subsequently covered by other debris. After glacial retreat, the ice in the block melts, leaving a depression in the landscape. These depressions or *kettles* fill with water, creating lakes and swamps. A relatively large kettle is shown in Figure 14-56a.

Sierra Nevada Mountain Range in California

Lake Tahoe

N

Figure 14-54   Landsat MSS band 3 (near-infrared) image of Lake Tahoe in the Sierra Nevada of California draped over Defense Mapping Agency Level-1 Digital Terrain Elevation Data (vertical exaggeration is 5x). In addition to Lake Tahoe, there are also several stair-stepped glaciated valleys in the southern portion of the image that contain lakes (courtesy USGS NALC database). A color version is found in Color Plate 14-10c.

A *drumlin* is a smooth, oval-shaped (elliptical) hill of glacial till material with its long axis parallel to the direction of flow of the former glacier. It is generally more blunt and steep on the glacier-facing end and more elongated down-glacier with a thinning tail. Drumlins often occur in clusters, referred to as drumlin swarms. Drumlins consist of unsorted gravel and sand mixed with some clay. There is much controversy as to the processes that produce drumlins. It is generally believed that a drumlin develops beneath active ice and is formed by high water pressure that molds the saturated till into a streamlined form. Drumlins may be 10 – 50 m high, 100 – 300 m wide and 2 – 3 km long. A small drumlin field in and around Grafton Lake in the Kejimkujik National Park of Canada is shown in Figure 14-56b.

Eolian Landforms

When wind erodes, transports, and/or deposits material, it is called *eolian* (aeolian) activity. Wind typically creates both erosional and depositional landforms in coastal areas and in deserts. Eolian landforms may develop when there is an ample supply of sand and enough wind velocity to move individual soil particles. Good sources of sand include coastal and lacustrine beaches, alluvial river deposits, and material from glacial till plains. Given a strong wind and ample supply of sand, the individual grains of sand bounce along the terrain through the process of *saltation* (Selby,

1989). Most sand grains moved during saltation only rise 1 – 2 cm above the ground while a few bounce 1 – 3 m. The repeated collision of particles causes them to become increasingly rounded and smooth (and slightly smaller), which makes it even easier for the particles to become entrained in the wind field. Also, some of the energy of saltating grains of sand is transferred directly to particles on the ground upon impact. This causes those grains to move forward slightly or creep. Extremely fine-grained silt smaller than 200 μm may be captured in the first 1 – 2 km of a turbulent atmosphere and moved great distances as dust. For example, approximately 10 cm of silt has been deposited in southern France from sources in Africa in the last 300 years. Thus, particles of sand are transported in the wind by suspension and saltation, and on the ground by creep.

Sometimes the blowing, saltating sand encounters vegetation or other phenomena that impede its forward progress, or wind velocity decreases, causing the sand particles to be deposited. When sufficient sand is accumulated, we have a *sand dune* which is defined as a mound, ridge, or hill of windblown sand-size particles. This mound of sand affects the flow of the air over the mound. Generally, the wind speed is greatest on the windward (upwind) side of the mound and decreases on the leeward (slipface), wind-shadow side (Figure 14-57a). The zone of reduced air velocity causes the sand to be deposited just beyond the top of the mound onto the leeward back-slope (slipface). Every size of

a. Lateral and medial moraines on Barnard Glacier, AK.

b. Lateral and terminal glacial moraines on Bylot Island, Baffin Island, NW Territories, Canada.

Figure 14-55 a) High-oblique aerial photograph of the Barnard Glacier, AK. Each of the tributary glaciers has lateral moraines. The confluence of the tributary glaciers creates medial moraines. The tributary glacier on the immediate right receives material from at least one other tributary glacier, as evidenced by the two medial moraines within the center of its flow (courtesy American Geographical Society Collection at the National Snow and Ice Data Center, Univ. of Colorado, Boulder). b) Low-oblique photograph of lateral and terminal glacial moraines on Bylot Island, Baffin Island, in the Canadian Northwest Territories.

sand grain has a different angle-of-repose, i.e., the maximum slope at which loose soil material remains stable. When this angle is reached for a particular sand (e.g., 30 – 40°), the sand particles slide down the leeward slipface causing the mound to gradually move forward in the direction of the prevailing wind.

Sand deposits may be shaped into a variety of bedforms including *ripples* (between 5 cm and 3 m apart and 0.1 – 5

a. An esker and kettle in Northern Manitoba, Canada.

b. Drumlins in Kejimkujik National Park, Canada.

Figure 14-56 a) Low-oblique photograph of a sinuous, gravel esker and oval-shaped kettle in the boreal forest of Northern Manitoba. The light-toned area to the left and right of the esker in the foreground are sand deltas deposited by the esker into the glacial lake that was formed as the glacier retreated. b) Low-oblique aerial photograph of several drumlins in the Kejimkujik National Park of Canada (courtesy Terrain Sciences Division, Geological Survey of Canada).

cm high), *dunes* (3 – 600 m apart and 0.1 – 20 m high), and *megadunes* (300 m – 3 km apart, 200 – 400 m high, and many kilometers long). Of course, small ripples may exist on individual sand dunes and megadunes. This discussion focuses on dunes and megadunes.

Depositional Eolian Landforms

Most dunes are composed of medium grain-size quartz sand; however, they may also be composed of volcanic ash, calcite, and even gypsum particles. There are a variety of depositional sand dune types that have characteristic developmental cycles and diagnostic shapes that can be recognized in aerial photography or other imagery. Scientists have compiled inventories of the major types of dunes found throughout the world and how the dunes appear in remotely sensed data. For example, see work by Rinker et al. (1991), Dokka et al. (1997), and Walker (1986; 1998).

The earliest significant work on the physics, origin, and evolution of eolian sand dunes was performed by Ralph Bagnold (1941), who worked in Egypt prior to World War II. He recognized two basic dune types: the crescentric dune, which he called "barchan," and the linear dune, which he called longitudinal, or "seif" (Arabic for "sword"). Unfortunately, his terminology was then applied to many other types of dunes, resulting in a bewildering terminology. This section utilizes the following dune classification system adapted from several sources (Walker, 1986; 1998; Rinker et al., 1991): crescentric, linear, dome, star, and parabolic.

All dune types may occur in three forms: simple, compound, and complex. *Simple* dunes are basic forms with a minimum number of slipfaces that define the geometric type. *Compound* dunes are large dunes on which smaller dunes of similar type and slipface orientation are superimposed. *Complex* dunes are combinations of two or more dune types (Walker, 1998).

Crescentric: The crescentric dune is the most common dune on Earth (and Mars). They are generally more wide than they are long. Based on the work of Bagnold (1941), these dunes are often referred to as barchan, barchanoid, or transverse dunes. Barchan crescentric dunes such as those shown in Figure 14-57b,c may be produced when 1) the prevailing wind comes from a uniform, consistent direction, 2) there is little vegetation present to anchor any portion of the dune, and 3) there is a relatively sparse supply of sand. The two horns of the crescent point downwind. The gentle windward slope is usually < 15°. The leeward, slipface side of the barchan dune is steepest with an angle-of-repose of approximately 34 – 40° for dry sand. Individual barchans or extremely large megabarchans commonly occur in elongate chains or trains that may merge with coalesced dunes in fields or ergs. Small barchans move several meters per year, but megabarchans move more slowly (Walker, 1998).

Linear: Linear (or longitudinal) dunes may be straight or irregularly sinuous. A low-oblique photograph of a linear dune in the south end of Soda Lake is shown in Figure 14-58a. These dunes may be many kilometers long and hundreds of meters high. They are generally composed of loose,

a. Sand movement on a dune.

b. Oblique view of crescentric dunes
(also called barchan or transverse dunes).

c. View of a single crescentric dune.

Figure 14-57 a) General characteristics of how sand is moved through saltation and creep on a sand dune. This illustration is keyed to the wind direction and features in (b). b) Several crescentric (barchan) dunes on the coast of Peru migrating from right to left, parallel with the wind direction (courtesy U.S. Geological Survey; Walker, 1998; and John McCauley). c) View of a single crescentric, barchanoid dune (courtesy Desert Processes Working Group; Rinker et al., 1991; Dokka et al., 1997).

well-sorted very fine to medium grain sand. Extremely straight, linear dunes are often referred to as *sand ridges,* while the linear, sinuous varieties are often called *seifs.* The length of a linear dune is much greater than its width. While the crescentric dune is the most commonly occurring dune, linear dunes cover more desert area, especially in central Australia, southern Africa, the Arabian Peninsula, and parts of the Sahara. The largest and tallest dunes in the world are found in the Namib Desert, Namibia, Africa. Some dunes are greater than 300 m high (Figure 14-59). Color Plate 14-11 depicts a portion of the vast Namibian sand dune landscape using Landsat 7 ETM and ASTER data draped over a digital elevation model.

The origin of linear dunes is controversial. Some suggest that the linear dunes are produced by unidirectional winds and that the alignment of the linear dunes is parallel with the wind direction. Others suggest that multidirectional winds

also play a role in their development. It may be that the length of the dune is enlarged when the wind blows in the predominant direction. The width of the dune is enlarged when the wind blows strongly at right angles to the prevailing wind direction for relatively brief periods of time. An extensive sea of linear dunes is shown in the Space Shuttle photograph of the Marzuq Desert of Libya in Figure 14-58b.

Dome Dunes: Oval or circular mounds of sand that generally lack a slipface are called *dome dunes.* They are relatively rare. Figure 14-58c depicts a stereopair of dome dunes in the Western Shield of Saudi Arabia. The domes are partly separated by interdune hollows that are enclosed by subsidiary dunes of hard-packed sand.

Star Dunes: Star dunes may develop where there is an ample amount of sand and the wind blows in many directions. They are also called pyramid or radial dunes. Three or

a. Oblique view of a linear (longitudinal) dune on Soda Lake.

b. Shuttle photograph of linear dunes in the Marzuq Desert, Libya.

c. Stereopair of dome dunes in the Western Shield of Saudi Arabia.

Figure 14-58 a) Example of a linear (longitudinal) dune. Note the physical obstruction causing the majority of sand to deviate from its course while some spills over the top (courtesy of Desert Processes Working Group; Dokka et al., 1997). b) Black-and-white version of Space Shuttle photograph STS054-152-189 obtained in January 1993, depicting an extensive sea of linear dunes. The dark areas are part of the Hammadat Marzuq (rocky desert) consisting of bedrock outcrops (courtesy NASA Johnson Space Center). c) Panchromatic stereopair of dome dunes in the Western Shield of Saudi Arabia (photography obtained by the Army Map Service; courtesy of Desert Processes Working Group; Rinker et al., 1991; Dokka et al., 1997).

more arms of sand typically radiate from a high central mound. The arms can vary in length, width, number, and shape, but each has a slipface. Star dunes accumulate in areas with multidirectional wind regimes. Star dunes tend to increase in height rather than migrating horizontally across the landscape, causing them to be one of the tallest dunes found. They may reach 200 – 300 m in the great deserts of the world. Star dunes also occur as secondary elements on or in combination with other dunes. A single star dune in the Namib Desert is shown in Figure 14-60a, and a Space Shuttle photograph of star dunes superimposed on linear dunes in Algeria is shown in Figure 14-60b.

Namib Desert, Namibia, Africa

Figure 14-59 Longitudinal dunes in the Namib Desert in Namib-Naukluft National Park recorded by the Landsat 7 ETM+ on August 12, 2000. These are the tallest dunes in the world, some > 300 m (almost 980 ft) (courtesy USGS and NASA). Refer to Color Plate 14-11a for a view of a portion of the same area recorded by the ASTER sensor.

Parabolic Dunes: A parabolic dune has a crescent, U-shape like a barchan dune *except* that its horns point in the opposite direction into the wind. They are also called U-shaped, blowout, or hairpin dunes. Parabolic dunes are created when wind is sufficiently intense to literally blow out the center of a dune but a relatively small amount of vegetation on either side causes the edges of the dune to be stabilized. Sometimes the horns of the parabolic dune remain attached to the vegetation for some time as the main body of the dune migrates downwind. When this occurs, the parabolic dune resembles a hairpin lying on its side instead of a neatly shaped barchan crescent. Parabolic dunes are always associated with vegetation — grasses, shrubs, and occasional trees that anchor the trailing arms. They usually form in areas with strong unidirectional winds.

Loess: As a continental glacier retreats, it may deposit glacial till in an outwash plain. Strong winds blowing for millennia over the outwash plain may extract the small, silt-size particles. These windborne, silt-size particles may be deposited downwind from the till plain as a layer of *loess* soil. For example, the thickest known deposit of loess is 335 m on the Loess Plateau of China. Loess accumulations in Europe and in the Americas resulting from the Wisconsin glaciation are generally 20 – 30 m deep. Loess is a very highly erodable soil that is usually just a few feet thick. Drainage channels in loess soil often have very steep banks. Dunes are not gener-ally formed on loess deposits, because of increased cohesion between the loess soil particles.

Erosional Eolian Landforms

Blowing sand can also produce impressive erosional features. The kinetic energy of wind acts upon the ground surface. Some of the energy is transferred to the ground as heat while some of it detaches soil particles from the ground and moves them along through saltation. When material is removed from a surface via wind, we have *deflation*. The continuous removal of both sand and silt from areas with an alluvial deposit leaves only a combination of coarse sand and pebbles and some clay material. This can result in the creation of a *desert pavement* landscape which is a sheetlike surface of rock fragments that remain after wind and water have removed the fine particles. Almost half of the Earth's desert surfaces are stony deflation zones. Desert pavement regions are called *regs* in Algeria, *serir* in Libya, and *gibbers* in Australia. These deflated regions may be very extensive in large deserts. A deflated basin is called a *blowout*. Sometimes a dark, shiny desert varnish is found on the desert pavement surfaces produced from manganese, iron oxides, hydroxides, and clay minerals.

The surfaces of rock bodies may also be eroded by abrasion and sandblasting by windborne particles. In extreme cases the wind may be so strong that it cuts channels in the less

a. Star dunes in the Namib Desert.

b. Space Shuttle photograph of the complex linear Tifernine Dunes in the Tassili N'Ajjer of southeastern Algeria superimposed with star dunes.

Figure 14-60 a) Large-scale aerial photograph of a star dune in the Namib desert. Each star dune consists of multiple arms each with their own slipfaces radiating from a central area. Star dunes are produced by winds coming from multiple directions (courtesy USGS; Walker, 1998). b) Black-and-white version of Space Shuttle photograph STS070-705-094 of the Tifernine Dunes of Algeria in July, 1995. Extremely long, linear dunes are present. The linear dunes have star dunes superimposed on portions of them. Resistant, weathered sandstone lies to the south and east and appears in dark tones. Several white playas are present in the image (courtesy NASA Johnson Space Center).

resistant rock, leaving residual streamlined spines called *yardangs,* which may be tens of meters high and kilometers long.

 References

ASTER, 2006, *Advanced Spaceborne Thermal Emission and Reflection Radiometer*, Washington: NASA, http://asterweb.jpl.nasa.gov/.

Baldi, P., Fabris, M., Marsella, M. and R. Monticelli, 2005, "Monitoring the Morphological Evolution of the Sciara del Fuoco during the 2002-2003 Stromboli Eruption using Multitemporal Photogrammetry," *ISPRS Journal of Photogrammetry & Remote Sensing*, 59:199–211.

Bagnold, R. A., 1941, *The Physics of Blown Sand and Desert Dunes*, London: Methuen, 265 p.

Belnap, J., 2003, "The World at your Feet: Desert Biological Soil Crusts," *Frontiers in Ecology and the Environment*, 1(5):181–189.

Bocco, G., Velazquez, A. and C. Siebe, 2005, "Using Geomorphologic Mapping to Strengthen Natural Resource Management in Developing Countries," *Catena*, 60:239–253.

Chabrillat, S., Goetz, A. F. H., Krosley, L. and H. W. Olsen, 2002, "Use of Hyperspectral Images in the Identification and Mapping of Expansive Clay Soils and the Role of Spatial Resolution," *Remote Sensing of Environment*, 82:431–445.

Clark, R. N., 1999, "Spectroscopy of Rocks and Minerals, and Principles of Spectroscopy," in *Manual of Remote Sensing – Geosciences,* A. Rencz (Ed.), NY: John Wiley.

Chen, J., Zhang, M. Y., Wang, L., Shimazaki, H. and M. Tamura, 2005, "A New Index for Mapping Lichen-dominated biological Soil Crusts in Desert Areas," *Remote Sensing of Environment*, 96:165–175.

Fernandes da Silva, P. C., Cripps, J. C. and S. M. Wise, 2005, "The Use of Remote Sensing Techniques and Empirical Tectonic Models for Inference of Geological Structures: Bridging from Regional to Local Scales," *Remote Sensing of Environment*, 96:18–36.

Dalton, J. B., Bove, D. J., Mladinich, C. S. and B. W. Rockwell, 2004, "Identification of Spectrally Similar Materials using the USGS Tetracorder Algorithm: the Calcite-epidote-chlorite Problem," *Remote Sensing of Environment*, 89:455–466.

Davis, B. A. and J. R. Jensen, 1998, "Remote Sensing of Mangrove Biophysical Characteristics," *Geocarto Intl.*, 13:55–64.

Dokka, R. K., Muyer, L., Rinker, J., Watts, J., Wakefield, G., McCauley, J. F. and C. Breed, 1997, *MDEP Surface Materials Mapping Group Database Design Document Version 2 (MDEI-2000)*.

Ellis, R. J. and P. W. Scott, 2004, "Evaluation of Hyperspectral Remote Sensing as a Means of Environmental Monitoring in the St. Austell China Clay (Kaolin) Region, Cornwall, UK," *Remote Sensing of Environment*, 93:118–130.

Farifteh, J., Farshad, A. and R. J. George, 2006, "Assessing Salt-affected Soils Using Remote Sensing, Solute Modelling, and Geophysics," *Geoderma*, 130:191–206.

Galvao, L. S., Almeida-Filho, R. and I. Vitorello, 2005, "Spectral Discrimination of Hydrothermally Altered Materials using ASTER Short-wave Infrared Bands: Evaluation in A Tropical Savannah Environment," *Intl. Journal of Applied Earth Observation and Geoinformation*, 7:107–114.

Hapke, B., 1993, *Theory of Reflectance and Emittance Spectroscopy*, New York, NY: Cambridge University Press.

Hoffer, R., 1978, "Biological and Physical Considerations," *Remote Sensing: The Quantitative Approach*, 241–251.

Hook, S. J., Dmochowski, J. E., Howard, K. A., Rowan, L. C., Karlstrom, K. E. and J. M. Stock, 2005, "Mapping Variations in Weight Percent Silica Measured from Multispectral Thermal Infrared Imagery," *Remote Sensing of Environment*, 95:273–289.

Hooper, D. M., Bursik, M. I. and F. H. Webb, 2003, "Application of High-resolution, Interferometric DEMs to Geomorphic Studies of Fault Scarps, Fish Lake Valley, Nevada-California, USA," *Remote Sensing of Environment*, 84:255–267.

Hunt, G. R., 1977, "Spectral Signatures of Particulate Minerals in the Visible and Near-Infrared," *Geophysics*, 42: 501–513.

James, A., 2006, correspondence, geomorphologist in the Dept. of Geography, Univ. of South Carolina, Columbia, SC.

JPL, 2006, *ASTER Spectral Library*, Pasadena: NASA Jet Propulsion Lab, http://speclib.jpl.nasa.gov/.

King, C., Baghdadi, N., Lecomte, V. and O. Cerdan, 2005, "The Application of Remote Sensing Data to Monitoring and Modelling Soil Erosion," *Catena*, 62:79–93.

Kruse, F. A., Kierein-Young, K. S. and J. W. Boardman, 1990, "Mineral Mapping at Cuprite, Nevada with A 63-channel Imaging Spectrometer," *Photogrammetric Engineering and Remote Sensing*, 56(1):83–92.

Loynachan, T. E., Brown, K. W., Cooper, T. H. and M. H. Milford, 1999, *Sustaining Our Soils and Society*, Alexandria: Soil Science Society of America and USDA, 66 p.

Lu, Z., Rykhus, R., Masterlark, T. and K. G. Dean, 2004, "Mapping Recent Lava Flows at Westdahl Volcano, Alaska, Using RADAR and Optical Satellite Imagery," *Remote Sensing of Environment*, 91:345–353.

Marsh, W. M. and J. Dozier, 1981, *Landscape: An Introduction to Physical Geography*, Reading: Addison Wesley, 637 p.

McCauley J. F., Schaber, G. G., Breed, C. S., Grolier, M. J., Haynes, C. V., Issawi, B., Elachi, C., and R. Blom, 1982, "Subsurface Valleys and Geo-archaeology of the Eastern Sahara Revealed by Shuttle Radar," *Science*, 218:1004–1020.

Metternicht, G. I. and J. A. Zinck, 2003, "Remote Sensing of Soil Salinity: Potentials and Constraints," *Remote Sensing of Environment*, 85:1–20.

Mojave Desert Ecosystem Program, 1998, *Mojave Desert Ecosystem Program Electronic Database*, Washington: Dept. of Defense and USDI, 8 CDs and maps.

Okin, G. S. and T. H. Painter, 2004, "Effect of Grain Size on Remotely Sensed Spectral Reflectance of Sandy Desert Surfaces," *Remote Sensing of Environment*, 89:272–280.

Pergola, N., Marchese, F. and V. Tramutoli, 2004, "Automated Detection of Thermal Features of Active Volcanoes by Means of Infrared AVHRR REcords," *Remote Sensing of Environment*, 93:311–327.

Petersen, G., 1999, correspondence, President of the Soil Science Society of America, Pennsylvania State University.

Rao, B., Sankar, T., Dwivedi, R., Thammappa, S., Venkataratnam, L., Sharma, R. and S. Das, 1995, "Spectral Behavior of

Salt-affected Soils," *Intl. Journal of Remote Sensing,* 16:2125–2136.

Rasher and Weaver, 1990, *Basic Photo Interpretation,* Washington: Soil Conservation Service, with photographs and slides.

Reddy, G. P. O., Maji, A. K. and K. S. Gajbhiye, 2004, "Drainage Morphometry and its Influence on Landform Characteristics in a Basaltic Terrain, Central India – A Remote Sensing and GIS Approach," *Intl. Journal of Applied Earth Observation and Geoinformation,* 6:1–16.

Rencz, A., (Ed.), 1999, *Manual of Remote Sensing – Geosciences,* NY: John Wiley.

Rinker, J. N., Breed, C. S., McCauley, J. F. and F. A. Cord, 1991, *Remote Sensing Field Guide - Desert,* Ft. Belvoir: U.S. Army Engineer Topographic Lab, 568 p., http://www.tec.army.mil/terrain/desert/coverpg.htm.

Rowan, L. C., Crowley, J. K., Schmidt, R. G., Ager, C. M. and J. C. Mars, 2000, "Mapping Hydrothermally Altered Rocks by Analyzing Hyperspectral Image (AVIRIS) Data of Forested Areas in the Southeastern United States," *Journal of Geochemical Exploration,* 68:145–166.

Salisbury, J. W., Walter, L. S., Vergo, N. and D. M. D'Aria, 1991, *Infrared (2.1 - 25 μm) Spectra of Minerals,* Baltimore: John Hopkins University Press, 267 p.

Selby, M. J., 1989, *Earth's Changing Surface: An Introduction to Geomorphology,* Oxford: Clarendon Press, 607 p.

Shelton, J. S., 1966, *Geology Illustrated,* San Francisco: W. H. Freeman, 434 p.

Short, N. M. and R. W. Blair, 1999, *Geomorphology from Space, A Global Overview of Regional Landforms,* NASA: Washington, http://daac.gsfc.nasa.gov//DAAC_DOCS/daac_ed.html.

Strain, P. and F. Engle, 1993, *Looking at Earth,* Atlanta: Turner Publishing, 304 p.

Swayze, G. A., Clark, R. N., Goetz, A. F. H., Livo, K. E. and S. S. Sutley, 1999, "AVIRIS 1998 Low Altitude Versus High Altitude Comparison Over Cuprite, Nevada," Denver: USGS, http://speclab.cr.usgs.gov/.

Trefil, J. and R. M. Hazen, 1995, *The Sciences: An Integrated Approach,* NY: John Wiley, 634 p.

USDA, 1998, *Keys to Soil Taxonomy,* Washington: USDA, 8th Ed., 371 p.

USDA, 2003, *Keys to Soil Taxonomy,* Washington: USDA, 9th Ed., 332 p., http://soils.usda.gov/technical/classification/tax_keys/keysweb.pdf.

USGS Speclab, 2006, *USGS Digital Splibo4 Spectral Library: 0.2 to 3.0 micrometers,* Washington: USGS, http://speclab.cr.usgs.gov/spectral.lib04/spectral-lib04.html.

Van der Meer, F., 1994, "The Extraction of Mineral Absorption Features from High-Spectral Resolution Data Using Non-Parametric Geostatistical Techniques," *Intl. Journal of Remote Sensing,* 15:2193–2214.

Walker, A. S., 1986, "Eolian Landforms," in Short, N. M. and R. W. Blair, *Geomorphology from Space, A Global Overview of Regional Landforms,* Washington: NASA, 447–520.

Walker, A. S., 1998, *Deserts: Geology and Resources,* Washington: USGS, http://pubs.usgs.gov/gip/deserts/contents.

Walsh, S. J., Butler, D. R.and G. P. Malanson, 1998, "An Overview of Scale, Pattern, Process Relationships in Geomorphology: A Remote Sensing and GIS Perspective," *Geomorphology,* 21:183–206.

Wassenaar, T., Andrieux, P., Baret, F., and J. M. Robbez-Masson, 2005, "Soil Surface Infiltration Capacity Classification based on the Bi-directional Reflectance Distribution Function Sampled by Aerial Photographs," *Catena,* 62:94–110.

Way, D. S. and J. R. Everett, 1997, "Chapter 3: Landforms and Geology," *Manual of Photographic Interpretation,* 2nd Ed., Bethesda: ASP&RS, 117–165.

Zribi, M., Baghdadi, N., Holah, N, Fafin, O. and C. Guerin, 2005, "Evaluation of A Rough Soil Surface Description with ASAR-ENVISAT Radar Data," *Remote Sensing of Environment,* 95:67–76.

In Situ Reflectance Measurement 15

I*n situ* spectral reflectance measurements obtained on the ground using a handheld spectroradiometer can be used to:

- gain insight into the spectral reflectance characteristics of selected materials;

- calibrate remote sensor data; and

- provide unique spectral data for improved information extraction using multispectral and hyperspectral remote sensor data.

In situ spectral reflectance data are often collected by scientists simply to gain insight into the spectral characteristics of Earth-surface materials (e.g., Bahiagrass). Additional knowledge about the material can be obtained by subjecting it to various treatments (e.g., application of different amounts of Nitrogen to Bahiagrass) and determining if the treatments result in different spectral reflectance patterns. The various treatments and their spectral reflectance characteristics can be monitored through time to obtain additional insight. Scientists often create *spectral reflectance libraries* of the materials found in their study area. Several public agencies share their spectral libraries via the Internet, e.g., U.S. Geological Survey Spectral Library #5 (Clark et al., 2003) and the ASTER Spectral Library (ASTER, 2006).

Remote sensing systems collect spectral reflectance data of the terrain in the visible and near-infrared (VNIR) and short-wavelength infrared (SWIR) portion of the electromagnetic spectrum (0.35 - 2.5 μm). Atmospheric scattering, absorption, and reflection often impact the energy reflected from the terrain and recorded by the remote sensing system (refer to Chapter 2). Fortunately, remote sensing–derived spectral reflectance data can be correlated with *in situ* percent reflectance data of the same phenomena to correct for atmospheric effects (Teillet et al., 2002). For example, a 3 × 3 m plot of Bahiagrass on the ground can be measured using a) a high spatial resolution airborne multispectral scanner, and b) a handheld spectroradiometer. The remote sensing data may be collected thousands of meters AGL. The *in situ* spectroradiometer measurement is typically collected approximately 1 m above the canopy. Numerous paired airborne and ground-based spectral reflectance measurements (e.g., 20 to 30) can be used to calibrate (adjust) the remote sensing-derived spectral reflectance measurements, to minimize the effects of atmospheric scattering, absorption, and reflection (Goetz, 2002). The result is a *calibrated* remote sensing dataset that often yields more accurate land cover or biophysical information (e.g., biomass, leaf-area-index) than an uncalibrated dataset.

Spectral Reflectance Measurement Using a Handheld Spectroradiometer

Reference	Target	
(*Spectralon* Plate)	(*e.g., Bahiagrass*)	Reflectance = Target / Reference

a.   b.   c.

Figure 15-1 A spectral reflectance curve of Bahiagrass obtained by dividing the target spectra (Bahiagrass in this example) by the reference spectra (e.g., a white Spectralon reference plate).

In situ spectral reflectance data from almost pure Earth surface materials (e.g., healthy full-canopy Bahiagrass, fresh asphalt, deep-clear water) can be used to locate endmembers (i.e., pure classes) for use during multispectral or hyperspectral data analysis. The endmembers are used to perform spectral unmixing which can be used to produce maps that reveal the proportion of materials within each pixel in the dataset (e.g., one pixel might contain 40% Bahiagrass, 30% asphalt, and 30% water). Jensen (2005) provides a discussion of linear spectral unmixing.

This chapter describes several parameters that should be considered when obtaining *in situ* spectral reflectance measurements of land cover materials using a handheld spectroradiometer.

Spectral Reflectance of a Material

The process of obtaining handheld spectral reflectance measurements is intriguing. To obtain the reflectance spectrum

of a target material, ρ_T, it is necessary to obtain two spectral measurements:

1. the amount of radiance reflected in a specific wavelength interval from a *reference* sample, L_r (e.g., a Spectralon plate);

2. the amount of radiance reflected from the target of interest, L_T, in the same wavelength interval (e.g., Bahiagrass).

The reflectance spectrum, ρ_T, is computed by dividing the spectral response of the target of interest by the spectral response of the reference sample:

$$\rho_T = \frac{L_T}{L_r} \times k . \qquad (15\text{-}1)$$

The correction factor (k) is the ratio of solar irradiance to reference plate exitance and ideally should be near 1 (McCoy, 2005). An example of reference spectra (Spectralon), target

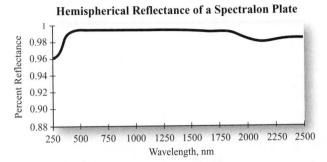

Figure 15-2 Spectral reflectance characteristics of a Spectralon plate over the spectral range 250 – 2500 nm measured in the laboratory under controlled artificial illumination. The reflectance curve does not exhibit the atmospheric absorption bands that are present when measurements are obtained in direct solar illumination as shown in Figure 15-1a.

spectra (Bahiagrass), and percent reflectance of Bahiagrass is demonstrated in Figure 15-1.

Reference Material

Spectralon is the most widely used reference material when obtaining *in situ* spectral reflectance measurements. It is a thermoplastic white (or gray) resin that has demonstrated the highest diffuse reflectance of any known material or coating for terrestrial remote sensing applications for both field and laboratory applications. Space-grade Spectralon combines high reflectance with an extremely lambertian reflectance profile, exhibiting reflectance values of >99% over the wavelength range from 400 to 1500 nm and >95% from 250 to 2500 nm (Labsphere, 2006). Spectral reflectance characteristics of a Spectralon plate measured in the laboratory are shown in Figure 15-2. Spectralon plates range in size from 2 × 2 × 0.56 in. up to 24 × 24 × 0.56 in. Most scientists use 10 × 10 or 12 × 12 in. plates mounted in a durable anodized aluminum frame. It is possible to purchase larger plates but they become difficult to handle in the field. They are also very expensive. The surface of the Spectralon plate should be kept clean from dust and debris. After heavy use it may be necessary to use water and very-fine sandpaper to restore the surface.

Each Spectralon plate is calibrated by the manufacturer. Calibration data from 250 to 2500 nm, every 50 nm is supplied with a Spectralon plate. Calibration data are traceable to national laboratory standards.

Illumination Considerations

When computing the reflectance spectrum of a material in the field or in the laboratory, it is important that the illumination characteristics of the *reference* material (e.g., Spectralon) and the target of interest (e.g., Bahiagrass) remain the same.

Any variation in the illumination characteristics between the time the reference spectra are collected and the time the target of interest spectra are collected results in error in the derived reflectance spectrum. That said, it is useful to review various scene illumination considerations.

In Situ Spectral Reflectance Measurement Using Solar Illumination

The solar energy reaching the Earth's surface and the dominant factors affecting the illumination were discussed in Chapter 2. They are revisited in Figure 15-3 for this handheld radiometry discussion. Water vapor impacts the incoming radiant flux the most (Gao and Goetz, 1990). Solar energy reaching the Earth's surface is essentially zero in the regions around 1400 nm and 1900 nm because atmospheric water vapor absorbs light in these regions over the long path length of the atmosphere. The greater the distance that the energy has to pass through the atmosphere (i.e., the path length), the greater the amount of absorption that takes place by the various atmospheric constituents. The amount of water vapor in the atmosphere changes diurnally and seasonally and can vary in space over short distances depending upon the characteristics of regional air masses.

Reflectance spectra obtained in the field using a handheld spectroradiometer are influenced by five types of solar illumination. The most dominant irradiance is direct solar illumination (E_o: path 1) as shown in Figure 15-3. It may contribute as much as 90% of the total energy illuminating the target under clear sky conditions. Seasonal solar elevation angle (θ) and atmospheric conditions (e.g., high water-vapor content) will affect the amount of radiant flux and its spectral characteristics illuminating the target. Figure 15-4 documents the solar irradiance that actually makes it to the Earth's surface after passing through the atmosphere.

Scattering in the atmosphere results in diffuse skylight illumination. The diffuse skylight usually accounts for 10–20% of the total energy illuminating the terrain. The spectral characteristics of diffuse skylight illumination are a function

In Situ Spectral Reflectance Measurement Using Solar Illumination

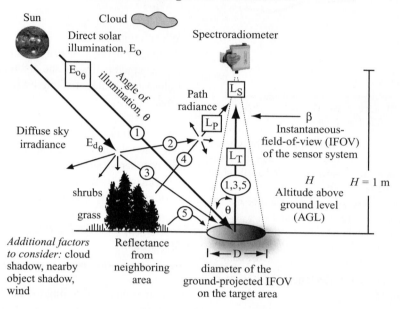

Figure 15-3 Spectroradiometer setup for obtaining *in situ* spectral reflectance measurements using solar illumination.

of the type and amount of atmospheric scattering and absorption taking place. Some of this diffuse sky illumination never reaches the target but is scattered into the remote sensing instrument field of view (E_d: path 2). Some of the diffuse sky irradiance reaches the target and is subsequently reflected into the remote sensing instrument field of view (E_d: path 3). Some irradiance on nearby terrain is reflected into the remote sensing instrument's field of view (4). Finally, the target of interest (e.g., Bahiagrass) may be illuminated by light scattered from nearby agriculture, grass, forest, or urban structures (5). Curtiss and Ustin (1988) found that as much as 20% of the illumination on cleared forest areas came from adjacent forest cover.

The irradiance that never reaches the ground (2) and the reflectance from nearby features on the ground (4) are considered to be noise and are called *path radiance* (L_P). Thus, the total radiance reaching the sensor (L_S) is a function of the atmospheric path radiance (L_P) and radiance reflected from the target of interest (L_T). Several atmospheric characteristics also affect illumination, including the presence of clouds and wind.

Clouds are visual evidence of water vapor in the atmosphere. Clouds can modify the spectral characteristics of the irradiance reaching the terrain. Therefore, one might think that it is necessary to have an absolutely clear sky devoid of

Solar Illumination Reaching the Earth's Surface

Figure 15-4 Solar illumination reaching the Earth's surface.

clouds in order to obtain reflectance spectra. This is not the case. As long as the reference spectrum (e.g., derived from a Spectralon plate) and the target spectrum (e.g., derived from Bahiagrass) are obtained under almost identical cloud-cover characteristics (i.e., taken in very rapid succession), then the subsequent ratioing will reduce the error introduced by atmospheric cloud and water vapor variability (clouds).

In Situ Spectral Reflectance Measurement Using Artificial Illumination

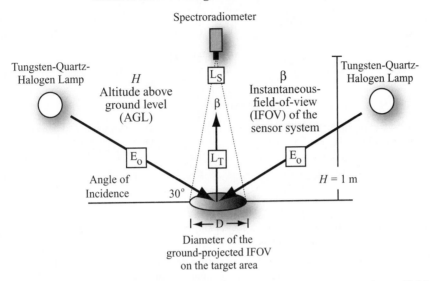

Figure 15-5 Spectroradiometer setup for obtaining *in situ* spectral reflectance measurements using artificial illumination in a lab.

If the *in situ* reflectance spectra are going to be used to calibrate satellite or aircraft remote sensing–derived spectra, then the *in situ* reflectance spectra should be collected at approximately the same time and under the same atmospheric conditions as the remote sensing mission.

Wind can play havoc when obtaining *in situ* reflectance spectra. Vegetation canopies are particularly susceptible because reflectance spectra for a target typically are not obtained in a single measurement. Rather, a Spectralon reference measurement is made, and then perhaps 10 or more target measurements are made. The 10 target spectral reflectance measurements are then averaged to yield a single target reflectance measurement that is divided by the reference spectra to yield a calibrated reflectance spectrum for the canopy under investigation. Ideally, the canopy remains completely still during the collection of the 10 target measurements. Ideally, the leaves are in the same position, and so are the shadows and the amount of bare soil seen by the spectroradiometer detectors through the optical instantaneous field of view (IFOV). Wind moves the leaves in the canopy, modifying the amount of vegetation, shadow, and bare soil measured within the constant IFOV of the spectroradiometer. Wind can even cause different structural parts of the canopy (e.g., stems, the underside of leaves) to be viewed during different scans. This introduces noise into the reflectance measurements.

In Situ Spectral Reflectance Measurement Using Artificial Illumination in the Laboratory

Sometimes it is useful to obtain the spectral reflectance characteristics of grass, leaves, or other land cover subjected to various treatments in a controlled laboratory environment. In a laboratory environment, the analyst has control of:

- illumination geometry based on the use of carefully positioned and calibrated lamps;
- spectroradiometer viewing geometry (usually pointing at nadir);
- environmental effects (e.g., there is no wind, and it does not rain in the laboratory);
- atmospheric effects (e.g., it is possible to control the temperature and water vapor in the lab).

A good configuration for indoor artificial illumination is shown in Figure 15-5. It consists of two 200 – 500 W Tungsten Quartz Halogen filament lamps (each approximately 3400 K color temperature) placed in aluminum reflector housings, oriented 30° from horizontal, and placed 1 m above the phenomena of interest (ASD, 2005). The power supply should be stable. The spectroradiometer optics are positioned to view the vegetation canopy at nadir. Ideally,

Figure 15-6 The spectrum of a fluorescent light.

Figure 15-7 The spectrum of a Tungsten-Quartz-Halogen lamp.

the sensor field of view is completely filled by the target materials of interest (e.g., Bahiagrass). Care must be exercised when using artificial illumination to collect target reflectance spectra, however, because extensive exposure to the Tungsten-Quartz-Halogen lamps can actually "cook" the vegetation. This can degrade the chlorophyll in the palisade parenchyma and cause plant water to be evaporated from plant spongy mesophyll. These conditions result in spurious vegetation reflectance spectra.

Why Tungsten–Quartz–Halogen Illumination? Many sources of light can be used when obtaining *in situ* spectra using a handheld spectroradiometer. For example, we could use commercial fluorescent lights powered by 60 cycle AC. Unfortunately, fluorescent lights produce several sharp energy peaks throughout the visible portion of the spectrum (Figure 15-6). They also provide little energy longer than 750 nm. Conversely, the energy spectrum of a DC powered Tungsten–Quartz–Halogen lamp produces energy that peaks at approximately 900 nm and yields energy well beyond 2500 nm (Figure 15-7) (ASD, 2005). The tungsten filament resides inside a quartz glass bulb containing halogen gas. This combination of materials emits a smooth energy curve. Because this energy is measured "closeup" in the lab, there are no atmospheric absorption features present in the reference or target spectra obtained in the lab. Also, there are no solar absorption features (Fraunhofer lines) discussed in Chapter 2.

Characteristics of the Spectroradiometer Used to Obtain the *In Situ* Reflectance Spectrum

The reflectance spectra to be measured are specified in the project experimental design. For example, perhaps we want to collect Bahiagrass reflectance spectra. There are several

other important characteristics that must be specified to obtain the desired spectra, including: general spectral region of interest, spectral resolution, spectral sampling interval, spectrometer design, data collection speed, viewing geometry, and field of view.

Spectral Region of Interest

If the goal is to obtain reflectance spectra for a particular material (e.g., Bahiagrass) and neighboring land cover types (e.g., bare soil, forest, asphalt), we may need to measure the spectral reflectance of the materials throughout the wavelength region 350 to 2500 nm. The most important absorption features associated with vegetation are summarized in Table 15-1 for the wavelength interval 430 nm to 2350 nm. Figure 15-8 depicts the spectral reflectance characteristics of healthy green vegetation from 400 nm to 2400 nm.

Strong chlorophyll absorption bands occur at 430 nm and 660 nm for chlorophyll *a* and 460 nm and 640 nm for chlorophyll *b*. When vegetation is stressed there is also noticeable movement in the transition region from near-infrared to red reflectance known as the *red edge position*. The greater the amount of stress, the greater the movement of the red-edge to shorter wavelengths (Carter et al., 1996). The short wavelength infrared (SWIR) from 1050 to 2500 nm provides valuable information on the amount of water in a plant (i.e., its relative turgidity) (Figure 15-8).

Numerous vegetation indices discussed in Chapter 11 require information in the chlorophyll absorption bands. The theoretical basis for "empirically-based" vegetation indices is derived from examination of typical spectral reflectance spectra of leaves (Figure 15-8). The reflected energy in the visible bands is low as a result of high absorption by photo-

Table 15-1. Foliar spectral absorption features (adapted from Curran, 1989; ASD, 2005).

Wavelength	Mechanism	Chemical(s)
0430	Electron transition	Chlorophyll *a*
0460	Electron transition	Chlorophyll *b*
0640	Electron transition	Chlorophyll *b*
0660	Electron transition	Chlorophyll *a*
0910	C-H stretch, 3rd overtone	Protein
0930	C-H stretch, 3rd overtone	Oil
0970	O-H stretch, 1nd overtone	Water, starch
0990	O-H stretch, 2nd overtone	Starch
1020	N-H stretch	Protein
1040	C-H stretch, C-H deformation	Oil
1120	C-H stretch, 2nd overtone	Lignin
1200	O-H bend, 1st overtone	Water, cellulose, starch, lignin
1400	O-H bend, 1st overtone	Water
1420	C-H stretch, C-H deformation	Lignin
1450	O-H stretch, 1st overtone, C-H stretch, C-H deformation	Starch, sugar, lignin, water
1490	O-H stretch, 1st overtone	Cellulose, sugar
1510	N-H stretch, 1st overtone	Protein, nitrogen
1530	O-H stretch, 1st overtone	Starch
1540	O-H stretch, 1st overtone	Starch, cellulose
1580	O-H stretch, 1st overtone	Starch, sugar
1690	C-H stretch, 1st overtone	Lignin, starch, protein, nitrogen
1780	C-H stretch, 1st overtone/O-H stretch/H-O-H deformation	Cellulose, sugar, starch
1820	O-H stretch/C-O stretch, 2nd overtone	Cellulose
1900	O-H stretch, C-O stretch	Starch
1930	Absorbed water	Water
1940	O-H stretch, O-H deformation	Water, lignin, protein, nitrogen, starch, cellulose
1960	O-H stretch/O-H bend	Sugar, starch
1980	N-H asymmetry	Protein
2000	O-H deformation, C-O deformation	Starch
2060	N-H bend, 2nd overtone/N-H bend/N-H stretch	Protein, nitrogen
2080	O-H stretch/O-H deformation	Sugar, starch
2100	O-H bend/C-O stretch/C-O-C stretch, 3rd overtone	Starch, cellulose
2130	N-H stretch	Protein
2180	N-H bend, 2nd overtone/C-H stretch/C-O stretch/C-N stretch	Protein
2240	C-H stretch	Protein
2250	O-H stretch, O-H deformation	Starch
2270	C-H stretch/O-H stretch, CH2 bend/CH2 stretch	Cellulose, sugar, starch
2280	C-H stretch/CH2 deformation	Starch, cellulose
2300	N-H stretch, C-0 stretch, C-H bend, 2nd overtone	Protein, nitrogen
2310	C-H bend, 2nd overtone	Oil
2320	C-H stretch/CH2 deformation	Starch
2340	C-H stretch/O-H deformation/C-H deformation/O-H stretch	Cellulose
2350	CH2 bend, 2nd overtone, C-H deformation, 2nd overtone	Cellulose, protein, nitrogen

synthetically active pigments with maximum sensitivity in the blue (470 nm) and red (670 nm) wavelengths. Most of the near-infrared radiation is scattered (reflected and transmitted) with very little absorption, in a manner dependent upon the structural properties of the canopy (LAI, leaf angle distribution, leaf morphology). As a result, the ***contrast*** between red and near-infrared responses is a sensitive measure of vegetation amount, with maximum red versus NIR differences occurring over a full canopy, and minimal contrast over targets with little or no vegetation. For low and medium amounts of vegetation, the contrast is a result of both red and NIR changes, while at higher amounts of vegetation, only the NIR contributes to increasing contrast as the red band becomes saturated due to chlorophyll absorption.

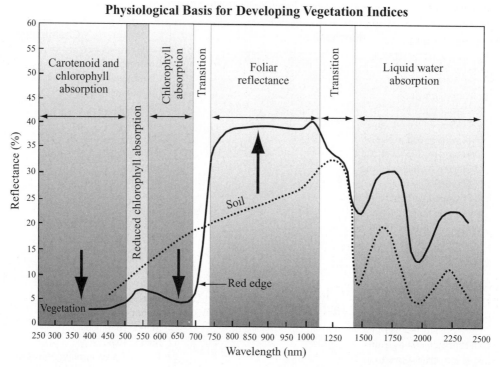

Figure 15-8 Physiological basis for developing vegetation indices in the region from 250 – 2400 nm.

The red versus NIR contrast can be quantified through the use of ratios (NIR / Red), differences (NIR – Red), weighted differences [NIR – (*k* * Red)], linear band combinations [(NIR – Red) / (NIR + Red)], or hybrid approaches of the above. Vegetation indexes are measures of this contrast and thus are integrative functions of canopy structural (e.g., percent cover, LAI) and physiological (e.g., pigments, photosynthesis) parameters (Huete et al., 2002). Numerous vegetation indices were summarized in Chapter 11.

Spectral Resolution

Spectral resolution is defined as the full-width-half-maximum (FWHM) of the spectroradiometer response to a monochromatic light source (ASD, 2005). The FWHM concept was introduced in Figure 1-6. Spectral resolution may also be defined as the narrowest spectral feature that can be resolved by a spectrometer. Many scientists are interested in obtaining reflectance spectra for selected land cover types subjected to various treatments in order to understand the spectral response to these treatments. If possible, it is ideal to use a handheld spectroradiometer that has a spectral resolution of 2 to 3 nm for the region 360 to 1000 nm, and 10 nm for the region 1000 nm to 2400 nm. Hyperspectral airborne remote sensing systems such as AVIRIS have a spectral res-

olution of about 10 nm (refer to Chapter 7 for information about AVIRIS).

Spectral Sampling Interval

Spectral sampling interval is the spacing between sample points in the reflectance spectrum. It is wise to use a handheld spectroradiometer that has a spectral sampling interval of 1.4 nm @ 350 – 1000 nm, and 2 nm @ 1000 – 2500 nm (ASD, 2005).

Spectrometer Design and Data Collection Speed

Some spectroradiometers used for *in situ* data collection record the spectral information more rapidly and accurately than others. The more rapidly the spectroradiometer collects an individual scan in the field, the higher the probability that additional spectra will be obtained before environmental conditions (e.g., cloud cover, wind) and sensor system conditions (e.g., viewing geometry) change. Remember, the goal is to hold as many environmental and sensor parameters as constant as possible while the target and reference scans are collected.

Silicon detectors provide the best performance for remote sensing investigations in the visible and near-infrared (VNIR) portion of the spectrum from 350 to 1000 nm. In the short-wavelength infrared (SWIR) region from 1000 – 2500 nm, the amount of solar energy available for measurement decreases. Therefore, more sensitive detectors are required to measure energy in the SWIR region (ASD, 2005). Typical detectors include two separate InGaAs (Indium gallium arsenide) photodiodes.

All spectroradiometers produce some noise. Dark-current (i.e., electronic or detector noise) can be measured by viewing a black reflectance target or by taking measurements with the optics covered. The dark-current signal can then be subtracted from subsequent spectral measurements. Most spectroradiometers include software for recording and subtracting dark-current.

Viewing Geometry and Field of View

Ideally, *in situ* and remote sensing–derived reflectance spectra are obtained by a spectroradiometer viewing the terrain at nadir. Most spectral investigations have problems dealing with the bidirectional reflectance distribution function (BRDF) effects introduced by the diurnal change in sun elevation and azimuth. Schill et al. (2004) documented how the spectral reflectance of Smooth Cordgrass (*Spartina alterniflora*) changed throughout the day due to changes in Sun altitude and zenith angles. The BRDF problems become even more acute when the sensor obtains data off-nadir. Therefore, whenever possible, *in situ* reflectance data should be obtained at nadir.

The actual terrain viewed by the spectroradiometer at nadir is a function of several factors. The diameter (D) of the ground-projected instantaneous field of view observed by the spectroradiometer optics is a function of the angular field of view (β) multiplied by the distance of the radiometer above the phenomena of interest (H):

$$D = \beta \times H. \qquad (15\text{-}2)$$

The diameter (D) of the ground-projected IFOV should be decided early in the project. In exploratory studies, D might be set at relatively small dimension to obtain spectra at the leaf, branch, and canopy spatial scale [e.g., D values ranging from just a few centimeters to 0.25 m (approximately 1 ft.)]. This would provide detailed information on vegetation, shadow, and soil characteristics typically encountered in a remote sensing investigation. Reflectance measurements from individual leaves and branches are best obtained in a controlled laboratory environment for reasons previously

Fields of View Associated with Various Fore-optics

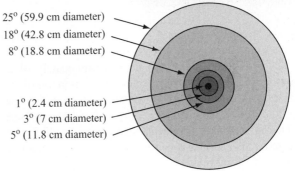

25° (59.9 cm diameter)
18° (42.8 cm diameter)
8° (18.8 cm diameter)

1° (2.4 cm diameter)
3° (7 cm diameter)
5° (11.8 cm diameter)

Figure 15-9 Fields of view using 1, 3, 5, 8, 18, and 25 degree fore-optics for a handheld spectroradiometer positioned 135 cm above the terrain of interest.

discussed (e.g., illumination variations caused by clouds, wind). Valuable information about the effects of the various treatments might be discovered in this manner. However, it should be noted that reflectance spectra derived from individual leaves and branches are usually of limited value when trying to perform matched-filtering or other types of image classification based on *in situ* reflectance spectra (Research Systems, Inc., 2004).

Conversely, if we are interested in using the reflectance spectrum for calibrating a specific type of remote sensor data (e.g., 224 bands of AVIRIS data at 3 × 3 m) or performing matched-filtering classification using these data, then it would be logical to adjust the diameter (D) of the ground-projected IFOV of the handheld spectroradiometer to be 3 m. We might need to collect the *in situ* spectra while standing on a ladder in order to obtain such a field of view.

Figure 15-9 depicts fields of view available using 1–, 3–, 5– 8–, 18–, and 25–degree fore-optics for a spectroradiometer held 135 cm above the target of interest. The 25–degree field of view fore-optics may allow the spectroradiometer to obtain data that is more representative of airborne hyperspectral remote sensing systems.

Candidate Spectroradiometers for In Situ Spectral Reflectance Data Collection

There are a variety of handheld spectroradiometers that may be used to collect the required *in situ* spectral reflectance information. Two of the most widely used spectroradiometers are produced by

- Geophysical Environmental Research (GER),

- Analytical Spectral Devices.

Spectroradiometers produced by these vendors are capable of obtaining the required spectral reflectance data over the spectral region from 400 – 2500 nm at approximately 10 nm resolution. Each is easy to use, stores information in a digital format, and can be customized to fulfill specific user requirements in the field (e.g., field of view, interfaces, backpacks). Spectroradiometer instrument specifications change rapidly, so it is necessary to contact the vendors to obtain current information.

 References

ASD, 2005, *Technical Guide*, 4th Ed., Boulder: Analytical Spectral Devices, http://www.asdi.com.

ASD, 2006, *FieldSpec Pro*, Boulder: Analytical Spectral Devices, http://www.asdi.com/products-spectroradiometers.asp.

ASTER, 2006, *ASTER Spectral Library*, Pasadena: ASTER Committee, http://speclib.jpl.nasa.gov/.

Carter, G. A., Cibula, W. G. and R. L. Miller, 1996, "Narrow-band Reflectance Imagery Compared with Thermal Imagery for Early Detection of Plant Stress," *Journal of Plant Physiology*, 148:515–522.

Clark, R. N., Swayze, G. A., Wise, R., Livo, K. E., Hoefen, T. M., Kokaly, R. F. and S. J. Sutley, 2003, *USGS Digital Spectral Library - Splib05a*, USGS Open File Report 03-395, Washington: USGS, http://speclab.cr.usgs.gov/spectral-lib.html.

Curran, P. J., 1989, "Remote Sensing of Foliar Chemistry," *Remote Sensing of Environment*, 30:271–278.

Curtiss, B. and S. L. Ustin, 1988, "Spectral Changes in Ponderosa Pine associated with Natural Ozone Exposure," *Forest Service Forest Response Program Annual Meeting*, Corpus Christi, February, pp 23–26.

Filippi, A., 1999, *Hyperspectral Image Classification Using a Batch Descending Fuzzy Learning Vector Quantization Artificial Neural Network: Vegetation Mapping at the John F. Kennedy Space Center*, Master's thesis, Columbia: Univ. of S.C., 276 p.

GER, 2006, *Field Portable Spectroradiometers*, Millbrook, NY: Group for Environmental Research, www.ger.com.

Goetz, S. J., 2002, "Recent Advances in Remote Sensing of Biophysical Variables: An Overview of the Special Issue," *Remote Sensing of Environment*, 79:145–146.

Jensen, J. R., 2005, *Introductory Digital Image Processing: A Remote Sensing Perspective*, 3rd Ed., Upper Saddle River: Prentice-Hall, 525 p.

Gao, B. C. and A. F. H. Goetz, 1990, "Column Atmospheric Water Vapor Retrievals from Airborne Imaging Spectrometer Data," *Journal of Geophysical Research - Atmospheres*, 95:3549–3564

Huete, A. R., Didan, K., Miura, T., Rodriguez, E. P., Gao, X. and G. Ferreira, 2002, "Overview of the Radiometric and Biophysical Performance of the MODIS Vegetation Indices," *Remote Sensing of Environment* 83:195–213.

Labsphere, 2006, *Spectralon*, North Sutton, NH, http://www.labsphere.com.

McCoy, R. M., 2004, *Field Methods in Remote Sensing*, NY: Guilford, 159 p.

Schill, S., Jensen, J. R. and G. Raber, 2004, "Bi-directional Reflectance Distribution Function of Smooth Cordgrass (*Spartina Alterniflora*)," *GIScience & Remote Sensing*, 41(2):116–135.

Teillet, P. M., Gauthier, R. P., Chichagov, A. and G. Fedosejevs, 2002, "Towards Integrated Earth Sensing: Advanced Technologies for In Situ Sensing in the Context of Earth Observation," *Canadian Journal of Remote Sensing*, 28(6):713-718.

Appendix - Sources of Remote Sensing Information

On-line Tutorials

Fundamentals of Remote Sensing, by Canada Center for Remote Sensing, www.ccrs.nrcan.gc.ca/resource/tutor/fundam/index_e.php

NASA Observatorium, http://observe.arc.nasa.gov/nasa/core.shtml.html

NASA Global Change Master Directory, http://gcmd.gsfc.nasa.gov/index.html

Remote Sensing Core Curriculum, by ICRSE, ASPRS, and NCGIA, http://www.r-s-c-c.org/

Remote Sensing Tutorial, by Nicholas M. Short, NASA Goddard Space Flight Center, http://rst.gsfc.nasa.gov/

Remote Sensing Tutorial, by the Virtual Hawaii, http://satftp.soest.hawaii.edu/space/hawaii/vfts/oahu/rem_sens_ex/rsex.spetral.1.html

Virtual Library: Remote Sensing, http://virtual.vtt.fi/space/rsvlib/

IAEGS Model Curriculum, by the Univ. of Mississippi Institute for Advanced Education in Geospatial Sciences, http://geoworkforce.olemiss.edu/overview.php

Societies

Alliance for Marine Remote Sensing – www.waterobserver.org

American Geophysical Union – www.agu.org/

American Soc. for Photogrammetry & Remote Sensing – www.asprs.org

Canadian Remote Sensing Soc. – www.casi.ca/ind-ex.php?pg=crss

European Assoc. Remote Sensing Labs – www.earsel.org/welcome.html

Federation of American Scientists; Image intelligence – www.fas.org/

IEEE Geoscience and Remote Sensing Society – www.grss-ieee.org/

Intl. Commission on Remote Sensing – http://hydrolab. arsusda.gov/~jritchie/

Intl. Soc. for Photogrammetry & Remote Sensing – www. isprs.org

National Stereoscopic Assoc. – www.stereoview. org/

Remote Sensing and Photogrammetry Soc. (UK) – www. rspsoc.org/

Remote Sensing Specialty Group, Assoc. Amer. Geographers – www.aagrssg.org/

SPIE Intl. Soc. For Optical Engineering – www.spie.org/

National Space Agencies

Canada Center Remote Sensing – www.ccrs.nrcan. gc.ca/

Centre National d'Etudes Spatiales – www.cnes.fr/

CSIRO-Scientific and Industrial Research for Australia – www.csiro.au/

European Space Agency – www.esa.int/

National Aeronautics & Space Administration – www. nasa.gov/

National Remote Sensing Agency India – www.nrsa. gov.in/

National Remote Sensing Agency Japan – www.nasda. go.jp/index_e.html

Journals

Advances in Space Research – www.elsevier.com/wps/ product/cws_home/644/

Atmospheric Research – www.elsevier.com/wps/product/ cws_home/503323

Backscatter – marine remote sensing: www.water observer.org/

Cartography & Geographic Information Science – www. acsm.net/cagis/cagisjr.html

Canadian Journal of Remote Sensing – www.casi.ca/ index.php

Computers & Geosciences – www.elsevier.com/locate/ compgeosci/

Earth Observation Magazine – www.eomonline.com/

Geocarto International – www.geocarto.com/geocarto. html

GeoInformatics – www.j-geoinfo.net/

Geospatial Solutions – www.geospatial-online.com/

GIScience & Remote Sensing – www.bellpub.com/msrs/

IEEE Transactions on Geoscience & Remote Sensing – http://ieeexplore.ieee.org/xpl/RecentIssue.jsp? punumber=36

Intl. Journal of Remote Sensing – www.tan-df.co.uk/ journals/online/0143-1161.asp

ISPRS Journal of Photogrammetry and Remote Sensing – www.elsevier.com/locate/isprsjprs

Pattern Recognition – http://science.direct.com

Remote Sensing of Environment – www.elsevier.com

Photogrammetric Engineering & Remote Sensing – www. asprs.org/publications/index.html

Stereo World – National Stereoscopic Association - www. stereoview.org/stereoworld.html

Aerial Photography Sources

Kite aerial photography – http://arch.ced.berkeley.edu/ kap/kaptoc.html

NHAP and NHAP – USGS, National High Altitude Photography (NHAP) & National Aerial Photography Program (NAPP), http://edc.usgs.gov/guides napp.html

USDA Photography – www.apfo.usda. gov/

Multispectral Remote Sensing Data Resources

ADEOS-II – NASDA Advanced Earth Observing Satellite-II, http://sharaku.eorc.jaxa.jp/ADEOS2/index.html

AMSR – NASA *Terra* Advanced Microwave Scanning Radiometer (onboard ADEOS-II), http://sharaku.eorc. jaxa.jp/AMSR/index_e.htm

ASTER – NASA *Terra* Advanced Spaceborne Thermal Emission and Reflection Radiometer, http://asterweb. jpl. nasa. gov/

AVHRR – NOAA Advanced Very High Resolution Radiometer, http://edc.usgs.gov/products/satellite/avhrr. html

AVHRR Pathfinder – http://podaac.jpl.nasa.gov/cgi-bin/ dcatalog/fam_summary.pl?sst+pfsst

AVIRIS – NASA Advanced Visible/infrared Imaging Spectrometer, http://aviris.jpl.nasa.gov/

CASI – Compact Airborne Spectrographic Imager, ITRES Research Limited of Canada, www.itres.com/

CERES – NASA *Terra* Clouds and the Earth's Radiant Energy System, http://asd-www.larc.nasa.gov/ceres/ ASDceres.html

CORONA, ARGON, LANYARD declassified images, http://edc.usgs.gov/products/satellite/declass1. html

DMSP – Defense Meteorological Satellite Program, www. ngdc.noaa.gov/dmsp/index.html

EO-1 – NASA New Millennium Program, Earth Observing-1, Advanced Land Imager, http://eo1.gsfc.nasa. gov/

GOES – NOAA/NASA Geostationary Operational Environ-mental Satellite, http://goes.gsfc.nasa.gov/

Google Earth – Google, Inc., http://earth.google.com/

I^3 (Information, Integration, Image, Inc.), www.i3.com/

IKONOS – GeoEye, Inc., www.GeoEye.com

IRS-1 A, B, C, D – Indian Remote Sensing System, Linear Imaging Self-scanning Sensor (LISS), http://www. space-imaging.com/products/irs/index.htm

LANDSAT MSS – Multispectral Scanner, http://edc.usgs. gov/products/satellite/mss.html

LANDSAT TM – Thematic Mapper, http://edc.usgs.gov/ products/satellite/tm.html

LANDSAT 7 (ETM+) – Landsat 7 Enhanced Thematic Mapper Plus, http://landsat7.usgs.gov/

Mercury, Gemini, Apollo, Skylab – www-pao.ksc.nasa. gov/history/mercury/mercury.htm

METEOSAT – http://oiswww.eumetsat.org/IDDS-cgi/ listImages.pl?m=bnw

MISR – NASA *Terra* Multi-angle Imaging Spectroradiometer, www-misr.jpl.nasa.gov/

MODIS – NASA *Terra* Moderate Resolution Imaging Spectrometer, http://modis.gsfc.nasa.gov/

MOPITT – NASA *Terra* Measurement of Pollution in the Troposhere, http://eosweb.larc.nasa.gov/PRODOCS/ mopitt/table_mopitt.html

MOS-1/MOS-1b – Marine Observation Satellite, http:// www.eoc.jaxa.jp/satellite/satdata/mos_e.html

OCTS – Ocean Color and Temperature Scanner, www. eoc.jaxa.jp/satellite/sendata/octs_e.html

OrbView-3 – GeoEye, Ind., www.GeoEye.com

QuickBird – DigitalGlobe, Inc., www.digital-globe.com/ about/quickbird.html

RESURS-01 – Russian satellite, http://sputnik.infospace. ru/resurs/engl/resurs.htm

SeaWiFS (OrbView-2) – GeoEye, Ind., www.Geo-Eye.com

SOHO – Solar and Heliospheric Observatory, http:// sohowww.nascom.nasa.gov/

Space Shuttle Photography – www.nasm.si.edu/ceps/rpif/ sspr.cfm

SPIN-2 TK-350 and KVR-1000 – U.S. coverage served on Microsoft's Terra Server (www.terraserver.com/ providers/SPIN2.asp)

SPOT 1- 5 – www.spot.com

SSM/I – NOAA/NASA Special Sensor Microwave/Imager, http://podaac.jpl.nasa.gov:2031/sensor_docs/ssmi.html

TOMS – Total Ozone Mapping Spectrometer, http://jwocky.gsfc.nasa.gov/

TOPEX/POSEIDON – http://topex-www.jpl.nasa.gov/

TRMM - NASA/NASDA Tropical Rainfall Measuring Mission, http://trmm.gsfc.nasa.gov/

UARS – NASA Upper Atmosphere Research Satellite, http://umpgal.gsfc.nasa.gov/Radar Sensor Systems

AIRSAR – NASA JPL SAR, http://airsar.jpl.nasa.gov/

ENVISAT-1 – European Space Agency, http://envisat.esa.int/

ERS-1, 2 – European Space Agency, http://earth.esa.int/ers/

JERS-1 – Japanese Earth Resource Satellite, www.nasda.go.jp/projects/sat/jers1/index_e.html

LightSAR – NASA Spaceborne Radar (proposed), http://southport.jpl.nasa.gov/lightsar/

RADARSAT – Canadian, www.rsi.ca/

SEASAT – http://southport.jpl.nasa.gov/scienceapps/seasat.html

SIR-A, B, C/X-SAR – NASA Space Shuttle Imaging Radar:

• SIR-A – http://southport.jpl.nasa.gov/scienceapps/sira.html

• SIR-B – http://southport.jpl.nasa.gov/scienceapps/sirb.html

• SIR-C/X-SAR – http://southport.jpl.nasa.gov/sir-c/

SLAR – NASA Side-Looking Airborne Radar, http://pubs.usgs.gov/of/2000/of00-006/htm/slar.htm

SRTM – Shuttle Radar Topography Mission, www2.jpl.nasa.gov/srtm/

STAR-3i – Sea Ice and Terrain Assessment Radar, Intermap Technologies, www.intermap.com/images/papers/Intermap_STAR-3i_upgrade_screen_v2.pdf

Virtual Earth – Microsoft Inc., www.local.live.com

On-line Remote Sensing Resources at the University of South Carolina

Dr. Jensen provides PowerPoint[TM] lecture materials containing non-copyrighted illustrations found in this book free of charge for educational purposes. The process can be initiated by visiting the University of South Carolina Center for GIS & Remote Sensing at

• www.cas.sc.edu/geog/cgisrs/.

Dr. Jensen also provides free access to remote sensing exercises and selected remote sensing datasets that he uses in the University of South Carolina "Remote Sensing of Environment" course. Please access the information at:

• www.cas.sc.edu/geog/Rsbook/index.html.

You are welcome to use the lecture materials and exercises and adapt them to your requirements. Several of the exercises require access to the ERDAS Imagine digital image processing system (or equivalent) to perform fundamental operations such as viewing a color-composite image and making measurements. In most instances, the imagery used in the exercises is in the public domain unless otherwise noted. If you use the remote sensing exercise materials, please send the author information about innovative uses, methodological refinements that could be made, and improved datasets that might be used.

Index